PERMUTATIONS

The number of permutations of n elements taken r at a time, where $r \leq n$, is

$$_nP_r = \frac{n!}{(n - r)!}.$$

COMBINATIONS

The number of combinations of n elements taken r at a time, where $r \leq n$, is

$$\binom{n}{r} = {}_nC_r = \frac{n!}{(n - r)!r!}.$$

BINOMIAL PROBABILITY

If p is the probability of success in a single trial of a binomial experiment, the probability of x successes and $n - x$ failures in n independent repeated trials of the experiment is

$$\binom{n}{x}p^x(1 - p)^{n-x}.$$

MEAN

The mean of the n numbers, $x_1, x_2, x_3, \ldots, x_n$, is

$$\bar{x} = \frac{x_1 + x_2 + \cdots + x_n}{n} = \frac{\Sigma(x)}{n}.$$

SAMPLE STANDARD DEVIATION

The standard deviation of a sample of n numbers, $x_1, x_2, x_3, \ldots, x_n$, with mean $\bar{x}$, is

$$s = \sqrt{\frac{\Sigma(x - \bar{x})^2}{n - 1}}.$$

BINOMIAL DISTRIBUTION

Suppose an experiment is a series of n independent repeated trials, where the probability of a success in a single trial is always p. Let x be the number of successes in the n trials. Then the probability that exactly x successes will occur in n trials is given by

$$\binom{n}{x}p^x(1 - p)^{n-x}.$$

The mean μ and variance σ^2 of a binomial distribution are, respectively,

$$\mu = np \quad \text{and} \quad \sigma^2 = np(1 - p).$$

The standard deviation is

$$\sigma = \sqrt{np(1 - p)}.$$

8th EDITION

Mathematics with Applications

In the Management, Natural, and Social Sciences

FINITE VERSION

8th EDITION

Mathematics with Applications

In the Management, Natural, and Social Sciences

FINITE VERSION

MARGARET L. LIAL

AMERICAN RIVER COLLEGE

THOMAS W. HUNGERFORD

CLEVELAND STATE UNIVERSITY

Addison
Wesley

Boston San Francisco New York
London Toronto Sydney Tokyo Singapore Madrid
Mexico City Munich Paris Cape Town Hong Kong Montreal

Publisher: Greg Tobin
Sponsoring Editor: William Hoffman
Executive Project Manager: Christine O'Brien
Assistant Editor: Stefanie Borge
Project Manager: Sandi Goldstein
Senior Production Supervisor: Karen Wernholm
Production Coordination: Elm Street Publishing Services, Inc.
Marketing Manager: Michael Boezi
Marketing Coordinator: Weslie Lewis
Media Producer: Ruth Berry
Software Editor/InterAct Math: Malcolm Litowitz
Software Editor/TestGen-EQ: David Malone
Manufacturing Buyer: Evelyn Beaton
Senior Designer: Barbara T. Atkinson
Text Design: Silvers Design
Cover Design: Leslie Haimes
Compositor: Nesbitt Graphics, Inc.
Illustrations: Techsetters, Inc.
Cover Photo: ©2002 Stone/Lonnie Duka

Photo Credits:
page 10 ©Chuck Savage/corbisstockmarket.com; **page 39** ©David Stoecklein/corbisstockmarket.com; **page 58** ©2002 PhotoDisc; **page 64** ©Mug Shots/corbisstockmarket.com; **page 95** ©Fritz Hoffmann/The Image Works; **page 99** AP/Wide World Photos; **page 107** ©Kathy Ferguson-Johnson/Photo Edit; **page 128** ©James L. Amos/CORBIS; **page 150** AP/Wide World Photos; **page 186** ©Reuters NewMedia Inc./CORBIS; **page 198** ©2002 Photo Disc; **page 215** ©Graham Neden/CORBIS; **page 225** Courtesy of NASA; **page 229** ©Brandon D. Cole/CORBIS; **page 266** ©Mug Shots/corbisstockmarket.com; **page 271** ©AP/Wide World Photos; **page 272** ©PhotoDisc; **page 309** ©Richard Hamilton Smith/CORBIS; **page 324** ©Owen Franken; **page 288** ©Lee Snider/CORBIS; **page 341** ©D. Boone/CORBIS; **page 359** ©Corbis Corporation/Tom Stewart, 2001; **page 384** ©Bob Rowan; Progressive Image/CORBIS; **page 409** ©PhotoDisc; **page 413** ©2002 PhotoDisc; **page 422** ©Gail Mooney/CORBIS; **page 438** ©Mary Boucher; **page 462** ©Corbis Corporation/Ariel Skelley, 2001; **page 479** ©Billy E. Barnes/PhotoEdit; **page 502** ©Mary Steinbacher/PhotoEdit; **page 507** ©Mug Shots/corbis-stockmarket.com; **page 523** ©Kevin R. Morris/CORBIS; **page 543** ©Bob Daemmrich/The Image Works; **page 553** ©2002 PhotoDisc; **page 582** ©V.C.L./FPG; **page 593** AP/Wide World Photos; **page 594** ©David Frazier/Stone

Library of Congress Cataloging-in-Publication Data

Lial, Margaret L.
  Mathematics with applications / Margaret L. Lial, Thomas W. Hungerford.—8th ed., finite version.
    p. cm.
  Includes index.
  ISBN 0-201-77003-2 (alk. paper)
  1. Mathematics.   I. Hungerford, Thomas W.   II. Title.
QA37.3.L55 2002b
510—dc21   2002066567

1 2 3 4 5 6 7 8 9 10—QWT—05040302

Contents

CHAPTER 9

Counting, Probability Distributions, and Further Topics in Probability 481

CHAPTER 10

Introduction to Statistics 545

Preface

Like its predecessors, this edition of *Mathematics with Applications, Finite Version* is an applications-oriented text for students in business, management, social sciences, and health sciences. The only prerequisite is a course in algebra. Chapter 1 provides a thorough review of basic algebra for those students who need it.

The book is written at a level appropriate for its intended audience. We have done our best to present sound mathematics in an understandable manner, proceeding from the familiar to new material and from concrete examples to general rules and formulas. There is an ongoing focus on real-world problem solving, and almost every section includes relevant contemporary applications.

KEY ORGANIZATIONAL AND CONTENT CHANGES

The table of contents for this edition is substantially the same as that of the previous edition. Organizational changes are:

- Chapters 3 and 4 (Functions and Graphs, Polynomial and Rational Functions) of the previous edition were combined into a single chapter (Chapter 3, Functions and Graphs).
- The order of topics in Chapters 1, 2, and 9 is slightly changed.
- Many minor changes have been made throughout the book to improve clarity and coverage.

The major content changes are:

- A much more complete discussion of linear models appears in Section 2.3, including (optional) material on linear regression. Other regression models are mentioned at appropriate points later in the text.
- Section 4.4 has been substantially rewritten to give students more guidance in handling a variety of different types of logarithmic and exponential equations.
- Section 9.5 has been rewritten to clarify the properties of regular Markov chains.
- A new table summarizing common geometric area and volume formulas has been added to Appendix B.

NEW AND ENHANCED FEATURES

BALANCED APPROACH   Multiple representations of a topic (symbolic, numerical, graphical, verbal) are given when appropriate. However, we do not believe that all representations are useful for all topics. So we are pedagogically selective in presenting alternative methods. Effective alternatives are discussed only when they are likely to increase student understanding.

TECHNOLOGY   It is assumed that all students have a calculator that will handle exponential and logarithmic functions. Beyond that, however,

the use of technology in this text is optional.

Examples and exercises that definitely require some sort of technology (graphing calculators, spreadsheets, or other computer programs) are marked with the icon , so instructors who want to omit these discussions and exercises can easily do so. They should note, however, that some examples that do *not* require technology may include some mention of it or a graphing calculator screen (for instance, one showing the matrix associated with a system of linear equations).

Instructors who routinely use technology in their courses—and expect their students to do the same—will find more than enough material here to satisfy their needs. Here are some of the features they may want to incorporate into their courses.

- *Examples and Exercises marked with* A number of examples show students how various features of graphing calculators and spreadsheets can be applied to the topics in this book. These include, but are not limited to, graphing functions, solving equations, row-reducing matrices, and finding linear regression models for data, statistical analysis. There are also many exercises marked for technology use.
- *Technology Tips* These are placed at appropriate points in the text to inform students of various features of their graphing calculator or spreadsheet programs and to guide them in using these tools.
- *Graphing Calculator Appendix* Part 1 is a brief introduction to the features of graphing calculators that are relevant to topics in the text. Part 2 contains eleven programs for carrying out various procedures discussed in the text (including financial functions, a RREF program for TI-82 and Casio, and a new two-stage method program for linear programming problems with mixed constraints).

  An outline of the Graphing Calculator Appendix (including the complete list of programs) is on page 597. In order to facilitate easy updating of material, the full Appendix is on the text's Web site: www.aw.com/MWA8, where it can be downloaded and printed. Printed copies of the Appendix can be made available; contact your local Addison-Wesley sales representative.

REAL-DATA EXAMPLES AND APPLICATIONS   Real-data exercises have long been a popular feature of this book. For this edition, almost all of these exercises have been updated and many new ones added. A significant number of real-data examples have also been introduced into the text.

CASE STUDIES   A number of new end-of-chapter case studies have been added and others updated. The case studies are contemporary real-world applications of some of the mathematics presented in the chapter. They provide an (at least partial) answer to that old question, "What's this stuff good for?"

FULL-COLOR GRAPHICS AND PHOTOGRAPHY   The new multicolor graphics in this edition not only make the book more attractive but are effectively used to enhance the mathematical exposition and to clarify many graphical presentations. Photographs now illustrate many of the applied exercises.

CONTINUING PEDAGOGICAL FEATURES

Popular features from earlier editions of *Mathematics with Applications* have been retained.

- Margin problems, keyed to the discussion in the text
- Extensive examples
- Boxed rules, definitions, and summaries
- "Cautions" that highlight common student difficulties and warn against errors
- An *Index of Applications* showing the wide variety of real-data applications in the text and allowing direct reference to particular topics
- Exercises keyed to the text
- Writing exercises (marked)
- Connection exercises that involve material from earlier sections (marked ⟳)
- Exercises from CPA exams to give interested students valuable exposure to such exams
- End-of-chapter summaries and review exercises

COURSE FLEXIBILITY

This book can be used for a variety of courses, including the following.

> **Finite Mathematics** (one semester or two quarters). Use as much of Chapters 1–4 as needed and then go into Chapters 5–10 as time permits and local needs require.
>
> **College Algebra with Applications** (one semester or quarter). Use Chapters 1–8, with Chapters 7 and 8 being optional.

Chapter interdependence is as follows.

	Chapter	*Prerequisite*
1	Algebra and Equations	None
2	Graphs, Lines, and Inequalities	Chapter 1

	Chapter	*Prerequisite*
3	Functions and Graphs	Chapters 1 and 2
4	Exponential and Logarithmic Functions	Chapter 3
5	Mathematics of Finance	Chapter 4
6	Systems of Linear Equations and Matrices	Chapters 1 and 2
7	Linear Programming	Chapters 3 and 6
8	Sets and Probability	None
9	Counting, Probability Distributions, and Further Topics in Probability	Chapter 8
10	Introduction to Statistics	Chapter 8

Contact your local Addison-Wesley sales representative to order a customized version of this text.

SUPPLEMENTS

For the Instructor

Instructor's Solutions Manual (ISBN 0-201-77009-1)
This manual contains detailed solutions to all text problems, suggested course outlines, and a chapter interdependence chart.

TestGen-EQ with QuizMaster (ISBN 0-201-79422-5)
Available on dual-platform Windows/Macintosh CD-ROM, TestGen enables instructors to build, edit, print, and administer tests using a computerized test bank of questions. The program also offers many options for organizing and displaying test banks and tests according to text chapter content. Tests can be printed or saved for online testing via a network or the Web, and the software can generate a variety of grading reports for tests and quizzes.

MyMathLab
MyMathLab is a complete online course available with this text and is ideal for helping students succeed in their course. This site offers a wide variety of resources, from dynamic multimedia—video clips, animations, audio explanations, and more—to an online grade book that allows students to track their performance on homework assignments and tests. With MyMathLab students can increase comprehension and get help with concepts they don't understand.

- The entire textbook is available online and is supplemented by multimedia content to help explain concepts.
- MyMathLab allows students to do practice work and to complete instructor-assigned tests and homework online. Based on these results, MyMathLab automatically builds an individual study plan that students can use to improve their skills.
- MyMathLab offers a full array of supplements such as work sheets, quizzes, and solution manuals and allows students to print them out and use them offline.
- With MyMathLab, students can learn and practice with tutorial exercises tied directly to the exercises in the textbook.

MyMathLab requires a student access code. For more information, go to www.mymathlab.com.

For the Student

Student's Solutions Manual (ISBN 0-201-77010-5)
This manual contains detailed, carefully worked-out solutions to all odd-numbered section exercises and all Chapter Review and Case Study exercises.

InterAct Tutorial Web Site
www.interactmath.com
This interactive tutorial Web site provides algorithmically generated practice exercises that correlate directly to the exercises in the text. A detailed worked-out example and guided solution accompany each practice exercise. The Web site recognizes student errors and provides feedback.

InterAct MathXL®
www.mathxl.com
The MathXL Web site provides diagnostic testing and tutorial help, all online, using InterAct Math tutorial software and TestGen-EQ testing software. Students can take chapter tests correlated to this textbook, receive individualized study plans based on those test results, work practice problems and receive tutorial instruction for areas in which they need improvement, and take further tests to gauge their progress. Instructors can customize tests and track all student test results, study plans, and practice work. The site is free when an access code is bundled with a new text.

Addison-Wesley Math Tutor Center
The Addison-Wesley Math Tutor Center is staffed by qualified mathematics instructors who provide students with tutoring on examples and exercises answered at the back of the textbook. Tutoring is available via toll-free telephone, fax, e-mail, or whiteboard technology—which allows tutors and students to actually see the problems worked while they "talk" in real time over the Internet. This service is available five days a week, seven hours a day. For more information, go to www.aw.com/tutorcenter. An access card is required.

ACKNOWLEDGMENTS

We thank the following instructors, who reviewed the manuscript and made many helpful suggestions for improvement.

Erol Barbut, *University of Idaho*
Jean Davis, *Southwest Texas State University*
Frank Fitzgerald, *Boston University*
Kim Gregor, *Delaware Technical and Community College*
Joseph Guthrie, *University of Texas at El Paso*
Carol M. Hurwitz, *Manhattan College*
Michael J. Kallaher, *Washington State University*
Jeff Lee, *Texas Tech University*
James Mazzarella, *Holy Family College*

Shannon Michaux, *University of Colorado at Colorado Springs*
Kathy O'Dell, *University of Alabama at Huntsville*
Charles Odion, *Houston Community College*
Steven A. Schonefeld, *Tri-State University*
Surinder Sehgal, *Ohio State University*
Calvin Shipley, *Henderson State University*
Bhushan Wadhwa, *Cleveland State University*
Thelma West, *University of Louisiana at Lafayette*
Hing-Sing Yu, *University of Texas at San Antonio*
Cathy Zucco-Teveloff, *Trinity College*

We also wish to thank our accuracy checkers, who did an excellent job of checking both text and exercise answers: Elka Block, Tim Mogill, Frank Purcell, Deana Richmond, Laurie Semarne, and Tom Wegleitner. A second (and very helpful) accuracy check of the Exercises was done by Bill Fellers and Cindy Trimble, who prepared the *Solutions Manual* for the text. Special thanks go to Elka Block and Frank Purcell, whose careful proofreading has helped eliminate errors from the text; to Paul Van Erden, who created an accurate and complete index for us; and to Becky Troutman, who carefully compiled the Index of Applications.

We want to thank the staff of Addison-Wesley for their assistance with and contributions to this book, particularly Greg Tobin, Christine O'Brien, Stefanie Borge, Karen Wernholm, Barbara Atkinson, Michael Boezi, Weslie Lewis, and Ruth Berry. Finally, we wish to express our appreciation to freelance project editor Sandi Goldstein and to Phyllis Crittenden of Elm Street Publishing Services, Inc.

MARGARET L. LIAL
THOMAS W. HUNGERFORD

To the Student

Several features of the text are designed to assist you in understanding the concepts and learning the mathematical procedures involved. To help you learn new concepts and reinforce your understanding, there are numerous *side problems* in the margin. They are referred to in the text by numbers in colored circles, such as ②. When you see that symbol, you should work the indicated problem in the margin before going on.

Technology (such as graphing calculators, spreadsheets, or other computer programs) is not required to use this book. However, there are a number of optional examples and exercises that use graphing calculators; they are marked with the icon ◣. There is also a *Graphing Calculator Appendix* that covers the basics of calculator use and provides some helpful programs for dealing with some of the topics in this book. The appendix can be found on the text's Web site: www.aw.com/MWA8. In addition, there are *Technology Tips* throughout the text that describe the proper menu or keys to be used in order to carry out a particular procedure on a graphing calculator.

The key to succeeding in this course is to remember that

mathematics is not a spectator sport.

You can't expect to learn mathematics without *doing* mathematics any more than you could learn to swim without getting wet. You have to take an active role, making use of all the resources at your disposal: your instructor, your fellow students, and this book.

There is no way that your instructor can possibly cover every aspect of a topic during class time. You simply won't develop the level of understanding you need to succeed unless you read the text carefully. In particular, you should read the text *before* starting the exercises. However, you can't read a math book the way you read a novel. You should have pencil, paper, and calculator handy to do the side problems, work out the statements you don't understand, and make notes on things to ask your fellow students and/or your instructor.

Finally, remember the words of the great Hillel: "The bashful do not learn." There is no such thing as a "dumb question" (assuming, of course, that you have read the book and your class notes and attempted the homework). Your instructor will welcome questions that arise from a serious effort on your part. So get your money's worth: Ask questions.

Chapter 1

Algebra and Equations

Mathematics is widely used in business and the social and biological sciences, from determining the most efficient production schedule for a factory to mapping the human genome. Because algebra and equations are the basic mathematical tools for many such applications, we begin by reviewing the fundamental ideas of algebra. Your success in this course will depend on knowing the algebraic skills presented here.

1.1 THE REAL NUMBERS

Only real numbers will be used in this book.* The names of the most common types of real numbers are as follows.

The Real Numbers	
Natural (counting) numbers	$1, 2, 3, 4, \ldots$
Whole numbers	$0, 1, 2, 3, 4, \ldots$
Integers	$\ldots, -3, -2, -1, 0, 1, 2, 3, \ldots$
Rational numbers	All numbers that can be written in the form p/q, where p and q are integers, with $q \neq 0$
Irrational numbers	Real numbers that are not rational

---

*Not all numbers are real numbers. An example of a number that is not a real number is $\sqrt{-1}$.

1

The relationships among these types of numbers are shown in Figure 1.1. Notice, for example, that the integers are also rational numbers and real numbers, but the integers are not irrational numbers.

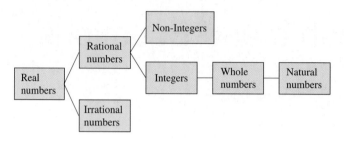

FIGURE 1.1

One example of an irrational number is π, the ratio of the circumference of a circle to its diameter. The number π can be approximated as $\pi \approx 3.14159$ ($\approx$ means "is approximately equal to"), but there is no rational number that is exactly equal to π.

1

Name all the types of numbers that apply to the following.

(a) -2

(b) $-5/8$

(c) $\pi/5$

Answers:

(a) Integer, rational, real

(b) Rational, real

(c) Irrational, real

▶**EXAMPLE 1**　What kind of number is each of the following?

(a) 6

The number 6 is a natural number, whole number, integer, rational number, and real number.

(b) 3/4

This number is rational and real.

(c) 3π

Because π is not a rational number, 3π is irrational and real.　◀ *

All real numbers can be written in decimal form. A rational number, when written in decimal form, is either a terminating decimal, such as .5 or .128, or a repeating decimal in which some block of digits eventually repeats forever, such as 1.3333. . . or 4.7234234234. . . .[†] Irrational numbers are decimals that neither terminate nor repeat.

The only real numbers that can be entered exactly into a calculator are those rational numbers that are terminating decimals of no more than 10 or 12 digits (depending on the calculator). Similarly, the answers produced by a calculator are often 10–12 digit decimal *approximations*—accurate enough for most applications. As a general rule, *you should not round off any numbers during a long calculator computation,* so that your final answer will be as accurate as possible. For convenience, however, we usually round off the final answer.

The important basic properties of the real numbers are as follows.

---

*The use of margin problems is explained in the "To the Student" section preceding this chapter.

†Some graphing calculators have a FRAC key that automatically converts some repeating decimals to fraction form. FRAC programs for other graphing calculators are in the Program Appendix.

Properties of the Real Numbers

For all real numbers, a, b, and c, the following properties hold true.

Commutative properties $a + b = b + a$ $\qquad ab = ba$

Associative properties $(a + b) + c = a + (b + c)$ $(ab)c = a(bc)$

Identity properties   There exists a unique real number 0, called the **additive identity,** such that

$$a + 0 = a \quad \text{and} \quad 0 + a = a.$$

There exists a unique real number 1, called the **multiplicative identity,** such that

$$a \cdot 1 = a \quad \text{and} \quad 1 \cdot a = a.$$

Inverse properties   For each real number a, there exists a unique real number $-a$, called the **additive inverse** of a, such that

$$a + (-a) = 0 \quad \text{and} \quad (-a) + a = 0.$$

If $a \neq 0$, there exists a unique real number $1/a$, called the **multiplicative inverse** of a, such that

$$a \cdot \frac{1}{a} = 1 \quad \text{and} \quad \frac{1}{a} \cdot a = 1.$$

Distributive property $a(b + c) = ab + ac$

The next five examples illustrate the properties listed in the preceding box.

② EXAMPLE 2   The commutative property says that the order in which you add or multiply two quantities doesn't matter.

(a) $(6 + x) + 9 = 9 + (x + 6)$

(b) $5 \cdot (9 \cdot 8) = (9 \cdot 8) \cdot 5$ ◀

EXAMPLE 3   When the associative property is used, the order of the numbers does not change, but the placement of parentheses does change.

(a) $4 + (9 + 8) = (4 + 9) + 8$

(b) $3(9x) = (3 \cdot 9)x$ ◀ ②

EXAMPLE 4   By the identity properties,

(a) $-8 + 0 = -8$ (b) $(-9) \cdot 1 = -9.$ ◀

② Name the property illustrated in each of the following examples.

(a) $(2 + 3) + 9$
  $= (3 + 2) + 9$
(b) $(2 + 3) + 9$
  $= 2 + (3 + 9)$
(c) $(2 + 3) + 9$
  $= 9 + (2 + 3)$
(d) $(4 \cdot 6)p = (6 \cdot 4)p$
(e) $4(6p) = (4 \cdot 6)p$

Answers:

(a) Commutative property
(b) Associative property
(c) Commutative property
(d) Commutative property
(e) Associative property

③

Name the property illustrated in each of the following examples.

(a) $2 + 0 = 2$

(b) $-\dfrac{1}{4} \cdot (-4) = 1$

(c) $-\dfrac{1}{4} + \dfrac{1}{4} = 0$

(d) $1 \cdot \dfrac{2}{3} = \dfrac{2}{3}$

Answers:

(a) Identity property

(b) Inverse property

(c) Inverse property

(d) Identity property

④

Use the distributive property to complete each of the following.

(a) $4(-2 + 5)$

(b) $2(a + b)$

(c) $-3(p + 1)$

(d) $(8 - k)m$

(e) $5x + 3x$

Answers:

(a) $4(-2) + 4(5) = 12$

(b) $2a + 2b$

(c) $-3p - 3$

(d) $8m - km$

(e) $(5 + 3)x = 8x$

TECHNOLOGY TIP To enter -8 on a calculator, use the negation key (labeled $(-)$ or $+/-$), *not* the subtraction key. On most one-line scientific calculators, key in 8 $+/-$. On graphing calculators or two-line scientific calculators, key in either $(-)$ 8 or $+/-$ 8.

▶**EXAMPLE 5**   By the inverse properties, the statements in parts (a) through (d) are true.

(a) $9 + (-9) = 0$

(b) $-15 + 15 = 0$

(c) $-8 \cdot \left(\dfrac{1}{-8}\right) = 1$

(d) $\dfrac{1}{\sqrt{5}} \cdot \sqrt{5} = 1$ ◀

NOTE   There is no real number x such that $0 \cdot x = 1$, so 0 has no multiplicative inverse. ③

▶**EXAMPLE 6**   By the distributive property,

(a) $9(6 + 4) = 9 \cdot 6 + 9 \cdot 4$

(b) $3(x + y) = 3x + 3y$

(c) $-8(m + 2) = (-8)(m) + (-8)(2) = -8m - 16$

(d) $(5 + x)y = 5y + xy.$ ◀

NOTE   As shown in Example 6(d), by the commutative property, the distributive property can also be written as $(a + b)c = ac + bc.$ ④

Order of Operations

We avoid possible ambiguity when working problems with real numbers by using the following **order of operations,** which has been agreed on as the most useful. This order of operations is used by computers and calculators.

Order of Operations

If parentheses or square brackets are present:

1. Work separately above and below any fraction bar.
2. Use the rules below within each set of parentheses or square brackets. Start with the innermost set and work outward.

If no parentheses are present:

1. Find all powers and roots, working from left to right.
2. Do any multiplications or divisions in the order in which they occur, working from left to right.
3. Do any additions or subtractions in the order in which they occur, working from left to right.

(5)

Evaluate the following if $m = -5$ and $n = 8$.

(a) $-2mn - 2m^2$

(b) $\dfrac{4(n - 5)^2 - m}{m + n}$

Answers:

(a) 30

(b) $\dfrac{41}{3}$

▶ **EXAMPLE 7** Use the order of operations to evaluate each expression if $x = -2, y = 5,$ and $z = -3$.

(a) $-4x^2 - 7y + 4z$

Use parentheses when replacing letters with numbers.

$$-4x^2 - 7y + 4z = -4(-2)^2 - 7(5) + 4(-3)$$
$$= -4(4) - 7(5) + 4(-3) = -16 - 35 - 12 = -63$$

(b) $\dfrac{2(x - 5)^2 + 4y}{z + 4} = \dfrac{2(-2 - 5)^2 + 4(5)}{-3 + 4}$

$$= \dfrac{2(-7)^2 + 20}{1}$$

$$= 2(49) + 20 = 118 \quad ◀ \; (5)$$

(6)

Use a calculator to evaluate the following.

(a) $4^2 \div 8 + 3^2 \div 3$

(b) $[-7 + (-9)](-4) - 8(3)$

(c) $\dfrac{-11 - (-12) - 4 \cdot 5}{4(-2) - (-6)(-5)}$

(d) $\dfrac{36 \div 4 \cdot 3 \div 9 + 1}{9 \div (-6) \cdot 8 - 4}$

Answers:

(a) 5

(b) 40

(c) $\dfrac{19}{38} = \dfrac{1}{2} = .5$

(d) $-\dfrac{1}{4} = -.25$

▶ **EXAMPLE 8** Use a calculator to evaluate

$$\dfrac{-9(-3) + (-5)}{3(-4) - 5(2)}$$

Use extra parentheses (shown here in blue) around the numerator and denominator when you enter the number in your calculator and be careful to distinguish the negation key from the subtraction key.

If you don't get -1 as the answer, then you are entering something incorrectly.

◀ **(6)**

Square Roots

There are two numbers whose square is 16, namely, 4 and -4. The positive one, 4, is called the **square root** of 16. Similarly, the square root of a nonnegative number d is defined to be the *nonnegative* number whose square is d; it is denoted $\sqrt{d}$. For instance,

$$\sqrt{36} = 6 \text{ because } 6^2 = 36 \quad \text{and} \quad \sqrt{1.44} = 1.2 \text{ because } (1.2)^2 = 1.44.$$

No negative number has a square root in the real numbers. For instance, there is no real number whose square is -4, so -4 has no square root.

Every nonnegative real number has a square root. Unless an integer is a perfect square (such as $64 = 8^2$), its square root is an irrational number. A calculator can be used to obtain a rational approximation of these square roots.

TECHNOLOGY TIP On one-line scientific calculators, $\sqrt{40}$ is entered as $40 \; \sqrt{}$. On graphing calculators and two-line scientific calculators, key in

$$\sqrt{} \; 40 \text{ ENTER (or } = \text{ or } \text{ EXE).}$$

⑦

Estimate each of the following.

(a) $\sqrt{73}$

(b) $\sqrt{22} + 3$

(c) Confirm your estimates in parts (a) and (b) with a calculator.

Answers:

(a) Between 8 and 9

(b) Between 7 and 8

(c) 8.5440; 7.6904

▶ **EXAMPLE 9** Estimate each of the following quantities. Verify your estimates with a calculator.

(a) $\sqrt{40}$

Since $6^2 = 36$ and $7^2 = 49$, $\sqrt{40}$ must be a number between 6 and 7. A typical calculator shows that $\sqrt{40} \approx 6.32455532$.

(b) $5\sqrt{7}$

$\sqrt{7}$ is between 2 and 3 because $2^2 = 4$ and $3^2 = 9$, so $5\sqrt{7}$ must be a number between $5 \cdot 2 = 10$ and $5 \cdot 3 = 15$. A calculator shows that $5\sqrt{7} \approx 13.22875656$. ◀ ⑦

CAUTION If c and d are positive real numbers, then $\sqrt{c + d}$ is *not* equal to $\sqrt{c} + \sqrt{d}$. For example, $\sqrt{9 + 16} = \sqrt{25} = 5$, but $\sqrt{9} + \sqrt{16} = 3 + 4 = 7$.

⑧

Draw a number line and graph the numbers $-4, -1, 0, 1, 2.5$, and $13/4$ on it.

Answer:

The Number Line

The real numbers can be illustrated geometrically with a diagram called a **number line.** Each real number corresponds to exactly one point on the line and vice-versa. A number line with several sample numbers located (or **graphed**) on it is shown in Figure 1.2. ◀ ⑧

FIGURE 1.2

When comparing the sizes of two real numbers, the following symbols are used.

Symbol	Read	Meaning
$a < b$	a is less than b.	a lies to the *left* of b on the number line.
$b > a$	b is greater than a.	b lies to the *right* of a on the number line.

Note that $a < b$ means the same thing as $b > a$. The inequality symbols are sometimes joined with the equal sign, as follows.

Symbol	Read	Meaning
$a \leq b$	a is less than or equal to b.	either $a < b$ or $a = b$
$b \geq a$	b is greater than or equal to a.	either $b > a$ or $b = a$

Only one part of an "either ... or" statement needs to be true for the entire statement to be considered true. So the statement $3 \leq 7$ is true because $3 < 7$, and the statement $3 \leq 3$ is true because $3 = 3$.

9

Write *true* or *false* for the following.

(a) $-9 \leq -2$

(b) $8 > -3$

(c) $-14 \leq -20$

Answers:

(a) True

(b) True

(c) False

▶ **EXAMPLE 10**   Write *true* or *false* for each of the following.

(a) $8 < 12$

This statement says that 8 is less than 12, which is true.

(b) $-6 > -3$

The graph in Figure 1.3 shows that -6 is to the *left* of -3. Thus, $-6 < -3$, and the given statement is false.

FIGURE 1.3

(c) $-2 \leq -2$

Because $-2 = -2$, this statement is true. ◀ **9**

TECHNOLOGY TIP If your graphing calculator has inequality symbols (usually located on the TEST menu), you can key in statements such as "5 < 12" or "−2 ≥ 3." When you press ENTER, the calculator will display 1 if the statement is true and 0 if it is false.

A number line can be used to draw the graph of a set of numbers, as shown in the next few examples.

10

Graph all real numbers x such that

(a) $-5 < x < 1$

(b) $4 < x < 7.$

Answers:

(a)

$\begin{array}{c} \xleftarrow{\quad(\underline{\qquad\qquad})\quad}\rightarrow \\ \;\;-5 \qquad\;\; 1 \end{array}$

(b)

$\begin{array}{c} \xleftarrow{\quad(\underline{\qquad\qquad})\quad}\rightarrow \\ \;\;4 \qquad\qquad 7 \end{array}$

▶ **EXAMPLE 11**   Graph all real numbers x such that $1 < x < 5$.

The graph includes all the real numbers between 1 and 5 and not just the integers. Graph these numbers by drawing a heavy line from 1 to 5 on the number line, as in Figure 1.4. Parentheses at 1 and 5 show that neither of these points belongs to the graph. ◀ **10**

FIGURE 1.4

A set that consists of all the real numbers between two points, such as $1 < x < 5$ in Example 11, is called an **interval**. A special notation called **interval notation** is used to indicate an interval on the number line. For example, the interval including all numbers x, where $-2 < x < 3$, is written as $(-2, 3)$. The parentheses indicate that the numbers -2 and 3 are *not* included. If -2 and 3 are to be included in the interval, square brackets are used, as in $[-2, 3]$. The chart below shows several typical intervals, where $a < b$.

Inequality	Interval Notation	Explanation
$a \leq x \leq b$	$[a, b]$	Both a and b are included.
$a \leq x < b$	$[a, b)$	a is included; b is not.
$a < x \leq b$	$(a, b]$	b is included; a is not.
$a < x < b$	(a, b)	Neither a nor b is included.

(11)

Graph all real numbers x in
the given interval.

(a) $(-\infty, 4]$

(b) $[-2, 1]$

Answers:

(a)

(b)

Interval notation is also used to describe sets such as the set of all numbers x, with $x \geq -2$. This interval is written $[-2, \infty)$. The set of all real numbers is written $(-\infty, \infty)$ in interval notation.

▶ **EXAMPLE 12** Graph the interval $[-2, \infty)$.

Start at -2 and draw a heavy line to the right, as in Figure 1.5. Use a square bracket at -2 to show that -2 itself is part of the graph. The symbol, ∞, read "infinity," *does not* represent a number. This notation simply indicates that *all* numbers greater than -2 are in the interval. Similarly, the notation $(-\infty, 2)$ indicates the set of all numbers x with $x < 2$. ◀ (11)

FIGURE 1.5

Absolute Value

Distance is always a nonnegative number. For example, the distance from 0 to -2 on a number line is 2, the same as the distance from 0 to 2. The **absolute value** of a number a is the distance on the number line from a to 0. Thus, the absolute value of both 2 and -2 is 2. We write the absolute value of the real number a as $|a|$. For example, the distance on the number line from 9 to 0 is 9, as is the distance from -9 to 0. (See Figure 1.6.) By definition, $|9| = 9$ and $|-9| = 9$.

Distance is 9   Distance is 9

-9 0   9

FIGURE 1.6

The facts that $|9| = 9$ and that $|-9| = 9 = -(-9)$ suggest the following algebraic definition of absolute value.

(12)

Find the following.

(a) $|-6|$

(b) $-|7|$

(c) $-|-2|$

(d) $|-3 - 4|$

(e) $|2 - 7|$

Answers:

(a) 6

(b) -7

(c) -2

(d) 7

(e) 5

Absolute Value

For any real number a,

$$|a| = a \qquad \text{if } a \geq 0$$
$$|a| = -a \qquad \text{if } a < 0.$$

Note the second part of the definition: for a negative number, say, -5, the negative of -5 is the positive number $-(-5) = 5$. Similarly, if a is any negative number, then $-a$ is a *positive* number. Thus, *for every real number a, $|a|$ is nonnegative.*

▶ **EXAMPLE 13** To evaluate $|8 - 9|$, you should first simplify the expression within the absolute value bars:

$$|8 - 9| = |-1| = 1.$$

Similarly, $-|-5 - 8| = -|-13| = -13$. ◀ (12)

1.1 Exercises

Label each of the following as true or false.

1. Every integer is a rational number.

2. No whole numbers are rational numbers.

Identify the properties that are illustrated in each of the following. Some may require more than one property. Assume all variables represent real numbers. (See Examples 2–6.)

3. $-7 + 0 = -7$

4. $3 + (-3) = (-3) + 3$

5. $0 + (-7) = -7 + 0$

6. $8 + (12 + 6) = (8 + 12) + 6$

7. $[5(-8)](-3) = 5[(-8)(-3)]$

8. $8(m + 4) = 8m + 8 \cdot 4$

9. How is the additive inverse property related to the additive identity property? the multiplicative inverse property to the multiplicative identity property?

10. Explain the distinction between the commutative and associative properties.

Evaluate each of the following if $p = -2$, $q = 4$, and $r = -5$. (See Examples 7 and 8.)

11. $-3(p + 5q)$

12. $2(q - r)$

13. $\dfrac{q + r}{q + p}$

14. $\dfrac{3q}{3p - 2r}$

Business   *Lenders are required to state the annual percentage rate (APR) for every loan, using the formula APR = 12r, where r is the monthly interest rate. Find the APR when*

15. $r = 1.5$

16. $r = 1.67$

Find the monthly interest rate when

17. $APR = 9$

18. $APR = 19.5$

Evaluate each expression, using the order of operations given in the text. (See Examples 7 and 8.)

19. $8 - 4^2 - (-12)$ 8-16 +12 = 4

20. $8 - (-4)^2 - (-12)$

21. $-(3 - 5) - [2 - (3^2 - 13)]$ 2- (9-13)

22. $\dfrac{2(3 - 7) + 4(8)}{4(-3) + (-3)(-2)}$

23. $\dfrac{2(-3) + 3/(-2) - 2/(-\sqrt{16})}{\sqrt{64} - 1}$

24. $\dfrac{6^2 - 3\sqrt{25}}{\sqrt{6^2 + 13}}$

State whether each of the following numbers is rational or irrational. If the number is irrational, approximate it to four decimal places.

25. 3π

26. $2/\pi$

27. $\sqrt{3}$

28. $\sqrt{4^2 - 3^2}$

Express each of the following statements in symbols, using $<$, $>$, $\leq$, or $\geq$.

29. 5 is less than 7.

30. -4 is greater than -9.

31. y is less than or equal to 8.3.

32. z is greater than or equal to -3.

33. t is positive.

34. c is at most 14.

Graph the following intervals on a number line. (See Examples 11 and 12.)

35. $(-8, -1)$ 36. $[-1, 10]$ 37. $[-2, 2)$

38. $(3, 7]$ 39. $(-2, 3]$ 40. $(-2, \infty)$

41. $(3, \infty)$ 42. $(-\infty, 5)$ 43. $(-\infty, -2]$

44. $[-4, \infty)$

Business   *The Consumer Price Index (CPI) tracks the cost of a typical sample of consumer goods. The table shows the percentage increase for each year in a ten-year period.**

Year	1991	1992	1993	1994	1995
% change in CPI	3.1	2.9	2.7	2.7	2.5

Year	1996	1997	1998	1999	2000
% change in CPI	3.3	1.7	1.6	2.7	3.4

Let r denote the yearly percentage increase in the CPI. Find the number of years during this period that r satisfied the given inequality.

45. $r > 2.7$ 46. $r < 2.7$ 47. $r \leq 2.7$

48. $r \geq 3$ 49. $r > 4$ 50. $r \leq 1.7$

---

*U.S. Bureau of Labor Statistics.

Health The Body Mass Index *(BMI)* is a number B that measures the relationship between a person's height H *(in inches)* and weight W *(in pounds)*. It is given by this formula:

$$B = \frac{.455W}{(.0254H)^2}. \quad ^*$$

Federal guidelines suggest that the desirable range for B is $19 \le B \le 25$.

(a) Find the BMI of the following athletes.

(b) Determine whether or not each athlete's BMI falls in the desirable range.

51. Steffi Graf (119 pounds; 5 ft 9 in)

52. Jackie Joyner-Kersee (153 pounds; 5 ft 10 in)

53. Shaquille O'Neal (300 pounds; 7 ft 1 in)

54. Deion Sanders (195 pounds; 6 ft 1 in)

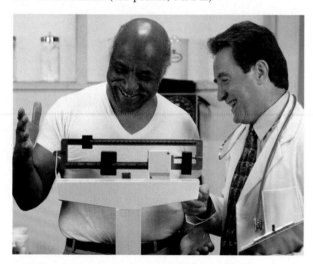

Physical Science The wind-chill factor is a measure of the cooling effect that the wind has on a person's skin. It calculates the equivalent cooling temperature if there were no wind.[†]

				Wind (mph)				
Calm	**5**	**10**	**15**	**20**	**25**	**30**	**35**	**40**
40	36	34	32	30	29	28	28	27
30	25	21	19	17	16	15	14	13
20	13	9	6	4	3	1	0	−1
10	1	−4	−7	−9	−11	−12	−14	−15
0	−11	−16	−19	−22	−24	−26	−27	−29
−10	−22	−28	−32	−35	−37	−39	−41	−43
−20	−34	−41	−45	−48	−51	−53	−55	−57
−30	−46	−53	−58	−61	−64	−67	−69	−71
−40	−57	−66	−71	−74	−78	−80	−82	−84

Temperature (°F) *(row labels in leftmost column)*

Suppose that we wish to determine the difference between two of these entries and we are interested only in the magnitude, or absolute value, of this difference. Then we subtract the two entries and find the absolute value. For example, the difference in wind-chill factors for wind at 20 miles per hour with a 20° temperature and wind at 30 miles per hour with a 10° temperature is $|-12° - 4°| = 16°$, or equivalently, $|4° - (-12°)| = 16°$.

Find the absolute value of the difference of the two indicated wind-chill factors.

55. Wind at 15 miles per hour with a 30° temperature and wind at 10 miles per hour with a −10° temperature

56. Wind at 20 miles per hour with a −20° temperature and wind at 5 miles per hour with a 30° temperature

57. Wind at 30 miles per hour with a −30° temperature and wind at 15 miles per hour with a −20° temperature

58. Wind at 40 miles per hour with a 40° temperature and wind at 25 miles per hour with a −30° temperature

Evaluate each of the following. (See Example 13.)

59. $|8| - |-4|$

60. $|-9| - |-12|$

61. $-|-4| - |-1 - 14|$

62. $-|6| - |-12 - 4|$

In each of the following problems, fill in the blank with either =, <, or >, so that the resulting statement is true.

63. $|5|$ _____ $|-5|$

64. $-|-4|$ _____ $|4|$

65. $|10 - 3|$ _____ $|3 - 10|$

66. $|6 - (-4)|$ _____ $|-4 - 6|$

67. $|-2 + 8|$ _____ $|2 - 8|$

68. $|3| \cdot |-5|$ _____ $|3(-5)|$

69. $|3 - 2|$ _____ $|3| - |2|$

70. $|5 - 1|$ _____ $|5| - |1|$

Write the expression without using absolute value.

71. $|a - 7|$ if $a < 7$

72. $|b - c|$ if $b \ge c$

73. In general, if a and b are any real numbers having the same sign (both negative or both positive), is it always true that $|a + b| = |a| + |b|$? Explain your answer.

74. If a and b are any two real numbers, is it always true that $|a - b| = |b - a|$? Explain your answer.

75. If a and b are any real numbers, is it always true that $|a + b| = |a| + |b|$? Explain your answer.

76. For which real numbers b does $|2 - b| = |2 + b|$? Explain your answer.

*Washington Post.

†Table from the Joint Action Group for Temperature Indices, 2001.

Social Science   *Use inequality symbols to rewrite each of the following statements, which are each based on an article in* The Sacramento Bee *newspaper.* Using x as the variable, describe what x represents in each exercise and then write an inequality. Example: At least 4000 foreign students attend the University of Southern California (USC). Let x represent the number of foreign students attending USC. Then x ≥ 4000.*

77. Foreign students contribute more than $1 billion annually to the California economy.

---

*"State colleges drawing foreigners despite cuts" by Lisa Lapin from *The Sacramento Bee,* December 2, 1992. Copyright, The Sacramento Bee, 1994. Reprinted by permission.

78. More than 60% of the international students come from Asian countries.

79. Less than 7.5% of foreign students now in the United States now come from Middle Eastern countries.

80. No more than 10% of the foreign students in the United States come from Japan.

81. California has more than 13% of all foreign students in the United States.

82. Foreign students must prove they have at least $22,000 in cash to spend here each year.

1.2 POLYNOMIALS

Polynomials are the fundamental tools of algebra and will play a central role in this course. In order to do polynomial arithmetic, you must first understand exponents. So we begin with them. You are familiar with the usual notation for squares and cubes:

$$5^2 = 5 \cdot 5 \quad \text{and} \quad 6^3 = 6 \cdot 6 \cdot 6.$$

We now extend this convenient notation to other cases.

> If n is a natural number and a is any real number, then
>
> $$a^n \quad \text{denotes the product} \quad a \cdot a \cdot a \cdots a \; (n \text{ factors}).$$
>
> The number a is the **base** and n is the **exponent.**

▶ **EXAMPLE 1** 4^6, which is read "four to the sixth," is the number

$$4 \cdot 4 \cdot 4 \cdot 4 \cdot 4 \cdot 4 = 4096.$$

Similarly, $(-5)^3 = (-5)(-5)(-5) = -125$ and

$$\left(\frac{3}{2}\right)^4 = \frac{3}{2} \cdot \frac{3}{2} \cdot \frac{3}{2} \cdot \frac{3}{2} = \frac{81}{16}. \quad ◀$$

▶ **EXAMPLE 2**   Use a calculator to approximate the following.

(a) $(1.2)^8$

Key in 1.2 and then use the x^y key (labeled $\wedge$ on some calculators); finally, key in the exponent 8. The calculator displays the (exact) answer 4.29981696.

(b) $\left(\frac{12}{7}\right)^{23}$

Don't compute 12/7 separately. Use parentheses and key in (12/7), followed by the x^y key and the exponent 23 to obtain the approximate answer 242,054.822.   ◀ ①

① Evaluate the following.

(a) 6^3

(b) 5^{12}

(c) 1^9

(d) $\left(\frac{7}{5}\right)^8$

Answers:

(a) 216

(b) 244,140,625

(c) 1

(d) 14.75789056

②

Evaluate the following.

(a) $3 \cdot 6^2$

(b) $5 \cdot 4^3$

(c) -3^6

(d) $(-3)^6$

(e) $-2 \cdot (-3)^5$

Answers:

(a) 108

(b) 320

(c) −729

(d) 729

(e) 486

CAUTION   A common error in using exponents occurs with expressions such as $4 \cdot 3^2$. The exponent of 2 applies only to the base 3, so that

$$4 \cdot 3^2 = 4 \cdot 3 \cdot 3 = 36.$$

On the other hand,

$$(4 \cdot 3)^2 = (4 \cdot 3)(4 \cdot 3) = 12 \cdot 12 = 144,$$

and so

$$4 \cdot 3^2 \neq (4 \cdot 3)^2.$$

Be careful to distinguish between expressions like -2^4 and $(-2)^4$.

$$-2^4 = -(2^4) = -(2 \cdot 2 \cdot 2 \cdot 2) = -16$$
$$(-2)^4 = (-2)(-2)(-2)(-2) = 16,$$

and so

$$-2^4 \neq (-2)^4. \ ②$$

By the definition of an exponent,

$$3^4 \cdot 3^2 = (3 \cdot 3 \cdot 3 \cdot 3)(3 \cdot 3) = 3^6.$$

This suggests the following property for the product of two powers of a number.

③

Simplify the following.

(a) $5^3 \cdot 5^6$

(b) $(-3)^4 \cdot (-3)^{10}$

(c) $(5p)^2 \cdot (5p)^8$

Answers:

(a) 5^9

(b) $(-3)^{14}$

(c) $(5p)^{10}$

If m and n are natural numbers and a is a real number, then

$$a^m \cdot a^n = a^{m+n}.$$

▶ **EXAMPLE 3**   Simplify the following.

(a) $7^4 \cdot 7^6 = 7^{4+6} = 7^{10}$

(b) $(-2)^3 \cdot (-2)^5 = (-2)^{3+5} = (-2)^8$

(c) $(3k)^2 \cdot (3k)^3 = (3k)^5$

(d) $(m + n)^2 \cdot (m + n)^5 = (m + n)^7$ ◀ ③

Polynomials

A polynomial is an algebraic expression like

$$5x^4 + 2x^3 + 6x, \quad 8m^3 + 9m^2 - 6m + 3, \quad 10p, \quad \text{or} \quad -9.$$

The letter used is called a *variable*. In formal terms, we define a polynomial as follows.

A **polynomial** in one variable x is an expression of the form

$$a_n x^n + a_{n-1} x^{n-1} + \cdots + a_1 x + a_0,$$

where n is a whole number, x is a variable, and $a_0, a_1, \ldots, a_n$ are real numbers (called the **coefficients** of the polynomial).

For example, the polynomial

$$8x^3 + 9x^2 - 6x + 3$$

is of the form

$$a_n x^n + a_{n-1} x^{n-1} + \cdots + a_1 x + a_0,$$

with $n = 3$, $a_n = a_3 = 8$, $a_{n-1} = a_2 = 9$, $a_1 = -6$, and $a_0 = 3$. Each of the expressions $8x^3$, $9x^2$, $-6x$, and 3 is called a **term** of the polynomial $8x^3 + 9x^2 - 6x + 3$. The coefficient a_0 of a polynomial (for instance, 3 in $8x^3 + 9x^2 - 6x + 3$) is called the **constant term.** Letters other than x may be used for the variable of a polynomial.

Only expressions that can be put in the form shown in the preceding box are polynomials. Consequently, the following expressions are *not* polynomials:

$$8x^3 + \frac{6}{x}, \quad \frac{9+x}{2-x}, \quad \text{and} \quad \frac{-p^2 + 5p + 3}{2p - 1}.$$

The **degree of a nonzero term** with only one variable is the exponent on the variable. For example, the term $9p^4$ has degree 4. The **degree of a polynomial** is the highest degree of any of its nonzero terms. Thus, the degree of $-p^2 + 5p + 3$ is 2. The **zero polynomial** consists of the constant term 0 and no other terms. No degree is assigned to the zero polynomial because it has no nonzero terms. A polynomial with two terms, such as $5x + 2$ or $x^3 + 7$, is called a **binomial,** and a polynomial with three terms, such as $3x^2 - 4x + 7$, is called a **trinomial.**

Addition and Subtraction

Two terms having the same variable with the same exponent are called **like terms;** other terms are called **unlike terms.** Polynomials can be added or subtracted by using the distributive property to combine like terms. Only like terms can be combined. For example,

$$12y^4 + 6y^4 = (12 + 6)y^4 = 18y^4$$

and

$$-2m^2 + 8m^2 = (-2 + 8)m^2 = 6m^2.$$

The polynomial $8y^4 + 2y^5$ has unlike terms, so it cannot be further simplified. The next example shows how to add and subtract polynomials by combining terms.

▶ **EXAMPLE 4**　Add or subtract as indicated.

(a) $(8x^3 - 4x^2 + 6x) + (3x^3 + 5x^2 - 9x + 8)$
　　Combine like terms.
$$(8x^3 - 4x^2 + 6x) + (3x^3 + 5x^2 - 9x + 8)$$
　　　Commutative property and associative property
$$= (8x^3 + 3x^3) + (-4x^2 + 5x^2) + (6x - 9x) + 8$$
$$= 11x^3 + x^2 - 3x + 8$$
　　　Distributive property

(b) $(-4x^4 + 6x^3 - 9x^2 - 12) + (-3x^3 + 8x^2 - 11x + 7)$
$$= -4x^4 + 3x^3 - x^2 - 11x - 5$$

Add or subtract.

(a) $(-2x^2 + 7x + 9)$
$+ (3x^2 + 2x - 7)$

(b) $(4x + 6) - (13x - 9)$

(c) $(9x^3 - 8x^2 + 2x)$
$- (9x^3 - 2x^2 - 10)$

Answers:

(a) $x^2 + 9x + 2$

(b) $-9x + 15$

(c) $-6x^2 + 2x + 10$

(c) $(2x^2 - 11x + 8) - (7x^2 - 6x + 2)$

Use the definition of subtraction: $a - b = a + (-b)$. Here a and b are polynomials; a is $2x^2 - 11x + 8$ and $-b$ is

$$-(7x^2 - 6x + 2) = -7x^2 + 6x - 2.$$

Now perform the subtraction.

$$(2x^2 - 11x + 8) - (7x^2 - 6x + 2)$$
$$= (2x^2 - 11x + 8) + (-7x^2 + 6x - 2)$$
$$= -5x^2 - 5x + 6 \quad \blacktriangleleft \quad \text{④}$$

Multiplication

The distributive property is also used to multiply polynomials. For example, the product of $8x$ and $6x - 4$ is found as follows.

$$8x(6x - 4) = 8x(6x) - 8x(4) \qquad \text{Distributive property}$$
$$= 48x^2 - 32x \qquad x \cdot x = x^2$$

⑤

Find the following products.

(a) $-6r(2r - 5)$

(b) $(8m + 3)(m^4 - 2m^2 + 6m)$

Answers:

(a) $-12r^2 + 30r$

(b) $8m^5 + 3m^4 - 16m^3$
$+ 42m^2 + 18m$

▶ **EXAMPLE 5** Find each product.

(a) $2p^3(3p^2 - 2p + 5) = 2p^3(3p^2) + 2p^3(-2p) + 2p^3(5)$
$$= 6p^5 - 4p^4 + 10p^3$$

(b) $(3k - 2)(k^2 + 5k - 4) = 3k(k^2 + 5k - 4) - 2(k^2 + 5k - 4)$
$$= 3k^3 + 15k^2 - 12k - 2k^2 - 10k + 8$$
$$= 3k^3 + 13k^2 - 22k + 8 \quad \blacktriangleleft \quad \text{⑤}$$

▶ **EXAMPLE 6** The product $(2x - 5)(3x + 4)$ can be found by using the distributive property twice.

$$(2x - 5)(3x + 4) = 2x(3x + 4) - 5(3x + 4)$$
$$= 2x \cdot 3x + 2x \cdot 4 + (-5) \cdot 3x + (-5) \cdot 4$$
$$= 6x^2 + \underbrace{8x - 15x}_{} - 20$$
$$= 6x^2 - \quad 7x \quad - 20 \quad \blacktriangleleft$$

Observe the pattern in the second line of Example 6 and its relationship to the terms being multiplied.

$$(2x - 5)(3x + 4) = 2x \cdot 3x + 2x \cdot 4 + (-5) \cdot 3x + (-5) \cdot 4$$

First terms

$(2x - 5)(3x + 4)$

Outside terms

$(2x - 5)(3x + 4)$

Inside terms

$(2x - 5)(3x + 4)$

Last terms

This pattern is easy to remember by using the acronym **FOIL** (First, Outside, Inside, Last). The FOIL method makes it easy to find products such as this one mentally, without the necessity of writing out the intermediate steps.

⑥

Use FOIL to find these products.

(a) $(5k - 1)(2k + 3)$

(b) $(7z - 3)(2z + 5)$

Answers:

(a) $10k^2 + 13k - 3$

(b) $14z^2 + 29z - 15$

▶ EXAMPLE 7

$$(3x + 2)(x + 5) = \underset{\underset{\text{First}}{\uparrow}}{3x^2} + \underset{\underset{\text{Outside}}{\uparrow}}{15x} + \underset{\underset{\text{Inside}}{\uparrow}}{2x} + \underset{\underset{\text{Last}}{\uparrow}}{10} = 3x^2 + 17x + 10$$

◀ ⑥

In business, the *revenue* from the sales of an item is given by

Revenue = (price per item) × (number of items sold).

The *cost* to manufacture and sell these items is given by

Cost = Fixed Costs + Variable Costs,

where the fixed costs include such things as buildings and machinery (which don't depend on how many items are made) and variable costs include such things as labor and materials (which vary, depending on how many items are made). Then

Profit = Revenue − Cost.

▶ EXAMPLE 8

Hot Rocks Music sells CDs for $8 each (wholesale) and can produce a maximum of 200,000 CDs. The variable cost of producing x thousand CDs is $3550x - 8x^2$ dollars, and the fixed costs for the manufacturing operation are $215,000. If x thousand CDs are manufactured and sold, find expressions for the revenue, cost, and profit.

If x thousand CDs are sold at $8 each, then

$$\text{Revenue} = \text{(price per item)} \times \text{(number of items sold)}$$
$$R = 8 \times 1000x = 8000x,$$

where $x \leq 200$ (because only 200,000 CDs can be made). The variable cost of making x thousand CDs is $3550x - 8x^2$, so that

$$\text{Cost} = \text{Fixed Costs} + \text{Variable Costs}$$
$$C = 215,000 + (3550x - 8x^2) \qquad (x \leq 200)$$

Therefore, the profit is given by

$$P = R - C = 8000x - (215,000 + 3550x - 8x^2)$$
$$= 8000x - 215,000 - 3550x + 8x^2$$
$$P = 8x^2 + 4450x - 215,000 \qquad (x \leq 200). \quad ◀ ⑦$$

⑦

Suppose revenue is given by $7x^2 - 3x$, fixed costs are $500, and variable costs are given by $3x^2 + 5x - 25$. Write an expression for

(a) Cost

(b) Profit.

Answers:

(a) $C = 3x^2 + 5x + 475$

(b) $P = 4x^2 - 8x - 475$

1.2 Exercises

Use a calculator to approximate these numbers. (See Examples 1 and 2.)

1. 12.1^7

2. $(-6.54)^{11}$

3. $(-18/7)^6$

4. $(7/9)^8$

5. Explain how the value of -3^2 differs from $(-3)^2$. Do -3^3 and $(-3)^3$ differ in the same way? Why or why not?

6. Describe the steps used to multiply 4^3 and 4^5. Is the product of 4^3 and 3^4 found in the same way? Explain.

Simplify each of the following. Leave answers with exponents. (See Example 3.)

7. $2^4 \cdot 2^3$

8. $3^8 \cdot 3^3$

9. $(-5)^2 \cdot (-5)^5$

10. $(-4)^4 \cdot (-4)^6$

11. $(2z)^5 \cdot (2z)^6$

12. $(6y)^3 \cdot (6y)^5$

Add or subtract as indicated. (See Example 4.)

13. $(3x^3 + 2x^2 - 5x) + (-4x^3 - x^2 + 8x)$

14. $(-2p^3 - 5p + 7) + (-4p^2 + 8p + 2)$

15. $(-4y^2 - 3y + 8) - (2y^2 - 6y - 2)$

16. $(7b^2 + 2b - 5) - (3b^2 + 2b - 6)$

17. $(2x^3 - 2x^2 + 4x - 3) - (2x^3 + 8x^2 - 1)$

18. $(3y^3 + 9y^2 - 11y + 8) - (-4y^2 + 10y - 6)$

Find each of the following. (See Examples 5–7.)

19. $-9m(2m^2 + 3m - 1)$

20. $2a(4a^2 - 6a + 3)$

21. $(3z + 5)(4z^2 - 2z + 1)$

22. $(2k + 3)(4k^3 - 3k^2 + k)$

23. $(6k - 1)(2k - 3)$

24. $(8r + 3)(r - 1)$

25. $(3y + 5)(2y - 1)$

26. $(5r - 3s)(5r + 4s)$

27. $(9k + q)(2k - q)$

28. $(.012x - .17)(.3x + .54)$

29. $(6.2m - 3.4)(.7m + 1.3)$

30. $2p - 3[4p - (3p + 1)]$

31. $5k - [k + (-3 + 5k)]$

32. $(3x - 1)(x + 2) - (2x + 5)^2$

Business *Find expressions for the Revenue, Cost, and Profit from selling x thousand items. (See Example 8.)*

	Item Price	Fixed Costs	Variable Costs
33.	$5.00	$150,000	$2250x$
34.	$7.95	$220,000	$4300x$
35.	$7.50	$300,000	$-3x^2 + 3480x - 325$
36.	$12.75	$450,000	$-4x^2 + 2880x - 295$

Business *The bar graph below shows the number (in millions) of U.S. households that are on-line, as reported by Jupiter Communications. The polynomial*

$$-.264x^2 + 6.84x + 8.9$$

gives a good approximation of the number of on-line households in year x , where $x = 0$ corresponds to 1995, $x = 1$ to 1996, and so on. For each year

(a) *use the bar graph to determine the number of on-line households;*

(b) *use the polynomial to determine the number of on-line households.*

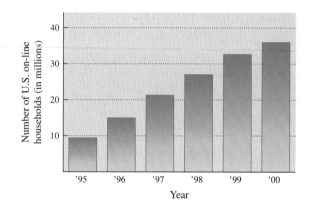

37. 1995 **38.** 1997 **39.** 1999 **40.** 2000

Assuming the polynomial approximation remains accurate, estimate the number of on-line households in each of these years.

41. 2002 **42.** 2003 **43.** 2004 **44.** 2005

45. **Physical Science** One of the most amazing formulas in all of ancient mathematics is the formula discovered by the Egyptians to find the volume of the frustum of a square pyramid, as shown in the figure. Its volume is given by $(1/3)h(a^2 + ab + b^2)$, where b is the length of the base, a is the length of the top, and h is the height.*

---

*H. A. Freebury, *A History of Mathematics*. (New York: MacMillan Company, 1968).

(a) When the Great Pyramid in Egypt was partially completed to a height h of 200 feet, b was 756 feet and a was 314 feet. Calculate its volume at this stage of construction.

(b) Try to visualize the figure if $a = b$. What is the resulting shape? Find its volume.

(c) Let $a = b$ in the Egyptian formula and simplify. Are the results the same?

46. Physical Science   Refer to the formula and the discussion in Exercise 45.

(a) Use the expression $(1/3)h(a^2 + ab + b^2)$ to determine a formula for the volume of a pyramid with a square base b and height h by letting $a = 0$.

(b) The Great Pyramid in Egypt had a square base of 756 feet and a height of 481 feet. Find the volume of the Great Pyramid. Compare it with the 273-foot-tall Louisiana Superdome, which has an approximate volume of 125 million cubic feet.*

(c) The Superdome covers an area of 13 acres. How many acres does the Great Pyramid cover? (*Hint:* 1 acre = 43,560 ft^2.)

---

*Louisiana Superdome (www.superdome.com).

47. Suppose one polynomial has degree 3 and another also has degree 3. Find all possible values for the degree of their

(a) sum

(b) difference

(c) product.

Business   *The following exercise is suitable for group work. A graphing calculator is needed.*

48. (a) Use the table feature to make a table of values for the profit function in Example 8, using $x = 0, 5, 10,...,$ 225. Use the table to answer the following questions.

(b) What is the profit or loss (= negative profit) when 20,000 CDs are sold? When 40,000 are sold? Use data given in Example 8 to explain these answers.

(c) Approximately how many CDs must be sold in order for the company to make a profit?

(d) What is the profit from selling 100,000 CDs? From 150,000 CDs? From 200,000 CDs?

(e) Explain why the profit amounts shown in the table for 205,000 CDs and beyond are meaningless.

1.3 FACTORING

The number 18 can be written as a product in several ways: $9 \cdot 2$, $(-3)(-6)$, $1 \cdot 18$, etc. The numbers in each product (9, 2, -3, etc.) are called **factors** and the process of writing 18 as a product of factors is called **factoring.** Thus, factoring is the reverse of multiplication.

Factoring of polynomials provides a means to simplify many expressions and to solve certain types of equations. As is the usual custom, factoring of polynomials in this book will be restricted to finding factors with *integer* coefficients (otherwise there is an infinite number of possible factors).

Greatest Common Factor

The algebraic expression $15m + 45$ is made up of two terms, $15m$ and 45. Each of these terms has 15 as a factor. In fact, $15m = 15 \cdot m$ and $45 = 15 \cdot 3$. By the distributive property,

$$15m + 45 = 15 \cdot m + 15 \cdot 3 = 15(m + 3).$$

Both 15 and $m + 3$ are factors of $15m + 45$. Since 15 divides into all terms of $15m + 45$ and is the largest number that will do so, it is called the **greatest common factor** for the polynomial $15m + 45$. The process of writing $15m + 45$ as $15(m + 3)$ is called **factoring out** the greatest common factor.

▶ **EXAMPLE 1** Factor out the greatest common factor.

(a) $12p - 18q$

Both $12p$ and $18q$ are divisible by 6, and

$$12p - 18q = 6 \cdot 2p - 6 \cdot 3q$$
$$= 6(2p - 3q).$$

(b) $8x^3 - 9x^2 + 15x$

Each of these terms is divisible by x.

$$8x^3 - 9x^2 + 15x = (8x^2) \cdot x - (9x) \cdot x + 15 \cdot x$$
$$= x(8x^2 - 9x + 15)$$

(c) $5(4x - 3)^3 + 2(4x - 3)^2$

The quantity $(4x - 3)^2$ is a common factor. Factoring it out gives

$$5(4x - 3)^3 + 2(4x - 3)^2 = (4x - 3)^2[5(4x - 3) + 2]$$
$$= (4x - 3)^2(20x - 15 + 2)$$
$$= (4x - 3)^2(20x - 13). \quad ◀ ①$$

①

Factor out the greatest common factor.

(a) $12r + 9k$

(b) $75m^2 + 100n^2$

(c) $6m^4 - 9m^3 + 12m^2$

(d) $3(2k + 1)^3 + 4(2k + 1)^4$

Answers:

(a) $3(4r + 3k)$

(b) $25(3m^2 + 4n^2)$

(c) $3m^2(2m^2 - 3m + 4)$

(d) $(2k + 1)^3(7 + 8k)$

Factoring Trinomials

Factoring is the opposite of multiplication. Because the product of two binomials is usually a trinomial, we can expect factorable trinomials (that have terms with no common factor) to have two binomial factors. Thus, factoring trinomials requires using FOIL backward.

▶ **EXAMPLE 2** Factor $x^2 + 9x + 18$.

We must find integers b and d such that

$$x^2 + 9x + 18 = (x + b)(x + d)$$
$$= x^2 + dx + bx + bd$$
$$= x^2 + (b + d)x + bd.$$

Since the constant coefficients on each side of the equation must be equal, we must have $bd = 18$; that is, b and d are factors of 18. Similarly, the coefficients of x must be the same, so that $b + d = 9$. The possibilities are summarized in this table.

②

Factor the following.

(a) $r^2 + 7r + 10$

(b) $x^2 + 4x + 3$

(c) $y^2 + 6y + 8$

Answers:

(a) $(r + 2)(r + 5)$

(b) $(x + 3)(x + 1)$

(c) $(y + 2)(y + 4)$

Factors b, d of 18	Sum $b + d$
$18 \cdot 1$	$18 + 1 = 19$
$9 \cdot 2$	$9 + 2 = 11$
$6 \cdot 3$	$6 + 3 = 9$

There's no need to list negative factors, such as $(-3)(-6)$, because their sum is negative. The table suggests that 6 and 3 will work. Verify that

$$(x + 6)(x + 3) = x^2 + 9x + 18. \quad ◀ ②$$

Factor the following.

(a) $x^2 - 4x + 3$

(b) $2y^2 - 5y + 2$

(c) $6z^2 - 13z + 6$

Answers:

(a) $(x - 3)(x - 1)$

(b) $(2y - 1)(y - 2)$

(c) $(3z - 2)(2z - 3)$

▶ EXAMPLE 3   Factor $4y^2 - 11y + 6$.

We must find integers a, b, c, and d such that
$$4y^2 - 11y + 6 = (ay + b)(cy + d)$$
$$= acy^2 + ady + bcy + bd$$
$$= acy^2 + (ad + bc)y + bd.$$

Since the coefficients of y^2 must be the same on both sides, we see that $ac = 4$. Similarly, the constant terms show that $bd = 6$. The positive factors of 4 are 4 and 1 or 2 and 2. Since the middle term is negative, we consider only negative factors of 6. The possibilities are -2 and -3 or -1 and -6. Now we try various arrangements of these factors until we find one that gives the correct coefficient of y.

$$(2y - 1)(2y - 6) = 4y^2 - \mathbf{14}y + 6 \qquad \text{Incorrect}$$
$$(2y - 2)(2y - 3) = 4y^2 - \mathbf{10}y + 6 \qquad \text{Incorrect}$$
$$(y - 2)(4y - 3) = 4y^2 - \mathbf{11}y + 6 \qquad \text{Correct}$$

The last trial gives the correct factorization.   ◀ ③

▶ EXAMPLE 4   Factor $6p^2 - 7pq - 5q^2$.

Again, we try various possibilities. The positive factors of 6 could be 2 and 3 or 1 and 6. As factors of -5 we have only -1 and 5 or -5 and 1. Try different combinations of these factors until the correct one is found.

$$(2p - 5q)(3p + q) = 6p^2 - \mathbf{13}pq - 5q^2 \qquad \text{Incorrect}$$
$$(3p - 5q)(2p + q) = 6p^2 - \mathbf{7}pq - 5q^2 \qquad \text{Correct}$$

Factor the following.

(a) $r^2 - 5r - 14$

(b) $3m^2 + 5m - 2$

(c) $6p^2 + 13pq - 5q^2$

Answers:

(a) $(r - 7)(r + 2)$

(b) $(3m - 1)(m + 2)$

(c) $(2p + 5q)(3p - q)$

Finally, $6p^2 - 7pq - 5q^2$ factors as $(3p - 5q)(2p + q)$.   ◀ ④

NOTE   In Examples 2–4, we chose positive factors of the positive first term. Of course, we could have used two negative factors, but the work is easier if positive factors are used.

The method shown above can be used to factor a **perfect square trinomial**, one that is the square of a binomial. The binomial can be predicted by observing the following patterns that always apply to a perfect square trinomial.

$$\left.\begin{array}{l} x^2 + 2xy + y^2 = (x + y)^2 \\ x^2 - 2xy + y^2 = (x - y)^2 \end{array}\right\} \text{Perfect square trinomials}$$

▶ EXAMPLE 5   Factor each trinomial.

(a) $16p^2 - 40pq + 25q^2$

Because $16p^2 = (4p)^2$ and $25q^2 = (5q)^2$, use the second pattern shown above with $4p$ replacing x and $5q$ replacing y to get

$$16p^2 - 40pq + 25q^2 = (4p)^2 - 2(\mathbf{4p})(\mathbf{5q}) + (\mathbf{5q})^2$$
$$= (4p - 5q)^2.$$

Factor each trinomial.

(a) $4m^2 + 4m + 1$

(b) $25z^2 - 80zt + 64t^2$

Answers:

(a) $(2m + 1)^2$

(b) $(5z - 8t)^2$

Make sure that the middle term of the trinomial being factored, $-40pq$ here, is twice the product of the two terms in the binomial $4p - 5q$.

$$-40pq = 2(4p)(-5q)$$

(b) $169x^2 + 104xy^2 + 16y^4 = (13x + 4y^2)^2$, since $2(13x)(4y^2) = 104xy^2$. ◀

Factoring Binomials

Three special factoring patterns are listed below. Each can be verified by multiplying on the right side of the equation. These formulas should be memorized.

$x^2 - y^2 = (x + y)(x - y)$	Difference of two squares
$x^3 - y^3 = (x - y)(x^2 + xy + y^2)$	Difference of two cubes
$x^3 + y^3 = (x + y)(x^2 - xy + y^2)$	Sum of two cubes

▶ **EXAMPLE 6**   Factor each of the following.

(a) $4m^2 - 9$

Notice that $4m^2 - 9$ is the difference of two squares, since $4m^2 = (2m)^2$ and $9 = 3^2$. Use the pattern for the difference of two squares, letting $2m$ replace x and 3 replace y. Then the pattern $x^2 - y^2 = (x + y)(x - y)$ becomes

$$4m^2 - 9 = (2m)^2 - 3^2$$
$$= (2m + 3)(2m - 3).$$

(b) $128p^2 - 98q^2$

First factor out the common factor of 2.

$$128p^2 - 98q^2 = 2(64p^2 - 49q^2)$$
$$= 2[(8p)^2 - (7q)^2]$$
$$= 2(8p + 7q)(8p - 7q)$$

(c) $x^2 + 36$

The *sum* of two squares usually cannot be factored. To see this, check some possibilities.

$$(x + 6)(x + 6) = (x + 6)^2 = x^2 + 12x + 36$$
$$(x + 4)(x + 9) = x^2 + 13x + 36$$

Any product of two binomials will always have a middle term unless it is the *difference* of two squares.

(d) $4z^2 + 12z + 9 - w^2$

Notice that the first three terms can be factored as a perfect square.

$$4z^2 + 12z + 9 - w^2 = (2z + 3)^2 - w^2$$

Factor the following.

(a) $9p^2 - 49$

(b) $y^2 + 100$

(c) $9r^2 + 12r + 4 - t^2$

(d) $81x^4 - 16y^4$

Answers:

(a) $(3p + 7)(3p - 7)$

(b) Cannot be factored

(c) $(3r + 2 + t)(3r + 2 - t)$

(d) $(9x^2 + 4y^2)(3x + 2y) \cdot$
$(3x - 2y)$

Written in this form, the expression is the difference of squares, which can be factored as

$$(2z + 3)^2 - w^2 = [(2z + 3) + w][(2z + 3) - w]$$
$$= (2z + 3 + w)(2z + 3 - w).$$

(e) $256k^4 - 625m^4$

Use the difference of two squares pattern twice, as follows:

$$256k^4 - 625m^4 = (16k^2)^2 - (25m^2)^2$$
$$= (16k^2 + 25m^2)(16k^2 - 25m^2)$$
$$= (16k^2 + 25m^2)(4k + 5m)(4k - 5m). \blacktriangleleft \ ⑥$$

▶ **EXAMPLE 7** Factor each of the following.

(a) $k^3 - 8$

Use the pattern for the difference of two cubes, since $8 = 2^3$, to get

$$k^3 - 8 = k^3 - 2^3 = (k - 2)(k^2 + 2k + 4).$$

(b) $m^3 + 125 = m^3 + 5^3 = (m + 5)(m^2 - 5m + 25)$

(c) $8k^3 - 27z^3 = (2k)^3 - (3z)^3 = (2k - 3z)(4k^2 + 6kz + 9z^2) \blacktriangleleft \ ⑦$

Factor the following.

(a) $a^3 + 1000$

(b) $z^3 - 64$

(c) $1000m^3 - 27z^3$

Answers:

(a) $(a + 10)(a^2 - 10a + 100)$

(b) $(z - 4)(z^2 + 4z + 16)$

(c) $(10m - 3z) \cdot$
$(100m^2 + 30mz + 9z^2)$

▶ **EXAMPLE 8** Factor each of the following.

(a) $12x^2 - 26x - 10$

Look first for a common factor. Here there is a common factor of 2: $12x^2 - 26x - 10 = 2(6x^2 - 13x - 5)$. Now try to factor $6x^2 - 13x - 5$. Possible factors of 6 are 3 and 2 or 6 and 1. The only factors of -5 are -5 and 1 or 5 and -1. Try various combinations. You should find the trinomial factors as $(3x + 1)(2x - 5)$. Thus,

$$12x^2 - 26x - 10 = 2(3x + 1)(2x - 5).$$

(b) $16a^2 - 100 - 48ac + 36c^2$

Factor out the common factor of 4 first.

$$16a^2 - 100 - 48ac + 36c^2 = 4[4a^2 - 25 - 12ac + 9c^2]$$
$$= 4[(4a^2 - 12ac + 9c^2) - 25] \quad \text{Rearrange terms and group.}$$
$$= 4[(2a - 3c)^2 - 25] \quad \text{Factor the trinomial.}$$
$$= 4(2a - 3c + 5)(2a - 3c - 5) \quad \text{Factor the difference of squares.}$$

Factor.

(a) $6x^2 - 27x - 15$

(b) $18 - 8xy - 2y^2 - 8x^2$

Answers:

(a) $3(2x + 1)(x - 5)$

(b) $2(3 - 2x - y)(3 + 2x + y)$

CAUTION Remember always to look first for a common factor. ◀ ⑧

1.3 Exercises

Factor out the greatest common factor in each of the following. (See Example 1.)

1. $12x^2 - 24x$

2. $5y - 25xy$

3. $r^3 - 5r^2 + r$

4. $t^3 + 3t^2 + 8t$

5. $6z^3 - 12z^2 + 18z$

6. $5x^3 + 35x^2 + 10x$

7. $3(2y - 1)^2 + 5(2y - 1)^3$

8. $(3x + 7)^5 - 2(3x + 7)^3$

9. $3(x + 5)^4 + (x + 5)^6$

10. $3(x + 6)^2 + 2(x + 6)^4$

Factor the polynomial. (See Example 2.)

11. $x^2 + 5x + 4$

12. $u^2 + 7u + 6$

13. $x^2 + 7x + 12$

14. $y^2 + 8y + 12$

15. $z^2 + 10z + 24$

16. $r^2 + 16r + 60$

Factor the polynomial. (See Examples 3 and 4.)

17. $2x^2 - 9x + 4$

18. $3w^2 - 8w + 4$

19. $15p^2 - 23p + 4$

20. $8x^2 - 14x + 3$

21. $4z^2 - 16z + 15$

22. $12y^2 - 29y + 15$

23. $6x^2 - 5x - 4$

24. $12z^2 + z - 1$

25. $10y^2 + 21y - 10$

26. $15u^2 + 4u - 4$

27. $6x^2 + 5x - 4$

28. $12y^2 + 7y - 10$

Factor each polynomial completely. Factor out the greatest common factor as necessary. (See Examples 2–6 and 8.)

29. $2a^2 + 3a - 5$

30. $6a^2 - 48a - 120$

31. $x^2 - 64$

32. $x^2 + 17xy + 72y^2$

33. $9p^2 - 24p + 16$

34. $3r^2 - r - 2$

35. $r^2 - 3rt - 10t^2$

36. $2a^2 + ab - 6b^2$

37. $m^2 - 6mn + 9n^2$

38. $8k^2 - 16k - 10$

39. $4p^2 - 9$

40. $8r^2 + r + 6$

41. $3x^2 - 24xz + 48z^2$

42. $9m^2 - 25$

43. $a^2 + 4ab + 5b^2$

44. $6y^2 - 11y - 7$

45. $-x^2 + 7x - 12$

46. $4y^2 + y - 3$

47. $3a^2 - 13a - 30$

48. $3k^2 + 2k - 8$

49. $21m^2 + 13mn + 2n^2$

50. $81y^2 - 100$

51. $20y^2 + 39yx - 11x^2$

52. $12s^2 + 11st - 5t^2$

53. $y^2 - 4yz - 21z^2$

54. $49a^2 + 9$

55. $121x^2 - 64$

56. $4z^2 + 56zy + 196y^2$

57. $5m^3(m^3 - 1)^2 - 3m^5(m^3 - 1)^3$

58. $9(x - 4)^5 - (x - 4)^3$

Factor each binomial. (See Example 7.)

59. $a^3 - 216$

60. $b^3 + 125$

61. $8r^3 - 27s^3$

62. $1000p^3 + 27q^3$

63. $64m^3 + 125$

64. $216y^3 - 343$

65. $1000y^3 - z^3$

66. $125p^3 + 8q^3$

67. When asked to factor $6x^4 - 3x^2 - 3$ completely, a student gave the following result:
$$6x^4 - 3x^2 - 3 = (2x^2 + 1)(3x^2 - 3).$$
Is this answer correct? Explain why.

68. When can the sum of two squares be factored? Give examples.

69. Explain why $(x + 2)^3$ is not the correct factorization of $x^3 + 8$ and give the correct factorization.

70. Describe how factoring and multiplication are related. Give examples.

1.4 RATIONAL EXPRESSIONS

We now consider **rational expressions,** such as

$$\frac{8}{x - 1}, \quad \frac{3x^2 + 4x}{5x - 6}, \quad \text{and} \quad \frac{2 + \dfrac{1}{y}}{y}.$$

Because rational expressions involve quotients, it is important to keep in mind values of the variables that make denominators 0. For example, 1 cannot be used as a

1

What values of the variable make each denominator equal 0?

(a) $\dfrac{5}{x - 3}$

(b) $\dfrac{2x - 3}{4x - 1}$

(c) $\dfrac{x + 2}{x}$

(d) Why do we need to determine these values?

Answers:

(a) 3

(b) 1/4

(c) 0

(d) Because division by 0 is undefined.

replacement for x in the first rational expression above, and 6/5 cannot be used in the second one, since these values make the respective denominators equal 0. **1**

Operations with Rational Expressions

The rules for operations with rational expressions are the usual rules for fractions.

Operations with Rational Expressions

For all mathematical expressions P, $Q \neq 0$, R, and $S \neq 0$,

(a) $\dfrac{P}{Q} = \dfrac{PS}{QS}$ Fundamental property

(b) $\dfrac{P}{Q} \cdot \dfrac{R}{S} = \dfrac{PR}{QS}$ Multiplication

(c) $\dfrac{P}{Q} + \dfrac{R}{Q} = \dfrac{P + R}{Q}$ Addition

(d) $\dfrac{P}{Q} - \dfrac{R}{Q} = \dfrac{P - R}{Q}$ Subtraction

(e) $\dfrac{P}{Q} \div \dfrac{R}{S} = \dfrac{P}{Q} \cdot \dfrac{S}{R}$, $R \neq 0$.   Division

The following examples illustrate these operations.

▶ **EXAMPLE 1** Write each of the following rational expressions in lowest terms (so that the numerator and denominator have no common factor with integer coefficients except 1 or -1).

(a) $\dfrac{12m}{-18}$

Both $12m$ and -18 are divisible by 6. By operation (a) above,

$$\frac{12m}{-18} = \frac{2m \cdot 6}{-3 \cdot 6}$$

$$= \frac{2m}{-3}$$

$$= -\frac{2m}{3}.$$

2

Write each of the following in lowest terms.

(a) $\dfrac{12k + 36}{18}$

(b) $\dfrac{15m + 30m^2}{5m}$

(c) $\dfrac{2p^2 + 3p + 1}{p^2 + 3p + 2}$

Answers:

(a) $\dfrac{2(k + 3)}{3}$ or $\dfrac{2k + 6}{3}$

(b) $3(1 + 2m)$ or $3 + 6m$

(c) $\dfrac{2p + 1}{p + 2}$

(b) $\dfrac{8x + 16}{4} = \dfrac{8(x + 2)}{4} = \dfrac{4 \cdot 2(x + 2)}{4} = \dfrac{2(x + 2)}{1} = 2(x + 2)$

The numerator, $8x + 16$, was factored so that the common factor could be identified. The answer could also be written as $2x + 4$, if desired.

(c) $\dfrac{k^2 + 7k + 12}{k^2 + 2k - 3} = \dfrac{(k + 4)(k + 3)}{(k - 1)(k + 3)} = \dfrac{k + 4}{k - 1}$ ◀ **2**

The values of k in Example 1(c) are restricted to $k \neq 1$ and $k \neq -3$. From now on, such restrictions will be assumed when working with rational expressions.

▶ **EXAMPLE 2**

(a) Multiply $\dfrac{2}{3} \cdot \dfrac{y}{5}$.

Use operation (b). Multiply the numerators and then the denominators.

$$\frac{2}{3} \cdot \frac{y}{5} = \frac{2 \cdot y}{3 \cdot 5} = \frac{2y}{15}$$

The result, $2y/15$, is in lowest terms.

(b) $\dfrac{3y + 9}{6} \cdot \dfrac{18}{5y + 15}$

Factor where possible.

$$\frac{3y + 9}{6} \cdot \frac{18}{5y + 15} = \frac{3(y + 3)}{6} \cdot \frac{18}{5(y + 3)}$$

$$= \frac{3 \cdot 18(y + 3)}{6 \cdot 5(y + 3)} \qquad \text{Multiply numerators and denominators.}$$

$$= \frac{3 \cdot 6 \cdot 3(y + 3)}{6 \cdot 5(y + 3)} \qquad 18 = 6 \cdot 3$$

$$= \frac{3 \cdot 3}{5} \qquad \text{Write in lowest terms.}$$

$$= \frac{9}{5}$$

③

Multiply.

(a) $\dfrac{3r^2}{5} \cdot \dfrac{20}{9r}$

(b) $\dfrac{y - 4}{y^2 - 2y - 8} \cdot \dfrac{y^2 - 4}{3y}$

Answers:

(a) $\dfrac{4r}{3}$

(b) $\dfrac{y - 2}{3y}$

(c) $\dfrac{m^2 + 5m + 6}{m + 3} \cdot \dfrac{m^2 + m - 6}{m^2 + 3m + 2}$

$$= \frac{(m + 2)(m + 3)}{m + 3} \cdot \frac{(m - 2)(m + 3)}{(m + 2)(m + 1)} \qquad \text{Factor.}$$

$$= \frac{(m + 2)(m + 3)(m - 2)(m + 3)}{(m + 3)(m + 2)(m + 1)} \qquad \text{Multiply.}$$

$$= \frac{(m - 2)(m + 3)}{m + 1} \qquad \text{Lowest terms}$$

$$= \frac{m^2 + m - 6}{m + 1} \qquad ◀ ③$$

▶ **EXAMPLE 3**

(a) Divide $\dfrac{8x}{5} \div \dfrac{11x^2}{20}$.

④

Divide.

(a) $\dfrac{5m}{16} \div \dfrac{m^2}{10}$

(b) $\dfrac{2y - 8}{6} \div \dfrac{5y - 20}{3}$

(c) $\dfrac{m^2 - 2m - 3}{m(m + 1)} \div \dfrac{m + 4}{5m}$

Answers:

(a) $\dfrac{25}{8m}$

(b) $\dfrac{1}{5}$

(c) $\dfrac{5(m - 3)}{m + 4}$

Invert the second expression and multiply [operation (e)].

$$\frac{8x}{5} \div \frac{11x^2}{20} = \frac{8x}{5} \cdot \frac{20}{11x^2} \qquad \text{Invert and multiply.}$$

$$= \frac{8x \cdot 20}{5 \cdot 11x^2} \qquad \text{Multiply.}$$

$$= \frac{32}{11x} \qquad \text{Lowest terms}$$

(b) $\dfrac{9p - 36}{12} \div \dfrac{5(p - 4)}{18}$

$$= \frac{9p - 36}{12} \cdot \frac{18}{5(p - 4)} \qquad \text{Invert and multiply.}$$

$$= \frac{9(p - 4)}{12} \cdot \frac{18}{5(p - 4)} \qquad \text{Factor.}$$

$$= \frac{27}{10} \qquad \text{Multiply and write in lowest terms.} \quad ◀ \quad ④$$

▶ **EXAMPLE 4**   Add or subtract as indicated.

(a) $\dfrac{4}{5k} - \dfrac{11}{5k}$

When two rational expressions have the same denominators, subtract by subtracting the numerators and keeping the common denominator. [See operation (d).]

$$\frac{4}{5k} - \frac{11}{5k} = \frac{4 - 11}{5k} = -\frac{7}{5k}$$

(b) $\dfrac{7}{p} + \dfrac{9}{2p} + \dfrac{1}{3p}.$

These three denominators are different; addition requires the same denominators. Find a common denominator, one which can be divided by p, $2p$, and $3p$. A common denominator here is $6p$. Rewrite each rational expression, using operation (a), with a denominator of $6p$. Then, using operation (c), add the numerators and keep the common denominator.

$$\frac{7}{p} + \frac{9}{2p} + \frac{1}{3p} = \frac{6 \cdot 7}{6 \cdot p} + \frac{3 \cdot 9}{3 \cdot 2p} + \frac{2 \cdot 1}{2 \cdot 3p} \qquad \text{Operation (a)}$$

$$= \frac{42}{6p} + \frac{27}{6p} + \frac{2}{6p}$$

$$= \frac{42 + 27 + 2}{6p} \qquad \text{Operation (c)}$$

$$= \frac{71}{6p}$$

(c) $\dfrac{k^2}{k^2 - 1} - \dfrac{2k^2 - k - 3}{k^2 + 3k + 2}$

Factor the denominators to find a common denominator.

$$\frac{k^2}{k^2 - 1} - \frac{2k^2 - k - 3}{k^2 + 3k + 2} = \frac{k^2}{(k + 1)(k - 1)} - \frac{2k^2 - k - 3}{(k + 1)(k + 2)}$$

The common denominator is $(k + 1)(k - 1)(k + 2)$. Write each fraction with the common denominator.

$$\frac{k^2}{(k + 1)(k - 1)} - \frac{2k^2 - k - 3}{(k + 1)(k + 2)}$$

$$= \frac{k^2(k + 2)}{(k + 1)(k - 1)(k + 2)} - \frac{(2k^2 - k - 3)(k - 1)}{(k + 1)(k - 1)(k + 2)}$$

$$= \frac{k^3 + 2k^2 - (2k^2 - k - 3)(k - 1)}{(k + 1)(k - 1)(k + 2)} \qquad \text{Subtract fractions.}$$

$$= \frac{k^3 + 2k^2 - (2k^3 - 3k^2 - 2k + 3)}{(k + 1)(k - 1)(k + 2)} \qquad \text{Multiply } (2k^2 - k - 3)(k - 1).$$

$$= \frac{k^3 + 2k^2 - 2k^3 + 3k^2 + 2k - 3}{(k + 1)(k - 1)(k + 2)} \qquad \text{Polynomial subtraction}$$

$$= \frac{-k^3 + 5k^2 + 2k - 3}{(k + 1)(k - 1)(k + 2)} \qquad \text{Combine terms.} \quad \blacktriangleleft \; \text{⑤}$$

---

⑤

Add or subtract.

(a) $\dfrac{3}{4r} + \dfrac{8}{3r}$

(b) $\dfrac{1}{m - 2} - \dfrac{3}{2(m - 2)}$

(c) $\dfrac{p + 1}{p^2 - p} - \dfrac{p^2 - 1}{p^2 + p - 2}$

Answers:

(a) $\dfrac{41}{12r}$

(b) $\dfrac{-1}{2(m - 2)}$

(c) $\dfrac{-p^3 + p^2 + 4p + 2}{p(p - 1)(p + 2)}$

Complex Fractions

Any quotient of two rational expressions is called a **complex fraction.** Complex fractions are simplified as follows.

▶ **EXAMPLE 5**   Simplify each complex fraction.

(a) $\dfrac{6 - \dfrac{5}{k}}{1 + \dfrac{5}{k}}$

Multiply both numerator and denominator by the common denominator k.

$$\frac{6 - \dfrac{5}{k}}{1 + \dfrac{5}{k}} = \frac{k\left(6 - \dfrac{5}{k}\right)}{k\left(1 + \dfrac{5}{k}\right)} \qquad \text{Multiply by } \frac{k}{k}.$$

$$= \frac{6k - k\left(\dfrac{5}{k}\right)}{k + k\left(\dfrac{5}{k}\right)} \qquad \text{Distributive property}$$

$$= \frac{6k - 5}{k + 5} \qquad \text{Simplify.}$$

(b) $\dfrac{\dfrac{a}{a+1}+\dfrac{1}{a}}{\dfrac{1}{a}+\dfrac{1}{a+1}}$

Multiply both numerator and denominator by the common denominator of all the fractions, in this case $a(a+1)$. Doing so gives

$$\frac{\dfrac{a}{a+1}+\dfrac{1}{a}}{\dfrac{1}{a}+\dfrac{1}{a+1}}=\frac{\left(\dfrac{a}{a+1}+\dfrac{1}{a}\right)a(a+1)}{\left(\dfrac{1}{a}+\dfrac{1}{a+1}\right)a(a+1)}$$

$$=\frac{a^2+(a+1)}{(a+1)+a}=\frac{a^2+a+1}{2a+1}.$$

As an alternative method of solution, first perform the indicated additions in the numerator and denominator and then divide.

$$\frac{\dfrac{a}{a+1}+\dfrac{1}{a}}{\dfrac{1}{a}+\dfrac{1}{a+1}}=\frac{\dfrac{a^2+1(a+1)}{a(a+1)}}{\dfrac{1(a+1)+1(a)}{a(a+1)}}=\frac{\dfrac{a^2+a+1}{a(a+1)}}{\dfrac{2a+1}{a(a+1)}}$$

$$=\frac{a^2+a+1}{a(a+1)}\cdot\frac{a(a+1)}{2a+1}=\frac{a^2+a+1}{2a+1}$$ ◀ ⑥

---

⑥

Simplify each complex fraction.

(a) $\dfrac{t-\dfrac{1}{t}}{2t+\dfrac{3}{t}}$

(b) $\dfrac{\dfrac{m}{m+2}+\dfrac{1}{m}}{\dfrac{1}{m}-\dfrac{1}{m+2}}$

Answers:

(a) $\dfrac{t^2-1}{2t^2+3}$

(b) $\dfrac{m^2+m+2}{2}$

1.4 Exercises

Write each of the following in lowest terms. Factor as necessary. (See Example 1.)

1. $\dfrac{8x^2}{40x}$

2. $\dfrac{27m}{81m^3}$

3. $\dfrac{20p^2}{35p^3}$

4. $\dfrac{18y^4}{27y^2}$

5. $\dfrac{5m+15}{4m+12}$

6. $\dfrac{10z+5}{20z+10}$

7. $\dfrac{4(w-3)}{(w-3)(w+3)}$

8. $\dfrac{-6(x+2)}{(x-4)(x+2)}$

9. $\dfrac{3y^2-12y}{9y^3}$

10. $\dfrac{15k^2+45k}{9k^2}$

11. $\dfrac{m^2-4m+4}{m^2+m-6}$

12. $\dfrac{r^2-r-6}{r^2+r-12}$

13. $\dfrac{x^2+3x-4}{x^2-1}$

14. $\dfrac{z^2-5z+6}{z^2-4}$

Multiply or divide as indicated in each of the following. Write all answers in lowest terms. (See Examples 2 and 3.)

15. $\dfrac{4p^3}{49}\cdot\dfrac{7}{2p^2}$

16. $\dfrac{24n^4}{6n^2}\cdot\dfrac{18n^2}{9n}$

17. $\dfrac{21a^5}{14a^3}\div\dfrac{8a}{12a^2}$

18. $\dfrac{2x^3}{6x^2}\div\dfrac{10x^2}{15x}$

19. $\dfrac{2a+b}{2c}\cdot\dfrac{15}{4(2a+b)}$

20. $\dfrac{4(x+2)}{w}\cdot\dfrac{3w}{8(x+2)}$

21. $\dfrac{15p-3}{6}\div\dfrac{10p-2}{3}$

22. $\dfrac{2k+8}{6}\div\dfrac{3k+12}{2}$

23. $\dfrac{9y-18}{6y+12}\cdot\dfrac{3y+6}{15y-30}$

24. $\dfrac{12r + 24}{36r - 36} \div \dfrac{6r + 12}{8r - 8}$

25. $\dfrac{4a + 12}{2a - 10} \div \dfrac{a^2 - 9}{a^2 - a - 20}$

26. $\dfrac{6r - 18}{9r^2 + 6r - 24} \cdot \dfrac{12r - 16}{4r - 12}$

27. $\dfrac{k^2 - k - 6}{k^2 + k - 12} \cdot \dfrac{k^2 + 3k - 4}{k^2 + 2k - 3}$

28. $\dfrac{n^2 - n - 6}{n^2 - 2n - 8} \div \dfrac{n^2 - 9}{n^2 + 7n + 12}$

29. In your own words, explain how to find the least common denominator for two fractions.

30. Describe the steps required to add three rational expressions. You may use an example to illustrate.

Add or subtract as indicated in each of the following. Write all answers in lowest terms. (See Example 4.)

31. $\dfrac{3}{5z} - \dfrac{2}{3z}$

32. $\dfrac{7}{4z} - \dfrac{5}{3z}$

33. $\dfrac{r + 2}{3} - \dfrac{r - 2}{3}$

34. $\dfrac{3y - 1}{8} - \dfrac{3y + 1}{8}$

35. $\dfrac{4}{x} + \dfrac{1}{3}$

36. $\dfrac{6}{r} - \dfrac{3}{4}$

37. $\dfrac{1}{m - 1} + \dfrac{2}{m}$

38. $\dfrac{8}{y + 2} - \dfrac{3}{y}$

39. $\dfrac{8}{3(a - 1)} + \dfrac{2}{a - 1}$

40. $\dfrac{5}{2(k + 3)} + \dfrac{2}{k + 3}$

41. $\dfrac{2}{5(k - 2)} + \dfrac{3}{4(k - 2)}$

42. $\dfrac{11}{3(p + 4)} - \dfrac{5}{6(p + 4)}$

43. $\dfrac{2}{x^2 - 2x - 3} + \dfrac{5}{x^2 - x - 6}$

44. $\dfrac{3}{m^2 - 3m - 10} + \dfrac{5}{m^2 - m - 20}$

45. $\dfrac{2y}{y^2 + 7y + 12} - \dfrac{y}{y^2 + 5y + 6}$

46. $\dfrac{-r}{r^2 - 10r + 16} - \dfrac{3r}{r^2 + 2r - 8}$

47. $\dfrac{3k}{2k^2 + 3k - 2} - \dfrac{2k}{2k^2 - 7k + 3}$

48. $\dfrac{4m}{3m^2 + 7m - 6} - \dfrac{m}{3m^2 - 14m + 8}$

In each of the following exercises, simplify the complex fraction. (See Example 5.)

49. $\dfrac{1 + \dfrac{1}{x}}{1 - \dfrac{1}{x}}$

50. $\dfrac{2 - \dfrac{2}{y}}{2 + \dfrac{2}{y}}$

51. $\dfrac{\dfrac{1}{x + h} - \dfrac{1}{x}}{h}$

52. $\dfrac{\dfrac{1}{(x + h)^2} - \dfrac{1}{x^2}}{h}$

53. Social Sciences   When an average of ten cars per minute are arriving at the entrance gate of an amusement park and the average number being admitted per minute is x (with $x > 10$), then the average waiting time in minutes for each car is given by

$$\frac{x^2 - 10x + 25}{x^3 - 15x^2 + 50x}.^*$$

(a) Reduce this fraction to lowest terms.

(b) Use the reduced fraction to determine the average waiting time *in seconds,* when $x = 11, 15, 20$.

54. Business   In Example 8 of Section 1.2, we saw that the cost C of producing x thousand CDs is given by

$$C = -8x^2 + 3550x + 215{,}000 \qquad (x \le 200)$$

(a) Write a rational expression in lowest terms that gives the average cost per CD when x thousand are produced. (*Hint:* The average cost is the total cost C divided by the number of CDs produced.)

(b) Find the average cost per CD for each of these production levels: 20,000, 50,000, and 150,000.

---

*L. Haefner, *Introduction to Transportation Systems,* New York: Holt, Rinehart and Winston, 1986.

1.5 EXPONENTS AND RADICALS

Exponents were introduced in Section 1.2. In this section the definition of exponents will be extended to include negative exponents and rational number exponents, such as 1/2 and 7/3.

Integer Exponents

In Section 1.2 we defined positive integer exponents and noted that $a^m \cdot a^n = a^{m+n}$ for positive integer values of m and n. Now we develop an analogous property for quotients. By definition,

$$\frac{6^5}{6^2} = \frac{6 \cdot 6 \cdot 6 \cdot 6 \cdot 6}{6 \cdot 6} = 6 \cdot 6 \cdot 6 = 6^3.$$

Because there are 5 factors of 6 in the numerator and 2 factors of 6 in the denominator, the quotient has $5 - 2 = 3$ factors of 6. In general, we can make the following statement.

> If a is a nonzero real number and m, n are positive integers with $m > n$, then
>
> $$\frac{a^m}{a^n} = a^{m-n}.$$

Next, we want to give a meaning to expressions such as 3^0. If the quotient property in the preceding box is to continue to be valid, we must define 3^0 in such a way that

$$\frac{3^5}{3^5} = 3^{5-5} = 3^0.$$

Since $3^5/3^5 = 1$, it is reasonable to define $3^0 = 1$, and similarly in the general case. The symbol 0^0 is undefined.

①

Evaluate the following.

(a) 17^0

(b) 30^0

(c) $(-10)^0$

(d) $-(12)^0$

Answers:

(a) 1

(b) 1

(c) 1

(d) -1

> ### Zero Exponent
>
> If a is any nonzero real number, then
>
> $$a^0 = 1.$$

▶ **EXAMPLE 1** Evaluate the following.

(a) $6^0 = 1$

(b) $(-9)^0 = 1$

(c) $-(4)^0 = -(1) = -1$ ◀ ①

The next step is to define negative integer exponents. If they are to be defined in such a way that the quotient rule given above remains valid, then we must have, for example,

$$\frac{3^2}{3^4} = 3^{2-4} = 3^{-2}.$$

However,

$$\frac{3^2}{3^4} = \frac{\cancel{3} \cdot \cancel{3}}{\cancel{3} \cdot \cancel{3} \cdot 3 \cdot 3} = \frac{1}{3^2},$$

which suggests that 3^{-2} should be defined to be $1/3^2$. Thus, we have the following definition of a negative exponent.

Negative Exponent

If n is a natural number, and if $a \neq 0$, then

$$a^{-n} = \frac{1}{a^n}.$$

---

②

Evaluate the following.

(a) 6^{-2}

(b) -6^{-3}

(c) -3^{-4}

(d) $\left(\dfrac{5}{8}\right)^{-1}$

(e) $\left(\dfrac{1}{2}\right)^{-4}$

(f) $\left(\dfrac{7}{3}\right)^{-2}$

Answers:

(a) $1/36$

(b) $-1/216$

(c) $-1/81$

(d) $8/5$

(e) 16

(f) $9/49$

▶ **EXAMPLE 2** Evaluate the following.

(a) $3^{-2} = \dfrac{1}{3^2} = \dfrac{1}{9}$

(b) $5^{-4} = \dfrac{1}{5^4} = \dfrac{1}{625}$

(c) $9^{-1} = \dfrac{1}{9^1} = \dfrac{1}{9}$

(d) $-4^{-2} = -\dfrac{1}{4^2} = -\dfrac{1}{16}$

(e) $\left(\dfrac{3}{4}\right)^{-1} = \dfrac{1}{\left(\dfrac{3}{4}\right)^1} = \dfrac{1}{\dfrac{3}{4}} = \dfrac{4}{3}$

(f) $\left(\dfrac{2}{3}\right)^{-3} = \dfrac{1}{\left(\dfrac{2}{3}\right)^3} = \dfrac{1}{\left(\dfrac{2^3}{3^3}\right)} = 1 \cdot \dfrac{3^3}{2^3} = \dfrac{3^3}{2^3} = \dfrac{27}{8}$ ◀

Parts (e) and (f) of Example 2 involve work with fractions that can lead to error. For a useful shortcut with such fractions, use the properties of division of rational numbers and the definition of a negative exponent to get

$$\left(\frac{a}{b}\right)^{-n} = \frac{1}{\left(\dfrac{a}{b}\right)^n} = \frac{1}{\left(\dfrac{a^n}{b^n}\right)} = 1 \cdot \frac{b^n}{a^n} = \frac{b^n}{a^n} = \left(\frac{b}{a}\right)^n. \quad ②$$

When keying in negative exponents on a calculator, be sure to use the negation key (labeled $(-)$ or $+/-$), not the subtraction key. Calculators normally display answers as decimals, as shown in Figure 1.7 for parts (a), (b), and (f) of Example 2. Some graphing calculators have a FRAC key that converts these decimals to fractions as shown in Figure 1.8.

FIGURE 1.7

FIGURE 1.8

TECHNOLOGY TIP The FRAC key is in the MATH menu of TI graphing calculators. A FRAC program for other graphing calculators is in the Program Appendix.
Fractions can be displayed on some graphing calculators by changing the number
display format (in the MODES menu) to "fraction" or "exact."

Roots and Rational Exponents

The definition of a^n will now be extended to include rational values of n, such as 1/2
and 5/4. In order to do this, we need some terminology.

There are two numbers whose square is 16, 4 and -4. As we saw in Section
1.1, the positive one, 4, is called the *square root* (or second root) of 16. Similarly,
there are two numbers whose fourth power is 16, 2 and -2. We call 2 the **fourth
root** of 16. This suggests the following generalization.

> If n is even, the **nth root of a** is the positive real number whose nth power is a.

All nonnegative numbers have nth roots for every natural number n, but *no negative
number has a real, even nth root*. For example, there is no real number whose
square is -16, so -16 has no square root.

We say that the **cube root** (or third root) of 8 is 2 because $2^3 = 8$. Similarly,
since $(-2)^3 = -8$, we say -2 is the cube root of -8. Again, we can generalize.

> If n is odd, the **nth root of a** is the real number whose nth power is a.

Every real number has an nth root for every *odd* natural number n.

We can now define rational exponents. If they are to have the same properties
as integer exponents, we want $a^{1/2}$ to be a number such that

$$(a^{1/2})^2 = a^{1/2} \cdot a^{1/2} = a^{1/2 + 1/2} = a^1 = a.$$

Thus, $a^{1/2}$ should be a number whose square is a and it is reasonable to *define $a^{1/2}$* to
be the square root of a (if it exists). Similarly, $a^{1/3}$ is defined to be the cube root of a
and we have the following definition.

Evaluate the following.

(a) $16^{1/2}$

(b) $16^{1/4}$

(c) $-256^{1/2}$

(d) $(-256)^{1/2}$

(e) $-8^{1/3}$

(f) $243^{1/5}$

Answers:

(a) 4

(b) 2

(c) -16

(d) Not a real number

(e) -2

(f) 3

If a is *a* real number and n is a positive integer, then

$a^{1/n}$ is defined to be the nth root of a (if it exists).

▶ **EXAMPLE 3**   Evaluate the following roots.

(a) $36^{1/2} = 6$ because $6^2 = 36$.

(b) $100^{1/2} = 10$ because $10^2 = 100$.

(c) $-(225^{1/2}) = -15$ because $15^2 = 225$.

(d) $625^{1/4} = 5$ because $5^4 = 625$.

(e) $(-1296)^{1/4}$ is not a real number, although $-1296^{1/4} = -6$ because $6^4 = 1296$.

(f) $(-27)^{1/3} = -3$ because $(-3)^3 = -27$.

(g) $-32^{1/5} = -2$ because $2^5 = 32$. ◀

A calculator can be used to evaluate expressions with fractional exponents. Whenever it's easy to do so, enter the fractional exponents in their equivalent decimal form. For instance, to find $625^{1/4}$, enter $625^{.25}$ on the calculator. When the decimal equivalent of a fraction is an infinitely repeating decimal, however, it is best to enter the fractional exponent directly using parentheses, as in Figure 1.9. If you omit the parentheses or use a shortened decimal approximation (such as .333 for 1/3), you will not get the correct answers because the calculator will not interpret these exponents as you intended, as shown in Figure 1.10.

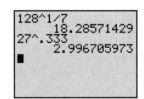

FIGURE 1.9   **FIGURE 1.10**

For more general rational exponents, the symbol $a^{m/n}$ should be defined so that the properties for exponents still hold. For example, by the product property, we want

$$(a^{1/3})^2 = a^{1/3} \cdot a^{1/3} = a^{1/3+1/3} = a^{2/3}$$

This suggests the following definition.

For all integers m and all positive integers n, and for all real numbers a for which $a^{1/n}$ is a real number,

$$a^{m/n} = (a^{1/n})^m.$$

4

Evaluate the following.

(a) $16^{3/4}$

(b) $25^{5/2}$

(c) $32^{7/5}$

(d) $100^{3/2}$

Answers:

(a) 8

(b) 3125

(c) 128

(d) 1000

▶ **EXAMPLE 4**   Evaluate the following.

(a) $27^{2/3} = (27^{1/3})^2 = 3^2 = 9$

(b) $32^{2/5} = (32^{1/5})^2 = 2^2 = 4$

(c) $64^{4/3} = (64^{1/3})^4 = 4^4 = 256$

(d) $25^{3/2} = (25^{1/2})^3 = 5^3 = 125$

◀

CAUTION   When the base is negative, as in $(-8)^{2/3}$, many calculators produce an error message. On such calculators, you should first compute $(-8)^{1/3}$ and then square the result; that is, compute $[(-8)^{1/3}]^2$.

Rational exponents were defined so that one of the familiar properties of exponents remains valid. In fact, all of the rules for exponents developed earlier for integer exponents are valid for rational exponents, as summarized below and illustrated in Examples 5 and 6.

Properties of Exponents

For any rational numbers m and n, and any real numbers a and b for which the following exist,

(a) $a^m \cdot a^n = a^{m+n}$ Product property

(b) $\dfrac{a^m}{a^n} = a^{m-n}$ Quotient property

(c) $(a^m)^n = a^{mn}$

(d) $(ab)^m = a^m \cdot b^m$ Power properties

(e) $\left(\dfrac{a}{b}\right)^m = \dfrac{a^m}{b^m}$

(f) $a^0 = 1$

(g) $a^{-n} = \dfrac{1}{a^n}$

(h) $\left(\dfrac{a}{b}\right)^{-n} = \left(\dfrac{b}{a}\right)^n.$

▶ **EXAMPLE 5**   Use the properties of exponents to simplify each of the following. Write answers with positive exponents.

(a) $7^{-4} \cdot 7^6 = 7^{-4+6} = 7^2$ Product property

(b) $\dfrac{9^{14}}{9^{-6}} = 9^{14-(-6)} = 9^{20}$ Quotient property

(c) $(2^{-3})^{-4} = 2^{(-3)(-4)} = 2^{12}$ Power property

⑤

Simplify the following.

(a) $9^6 \cdot 9^{-4}$

(b) $\dfrac{8^7}{8^{-3}}$

(c) $(13^4)^{-3}$

(d) $6^{2/5} \cdot 6^{3/5}$

(e) $\dfrac{8^{2/3} \cdot 8^{-4/3}}{8^2}$

Answers:

(a) 9^2

(b) 8^{10}

(c) $1/13^{12}$

(d) 6

(e) $1/8^{8/3}$ or $1/2^8$

(d) $\dfrac{27^{1/3} \cdot 27^{5/3}}{27^3} = \dfrac{27^{1/3+5/3}}{27^3}$ Product property

$= \dfrac{27^2}{27^3} = 27^{2-3}$ Quotient property

$= 27^{-1} = \dfrac{1}{27}$ Definition of negative exponent ◀ ⑤

You can use a calculator to check computations, such as those in Example 5, by computing the left and right sides separately and confirming that the answers are the same in each case. Figure 1.11 shows this for part (a) of Example 5.

```
7^-4*7^6
              49
7^2
              49
```

FIGURE 1.11

⑥

Simplify the following. Give answers with only positive exponents. Assume all variables represent positive real numbers.

(a) $\dfrac{(t^{-1})^2}{t^{-5}}$

(b) $\dfrac{(3z)^{-1}z^4}{z^2}$

(c) $3x^{1/4} \cdot 5x^{5/4}$

(d) $\left(\dfrac{2k^{1/3}}{p^{5/4}}\right)^2 \cdot \left(\dfrac{4k^{-2}}{p^5}\right)^{3/2}$

(e) $a^{5/8}(2a^{3/8} + a^{-1/8})$

Answers:

(a) t^3

(b) $z/3$

(c) $15x^{3/2}$

(d) $32/(p^{10}k^{7/3})$

(e) $2a + a^{1/2}$

▶ **EXAMPLE 6**   Simplify each expression. Give answers with only positive exponents. Assume all variables represent positive real numbers.

(a) $\dfrac{(m^3)^{-2}}{m^4} = \dfrac{m^{-6}}{m^4} = m^{-6-4} = m^{-10} = \dfrac{1}{m^{10}}$

(b) $6y^{2/3} \cdot 2y^{-1/2} = 12y^{2/3-1/2} = 12y^{1/6}$

(c) $\left(\dfrac{3m^{5/6}}{y^{3/4}}\right)^2 = \dfrac{3^2(m^{5/6})^2}{(y^{3/4})^2} = \dfrac{3^2m^{5/3}}{y^{3/2}} = \dfrac{9m^{5/3}}{y^{3/2}}$

(d) $m^{2/3}(m^{7/3} + 2m^{1/3}) = (m^{2/3+7/3} + 2m^{2/3+1/3}) = m^3 + 2m$ ◀ ⑥

Radicals

The nth root of a was denoted above as $a^{1/n}$. An alternative notation for nth roots uses the **radical** symbol $\sqrt[n]{\ }$.

If n is an even natural number and $a \geq 0$, or if n is an odd natural number,

$$\sqrt[n]{a} = a^{1/n}.$$

In the radical expression $\sqrt[n]{a}$, a is called the **radicand** and n is called the **index**. When $n = 2$, the familiar square root symbol $\sqrt{a}$ is used instead of $\sqrt[2]{a}$.

⑦

Simplify.

(a) $\sqrt[3]{27}$

(b) $\sqrt[4]{625}$

(c) $\sqrt[6]{64}$

(d) $\sqrt[3]{\dfrac{64}{125}}$

Answers:

(a) 3

(b) 5

(c) 2

(d) 4/5

▶ **EXAMPLE 7**   Simplify the following.

(a) $\sqrt[4]{16} = 16^{1/4} = 2$

(b) $\sqrt[5]{-32} = -2$

(c) $\sqrt[3]{1000} = 10$

(d) $\sqrt[6]{\dfrac{64}{729}} = \left(\dfrac{64}{729}\right)^{1/6} = \dfrac{64^{1/6}}{729^{1/6}} = \dfrac{2}{3}$ ◀ ⑦

The symbol $a^{m/n}$ also can be written in an alternative notation using radicals.

For all rational numbers m/n and all real numbers a for which $\sqrt[n]{a}$ exists,

$$a^{m/n} = (\sqrt[n]{a})^m \quad \text{or} \quad a^{m/n} = \sqrt[n]{a^m}.$$

Notice that $\sqrt[n]{x^n}$ cannot be written simply as x when n is even. For example, if $x = -5$,

$$\sqrt{x^2} = \sqrt{(-5)^2} = \sqrt{25} = 5 \neq x.$$

However, $|-5| = 5$, so that $\sqrt{x^2} = |x|$ when x is -5. This is true in general.

For any real number a and any natural number n,

$$\sqrt[n]{a^n} = |a| \text{ if } n \text{ is even}$$

and

$$\sqrt[n]{a^n} = a \text{ if } n \text{ is odd.}$$

To avoid this difficulty that $\sqrt[n]{a^n}$ is not necessarily equal to a, we shall assume that all variables in radicands represent only nonnegative numbers, as they usually do in applications.

The properties of exponents can be written with radicals as shown below.

For all real numbers a and b and positive integers n for which all indicated roots exist,

(a) $\sqrt[n]{a} \cdot \sqrt[n]{b} = \sqrt[n]{ab}$ (b) $\dfrac{\sqrt[n]{a}}{\sqrt[n]{b}} = \sqrt[n]{\dfrac{a}{b}} \quad (b \neq 0).$

⑧

Simplify.

(a) $\sqrt{3} \cdot \sqrt{27}$

(b) $\sqrt{\dfrac{3}{49}}$

(c) $\sqrt{25 - 4}$

(d) $\sqrt{25} - \sqrt{4}$

Answers:

(a) 9

(b) $\dfrac{\sqrt{3}}{7}$

(c) $\sqrt{21}$

(d) 3

▶ EXAMPLE 8   Simplify the following.

(a) $\sqrt{6} \cdot \sqrt{54} = \sqrt{6 \cdot 54} = \sqrt{324} = 18$

Alternatively, simplify $\sqrt{54}$ first.

$$\sqrt{6} \cdot \sqrt{54} = \sqrt{6} \cdot \sqrt{9 \cdot 6}$$
$$= \sqrt{6} \cdot 3\sqrt{6} = 3 \cdot 6 = 18$$

(b) $\sqrt{\dfrac{7}{64}} = \dfrac{\sqrt{7}}{\sqrt{64}} = \dfrac{\sqrt{7}}{8}$

(c) $\sqrt{9 + 16} = \sqrt{25} = 5$

When using a calculator to evaluate radical expressions, be sure to use parentheses. To compute $\sqrt{9 + 16}$ on a calculator, for example, you must enter it as $\sqrt{}(9 + 16)$ in order to get the correct answer 5. If you enter $\sqrt{}\ 9 + 16$, most calculators will interpret it as $3 + 16 = 19$, as shown in Figure 1.12.   ◀ ⑧

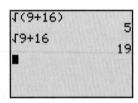

FIGURE 1.12

Multiplying radical expressions is much like multiplying polynomials.

⑨

Multiply.

(a) $(\sqrt{5} - \sqrt{2})(3 + \sqrt{2})$

(b) $(\sqrt{3} + \sqrt{7})(\sqrt{3} - \sqrt{7})$

Answers:

(a) $3\sqrt{5} + \sqrt{10} - 3\sqrt{2} - 2$

(b) -4

▶ EXAMPLE 9   Multiply the following.

(a) $(\sqrt{2} + 3)(\sqrt{8} - 5) = \sqrt{2}(\sqrt{8}) - \sqrt{2}(5) + 3\sqrt{8} - 3(5)$ FOIL
$$= \sqrt{16} - 5\sqrt{2} + 3(2\sqrt{2}) - 15$$
$$= 4 - 5\sqrt{2} + 6\sqrt{2} - 15$$
$$= -11 + \sqrt{2}$$

(b) $(\sqrt{7} - \sqrt{10})(\sqrt{7} + \sqrt{10}) = (\sqrt{7})^2 - (\sqrt{10})^2$
$$= 7 - 10 = -3$$ ◀ ⑧

Rationalizing Denominators and Numerators

Before the invention of calculators, it was customary to **rationalize the denominators** of fractions (that is, write equivalent fractions with no radicals in the denominator) because this made many computations easier. Although there is no longer a computational reason, rationalizing denominators (and sometimes numerators) is still used today to simplify expressions and to derive useful formulas.

▶ **EXAMPLE 10**   Rationalize each denominator.

(a) $\dfrac{4}{\sqrt{3}}$

The key is to multiply by 1, with 1 written as a radical fraction:

$$\frac{4}{\sqrt{3}} = \frac{4}{\sqrt{3}} \cdot 1 = \frac{4}{\sqrt{3}} \cdot \frac{\sqrt{3}}{\sqrt{3}} = \frac{4\sqrt{3}}{3}.$$

(b) $\dfrac{1}{3 - \sqrt{2}}$

The same technique works here, using $1 = \dfrac{3 + \sqrt{2}}{3 + \sqrt{2}}$:

$$\frac{1}{3 - \sqrt{2}} = \frac{1}{3 - \sqrt{2}} \cdot 1 = \frac{1}{3 - \sqrt{2}} \cdot \frac{3 + \sqrt{2}}{3 + \sqrt{2}} = \frac{3 + \sqrt{2}}{(3 - \sqrt{2})(3 + \sqrt{2})}$$

$$= \frac{3 + \sqrt{2}}{9 - 2} = \frac{3 + \sqrt{2}}{7}. \quad ◀ \quad ⑩$$

▶ **EXAMPLE 11**   Rationalize the numerator of $\dfrac{2 + \sqrt{5}}{1 + \sqrt{3}}$.

Multiply the fraction by $1 = \dfrac{2 - \sqrt{5}}{2 - \sqrt{5}}$:

$$\frac{2 + \sqrt{5}}{1 + \sqrt{3}} = \frac{2 + \sqrt{5}}{1 + \sqrt{3}} \cdot \frac{2 - \sqrt{5}}{2 - \sqrt{5}} = \frac{4 - 5}{2 - \sqrt{5} + 2\sqrt{3} - \sqrt{3}\sqrt{5}}$$

$$= \frac{-1}{2 - \sqrt{5} + 2\sqrt{3} - \sqrt{15}}. \quad ◀$$

⑩

Rationalize the denominator.

(a) $\dfrac{2}{\sqrt{5}}$

(b) $\dfrac{1}{2 + \sqrt{3}}$

Answers:

(a) $\dfrac{2\sqrt{5}}{5}$

(b) $2 - \sqrt{3}$

1.5 Exercises

Evaluate each expression. Write all answers without exponents. (See Examples 1 and 2.)

1. 5^0

2. 8^0

3. 6^{-1}

4. 10^{-3}

5. 2^{-5}

6. -4^{-3}

7. -7^{-4}

8. $\left(\dfrac{1}{3}\right)^{-2}$

9. $\left(\dfrac{1}{6}\right)^{-3}$

10. $\left(\dfrac{2}{5}\right)^{-4}$

11. $\left(\dfrac{4}{3}\right)^{-2}$

12. Explain why $-2^{-4} = -1/16$, but $(-2)^{-4} = 1/16$.

Evaluate each expression. Write all answers without exponents. Write decimal answers to the nearest tenth. (See Examples 3 and 4.)

13. $49^{1/2}$

14. $8^{1/3}$

15. $(7.51)^{1/4}$

16. $(68.93)^{1/5}$

17. $27^{2/3}$

18. $24^{3/2}$

19. $-64^{2/3}$

20. $-64^{3/2}$

21. $(8/27)^{-4/3}$

22. $(27/64)^{-1/3}$

Simplify each expression. Write all answers using only positive exponents. (See Example 5.)

23. $\dfrac{4^{-2}}{4^3}$

24. $\dfrac{9^{-4}}{9^{-3}}$

25. $4^{-3} \cdot 4^6$

26. $5^{-9} \cdot 5^{10}$

27. $8^{2/3} \cdot 8^{-1/3}$

28. $12^{-3/4} \cdot 12^{1/4}$

29. $\dfrac{8^9 \cdot 8^{-7}}{8^{-3}}$

30. $\dfrac{5^{-4} \cdot 5^6}{5^{-1}}$

31. $\dfrac{9^{-5/3}}{9^{2/3} \cdot 9^{-1/5}}$

32. $\dfrac{3^{5/3} \cdot 3^{-3/4}}{3^{-1/4}}$

Simplify each expression. Assume all variables represent positive real numbers. Write answers with only positive exponents. (See Example 6.)

33. $\dfrac{z^5 \cdot z^2}{z^4}$

34. $\dfrac{k^6 \cdot k^9}{k^{12}}$

35. $\dfrac{2^{-1}(p^{-1})^3}{2p^{-4}}$

36. $\dfrac{(5x^3)^{-2}}{x^4}$

37. $(q^{-5}r^2)^{-1}$

38. $(2y^2z^{-2})^{-3}$

39. $(2p^{-1})^3 \cdot (5p^2)^{-2}$

40. $(5^{-1}m^2)^{-3} \cdot (3m^{-2})^4$

41. $(2p)^{1/2} \cdot (2p^3)^{1/3}$

42. $(5k^2)^{3/2} \cdot (5k^{1/3})^{3/4}$

43. $p^{2/3}(2p^{1/3} + 5p)$

44. $2z^{1/2}(3z^{-1/2} + z^{1/2})$

Match the rational exponent expression in Column I with the equivalent radical expression in Column II. Assume that x is not zero.

I	II
45. $(-3x)^{1/3}$	**(a)** $\dfrac{3}{\sqrt[3]{x}}$
46. $-3x^{1/3}$	**(b)** $-3\sqrt[3]{x}$
47. $(-3x)^{-1/3}$	**(c)** $\dfrac{1}{\sqrt[3]{3x}}$
48. $-3x^{-1/3}$	**(d)** $\dfrac{-3}{\sqrt[3]{x}}$
49. $(3x)^{1/3}$	**(e)** $3\sqrt[3]{x}$
50. $3x^{-1/3}$	**(f)** $\sqrt[3]{-3x}$
51. $(3x)^{-1/3}$	**(g)** $\sqrt[3]{3x}$
52. $3x^{1/3}$	**(h)** $\dfrac{1}{\sqrt[3]{-3x}}$

Simplify each of the following. (See Examples 7–9.)

53. $\sqrt[3]{64}$

54. $\sqrt[6]{64}$

55. $\sqrt[4]{625}$

56. $\sqrt[5]{-243}$

57. $\sqrt[7]{-128}$

58. $\sqrt{44} \cdot \sqrt{11}$

59. $\sqrt[3]{81} \cdot \sqrt[3]{9}$

60. $\sqrt{49 - 16}$

61. $\sqrt{81 - 4}$

62. $\sqrt{49} - \sqrt{16}$

63. $\sqrt{81} - \sqrt{4}$

64. $(\sqrt{2} + 3)(\sqrt{2} - 3)$

65. $(\sqrt{5} + \sqrt{2})(\sqrt{5} - \sqrt{2})$

Rationalize the denominator of each of the following. (See Example 10.)

66. $\dfrac{3}{1 - \sqrt{2}}$

67. $\dfrac{2}{1 + \sqrt{5}}$

68. $\dfrac{4 - \sqrt{2}}{2 - \sqrt{2}}$

69. $\dfrac{\sqrt{3} - 1}{\sqrt{3} - 2}$

Rationalize the numerator of each of the following. (See Example 11.)

70. $\dfrac{1 + \sqrt{7}}{2 - \sqrt{3}}$

71. $\dfrac{2 - \sqrt{3}}{2 + \sqrt{3}}$

72. What is wrong with the statement $\sqrt[3]{4} \cdot \sqrt[3]{4} = 4$?

The following exercises are applications of exponentiation and radicals.

73. Business The theory of economic lot size shows that, under certain conditions, the number of units to order to minimize total cost is

$$x = \sqrt{\dfrac{kM}{f}}.$$

Here k is the cost to store one unit for one year, f is the (constant) setup cost to manufacture the product, and M is the total number of units produced annually. (See Section 12.3.) Find x for the following values of f, k, and M.
   (a) $k = \$1$, $f = \$500$, $M = 100,000$
   (b) $k = \$3$, $f = \$7$, $M = 16,700$
   (c) $k = \$1$, $f = \$5$, $M = 16,800$

74. Social Science The threshold weight T for a person is the weight above which the risk of death increases greatly. One researcher found that the threshold weight in pounds for men aged 40–49 is related to height in inches by the equation $h = 12.3T^{1/3}$. What height corresponds to a threshold of 216 pounds for a man in this age group?

Natural Science *The following exercises refer to the table of wind-chill factors that precedes Exercises 55–58 of Section 1.1.*

75. The equation for calculating the wind-chill C was revised in 2001. Before that time, the following equation could be used to approximate the wind-chill:

$$C = 91.4 - (91.4 - T)(.478 + .301\sqrt{v} - .02v),$$

where T is the temperature and v is the wind velocity. Use this equation to find C when
   (a) $T = -10$ and $v = 30$
   (b) $T = -40$ and $v = 5$.
   (c) How do the old wind-chill factors in parts (a) and (b) compare with the new ones given by the table?

76. Compare the wind-chill factors given by the old equation in Exercise 75 and the current table for a temperature of 0° and wind velocities of 10, 20, 30, and 40 mph. Under these conditions, how do the two formulas compare?

Business   *The Consumer Price Index (CPI) in U.S. cities stood at 100 in 1982–1984. In recent years it can be approximated by*

$$161.44x^{.0442} \qquad (1 \leq x \leq 10),$$

where x = 1 corresponds to 1998, x = 2 to 1999, and so on.* Approximate the CPI in the following years.

77. First half of 1999 ($x = 2$)

78. Last half of 2000 ($x = 3.5$)

79. Last half of 2003   **80.** First half of 2004

Health   *According to data from the National Organ Procurement and Transplantation Network, the number of kidney transplants (in thousands) in year x can be approximated by*

$$8.064(1.032^{x}) \qquad (3 \leq x \leq 18),$$

where x = 3 corresponds to 1988, x = 4 to 1989, and so on. Find the approximate number of kidney transplants for each year.

81. 1990   **82.** 1996   **83.** 1999

84. 2001   **85.** 2003

86. According to the United Network for Organ Sharing, approximately 46,800 people were waiting for a kidney transplant in 2000. About how many of these people did *not* get a transplant that year?

---

*U.S. Bureau of Labor.

87. Business   Based on data from *Ward's Communications,* the number of sport utility vehicles (SUVs) sold each year (in millions) can be approximated by

$$\frac{12.8}{1 + 14.97(.838^{x})} \qquad (1 \leq x \leq 14),$$

where $x = 1$ corresponds to 1991, $x = 2$ to 1992, and so on.

(a) Approximately how many SUVs were sold in 1996? In 1999? In 2002?

(b) Assume that this formula remains valid over the next forty years (probably unlikely). Use the table feature of a graphing calculator to determine when SUV sales would reach 12,000,000. According to this approximation, would sales ever reach 13,000,000?

1.6 FIRST-DEGREE EQUATIONS

An **equation** is a statement that two mathematical expressions are equal; for example,

$$5x - 3 = 13, \qquad 8y = 4, \qquad -3p + 5 = 4p - 8.$$

The letter in each equation is called the variable. This section concentrates on **first-degree equations,** which are equations that involve only constants and the first power of the variable. All of the equations displayed above are first-degree equations, but neither of the following equations is first-degree: $2x^2 = 5x + 6$ (variable has exponent greater than 1) and $\sqrt{x + 2} = 4$ (because the variable is under the radical).

Is -4 a solution of the following equations?

(a) $3x + 5 = -7$

(b) $2x - 3 = 5$

(c) Is there more than one solution of the equation in part (a)?

Answers:

(a) Yes

(b) No

(c) No

A **solution** of an equation is a number that can be substituted for the variable in the equation to produce a true statement. For example, substituting the number 9 for x in the equation $2x + 1 = 19$ gives

$$2x + 1 = 19$$
$$2(9) + 1 \overset{?}{=} 19 \qquad \text{Let } x = 9.$$
$$18 + 1 = 19. \qquad \text{True}$$

This true statement indicates that 9 is a solution of $2x + 1 = 19$. ①

The following properties are used to solve equations.

Properties of Equality

1. The same number may be added to or subtracted from both sides of an equation:

$$\text{If } a = b, \text{ then } a + c = b + c \quad \text{and} \quad a - c = b - c.$$

2. Both sides of an equation may be multiplied or divided by the same nonzero number:

$$\text{If } a = b \text{ and } c \neq 0, \text{ then } ac = bc \quad \text{and} \quad \frac{a}{c} = \frac{b}{c}.$$

▶ **EXAMPLE 1**   Solve the equation $5x - 3 = 12$.

Using the first property of equality, add 3 to both sides. This isolates the term containing the variable on one side of the equation.

$$5x - 3 = 12$$
$$5x - 3 + 3 = 12 + 3 \qquad \text{Add 3 to both sides.}$$
$$5x = 15$$

Now arrange for the coefficient of x to be 1 by using the second property of equality.

$$5x = 15$$
$$\frac{5x}{5} = \frac{15}{5} \qquad \text{Divide both sides by 5.}$$
$$x = 3$$

②

Solve the following.

(a) $3p - 5 = 19$

(b) $4y + 3 = -5$

(c) $-2k + 6 = 2$

Answers:

(a) 8

(b) -2

(c) 2

The solution of the original equation, $5x - 3 = 12$, is 3. Check the solution by substituting 3 for x in the original equation. ◀ ②

▶ **EXAMPLE 2** Solve $2k + 3(k - 4) = 2(k - 3)$.

First, simplify the equation by using the distributive property on the left-side term $3(k - 4)$ and right-side term $2(k - 3)$:

$$2k + 3(k - 4) = 2(k - 3)$$
$$2k + 3k - 12 = 2(k - 3) \qquad \text{Distributive property}$$
$$2k + 3k - 12 = 2k - 6 \qquad \text{Distributive property}$$
$$5k - 12 = 2k - 6 \qquad \text{Collect like terms on left side.}$$

One way to proceed is to add $-2k$ to both sides.

$$5k - 12 + (-2k) = 2k - 6 + (-2k) \qquad \text{Add } -2k \text{ to both sides.}$$
$$3k - 12 = -6$$
$$3k - 12 + 12 = -6 + 12 \qquad \text{Add 12 to both sides.}$$
$$3k = 6$$
$$\frac{1}{3}(3k) = \frac{1}{3}(6) \qquad \text{Multiply both sides by } \frac{1}{3}.$$
$$k = 2$$

The solution is 2. Check this result by substituting 2 for k in the original equation. ◀ ③

③

Solve the following.

(a) $3(m - 6) + 2(m + 4)$
$= 4m - 2$

(b) $-2(y + 3) + 4y$
$= 3(y + 1) - 6$

Answers:

(a) 8

(b) −3

▶ **EXAMPLE 3** Use a calculator to solve $42.19x + 121.34 = 16.83x + 19.15$.

To avoid round-off errors in the intermediate steps, do all the algebra first, without using the calculator.

$$42.19x = 16.83x + 19.15 - 121.34 \qquad \text{Subtract 121.34 from both sides.}$$

$$42.19x - 16.83x = 19.15 - 121.34 \qquad \text{Subtract } 16.83x \text{ from both sides.}$$

$$(42.19 - 16.83)x = 19.15 - 121.34 \qquad \text{Distributive property}$$

$$x = \frac{19.15 - 121.34}{42.19 - 16.83} \qquad \text{Divide both sides by } (42.19 - 16.83).$$

Now use a calculator, being careful to put parentheses around numerator and denominator, and determine that $x \approx -4.0296$, as shown in Figure 1.13. Because this answer is approximate, it may not check exactly when substituted in the original equation. ◀

FIGURE 1.13

The next three examples show how to simplify the solution of first-degree equations involving fractions. We solve these equations by multiplying both sides of the equation by a *common denominator*. This step will eliminate the fractions.

▶ **EXAMPLE 4** Solve $\dfrac{r}{10} - \dfrac{2}{15} = \dfrac{3r}{20} - \dfrac{1}{5}$.

Here the denominators are 10, 15, 20, and 5. Each of these numbers can be divided into 60; therefore, 60 is a common denominator. Multiply both sides of the equation by 60.

$$60\left(\frac{r}{10} - \frac{2}{15}\right) = 60\left(\frac{3r}{20} - \frac{1}{5}\right)$$

$$60\left(\frac{r}{10}\right) - 60\left(\frac{2}{15}\right) = 60\left(\frac{3r}{20}\right) - 60\left(\frac{1}{5}\right) \qquad \text{Distributive property}$$

$$6r - 8 = 9r - 12$$

$$6r - 8 + (-6r) + 12 = 9r - 12 + (-6r) + 12 \qquad \text{Add } -6r \text{ and 12 to both sides.}$$

$$4 = 3r$$

$$r = \frac{4}{3} \qquad \text{Multiply both sides by 1/3.}$$

Check this solution in the original equation. ◀ ④

④

Solve the following.

(a) $\dfrac{x}{2} - \dfrac{x}{4} = 6$

(b) $\dfrac{2x}{3} + \dfrac{1}{2} = \dfrac{x}{4} - \dfrac{9}{2}$

Answers:

(a) 24

(b) −12

CAUTION Multiplying *both* sides of the *equation* by a number to eliminate fractions is valid. But multiplying a single fraction to simplify it is not valid. For instance, multiplying $\dfrac{3x}{8}$ by 8 *changes* it to $3x$, which is *not equal* to $\dfrac{3x}{8}$.

▶ **EXAMPLE 5** Solve $\dfrac{4}{3(k+2)} - \dfrac{k}{3(k+2)} = \dfrac{5}{3}$.

Multiply both sides of the equation by the common denominator $3(k+2)$. Here $k \neq -2$, since $k = -2$ would give a 0 denominator, making the fraction undefined.

$$3(k+2) \cdot \frac{4}{3(k+2)} - 3(k+2) \cdot \frac{k}{3(k+2)} = 3(k+2) \cdot \frac{5}{3}$$

Simplify each side and solve for k.

$$4 - k = 5(k+2)$$

$$4 - k = 5k + 10 \qquad \text{Distributive property}$$

$$4 - k + k = 5k + 10 + k \qquad \text{Add } k \text{ to both sides.}$$

$$4 = 6k + 10$$

$$4 + (-10) = 6k + 10 + (-10) \qquad \text{Add } -10 \text{ to both sides.}$$

$$-6 = 6k$$

$$-1 = k \qquad \text{Multiply both sides by } \frac{1}{6}.$$

5

Solve the equation.

$$\frac{5p + 1}{3(p + 1)} =$$
$$\frac{3p - 3}{3(p + 1)} + \frac{9p - 3}{3(p + 1)}$$

Answer:

1

The solution is -1. Substitute -1 for k as a check.

$$\frac{4}{3(-1 + 2)} - \frac{-1}{3(-1 + 2)} \stackrel{?}{=} \frac{5}{3}$$

$$\frac{4}{3} - \frac{-1}{3} \stackrel{?}{=} \frac{5}{3}$$

$$\frac{5}{3} = \frac{5}{3}$$

The check shows that -1 is the solution.   ◀ **5**

CAUTION   Whenever you multiply an equation by an expression involving the variable, as in Example 5, it is *essential* to check the solution, as the next example illustrates.

6

Solve each equation.

(a) $\dfrac{3p}{p + 1} = 1 - \dfrac{3}{p + 1}$

(b) $\dfrac{8y}{y - 4} = \dfrac{32}{y - 4} - 3.$

Answer:

Neither equation has a solution.

▶ **EXAMPLE 6**   Solve $\dfrac{3x - 4}{x - 2} = \dfrac{x}{x - 2}.$

Multiplying both sides by $x - 2$ produces

$$3x - 4 = x$$
$$2x - 4 = 0 \qquad \text{Subtract } x \text{ from both sides.}$$
$$2x = 4 \qquad \text{Add 4 to both sides.}$$
$$x = 2 \qquad \text{Divide both sides by 2.}$$

Substituting 2 for x in the original equation produces fractions with 0 denominators. Since division by 0 is not defined, the original equation has no solution.   ◀ **6**

Sometimes an equation with several variables must be solved for one of the variables. This process is called **solving for a specified variable.**

▶ **EXAMPLE 7**   Solve for x: $3(ax - 5a) + 4b = 4x - 2.$

Use the distributive property to get

$$3ax - 15a + 4b = 4x - 2.$$

7

Solve for x.

(a) $2x - 7y = 3xk$

(b) $8(4 - x) + 6p$
$= -5k - 11yx$

Answers:

(a) $x = \dfrac{7y}{2 - 3k}$

(b) $x = \dfrac{5k + 32 + 6p}{8 - 11y}$

Treat x as the variable, the other letters as constants. Get all terms with x on one side of the equation and all terms without x on the other side.

$$3ax - 4x = 15a - 4b - 2 \qquad \text{Isolate terms with } x \text{ on the left.}$$
$$(3a - 4)x = 15a - 4b - 2 \qquad \text{Distributive property}$$
$$x = \frac{15a - 4b - 2}{3a - 4} \qquad \text{Multiply by } \frac{1}{3a - 4}.$$

The final equation is solved for x, as required.   ◀ **7**

Absolute Value Equations

Recall from Section 1.1 that the absolute value of a number a is either a or $-a$, whichever one is positive. For instance, $|4| = 4$ and $|-7| = -(-7) = 7$.

▶ **EXAMPLE 8**   Solve $|x| = 3$.

Since $|x|$ is either x or $-x$, the equation says that

$$x = 3 \quad \text{or} \quad -x = 3$$
$$x = -3.$$

The solutions of $|x| = 3$ are 3 and -3. ◀

⑧

Solve each equation.

(a) $|y| = 9$

(b) $|r + 3| = 1$

(c) $|2k - 3| = 7$

Answers:

(a) 9, -9

(b) $-2, -4$

(c) 5, -2

▶ **EXAMPLE 9**   Solve $|p - 4| = 2$.

Since $|p - 4|$ is either $p - 4$ or $-(p - 4)$, we have

$$p - 4 = 2 \quad \text{or} \quad -(p - 4) = 2$$
$$p = 6 \qquad\qquad -p + 4 = 2$$
$$-p = -2$$
$$p = 2,$$

so that 6 and 2 are possible solutions. Checking them in the original equation shows that both are solutions. ◀ ⑧

⑨

Solve each equation.

(a) $|r + 6| = |2r + 1|$

(b) $|5k - 7| = |10k - 2|$

Answers:

(a) 5, $-7/3$

(b) $-1, 3/5$

▶ **EXAMPLE 10**   Solve $|4m - 3| = |m + 6|$.

The quantities in absolute value bars must either be equal or be negatives of one another to satisfy the equation. That is,

$$4m - 3 = m + 6 \quad \text{or} \quad 4m - 3 = -(m + 6)$$
$$3m = 9 \qquad\qquad 4m - 3 = -m - 6$$
$$m = 3 \qquad\qquad 5m = -3$$
$$m = -\frac{3}{5}.$$

Check that the solutions for the original equation are 3 and $-3/5$. ◀

Applications

One of the main reasons for learning mathematics is to be able to use it to solve practical problems. There are no hard and fast rules for dealing with real-world applications, except perhaps to use common sense. However, you will find it much easier to deal with such problems if you don't try to do everything at once. After reading the problem carefully, attack it in stages, as suggested in the following guidelines.

Solving Applied Problems

Step 1   Decide on the unknown. Name it with some variable that you *write down*. Many students try to skip this step. They are eager to get on with the writing of the equation. But this is an important step. If you don't know what the variable represents, how can you write a meaningful equation or interpret a result?

Step 2 Decide on a variable expression to represent any other unknowns in the problem. For example, if x represents the width of a rectangle, and you know that the length is one more than twice the width, then *write down* that the length is $1 + 2x$.

Step 3 Draw a sketch or make a chart, if appropriate, showing the information given in the problem.

Step 4 Using the results of Steps 1–3, write an equation that expresses a condition that must be satisfied.

Step 5 Solve the equation.

Step 6 Check the solution in the words of the *original problem,* not just in the equation you have written.

The following examples illustrate this approach.

▶ **EXAMPLE 11** A financial manager has $14,000 to invest for her company. She plans to invest part of the money in tax-free bonds at 6% interest and the remainder in taxable bonds at 9%. She wants to earn $1005 per year in interest from the investments. Find the amount she should invest at each rate.

Step 1 Let x represent the amount to be invested at 6%.

Step 2 After x dollars are invested, the remaining amount is $14,000 - x$ dollars, which is to be invested at 9%.

Step 3 Interest for one year is given by rate $\times$ amount invested. For instance, 6% of x dollars is $.06x$. The given information is summarized in this chart.

Investment	Amount Invested	Interest Rate	Interest Earned in 1 Year
Tax-free bonds	x	6% = .06	$.06x$
Taxable bonds	$14,000 - x$	9% = .09	$.09(14,000 - x)$
Totals	14,000		1005

Step 4 Because the total interest is to be $1005, the last column of the tables shows that

$$.06x + .09(14,000 - x) = 1005.$$

An investor owns two pieces of property. One, worth twice as much as the other, returns 6% in annual interest, while the other returns 4%. Find the value of each piece of property if the total annual interest earned is $8000.

Answers:

6% return: $100,000; 4% return: $50,000

Step 5 Solve this equation.

$$.06x + .09(14{,}000 - x) = 1005$$
$$.06x + .09(14{,}000) - .09x = 1005$$
$$.06x + 1260 - .09x = 1005$$
$$-.03x = -255$$
$$x = 8500.$$

The manager should invest $8500 at 6% and $14,000 − $8500 = $5500 at 9%.

Step 6 Check these results in the original problem. If $8500 is invested at 6%, the interest is .06(8500) = $510. If $5500 is invested at 9%, the interest is .09(5500) = $495. So the total interest is $510 + $495 = $1005 as required. ◀ ⑩

▶**EXAMPLE 12** Chuck travels 80 kilometers in the same time that Mary travels 180 kilometers. Mary travels 50 kilometers per hour faster than Chuck. Find the speed of each person.

Step 1 Use x to represent Chuck's speed and $x + 50$ to represent Mary's speed, which is 50 kilometers per hour faster than Chuck's.

Steps 2 and 3 Constant rate problems of this kind require the distance formula

$$d = rt,$$

where d is the distance traveled in t hours at a constant rate of speed r. The distance traveled by each person is given, along with the fact that the time traveled by each person is the same. Solve the formula $d = rt$ for t.

$$d = rt$$
$$\frac{1}{r} \cdot d = \frac{1}{r} \cdot rt$$
$$\frac{d}{r} = t$$

For Chuck, $d = 80$ and $r = x$, giving $t = 80/x$. For Mary, $d = 180$, $r = x + 50$, and $t = 180/(x + 50)$. Use these facts to complete a chart, which organizes the information given in the problem.

	d	r	t
Chuck	80	x	$\dfrac{80}{x}$
Mary	180	$x + 50$	$\dfrac{180}{x + 50}$

Step 4 Because both people traveled for the *same time*, the equation is

$$\frac{80}{x} = \frac{180}{x + 50}.$$

11

(a) Tom and Dick are in a run for charity. Tom runs at 7 mph and Dick runs at 5 mph. If they start at the same time, how long will it be until they are 1/2 mile apart?
(b) In part (a), suppose the run has a staggered start. If Dick starts first, and Tom starts 10 minutes later, how long will it be until they are neck and neck?

Answers:

(a) 15 minutes (1/4 hour)

(b) Tom has run 25 minutes.

Step 5 Multiply both sides of the equation by $x(x + 50)$.

$$x(x + 50)\,\frac{80}{x} = x(x + 50)\,\frac{180}{x + 50}$$

$$80(x + 50) = 180x$$
$$80x + 4000 = 180x$$
$$4000 = 100x$$
$$40 = x$$

Step 6 Since x represents Chuck's speed, Chuck went 40 kilometers per hour. Mary's speed is $x + 50$, or $40 + 50 = 90$ kilometers per hour. Check these results in the words of the original problem. ◀ **11**

▶ **EXAMPLE 13**   An oil company distributor needs to fill orders for 89 octane gas but only has 87 and 93 octane gas on hand. How much of each type should be mixed together to produce 54,000 gallons of 89 octane gas?

The octane rating of a gasoline indicates that it has the same antiknock qualities as a standard fuel made of heptane and isooctane. The octane rating is the percentage of isooctane in the standard fuel.* So we assume that standard fuels are being mixed. Let x be the amount of 87 octane gas. Then $54,000 - x$ is the amount of 93 octane gas. We can summarize the relevant information in a chart.

Type of Gas	Quantity	% Isooctane	Amount of Isooctane
87 octane	x	87%	$.87x$
93 octane	$54,000 - x$	93%	$.93(54,000 - x)$
Mixture	54,000	89%	$.89(54,000)$

The amount of isooctane satisfies this equation:

(Amount in 87 octane) + (Amount in 93 octane) = Amount in mixture
$$.87x + .93(54,000 - x) = .89(54,000)$$

Use the distributive property on the left side and multiply on the right side:

$$.87x + 50,220 - .93x = 48,060$$
$$-.06x = -2160$$
$$x = \frac{-2160}{-.06} = 36,000$$

12

How much 89 octane gas and how much 94 octane gas are needed to produce 1000 gallons of 91 octane gas?

Answer:

600 gallons of 89 octane; 400 gallons of 94 octane

So the distributor should mix 36,000 gallons of 87 octane gas with $54,000 - 36,000 = 18,000$ gallons of 93 octane gas to obtain 54,000 gallons of 89 octane gas ◀ **12**

*This is one of several possible ways of determining octane ratings.

1.6 Exercises

Solve each equation. (See Examples 1–6.)

1. $3x + 5 = 20$

2. $4 - 5y = 9$

3. $.6k - .3 = .5k + .4$

4. $2.5 + 5.04m = 8.5 - .06m$

5. $2a - 1 = 3(a + 1) + 7a + 5$

6. $3(k - 2) - 6 = 4k - (3k - 1)$

7. $2[x - (3 + 2x) + 9] = 2x + 4$

8. $-2[4(k + 2) - 3(k + 1)] = 14 + 2k$

9. $\dfrac{3x}{5} - \dfrac{4}{5}(x + 1) = 2 - \dfrac{3}{10}(3x - 4)$

10. $\dfrac{4}{3}(x - 2) - \dfrac{1}{2} = 2\left(\dfrac{3}{4}x - 1\right)$

11. $\dfrac{5y}{6} - 8 = 5 - \dfrac{2y}{3}$

12. $\dfrac{x}{2} - 3 = \dfrac{3x}{5} + 1$

13. $\dfrac{m}{2} - \dfrac{1}{m} = \dfrac{6m + 5}{12}$

14. $-\dfrac{3k}{2} + \dfrac{9k - 5}{6} = \dfrac{11k + 8}{k}$

15. $\dfrac{4}{x - 3} - \dfrac{8}{2x + 5} + \dfrac{3}{x - 3} = 0$

16. $\dfrac{5}{2p + 3} - \dfrac{3}{p - 2} = \dfrac{4}{2p + 3}$

17. $\dfrac{3}{2m + 4} = \dfrac{1}{m + 2} - 2$

18. $\dfrac{8}{3k - 9} - \dfrac{5}{k - 3} = 4$

Use a calculator to solve each of the following equations. Round your answer to the nearest hundredth. (See Example 3.)

19. $9.06x + 3.59(8x - 5) = 12.07x + .5612$

20. $-5.74(3.1 - 2.7p) = 1.09p + 5.2588$

21. $\dfrac{2.63r - 8.99}{1.25} - \dfrac{3.90r - 1.77}{2.45} = r$

22. $\dfrac{8.19m + 2.55}{4.34} - \dfrac{8.17m - 9.94}{1.04} = 4m$

Solve each equation for x. (See Example 7).

23. $4(a - x) = b - a + 2x$

24. $(3a + b) - bx = a(x - 2)$ $(a \neq -b)$

25. $5(b - x) = 2b + ax$ $(a \neq -5)$

26. $bx - 2b = 2a - ax$

Solve each equation for the specified variable. Assume all denominators are nonzero. (See Example 7.)

27. $PV = k$ for V

28. $i = prt$ for p

29. $V = V_0 + gt$ for g

30. $S = S_0 + gt^2 + k$ for g

31. $A = \dfrac{1}{2}(B + b)h$ for B

32. $C = \dfrac{5}{9}(F - 32)$ for F

Solve each equation. (See Examples 8–10.)

33. $|2h + 1| = 5$

34. $|4m - 3| = 12$

35. $|6 - 2p| = 10$

36. $|-5x + 7| = 15$

37. $\left|\dfrac{5}{r - 3}\right| = 10$

38. $\left|\dfrac{3}{2h - 1}\right| = 4$

39. $|3y - 2| = |4y + 5|$

40. $|1 - 3z| = |z + 2|$

Solve the following applied problems.

Business The approximate annual interest rate of a loan paid off with monthly payments is given by

$$A = \dfrac{24f}{b(p + 1)},$$

where f is the finance charge on the loan, p is the total number of payments, and b is the original balance of the loan. Use the formula to find the requested value in the following. Round A to the nearest percent and round other variables to the nearest whole numbers. (This formula is not accurate enough for the requirements of federal law.)

41. $f = \$800$, $b = \$4000$, $p = 36$; find A

42. $A = 5\%$, $b = \$1500$, $p = 24$; find f

43. $A = 6\%$, $f = \$370$, $p = 36$; find b

44. $A = 10\%$, $f = \$490$, $p = 48$; find b

Social Science The population P of the United States (in millions) can be approximated by the equation

$$P = 2.275x + 203.45,$$

where x is the number of years since 1970. Determine the year in which the United States has the given population.

45. 214,825,000

46. 248,950,000

47. 260,325,000

48. 283,075,000

Health   According to data from the U.S. Department of Health and Human Services, the total health care expenditures E in the United States (in billions of dollars) can be approximated by

$$E = 55.99x + 702.76,$$

where x is the number of years since 1990. Determine the year in which health care expenditures were at the given level.

49. $1038.7 billion

50. $1150.7 billion

51. $1318.7 billion

52. $1374.6 billion

Business   When a loan is paid off early, a portion of the finance charge must be returned to the borrower. By one method of calculating the finance charge (called the rule of 78), the amount of unearned interest (finance charge to be returned) is given by

$$u = f \cdot \frac{n(n + 1)}{q(q + 1)},$$

where u represents unearned interest, f is the original finance charge, n is the number of payments remaining when the loan is paid off, and q is the original number of payments. Find the amount of the unearned interest in each of the following.

53. Original finance charge = $800, loan scheduled to run 36 months, paid off with 18 payments remaining

54. Original finance charge = $1400, loan scheduled to run 48 months, paid off with 12 payments remaining

55. *Natural Science*   The excess lifetime cancer risk R is a measure of the likelihood that an individual will develop cancer from a particular pollutant. For example, if $R = .01$, then a person has a 1% increased chance of developing cancer during a lifetime. The value of R for formaldehyde can be calculated using the equation $R = kd$, where k is a constant and d is the daily dose in parts per million. The constant k for formaldehyde can be calculated using the formula $k = .132(\frac{B}{W})$, where B is the total number of cubic meters of air a person breathes in one day and W is a person's weight in kilograms.*

  (a) Find k for a person that breathes in 20 cubic meters of air per day and weighs 75 kg.

  (b) Mobile homes in Minnesota were found to have a mean daily dose d of .42 parts per million. Calculate R.[†]

  (c) For every 5000 people, how many cases of cancer could be expected each year from these levels of formaldehyde? Assume an average life expectancy of 72 years.

---

* A. Hines, T. Ghosh, S. Layalka, and R.Warder, *Indoor Air Quality and Control* Prentice Hall, 1993 (TD 883.1.I476 1993).

†I. Ritchie, and R. Lehnen, "An Analysis of Formaldehyde Concentration in Mobile and Conventional Homes" *Journal of Environmental Health* 47:300–305.

Business   Solve the following investment problems. (See Example 11.)

56. Joe Gonzalvez received $52,000 profit from the sale of some land. He invested part at 5% interest and the rest at 4% interest. He earned a total of $2290 interest per year. How much did he invest at 5%?

57. Weijen Luan invests $20,000 received from an insurance settlement in two ways: some at 6% and some at 4%. Altogether, she makes $1040 per year interest. How much is invested at 4%?

58. Maria Martinelli bought two plots of land for a total of $120,000. On the first plot, she made a profit of 15%. On the second, she lost 10%. Her total profit was $5500. How much did she pay for each piece of land?

59. Suppose $20,000 is invested at 5%. How much additional money must be invested at 4% to produce a yield of 4.8% on the entire amount invested?

Solve the following applied problems. (See Example 12.)

60. A plane flies nonstop from New York to London, cities which are about 3500 miles apart. After one hour and six minutes in the air, the plane passes over Halifax, Nova Scotia, which is 600 miles from New York. Estimate the flying time from New York to London.

61. On vacation, Le Hong averaged 50 mph traveling from Denver to Minneapolis. Returning by a different route that covered the same number of miles, he averaged 55 mph. What is the distance between the two cities if his total traveling time was 32 hours?

62. Russ and Janet are running in the Apple Hill Fun Run. Russ runs at 7 mph, Janet at 5 mph. If they start at the same time, how long will it be before they are 2/3 mile apart?

Natural Science   Using the same assumptions about octane ratings as in Example 13, solve the following problems.

63. How many liters of 94 octane gasoline should be mixed with 200 liters of 99 octane gasoline to get a mixture that is 97 octane?

64. A service station has 92 octane and 98 octane gasoline. How many liters of each gasoline should be mixed to provide 12 liters of 96 octane gasoline for a chemistry experiment?

Solve the following applied problems.

65. *Business*   A major car rental firm charges $45.56 a day with unlimited mileage. A discount firm offers a similar car for $20 a day plus 18 cents per mile. How far must you drive in a day in order for the cost to be the same at both firms?

66. Natural Science A car radiator contains 8 quarts of fluid, 40% of which is antifreeze. How much fluid should be drained and replaced with pure (100%) antifreeze in order that the new mixture is 60% antifreeze?

67. Transportation In Massachusetts speeding fines are determined by the equation

$$y = 10(x - 65) + 50, \qquad x \geq 65,$$

where y is the amount of the fine (in dollars) for driving x miles per hour. If Paul was fined $100 for speeding, how fast was he driving?

68. Health Based on data from the U.S. Office of Management and Budget, Medicare costs C (in billions of dollars) in year x can be approximated by $C = 18x - 35,753$. In what year were Medicare costs approximately 283 billion dollars?

69. Finance Jack borrowed his father's luxury car and promised to return it with a full tank of premium gas, which costs $1.49 per gallon. From experience, he knows that he needs 15.5 gallons. But he has only $22 (and no credit card), which isn't enough. He decides to get as much premium as possible and fill the remainder of the tank with regular gas, which costs $1.29 per gallon. To the nearest thousandth of a gallon, how much of each type of gas should he get?

Geometry *Solve each of these geometry problems. (Hint: In each case, draw an appropriate figure and label its parts.)*

70 The length of a rectangular label is 3 centimeters less than twice the width. The perimeter is 54 centimeters. Find the width.

71. A puzzle piece in the shape of a triangle has a perimeter of 30 centimeters. Two sides of the triangle are each twice as long as the shortest side. Find the length of the shortest side.

72. A triangle has a perimeter of 27 centimeters. One side is twice as long as the shortest side. The third side is seven centimeters longer than the shortest side. Find the length of the shortest side.

73. A closed recycling bin is in the shape of a rectangular box. Find the height of the bin if its length is 18 feet, its width is 8 feet, and its surface area is 496 square feet.

1.7 QUADRATIC EQUATIONS

An equation that can be written in the form

$$ax^2 + bx + c = 0,$$

where a, b, and c are real numbers with $a \neq 0$, is called a **quadratic equation.** For example, each of

$$2x^2 + 3x + 4 = 0, \qquad x^2 = 6x - 9, \qquad 3x^2 + x = 6, \qquad \text{and} \qquad x^2 = 5$$

is a quadratic equation. A solution of an equation that is a real number is said to be a **real solution** of the equation.

One method of solving quadratic equations is based on the following property of real numbers.

Zero-Factor Property

If a and b are real numbers, with $ab = 0$, then $a = 0$ or $b = 0$ or both.

▶ **EXAMPLE 1** Solve the equation $(x - 4)(3x + 7) = 0$.

By the zero-factor property, the product $(x - 4)(3x + 7)$ can equal 0 only if at least one of the factors equals 0. That is, the product equals zero only if $x - 4 = 0$

①

Solve the following equations.

(a) $(y - 6)(y + 2) = 0$

(b) $(5k - 3)(k + 5) = 0$

(c) $(2r - 9)(3r + 5)(r + 3)$
$= 0$

Answers:

(a) 6, −2

(b) 3/5, −5

(c) 9/2, −5/3, −3

or $3x + 7 = 0$. Solving each of these equations separately will give the solutions of the original equation.

$$x - 4 = 0 \quad \text{or} \quad 3x + 7 = 0$$
$$x = 4 \quad \text{or} \quad 3x = -7$$
$$x = -\frac{7}{3}$$

The solutions of the equation $(x - 4)(3x + 7) = 0$ are 4 and −7/3. Check these solutions by substitution in the original equation. ◀ ①

▶ **EXAMPLE 2**   Solve $6r^2 + 7r = 3$.

Rewrite the equation as

$$6r^2 + 7r - 3 = 0.$$

Now factor $6r^2 + 7r - 3$ to get

$$(3r - 1)(2r + 3) = 0.$$

By the zero-factor property, the product $(3r - 1)(2r + 3)$ can equal 0 only if

$$3r - 1 = 0 \quad \text{or} \quad 2r + 3 = 0.$$

Solving each of these equations separately gives the solutions of the original equation.

$$3r = 1 \quad \text{or} \quad 2r = -3$$
$$r = \frac{1}{3} \qquad r = -\frac{3}{2}$$

Verify that both 1/3 and −3/2 are solutions by substituting them in the original equation. ◀ ②

An equation such as $x^2 = 5$ has two solutions, $\sqrt{5}$ and $-\sqrt{5}$. The same idea is true in general.

②

Solve each equation by factoring.

(a) $y^2 + 3y = 10$

(b) $2r^2 + 9r = 5$

(c) $4k^2 = 9k$

Answers:

(a) 2, −5

(b) 1/2, −5

(c) 9/4, 0

Square-Root Property

If $b > 0$, then the solutions of $x^2 = b$ are $\sqrt{b}$ and $-\sqrt{b}$.

The two solutions are sometimes abbreviated $\pm\sqrt{b}$.

▶ **EXAMPLE 3**   Solve each equation.

(a) $m^2 = 17$

By the square-root property, the solutions are $\sqrt{17}$ and $-\sqrt{17}$, abbreviated $\pm\sqrt{17}$.

③

Solve each equation by using the square-root property.

(a) $p^2 = 21$

(b) $(m + 7)^2 = 15$

(c) $(2k - 3)^2 = 5$

Answers:

(a) $\pm\sqrt{21}$

(b) $-7 \pm \sqrt{15}$

(c) $(3 \pm \sqrt{5})/2$

(b) $(y - 4)^2 = 11$

Use a generalization of the square-root property, working as follows:

$$(y - 4)^2 = 11$$
$$y - 4 = \sqrt{11} \quad \text{or} \quad y - 4 = -\sqrt{11}$$
$$y = 4 + \sqrt{11} \qquad\qquad y = 4 - \sqrt{11}.$$

Abbreviate the solutions as $4 \pm \sqrt{11}$. ◀ ③

As suggested by Example 3(b), it is easy to solve any quadratic equation that can be written in the form $(x + d)^2 = k$. Fortunately, there is a simple algebraic technique that enables us to write any quadratic equation in this form.

▶**EXAMPLE 4**   Solve $4x^2 - 24x + 19 = 0$.

First, rearrange the equation so that the x terms are on the left and the constants on the right:

$$4x^2 - 24x = -19.$$

Now divide both sides by 4 so that 1 is the coefficient of x^2:

$$x^2 - 6x = -\frac{19}{4}.$$

Now take half the coefficient of x (namely, $-6/2 = -3$) and square it, obtaining 9. (The reason for this is explained below). Add 9 to both sides and factor the left side:

$$x^2 - 6x + 9 = -\frac{19}{4} + 9$$
$$(x - 3)^2 = \frac{17}{4}.$$

Use the square-root property to complete the solution.

$$x - 3 = \pm\sqrt{\frac{17}{4}} = \pm\frac{\sqrt{17}}{2}$$

$$x = 3 \pm \frac{\sqrt{17}}{2} = \frac{6 \pm \sqrt{17}}{2}$$

The two solutions are $\dfrac{6 + \sqrt{17}}{2}$ and $\dfrac{6 - \sqrt{17}}{2}$. ◀

To see why the technique used in Example 4 always works, look at this multiplication:

$$\left(x + \frac{b}{2}\right)^2 = x^2 + 2\left(\frac{b}{2}\right)x + \left(\frac{b}{2}\right)^2 = x^2 + bx + \left(\frac{b}{2}\right)^2.$$

If you read it from right to left, it shows that the expression $x^2 + bx$ can be made into a perfect square by adding the square of half of the coefficient of x, namely,

$\left(\dfrac{b}{2}\right)^2$. That's exactly what was done in Example 4 (where $b = -6$). This technique is called **completing the square.**

The method of completing the square can be used on the general quadratic equation,

$$ax^2 + bx + c = 0 \qquad (a \neq 0),$$

to convert it to one whose solutions can be found by the square-root property. This will give a general formula for solving any quadratic equation. Going through the necessary algebra produces the following important result, which you should memorize.

Quadratic Formula

The solutions of the quadratic equation $ax^2 + bx + c = 0$, where $a \neq 0$, are given by

$$x = \dfrac{-b \pm \sqrt{b^2 - 4ac}}{2a}.$$

CAUTION When using the quadratic formula, remember that the equation must be in the form $ax^2 + bx + c = 0$. Also, notice that the fraction bar in the quadratic formula extends under *both* terms in the numerator. Be sure to add $-b$ to $\pm\sqrt{b^2 - 4ac}$ *before* dividing by $2a$.

▶ **EXAMPLE 5**   Solve $x^2 + 1 = 4x$.

First add $-4x$ to both sides to get 0 alone on the right side.

$$x^2 - 4x + 1 = 0$$

Now identify the letters a, b, and c. Here $a = 1$, $b = -4$, and $c = 1$. Substitute these numbers into the quadratic formula

$$x = \dfrac{-(-4) \pm \sqrt{(-4)^2 - 4(1)(1)}}{2(1)}$$

$$= \dfrac{4 \pm \sqrt{16 - 4}}{2}$$

$$= \dfrac{4 \pm \sqrt{12}}{2}$$

$$= \dfrac{4 \pm 2\sqrt{3}}{2} \qquad \sqrt{12} = \sqrt{4 \cdot 3} = \sqrt{4} \cdot \sqrt{3} = 2\sqrt{3}$$

$$= \dfrac{2(2 \pm \sqrt{3})}{2} \qquad \text{Factor } 4 \pm 2\sqrt{3}.$$

$$x = 2 \pm \sqrt{3}.$$

The $\pm$ sign represents the two solutions of the equation. First use $+$ and then use $-$ to find each of the solutions: $2 + \sqrt{3}$ and $2 - \sqrt{3}$. ◀ ④

④

Use the quadratic formula to solve each equation.

(a) $x^2 - 2x = 2$

(b) $u^2 - 6u + 4 = 0$

Answers:

(a) $x = 1 + \sqrt{3}$ or $1 - \sqrt{3}$

(b) $u = 3 + \sqrt{5}$ or $3 - \sqrt{5}$

Example 5 shows that the quadratic formula produces exact solutions. In many real-world applications, however, you must use a calculator to find decimal approximations of the solutions. The approximate solutions in Example 5 are

$$x = 2 + \sqrt{3} \approx 3.732050808 \quad \text{and} \quad x = 2 - \sqrt{3} \approx .2679491924,$$

as shown in Figure 1.14.

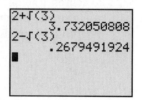

FIGURE 1.14

▶ **EXAMPLE 6** Use the quadratic formula and a calculator to solve

$$3.2x^2 + 15.93x - 7.1 = 0.$$

Compute $\sqrt{b^2 - 4ac} = \sqrt{15.93^2 - 4(3.2)(-7.1)} \approx 18.56461419$ and store the result in memory. This number may be inserted in any computation by using the memory recall key, which we shall denote here by M (it may be labeled differently on your calculator). Figure 1.15 shows that the approximate solutions are

$$x = \frac{-15.93 + M}{2 \cdot 3.2} \approx .4116584671 \quad \text{and} \quad x = \frac{-15.93 - M}{2 \cdot 3.2} \approx -5.389783467.$$

FIGURE 1.15

⑤

Find approximate solutions for $5.1x^2 - 3.3x - 240.624 = 0$.

Answer:

$x = 7.2$ or $x \approx -6.5529$

Note the use of parentheses in Figure 1.15—omitting the parentheses leads to wrong answers. Also remember that these answers are *approximations*, so they may not check exactly when substituted in the original equation. ◀ ⑤

TECHNOLOGY TIP You can approximate the solutions of quadratic equations on a graphing calculator by using a quadratic formula program (see the Program Appendix) or using a built-in quadratic equation solver if your calculator has one. Then you need only enter the coefficients a, b, c to obtain the approximate solutions. See your instruction manual for details.

▶**EXAMPLE 7**   Solve $9x^2 - 30x + 25 = 0$.

Applying the quadratic formula with $a = 9$, $b = -30$, and $c = 25$, we have

$$x = \frac{-(-30) \pm \sqrt{(-30)^2 - 4(9)(25)}}{2(9)}$$

$$= \frac{30 \pm \sqrt{900 - 900}}{18} = \frac{30 \pm 0}{18} = \frac{30}{18} = \frac{5}{3}.$$

Therefore, the given equation has only one real solution. The fact that the solution is a rational number indicates that this equation could have been solved by factoring. ◀

▶**EXAMPLE 8**   Solve $x^2 - 6x + 10 = 0$.

Apply the quadratic formula with $a = 1$, $b = -6$, and $c = 10$.

$$x = \frac{-(-6) \pm \sqrt{(-6)^2 - 4(1)(10)}}{2(1)}$$

$$= \frac{6 \pm \sqrt{36 - 40}}{2}$$

$$= \frac{6 \pm \sqrt{-4}}{2}$$

Since no negative number has a square root in the real number system, $\sqrt{-4}$ is not a real number. Hence the equation has no real solutions. ◀ ⑥

Examples 5–8 show that the number of solutions of the quadratic equation $ax^2 + bx + c = 0$ is determined by $b^2 - 4ac$, the quantity under the radical, which is called the **discriminant** of the equation. ⑦

---

⑥

Solve each equation.

(a) $9k^2 - 6k + 1 = 0$

(b) $4m^2 + 28m + 49 = 0$

(c) $2x^2 - 5x + 5 = 0$

Answers:

(a) 1/3

(b) −7/2

(c) No real solutions

---

⑦

Use the discriminant to determine the number of real solutions of each equation.

(a) $x^2 + 8x + 3 = 0$

(b) $2x^2 + x + 3 = 0$

(c) $x^2 - 194x + 9409 = 0$

Answers:

(a) 2

(b) 0

(c) 1

The Discriminant

The equation $ax^2 + bx + c = 0$ has either two, or one, or no real solutions.

If $b^2 - 4ac > 0$, there are two real solutions. (*Examples 5 and 6*)

If $b^2 - 4ac = 0$, there is one real solution. (*Example 7*)

If $b^2 - 4ac < 0$, there are no real solutions. (*Example 8*)

Applications

Quadratic equations arise in a variety of settings, as illustrated in the following examples. Example 9 depends on the following useful fact from geometry.

The Pythagorean Theorem

In a right triangle with legs of lengths a and b and hypotenuse of length c,

$$a^2 + b^2 = c^2.$$

▶**EXAMPLE 9**   The size of a computer monitor is the diagonal measurement of its screen. The height of the screen is approximately 3/4 of its width. Kathy claims that John's 14-inch monitor has less than half the viewing area of her 21-inch monitor. John says, "No way! 14/21 is 2/3, so my monitor has 2/3 of the viewing area of yours." Who is right?

First, find the area of Kathy's screen. Let x be its width. Then its height is 3/4 of x, that is, .75x, as shown in Figure 1.16.

FIGURE 1.16

⑧

Let x be the width of John's screen in Example 9.

(a) Write an equation that expresses the relationship between the width, height, and diagonal of the screen.

(b) Find the dimensions of the screen.

(c) Find its area.

Answer:

(a) $x^2 + (.75x)^2 = 14^2$

(b) 11.2 by 8.4 inches

(c) 94.08 square inches

By the Pythagorean theorem,

$$x^2 + (.75x)^2 = 21^2$$
$$x^2 + .5625x^2 = 441 \qquad \text{Expand } (.75x)^2$$
$$(1 + .5625)x^2 = 441 \qquad \text{Distributive property}$$
$$1.5625x^2 = 441$$
$$x^2 = 282.24 \qquad \text{Divide both sides by 1.5625}$$
$$x = \pm\sqrt{282.24} \qquad \text{Square root property}$$
$$x = \pm16.8$$

We can ignore the negative solution since x is a width. Thus, the width is 16.8 inches and the height is .75x = .75(16.8) = 12.6 inches, so the area is

$$\text{Area} = \text{width} \times \text{height} = 16.8 \times 12.6 = 211.68 \text{ square inches.}$$

Next, find the area of John's screen by working Problem ⑧ at the side, which shows that John's screen has an area of 94.08 square inches. So, who is right?   ◀

▶**EXAMPLE 10**   A landscape architect wants to make an exposed gravel border of uniform width around a small shed behind a company plant. The shed is 10 feet by 6 feet. He has enough gravel to cover 36 square feet. How wide should the border be?

A sketch of the shed with border is given in Figure 1.17. Let x represent the width of the border. Then the width of the large rectangle is $6 + 2x$ and its length is $10 + 2x$.

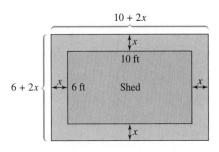

FIGURE 1.17

We must write an equation relating the given areas and dimensions. The area of the large rectangle is $(6 + 2x)(10 + 2x)$. The area occupied by the shed is $6 \cdot 10 = 60$. The area of the border is found by subtracting the area of the shed from the area of the large rectangle. This difference should be 36 square feet, giving the equation

$$(6 + 2x)(10 + 2x) - 60 = 36.$$

Solve this equation with the following sequence of steps.

$$60 + 32x + 4x^2 - 60 = 36$$
$$4x^2 + 32x - 36 = 0$$
$$x^2 + 8x - 9 = 0$$
$$(x + 9)(x - 1) = 0$$

The solutions are -9 and 1. The number -9 cannot be the width of the border, so the solution is to make the border 1 foot wide. ◀

In some applications it may be necessary to solve an equation in several variables for a specific variable.

▶ **EXAMPLE 11** Solve $v = mx^2 + x$ for x. (Assume m and v are positive.)

The equation is quadratic in x because of the x^2 term. Use the quadratic formula, first writing the equation in standard form.

$$v = mx^2 + x$$
$$0 = mx^2 + x - v$$

Let $a = m$, $b = 1$, and $c = -v$. Then the quadratic formula gives

$$x = \frac{-1 \pm \sqrt{1^2 - 4(m)(-v)}}{2m}$$

$$x = \frac{-1 \pm \sqrt{1 + 4mv}}{2m}.$$ ◀

The length of a picture is 2 inches more than the width. It is mounted on a mat that extends 2 inches beyond the picture on all sides. What are the dimensions of the picture if the area of the mat is 99 square inches?

Answer:

5 inches by 7 inches

Solve each of the following equations for the indicated variable. Assume all variables are positive.

(a) $k = mp^2 - bp$ for p

(b) $r = \dfrac{APk^2}{3}$ for k

Answers:

(a) $p = \dfrac{b \pm \sqrt{b^2 + 4mk}}{2m}$

(b) $k = \pm\sqrt{\dfrac{3r}{AP}}$ or $\dfrac{\pm\sqrt{3rAP}}{AP}$

1.7 Exercises

Use factoring to solve each equation. (See Examples 1 and 2.)

1. $(x + 3)(x - 12) = 0$
2. $(p - 16)(p - 5) = 0$
3. $x(x + 5) = 0$
4. $x^2 - 2x = 0$
5. $3z^2 = 6z$
6. $x^2 - 81 = 0$
7. $y^2 + 15y + 56 = 0$
8. $k^2 - 4k - 5 = 0$
9. $2x^2 = 5x - 3$
10. $2 = 12z^2 + 5z$
11. $6r^2 + r = 1$
12. $3y^2 = 16y - 5$
13. $2m^2 + 20 = 13m$
14. $10a^2 + 17a + 3 = 0$
15. $m(m - 7) = -10$
16. $z(2z + 7) = 4$
17. $9x^2 - 16 = 0$
18. $25y^2 - 64 = 0$
19. $16x^2 - 16x = 0$
20. $12y^2 - 48y = 0$

Solve each equation by using the square-root property. (See Example 3.)

21. $(r - 2)^2 = 7$
22. $(b + 5)^2 = 8$
23. $(4x - 1)^2 - 20$
24. $(3t + 5)^2 = 11$

Use the quadratic formula to solve each equation. If the solutions involve square roots, give both the exact and approximate solutions. (See Examples 5–8.)

25. $2x^2 + 5x + 1 = 0$
26. $3x^2 - x - 7 = 0$
27. $4k^2 + 2k = 1$
28. $r^2 = 3r + 5$
29. $5y^2 + 6y = 2$
30. $2z^2 + 3 = 8z$
31. $6x^2 + 6x + 5 = 0$
32. $3a^2 - 2a + 2 = 0$
33. $2r^2 - 7r + 5 = 0$
34. $8x^2 = 8x - 3$
35. $2x^2 - 7x + 30 = 0$
36. $3k^2 + k = 6$
37. $1 + \dfrac{7}{2a} = \dfrac{15}{2a^2}$
38. $5 - \dfrac{4}{k} - \dfrac{1}{k^2} = 0$

Use the discriminant to determine the number of real solutions of each equation. You need not solve the equations.

39. $25t^2 + 49 = 70t$
40. $9z^2 - 12z = 1$
41. $13x^2 + 24x - 6 = 0$
42. $22x^2 + 19x + 5 = 0$

Use a calculator and the quadratic formula to find approximate solutions of the equation. (See Example 6.)

43. $4.42x^2 - 10.14x + 3.79 = 0$
44. $3x^2 - 82.74x + 570.4923 = 0$
45. $7.63x^2 + 2.79x = 5.32$
46. $8.06x^2 + 25.8726x = 25.047256$

Solve the following problems. (See Example 3.)

47. **Physical Science**　When wind resistance is ignored, a falling object travels a distance of d feet in t seconds, where $d = 16t^2$.

(a) How long does it take for the object to fall 36 feet?

(b) How long does it take for the object to fall five times farther than in part (a)?

48. **Transportation**　According to the Federal Aviation Administration, the maximum recommended taxiing speed x (in miles per hour) for a plane on a curved runway exit is given by $R = .5x^2$, where R is the radius of the curve (in feet). Find the maximum taxiing speed for planes on such exits when the radius of the exit is

(a) 450 ft　　　　(b) 615 ft　　　　(c) 970 ft.

Solve the following problems. (See Example 6.)

49. **Social Science**　The number of drivers who die in automobile accidents is related to the ages of the drivers, with teenagers and the elderly having the worst records. According to data from the National Highway Traffic Safety Administration, the driver fatality rate D per 1000 licensed drivers every 100 million miles can be approximated by the equation $D = .0031x^2 - .291x + 7.1$, where x is the age of the driver.

(a) For what ages is the driver fatality rate about 1 death per 1000?

(b) For what ages is the rate three times greater than in part (a)?

50. **Social Science**　According to data from the U.S. Census Bureau, the population P of Cleveland, Ohio (in thousands), in year x can be approximated by $P = -.172x^2 + 16.82x + 487.62$, where $x = 0$ corresponds to 1900. In what year after 1940 was the population 877,620?

51. **Business**　According to data from the International Civil Aviation Organization, the number N of airline passengers (in millions) in year x can be approximated by the equation $N = .68x^2 + 3.8x + 24$, where $x = 0$ corresponds to 1950. In what year did the number of passengers first reach

(a) one billion?　　　　(b) two billion?

(Recall that one billion is 1000 million.)

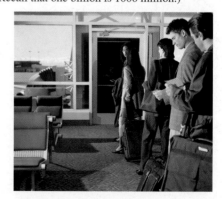

52. Social Science According to data from the U.S. Department of Health and Human Services, the cumulative number N of reported cases of AIDS in the United States in year x can be approximated by the equation $N = 3362.1x^2 - 17{,}270.3x + 24{,}043$, where $x = 0$ corresponds to 1980. In what year did the total reach 550,000?

Solve the following problems. (See Examples 9 and 10.)

53. Transportation Two trains leave the same city at the same time, one going north and the other east. The eastbound train travels 20 mph faster than the northbound one. After 5 hours the trains are 300 miles apart. Determine the speed of each train, using these steps.

(a) Let x denote the speed of the northbound train. Express the speed of the eastbound train in terms of x.

(b) Write expressions that give the distance traveled by each train after 5 hours.

(c) Use part (b) and the fact that the trains are 300 miles apart after 5 hours to write an equation. (A diagram of the situation may help.)

(d) Solve the equation and determine the speeds of the trains.

54. Chris and Josh have received walkie-talkies for Christmas. If they leave from the same point at the same time, Chris walking north at 2.5 mph and Josh walking east at 3 mph, how long will they be able to talk to each other if the range of the walkie-talkies is 4 miles? Round your answer to the nearest minute.

55. An ecology center wants to set up an experimental garden. It has 300 meters of fencing to enclose a rectangular area of 5000 square meters. Find the length and width of the rectangle as follows.

(a) Let $x =$ the length and write an expression for the width.

(b) Write an equation relating the length, width, and area, using the result of part (a).

(c) Solve the problem.

56. A landscape architect has included a rectangular flower bed measuring 9 feet by 5 feet in her plans for a new building. She wants to use two colors of flowers in the bed, one in the center and the other for a border of the same width on all four sides. If she can get just enough plants to cover 24 square feet for the border, how wide can the border be?

57. Joan wants to buy a rug for a room that is 12 feet by 15 feet. She wants to leave a uniform strip of floor around the rug. She can afford 108 square feet of carpeting. What dimensions should the rug have?

58. In 1991 Rick Mears won the (500 mile) Indianapolis 500 race. His speed (rate) was 100 mph (to the nearest mph) faster than that of the 1911 winner, Ray Harroun. Mears completed the race in 3.74 hours less time than Harroun. Find Mears's rate to the nearest whole number.

59. Physical Science If a ball is thrown upward with an initial velocity of 64 feet per second, then its height after t seconds is $h = 64t - 16t^2$. In how many seconds will the ball reach

(a) 64 ft?   (b) 28 ft?

(c) Why are two answers possible?

60. Physical Science A particle moves horizontally with its distance (in centimeters) from a starting point at t seconds given by $d = 11t^2 - 10t$.

(a) How long will it take before the particle returns to the starting point?

(b) When will the particle be 100 centimeters from the starting point?

Solve each of the following equations for the indicated variable. Assume all denominators are nonzero and that all variables represent positive real numbers. (See Example 11.)

61. $S = \dfrac{1}{2}gt^2$ for t

62. $a = \pi r^2$ for r

63. $L = \dfrac{d^4 k}{h^2}$ for h

64. $F = \dfrac{kMv^2}{r}$ for v

65. $P = \dfrac{E^2 R}{(r + R)^2}$ for R

66. $S = 2\pi rh + 2\pi r^2$ for r

67. Solve the equation $z^4 - 2z^2 = 15$ as follows.

(a) Let $x = z^2$ and write the equation in terms of x.

(b) Solve the new equation for x.

(c) Set z^2 equal to each positive answer in part (b) and solve the resulting equation.

Solve each of the following equations. (See Exercise 67.)

68. $6p^4 = p^2 + 2$

69. $2q^4 + 3q^2 - 9 = 0$

70. $4a^4 = 2 - 7a^2$

71. $z^4 - 3z^2 - 1 = 0$

72. $2r^4 - r^2 - 5 = 0$

CHAPTER 1 **SUMMARY**

Key Terms and Symbols

1.1 $\approx$ is approximately equal to
π pi
$|a|$ absolute value of a
real number
natural (counting) number
whole number
integer
rational number
irrational number
properties of real numbers
additive inverse
multiplicative inverse
order of operations
square roots
number line
interval
interval notation
absolute value

1.2 a^n a to the power n
exponent or power
base
polynomial

variable
coefficient
term
constant term
degree of a nonzero term
degree of a polynomial
zero polynomial
binomial
trinomial
like terms
FOIL
revenue
fixed cost
variable cost
profit

1.3 factor
factoring
greatest common factor
perfect square trinomial
difference of two squares
sum and difference of two cubes

1.4 rational expression
operations with rational
  expressions
complex fraction

1.5 $a^{1/n}$ nth root of a
$\sqrt{a}$ square root of a
$\sqrt[n]{a}$ nth root of a
zero exponent
properties of exponents
radical
radicand
index
rationalizing the denominator
rationalizing the numerator

1.6 first-degree equation
solution of an equation
properties of equality
solving for a specified variable

1.7 quadratic equation
real solution
zero-factor property
square-root property
completing the square
quadratic formula
discriminant
Pythagorean theorem

Key Concepts

Factoring

$x^2 + 2xy + y^2 = (x + y)^2$ $x^3 - y^3 = (x - y)(x^2 + xy + y^2)$

$x^2 - 2xy + y^2 = (x - y)^2$ $x^3 + y^3 = (x + y)(x^2 - xy + y^2)$

$x^2 - y^2 = (x + y)(x - y)$

Rules for Radicals

Let a and b be real numbers, n be a positive integer, and m be any integer for which the following exist.

$a^{m/n} = \sqrt[n]{a^m} = (\sqrt[n]{a})^m$ $\sqrt[n]{a^n} = |a|$ if n is even $\sqrt[n]{a^n} = a$ if n is odd

$\sqrt[n]{a} \cdot \sqrt[n]{b} = \sqrt[n]{ab}$ $\dfrac{\sqrt[n]{a}}{\sqrt[n]{b}} = \sqrt[n]{\dfrac{a}{b}}$ $(b \neq 0)$

Rules for Exponents

Let a, b, r, and s be any real numbers for which the following exist.

$$a^{-r} = \frac{1}{a^r} \qquad a^0 = 1 \qquad \left(\frac{a}{b}\right)^r = \frac{a^r}{b^r}$$

$$a^r \cdot a^s = a^{r+s} \qquad (a^r)^s = a^{rs} \qquad a^{1/r} = \sqrt[r]{a}$$

$$\frac{a^r}{a^s} = a^{r-s} \qquad (ab)^r = a^r b^r \qquad \left(\frac{a}{b}\right)^{-r} = \left(\frac{b}{a}\right)^r$$

Absolute Value

Assume a and b are real numbers with $b > 0$.

The solutions of $|a| = b$ or $|a| = |b|$ are $a = b$ or $a = -b$.

Quadratic Equations

Facts needed to solve quadratic equations (in which a, b, and c are real numbers):

Factoring: If $ab = 0$, then $a = 0$ or $b = 0$ or both.

Square-Root Property: If $b > 0$, then the solutions of $x^2 = b$ are $\sqrt{b}$ and $-\sqrt{b}$.

Quadratic Formula: The solutions of $ax^2 + bx + c$ ($a \neq 0$) are

$$x = \frac{-b \pm \sqrt{b^2 - 4ac}}{2a}.$$

Discriminant: There are two real solutions if $b^2 - 4ac > 0$, one real solution if $b^2 - 4ac = 0$, and no real solutions if $b^2 - 4ac < 0$.

Chapter 1 Review Exercises

Name the numbers from the list -12, -6, $-9/10$, $-\sqrt{7}$, $-\sqrt{4}$, 0, $1/8$, $\pi/4$, 6, $\sqrt{11}$ *that are*

1. whole numbers

2. integers

3. rational numbers

4. irrational numbers

Identify the properties of real numbers that are illustrated in each of the following.

5. $7[(-3)4] = 7[4(-3)]$

6. $9(4 + 5) = (4 + 5)9$

7. $6(x + y - 3) = 6x + 6y + 6(-3)$

8. $11 + (8 + 3) = (11 + 8) + 3$

Express each statement in symbols.

9. x is at least 6.

10. x is negative.

Write the following numbers in numerical order from smallest to largest.

11. $-7, -3, 8, \pi, -2, 0$

12. $\dfrac{5}{6}, \dfrac{1}{2}, -\dfrac{2}{3}, -\dfrac{5}{4}, -\dfrac{3}{8}$

13. $|6 - 4|, -|-2|, |8 + 1|, -|3 - (-2)|$

14. $\sqrt{7}, -\sqrt{8}, -|\sqrt{16}|, |-\sqrt{12}|$

Write without absolute value bars.

15. $-|-6| + |3|$

16. $|-5| + |-9|$

17. $7 - |-8|$

18. $|-2| - |-7 + 3|$

Graph each of the following on a number line.

19. $x \geq -3$

20. $-4 < x \leq 6$

21. $x < -2$

22. $x \leq 1$

Use the order of operations to simplify.

23. $(-6 + 2 \cdot 5)(-2)$

24. $-4(-7 - 9 \div 3)$

25. $\dfrac{-8 + (-6)(-3) \div 9}{6 - (-2)}$

26. $\dfrac{20 \div 4 \cdot 2 \div 5 - 1}{-9 - (-3) - 12 \div 3}$

Perform each of the indicated operations.

27. $(3x^4 - x^2 + 5x) - (-x^4 + 3x^2 - 8x)$

28. $(-8y^3 + 8y^2 - 3y) - (2y^3 - 4y^2 - 10)$

29. $-2(q^4 - 3q^3 + 4q^2) + 4(q^4 + 2q^3 + q^2)$

30. $5(3y^4 - 4y^5 + y^6) - 3(2y^4 + y^5 - 3y^6)$

31. $(5z + 2)(3z - 2)$ **32.** $(8p - 4)(5p + 3)$

33. $(4k - 3h)(4k + 3h)$ **34.** $(2r - 5y)(2r + 5y)$

35. $(6x + 3y)^2$ **36.** $(2a - 5b)^2$

Factor as completely as possible.

37. $2kh^2 - 4kh + 5k$ **38.** $2m^2n^2 + 6mn^2 + 16n^2$

39. $3a^4 + 13a^3 + 4a^2$ **40.** $24x^3 + 4x^2 - 4x$

41. $10y^2 - 11y + 3$ **42.** $8q^2 + 3m + 4qm + 6q$

43. $4a^2 - 20a + 25$ **44.** $36p^2 + 12p + 1$

45. $144p^2 - 169q^2$ **46.** $81z^2 - 25x^2$

47. $8y^3 - 1$ **48.** $125a^3 + 216$

Perform each operation.

49. $\dfrac{4x}{5} \cdot \dfrac{35x}{12}$

50. $\dfrac{5k^2}{24} - \dfrac{75k}{36}$

51. $\dfrac{c^2 - 3c + 2}{2c(c - 1)} \div \dfrac{c - 2}{8c}$

52. $\dfrac{p^3 - 2p^2 - 8p}{3p(p^2 - 16)} \div \dfrac{p^2 + 4p + 4}{9p^2}$

53. $\dfrac{2y - 10}{5y} \cdot \dfrac{20y - 25}{12}$

54. $\dfrac{m^2 - 2m}{15m^3} \cdot \dfrac{5}{m^2 - 4}$

55. $\dfrac{2m^2 - 4m + 2}{m^2 - 1} \div \dfrac{6m + 18}{m^2 + 2m - 3}$

56. $\dfrac{x^2 + 6x + 5}{4(x^2 + 1)} \cdot \dfrac{2x(x + 1)}{x^2 - 25}$

57. $\dfrac{6}{15z} + \dfrac{2}{3z} - \dfrac{9}{10z}$

58. $\dfrac{5}{y - 2} - \dfrac{4}{y}$

59. $\dfrac{2}{5q} + \dfrac{10}{7q}$

60. Give two ways to evaluate $125^{2/3}$ and then compare them. Which do you prefer? Why?

Simplify each of the following. Write all answers without negative exponents. Assume all variables represent positive real numbers.

61. 5^{-3} **62.** 10^{-2}

63. -7^0 **64.** -3^{-1}

65. $\left(-\dfrac{6}{5}\right)^{-2}$ **66.** $\left(\dfrac{2}{3}\right)^{-3}$

67. $4^6 \cdot 4^{-3}$ **68.** $7^{-5} \cdot 7^{-1}$

69. $\dfrac{9^{-4}}{9^{-3}}$ **70.** $\dfrac{6^{-2}}{6^3}$

71. $\dfrac{9^4 \cdot 9^{-5}}{(9^{-2})^2}$ **72.** $\dfrac{k^4 \cdot k^{-3}}{(k^{-2})^{-3}}$

73. $4^{-1} + 2^{-1}$ **74.** $3^{-2} + 3^{-1}$

75. $125^{2/3}$ **76.** $128^{3/7}$

77. $9^{-5/2}$ **78.** $\left(\dfrac{144}{49}\right)^{-1/2}$

79. $\dfrac{5^{1/3} \cdot 5^{1/2}}{5^{3/2}}$ **80.** $\dfrac{2^{3/4} \cdot 2^{-1/2}}{2^{1/4}}$

81. $(3a^2)^{1/2} \cdot (3^2a)^{3/2}$ **82.** $(4p)^{2/3} \cdot (2p^3)^{3/2}$

Simplify each of the following.

83. $\sqrt[3]{27}$ **84.** $\sqrt[6]{-64}$

85. $\sqrt{24}$ **86.** $\sqrt{63}$

87. $\sqrt[3]{54p^3q^5}$ **88.** $\sqrt[4]{64a^5b^3}$

89. $\sqrt{\dfrac{5n^2}{6m}}$ **90.** $\sqrt{\dfrac{3x^3}{2z}}$

91. $2\sqrt{3} - 5\sqrt{12}$

92. $8\sqrt{7} + 2\sqrt{28}$

93. $(\sqrt{5} - 1)(\sqrt{5} + 1)$

94. $(\sqrt{7} - \sqrt{3})(\sqrt{7} + \sqrt{3})$

Rationalize each denominator.

95. $\dfrac{\sqrt{2}}{1 + \sqrt{3}}$

96. $\dfrac{4 + \sqrt{2}}{4 - \sqrt{5}}$

Social Science   *In our system of government, the president is elected by the electoral college and not by individual voters. Because of this, smaller states have a greater voice in the selection of a president than they otherwise would have. Two political scientists have studied the problems of campaigning for president under the current system and have concluded that candidates should allot their money according to the formula*

$$\begin{array}{l}\text{Amount for} \\ \text{large state}\end{array} = \left(\dfrac{E_{large}}{E_{small}}\right)^{3/2} \times \begin{array}{l}\text{amount for} \\ \text{small state.}\end{array}$$

Here E_{large} represents the electoral vote of the large state, and E_{small} represents the electoral vote of the small state. Find the amount that should be spent in each of the following larger states if \$1,000,000 is spent in the small state and the following statements are true.

97. The large state has 48 electoral votes, and the small state has 3.

98. The large state has 36 electoral votes, and the small state has 4.

99. 6 votes in a small state; 28 in a large

100. 9 votes in a small state; 32 in a large

Solve each equation.

101. $2x - 5(x - 4) = 3x + 2$

102. $5y + 6 = -2(1 - 3y) + 4$

103. $\dfrac{2z}{5} - \dfrac{4z - 3}{10} = \dfrac{-z + 1}{10}$

104. $\dfrac{p}{p + 2} - \dfrac{3}{4} = \dfrac{2}{p + 2}$

105. $\dfrac{2m}{m - 3} = \dfrac{6}{m - 3} + 4$

106. $\dfrac{15}{k + 5} = 4 - \dfrac{3k}{k + 5}$

Solve for x.

107. $5ax - 1 = x$

108. $6x - 3y = 4bx$

109. $\dfrac{2x}{3 - c} = ax + 1$

110. $b^2x - 2x = 4b^2$

Solve each equation.

111. $|m - 3| = 9$

112. $|6 - x| = 12$

113. $\left|\dfrac{2 - y}{5}\right| = 8$

114. $|4k + 1| = |6k - 3|$

115. Business   The number of private-label items in grocery stores is steadily growing. The number P of private-label items available (in thousands) in year x can be approximated by $P = 2.4x + 26.4$, where $x = 0$ corresponds to 1990.* Find the year in which the number of private-label items was
 (a) 48,000   **(b)** 40,800.

116. Business   A laser printer is on sale for 15% off. The sale price is $306. What was the original price?

117. Health   The percentage M of people who retire early and receive medical coverage from their employer in year x can be approximated by $M = -1.9x + 50.8$, where $x = 0$ corresponds to 1990.† Find the year in which the percentage of early retirees receiving employer sponsored medical coverage was
 (a) 44.8%   **(b)** 30.8%.

118. Finance   Ellen borrowed $500 from a credit union at 12% annual interest and got $250 in cash with her credit card at 18% annual interest. What single rate of interest on $750 results in the same total amount of interest?

119. Finance   A real estate firm invests $100,000 proceeds from a sale in two ways. The first portion is invested in a shopping center that provides an annual return of 8%. The rest is invested in a small apartment building with an annual return of 5%. The firm wants annual income of $6800 from these investments. How much should be put into each investment?

120. Business   To make a special mix for Valentine's Day, the owner of a candy store wants to combine chocolate hearts which sell for $5 per pound with candy kisses which sell for $3.50 per pound. How many pounds of each kind should be used to get 30 pounds of a mix that can be sold for $4.50 per pound?

Determine the number of real solutions of each quadratic equation.

121. $x^2 - 6x = 4$

122. $-3x^2 + 5x + 2 = 0$

123. $4x^2 - 12x + 9 = 0$

124. $5x^2 + 2x + 1 = 0$

125. $x^2 + 3x + 5 = 0$

Find all real solutions of each equation.

126. $(b + 7)^2 = 5$ **127.** $(2p + 1)^2 = 7$

128. $2p^2 + 3p = 2$ **129.** $2y^2 = 15 + y$

130. $x^2 - 2x = 2$ **131.** $r^2 + 4r = 1$

132. $2m^2 - 12m = 11$ **133.** $9k^2 + 6k = 2$

134. $2a^2 + a - 15 = 0$ **135.** $12x^2 = 8x - 1$

136. $2q^2 - 11q = 21$ **137.** $3x^2 + 2x = 16$

138. $6k^4 + k^2 = 1$ **139.** $21p^4 = 2 + p^2$

140. $2x^4 = 7x^2 + 15$

141. $3m^4 + 20m^2 = 7$

142. $3 = \dfrac{13}{z} + \dfrac{10}{z^2}$

Solve each equation for the specified variable.

143. $p = \dfrac{E^2R}{(r + R)^2}$ for r

144. $p = \dfrac{E^2R}{(r + R)^2}$ for E

145. $K = s(s - a)$ for s

146. $kz^2 - hz - t = 0$ for z

---

*Datamonitor.

†William M. Mercer, Inc.

147. **Energy** The demand D for electricity in California (in thousands of gigawatt-hours) in year x can be approximated by $D = .74x^2 - 2.24x + 242.4$ $(0 \le x \le 10)$, where $x = 0$ corresponds to 1990.*
    (a) In what year was demand about 294,000 gigawatt-hour?
    (b) Assuming this model remains valid for the current decade, in what year will demand reach 396,000 gigawatt-hours?

148. **Business** A landscaper wants to put a cement walk of uniform width around a rectangular garden that measures 24 by 40 feet. She has enough cement to cover 740 square feet. To the nearest tenth of a foot, how wide should the walk be in order to use up all the cement?

149. **Business** A recreation director wants to fence off a rectangular playground beside an apartment building. The building forms one boundary, so she needs to fence only the other three sides. The area of the playground is to be 11,250 square meters. She has enough material to build 325 meters of fence. Find the length and width of the playground.

150. **Transportation** Two cars leave an intersection at the same time. One travels north and the other heads west traveling 10 mph faster. After 1 hour they are 50 miles apart. What were their speeds?

*Cambridge Energy Research Associates.

C a s e 1

Consumers Often Defy Common Sense*

Imagine two refrigerators in the appliance section of a department store. One sells for $700 and uses $85 worth of electricity a year. The other is $100 more expensive but costs only $25 a year to run. Given that either refrigerator should last at least 10 years without repair, consumers would overwhelmingly buy the second model, right?

Well, not exactly. Many studies by economists have shown that in a wide range of decisions about money—from paying taxes to buying major appliances—consumers consistently make decisions that defy common sense.

In some cases—as in the refrigerator example—this means that people are generally unwilling to pay a little more money up front to save a lot of money in the long run. At times, psychological studies have shown, consumers appear to assign entirely whimsical values to money, values that change depending on time and circumstances.

In recent years, these apparently irrational patterns of human behavior have become a subject of intense interest among economists and psychologists, both for what they say

*"Consumers often defy common sense" by Malcolm Gladwell from *The Washington Post*. Copyright © 1990. The Washington Post. Reprinted with permission.

about the way the human mind works and because of their implications for public policy.

How, for example, can the United States move toward a more efficient use of electricity if so many consumers refuse to buy energy-efficient appliances even when such a move is in their own best interest?

At the heart of research into the economic behavior of consumers is a concept known as the discount rate. It is a measure of how consumers compare the value of a dollar received today with one received tomorrow.

Consider, for example, if you won $1000 in a lottery. How much more money would officials have to give you before you would agree to postpone cashing the check for a year?

Some people might insist on at least another $100, or 10 percent, since that is roughly how much it would take to make up for the combined effects of a year's worth of inflation and lost interest.

But the studies show that someone who wants immediate gratification might not be willing to postpone receiving the $1000 for 20 percent or 30 percent or even 40 percent more money.

In the language of economists, this type of person has a high discount rate: He or she discounts the value of $1000 so much over a year that it would take hundreds of extra dollars to make waiting as attractive as getting the money immediately.

Of the two alternatives, waiting a year for more money is clearly more rational than taking the check now. Why would people turn down $1400 dollars next year in favor of $1000 today? Even if they needed the $1000 immediately, they would be better off borrowing it from a bank, even at 20 percent or even 30 percent interest. Then, a year later, they could pay off the loan—including the interest—with the $1400 and pocket the difference.

The fact is, however, that economists find numerous examples of such high discount rates implicit in consumer behavior.

While consumers were very much aware of savings to be made at the point of purchase, they so heavily discounted the value of monthly electrical costs that they would pay over the lifetime of their dryer or freezer that they were oblivious to the potential for greater savings.

Gas water heaters, for example, were found to carry an implicit discount rate of 100 percent. This means that in deciding which model was cheapest over the long run, consumers acted as if they valued a $100 gas bill for the first year as if it were really $50. Then, in the second year, they would value the next $100 gas bill as if it were really worth $25, and so on through the life of the appliance.

Few consumers actually make this formal calculation, of course. But there are clearly bizarre behavioral patterns in evidence.

Some experiments, for example, have shown that the way in which consumers make decisions about money depends a great deal on how much is at stake. Few people are willing to give up $10 now for $15 next year. But they are if the choice is between $100 now and $150 next year, a fact that would explain why consumers appear to care less about many small electricity bills—even if they add up to a lot—than one big initial outlay.

Exercises

1. Suppose a refrigerator that sells for $700 costs $85 a year for electricity. Write an expression for the cost to buy and run the refrigerator for x years.
2. Suppose another refrigerator costs $1000 and $25 a year for electricity. Write an expression for the total cost for this refrigerator over x years.
3. Over 10 years which refrigerator costs the most? By how much?
4. In how many years will the the total costs for the two refrigerators be equal?

Chapter 2

Graphs, Lines, and Inequalities

Graphical representations of data are commonly used in business, health sciences, and social sciences. Such data is often used to construct mathematical models useful for analyzing the current situation and predicting future performance. Lines, equations, and inequalities play an important role in developing such models. This chapter presents both algebraic and graphical methods for dealing with these topics.

2.1 GRAPHS

Just as the number line associates the points on a line with real numbers, a similar construction in two dimensions associates points in the plane with *ordered pairs* of real numbers. A **Cartesian coordinate system,** as shown in Figure 2.1, consists of a

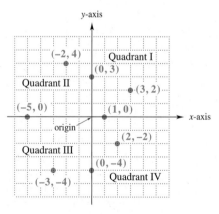

FIGURE 2.1

horizontal number line (usually called the **x-axis**) and a vertical number line (usually called the **y-axis**). The point where the number lines meet is called the **origin.** Each point in a Cartesian coordinate system is labeled with an **ordered pair** of real numbers, such as $(-2, 4)$ or $(3, 2)$. Several points and their corresponding ordered pairs are shown in Figure 2.1.

For the point labeled $(-2, 4)$, for example, -2 is the **x-coordinate** and 4 is the **y-coordinate.** You can think of these coordinates as directions telling you how to move to this point from the origin: you go 2 horizontal units to the left (x-coordinate) and 4 vertical units upward (y-coordinate). From now on, instead of referring to "the point labeled by the ordered pair $(-2, 4)$," we will say "the point $(-2, 4)$."

①

The x-axis and the y-axis divide the plane into four parts, or **quadrants,** which are numbered as shown in Figure 2.1. The points on the coordinate axes belong to no quadrant.

Equations and Graphs

A **solution of an equation** in two variables, such as

$$y = -2x + 3$$

or

$$y = x^2 + 7x - 2,$$

is an ordered pair of numbers such that the substitution of the first number for x and the second number for y produces a true statement.

▶ **EXAMPLE 1**   Which of the following are solutions of $y = -2x + 3$?

(a) $(2, -1)$

This is a solution of $y = -2x + 3$ because "$-1 = -2 \cdot 2 + 3$" is a true statement.

(b) $(4, 7)$

Since $-2 \cdot 4 + 3 = -5$ and not 7, the ordered pair $(4, 7)$ is not a solution of $y = -2x + 3$. ◀ ②

Equations in two variables, such as $y = -2x + 3$, typically have an infinite number of solutions. To find one, choose a number for x and then compute the value of y that produces a solution. For instance, if $x = 5$, then $y = -2 \cdot 5 + 3 = -7$, so that the pair $(5, -7)$ is a solution of $y = -2x + 3$. Similarly, if $x = 0$, then $y = -2 \cdot 0 + 3 = 3$, so that $(0, 3)$ is also a solution.

The **graph** of an equation in two variables is the set of points in the plane whose coordinates (ordered pairs) are solutions of the equation. Thus, the graph of an equation is a picture of its solutions. Since a typical equation has infinitely many solutions, its graph has infinitely many points.

---

①
Locate $(-1, 6)$, $(-3, -5)$, $(4, -3)$, $(0, 2)$, and $(-5, 0)$ on a coordinate system.

Answer:

②
Which of the following are solutions of $y = x^2 + 7x - 2$?

(a) $(1, 6)$

(b) $(-2, -20)$

(c) $(-1, -8)$

Answer:

(a) and (c)

③

Graph $x = 5y$.

Answer:

▶ **EXAMPLE 2**   Sketch the graph of $y = -2x + 5$.

Since we cannot plot infinitely many points, we construct a table of y-values for a reasonable number of x-values, plot the corresponding points, and make an "educated guess" about the rest. The table of values and points in Figure 2.2 suggests that the graph is a straight line, as shown in Figure 2.3. ◀ ③

x	-1	0	2	4	5
$-2x + 5$	7	5	1	-3	-5

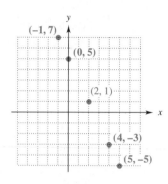

| **FIGURE 2.2** | **FIGURE 2.3** |

An **x-intercept** of a graph is the x-coordinate of a point where the graph intersects the x-axis (the y-coordinate of this point is 0 since it's on the axis). Consequently, to find the x-intercepts of the graph of an equation, set $y = 0$ and solve for x. For instance, in Example 2, the x-intercept of the graph of $y = -2x + 5$ (see Figure 2.3) is found by setting $y = 0$ and solving for x.

$$0 = -2x + 5$$
$$2x = 5$$
$$x = \frac{5}{2}$$

Similarly, a **y-intercept** of a graph is the y-coordinate of a point where the graph intersects the y-axis (the x-coordinate of this point is 0—why?) The y-intercepts are found by setting $x = 0$ and solving for y. For example, the graph of $y = -2x + 5$ in Figure 2.3 has y-intercept 5. ④

④

Find the x- and y-intercepts of the graphs of these equations.

(a) $3x + 4y = 12$

(b) $5x - 2y = 8$

Answers:

(a) x-intercept 4,
      y-intercept 3

(b) x-intercept 8/5,
      y-intercept -4

▶ **EXAMPLE 3**   Find the x- and y-intercepts of the graph of $y = 4 - x^2$ and sketch the graph.

Setting $x = 0$ in $y = 4 - x^2$, we see that the y-intercept is $y = 4$. Similarly, setting $y = 0$ gives $x^2 = 4$. Thus the x-intercepts are $x = 2$ and $x = -2$. Now make a table, being sure to take both positive and negative values for x, and plot the corresponding points, as in Figure 2.4. These points suggest that the entire graph looks like Figure 2.5. ◀

x	$4 - x^2$
-3	-5
-2	0
-1	3
0	4
1	3
2	0
3	-5

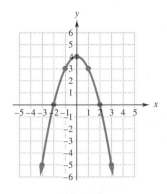

FIGURE 2.4   **FIGURE 2.5**

▶ **EXAMPLE 4**   Sketch the graph of $y = \sqrt{x}$.

Notice that x can take only nonnegative values (since the square root of a negative number is not defined) and that the corresponding value of y is also nonnegative. Hence all the points on the graph will lie in the first quadrant. Computing some typical values, we obtain the graph in Figure 2.6.   ◀

x	$\sqrt{x}$
0	0
1	1
4	2
6	$\sqrt{6} \approx 2.45$
9	3
11	$\sqrt{11} \approx 3.32$

FIGURE 2.6

Example 2 shows that the x-intercept of the graph of $y = -2x + 5$ is the solution of $-2x + 5 = 0$. Example 3 shows that the x-intercepts of the graph of $y = 4 - x^2$ are the solutions of $4 - x^2 = 0$. Similar facts hold in the general case.

Intercepts and Equations

The real solutions of a one-variable equation of the form

expression in x = 0

are the x-intercepts of the graph of the two-variable equation

y = expression in x.

Technology and Graphs

A graphing calculator or computer graphing program follows essentially the same procedure used when graphing by hand: the calculator selects a large number of x-values (95 or more), equally spaced along the x-axis, and plots the corresponding points, simultaneously connecting them with line segments. Calculator-generated graphs are generally quite accurate, although they may not appear as smooth as hand-drawn ones. The next example illustrates the basics of graphing on a graphing calculator (computer graphing software operates similarly). For a full discussion, see the Graphing Calculator Appendix.

EXAMPLE 5 Use a graphing calculator to sketch the graph of the equation $2x^3 - 2y - 10x + 2 = 0$.

First, set the **viewing window**—the portion of the coordinate plane that will appear on the screen. Press the WINDOW key (labeled RANGE or PLOT-SETUP on some calculators) and enter the appropriate numbers, as in Figure 2.7 (which shows a T1-83; other calculators are similar). Then the calculator will display the portion of the plane shown in Figure 2.8, that is, the points (x, y) with $-9 \le x \le 9$ and $-6 \le y \le 6$.

FIGURE 2.7   FIGURE 2.8

In Figure 2.7 we have set Xscl = 2 and Yscl = 1, which means the **tick marks** on the x-axis are two units apart and the tick marks on the y-axis are one unit apart (Figure 2.8).

Second, enter the equation to be graphed in the equation memory. To do this, you must first solve the equation for y (because a calculator only accepts equations of the form $y = $ expression in x):

$$2y = 2x^3 - 10x + 2$$
$$y = x^3 - 5x + 1.$$

Now press the Y= key (labeled SYMB on some calculators) and enter the equation, using the "variable key" for x, (This key is labeled X,T,θ, n or X,θ,T or x-VAR, depending on the calculator.) Figure 2.9 shows the equation entered on a TI-83; other calculators are similar. Now press GRAPH (or PLOT or DRW on some calculators) and obtain Figure 2.10.

FIGURE 2.9   FIGURE 2.10

Use a graphing calculator to graph $y = 18x - 3x^3$ in the following viewing windows.

(a) $-10 \le x \le 10$ and $-10 \le y \le 10$ with Xscl = 1, Yscl = 1

(b) $-5 \le x \le 5$ and $-20 \le y \le 20$ with Xscl = 1, Yscl = 5

Answers:

(a)

(b)

Finally, if necessary, change the viewing window to obtain a more readable graph. It is difficult to see the y-intercept in Figure 2.10, so press WINDOW and change the viewing window (Figure 2.11); then press GRAPH to obtain Figure 2.12 in which the y-intercept at $y = 1$ is clearly shown. (It isn't necessary to reenter the equation.) ◀ ⑤

FIGURE 2.11

FIGURE 2.12

▶ **EXAMPLE 6**   Use a graphing calculator to solve the equation $x^3 - 5x + 1 = 0$. The box on page 69 shows that the solutions of the equation are the x-intercepts of the graph of $y = x^3 - 5x + 1$ (Figure 2.12). To determine these x-intercepts as accurately as possible, use the **graphical root finder** on a graphing calculator. Look in the CALC or GRAPH MATH or PLOT FCN or G-SOLVE menu for "root," "zero," or "x-intercept." Check your instruction manual for the proper syntax. A typical root finder (Figure 2.13) shows that two of the solutions (x-intercepts) are $x \approx .2016$ and $x \approx 2.1284$. For the third solution, see Problem ⑥ at the side. ◀

Use graphical root finder to approximate the third solution of the equation in Example 6.

Answer:

$x \approx -2.330059$

FIGURE 2.13

Graph Reading

Information is often given in graphical form, so you must be able to read and interpret graphs, that is, to translate graphical information into statements in English.

▶ **EXAMPLE 7**   At various locations, the National Weather Service continuously records the temperature in graphical form. The results for March 20, 2001, in Cleveland, Ohio, are displayed in Figure 2.14 on the next page. The first coordinate of each point on the graph represents the time (measured in hours after midnight) and the second coordinate, the temperature at that time.

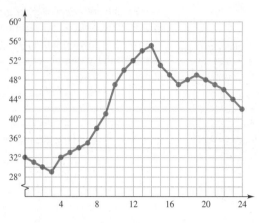

FIGURE 2.14

(a) What was the temperature at 9 A.M. and 6 P.M.?

The point (9, 41) is on the graph, which means that the temperature at 9 A.M. was 41°. Now 6 P.M. is 18 hours after midnight and the point (18, 48) is on the graph. So the temperature at 6 P.M. was 48°.

(b) At what times during the day was the temperature below 50°?

Look for the points whose second coordinates are less than 50, that is, points that lie below the horizontal line through 50°. The first coordinates of these points are the times when the temperature was below 50°. Figure 2.14 shows that these are the points with first coordinates less than 11 or greater than about 15.5. Since 15.5 hours corresponds to 3:30 P.M., the temperature was below 50° from midnight to 11 A.M. and from approximately 3:30 P.M. to midnight.   ◀ ⑦

The next example deals with the basic business relationship that was introduced in Section 1.2:

$$\text{Profit} = \text{Revenue} - \text{Cost}.$$

⑦

In Example 7, what are the highest and lowest temperatures during the day? When did they occur?

Answer:

Highest was 55° at 2 P.M.
Lowest was 29° at 3 A.M.

▶ **EXAMPLE 8**   Monthly revenue and costs for the Fenton Cell Phone Company are determined by the number *t* of phones produced and sold, as shown in Figure 2.15.

FIGURE 2.15

(a) How many phones should be produced each month if the company is to make a profit (assuming all phones produced are sold)?

Profit is Revenue − Cost, so the company makes a profit whenever revenue is greater than cost, that is, when the revenue graph is above the cost graph. Figure 2.15 shows that this occurs between $t = 12$ and $t = 48$, that is, when 12,000 to 48,000 phones are produced. If the company makes fewer than 12,000 phones, it will lose money (costs will be greater than revenue). They also lose money by making more than 48,000 phones (one reason might be that high production levels require large amounts of overtime pay, which drives costs up too much).

(b) Is it more profitable to make 40,000 or 44,000 phones?

On the revenue graph, the point with first coordinate 40 has second coordinate approximately 3.7, meaning that the revenue from 40,000 phones is about 3.7 million dollars. The point with first coordinate 40 on the cost graph is (40, 2), meaning that the cost of producing 40,000 phones is 2 million dollars. Therefore, the profit on 40,000 phones is about $3.7 - 2 = 1.7$ million dollars. For 44,000 phones, we have the approximate points (44, 4) on the revenue graph and (44, 3) on the cost graph. So the profit on 44,000 phones is $4 - 3 = 1$ million dollars. Consequently, it is more profitable to make 40,000 phones. ◀ ⑧

⑧

In Example 8, find the profit from making

(a) 32,000 phones;

(b) 4000 phones.

Answers:

(a) About $1,200,000 (rounded)

(b) About −$500,000 (that is a loss of $500,000)

2.1 Exercises

Determine whether the given ordered pair is a solution of the given equation. (See Example 1.)

1. $(1, -2)$; $3x - y - 5 = 0$

2. $(2, -1)$; $x^2 + y^2 - 6x + 8y = -15$

3. $(3, 4)$; $(x - 2)^2 + (y + 5)^2 = 4$

4. $(1, -1)$; $\dfrac{x^2}{2} + \dfrac{y^2}{3} = 1$

Sketch the graph of each of these equations. (See Example 2.)

5. $4y + 3x = 12$ **6.** $2x + 7y = 14$

7. $8x + 3y = 12$ **8.** $9y - 4x = 12$

9. $x = 2y + 3$ **10.** $x - 3y = 0$

List the x-intercepts and y-intercepts of each graph.

11.

12.

13.

14.

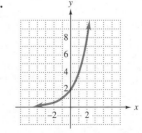

Find the x-intercepts and y-intercepts of the graph of each equation. You need not sketch the graph. (See Example 3.)

15. $3x + 4y = 24$ **16.** $x - 2y = 3$

17. $2x - 3y = 6$ **18.** $3x + y = 4$

19. $y = x^2 - 9$ **20.** $y = x^2 + 4$

Sketch the graph of the equation. (See Examples 2–4.)

21. $y = x^2$ **22.** $y = x^2 + 2$

23. $y = x^2 - 3$ **24.** $y = 2x^2$

25. $y = x^3$ **26.** $y = x^3 - 3$

27. $y = x^3 + 1$ **28.** $y = x^3/2$

29. $y = \sqrt{x + 2}$ **30.** $y = \sqrt{x - 2}$

31. $y = \sqrt{4 - x^2}$ **32.** $y = \sqrt{9 - x^2}$

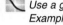 *Use a graphing calculator to find the graph of the equation. (See Example 5.)*

33. $y = x^2 + x + 1$ **34.** $y = 2 - x - x^2$

35. $y = (x - 3)^3$ **36.** $y = x^3 + 2x^2 + 2$

37. $y = x^3 - 3x^2 + x - 1$ **38.** $y = x^4 - 5x^2 - 2$

 Use a graphing calculator for Exercises 39–40.

39. Graph $y = x^4 - 2x^3 + 2x$ in a window with $-3 \le x \le 3$. Is the "flat" part of the graph near $x = 1$ really a horizontal line segment? (*Hint:* Use trace to move along the "flat" part and watch the y-coordinates. Do they remain the same [as they should on a horizontal segment]?)

40. (a) Graph $y = x^4 - 2x^3 + 2x$ in the **standard window** (the one with $-10 \le x \le 10$ and $-10 \le y \le 10$). Use the trace feature to approximate the coordinates of the lowest point on the graph.

    (b) Use a minimum finder to obtain an accurate approximation of the lowest point. How does this compare with your answer in part (a)?

 Use a graphing calculator to approximate all real solutions of the equation. (See Example 6.)

41. $x^3 - 3x^2 + 5 = 0$

42. $x^3 + x - 1 = 0$

43. $2x^3 - 4x^2 + x - 3 = 0$

44. $6x^3 - 5x^2 + 3x - 2 = 0$

45. $x^5 - 6x + 6 = 0$

46. $x^3 - 3x^2 + x - 1 = 0$

 Use a graphing calculator to work Exercises 47–50.

47. Physical Science   The surface area S of the right circular cone in the figure is given by $S = \pi r\sqrt{r^2 + h^2}$. What radius should be used to produce a cone of height 5 inches and surface area 100 square inches?

48. Physical Science   The surface area of the right square pyramid in the figure is given by $S = b\sqrt{b^2 + 4h^2}$. If the pyramid has height 10 feet and surface area 100 square feet, what is the length of a side b of its base?

49. Health   According to data from the National Cancer Institute, the rate of lung cancer cases per 100,000 women in the year x can be approximated by the equation $y = -.00075x^3 - .0083x^2 + 1.565x + 21.23$, where $x = 0$ corresponds to 1975. In what year did the female lung cancer rate first reach 40 per 100,000?

50. Natural Science   Carbon monoxide combines with the hemoglobin of the blood to form carboxyhemoglobin (COHb), which reduces transport of oxygen to tissue. A 4% to 6% COHb level in the blood (typical for smokers) can cause symptoms such as blood flow alterations, visual impairment, and poorer vigilance ability. When $50 \le x \le 100$, the equation $T = .00787x^2 - 1.528x + 75.89$ approximates the exposure time T (in hours) necessary to reach this 4% to 6% level, where x is the amount of carbon monoxide present in the air in parts per million (ppm).*

    (a) A kerosene heater or a room full of smokers is capable of producing 50 ppm of carbon monoxide. How long would it take for a nonsmoking person to start feeling the above symptoms?

    (b) What carbon monoxide concentration will cause a person to reach a 4% to 6% COHb level in three hours?

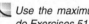 *Use the maximum/minimum finder on a graphing calculator to do Exercises 51 and 52.*

51. Social Science   The number N of boys in the United States involved in high school sports (in millions) in year x can be approximated by $N = -.00015x^3 + .01x^2 - .16x + 4.1$, where $x = 0$ corresponds to 1975. Between 1975

---

*U.S. Department of Energy

and 2000, when were the fewest boys involved in sports? How many were involved then?

52. **Health** The number y of AIDS cases (in thousands) diagnosed in year x can be approximated by

$$y = .21x^3 - 4.6x^2 + 21.9x + 50 \qquad (0 \le x \le 9),$$

where $x = 0$ corresponds to 1990. When were the most cases diagnosed?

Physical Science *The temperature graphs for Fargo and Seattle on the same day are shown in the figure. Use them to do the following exercises. (See Example 7.)*

53. Approximately when did the temperature first reach 60° in Fargo? In Seattle?

54. At what times during the day did the two cities have the same temperature?

55. At what times during the day was it warmer in Fargo than in Seattle?

56. Was there any time when it was at least 10° warmer in Seattle than in Fargo?

Business *Use the revenue and cost graphs for the Fenton Cell Phone Company in Example 8 to do Exercises 57–60.*

57. Find the approximate cost of manufacturing the given number of phones.
    (a) 20,000   (b) 36,000
    (c) 48,000

58. Find the approximate revenue from selling the given number of phones.
    (a) 12,000   (b) 24,000
    (c) 36,000

59. Find the approximate profit from manufacturing the given number of phones.
    (a) 20,000   (b) 28,000
    (c) 36,000

60. The company must replace its aging machinery with better, but much more expensive, machines. In addition, raw material prices increase, so that monthly costs go up by $250,000. Owing to competitive pressure, phone prices cannot be increased, so revenue remains the same. Under these new circumstances, find the approximate profit from manufacturing the given number of phones.
    (a) 20,000   (b) 36,000
    (c) 40,000

Physical Science *According to an article in the December 1994 issue of* Scientific American, *the coast-down time for a typical 1993 car as it drops 10 miles per hour from an initial speed depends on variations from the standard condition (automobile in neutral; average drag and tire pressure). The accompanying graph illustrates some of these conditions with coast-down time in seconds and initial speed in miles per hour.*

Use the graph to answer the following questions.

61. What is the approximate coast-down time in fifth gear if the initial speed is 40 miles per hour?

62. For what speed is the coast-down time the same for the conditions of high air drag and hard tires?

63. **Business** Sandi Goldstein takes out a 30-year mortgage on which her monthly payment is $850. During the early years of the mortgage, most of each payment is for interest and the rather small remainder for principal. As time goes on the portion of each payment that goes for interest decreases, while the portion for principal increases, as shown in the accompanying graph.

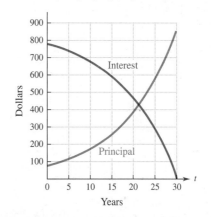

    (a) Approximately how much of the $850 monthly payment goes for interest in year 5? In year 15? In year 25?
    (b) In what year will the monthly payment be equally divided between interest and principal?

Business   *The accompanying graph shows non-agricultural employment (in thousands) in the United States during recent years.* Use it to answer the following questions.

64. How many people were employed at the beginning of 1995? At the beginning of 2000?

65. During what period was employment below 112,000,000?

66. Approximately when was employment the lowest?

U.S. Nonagricultural Employment
Seasonally Adjusted (in thousands)

*U.S. Bureau of Labor Statistics; data for 2000–2002 are projections.

Business   *The Consumer Confidence Index reflects people's feelings about their employment opportunities and income prospects. The accompanying graph from the* Seattle Post-Intelligencer *shows the index over a five-year period. Use it to answer the following questions.*

67. How did the Consumer Confidence Index vary during 1997 and 1998?

68. What was the lowest level of consumer confidence during 1998 and 1999? The highest level in these years?

69. During what periods was the index below 120?

Consumer Confidence Index
Seasonally Adjusted

Note: The index compares results with its base year, 1985, when it stood at 100.
Source: The Conference Board   THE ASSOCIATED PRESS

2.2 EQUATIONS OF LINES

Straight lines, which are the simplest graphs, play an important role in a wide variety of applications. They are considered here from both a geometric and an algebraic point of view.

The key geometric feature of a nonvertical straight line is how steeply it rises or falls as you move from left to right. The "steepness" of a line can be represented numerically by a number called the *slope* of the line.

To see how slope is defined, start with Figure 2.16, which shows a line passing through the two different points $(x_1, y_1) = (-3, 5)$ and $(x_2, y_2) = (2, -4)$. The difference in the two x-values,

$$x_2 - x_1 = 2 - (-3) = 5,$$

in this example is called the **change in x.** The Greek letter Δ (delta) is used to denote change. The symbol Δx (read "delta x") represents the change in x. In the same way, Δy represents the **change in y.** In this example,

$$\Delta y = y_2 - y_1 = -4 - 5 = -9.$$

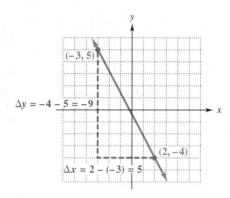

FIGURE 2.16

> The **slope** of the line through the two points (x_1, y_1) and (x_2, y_2), where $x_1 \neq x_2$, is defined as the quotient of the change in y and the change in x:
>
> $$\textbf{slope} = \frac{\textbf{change in } y}{\textbf{change in } x} = \frac{\Delta y}{\Delta x} = \frac{y_2 - y_1}{x_2 - x_1}.$$

The slope of the line in Figure 2.16 is

$$\text{slope} = \frac{\Delta y}{\Delta x} = \frac{-4 - 5}{2 - (-3)} = -\frac{9}{5}.$$

Using similar triangles from geometry, it can be shown that the slope is independent of the choice of points on the line. That is, the same value of the slope will be obtained for *any* choice of two different points on the line.

▶ **EXAMPLE 1**   Find the slope of the line through the points $(-7, 6)$ and $(4, 5)$.

Let $(x_1, y_1) = (-7, 6)$ and $(x_2, y_2) = (4, 5)$. Use the definition of slope.

$$\text{slope} = \frac{\Delta y}{\Delta x} = \frac{y_2 - y_1}{x_2 - x_1} = \frac{5 - 6}{4 - (-7)} = \frac{-1}{11} = -\frac{1}{11}$$

The slope can also be found by letting $(x_1, y_1) = (4, 5)$ and $(x_2, y_2) = (-7, 6)$. In that case,

$$\text{slope} = \frac{6 - 5}{-7 - 4} = \frac{1}{-11} = -\frac{1}{11},$$

the same answer.   ◀ ①

①

Find the slope of the line through

(a) $(6, 11), (-4, -3)$;

(b) $(-3, 5), (-2, 8)$.

Answers:

(a)   7/5

(b)   3

CAUTION   When finding the slope of a line, be careful to subtract the *x*-values and the *y*-values in the same order. For example, with the points $(4, 3)$ and $(2, 9)$, if you use $9 - 3$ for the numerator, you must use $2 - 4$ (*not* $4 - 2$) for the denominator.

▶ **EXAMPLE 2**   Find the slope of the horizontal line in Figure 2.17.

Every point on the line has the same *y*-coordinate, -5. Choose any two of them to compute the slope, say, $(x_1, y_1) = (-3, -5)$ and $(x_2, y_2) = (2, -5)$:

$$\text{slope} = \frac{-5 - (-5)}{2 - (-3)}$$

$$= \frac{0}{5}$$

$$= 0. \quad ◀$$

FIGURE 2.17   **FIGURE 2.18**

▶ **EXAMPLE 3**   What is the slope of the vertical line in Figure 2.18?

Every point on the line has the same *x*-coordinate, 4. If you attempt to compute the slope with two of these points, say $(x_1, y_1) = (4, -2)$ and $(x_2, y_2) = (4, 1)$, you obtain

$$\text{slope} = \frac{1 - (-2)}{4 - 4}$$

$$= \frac{3}{0}.$$

Division by 0 is not defined, so the slope of this line is undefined.   ◀

The arguments used in Examples 2 and 3 work in the general case and lead to the following conclusion.

The slope of every horizontal line is 0.

The slope of every vertical line is undefined.

Slope-Intercept Form

Slope can be used to develop an algebraic description of nonvertical straight lines. Assume that a line with slope *m* has *y*-intercept *b*, so that it goes through the point

$(0, b)$ (see Figure 2.19). Let (x, y) be any point on the line other than $(0, b)$. Using the definition of slope with the points $(0, b)$ and (x, y) gives

$$m = \frac{y - b}{x - 0}$$

$$m = \frac{y - b}{x}$$

$mx = y - b$ Multiply both sides by x.

$y = mx + b$ Add b to both sides. Reverse the equation.

In other words, the coordinates of any point on the line satisfy the equation $y = mx + b$.

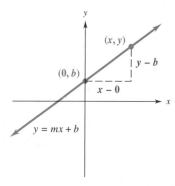

FIGURE 2.19

②

Find an equation for the line with

(a) y-intercept -3 and slope 2/3;

(b) y-intercept 1/4 and slope $-3/2$.

Answers:

(a) $y = \frac{2}{3}x - 3$

(b) $y = -\frac{3}{2}x + \frac{1}{4}$

Slope-Intercept Form

If a line has slope m and y-intercept b, then it is the graph of the equation

$$y = mx + b.$$

This equation is called the **slope-intercept form** of the equation of the line.

▶ **EXAMPLE 4** Find an equation for the line with y-intercept 7/2 and slope $-5/2$.

Use the slope-intercept form with $b = 7/2$ and $m = -5/2$.

$$y = mx + b$$

$$y = -\frac{5}{2}x + \frac{7}{2}$$ ◀ ②

▶ **EXAMPLE 5** Find the equation of the horizontal line with y-intercept 3.

The slope of the line is 0 (why?) and its y-intercept is 3, so its equation is

$$y = mx + b$$
$$y = 0x + 3$$
$$y = 3$$ ◀

The argument in Example 5 also works in the general case.

Find the slope and y-intercept for

(a) $x + 4y = 6$;

(b) $3x - 2y = 1$.

Answers:

(a) Slope $-1/4$; y-intercept $3/2$

(b) Slope $3/2$; y-intercept $-1/2$

If k is a constant, then the graph of the equation $y = k$ is the horizontal line with y-intercept k.

▶ **EXAMPLE 6**   Find the slope and y-intercept for each of the following lines.

(a) $5x - 3y = 1$
Solve for y.

$$5x - 3y = 1$$
$$-3y = -5x + 1$$
$$y = \frac{5}{3}x - \frac{1}{3}$$

This equation is in the form $y = mx + b$, with $m = 5/3$ and $b = -1/3$. So the slope is $5/3$ and the y-intercept is $-1/3$.

(b) $-9x + 6y = 2$
Solve for y.

$$-9x + 6y = 2$$
$$6y = 9x + 2$$
$$y = \frac{3}{2}x + \frac{1}{3}$$

The slope is $3/2$ (coefficient of x) and the y-intercept is $1/3$. ◀ ③

(a) List the slopes of the following lines:

E: $y = -.3x$,   F: $y = -x$,
G: $y = -2x$,   H: $y = -5x$.

(b) Graph all four lines on the same set of axes.

(c) How are the slopes of the lines related to their steepness?

Answers:

(a) Slope E $= -.3$; slope F $= -1$; slope G $= -2$; slope H $= -5$.

(b)

(c) The larger the slope in absolute value, the more steeply the line falls from left to right.

The slope-intercept form can be used to show how the slope measures the steepness of a line. Consider the straight lines A, B, C, and D given by the following equations; each has y-intercept 0 and slope as indicated.

A: $y = .5x$,   B: $y = x$,   C: $y = 3x$,   D: $y = 7x$
Slope .5   Slope 1   Slope 3   Slope 7

For these lines, Figure 2.20 shows that the bigger the slope, the more steeply the line rises from left to right. ④

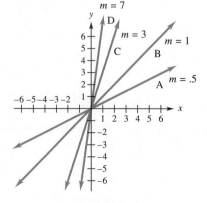

FIGURE 2.20

⑤

Graph the lines and label the intercepts.

(a) $3x + 4y = 12$

(b) $5x - 2y = 8$

Answers:

(a)

(b)

The preceding discussion and Problem 4 in the margin may be summarized as follows.

Direction of Line (moving from left to right)	Slope
Upward	**Positive** (larger for steeper lines)
Horizontal	**0**
Downward	**Negative** (larger in absolute value for steeper lines)
Vertical	**Undefined**

▶ **EXAMPLE 7**   Sketch the graph of $x + 2y = 5$ and label the intercepts.

Find the x-intercept by setting $y = 0$ and solving for x.

$$x + 2 \cdot 0 = 5$$
$$x = 5$$

The x-intercept is 5 and $(5, 0)$ is on the graph. The y-intercept is found similarly, by setting $x = 0$ and solving for y.

$$0 + 2y = 5$$
$$y = 5/2$$

The y-intercept is 5/2 and $(0, 5/2)$ is on the graph. The points $(5, 0)$ and $(0, 5/2)$ can be used to sketch the graph (Figure 2.21).   ◀ ⑤

FIGURE 2.21

TECHNOLOGY TIP   To graph a linear equation on a graphing calculator, you must first put the equation in slope-intercept form $y = mx + b$ so that it can be entered in the equation memory (called the Y= list on some calculators). Vertical lines cannot be graphed on most calculators.

Slopes of Parallel and Perpendicular Lines

We shall assume the following facts without proof. The first one is a consequence of the fact that slope measures steepness and parallel lines have the same steepness.

Two nonvertical lines are parallel whenever they have the same slope.

Two nonvertical lines are perpendicular whenever the product of their slopes is -1.

▶ **EXAMPLE 8** Determine whether each of the following pairs of lines are *parallel, perpendicular,* or *neither.*

(a) $2x + 3y = 5$ and $4x + 5 = -6y$

Put each equation in slope-intercept form by solving for y.

$$3y = -2x + 5 \qquad -6y = 4x + 5$$

$$y = -\frac{2}{3}x + \frac{5}{3} \qquad y = -\frac{2}{3}x - \frac{5}{6}$$

In each case the slope (coefficient of x) is $-2/3$, so the lines are parallel.

(b) $3x = y + 7$ and $x + 3y = 4$

Put each equation in slope-intercept form to determine its slope.

$$3x = y + 7 \qquad\qquad 3y = -x + 4$$

$$y = 3x - 7 \qquad\qquad y = -\frac{1}{3}x + \frac{4}{3}$$

$$\text{slope } 3 \qquad\qquad \text{slope } -1/3$$

Since $3(-1/3) = -1$, these lines are perpendicular.

(c) $x + y = 4$ and $x - 2y = 3$

Verify that the slope of the first line is -1 and the slope of the second is $1/2$. The slopes are not equal and their product is not -1, so the lines are neither parallel nor perpendicular. ◀ ⑥

⑥

Tell if the lines in each of the following pairs are *parallel, perpendicular,* or *neither.*

(a) $x - 2y = 6$ and
$2x + y = 5$

(b) $3x + 4y = 8$ and
$x + 3y = 2$

(c) $2x - y = 7$ and
$2y = 4x - 5$

Answers:

(a) Perpendicular

(b) Neither

(c) Parallel

TECHNOLOGY TIP Perpendicular lines may not appear perpendicular on a graphing calculator, unless you use a *square window*—one in which a one-unit segment on the *y*-axis is the same length as a one-unit segment on the *x*-axis. To obtain such a window on most calculators, use a viewing window in which the *y*-axis is about 2/3 as long as the *x*-axis. The SQUARE (or ZSQUARE) key in the ZOOM menu will change the current window to a square window by automatically adjusting the length of one of the axes.

Point-Slope Form

The slope-intercept form of the equation of a line is usually the most convenient for graphing and for understanding how slopes and lines are related. However, it is not always the best way to *find* the equation of a line. In many situations (particularly in calculus), the slope and a point on the line are known and you must find the equation of the line. In such cases, the best method is to use the *point-slope form,* which we now explain.

Suppose a line has slope m and that (x_1, y_1) is a point on the line. Let (x, y) represent any other point on the line. Since m is the slope, then, by the definition of slope,

$$\frac{y - y_1}{x - x_1} = m.$$

Multiplying both sides by $x - x_1$ shows that

$$y - y_1 = m(x - x_1).$$

Point-Slope Form

If a line has slope m and passes through the point (x_1, y_1), then

$$y - y_1 = m(x - x_1)$$

is the **point-slope form** of the equation of the line.

▶ **EXAMPLE 9** Find the equation of the line satisfying the given conditions.

(a) Slope 2; the point $(5, 3)$ is on the line.

Use the point-slope form with $m = 2$ and $(x_1, y_1) = (5, 3)$. Substitute $x_1 = 5$, $y_1 = 3$, and $m = 2$ into the point-slope form of the equation.

$$y - y_1 = m(x - x_1)$$
$$y - 3 = 2(x - 5).$$

For some purposes, this form of the equation is fine; in other cases, you may want to rewrite it in the slope-intercept form.

(b) Slope -3; the point $(-4, 1)$ is on the line.

Use the point-slope form with $m = -3$ and $(x_1, y_1) = (-4, 1)$.

$$y - y_1 = m(x - x_1)$$
$$y - 1 = -3[x - (-4)] \qquad \text{Point-slope form}$$

Using algebra we obtain the slope-intercept form of this equation.

$$y - 1 = -3(x + 4)$$
$$y - 1 = -3x - 12 \qquad \text{Distributive property}$$
$$y = -3x - 11 \qquad \text{Slope-intercept form} \quad \blacktriangleleft \quad \text{⑦}$$

The point-slope form can also be used to find an equation of a line given two different points on the line. The procedure is shown in the next example.

▶ **EXAMPLE 10** Find an equation of the line through $(5, 4)$ and $(-10, -2)$.

Begin by using the definition of slope to find the slope of the line that passes through the two points.

$$\text{slope} = m = \frac{-2 - 4}{-10 - 5} = \frac{-6}{-15} = \frac{2}{5}$$

7

Find both the point-slope and the slope-intercept forms of the equation of the line having the given slope and passing through the given point.

(a) $m = -3/5$, $(5, -2)$

(b) $m = 1/3$, $(6, 8)$

Answers:

(a) $y + 2 = -\dfrac{3}{5}(x - 5)$;

$y = -\dfrac{3}{5}x + 1.$

(b) $y - 8 = \dfrac{1}{3}(x - 6)$;

$y = \dfrac{1}{3}x + 6.$

Find an equation of the line through

(a) $(2, 3)$ and $(-4, 6)$;

(b) $(-8, 2)$ and $(3, -6)$.

Answers:

(a) $2y = -x + 8$

(b) $11y = -8x - 42$

Use $m = 2/5$ and either of the given points in the point-slope form. If $(x_1, y_1) = (5, 4)$, then

$$y - y_1 = m(x - x_1)$$

$$y - 4 = \frac{2}{5}(x - 5) \qquad \text{Let } y_1 = 4, \ m = \frac{2}{5}, \ x_1 = 5.$$

$$5(y - 4) = 2(x - 5) \qquad \text{Multiply both sides by 5.}$$

$$5y - 20 = 2x - 10 \qquad \text{Distributive property}$$

$$5y = 2x + 10.$$

Check that the result is the same when $(x_1, y_1) = (-10, -2)$. ◀

Vertical Lines

The equation forms developed above do not apply to vertical lines because slope is not defined for such lines. However, vertical lines can easily be described as graphs of equations.

▶ **EXAMPLE 11**   Find the equation whose graph is the vertical line in Figure 2.22.

Every point on the line has x-coordinate -1 and hence has the form $(-1, y)$. Thus every point is a solution of the equation $x + 0y = -1$, which is usually written simply as $x = -1$. Note that -1 is the x-intercept of the line. ◀

The argument in Example 11 also works in the general case.

If k is a constant, then the graph of the equation $x = k$ is the vertical line with x-intercept k.

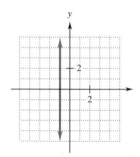

FIGURE 2.22

Linear Equations

An equation in two variables whose graph is a straight line is called a **linear equation.** Linear equations have a variety of forms, as summarized in the following table. The last form listed is called the **general form** of a linear equation, and every linear equation can be written in this form. For example, $y = 4x - 5$ can be written in general forms as $4x - y = 5$ and $x = 6$ can be written in general form as $x + 0y = 6$.

Equation	Description
$x = k$	**Vertical line,** x-intercept k, no y-intercept, undefined slope
$y = k$	**Horizontal line,** y-intercept k, no x-intercept, slope 0
$y = mx + b$	**Slope-intercept form,** slope m, y-intercept b
$y - y_1 = m(x - x_1)$	**Point-slope form,** slope m, the line passes through (x_1, y_1).
$ax + by = c$	**General form.** If $a \neq 0$ and $b \neq 0$, the line has x-intercept c/a and y-intercept c/b.

Applications

Many relationships are linear, or almost linear, so that they can be approximated by linear equations.

▶ **EXAMPLE 12** In an experiment testing a person's reaction time y (in seconds) after undergoing x hours of stressful activity, the linear equation $y = .1957x + .1243$ was found to be a good approximation of the relationship between stress and reaction time during the first five hours.

(a) Assuming that reaction times continue to follow this pattern, what would be the approximate reaction time after eight and a half hours?

Substitute $x = 8.5$ in the equation and use a calculator to compute y.

$$y = .1957x + .1243$$
$$y = .1957(8.5) + .1243 = 1.78775$$

So the reaction time is approximately 1.8 seconds.

(b) When the reaction time is 1.5 seconds, how long has the subject been undergoing stressful activity?

Substitute $y = 1.5$ in the equation and solve for x.

$$1.5 = .1957x + .1243$$
$$1.5 - .1243 = .1957x$$
$$x = \frac{1.5 - .1243}{.1957} \approx 7.0296$$

Hence, the stressful activity has been going on for slightly more than 7 hours. ◀

▶ **EXAMPLE 13** According to data from the College Board, the average cost of tuition and fees in public four-year colleges was \$2966 in the fall of 1996 and grew in an approximately linear fashion to \$3356 in the fall of 1999.

(a) Find a linear equation for this data.

Measure time along the x-axis and cost along the y-axis. Then the x-coordinate of each point is a year and the y-coordinate is the average cost of tuition and fees in that year. For convenience, let $x = 0$ correspond to 1990, so that $x = 6$ is 1996 and $x = 9$ is 1999. Then the given data points are (6, 2966) and (9, 3356). The slope of the line joining these points is

$$\frac{3356 - 2966}{9 - 6} = \frac{390}{3} = 130.$$

Using the point (6, 2966), we obtain the equation of this line

$$y - 2966 = 130(x - 6)$$
$$y - 2966 = 130x - 780$$
$$y = 130x + 2186$$

(b) Use this equation to estimate the average cost of tuition and fees in 1998.

Letting $x = 8$, we see that tuition and fees were

$$y = 130(8) + 2186 = \$3226.$$

(c) Assuming the equation remains valid in later years, estimate the average cost of tuition and fees in 2002.

When $x = 12$, then $y = 130(12) + 2186 = \$3746.$ ◀ ⑨

The average cost of tuition and fees in private four-year colleges was \$12,824 in 1996 and \$15,380 in 1999.

(a) Let $x = 0$ correspond to 1990 and find a linear equation for this data.

(b) Estimate the average cost in 2001.

Answers:

(a) $y = 852x + 7712$

(b) \$17,084

2.2 Exercises

Find the slope of the line, if it is defined. (See Examples 1–3.)

1. Through $(2, 5)$ and $(0, 6)$
2. Through $(9, 0)$ and $(12, 15)$
3. Through $(-4, 7)$ and $(3, 0)$
4. Through $(-5, -2)$ and $(-4, 11)$
5. Through the origin and $(-4, 6)$
6. Through the origin and $(8, -2)$
7. Through $(-1, 4)$ and $(-1, 8)$
8. Through $(-3, 5)$ and $(2, 5)$

Find an equation of the line with the given y-intercept and slope m. (See Examples 4 and 5.) $y = mx + b$

9. $5, m = 3$
10. $-3, m = -7$
11. $1.5, m = -2.3$
12. $-4.5, m = 1.5$
13. $4, m = -3/4$
14. $-3, m = 2/3$

Find the slope and the y-intercept of the line whose equation is given. (See Example 6.) $y - y_1 = m(x - x_1)$

15. $2x - y = 7$
16. $x + 2y = 7$
17. $6x = 2y + 4$
18. $4x + 3y = 12$
19. $6x - 9y = 14$
20. $4x + 2y = 0$
21. $2x - 3y = 0$
22. $y = 5$
23. $x = y - 5$ $y = y_1 + m(x - x_1)$

24. On one graph, sketch six straight lines that meet at a single point and satisfy this condition: one line has slope 0, two lines have positive slope, two lines have negative slope, and one line has undefined slope.

25. For which of the line segments in the figure is the slope
    (a) largest? (b) smallest?
    (c) largest in absolute value? (d) closest to 0?

26. Match each equation with the line at the top of the next column that most closely resembles its graph. (*Hint:* Consider the signs of m and b in the slope-intercept form.)
    (a) $y = 3x + 2$ (b) $y = -3x + 2$
    (c) $y = 3x - 2$ (d) $y = -3x - 2$

A   B

C   D

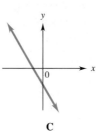

Sketch the graph of the equation and label its intercepts. (See Example 7.)

27. $2x - y = -2$
28. $2y + x = 4$
29. $2x + 3y = 4$
30. $-5x + 4y = 3$
31. $4x - 5y = 2$
32. $3x + 2y = 8$

Determine whether each pair of lines is parallel, perpendicular, *or* neither. *(See Example 8.)*

33. $4x - 3y = 6$ and $3x + 4y = 8$
34. $2x - 5y = 7$ and $15y - 5 = 6x$
35. $3x + 2y = 8$ and $6y = 5 - 9x$
36. $x - 3y = 4$ and $y = 1 - 3x$
37. $4x = 2y + 3$ and $2y = 2x + 3$
38. $2x - y = 6$ and $x - 2y = 4$

39. (a) Find the slope of each side of the triangle with vertices $(9, 6)$, $(-1, 2)$, and $(1, -3)$.
    (b) Is this triangle a right triangle? (*Hint:* Are two sides perpendicular?)

40. (a) Find the slope of each side of the quadrilateral with vertices $(-5, -2)$, $(-3, 1)$, $(3, 0)$, and $(1, -3)$.
    (b) Is this quadrilateral a parallelogram (opposite sides parallel)?

Find an equation of the line that passes through the given point and has the given slope. (See Examples 9 and 11.)

41. $(-1, 2), m = -2/3$
42. $(-4, -3), m = 5/4$
43. $(-2, -2), m = 2$

44. $(-2, 3), m = -1/2$

45. $(8, 2), m = 0$

46. $(2, -4), m = 0$

47. $(6, -5)$, undefined slope

48. $(-8, 9)$, undefined slope $X = 0$

Find an equation of the line that passes through the given points. (See Example 10.)

49. $(-1, 1)$ and $(2, 5)$ **50.** $(2, 5)$ and $(0, 6)$

51. $(1, 2)$ and $(3, 7)$ **52.** $(-1, -2)$ and $(2, -1)$

Find an equation of the line satisfying the given conditions.

53. Through the origin with slope 7

54. Through the origin and horizontal

55. Through $(5, 8)$ and vertical

56. Through $(7, 11)$ and parallel to $y = 6$

57. Through $(3, 4)$ and parallel to $4x - 2y = 5$

58. Through $(6, 8)$ and perpendicular to $y = 2x - 3$

59. x-intercept 5; y-intercept -5

60. Through $(-5, 2)$ and parallel to the line through $(1, 2)$ and $(4, 3)$

61. Through $(-1, 3)$ and perpendicular to the line through $(0, 1)$ and $(2, 3)$

62. y-intercept 3 and perpendicular to $2x - y + 6 = 0$

Business *The lost value of equipment over a period of time is called* depreciation. *The simplest method for calculating depreciation is* straight-line depreciation. *The annual straight-line depreciation of an item that cost x dollars with a useful life of n years is* $D = (1/n)x$. *Find the depreciation for items with the following characteristics.*

63. Cost: $12,482; life 10 yr

64. Cost: $39,700; life 12 yr

65. Cost: $145,000; life 28 yr

66. **Business** Ral Corp. has an incentive compensation plan under which a branch manager receives 10% of the branch's income after deduction of the bonus but before deduction of income tax.* Branch income for 1988 before the bonus and income tax was $165,000. The tax rate was 30%. The 1988 bonus amounted to

   (a) $12,600   **(b)** $15,000

   (c) $16,500   **(d)** $18,000.

Work the following applied problems. (See Examples 12 and 13.)

67. **Transportation** According to data from the U.S. Department of Transportation, the total number of pas-

senger miles (in millions) traveled on all forms of transportation (air, train, bus, etc.) in year x can be approximated by the equation

$$y = .12x + 3.29$$

where $x = 0$ corresponds to 1990. Find the approximate number of passenger miles in each of these years:

   (a) 1997   **(b)** 2001   **(c)** 2003.

   (d) In what year will the total passenger miles reach 5 million?

68. **Natural Sciences** Deer ticks cause concern because they can carry Lyme disease. One study found a relationship between the density of acorns produced in the fall and the density of deer tick larvae the following spring.* The relationship can be approximated by the linear equation

$$y = 34x + 230,$$

where x is the number of acorns per square meter in the fall and y is the number of deer tick larvae per 400 square meters the following spring. According to this formula, approximately how many acorns per square meter would result in 1000 deer tick larvae per 400 square meters?

69. **Business** According to data from the U.S. Department of Commerce, movie-theater box-office receipts (in millions of dollars) in year x can be approximated by

$$y = 494.65x + 2971.03,$$

where $x = 0$ corresponds to 1995. Assume this equation remains valid until 2008.

   (a) Estimate the box-office receipts in 2003.

   (b) In what year will the box-office receipts reach $8,000,000,000?

70. **Business** The Consumer Price Index (CPI) measures the cost of a common bundle of goods and services bought by the average consumer on a regular basis. It is scaled so that the CPI for 1982 is 100. For recent years, the CPI can be approximated by $y = 3.913x + 132.25$, where $x = 0$ corresponds to 1990.[†]

   (a) To the nearest whole number, what was the CPI in 1999?

   (b) If this trend continues, when will the CPI be 187.032?

71. **Business** An article in the *Economist* (March 31, 2001) states that worldwide automobile production was about 54 million vehicles in 2000 and is expected to be 63 million vehicles in 2008.

   (a) Let the x-axis denote time and the y-axis the number of vehicles produced (in millions). Let $x = 0$ corre-

---

*Uniform CPA Examination, May 1989, American Institute of Certified Public Accountants.

*Science, vol. 281, no. 5375, July 17, 1998, pp. 350–351.

†U.S. Bureau of Labor Statistics.

spond to the year 2000. Fill in the blanks: the given data is represented by the points (_____, 54) and (8, _____).

(b) Find the equation determined by the two points in part (a).

(c) Use the linear equation in part (b) to estimate the number of vehicles produced in 2002.

(d) In what year will vehicle production first reach 60 million?

72. Social Sciences   In 1974, a total of 86,821 people from other countries immigrated to the state of California. In 1996, the number of immigrants was 199,483.*

(a) If the change in foreign immigration to California is considered to be linear, write an equation expressing the number y, of immigrants, in terms of the number x of years after 1974.

(b) Use your result in part (a) to predict the foreign immigration to California in the year 2010.

73. Social Sciences   The number of unmarried couples in the United States who are living together has been rising at a roughly linear rate in recent years. The number of cohabitating adults was 1.1 million in 1977 and 4.9 million in 1997.[†]

(a) Write an equation expressing the number y of cohabitating adults (in millions) in terms of the number x of years after 1977.

(b) Use your result in part (a) to predict the number of cohabitating adults in the year 2010.

74. Social Sciences   The percentage of college students who are age 35 and older has been increasing at roughly a linear rate. In 1972 the percentage was 9%, and in 1998 it was 17%.[‡]

(a) Find an equation giving the percentage of college students age 35 and older in terms of time t, where t represents the number of years since 1970.

(b) If this linear trend continues, what percentage of college students will be 35 and over in 2010?

(c) If this linear trend continues, in what year will the percentage of college students 35 and over reach 31%?

75. Physical Sciences   In 2001, the Intergovernmental Panel on Climate Change predicted that the average temperature on the earth would rise about .3°C per decade in the coming century.[§] Let t measure the time in years since 1970, when the average global temperature was 15°C.

(a) Find a linear equation giving the average global temperature in degrees Celsius in terms of t, the number of years since 1970.

(b) Scientists have estimated that the sea level will rise by 65 cm if the average global temperature rises to 19°C. According to your answer to part (a), when would this occur?

76. Health   For you to achieve the maximum benefit for the heart when exercising, your heart rate (in pulse beats per minute) should be in the target heart rate zone. The lower limit of this zone is found by taking 65% of the difference between 220 and your age. The upper limit is found by using 85%.*

(a) Find linear equations for the upper and lower limits (U and L) in terms of age x.

(b) What is the target heart rate zone for a 20 year old?

(c) What is the target heart rate zone for a 40 year old?

(d) Two women in an aerobics class stop to take their pulses and are surprised to find that they have the same pulse. One woman is 48 years older than the other and is working at the upper limit of her target heart rate zone. The younger woman is working at the lower limit of her target heart rate zone. What are the ages of the two women, and what is their pulse?

(e) Run for 10 minutes, take your pulse, and see if it in your target heart rate zone. (After all, this is listed as an exercise!)

77. Physical Science   The graph shows the winning times (in minutes) at the Olympic Games for the 5000-meter run, together with a linear approximation of the data.[†]

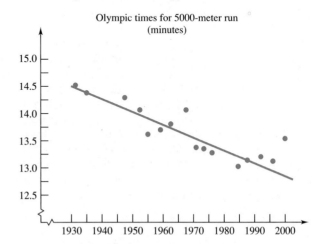

Olympic times for 5000-meter run (minutes)

*Legal Immigration to California in Federal Fiscal Year 1996, State of California Demographic Research Unit, June 1999.

†New York Times, February 15, 2000, p. F8.

‡New York Times, August 17, 1994, p. B9, and January 9, 2000, p. 4A.

§IPCC, Climatic Change 2001: the Scientific Basis.

*www.drkoop.com

†United States Olympic Committee.

(a) The equation for the linear approximation is $y = -.0221x + 57.14$. What does the slope of this line represent? Why is the slope negative?

(b) Can you think of any reason why there are no data points for the years 1940 and 1944?

78. Business   The graph at the right shows the number of U.S. radio stations on the air along with a linear approximation of the data.*

(a) Use the data points (1950, 2773) and (1999, 12913) to estimate the slope of the line shown. Interpret this number.

(b) Discuss the accuracy of the linear approximation.

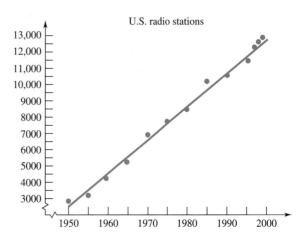

U.S. radio stations

*National Association of Broadcasters.

2.3 LINEAR MODELS

In business and science, it is often necessary to make judgments based on data from the past. For instance, a stock analyst might use a company's profits in previous years to estimate next year's profits. A life insurance company might look at life expectancies of people born in various years to predict how much money it should expect to pay out next year.

In such situations, the available data is used to construct a mathematical model, such as an equation or a graph, which is used to approximate the likely outcome in cases where complete data is not available. In this section, we consider applications in which the data can be modeled by a linear equation.

The simplest way to construct a linear model is to use the line determined by two of the data points, as illustrated in the following example.

▶ **EXAMPLE 1**   The profits of the General Electric Company (GE) (in billions of dollars) over a five-year period are shown in the table.*

Year	1996	1997	1998	1999	2000
Profit	7	8	9	11	13

(a) Let $x = 1$ correspond to 1996 and plot the points (x, y), where x is the year and y the profit.

*GE Annual Report 2000, March 2001. The profit figures (net earnings) have been rounded to the nearest billion.

The data points are $(1, 7)$, $(2, 8)$, $(3, 9)$, $(4, 11)$, and $(5, 13)$, as shown in Figure 2.23.

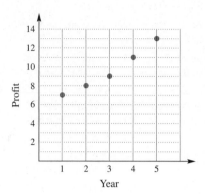

FIGURE 2.23

The data points are approximately linear, so a line should be a reasonable model.

(b) Use the data points $(1, 7)$ and $(3, 9)$ to find a line that models the data.

The slope of the line through $(1, 7)$ and $(3, 9)$ is $\dfrac{9 - 7}{3 - 1} = \dfrac{2}{2} = 1$. Using the point $(1, 7)$, we find that the point-slope equation of this line is

$$y - 7 = 1(x - 1)$$
$$y = x + 6.$$

This line and the data points are shown in Figure 2.24. Although the line fits the first three data points perfectly, it underestimates profits in later years.

FIGURE 2.24

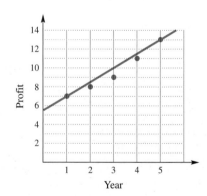

FIGURE 2.25

(c) Use the points $(1, 7)$ and $(5, 13)$ to find another line that models this data.

The slope is $\dfrac{13 - 7}{5 - 1} = \dfrac{6}{4} = \dfrac{3}{2} = 1.5$ and the equation is

$$y - 7 = 1.5(x - 1)$$
$$y = 1.5x + 5.5.$$

This line and the data points are graphed in Figure 2.25. Although this line passes through only two of the data points, it appears to fit the data reasonably well overall.

(d) Use the model in part (c) to estimate the profits in 2001.

Substitute $x = 6$ (corresponding to 2001) in the equation:

$$y = 1.5x + 5.5$$
$$y = 1.5(6) + 5.5 = 14.5 \text{ billion.}$$ ◀ ①

Unlike the situation in Example 1, it isn't always possible to determine visually which of several lines best fits the data. For a given linear model, however, we can compute the difference between the actual amount p and the amount y given by the line. If the data point is (x, p) and the corresponding point on the line is (x, y), then the difference $p - y$ measures the error in the model for that particular value of x. The number $p - y$ is called a **residual.** As shown in Figure 2.26, the residual $p - y$ is the vertical distance from the data point to the line (positive when the data point is above the line, negative when it is below the line, and 0 when it is on the line).

One way to determine how well a line fits the data points might be to compute the sum of its residuals, that is, the sum of the individual errors. Unfortunately, however, two different lines might each have 0 as the sum of their residuals, which doesn't enable us to decide which one is the better fit. (See Exercise 11 at the end of this section for an example of this possibility.) To avoid this difficulty, the sum of the *squares* of the residuals is used to measure how well a line fits the data points. When the sum of the squares is used, a smaller sum means a smaller error (and, hence, a better fit). The error is 0 only when *all* the data points lie on the line (a perfect fit).

①

Use the points (2, 8) and (4, 11) to find another linear model for the data in Example 1.

Answer:

$y = 1.5x + 5$

FIGURE 2.26

▶ **EXAMPLE 2** In Example 1, two linear models for GE's profits were constructed: $y = x + 6$ and $y = 1.5x + 5.5$. For each model, determine the five residuals, the square of each residual, and the sum of the squared residuals.

The information for each model is summarized in the following tables.

$y = x + 6$

Data Point (x, p)	Model Point (x, y)	Residual $p - y$	Squared Residual $(p - y)^2$
(1, 7)	(1, 7)	0	0
(2, 8)	(2, 8)	0	0
(3, 9)	(3, 9)	0	0
(4, 11)	(4, 10)	1	1
(5, 13)	(5, 11)	2	4
			sum: 5

$y = 1.5x + 5.5$

Data Point (x, p)	Model Point (x, y)	Residual $p - y$	Squared Residual $(p - y)^2$
(1, 7)	(1, 7)	0	0
(2, 8)	(2, 8.5)	−.5	.25
(3, 9)	(3, 10)	−1	1
(4, 11)	(4, 11.5)	−.5	.25
(5, 13)	(5, 13)	0	0
			sum: 1.5

In side problem 1, you showed that $y = 1.5x + 5$ is also a linear model for GE's profits. Use this line and the data in Example 1 to find

(a) the residuals;

(b) the sum of the squares of the residuals

(c) Does this line fit the data better than the two lines in Example 2?

Answers:

(a) $.5, 0, -.5, 0, .5$

(b) $.75$

(c) Yes, because the sum of the squares of its residuals is smaller.

These results confirm what we saw visually in Example 1. The line $y = 1.5x + 5.5$ is a better fit for the data because the sum of the squares of its residuals is smaller than the sum for $y = x + 6$. ◀ ②

Linear Regression (Optional)*

Mathematical techniques that are introduced in Chapter 14 can be used to prove the following result.

> For any set of data points, there is one and only one line for which the sum of the squares of the residuals is as small as possible.

This *line of best fit* is called the **least squares regression line,** and the computational process for finding its equation is called **linear regression.** The linear regression formulas are quite complicated and require a large amount of computation. Fortunately, most graphing calculators and spreadsheet programs can do linear regression quickly and easily.

▶ **EXAMPLE 3**   Recall that profits of GE (in billions of dollars) were as follows.

Year	1996	1997	1998	1999	2000
Profit	7	8	9	11	13

Use a graphing calculator to do the following.

(a) Plot the data points, with $x = 1$ corresponding to 1996.[†]

The data points are $(1, 7)$, $(2, 8)$, $(3, 9)$, $(4, 11)$, and $(5, 13)$. Press STAT EDIT to bring up the statistics editor. Enter the x-coordinates as a list L_1 and the corresponding y-coordinates as a list L_2, as shown in Figure 2.27. To plot the data points, go to the STAT PLOT menu, choose a plot (here it's Plot 1), choose ON, and enter the lists L_1 and L_2, as shown in Figure 2.28. Then set the viewing window as usual and press GRAPH to produce Figure 2.29.

FIGURE 2.27

FIGURE 2.28

FIGURE 2.29

---

*Examples 3–6 require either a graphing calculator or a spreadsheet program.

[†]The process outlined here works for most TI graphing calculators. Other graphing calculators and spreadsheet programs operate similarly, but check your instruction manual, or see the Graphing Calculator Appendix, which is available at www.aw.com/MWA8.

Use the least squares regression line $y = 1.5x + 5.1$ and the data points in Example 3 to find

(a)   the residuals;

(b)   the sum of the squares of the residuals.

(c)   How does this line compare to those in Example 1 and side problem 2?

Answers:

(a) $.4, -.1, -.6, -.1, .4$

(b) $.7$

(c)   It fits the data best because the sum of the squares of its residuals is smallest.

(b) Find the least squares regression line for this data.

Go to the STAT CALC menu and choose LIN REG, which returns you to the home screen. As shown in Figure 2.30, enter the list names and the place where the equation of the regression line should be stored (here Y_1 is chosen; it's on the VARS Y-VARS FUNCTION menu); then press ENTER. Figure 2.31 shows that the equation of the regression line is

$$y = 1.5x + 5.1.$$

(The number r in Figure 2.31 will be discussed below.)

FIGURE 2.30   **FIGURE 2.31**

(c) Graph the data points and the regression line on the same screen.

Press GRAPH to see the line plotted with the data points (Figure 2.32).

FIGURE 2.32

NOTE   The mathematical techniques used to find the least squares regression line are discussed and illustrated with a simple example in the case at the end of Chapter 14. These techniques require multivariable calculus.

▶ **EXAMPLE 4**   The table shows the number of accidental deaths per 100,000 population in the United States over the past century.*

Year	Death Rate	Year	Death Rate
1910	84.4	1960	52.1
1920	71.2	1970	56.2
1930	80.5	1980	46.5
1940	73.4	1990	36.9
1950	60.3	1998	34.1

*U.S. National Center for Health Statistics.

(a) Let $x = 0$ correspond to 1900. Use a graphing calculator or spreadsheet program to find the least squares regression line that models this data.

The data points are $(10, 84.4)$, $(20, 71.2)$, ..., $(98, 34.1)$. Enter the x-coordinates as list L_1 and the corresponding y-coordinates as list L_2 in a graphing calculator. Then find the regression line, as in Figure 2.33. Rounding the coefficients, the equation of the line is $y = -.565x + 90.525$.

FIGURE 2.33

FIGURE 2.34

④

Using only the data from 1960 and later in Example 4, find the equation of the least squares regression line. Round the coefficients to three decimal places.

Answers:

$y = -.575x + 90.933$

(b) Plot the data points and the regression line on the same screen.
See Figure 2.34, which shows that the line is a reasonable model for the data.

(c) Use the regression line to predict the number of accidental deaths in 2004.
2004 corresponds to $x = 104$. Substitute $x = 104$ in the regression line equation:

$$y = -.565(104) + 90.525 \approx 31.8.$$

The model predicts that there will be approximately 31.8 accidental deaths per 100,000 population in 2004. ◄ ④

◣ Correlation

Although the "best fit" line can always be found by linear regression, it may not be a good model. For instance, if the data points are widely scattered, no straight line will model the data accurately. When the linear regression line was computed in Examples 3 and 4, the calculator also displayed a number r and its square (Figures 2.31 and 2.33). This number is called the **coefficient of correlation.** It measures how closely the data points fit the regression line and thus indicates how good the regression line is for predictive purposes.

The coefficient of correlation r is always between -1 and 1. When $r = \pm 1$, the data points all lie on the regression line (a perfect fit). When the absolute value of r is close to 1, the line fits the data quite well, and when r is close to 0, the line is a poor fit for the data (but some other curve might be a good fit). Figure 2.35 shows how the value of r varies, depending on the pattern of the data points.

FIGURE 2.35

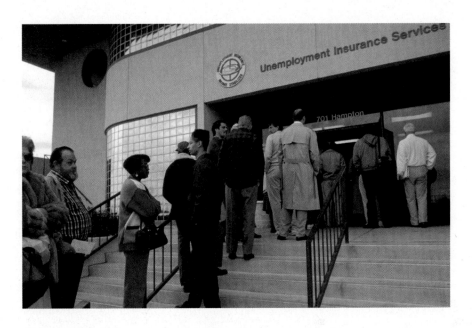

▶ **EXAMPLE 5**   The number of unemployed people in the U.S. labor force (in millions) in recent years is shown in the table.*

Year	Unemployed	Year	Unemployed	Year	Unemployed
1987	7.425	1992	9.613	1997	6.739
1988	6.701	1993	8.940	1998	6.210
1989	6.528	1994	7.996	1999	5.880
1990	7.047	1995	7.404		
1991	8.628	1996	7.236		

⑤

Using only the data from 1992 and later in Example 5, find

(a) the equation of the least squares regression line;

(b) the coefficient of correlation.

(c) How well does this line fit the data?

Answers:

(a) $y \approx -.52x + 15.57$
(b) $r \approx -.98$
(c) It fits very well because $|r|$ is very close to 1.

Determine whether or not a linear equation is a good model for this data.

Let $x = 0$ correspond to 1980 and plot the data points (7, 7.425), (8, 6.701), etc., either by hand or using a graphing calculator, as in Figure 2.36. They do not form a linear pattern (unemployment tends to rise and fall). Alternatively, you could compute the regression equation for the data, as in Figure 2.37. The coefficient of correlation is $r \approx -.278$, a number quite close to 0, which indicates that the regression line is a poor fit for the data. Therefore, a linear equation is not a good model for this data.   ◀ ⑤

FIGURE 2.36

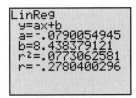

FIGURE 2.37

*U.S. Department of Labor, Bureau of Labor Statistics.

▶ **EXAMPLE 6**   Enrollment projections (in millions) for all U.S. colleges and universities (including two-year institutions) in selected years are shown in the table.*

Year	1995	1998	1999	2000	2001	2002	2003
Enrollment	14.262	14.608	14.881	15.072	15.158	15.168	15.262

```
LinReg
y=ax+b
a=.1313585526
b=13.63980263
r²=.9486510077
r=.9739871702
```

FIGURE 2.38

(a) Let $x = 0$ correspond to 1990. Use a graphing calculator or spreadsheet program to find a linear model for the data and determine how well it fits the data points.

The least squares regression line (with coefficients rounded) is

$$y = .1314x + 13.6398,$$

as shown in Figure 2.38. The coefficient of correlation is $r \approx .97$, which is very close to 1, so this line fits the data very well.

(b) Predict the enrollment in 2005.

Let $x = 15$ (corresponding to 2005) in the regression equation:

$$y = .1314(15) + 13.6398 = 15.6108.$$

Therefore, the enrollment in 2005 will be approximately 15,610,800 students.

(c) In what year will enrollment reach 16 million?

Let $y = 16$ and solve the regression equation for x.

$$16 = .1314x + 13.6398$$
$$2.3602 = .1314x$$
$$x \approx 17.96$$

Since these enrollment figures change once a year, use the nearest integer value for x, namely, 18. So enrollment will reach 16 million in 2008. ◀

---

*As of fall of each year. *Statistical Abstract of the United States*, 2000.

2.3 Exercises

1. *Physical Science*   The table shows equivalent Fahrenheit and Celsius temperatures.

Degrees Fahrenheit	32	68	104	140	176	212
Degrees Celsius	0	20	40	60	80	100

(a) Choose any two data points and use them to construct a linear equation that models the data, with x being Fahrenheit and y Celsius.

(b) Use the model in part (a) to find the Celsius temperature corresponding to

50° Fahrenheit and 75° Fahrenheit.

Physical Science   *Use the linear equation derived in Exercise 1 to work the following problems.*

2. Convert each temperature.
   (a) 58°F to Celsius   **(b)** 50°C to Fahrenheit
   (c) −10°C to Fahrenheit   **(d)** −20°F to Celsius

3. According to the *World Almanac and Book of Facts*, 2001, Venus is the hottest planet, with a surface temperature of 864° Fahrenheit. What is this temperature in Celsius?

4. Find the temperature at which Celsius and Fahrenheit temperatures are numerically equal.

In each of the following problems, assume the data can be modeled by a straight line. Use two data points to find such a line and then answer the question. (See Example 1.)

5. Business   The Consumer Price Index (CPI), which measures the cost of a typical package of consumer goods, stood at 130.7 in 1990 and 170.7 in 2000. Let $x = 0$ correspond to 1990 and estimate the CPI in 1995 and 2003.

6. Social Science   The average monthly family payment for Temporary Assistance to Needy Families was \$473 in 1997 and \$589 in 1999.* Let $x = 0$ correspond to 1990 and estimate the average monthly payment in 2004.

7. Business   The United States is China's largest export market. Imports from China grew from about 46 billion dollars in 1995 to 82 billion dollars in 1999. Let $x = 0$ correspond to 1990 and estimate the amount of imports in 2002.

8. Physical Science   Suppose a baseball is thrown at 85 miles per hour. The ball will travel 320 feet when hit by a bat swung at 50 miles per hour and will travel 440 feet when hit by a bat swung at 80 miles per hour. Let y be the number of feet traveled by the ball when hit by a bat swung at x miles per hour. (*Note:* This is valid for $50 \leq x \leq 90$, where the bat is 35 inches long, weighs 32 ounces, and strikes a waist-high pitch so the place of the swing lies at 10° from the diagonal.)[†] How much farther will a ball travel for each mile per hour increase in the speed of the bat?

9. Natural Science   The amount of tropical rain forests in Central America decreased from 130,000 square miles to about 80,000 square miles from 1969 to 1985. Let y be the amount (in ten thousands of square miles) x years after 1965. How large were the rain forests in the year 1997?

10. Physical Science   Ski resorts require large amounts of water in order to make snow. Snowmass Ski Area in Colorado plans to pump at least 1120 gallons of water per minute for at least 12 hours a day from Snowmass Creek between mid-October and late December.[‡] Environmentalists are concerned about the effects on the ecosystem. Find the minimum amount of water pumped in 30 days. (*Hint:* Let y be the total number of gallons pumped x days after pumping begins. Note that $(0, 0)$ is on the graph of the equation.)

In each of the following problems, two linear models are given for the data. For each model,
   (a) Find the residuals and their sum.
   (b) Find the sum of the squares of the residuals.
   (c) Decide which model is the better fit. (See Example 2.)

11. Business   The weekly amount spent on advertising and the weekly sales revenue of a small store over a five-week period are shown in the table.

Advertising Expenditure x (in hundreds of dollars)	1	2	3	4	5
Sales Revenue y (in thousands of dollars)	2	2	3	3	5

Two equations that model the data are $y = .5x + 1.5$ and $y = x$.

12. Business   The table shows the total outstanding consumer debt (excluding home mortgages) in billions of dollars in selected years.*

Year	1980	1985	1990	1995	1999
Consumer debt	349	585	789	1096	1395

Let $x = 0$ correspond to 1980. Two equations that model the data are $y = 54x + 314$ and $y = 55x + 350$.

In each of the following problems, determine whether or not a straight line is a good model for the data. You may do this either visually, by plotting the data points, or analytically, by finding the coefficient of correlation for the least squares regression line. (See Examples 5 and 6.)

13. Natural Science   The table shows the average monthly temperature (in degrees Fahrenheit) in Cleveland, Ohio, based on data from 1961 to 1990.[†] Let $x = 2$ correspond to February, $x = 4$ to April, etc.

Month	Feb	April	June	Aug	Oct	Dec
Temperature	27.3	47.5	67.5	70.3	52.7	30.9

14. Health   The table shows the number of deaths per 100,000 people from heart disease in selected years. Let $x = 0$ correspond to 1960.

Year	1960	1970	1980	1990	1999
Deaths	522	496	436	368	348

*Office of Planning, Research, and Evaluation, Administration for Children and Families, U.S. Department of Health and Human Services.

†Robert K. Adair, *The Physics of Baseball* (HarperCollins, 1990).

‡York Snow, Inc.

*Federal Reserve Bulletin.

†National Climatic Data Center.

In the following exercises, find the required linear model as follows. If you do not have a graphing calculator or spreadsheet program, use the first and last data points to determine a line. If you do have a graphing calculator or spreadsheet program, find the least squares regression line. (See Examples 1, 3, 4, and 6.)

15. Health   Use the data on death from heart disease in Exercise 14.
   (a) Find a linear model for the data, with $x = 0$ corresponding to 1960.
   (b) Estimate the number of deaths from heart disease in 2005.

16. Social Science   The table shows how poverty level income cutoffs (in dollars) for a family of four have changed over time (in large part because of inflation).*

Year	1970	1975	1980	1985	1990	1995
Income	3968	5500	8414	10,989	13,359	15,569

   (a) Find a linear model for the data, with $x = 0$ corresponding to 1970 and income measured in thousands of dollars (so that $3968 is 3.968, etc.).
   (b) What was the approximate poverty level in 2000? In 2003?

17. Physical Science   While shopping for an air conditioner, Adam Bryer consulted the following table giving a machine's BTUs and the square footage (ft²) that it would cool.†

ft² (x)	BTUs (y)
150	5000
175	5500
215	6000
250	6500
280	7000
310	7500
350	8000
370	8500
420	9000
450	9500

   (a) Find a linear model for the data.
   (b) To check the fit of the data to the line, use the results from part (a) to find the BTUs required to cool a room of 150 ft², 280 ft², and 420 ft². How well do the actual data agree with the predicted values?

   (c) Suppose Adam's room measures 235 ft². Use the results from part (a) to decide how many BTUs it requires. If air conditioners are available only with the BTU choices in the table, which should Adam choose?

18. Business   The number of new businesses started each year (in thousands) during the late 1990s is shown in the table.*

Year	1995	1996	1997	1998	1999
Starts	168	170	167	155	151

   (a) Find a linear model for the data, with $x = 0$ corresponding to 1990.
   (b) Estimate the number of business starts in 2003 and 2005.

The following exercises require a graphing calculator or spreadsheet program. (See Examples 3, 4, and 6.)

19. Business   The graph shows Intel's net revenues (in billions of dollars) from 1991 to 2000.†

   (a) List the data points, with $x = 0$ corresponding to 1990.
   (b) Find the least squares regression line that models the data.
   (c) If this trend continues, what will Intel's net revenues be in 2003 and 2005?

20. Natural Science   Biologists have observed a linear relationship between the temperature and the frequency with which a cricket chirps. The following data were measured for the striped ground cricket.‡

Statistical Abstracts of the United States 2000; figures rounded to nearest thousand.

†Reprinted by permission of Intel Corporation, Copyright Intel Corporation 2002.

‡Reprinted by permission of the publishers from *The Songs of Insects* by George W. Pierce. Cambridge, Mass.: Harvard University Press, Copyright © 1948 by the President and Fellows of Harvard College.

*U.S. Bureau of Census, *Current Population Reports.*

†Morris Carey and James Carey, "On the House," *Sacramento Bee,* July 29, 2000.

Temperature °F (x)	Chirps per second (y)
88.6	20.0
71.6	16.0
93.3	19.8
84.3	18.4
80.6	17.1
75.2	15.5
69.7	14.7
82.0	17.1
69.4	15.4
83.3	16.2
79.6	15.0
82.6	17.2
80.6	16.0
83.5	17.0
76.3	14.4

(a) Find the least squares regression line that models the data.

(b) Use the results of part (a) to determine how many chirps per second you would expect to hear from the striped ground cricket if the temperature were 73°F.

(c) Use the results of part (a) to determine what the temperature is when the striped ground crickets are chirping at a rate of 18 times per second.

(d) Find the coefficient of correlation.

21. Social Science   Over a fifteen-year period beginning in 1980, the average major league baseball game got longer and longer, as shown in the table.*

Year	Average Completion Time
1981	2:33
1982	2:34
1983	2:36
1984	2:35
1985	2:40
1986	2:44
1987	2:48
1988	2:45
1989	2:46
1990	2:48
1991	2:49
1992	2:49
1993	2:48
1994	2:54
1995	2:57

(a) Find the least squares regression line that models the data, with $x = 0$ corresponding to 1980 and y being the number of minutes beyond 2 hours (for instance, 2:36 corresponds to $y = 36$).

(b) If the trend in the data continues, when will the average completion time be 3 hours and 15 minutes?

22. Natural Science   The following table shows the men's and women's world records (in seconds) in the 800-meter run.*

Year	Men's Record	Women's Record
1905	113.4	—
1915	111.9	—
1925	111.9	144.0
1935	109.7	135.6
1945	106.6	132.0
1955	105.7	125.0
1965	104.3	118.0
1975	103.7	117.48
1985	101.73	113.28
1995	101.11	113.28

(a) Find the equation for the least squares line for the men's record (y) in terms of the year (x). Use 5 for 1905, 15 for 1915, and so on.

(b) Find the equation for the least squares line for the women's record.

(c) Suppose the men's and women's records continue to improve as predicted by the equations found in parts (a) and (b). In what year will the women's record catch up with the men's record? Do you believe that will happen? Why or why not?

(d) Calculate the coefficient of correlation for both the men's and the women's record. What do these numbers tell you?

23. Health   The table shows U.S. Health Care Financing Administration projections of personal health care expenditures (in billions of dollars).

Year	1990	1997	1999	2000	2001	2002
Expenditures	614.7	969.0	1078.3	1150.9	1227.3	1310.0

(a) Find the least squares regression line that models the data, with $x = 0$ corresponding to 1990.

(b) Does this model fit the data well? Why?

(c) Estimate health care expenditures in 2005.

*Brian J. Whipp and Susan A. Ward, "Will Women Soon Outrun Men?" *Nature*, vol. 355 January 2, 1992), p. 25. The data are from Peter Mathews, *Track and Field Athletics: The Records* (Guinness, 1986) pp. 11, 44, and from Robert W. Schutz and Yuanlong Liu in *Statistics in Sport*, edited by Jay Bennett (Arnold, 1998), p. 189.

(d) If this trend continues, when will expenditures reach 2000 billion dollars?

24. Health   Employees of large and medium-sized businesses who want to be covered under company-sponsored health insurance plans are often required to pay part of the premium. A decade-long survey showed that participation rates declined as the employee contribution increased, as summarized in the following table.*

Average Monthly Employee Contribution ($)	62	69.50	76	84.50
Percentage of Participating Employees	83	82	77	76

(a) Find the least squares regression line that models the data, with x being the employee contribution and y the percentage of participating employees.

(b) Does this model fit the data well? Why?

*U.S. Bureau of Labor Statistics.

(c) Estimate the participation rate when the monthly contribution is $100.

(d) At what contribution rate will participation fall below 50%?

25. Business   Total corporate profits before taxes (in billions of dollars) of U.S. companies during the 1990s are shown in the table.*

Year	1990	1994	1995	1996	1997	1998	1999
Profit	402	573	669	726	796	782	849

(a) Find the least squares regression line that models the data, with $x = 0$ corresponding to 1990.

(b) Does this model fit the data well? Why?

(c) If this trend continues, what will corporate profits be in 2004?

(d) If this trend continues, when will corporate profits reach $1475 billion?

*U.S. Bureau of Economic Analysis.

2.4 LINEAR INEQUALITIES

An **inequality** is a statement that one mathematical expression is greater than (or less than) another. Inequalities are very important in applications. For example, a company wants revenue to be *greater than* costs and must use *no more than* the total amount of capital or labor available.

Inequalities may be solved using algebraic or geometric methods. In this section we shall concentrate on algebraic methods for solving **linear inequalities,** such as

$$4 - 3x \le 7 + 2x \quad \text{and} \quad -2 < 5 + 3m < 20,$$

and absolute value inequalities, such as $|x - 2| < 5$. The following properties are the basic algebraic tools for working with inequalities.

Properties of Inequality

For real numbers a, b, and c,

(a)   if $a < b$, then $a + c < b + c$

(b)   if $a < b$, and if $c > 0$, then $ac < bc$

(c)   if $a < b$, and if $c < 0$, then $ac > bc$.

Throughout this section, definitions are given only for $<$, but they are equally valid for $>$, $\le$, or $\ge$.

(a) First multiply both sides of $-6 < -1$ by 4, and then multiply both sides of $-6 < -1$ by -7.

(b) First multiply both sides of $9 \geq -4$ by 2 and then multiply both sides of $9 \geq -4$ by -5.

(c) First add 4 to both sides of $-3 < -1$, and then add -6 to both sides of $-3 < -1$.

Answers:

(a) $-24 < -4$; $42 > 7$

(b) $18 \geq -8$; $-45 \leq 20$

(c) $1 < 3$; $-9 < -7$

CAUTION Pay careful attention to part (c): if both sides of an inequality are multiplied by a negative number, the direction of the inequality symbol must be reversed. For example, starting with the true statement $-3 < 5$ and multiplying both sides by the positive number 2 gives

$$-3 \cdot 2 < 5 \cdot 2,$$

or

$$-6 < 10,$$

still a true statement. On the other hand, starting with $-3 < 5$ and multiplying both sides by the negative number -2 gives a true result only if the direction of the inequality symbol is reversed:

$$-3(-2) > 5(-2)$$
$$6 > -10. \quad \text{①}$$

▶ **EXAMPLE 1**   Solve $3x + 5 > 11$. Graph the solution.

First, add -5 to both sides.

$$3x + 5 + (-5) > 11 + (-5)$$
$$3x > 6$$

Now multiply both sides by 1/3.

$$\frac{1}{3}(3x) > \frac{1}{3}(6)$$
$$x > 2$$

(Why was the direction of the inequality symbol not changed?) In interval notation (introduced in Section 1.1), the solution is the interval $(2, \infty)$, which is graphed on the number line in Figure 2.39. The parenthesis at 2 shows that 2 is not included in the solution.

FIGURE 2.39

As a partial check, note that 0, which is not part of the solution, makes the inequality false, while 3, which is part of the solution, makes it true.

$$3(0) + 5 \overset{?}{>} 11 \qquad\qquad 3(3) + 5 \overset{?}{>} 11$$
$$5 > 11 \quad \text{False} \qquad\qquad 14 > 11 \quad \text{True} \quad \text{◀} \quad \text{②}$$

②

Solve these inequalities. Graph each solution.

(a) $5z - 11 < 14$

(b) $-3k \leq -12$

(c) $-8y \geq 32$

Answers:

(a) $z < 5$

(b) $k \geq 4$

(c) $y \leq -4$

▶ **EXAMPLE 2**   Solve $4 - 3x \leq 7 + 2x$.

Add -4 to both sides.

$$4 - 3x + (-4) \leq 7 + 2x + (-4)$$
$$-3x \leq 3 + 2x$$

Solve these inequalities. Graph each solution.

(a) $8 - 6t \geq 2t + 24$

(b) $-4r + 3(r + 1) < 2r$

Answers:

(a) $t \leq -2$

(b) $r > 1$

Add $-2x$ to both sides. (Remember that *adding* to both sides never changes the direction of the inequality symbol.)

$$-3x + (-2x) \leq 3 + 2x + (-2x)$$
$$-5x \leq 3$$

Multiply both sides by $-1/5$. Since $-1/5$ is negative, change the direction of the inequality symbol.

$$-\frac{1}{5}(-5x) \geq -\frac{1}{5}(3)$$
$$x \geq -\frac{3}{5}$$

Figure 2.40 shows a graph of the solution $[-3/5, \infty)$. The bracket in Figure 2.40 shows that $-3/5$ is included in the solution. ◀ ③

FIGURE 2.40

Solve each of the following. Graph each solution.

(a) $9 < k + 5 < 13$

(b) $-6 \leq 2z + 4 \leq 12$

Answers:

(a) $4 < k < 8$

(b) $-5 \leq z \leq 4$

▶ **EXAMPLE 3**   Solve $-2 < 5 + 3m < 20$. Graph the solution.

The inequality $-2 < 5 + 3m < 20$ says that $5 + 3m$ is *between* -2 and 20. We can solve this inequality with an extension of the properties given above. Work as follows, first adding -5 to each part.

$$-2 + (-5) < 5 + 3m + (-5) < 20 + (-5)$$
$$-7 < 3m < 15$$

Now multiply each part by $1/3$.

$$-\frac{7}{3} < m < 5$$

A graph of the solution, $(-7/3, 5)$, is given in Figure 2.41. ◀ ④

FIGURE 2.41

▶ **EXAMPLE 4**   The formula for converting from Celsius to Fahrenheit temperature is

$$F = \frac{9}{5}C + 32.$$

What Celsius temperature range corresponds to the range from $32°F$ to $77°F$?

The Fahrenheit temperature range is $32 < F < 77$. Since $F = (9/5)C + 32$,

$$32 < \frac{9}{5}C + 32 < 77.$$

Solve the inequality for C.

$$32 < \frac{9}{5}C + 32 < 77$$

$$0 < \frac{9}{5}C < 45$$

$$\frac{5}{9} \cdot 0 < \frac{5}{9} \cdot \frac{9}{5}C < \frac{5}{9} \cdot 45$$

$$0 < C < 25$$

The corresponding Celsius temperature range is 0°C to 25°C. ◀ ⑤

⑤

In Example 4, what Celsius temperatures correspond to 5°F to 95°F?

Answer:

-15°C to 35°C

A product will break even, or produce a profit, only if the revenue R from selling the product at least equals the cost C of producing it, that is, if $R \geq C$.

▶ **EXAMPLE 5** A company analyst has determined that the cost to produce and sell x units of a certain product is $C = 20x + 1000$. The revenue for that product is $R = 70x$. Find the values of x for which the company will break even or make a profit on the product.

Solve the inequality $R \geq C$.

$$R \geq C$$
$$70x \geq 20x + 1000 \qquad \text{Let } R = 70x, \; C = 20x + 1000.$$
$$50x \geq 1000$$
$$x \geq 20$$

The company must produce and sell 20 items to break even and more than 20 to make a profit. ◀

Absolute Value Inequalities

You may wish to review the definition of absolute value in Section 1.1 before reading the following examples, which show how to solve inequalities involving absolute values.

▶ **EXAMPLE 6** Solve each inequality.

(a) $|x| < 5$

Because absolute value gives the distance from a number to 0, the inequality $|x| < 5$ is true for all real numbers whose distance from 0 is less than 5. This includes all numbers between -5 and 5, or numbers in the interval $(-5, 5)$. A graph of the solution is shown in Figure 2.42.

FIGURE 2.42

Solve each inequality. Graph each solution.

(a) $|x| \leq 1$

(b) $|y| \geq 3$

Answers:

(a) $[-1, 1]$

(b) All numbers in

$(-\infty, -3]$ or $[3, \infty)$

(b) $|x| > 5$

In a similar way, the solution of $|x| > 5$ is given by all those numbers whose distance from 0 is *greater* than 5. This includes the numbers satisfying $x < -5$ or $x > 5$. A graph of the solution, all numbers in

$$(-\infty, -5) \quad \text{or} \quad (5, \infty),$$

is shown in Figure 2.43. ◀ ⑥

FIGURE 2.43

The examples above suggest the following generalizations.

Assume a and b are real numbers with b positive.

1. Solve $|a| = b$ by solving $a = b$ or $a = -b$.
2. Solve $|a| < b$ by solving $-b < a < b$.
3. Solve $|a| > b$ by solving $a < -b$ or $a > b$.

▶ **EXAMPLE 7**   Solve $|x - 2| < 5$.

Replace a with $x - 2$ and b with 5 in property (2) in the box above. Now solve $|x - 2| < 5$ by solving the inequality

$$-5 < x - 2 < 5.$$

Add 2 to each part, getting the solution

$$-3 < x < 7,$$

which is graphed in Figure 2.44. ◀ ⑦

FIGURE 2.44

▶ **EXAMPLE 8**   Solve $|2 - 7m| - 1 > 4$.

First add 1 to both sides.

$$|2 - 7m| > 5$$

Now use property (3) from the box above to solve $|2 - 7m| > 5$ by solving the inequality

$$2 - 7m < -5 \quad \text{or} \quad 2 - 7m > 5.$$

⑦

Solve each inequality. Graph each solution.

(a) $|p + 3| < 4$

(b) $|2k - 1| \leq 7$

Answers:

(a) $(-7, 1)$

(b) $[-3, 4]$

Solve each part separately.

$$-7m < -7 \quad \text{or} \quad -7m > 3$$

$$m > 1 \quad \text{or} \quad m < -\frac{3}{7}$$

The solution, all numbers in $\left(-\infty, -\frac{3}{7}\right)$ or $(1, \infty)$, is graphed in Figure 2.45. ◀ ⑧

FIGURE 2.45

Solve each inequality. Graph each solution.

(a) $|y - 2| > 5$

(b) $|3k - 1| \geq 2$

(c) $|2 + 5r| - 4 \geq 1$

Answers:

(a) All numbers in $(-\infty, -3)$ or $(7, \infty)$

(b) All numbers in $\left(-\infty, -\frac{1}{3}\right]$ or $[1, \infty)$

(c) All numbers in $\left(-\infty, -\frac{7}{5}\right]$ or $\left[\frac{3}{5}, \infty\right)$

▶ EXAMPLE 9   Solve $|2 - 5x| \geq -4$.

The absolute value of a number is always nonnegative. Therefore, $|2 - 5x| \geq -4$ is always true; so the solution is the set of all real numbers. Note that the inequality $|2 - 5x| \leq -4$ has no solution, because the absolute value of a quantity can never be less than a negative number. ◀ ⑨

Absolute value inequalities can be used to indicate how far a number may be from a given number. The next example illustrates this use of absolute value.

▶ EXAMPLE 10   Write each statement using absolute value.

(a) k is at least 4 units from 1.

If k is at least 4 units from 1, then the distance from k to 1 is greater than or equal to 4. See Figure 2.46(a). Write this statement using absolute value as follows:

$$|k - 1| \geq 4.$$

FIGURE 2.46

(b) p is within 2 units of 5.

This statement means that the distance between p and 5 must be less than or equal to 2. See Figure 2.46(b). Using absolute value notation, the statement is written as

$$|p - 5| \leq 2. \quad ◀ \quad ⑩$$

Solve each inequality.

(a) $|5m - 2| > -1$

(b) $|2 + 3a| < -3$

(c) $|6 + r| > 0$

Answers:

(a) All real numbers

(b) No solution

(c) All real numbers except -6

Write each statement using absolute value.

(a) m is at least 3 units from 5.

(b) t is within .01 of 4.

Answers:

(a) $|m - 5| \geq 3$

(b) $|t - 4| \leq .01$

2.4 Exercises

1. Explain how to determine whether a parenthesis or a bracket is used when graphing the solution of a linear inequality.

2. The three-part inequality $p < x < q$ means "p is less than x and x is less than q." Which one of the following inequalities is not satisfied by any real number x? Explain why.
   (a) $-3 < x < 5$ (b) $0 < x < 4$
   (c) $-7 < x < -10$ (d) $-3 < x < -2$

Solve each inequality and graph each solution. (See Examples 1–3.)

3. $-8k \leq 32$ 4. $-6a \leq 36$

5. $-2b > 0$ 6. $6 - 6z < 0$

7. $3x + 4 \leq 12$ 8. $2y - 5 < 9$

9. $-4 - p \geq 3$ 10. $5 - 3r \leq -4$

11. $7m - 5 < 2m + 10$

12. $6x - 2 > 4x - 8$

13. $m - (4 + 2m) + 3 < 2m + 2$

14. $2p - (3 - p) \leq -7p - 2$

15. $-2(3y - 8) \geq 5(4y - 2)$

16. $5r - (r + 2) \geq 3(r - 1) + 5$

17. $3p - 1 < 6p + 2(p - 1)$

18. $x + 5(x + 1) > 4(2 - x) + x$

19. $-7 < y - 2 < 4$

20. $-3 < m + 6 < 2$

21. $8 \leq 3r + 1 \leq 13$

22. $-6 < 2p - 3 \leq 5$

23. $-4 \leq \dfrac{2k - 1}{3} \leq 2$

24. $-1 \leq \dfrac{5y + 2}{3} \leq 4$

25. $\dfrac{3}{5}(2p + 3) \geq \dfrac{1}{10}(5p + 1)$

26. $\dfrac{8}{3}(z - 4) \leq \dfrac{2}{9}(3z + 2)$

In the following exercises, write a linear inequality that describes the given graph.

27.

28.

29.

30.

31. Natural Science   Federal guidelines require drinking water to have fewer than .050 milligram per liter of lead. A test using 21 samples of water in a midwestern city found that the average amount of lead in the samples was .040 milligram per liter. All samples had lead content within 5% of the average.
    (a) Select a variable and write down what it represents.
    (b) Write a three-part inequality to express the sample results.
    (c) Did all the samples meet the federal requirement?

32. Natural Science   Exposure to radon gas is a known lung cancer risk. According to the Environmental Protection Agency (EPA), the individual lifetime excess cancer risk R for radon exposure is between .0015 and .006, where $R = .01$ represents a 1% increase in risk of developing cancer.*
    (a) Write the preceding information as an inequality.
    (b) Determine the range of individual annual risk by dividing R by an average life expectancy of 75 years.

33. Finance   The table below shows the federal income tax for a single person in 2002.

If Taxable Income Is: Over	But Not Over	The Tax Is:	Of the Amount Over
$0	$6,000	— 10%	$0
6,000	27,950	$600.00 + 15%	6,000
27,950	67,700	3,892.50 + 27%	27,950
67,700	141,250	14,625.00 + 30%	67,700
141,250	307,050	36,690.00 + 35%	141,250
307,050	—	94,720.00 + 38.6%	307,050

   (a) Let x denote the taxable income. Write each of the six income ranges in the table as an inequality.
   (b) Let T denote the income tax. Write an inequality that gives the tax range in dollars for each of the six income ranges in the table.

Indoor-Air-Assessment: A Review of Indoor Air Quality Risk Characterization Studies, Report No. EPA/600/8-90/044, Environmental Protection Agency, 1991.

Solve each inequality. Graph each solution. (See Examples 6–9.)

34. $|p| > 7$

35. $|m| < 1$

36. $|r| \leq 4$

37. $|a| < -2$

38. $|b| > -5$

39. $|2x + 5| < 3$

40. $\left| x - \dfrac{1}{2} \right| < 2$

41. $|3z + 1| \geq 7$

42. $|8b + 5| \geq 7$

43. $\left| 5x + \dfrac{1}{2} \right| - 2 < 5$

44. $\left| x + \dfrac{2}{3} \right| + 1 < 4$

Physical Science *The given inequality describes the monthly average temperature T in degrees Fahrenheit in the given location. What range of temperatures corresponds to the inequality?*

45. $|T - 49| \leq 20$; Santa Fe, New Mexico

46. $|T - 43| \leq 24$; Marquette, Michigan

47. $|T - 62| \leq 19$; Memphis, Tennessee

48. $|T - 10| \leq 36$; Chesterfield Inlet, Canada

49. Natural Science Human beings emit carbon dioxide when they breathe. In one study, the emission rates of carbon dioxide by college students were measured both during lectures and exams. The average individual rate R_L (in grams per hour) during a lecture class satisfied the inequality $|R_L - 26.75| \leq 1.42$, whereas during an exam the rate R_E satisfied the inequality $|R_E - 38.75| \leq 2.17$.[*]
   (a) Find the range of values for R_L and R_E.
   (b) The class had 225 students. If T_L and T_E represent the total amounts of carbon dioxide (in grams) emitted during a one-hour lecture and one-hour exam respectively, write inequalities that describe the ranges for T_L and T_E.

50. Social Science When administering a standard intelligence test, we expect about 1/3 of the scores to be more than 12 units above 100 or more than 12 units below 100. Describe this situation by writing an absolute value inequality.

Work the following problems. (See Example 4.)

51. Business Projections from the U.S. Department of Commerce show that the number of Internet users (in millions) is approximately $11.2x - 49.5$, where $x = 0$ corresponds to 1990. During what years was the number of Internet users between 52 and 97 million?

52. Business Based on data from the Federal Communications Commission, the number of international long-distance calls (in millions) placed in the United States in year x is approximately $452.1x + 884.9$, where $x = 0$ corresponds to 1990. During what years was the number of such calls between 4,050,000,000 and 6,300,000,000?

53. Health At age x, with $0 \leq x \leq 80$, the average remaining life expectancy of a male is approximately $-.8x + 70.5$. For what age range is life expectancy between 54.5 and 42.5 years?

54. Health According to data from the U.S. Census Bureau the percentage of persons not covered by health insurance in year x is approximately $.3x + 13.9$, where $x = 0$ corresponds to 1990. During what years was the percentage of uninsured individuals between 15.4 and 16.9?

55. Transportation The cost of a taxi in New York City is $2 plus 30 cents for each 1/5 of a mile. How far could you travel for at least $5 but no more than $11?

56. Transportation Anne Kelly went to a conference in Montreal for a week. She decided to rent a car and checked with two rental firms. Avery wanted $56 per day, with no mileage fee. Hart wanted $216 per week and $.28 per mile (or part of a mile). How many miles must she drive before the Avery car is the better deal?

Business *In Exercises 57–62, find all values of x where the following products will at least break even. (See Example 5.)*

57. The cost to produce x units of wire is $C = 50x + 5000$, while the revenue is $R = 60x$.

58. The cost to produce x units of squash is $C = 100x + 6000$, while the revenue is $R = 500x$.

59. $C = 85x + 900$; $R = 105x$

60. $C = 70x + 500$; $R = 60x$

61. $C = 1000x + 5000$; $R = 900x$

62. $C = 2500x + 10,000$; $R = 102,500x$

Write each of the following statements using absolute value. (See Example 10.)

63. x is within 4 units of 2.

64. m is no more than 8 units from 9.

65. z is no less than 2 units from 12.

66. p is at least 5 units from 9.

---

[*]T. C. Wang, *ASHRAE Transactions* 81 (Part 1), 32 (1975).

2.5 POLYNOMIAL AND RATIONAL INEQUALITIES

This section deals with the solution of polynomial and rational inequalities, such as

$$r^2 + 3r - 4 \geq 0, \qquad x^3 - x \leq 0, \qquad \text{and} \qquad \frac{2x - 1}{3x + 4} < 5.$$

We shall concentrate on algebraic solution methods, but to understand why these methods work, we must first look at such inequalities from a graphical point of view.

FIGURE 2.47

▶ **EXAMPLE 1** Use the graph of $y = x^3 - 5x^2 + 2x + 8$ in Figure 2.47 to solve each of the following inequalities.

(a) $x^3 - 5x^2 + 2x + 8 > 0$

Each point on the graph has coordinates of the form $(x, x^3 - 5x^2 + 2x + 8)$. The number x is a solution of the inequality exactly when the second coordinate of this point is positive, that is, when the point lies *above* the x-axis. So to solve the inequality, we need only find the first coordinates of points on the graph that are above the x-axis. This information can be read from Figure 2.47. The graph is above the x-axis when $-1 < x < 2$ and also when $x > 4$. Therefore, the solutions of the inequality are all numbers x in the intervals $(-1, 2)$ or $(4, \infty)$.

(b) $x^3 - 5x^2 + 2x + 8 < 0$

The number x is a solution of the inequality exactly when the second coordinate of the point $(x, x^3 - 5x^2 + 2x + 8)$ on the graph is negative, that is, when the point lies *below* the x-axis. Figure 2.47 shows that the graph is below the x-axis when $x < -1$ and also when $2 < x < 4$. Hence, the solutions are all numbers x in the intervals $(-\infty, -1)$ or $(2, 4)$. ◀

The solution process in Example 1 depends only on knowing the graph and its x-intercepts (that is, the points where the graph intersects the x-axis). This information can often be obtained algebraically, without doing any graphing, as illustrated in the next example.

▶ **EXAMPLE 2** Solve each of the following quadratic inequalities.

(a) $x^2 - x < 12$

First, rewrite the inequality as $x^2 - x - 12 < 0$. Now we don't know what the graph of $y = x^2 - x - 12$ looks like, but we can still find its x-intercepts by solving the equation

$$x^2 - x - 12 = 0$$
$$(x + 3)(x - 4) = 0.$$
$$x + 3 = 0 \quad \text{or} \quad x - 4 = 0$$
$$x = -3 \qquad\qquad x = 4$$

These numbers divide the x-axis (number line) into three regions, as indicated in Figure 2.48.

FIGURE 2.48

In each region, the graph of $y = x^2 - x - 12$ is an unbroken curve, so it will be entirely above or entirely below the axis. It can only pass from above to below the x-axis at the x-intercepts. To see whether the graph is above or below the x-axis when x is in region A, choose a value of x in region A, say $x = -5$, and substitute this in the equation.

$$y = x^2 - x - 12 = (-5)^2 - (-5) - 12 = 18$$

Therefore, the point $(-5, 18)$ is on the graph. Since its y-coordinate 18 is positive, this point lies above the x-axis, and hence the entire graph lies above the x-axis in region A.

Similarly, we can choose a value of x in region B, say $x = 0$. Then

$$y = x^2 - x - 12 = 0^2 - 0 - 12 = -12,$$

so that $(0, -12)$ is on the graph. Since this point lies below the x-axis (why?), the entire graph in region B must be below the x-axis. Finally, in region C, let $x = 5$. Then $y = 5^2 - 5 - 12 = 8$, so that $(5, 8)$ is on the graph and the entire graph in region C lies above the x-axis. We can summarize the results as follows.

Interval	$x < -3$	$-3 < x < 4$	$x > 4$
Test value in interval	-5	0	5
Value of $x^2 - x - 12$	18	-12	8
Graph	above x-axis	below x-axis	above x-axis
Conclusion	$x^2 - x - 12 > 0$	$x^2 - x - 12 < 0$	$x^2 - x - 12 > 0$

Solve each inequality. Graph the solution on the number line.

(a) $x^2 + 2x - 3 < 0$

(b) $2p^2 + 3p - 2 < 0$

Answers:

(a) $(-3, 1)$

(b) $(-2, 1/2)$

The last row shows that the only region where $x^2 - x - 12 < 0$ is region B, so the solutions of the inequality are all numbers x with $-3 < x < 4$, that is, the interval $(-3, 4)$, as shown in the number line graph in Figure 2.49.

FIGURE 2.49

(b) $x^2 - x - 12 > 0$

Use the chart in part (a). The last row shows that $x^2 - x - 12 > 0$ only when x is in region A or region C. Hence, the solutions of the inequality are all numbers x with $x < -3$ or $x > 4$, that is, all numbers in the intervals $(-\infty, -3)$ or $(4, \infty)$.

▶ **EXAMPLE 3** Solve the quadratic inequality $r^2 + 3r \geq 4$.

First rewrite the inequality so that one side is 0.

$$r^2 + 3r \geq 4$$
$$r^2 + 3r - 4 \geq 0 \qquad \text{Add } -4 \text{ to both sides.}$$

Now solve the corresponding equation.

$$r^2 + 3r - 4 = 0$$
$$(r - 1)(r + 4) = 0$$
$$r = 1 \quad \text{or} \quad r = -4$$

These numbers separate the number line into three regions, as shown in Figure 2.50. Test a number from each region.

Let $x = -5$ from region **A**: $(-5)^2 + 3(-5) - 4 = 6 > 0$.
Let $x = 0$ from region **B**: $(0)^2 + 3(0) - 4 = -4 < 0$.
Let $x = 2$ from region **C**: $(2)^2 + 3(2) - 4 = 6 > 0$.

We want the inequality to be positive or 0. The solution includes numbers in region A and in region C, as well as -4 and 1, the endpoints. The solution, which includes all numbers in the intervals $(-\infty, -4]$ or $[1, \infty)$, is graphed in Figure 2.50. ◀ ②

FIGURE 2.50

▶ **EXAMPLE 4** Solve $q^3 - 4q > 0$.

Solve the corresponding equation by factoring.

$$q^3 - 4q = 0$$
$$q(q^2 - 4) = 0$$
$$q(q + 2)(q - 2) = 0$$
$$q = 0 \quad \text{or} \quad q + 2 = 0 \quad \text{or} \quad q - 2 = 0$$
$$q = 0 \qquad \qquad q = -2 \qquad \qquad q = 2$$

These three numbers separate the number line into the four regions shown in Figure 2.51. Test a number from each region.

A: If $q = -3$, $(-3)^3 - 4(-3) = -15 < 0$.
B: If $q = -1$, $(-1)^3 - 4(-1) = 3 > 0$.
C: If $q = 1$, $(1)^3 - 4(1) = -3 < 0$.
D: If $q = 3$, $(3)^3 - 4(3) = 15 > 0$.

The numbers that make the polynomial positive are in the intervals

$$(-2, 0) \quad \text{or} \quad (2, \infty),$$

as graphed in Figure 2.51. ◀ ③

**② **

Solve each inequality. Graph each solution.

(a) $k^2 + 2k - 15 \geq 0$

(b) $3m^2 + 7m \geq 6$

Answers:

(a) All numbers in $(-\infty, -5]$ or $[3, \infty)$

(b) All numbers in $(-\infty, -3]$ or $[2/3, \infty)$

**③ **

Solve each inequality. Graph each solution.

(a) $m^3 - 9m > 0$

(b) $2k^3 - 50k \leq 0$

Answers:

(a) All numbers in $(-3, 0)$ or $(3, \infty)$

(b) All numbers in $(-\infty, -5]$ or $[0, 5]$

Region A	Region B	Region C	Region D
$q < -2$	$-2 < q < 0$	$0 < q < 2$	$q > 2$

FIGURE 2.51

A graphing calculator can be used to solve inequalities, without the need to evaluate at a test number in each interval. It is also useful for finding approximate solutions when the x-intercepts of the graph cannot be found algebraically.

 ▶ **EXAMPLE 5** Use a graphing calculator to solve $x^3 - 5x^2 + x + 6 > 0$.

−.9254   1.4481   4.4774

FIGURE 2.52

Begin by graphing $y = x^3 - 5x^2 + x + 6$ (Figure 2.52). Find the x-intercepts by solving $x^3 - 5x^2 + x + 6 = 0$. Since this cannot be readily done algebraically, use the graphical root finder to determine that the solutions (x-intercepts) are approximately $-.9254$, 1.4481, and 4.4774.

The graph is above the x-axis when $-.9254 < x < 1.4481$ and also when $x > 4.4774$. Therefore, the approximate solutions of the inequality are all numbers in the intervals $(-.9254, 1.4481)$ or $(4.4774, \infty)$. ◀ ④

(4)

Use graphical methods to find approximate solutions of these inequalities.

(a) $x^2 - 6x + 2 > 0$

(b) $x^2 - 6x + 2 < 0$

Answers:

(a) All numbers in $(-\infty, .3542)$ or $(5.6458, \infty)$

(b) All numbers in $(.3542, 5.6458)$

Rational Inequalities

Inequalities with quotients of algebraic expressions are called **rational inequalities.** These inequalities can be solved in much the same way as polynomial inequalities.

▶ **EXAMPLE 6** Solve the rational inequality $\dfrac{5}{x + 4} \geq 1$.

Write an equivalent inequality with one side equal to 0.

$$\frac{5}{x + 4} \geq 1$$

$$\frac{5}{x + 4} - 1 \geq 0$$

Write the left side as a single fraction.

$$\frac{5}{x + 4} - \frac{x + 4}{x + 4} \geq 0 \qquad \text{Get a common denominator.}$$

$$\frac{5 - (x + 4)}{x + 4} \geq 0 \qquad \text{Subtract fractions.}$$

$$\frac{5 - x - 4}{x + 4} \geq 0 \qquad \text{Distributive property}$$

$$\frac{1 - x}{x + 4} \geq 0$$

The quotient can change sign only at places where the denominator is 0 or the numerator is 0. (In graphical terms, these are the only places where the graph of $y = \dfrac{1 - x}{x + 4}$ can change from above the x-axis to below.) This happens when

$$1 - x = 0 \quad \text{or} \quad x + 4 = 0$$
$$x = 1 \quad \text{or} \quad x = -4.$$

As in the earlier examples, the numbers -4 and 1 divide the x-axis into three regions. Test a number from each of these regions:

$$x < -4 \qquad \text{Let } x = -5: \quad \frac{1 - (-5)}{-5 + 4} = -6 < 0.$$

$$-4 < x < 1 \qquad \text{Let } x = 0: \quad \frac{1 - 0}{0 + 4} = \frac{1}{4} > 0.$$

$$x > 1 \qquad \text{Let } x = 2: \quad \frac{1 - 2}{2 + 4} = -\frac{1}{6} < 0.$$

The test shows that numbers in $(-4, 1)$ satisfy the inequality. With a quotient, the endpoints must be considered individually to make sure that no denominator is 0. In this inequality, -4 makes the denominator 0, while 1 satisfies the given inequality. Write the solution as $(-4, 1]$. ◀

CAUTION As suggested by Example 6, be very careful with the endpoints of the intervals in the solution of rational inequalities. ⑤

▶ **EXAMPLE 7** Solve $\dfrac{2x - 1}{3x + 4} < 5$.

Write an equivalent inequality with 0 on one side. Begin by subtracting 5 on both sides and combining the terms on the left into a single fraction.

$$\frac{2x - 1}{3x + 4} < 5$$

$$\frac{2x - 1}{3x + 4} - 5 < 0 \qquad \text{Get 0 on one side.}$$

$$\frac{2x - 1 - 5(3x + 4)}{3x + 4} < 0 \qquad \text{Subtract.}$$

$$\frac{-13x - 21}{3x + 4} < 0 \qquad \text{Combine terms.}$$

Set the numerator and denominator each equal to 0 and solve the two equations.

$$-13x - 21 = 0 \quad \text{or} \quad 3x + 4 = 0$$

$$x = -\frac{21}{13} \quad \text{or} \quad x = -\frac{4}{3}$$

Use the values $-21/13$ and $-4/3$ to divide the number line into three intervals. Test a number from each interval in the inequality. The quotient is negative for numbers in $(-\infty, -21/13)$ or $(-4/3, \infty)$. Neither endpoint satisfies the given inequality. ◀

---

⑤

Solve each inequality.

(a) $\dfrac{3}{x - 2} \geq 4$

(b) $\dfrac{p}{1 - p} < 3$

(c) Why is 2 excluded from the solution in part (a)?

Answers:

(a) $(2, 11/4]$

(b) All numbers in $(-\infty, 3/4)$ or $(1, \infty)$

(c) When $x = 2$, the fraction is undefined.

⑥

Solve each rational inequality.

(a) $\dfrac{3y - 2}{2y + 5} < 1$

(b) $\dfrac{3c - 4}{2 - c} \geq -5$

Answers:

(a) $(-5/2, 7)$

(b) All numbers in $(-\infty, 2)$ or $[3, \infty)$

CAUTION In problems like Examples 6 and 7, we cannot begin by simply multiplying both sides by the denominator to simplify the inequality, because we do not know whether the variable denominator is positive or negative. ⑥

TECHNOLOGY TIP Rational inequalities can also be solved graphically. In Example 7, for instance, after writing the inequality with a single fraction on the left and determining the values that make the numerator and denominator 0, graph the left side. The graph of $y = \dfrac{-13x - 21}{3x + 4}$ in Figure 2.53 contains an erroneous vertical line segment at $x = -4/3$, but you can still read the necessary information: The graph is below the x-axis to the left of $-21/13$ and to the right of $-4/3$.

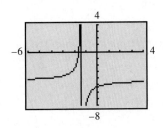

FIGURE 2.53

2.5 Exercises

Solve each of these quadratic inequalities. Graph the solutions on the number line. (See Examples 2 and 3.)

1. $(x + 5)(2x - 3) \leq 0$

2. $(5y - 1)(y + 4) > 0$

3. $r^2 + 4r > -3$

4. $z^2 + 6z < -8$

5. $4m^2 + 7m - 2 \leq 0$

6. $6p^2 - 11p + 3 \geq 0$

7. $4x^2 + 3x - 1 > 0$

8. $3x^2 - 5x > 2$

9. $x^2 \leq 25$

10. $y^2 \geq 4$

11. $p^2 - 16p > 0$

12. $r^2 - 9r < 0$

Solve these inequalities. (See Example 4.)

13. $x^3 - 9x \geq 0$

14. $p^3 - 25p \leq 0$

15. $(x + 6)(x + 1)(x - 4) \geq 0$

16. $(2x + 5)(x^2 - 1) \leq 0$

17. $(x + 4)(x^2 - 2x - 3) < 0$

18. $x^3 - 2x^2 - 3x \geq 0$

19. $6k^3 - 5k^2 < 4k$

20. $2m^3 + 7m^2 > 4m$

21. A student solved the inequality $p^2 < 16$ by taking the square root of both sides to get $p < 4$. She wrote the solution as $(-\infty, 4)$. Is her solution correct?

Use a graphing calculator to solve these inequalities. (See Example 5.)

22. $3x + 7 < 2x^2$

23. $.5x^2 - 1.2x < .1$

24. $3.1x^2 - 7.4x + 3.2 > 0$

25. $x^3 - 2x^2 - 5x + 7 \geq 2x + 1$

26. $x^4 - 6x^3 + 2x^2 < 5x - 2$

27. $2x^4 + 3x^3 < 2x^2 + 4x - 2$

28. $x^5 + 5x^4 > 4x^3 - 3x^2 - 2$

Solve the following rational inequalities. (See Examples 6 and 7.)

29. $\dfrac{r - 3}{r - 1} \geq 0$

30. $\dfrac{z + 6}{z + 3} > 1$

31. $\dfrac{a - 2}{a - 5} < -1$

32. $\dfrac{1}{3k - 5} < \dfrac{1}{3}$

33. $\dfrac{1}{p - 2} < \dfrac{1}{3}$

34. $\dfrac{7}{k + 2} \geq \dfrac{1}{k + 2}$

35. $\dfrac{5}{p + 1} > \dfrac{12}{p + 1}$

36. $\dfrac{x^2 - 4}{x} > 0$

37. $\dfrac{x^2 - x - 6}{x} < 0$

38. $\dfrac{x^2 + x - 2}{x^2 - 2x - 3} < 0$

 Use a graphing calculator to solve these inequalities. You may have to approximate the roots of the numerators or denominators.

39. $\dfrac{2x^2 + x - 1}{x^2 - 4x + 4} \leq 0$

40. $\dfrac{x^3 - 3x^2 + 5x - 29}{x^2 - 7} > 3$

Work these problems.

41. **Business**   An analyst has found that his company's profits, in hundreds of thousands of dollars, are given by $P = 3x^2 - 35x + 50$, where x is the amount, in hundreds of dollars, spent on advertising. For what values of x does the company make a profit?

42. **Business**   The commodities market is very unstable; money can be made or lost quickly on investments in soybeans, wheat, and so on. Suppose that an investor kept track of her total profit, P, at time t, in months, after she began investing, and she found that $P = 4t^2 - 29t + 30$. Find the time intervals where she has been ahead.

43. **Business**   The manager of a large apartment complex has found that the profit is given by $P = -x^2 + 250x - 15{,}000$, where x is the number of apartments rented. For what values of x does the complex produce a profit?

44. **Physical Science**   A physicist has found that the velocity of a moving particle is given by $2t^2 - 5t - 12$, where t is time in seconds since he began his observations. (Here t can be positive or negative; think of t seconds before his observations began.) Find the time intervals in which the velocity has been negative.

45. **Physical Science**   A projectile is fired from ground level. After t seconds its height above the ground is $220t - 16t^2$ feet. For what time period is the projectile at least 624 feet above the ground?

46. **Social Science**   According to data from the FBI Uniform Crime Report, the number of violent crimes per 100,000 inhabitants between 1984 and 1999 was approximately $-3.8x^2 + 87.4x + 233.2$, where $x = 0$ corresponds to 1980. During what years was the rate 712 or greater?

CHAPTER 2   SUMMARY

Key Terms and Symbols*

2.1
Cartesian coordinate system
x-axis
y-axis
origin
ordered pair
x-coordinate
y-coordinate
quadrant
solution of an equation
graph

x-intercept
y-intercept
[viewing window]
[graphical root finder]
[standard window]
graph reading

2.2
Δx change in x
Δy change in y
slope
slope-intercept form
point-slope form
linear equations
general form

2.3
linear models
residual
[least squares regression line]
[linear regression]
[coefficient of correlation]

2.4
linear inequality
properties of inequality
absolute value inequality

2.5
polynomial inequality
algebraic solution methods
[graphical solution methods]
rational inequality

*Terms in brackets deal with material in which a graphing calculator or other technology is used.

Key Concepts

The **slope** of the line through the points (x_1, y_1) and (x_2, y_2), where $x_1 \neq x_2$ is
$$m = \frac{y_2 - y_1}{x_2 - x_1}.$$
The line with equation $y = mx + b$ has slope m and y-intercept b.
The line with equation $y - y_1 = m(x - x_1)$ has slope m and goes through (x_1, y_1).
The line with equation $ax + by = c$ (with $a \neq 0$, $b \neq 0$) has x-intercept c/a and y-intercept c/b.
The line with equation $x = k$ is vertical with x-intercept k, no y-intercept, and undefined slope.
The line with equation $y = k$ is horizontal, with y-intercept k, no x-intercept, and slope 0.
Nonvertical **parallel lines** have the same slope, and **perpendicular lines,** if neither is vertical, have slopes with a product of -1.

Chapter 2 Review Exercises

Which of the ordered pairs $(-2, 3)$, $(0, -5)$, $(2, -3)$, $(3, -2)$, $(4, 3)$, $(7, 2)$ *are solutions of the given equation?*

1. $y = x^2 - 2x - 5$ **2.** $x - y = 5$

Sketch the graph of each equation.

3. $5x - 3y = 15$ **4.** $2x + 7y - 21 = 0$

5. $y + 3 = 0$ **6.** $y - 2x = 0$

7. $y = .25x^2 + 1$ **8.** $y = \sqrt{x} + 4$

9. The following temperature graph was recorded in Bratenahl, Ohio.
   (a) At what times during the day was the temperature over 55°?
   (b) When was the temperature below 40°?

10. Greenville, South Carolina, is 500 miles south of Bratenahl, Ohio, and its temperature is 7° higher all day long (see the graph in Exercise 9). At what time was the temperature in Greenville the same as the temperature at noon in Bratenahl?

11. In your own words, define the slope of a line.

In Exercises 12–21, find the slope of the line.

12. Through $(-1, 4)$ and $(2, 3)$

13. Through $(5, -3)$ and $(-1, 2)$

14. Through $(7, -2)$ and the origin

15. Through $(8, 5)$ and $(0, 3)$

16. $2x + 3y = 30$

17. $4x - y = 7$

18. $x + 5 = 0$

19. $y = 3$

20. Parallel to $3x + 8y = 0$

21. Perpendicular to $x = 3y$

22. Graph the line through $(0, 5)$ with slope $m = -2/3$.

23. Graph the line through $(-4, 1)$ with $m = 3$.

24. What information is needed to determine the equation of a line?

Find an equation for each of the following lines.

25. Through $(5, -1)$, slope $2/3$

26. Through $(8, 0)$, slope $-1/4$

27. Through $(5, -2)$ and $(1, 3)$

28. Through $(2, -3)$ and $(-3, 4)$

29. Undefined slope, through $(-1, 4)$

30. Slope 0, through $(-2, 5)$

31. x-intercept -3, y-intercept 5

32. x-intercept $-2/3$, y-intercept $1/2$

33. Health According to the National Center for Health Statistics, 25.3% of adults were smokers in 1990. By 1998, only 24% smoked.

(a) Assuming the decline in smoking is linear, write an equation that gives the percentage of smokers in year x, with $x = 0$ corresponding to 1990.

(b) Is the slope of the line positive or negative? Why?

(c) Estimate the percentage of smokers in 2006.

34. Social Science In 1980 the percent of children living with two parents was 77 and in 1998 it was 68.*

(a) Assuming the decrease was linear, write an equation that relates the percent y and year x, with 1980 corresponding to $x = 0$.

(b) Graph the equation for the years 1990 to 2005.

(c) Estimate the number of children living in two-parent families in 2005.

35. Business The table shows the average hourly earnings of U.S. production workers.[†]

Year	1991	1993	1995	1997	1999
Earnings	10.32	10.83	11.43	12.28	13.24

(a) Use the first and last data points to find a linear model for the data, with $x = 0$ corresponding to 1990.

(b) If you have access to appropriate technology, find the least squares regression line for the data.

(c) Use the models from part (a) and/or (b) to estimate the hourly earnings in 1996. The actual average in 1996 was $11.81. How far off is the model?

(d) Use the model from part (a) and/or (b) to estimate the hourly earnings in 2004.

36. Finance The table shows the number of Automatic Teller Machine (ATM) transactions (in billions) in selected years.[‡]

Year	1985	1990	1994	1995
Transactions	3.5656	5.751	8.454	9.689

Year	1996	1997	1998	1999
Transactions	10.684	10.980	11.160	10.889

(a) Use the first and last data points to find a linear model for the data, with $x = 0$ corresponding to 1980.

(b) If you have access to appropriate technology, find the least squares regression line for the data.

(c) Use the model from part (a) and/or (b) to estimate the number of ATM transactions in 2003.

37. Transportation The table, based on data from the American Automobile Manufacturers Association, shows the average expenditure for a new car in various years.

Year	1992	1993	1994	1995
Expenditure	16,336	16,871	17,903	17,959

Year	1996	1997	1998	1999
Expenditure	18,777	19,551	20,849	21,022

(a) Let $x = 0$ correspond to 1990 and find the least squares regression line for the data.

(b) What is the coefficient of correlation? Does it indicate that the line is a good fit for the data?

(c) Estimate the expenditure on a new car in 2004.

38. Business The following table, from the U.S. Federal Deposit Insurance Corporation, shows the net after-tax income (in billions of dollars) of insured commercial banks.

Year	1990	1994	1995	1996
Income	16	44.6	48.7	52.4

Year	1997	1998	1999
Income	59.2	61.8	71.7

(a) Let $x = 0$ correspond to 1990 and find the least squares regression line for the data.

(b) What is the coefficient of correlation? Does it indicate that the line is a good fit for the data?

(c) Estimate the income for the year 2001.

Solve each inequality.

39. $-6x + 3 < 2x$

40. $12z \geq 5z - 7$

41. $2(3 - 2m) \geq 8m + 3$

42. $6p - 5 > -(2p + 3)$

43. $-3 \leq 4x - 1 \leq 7$

44. $0 \leq 3 - 2a \leq 15$

*U.S. Census Bureau.

†U.S. Bureau of Labor Statistics.

‡*Statistical Abstract of the United States*, 2000.

45. $|b| \leq 8$

46. $|a| > 7$

47. $|2x - 7| \geq 3$

48. $|4m + 9| \leq 16$

49. $|5k + 2| - 3 \leq 4$

50. $|3z - 5| + 2 \geq 10$

51. Natural Science   Dr. Ryan has found that, over the years, 95% of the babies he has delivered have weighed y pounds, where $|y - 7.5| \leq 2$. What range of weights corresponds to this inequality?

52. Natural Science   The number of milligrams of a certain substance per liter in drinking water samples all tested within .05 of 40 milligrams per liter. Write this information as an inequality, using absolute value.

53. Business   Total energy production in the Asia-Pacific region was 415 million metric tons of oil equivalent (Mtoe) in 1990 and 645 Mtoe in 2000 and is increasing linearly.*
   (a) Find the linear equation that gives the energy production y in year x, with $x = 0$ corresponding to 1990.
   (b) When is energy production less than 599 Mtoe?
   (c) When is energy production more than 714 Mtoe?

54. Business   One car rental firm charges $75 for a weekend rental (Friday afternoon through Monday morning) with unlimited mileage. A second firm charges $50 plus 5 cents per mile. For what range of miles driven is the second firm cheaper?

Solve each inequality.

55. $r^2 + r - 6 < 0$

56. $y^2 + 4y - 5 \geq 0$

57. $2z^2 + 7z \geq 15$

58. $3k^2 \leq k + 14$

59. $(x - 3)(x^2 + 7x + 10) \leq 0$

60. $(x + 4)(x^2 - 1) \geq 0$

61. $\dfrac{m + 2}{m} \leq 0$

62. $\dfrac{q - 4}{q + 3} > 0$

63. $\dfrac{5}{p + 1} > 2$

64. $\dfrac{6}{a - 2} \leq -3$

65. $\dfrac{2}{r + 5} \leq \dfrac{3}{r - 2}$

66. $\dfrac{1}{z - 1} > \dfrac{2}{z + 1}$

67. Social Science   During the last century, the population of Cleveland, Ohio (in thousands), in year x was approximately $.06x^2 - 12.5x + 926$, where $x = 0$ corresponds to 1950. During what years since 1950 was the population at least 500,000?

---

*International Energy Agency, *Global Energy: The Changing Outlook*, 1992.

Case 2

Using Extrapolation to Predict Life Expectancy

One reason for developing a mathematical model is to make predictions. If your model is a least squares line, you can predict the y value corresponding to some new x by substituting this x into an equation of the form $Y = mx + b$. (We use a capital Y to remind us that we're getting a predicted value rather than an actual data value.) Data analysts distinguish two very different kinds of prediction, *interpolation* and *extrapolation*. An interpolation uses a new x inside the x range of your original data. For example, if you have inflation data at five-year intervals from 1950 to 2000, estimating the rate of inflation in 1957 is an interpolation problem. But if you use the same data to estimate what the inflation rate was in 1920, or what it will be in 2020, you are extrapolating.

In general, interpolation is much safer than extrapolation, because data that are approximately linear over a short interval may be nonlinear over a larger interval. One way to detect nonlinearity is to look at *residuals,* which are the differences between the actual data values and the values predicted by the line of best fit. Here is a simple example:

FIGURE 1

The regression equation for the best-fit line in Figure 1 is $Y = 3.431 + 1.334x$. Since the r value for this regression line is 0.93, our linear model fits the data very well. But we might notice that the predictions are a bit low at the ends and high in the middle. We can get a better look at this pattern by plotting the residuals. To find them, we put each value of the independent variable into the regression equation, calculate the predicted value Y, and subtract it from the actual y value. The residual plot in Figure 2 has the vertical axis rescaled to exaggerate the pattern. The resid-

uals indicate that our data have a nonlinear U-shaped component that isn't captured by the linear fit. Extrapolating from this data set is probably not a good idea; our linear prediction for the value of y when x is 10 may be much too low.

FIGURE 2

Exercises

The following table gives the life expectancy at birth of females born in the United States in various years from 1950 to 1995.*

Year of Birth	Life Expectancy in Years
1950	71.3
1960	73.1
1970	74.7
1980	77.4
1985	78.2
1990	78.8
1995	78.9

1. If you have appropriate technology, verify that the least squares regression line that models this data (with coeffi-

---

**Health, United States, 1998,* Centers for Disease Control.

cients rounded) is $Y = .1827x + 62.2251$, where Y is the life expectancy of a woman born in year x and $x = 0$ corresponds to 1900.

2. Use the regression equation in Exercise 1 to guess a value for the life expectancy of females born in 1900.

3. Compare your answer with the actual life expectancy for females born in 1900, which was 48.3 years. Are you surprised?

4. Find the life expectancy (rounded to one decimal place) predicted by the regression equation for each year in the table and subtract it from the actual value in the second column. This gives you a table of residuals. Plot your residuals as points on a graph.

5. Now look at the residuals as a fresh data set and see if you can sketch the graph of a smooth function that fits the residuals well. How easy do you think it will be to predict the life expectancy at birth of females born in 2010?

6. What will happen if you try linear regression on the *residuals*? If you're not sure, use technology to find the regression equation for the residuals. Why does this result make sense?

7. Since most of the females born in 1985 are still alive, how did the Public Health Service come up with a life expectancy of 78.2 years for these women?

Chapter 3

Functions and Graphs

Functions are an extremely useful way of describing many real-world situations in which the value of one quantity varies with, depends on, or determines the value of another. In this chapter you will be introduced to functions, learn how to use functional notation, develop skill in constructing and interpreting the graphs of functions, and, finally, learn to apply this knowledge in a variety of situations.

3.1 FUNCTIONS

To understand the origin of the concept of function, we consider some "real life" situations in which one numerical quantity depends on, corresponds to, or determines another.

▶ **EXAMPLE 1**   The amount of income tax you pay depends on the amount of your income. The way in which the income determines the tax is given by the tax law. ◀

▶ **EXAMPLE 2**   The weather bureau records the temperature over a 24-hour period in the form of a graph (see Figure 3.1). The graph shows the temperature that corresponds to each given time. ◀

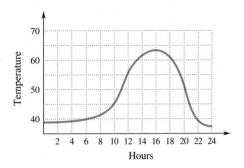

FIGURE 3.1

▶ **EXAMPLE 3**   Suppose a rock is dropped straight down from a high point. From physics we know that the distance traveled by the rock in t seconds is $16t^2$ feet. So the distance depends on the time.   ◀

The first common feature shared by these examples is that each involves two sets of numbers, which we can think of as a set of inputs and a set of outputs.

	Set of Inputs	*Set of Outputs*
Example 1	All incomes	All tax amounts
Example 2	Hours since midnight	Temperatures during the day
Example 3	Seconds elapsed after dropping the rock	Distances the rock travels

The second common feature is that in each example there is a definite *rule* by which each input determines an output. In Example 1 the rule is given by the tax law, which specifies how each income (input) determines a tax amount (output). Similarly, the rule is given by the time/temperature graph in Example 2 and by the formula (distance $= 16t^2$) in Example 3.

Each of these examples could be represented by an idealized calculator that has a single operation key and can receive or display any real number. When a number is entered (*input*), and the "rule key" is pressed, an answer is displayed (*output*). (See Figure 3.2.) The formal definition of function has these same common features (input/rule/output), with a slight change of terminology.

FIGURE 3.2

> A **function** consists of a set of inputs called the **domain,** a set of outputs called the **range,** and a rule by which each input determines *exactly one* output.

The domain in Example 1 consists of all possible income amounts; the rule is given by the tax law, and the range consists of all possible tax amounts. In Example 2 the domain is the set of hours in the day (that is, all real numbers from 0 to 24); the rule is given by the time/temperature graph, which shows the temperature at each time. The graph also shows that the range (the temperatures that actually occur during the day) includes all the numbers from 38 to 63.

Be sure that you understand the phrase "exactly one output" in the definition of the rule of a function. In Example 2, for instance, each time of day (input) determines exactly one temperature (output)—you can't have two different temperatures at the same time. However, it is quite possible to have the same temperature (output) at two different times (inputs). In other words:

> **In a function, each input produces a single output,**
> **but different inputs may produce the same output.**

Do the following define functions?

(a) The correspondence defined by the rule $y = x^2 + 5$, where x is the input and y is the output.

(b) Enter a nonzero number in a calculator and press the $1/x$ key.

(c) The correspondence between a computer, x, and several users of the computer, y

Answers:

(a) Yes

(b) Yes

(c) No

▶ **EXAMPLE 4**   Which of the following rules describe functions?

(a) Use the optical reader at the checkout counter of the supermarket to convert codes to prices.

   For each code, the reader produces exactly one price, so this is a function.

(b) Enter a number in a calculator and press the x^2 key.

   This is a function because the calculator produces just one number x^2 for each number x that is entered.

(c) Assign to each number x the number y given by this table.

x	1	1	2	2	3	3
y	3	−3	5	−5	8	−8

Since at least one x-value corresponds to more than one y-value, this table does not define a function.

(d) Assign to each number x the number y given by this equation: $y = 3x - 5$.

   Because the equation determines a unique value of y for each value of x, it defines a function. ◀ ①

The equation $y = 3x - 5$ in part (d) of Example 4 defines a function, with x as input and y as output, because each value of x determines a *unique* value of y. In such a case, the equation is said to *define y as a function of x.*

2

Do the following define y as a function of x?

(a) $y = -6x + 1$

(b) $y = x^2$

(c) $x = y^2 - 1$

(d) $y < x + 2$

Answers:

(a) Yes

(b) Yes

(c) No

(d) No

▶ **EXAMPLE 5**   Decide whether each of the following equations defines y as a function of x.

(a) $y = -4x + 11$

For a given value of x, calculating $-4x + 11$ produces exactly one value of y. (For example, if $x = -7$, then $y = -4(-7) + 11 = 39$.) Because one value of the input variable leads to exactly one value of the output variable, $y = -4x + 11$ defines y as a function of x.

(b) $y^2 = x$

Suppose $x = 36$. Then $y^2 = x$ becomes $y^2 = 36$, from which $y = 6$ or $y = -6$. Since one value of x can lead to two values of y, $y^2 = x$ does *not* define y as a function of x. ◀ **2**

Almost all the functions in this book are defined by formulas or equations, as in part (a) of Example 5. The domain of such a function is determined by the following *agreement on domains*.

> Unless otherwise stated, assume that the domain of any function defined by a formula or an equation is the largest set of real numbers that are meaningful replacements for the input variable.

For example, suppose

$$y = \frac{-4x}{2x - 3}.$$

Any real number can be used for x except $x = 3/2$, which makes the denominator equal 0. By the agreement on domains, the domain of this function is the set of all real numbers except 3/2.

3

Give the domain.

(a) $y = 3x + 1$

(b) $y = x^2$

(c) $y = \sqrt{-x}$

Answers:

(a) $(-\infty, \infty)$

(b) $(-\infty, \infty)$

(c) $(-\infty, 0]$

▶ **EXAMPLE 6**   Each of the following equations defines y as a function of x. Find the domain of each function.

(a) $y = x^4$

Any number can be raised to the fourth power, so the domain is the set of all real numbers, which is sometimes written as $(-\infty, \infty)$.

(b) $y = \sqrt{6 - x}$

For y to be a real number, $6 - x$ must be nonnegative. This happens only when $6 - x \geq 0$, or $6 \geq x$, making the domain the interval $(-\infty, 6]$.

(c) $y = \dfrac{1}{x + 3}$

Because the denominator cannot be 0, $x \neq -3$ and the domain consists of all numbers in the intervals

$$(-\infty, -3) \quad \text{or} \quad (-3, \infty). \quad ◀ \; \text{③}$$

CAUTION Notice the difference between the rule $y = \sqrt{6 - x}$ in Example 6(b) and the rule $y^2 = 6 - x$. For a particular x-value, say 2, the radical expression in the first rule represents a single positive number. The second rule, however, when $x = 2$, produces *two* numbers, $y = 2$ or $y = -2$.

Functional Notation

In actual practice, functions are seldom presented in the style of domain, rule, range, as they have been here. Functions are usually denoted by a letter (f is frequently used). If x is an input, then $f(x)$ denotes the output number that the function f produces from the input x. The symbol $f(x)$ is read "f of x." The rule is usually given by a formula, such as $f(x) = \sqrt{x^2 + 1}$. This formula can be thought of as a set of directions.

Name of function   Input number

$$f(x) = \sqrt{x^2 + 1}$$

Output number   Directions that tell you what to do with input x in order to produce the corresponding output $f(x)$; namely, "square it, add 1, and take the square root of the result."

For example, to find $f(3)$ (the output number produced by the input 3), simply replace x by 3 in the formula:

$$f(3) = \sqrt{3^2 + 1}$$
$$= \sqrt{10}.$$

Similarly, replacing x by -5 and 0 shows that

$$f(-5) = \sqrt{(-5)^2 + 1} \qquad \text{and} \qquad f(0) = \sqrt{0^2 + 1}$$
$$= \sqrt{26} \qquad\qquad\qquad = 1.$$

These directions can be applied to any quantities, such as $a + b$ or c^4 (where a, b, c are real numbers). Thus, to compute $f(a + b)$, the output corresponding to input $a + b$, we square the input [obtaining $(a + b)^2$], add 1 [obtaining $(a + b)^2 + 1$], and take the square root of the result:

$$f(a + b) = \sqrt{(a + b)^2 + 1}$$
$$= \sqrt{a^2 + 2ab + b^2 + 1}.$$

Similarly, the output $f(c^4)$ corresponding to the input c^4 is computed by squaring the input $[(c^4)^2]$, adding 1 $[(c^4)^2 + 1]$, and taking the square root of the result:

$$f(c^4) = \sqrt{(c^4)^2 + 1}$$
$$= \sqrt{c^8 + 1}.$$

▶ **EXAMPLE 7**   Let $g(x) = -x^2 + 4x - 5$. Find each of the following.

(a) $g(-2)$
Replace x with -2.

$$g(-2) = -(-2)^2 + 4(-2) - 5$$
$$= -4 - 8 - 5$$
$$= -17$$

(b) $g(x + h)$
Replace x by the quantity $x + h$ in the rule of g.

$$g(x + h) = -(x + h)^2 + 4(x + h) - 5$$
$$= -(x^2 + 2xh + h^2) + (4x + 4h) - 5$$
$$= -x^2 - 2xh - h^2 + 4x + 4h - 5$$

(c) $g(x + h) - g(x)$
Use the result from part (b) and the rule for $g(x)$.

$$g(x + h) - g(x) = (-x^2 - 2xh - h^2 + 4x + 4h - 5) - (-x^2 + 4x - 5)$$
$$= -2xh - h^2 + 4h$$

(d) $\dfrac{g(x + h) - g(x)}{h}$ (assuming $h \neq 0$)

The numerator was found in part (c). Divide it by h as follows.

$$\frac{g(x + h) - g(x)}{h} = \frac{-2xh - h^2 + 4h}{h}$$
$$= \frac{h(-2x - h + 4)}{h}$$
$$= -2x - h + 4 \blacktriangleleft$$

The quotient found in Example 7(d)

$$\frac{g(x + h) - g(x)}{h}$$

is important in calculus, and we will see it again in Chapter 11. ④

④

Let $f(x) = 5x^2 - 2x + 1$. Find the following.

(a) $f(1)$
(b) $f(3)$
(c) $f(1 + 3)$
(d) $f(1) + f(3)$
(e) $f(m)$
(f) $f(x + h) - f(x)$

Answers:

(a) 4
(b) 40
(c) 73
(d) 44
(e) $5m^2 - 2m + 1$
(f) $10xh + 5h^2 - 2h$

CAUTION   Functional notation is *not* the same as ordinary algebraic notation. You cannot simplify an expression such as $f(x + h)$ by writing $f(x) + f(h)$. To see why, consider the answers to problems 4(c) and (d) at the side, which show that

$$f(1 + 3) \neq f(1) + f(3).$$

▶ **EXAMPLE 8**   Suppose the projected sales (in thousands of dollars) of a small company over the next ten years are approximated by the function

$$S(x) = .08x^4 - .04x^3 + x^2 + 9x + 54.$$

A developer estimates that the total cost of building x large apartment complexes in a year is approximated by

$$A(x) = x^2 + 80x + 60,$$

where $A(x)$ represents the cost in hundred thousands of dollars. Find the cost of building

(a) 4 complexes;

(b) 10 complexes.

Answers:

(a) $39,600,000

(b) $96,000,000

(a) **What are the projected sales this year?**
The current year corresponds to $x = 0$ and the sales this year are given by $S(0)$. Substituting 0 for x in the rule of S, we see that $S(0) = 54$. So the current projected sales are $54,000.

(b) **What will sales be in three years?**
The sales three years from now are given by $S(3)$, which can be computed by hand or with a calculator:

$$S(x) = .08x^4 - .04x^3 + x^2 + 9x + 54$$
$$S(3) = .08(3)^4 - .04(3)^3 + (3)^2 + 9(3) + 54 \qquad \text{Let } x = 3.$$
$$= 95.4,$$

which means that sales are projected to be $95,400. ◀ ⑤

If you have a graphing calculator, you should learn to use its table feature to evaluate functions.

⟋▶EXAMPLE 9   Use the table feature of a graphing calculator to find the projected sales of the company in Example 8 for years 4 through 10.
Enter the sales equation $y = .08x^4 - .04x^3 + x^2 + 9x + 54$ in the equation memory of the calculator (often called the Y= list). Check your instruction manual for how to set the table to start at $x = 4$ and go through $x = 10$. Then display the table, as in Figure 3.3. It shows that sales are projected to rise from $123,920 in year 4 to $1,004,000 in year 10. ◀

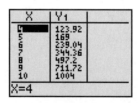

FIGURE 3.3

3.1 Exercises

Which of the following rules define y as a function of x? (See Examples 1–5.)

1.

x	3	2	1	0	−1	−2	−3
y	9	4	1	0	1	4	9

2.

x	9	4	1	0	1	4	9
y	3	2	1	0	−1	−2	−3

3. $y = x^3$

4. $y = \sqrt{x - 1}$

5. $x = |y + 2|$

6. $x = y^2 + 3$

7. $y = \dfrac{-1}{x - 1}$

8. $y = \dfrac{4}{2x + 3}$

State the domain of each function. (See Example 6.)

9. $f(x) = 4x - 1$ **10.** $f(x) = 2x + 7$

11. $f(x) = x^4 - 1$ **12.** $f(x) = (2x + 5)^2$

13. $f(x) = \sqrt{-x} + 3$ **14.** $f(x) = \sqrt{4 - x}$

15. $f(x) = \dfrac{1}{x - 1}$ **16.** $f(x) = \dfrac{1}{x - 3}$

17. $f(x) = |5 - 4x|$ **18.** $f(x) = |-x - 6|$

For each of the following functions, find

(a) $f(4)$ **(b)** $f(-3)$ **(c)** $f(2.7)$ **(d)** $f(-4.9)$.

(See Example 7.)

19. $f(x) = 6$ **20.** $f(x) = 0$

21. $f(x) = 2x^2 + 4x$ **22.** $f(x) = x^2 - 2x$

23. $f(x) = \sqrt{x + 3}$ **24.** $f(x) = \sqrt{5 - x}$

25. $f(x) = x^3 - 6.2x^2 + 4.5x - 1$

26. $f(x) = x^4 + 5.5x^2 - .3x$

27. $f(x) = |x^2 - 6x - 4|$

28. $f(x) = |x^3 - x^2 + x - 1|$

29. $f(x) = \dfrac{\sqrt{x - 1}}{x^2 - 1}$

30. $f(x) = \sqrt{-x} + \dfrac{2}{x + 1}$

If you have a graphing calculator with table-making ability, display a table showing the (approximate) values of the given function at 3.5, 3.9, 4.3, 4.7, 5.1, and 5.5. (See Example 9.)

31. $g(x) = 3x^4 - x^3 + 2x$

32. $f(x) = \sqrt{x^2 - 2.4x + 8}$

For each of the following functions, find

(a) $f(p)$ **(b)** $f(-r)$ **(c)** $f(m + 3)$.

(See Example 7.)

33. $f(x) = 5 - x$ **34.** $f(x) = 3x + 7$

35. $f(x) = \sqrt{4 - x}$ **36.** $f(x) = \sqrt{-2x}$

37. $f(x) = x^3 + 1$ **38.** $f(x) = 2 - x^3$

39. $f(x) = \dfrac{3}{x - 1}$ **40.** $f(x) = \dfrac{-1}{2 + x}$

For each of the following functions find

$$\dfrac{f(x + h) - f(x)}{h}.$$

(See Example 7.)

41. $f(x) = 2x - 4$ **42.** $f(x) = 2 - 3x$

43. $f(x) = x^2 + 1$ **44.** $f(x) = x^2 - x$

Use a calculator to work these exercises. (See Example 8.)

45. Health   The table contains incidence ratios by age for death from coronary heart disease (CHD) and lung cancer

(LC) when comparing smokers (21–39 cigarettes per day) to nonsmokers.*

Age	CHD	LC
55–64	1.9	10
65–74	1.7	9

The incidence ratio of 10 means that smokers are 10 times more likely than nonsmokers to die of lung cancer between the ages of 55 and 64. If the incidence ratio is x, then the percent P (expressed as a decimal) of deaths caused by smoking is given by the function $P(x) = \dfrac{x - 1}{x}$.

(a) What is the percent of lung cancer deaths that can be attributed to smoking between the ages of 65 and 74?

(b) What is the percent of coronary heart disease deaths that can be attributed to smoking between the ages of 55 and 64?

46. Transportation   The number of fliers on commuter airlines (10- to 30-seat planes) between 1975 and 2010 is approximated by $g(x) = .0138x^2 - .172x + 1.4$ (where $x = 0$ corresponds to 1975 and $g(x)$ is in millions).[†]

(a) How many fliers were there in 1975 and 1994?

(b) How many fliers are projected for 2006?

47. Social Science   The number of Americans (in thousands) that are or are expected to be over 100 years old in year x can be approximated by the function $h(x) = .4018x^2 + 2.039x + 50$ (where $x = 0$ corresponds to 1994).[‡]

(a) How many Americans were over 100 in 1994? In 1996?

(b) Predict the number of Americans that will be over 100 years old in the year 2008.

48. Natural Science   High concentrations of zinc ions in the water are lethal to rainbow trout. The function

$$f(x) = \left(\dfrac{x}{1960}\right)^{-.833}$$

gives the approximate average survival time (in minutes) for trout exposed to x milligrams per liter (mg/L) of zinc ions.[§] Find the survival time (to the nearest minute) for the following concentrations of zinc ions.

(a) 110   **(b)** 525   **(c)** 1960   **(d)** 4500

*Walker, A., *Observations and Inference: An Introduction to the Methods of Epidemiology,* (Newton Lower Falls, MA: Epidemiology Resources, Inc., 1991).

†Based on data from the Federal Aviation Administration (in *USA Today*, March 27, 1995).

‡Based on data from the U.S. Census Bureau.

§C. Mason, *Biology of Freshwater Pollution.*

Use the table feature of a graphing calculator to do these exercises. (See Example 9.)

49. Finance Based on data from the U.S. Department of Education, federal student financial assistance (Pell grants, Perkins loans, work-study funds, etc.) is modeled by the function

$$g(x) = .104x^3 - 2.34x^2 + 18.23x - 9.6 \quad (4 \le x \le 10),$$

where $x = 0$ corresponds to 1990 and $g(x)$ is in billions of dollars.
(a) List the values of $g(x)$ for $x = 4, 5, \ldots, 10$.
(b) Explain why this model is not likely to be accurate after 2001. (*Hint:* find $g(15)$ and $g(18)$; do these numbers seem plausible?)

50. Natural Science The average monthly temperature in Austin, Texas, can be approximated by the function

$$f(x) = .03x^4 - .716x^3 + 4.38x^2 - 1.07x + 49$$
$$(0 \le x < 12),$$

where $x = 0$ corresponds to January 1, $x = 2$ to February 1, etc. List the average temperature on the first day of each month from April through October.

Work these problems.

51. Transportation The distance from Chicago to Seattle is approximately 2000 miles. A plane flying directly to Seattle passes over Chicago at noon. If the plane travels at 475 mph, find the rule of the function $f(t)$ that gives the distance of the plane from Seattle at time t hours (with $t = 0$ corresponding to noon).

52. Business A pretzel factory has daily fixed costs of $1800. In addition, it costs 50 cents to produce each bag of pretzels. A bag of pretzels sells for $1.20.
(a) Find the rule of the cost function $c(x)$ that gives the total daily cost of producing x bags of pretzels.
(b) Find the rule of the revenue function $r(x)$ that gives the daily revenue from selling x bags of pretzels.
(c) Find the rule of the profit function $p(x)$ that gives the daily profit from x bags of pretzels.

53. Business A bicycle factory has weekly fixed costs of $36,000. The material and labor costs for each bike are $125.
(a) Express the total weekly cost $C(x)$ as a function of the number x of bicycles that are made.
(b) What is the cost *per bicycle* of producing 600 bikes in a week?

54. Geometry Find an equation that expresses the area y of a square as a function of its
(a) side x (b) diagonal d

3.2 GRAPHS OF FUNCTIONS

The **graph** of a function $f(x)$ is defined to be the graph of the *equation* $y = f(x)$. For example, the graph of $f(x) = x^3 + x + 5$ is the graph of the equation $y = x^3 + x + 5$. Thus, the graph consists of all points in the coordinate plane whose coordinates are of the form $(x, x^3 + x + 5)$, that is, all points $(x, f(x))$. The same thing is true in the general case: Each point on the graph of a function f is an ordered pair whose first coordinate is an input number from the domain of f and whose second coordinate is the corresponding output number.

We begin with the simplest function graphs. In Section 2.2 we saw that the graph of an equation of the form $y = ax + b$ is a straight line. This equation also defines y as a function of x, which leads to the following terminology.

A **linear function** is a function whose rule can be written in the form

$$f(x) = ax + b$$

for some constants a and b.

▶ **EXAMPLE 1**   The graph of the linear function $g(x) = .5x - 3$ is the graph of the equation $y = .5x - 3$. So the graph is a straight line with slope .5 and y-intercept -3, as shown in Figure 3.4. ◀

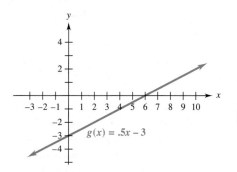

FIGURE 3.4

Piecewise Linear Functions

We now consider functions whose graphs consist of straight line segments. Such functions are called **piecewise linear functions** and are typically defined with different equations for different parts of the domain.

▶ **EXAMPLE 2**   Graph the following function.

$$f(x) = \begin{cases} x + 1 & \text{if } x \le 2 \\ -2x + 7 & \text{if } x > 2 \end{cases}$$

When $x \le 2$, the graph consists of the part of the line $y = x + 1$ that lies to the left of $x = 2$. When $x > 2$, the graph consists of the part of the line $y = -2x + 7$ that lies to the right of $x = 2$. These line segments can be graphed by plotting the points determined by the following tables.

$x \le 2$

x		-2	0	2
$y = x + 1$		-1	1	3

$x > 2$

x		2	3	4
$y = -2x + 7$		3	1	-1

Note that even though 2 is not in the interval $x > 2$, we found the ordered pair for that endpoint because the graph will extend right up to that point. Because this endpoint (2, 3) agrees with the endpoint for the interval $x \leq 2$, the two parts of the graph are joined at that point as shown in Figure 3.5. ◀

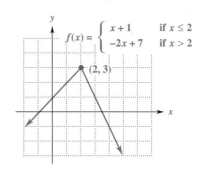

$$f(x) = \begin{cases} x + 1 & \text{if } x \leq 2 \\ -2x + 7 & \text{if } x > 2 \end{cases}$$

FIGURE 3.5

Graph each function.

(a) $f(x) = 2 - |x|$

(b) $f(x) = |5x - 7|$

Answers:

(a)

(0, 2)

(b)

$\left(\frac{7}{5}, 0\right)$

CAUTION In Example 2, notice that we did not graph the entire lines but only those portions with domain as given. Graphs of these functions should *not* be two intersecting lines.

▶ **EXAMPLE 3** Graph the **absolute value function,** whose rule is $f(x) = |x|$.

The function f is a piecewise linear function because from the definition of $|x|$,

$$f(x) = \begin{cases} x & \text{if } x \geq 0 \\ -x & \text{if } x < 0. \end{cases}$$

So the right half of the graph (that is, where $x \geq 0$) will consist of a portion of the line $y = x$. It can be graphed by plotting two points, say, (0, 0) and (1, 1). The left half of the graph (where $x < 0$) will consist of a portion of the line $y = -x$, which can be graphed by plotting (−2, 2) and (−1, 1), as shown in Figure 3.6. ◀ ①

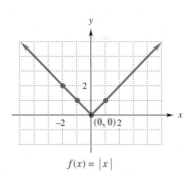

$f(x) = |x|$

FIGURE 3.6

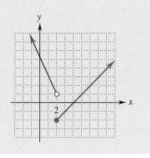

Graph $f(x)$, where

$$f(x) = \begin{cases} -2x + 5 & \text{if } x < 2 \\ x - 4 & \text{if } x \geq 2. \end{cases}$$

Answer:

▶ **EXAMPLE 4**   Graph the function

$$f(x) = \begin{cases} x - 2 & \text{if } x \leq 3 \\ -x + 8 & \text{if } x > 3. \end{cases}$$

The ordered pairs $(-2, -4)$, $(0, -2)$, and $(3, 1)$ satisfy $y = x - 2$, and the pairs $(4, 4)$ and $(6, 2)$ satisfy $y = -x + 8$. Use them to graph the two straight line segments that make up the graph of the function. The point $(3, 5)$, where the right half of the graph begins, is an endpoint, but is *not* part of the graph, as indicated by the open circle in Figure 3.7.   ◀ ②

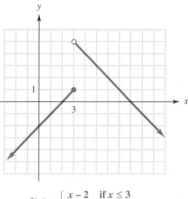

$$f(x) = \begin{cases} x - 2 & \text{if } x \leq 3 \\ -x + 8 & \text{if } x > 3 \end{cases}$$

FIGURE 3.7

T E C H N O L O G Y   T I P   To graph most piecewise linear functions on a graphing calculator, you must use a special syntax. For example, on TI and HP-39 calculators, the best way to obtain the graph in Example 4 is to graph two separate equations on the same screen:

$$y_1 = (x - 2)/(x \leq 3) \quad \text{and} \quad y_2 = (-x + 8)/(x > 3);$$

the inequality symbols are in the TEST (or CHAR) menu. However, most calculators will graph absolute value functions directly. To graph $f(x) = |x + 2|$, for instance, graph the equation $y = \text{abs}(x + 2)$. "Abs" (for absolute value) is on the keyboard or in the MATH menu.

Graphs used in business and the social sciences are often graphs of piecewise linear functions. It is just as important to know how to *read* such graphs as it is to construct them.

▶ **EXAMPLE 5** Figure 3.8 shows the graphs of three functions. Each function is the fatal crash rate per 10,000 drivers for a group of Ohio drivers.*

FIGURE 3.8

(a) Describe the fatal crash record for 16-year-old drivers and for drivers at ages 25–54. How do they compare?

The graph suggests that fatal car crashes of 16-year-old drivers have followed a cyclic pattern, dropping significantly from 1988 to 1992, rising from 1992 to 1995, and dropping again in recent years (with a slight rise in 1998), with the rate generally being between 8 and 12. The graph for drivers ages 25–54 is almost a horizontal line, indicating a virtually constant fatality rate of about 2 per 10,000. Thus, the 16-year-old rate is about 4 to 6 times higher than the rate for older drivers, depending on the year.

(b) What were the worst and best years for 16- and 17-year-old drivers?

Sixteen-year-olds had their worst year in 1995, where their graph peaks, and their best year in 1992. The worst year for 17 year olds was 1989, and their best during this period was 1996. ◀ ③

③

In Example 5, describe the fatal crash record of 17-year-old drivers. How does it compare to that of the other two groups?

Answer:

The crash rate shows a general downward trend since 1989 (with a very slight rise from 1991 to 1994 and a notable rise from 1996 to 1997). Their fatality rate is significantly lower than the 16-year-old rate, but two to three times that of older drivers.

Step Functions

The **greatest integer function,** usually written $f(x) = [x]$, is defined by saying that $[x]$ denotes the largest integer that is less than or equal to x. For example, $[8] = 8$, $[7.45] = 7$, $[\pi] = 3$, $[-1] = -1$, $[-2.6] = -3$, and so on.

④

Graph $y = [\frac{1}{2}x + 1]$.

Answer:

▶ **EXAMPLE 6** Graph $f(x) = [x]$.

If $-1 \le x < 0$, then $[x] = -1$. If $0 \le x < 1$, then $[x] = 0$. If $1 \le x < 2$, then $[x] = 1$, and so on. Thus, the graph, as shown in Figure 3.9, consists of a series of horizontal line segments. In each one, the left endpoint is included and the right endpoint is excluded. The shape of the graph is the reason that this function is called a **step function.** ◀ ④

*Ohio Department of Public Safety.

FIGURE 3.9

⑤

Assume that the post office charges 80¢ per ounce, or fraction of an ounce, to mail a letter to Europe. Graph the ordered pairs (ounces, cost).

Answer:

▶**EXAMPLE 7** An overnight delivery service charges $25 for a package weighing up to 2 pounds. For each additional pound or fraction of a pound there is an additional charge of $3. Let $D(x)$ represent the cost to send a package weighing x pounds. Graph $D(x)$ for x in the interval (0, 6].

For x in the interval (0, 2], $y = 25$. For x in (2, 3], $y = 25 + 3 = 28$. For x in (3, 4], $y = 28 + 3 = 31$, and so on. The graph, which is that of a step function, is shown in Figure 3.10. ◀ ⑤

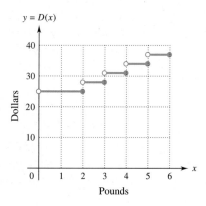

FIGURE 3.10

TECHNOLOGY TIP On most graphing calculators the greatest integer function is denoted INT or FLOOR (look on the MATH menu or its NUM submenu). Casio calculators use INTG for the greatest integer function and INT for a different function. When graphing these functions, put your calculator in "dot" graphing mode rather than the usual "connected" mode to avoid erroneous vertical line segments in the graph.

Other Functions

The graphs of many functions do not consist only of straight line segments. As a general rule when graphing functions by hand, you should follow the procedure introduced in Section 2.1 and summarized here.

Graphing a Function by Point Plotting

1. Determine the domain of the function.
2. Select a few numbers in the domain of f (include both negative and positive ones when possible) and compute the corresponding values of $f(x)$.
3. Plot the points $(x, f(x))$ computed in step two. Use these points and any other information you may have about the function to make an "educated guess" about the shape of the entire graph.
4. Unless you have information to the contrary, assume that the graph is continuous (unbroken) wherever it is defined.

6

Graph $f(x) = \sqrt{4 - x}$.

Answer:

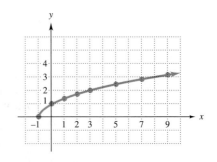

This method was used to find the graphs of the functions $f(x) = 4 - x^2$ and $g(x) = \sqrt{x}$ in Examples 3 and 4 of Section 2.1. Here are some more examples.

▶ **EXAMPLE 8**   Graph $g(x) = \sqrt{x + 1}$.

Because the rule of the function is defined only when $x + 1 \geq 0$ (that is, when $x \geq -1$), the domain of g is the interval $[-1, \infty)$. Use a calculator to get a table of ordered pairs, such as the one in Figure 3.11. Plot the points and connect them in order (as x increases) to get the graph in Figure 3.11.   ◀   6

x	$g(x) = \sqrt{x + 1}$
−1	0
0	1
1	1.414
2	1.732
3	2
5	2.449
7	2.828
9	3.162

FIGURE 3.11

▶ **EXAMPLE 9**   Graph the function whose rule is $f(x) = 2 - x^3/5$.

Make a table of values and plot the corresponding points. They suggest the graph in Figure 3.12.   ◀

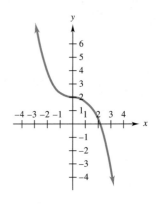

x	$f(x) = 2 - x^3/5$
-4	14.8
-3	7.4
-2	3.6
-1	2.2
0	2
1	1.8
2	.4
3	-3.4
4	-10.8

FIGURE 3.12

Many functions that arise in real-world applications are based on data and given in graphical form, so it is important to be able to read such graphs.

▶ **EXAMPLE 10**   Figure 3.13 is the graph of the function h whose rule is $h(x) = $ Household debt payment as a percentage of after-tax income in year x.*

FIGURE 3.13

(a) Find the function values $h(1992)$, and $h(2000)$ and interptet them.

The point (1992, 12) is on the graph, which means that $h(1992) = 12$. Similarly, $h(2000) = 14.3$ because the point (2000, 14.3) is on the graph.[†] In early 1992, debt payment was 12% of after-tax income, and in early 2000 it was 14.3%.

(b) During what period was debt payment below 13%?

Look for points on the graph whose second coordinates are less than 13, that is, points below the horizontal line through 13. These points represent the period from late 1990 to mid-1995.

(c) During what period was debt payment above 13.5%?

*Federal Reserve Board.

[†]Since this small graph cannot be read with perfect accuracy, these results are approximate.

Look for points whose second coordinates are greater than 13.5 (above the horizontal line through 13.5). They occur during two periods: from 1986 to the beginning of 1990 and since the first quarter of 1998. ◄

The Vertical Line Test

The following fact distinguishes function graphs from other graphs.

Vertical Line Test

No vertical line intersects the graph of a function more than once.

In other words, if a vertical line intersects a graph at more than one point, the graph is not the graph of a function. To see why this is true, consider the graph in Figure 3.14. The vertical line $x = 3$ intersects the graph at two points. If this were the graph of a function f, this would mean that $f(3) = 2$ (because (3, 2) is on the graph) and that $f(3) = -1$ (because (3, -1) is on the graph). This is impossible because a *function* can have only one value when $x = 3$ (each input determines exactly one output). Therefore, this cannot be the graph of a function. A similar argument works in the general case.

FIGURE 3.14

▶ EXAMPLE 11   Which of the graphs in Figure 3.15 are the graphs of functions?

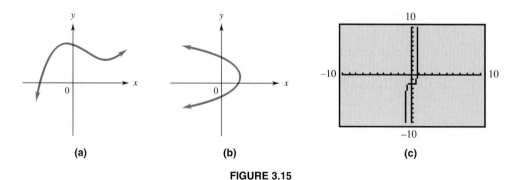

(a)   (b)   (c)

FIGURE 3.15

Every vertical line intersects graph (a) in at most one point, so this graph is the graph of a function. Many vertical lines (including the *y*-axis) intersect graph (b) twice, so it is not the graph of a function.

Graph (c) appears to fail the Vertical Line Test near $x = 1$ and $x = -1$, indicating this is not the graph of a function. But this is misleading because of the low resolution of a calculator screen. The table in Figure 3.16 and the very narrow segment of the graph in Figure 3.17 show that the graph actually moves to the right as it moves upward. A similar fact is true near $x = -1$. So this graph *does* pass the Vertical Line Test and *is* the graph of a function. (Its rule is $f(x) = 15x^{11} - 2$.) The moral of this story is that you can't always trust a graphing calculator image. When in doubt, try other viewing windows or a table to see what's really going on. ◀ ⑦

(7)

Find a viewing window that indicates the actual shape of the graph of the function $f(x) = 15x^{11} - 2$ of Example 10 near the point $(-1, -17)$.

Answer:

There are many correct answers, including,
$-1.4 \le x \le -.6$ and
$-30 \le y \le 0$.

FIGURE 3.16

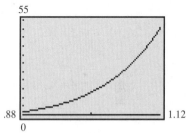

FIGURE 3.17

3.2 Exercises

Graph each function. (See Examples 1–4.)

1. $f(x) = -.5x + 2$

2. $g(x) = 3 - x$

3. $f(x) = \begin{cases} x + 2 & \text{if } x \le 1 \\ 3 & \text{if } x > 1 \end{cases}$

4. $g(x) = \begin{cases} 2x - 1 & \text{if } x < 0 \\ -1 & \text{if } x \ge 0 \end{cases}$

5. $y = \begin{cases} 3 - x & \text{if } x \le 0 \\ 2x + 3 & \text{if } x > 0 \end{cases}$

6. $y = \begin{cases} x + 5 & \text{if } x \le 1 \\ 2 - 3x & \text{if } x > 1 \end{cases}$

7. $f(x) = \begin{cases} |x| & \text{if } x \le 2 \\ -x & \text{if } x > 2 \end{cases}$

8. $g(x) = \begin{cases} -|x| & \text{if } x \le 1 \\ 2x & \text{if } x > 1 \end{cases}$

9. $f(x) = |x - 4|$

10. $g(x) = |4 - x|$

11. $f(x) = |3 - 4x|$

12. $g(x) = -|x|$

13. $y = -|x - 1|$

14. $f(x) = |x| - 2$

15. $y = |x| + 3$

16. $|x| + |y| = 1$ (*Hint:* This is not the graph of a function but is made up of four straight line segments. Find them by using the definition of absolute value in these four cases: $x \ge 0$ and $y \ge 0$; $x \ge 0$ and $y < 0$; $x < 0$ and $y \ge 0$; $x < 0$ and $y < 0$.)

Graph each of the following functions. (See Examples 6 and 7.)

17. $f(x) = [x - 3]$

18. $g(x) = [x + 2]$

19. $g(x) = [-x]$

20. $f(x) = -[x]$

21. $f(x) = [x] + [-x]$ (The graph contains horizontal segments but is *not* a horizontal line.)

22. Assume that postage rates are 37¢ for the first ounce, plus 23¢ for each additional ounce, and that each letter carries one 37¢ stamp and as many 23¢ stamps as necessary. Graph the *postage stamp function* whose rule is

$p(x) =$ the number of stamps on a letter weighing x ounces.

Graph each function. (See Examples 8 and 9.)

23. $f(x) = 3 - 2x^2$

24. $g(x) = 2 - x^2$

25. $h(x) = x^3/10 + 2$

26. $f(x) = x^3/20 - 3$

27. $g(x) = \sqrt{-x}$ **28.** $h(x) = \sqrt{x} - 1$

29. $f(x) = \sqrt[3]{x}$ **30.** $g(x) = \sqrt[3]{x} - 4$

Which of these are graphs of functions? (See Example 11.)

31.

32.

33.

34.

35.

36.

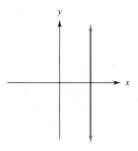

Work the following problems. (See Examples 3–5.)

37. Finance   The maximum allowable yearly contribution to an individual retirement account (IRA) was $2000 from 1981 to 2001 and has steadily increased since then. When the maximum is adjusted for inflation, however, the picture is a little different. The approximate maximum IRA contribution in 1981 dollars in year x is given by the function

$$f(x) = \begin{cases} -50x + 2050 & \text{if } 1 \le x < 22 \\ 84x - 348 & \text{if } 22 \le x < 29 \end{cases}$$,

where $x = 0$ corresponds to 1980.

(a) Graph this function.

(b) What does the graph say about inflation from 1981 to 2001?

38. Finance   The Connecticut state income tax in 2001 was given by the function

$$g(x) = \begin{cases} .03x & \text{if } 0 \le x \le 20{,}000 \\ .045x - 300 & \text{if } x > 20{,}000 \end{cases}$$,

where x is the income in dollars. Graph this function for incomes from 0 to $40,000.

39. Natural Science   The snow depth in Michigan's Isle Royale National Park varies throughout the winter. In a typical winter, the snow depth in inches is approximated by the following function.

$$f(x) = \begin{cases} 6.5x & \text{if } 0 \le x \le 4 \\ -5.5x + 48 & \text{if } 4 < x \le 6 \\ -30x + 195 & \text{if } 6 < x \le 6.5 \end{cases}$$

Here, x represents the time in months with $x = 0$ representing the beginning of October, $x = 1$ representing the beginning of November, and so on.

(a) Graph $f(x)$.

(b) In what month is the snow deepest? What is the deepest snow depth?

(c) In what months does the snow begin and end?

40. Natural Science   A factory begins emitting particulate matter into the atmosphere at 8 A.M. each workday, with the emissions continuing until 4 P.M. The level of pollutants, $P(t)$, measured by a monitoring station 1/2 mile away is approximated as follows, where t represents the number of hours since 8 A.M.

$$P(t) = \begin{cases} 75t + 100 & \text{if } 0 \le t \le 4. \\ 400 & \text{if } 4 < t < 8 \\ -100t + 1200 & \text{if } 8 \le t \le 10 \\ -\dfrac{50}{7}t + \dfrac{1900}{7} & \text{if } 10 < t < 24 \end{cases}$$

Find the level of pollution at

(a) 9 A.M. (b) 11 A.M. (c) 5 P.M.

(d) 7 P.M. (e) Midnight.

(f) Graph $y = P(t)$.

(g) From the graph in part (f), at what time(s) is the pollution level highest? lowest?

41. Health The table shows the consumer price index (CPI) for medical care in selected years.*

Year	Medical CPI
1950	15.1
1975	47.5
1999	250.6

(a) Let $x = 0$ correspond to 1950. Find the rule of a piecewise linear function that models this data, that is, a piecewise linear function f with $f(0) = 15.1$, $f(25) = 47.5$, and $f(49) = 250.6$. Round all coefficients in your final answer to one decimal place.

(b) Graph the function f for $0 \le x \le 50$

(c) Use the function to estimate the medical CPI in 1995.

(d) Assume that this model remains accurate after 2000 and estimate the medical CPI in 2004.

42. Business The graph (from the Bureau of Public Debt) shows the federal debt from 1990 to 2000 (in billions of dollars).

(a) Let $x = 0$ correspond to 1990. Find the rule of a linear function g that passes through the two points cor-

responding to 1990 and 2000. Draw the graph of g in the preceding figure.

(b) Is the function g a good model for the federal debt? Why?

Finance *The figure shows the annual percentage change in consumer price indexes (CPI). Use it for Exercises 43 and 44. (See Examples 5 and 10.)*

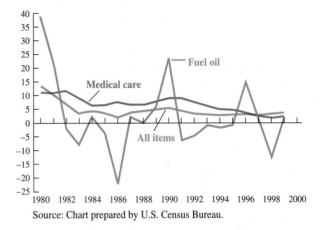

Source: Chart prepared by U.S. Census Bureau.

43. Answer each of these questions and explain how you obtained the answer from the graph.

(a) Was there any period between 1980 and 1998 when the CPI for all items was decreasing?

(b) During what years was the CPI for fuel oil increasing or constant?

(c) During what years was the CPI for fuel oil decreasing? How can you tell from the graph?

44. (a) When was the fuel oil CPI decreasing at the fastest rate?

(b) After 1982, when was the fuel oil CPI increasing at the fastest rate?

(c) Between 1985 and 1995, when were the medical care CPI and the fuel oil CPI increasing at the same rate? How can you tell from the graph?

45. Natural Science The table gives estimates of the percent of ozone change from 1985 for several years in the future, if chlorofluorocarbon production is reduced 80% globally.

Year	Percent
1985	0
2005	1.5
2025	3
2065	4
2085	5

---

*U.S. Bureau of Labor Statistics.

(a) Plot these ordered pairs on a grid.

(b) The points from part (a) should lie approximately on a straight line. Use the pairs (2005, 1.5) and (2085, 5) to write an equation of the line.

(c) Letting $f(x)$ represent the percent of ozone change and x represent the year, write your equation from part (b) as a rule that defines a function.

(d) Find $f(2065)$. Does it agree fairly closely with the number in the table that corresponds to 2065? Do you think the expression from part (c) describes this function adequately?

46. Business   The graph below shows the number of cellular telephone accounts (in millions) since 1989.

(a) Is this the graph of a function?

(b) What does the domain represent?

(c) Estimate the range.

Business   *Work these problems. (See Examples 6, 7, and 11.)*

47. Whenever postage rates change, some newspaper publishes a graph like this one, which shows the price of a first-class stamp from 1982 to 2003.

(a) Let f be the function whose rule is

$f(x)$ = cost of a first-class stamp in year x.

Find $f(1990)$ and $f(2000)$.

(b) Explain why the graph in the figure is not the graph of the *function f*. What must be done to the figure to make it an accurate graph of the function f?

48. A chain-saw rental firm charges $7 per day or fraction of a day to rent a saw, plus a fixed fee of $4 for resharpening

the blade. Let $S(x)$ represent the cost of renting a saw for x days. Find each of the following.

(a) $S\left(\dfrac{1}{2}\right)$ **(b)** $S(1)$ **(c)** $S\left(1\dfrac{1}{4}\right)$ **(d)** $S\left(3\dfrac{1}{2}\right)$

(e) What does it cost to rent a saw for $4\dfrac{9}{10}$ days?

(f) A portion of the graph of $y = S(x)$ is shown here. Explain how the graph could be continued.

(g) What is the domain variable?

(h) What is the range variable?

(i) Write a sentence or two explaining what part (c) and its answer represent.

(j) We have left $x = 0$ out of the graph. Discuss why it should or shouldn't be included. If it were included, how would you define $S(0)$?

49. The charge to rent a Haul-It-Yourself Trailer is $25 plus $2 per hour or portion of an hour. Find the cost to rent a trailer for

(a) 2 hours;   **(b)** 1.5 hours;   **(c)** 4 hours;

(d) 3.7 hours.

(e) Graph the ordered pairs (hours, cost).

50. A delivery company charges $3 plus 50¢ per mile or part of a mile. Find the cost for a trip of

(a) 3 miles;   **(b)** 3.4 miles;   **(c)** 3.9 miles;

(d) 5 miles.

(e) Graph the ordered pairs (miles, cost).

(f) Is this a function?

Work these problems.

51. Natural Science   A laboratory culture contains about one million bacteria at midnight. The culture grows very rapidly until noon, when a bactericide is introduced and the bacteria population plunges. By 4 P.M. the bacteria have adapted to the bactericide and the culture slowly increases in population until 9 P.M. when the culture is accidentally destroyed by the clean-up crew. Let $g(t)$ denote the bacteria population at time t (with $t = 0$ corresponding to midnight) and draw a plausible graph of the function g. (Many correct answers are possible.)

52. Transportation   A plane flies from Austin, Texas, to Cleveland, Ohio, a distance of 1200 miles. Let f be the function whose rule is

$f(t)$ = distance (in miles) from Austin at time t hours,

with $t = 0$ corresponding to the 4 P.M. takeoff. In each of the following, draw a plausible graph of f under the given circumstances. (There are many correct answers for each part.)

(a) The flight is nonstop and takes between 3.5 and 4 hours.

(b) Bad weather forces the plane to land in Dallas (about 200 miles from Austin) at 5 P.M., remain overnight, and leave at 8 A.M. the next morning, flying nonstop to Cleveland.

(c) The plane flies nonstop, but due to heavy traffic it must fly in a holding pattern for an hour over Cincinnati (about 200 miles from Cleveland) and then go on to Cleveland.

3.3 APPLICATIONS OF LINEAR FUNCTIONS

Most of this section deals with the basic business relationships that were introduced in Section 1.2:

$$\textbf{Revenue} = \textbf{(Price per item)} \times \textbf{(Number of items)}$$
$$\textbf{Cost} = \textbf{Fixed Costs} + \textbf{Variable Costs}$$
$$\textbf{Profit} = \textbf{Revenue} - \textbf{Cost}$$

The examples will use only linear functions, but the methods presented here also apply to more complicated functions.

Cost Analysis

Recall that the fixed costs are for such things as buildings, machinery, real estate taxes, and product design. Within broad limits, the fixed cost is constant for a particular product and does not change as more items are made. The variable costs are for labor, materials, shipping, and so on. The variable costs depend on the number of items made.

If $C(x)$ is the cost of making x items, then the fixed cost (the cost which occurs even when no items are produced) can be found by letting $x = 0$. For example, for the cost function $C(x) = 45x + 250,000$, the fixed cost is

$$C(0) = 45(0) + 250,000 = \$250,000.$$

In this case, the variable cost of making x items is $45x$, that is, \$45 per item manufactured.

▶ **EXAMPLE 1** An anticlot drug can be made for \$10 per unit. The total cost to produce 100 units is \$1500.

(a) Assume that the cost function is linear and find its rule.

Since the cost function $C(x)$ is linear, its rule is of the form $C(x) = mx + b$. We are given that m (the cost per item) is 10, so the rule is $C(x) = 10x + b$. To find b use the fact that it costs \$1500 to produce 100 units, which means that

$$C(100) = 1500$$
$$10(100) + b = 1500 \qquad C(x) = 10x + b$$
$$1000 + b = 1500$$
$$b = 500.$$

So the rule of the cost function is $C(x) = 10x + 500$.

(b) What are the fixed costs?

The fixed costs are $C(0) = 10(0) + 500 = \$500.$ ◀

If $C(x)$ is the total cost to produce x items, then the **average cost** per item is given by

$$\overline{C}(x) = \frac{C(x)}{x}.$$

As more and more items are produced, the average cost per item typically decreases.

▶ **EXAMPLE 2** Find the average cost to produce 100 and 1000 units of the anticlot drug in Example 1.

The cost function is $C(x) = 10x + 500$, so the average cost of 100 units is

$$\overline{C}(100) = \frac{C(100)}{100} = \frac{10(100) + 500}{100} = \frac{1500}{100} = \$15.00 \text{ per unit.}$$

The average cost of 1000 units is

$$\overline{C}(1000) = \frac{C(1000)}{1000} = \frac{10(1000) + 500}{1000} = \frac{10500}{1000} = \$10.50 \text{ per unit.} \quad ◀ \quad ②$$

Rates of Change

The rate at which a quantity (such as revenue or profit) is changing can be quite important. For instance, if a company determines that the rate of change of its revenue is decreasing, then sales growth is slowing down, a trend that may require a response.

The rate of change of a linear function is easily determined. For example, suppose $f(x) = 3x + 5$ and consider the following table of values. The table shows that each time x changes by 1, the corresponding value of $f(x)$ changes by 3.

x	$f(x) = 3x + 5$
1	8
2	11
3	14
4	17
5	20

Thus, the rate of change of $f(x) = 3x + 5$ with respect to x is 3, which is the slope of the line $y = 3x + 5$. The same thing happens for any linear function:

The rate of change of a linear function $f(x) = mx + b$ is the slope m.

In particular, the rate of change of a linear function is constant. Rates of change of nonlinear functions, which are more complicated (and not constant), will be studied in Section 11.2.

In economics, the rate of change of the cost function is called the **marginal cost.** Marginal cost is important to management in making decisions in such areas as cost control, pricing, and production planning. When the cost function is linear, say, $C(x) = mx + b$, then the marginal cost is the number m (the slope of the graph of C). Marginal cost can also be thought of as the cost of producing one more item, as the next example demonstrates.

The cost in dollars to produce x kilograms of chocolate candy is given by $C(x)$, where

$$C(x) = 3.5x + 800.$$

Find each of the following.

(a) The fixed cost

(b) The total cost for 12 kilograms

(c) The marginal cost of the 40th kilogram

(d) The marginal cost per kilogram

Answers:

(a) $800

(b) $842

(c) $3.50

(d) $3.50

▶ **EXAMPLE 3** An electronics company manufactures handheld PCs. The cost function for one of their models is $C(x) = 160x + 750,000$.

(a) What are the fixed costs for this product?

The fixed costs are $C(0) = 160(0) + 750,000 = \$750,000$.

(b) What is the marginal cost?

The slope of $C(x) = 160x + 750,000$ is 160, so the marginal cost is $160.

(c) After 50,000 units have been produced, what is the cost of producing one more?

The cost of producing 50,000 is

$$C(50,000) = 160(50,000) + 750,000 = \$8,750,000.$$

The cost of 50,001 units is

$$C(50,001) = 160(50,001) + 750,000 = \$8,750,160.$$

The cost of the additional unit is the difference

$$C(50,001) - C(50,000) = 8,750,160 - 8,750,000 = \$160.$$

Thus, the cost of one more item is the marginal cost. ◀

Example 3 is typical of the general case, as summarized here.

In a **linear cost function** $C(x) = mx + b$, m represents the marginal cost and b the fixed cost. The marginal cost is the cost of producing one more item.

Conversely, if the fixed cost is b and the marginal cost is always the same constant m, then the cost function for producing x items is $C(x) = mx + b$.

Break-Even Analysis

A typical company must analyze its cost and the potential market for its product to determine when (or if) it will make a profit.

▶ **EXAMPLE 4** A company manufactures a particular model of DVD player that sells to retailers for $168. The cost of making x of these DVD players is given by the function $C(x) = 118x + 800,000$.

(a) Find the function R that gives the revenue from selling x players.

Since revenue is the product of the price per item and the number of items, $R(x) = 168x$.

(b) What is the revenue from selling 40,00 players?

Evaluate the revenue function R at 40,000:

$$R(40,000) = 168(40,000) = \$6,720,000.$$

(c) Find the profit function P.

Since profit = revenue − cost,

$$P(x) = R(x) - C(x) = 168x - (118x + 800,000) = 50x - 800,000.$$

(d) What is the profit from selling 10,000 players?

Evaluate the profit function at 10,000:

$$P(10,000) = 50(10,000) - 800,000 = -300,000,$$

that is, a loss of $300,000. ◀

A company can make a profit only if the revenue on a product exceeds the cost of manufacturing it. The number of units at which revenue equals cost (that is, profit is 0) is the **break-even point.**

For a certain magazine, the cost equation is $c = .70x + 1200$, where x is the number of magazines sold. The magazine sells for $1 per copy. Find the break-even point.

Answer:

4000 magazines

▶ **EXAMPLE 5** Find the break-even point for the company in Example 4.

The company will break even when revenue equals cost, that is, when

$$R(x) = C(x)$$
$$168x = 118x + 800,000$$
$$50x = 800,000$$
$$x = 16,000$$

The company breaks even by selling 16,000 DVD players. The graphs of the revenue and cost functions and the break-even point (where $x = 16,000$) are shown in Figure 3.18. The company must sell more than 16,000 players ($x > 16,000$) in order to make a profit. ◀ ④

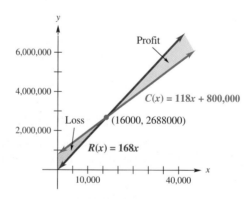

FIGURE 3.18

TECHNOLOGY TIP The break-even point in Example 5 can be found on a graphing calculator by graphing the cost and revenue functions on the same screen and using the calculator's intersection finder, as shown in Figure 3.19. Depending on the calculator, the intersection finder is in the CALC or G-SOLVE menu or in the MATH or FCN submenu of the GRAPH menu.

FIGURE 3.19

Supply and Demand

The supply of and demand for an item are usually related to its price. Producers will supply large numbers of the item at a high price, but consumer demand will be low. As the price of the item decreases, consumer demand increases, but producers are less willing to supply large numbers of the item. The curves that show the quantity that will be supplied at a given price and the quantity that will be demanded at a given price are called **supply and demand curves.** In supply and demand problems, we use p for price and q for quantity. We will discuss the economic concepts of supply and demand in more detail in later chapters.

▶ **EXAMPLE 6** Bill Cornett, an economist, has studied the supply and demand for aluminum siding and has determined that price per unit,* p, and the quantity demanded, q, are related by the linear equation

$$p = 60 - \frac{3}{4}q.$$

(a) Find the demand at a price of $40 per unit.

Let $p = 40$.

$$p = 60 - \frac{3}{4}q$$

$$40 = 60 - \frac{3}{4}q \qquad \text{Let } p = 40.$$

$$-20 = -\frac{3}{4}q \qquad \text{Add } -60 \text{ on both sides.}$$

$$\frac{80}{3} = q \qquad \text{Multiply both sides by } -\frac{4}{3}.$$

At a price of $40 per unit, 80/3 (or $26\frac{2}{3}$) units will be demanded.

(b) Find the price if the demand is 32 units.

Let $q = 32$.

$$p = 60 - \frac{3}{4}q$$

$$p = 60 - \frac{3}{4}(32) \qquad \text{Let } q = 32.$$

$$p = 60 - 24$$

$$p = 36$$

With a demand of 32 units, the price is $36.

(c) Graph $p = 60 - \frac{3}{4}q$.

It is customary to use the horizontal axis for quantity q and the vertical axis for price p. In part (a) we saw that 80/3 units would be demanded at a price of $40 per unit; this gives the ordered pair (80/3, 40). Part (b) shows that with a demand of 32

---

*An appropriate unit here might be, for example, one thousand square feet of siding.

Suppose price and demand are related by $p = 100 - 4q$.

(a) Find the price if the demand is 10 units.

(b) Find the demand if the price is $80.

(c) Write the corresponding ordered pairs.

Answers:

(a) $60

(b) 5 units

(c) (10, 60); (5, 80)

units, the price is $36, which gives the ordered pair (32, 36). Using the points (80/3, 40) and (32, 36) to get the demand graph shown in Figure 3.20. Only the portion of the graph in Quadrant I is shown, because supply and demand are meaningful only for positive values of p and q. (5)

FIGURE 3.20

(d) From Figure 3.20, at a price of $30, what quantity is demanded?

Price is located on the vertical axis. Look for 30 on the p-axis and read across to where the line $p = 30$ crosses the demand graph. As the graph shows, this occurs where the demand is 40.

(e) At what price will 60 units be demanded?

Quantity is located on the horizontal axis. Find 60 on the q-axis and read up to where the vertical line $q = 60$ crosses the demand graph. This occurs where the price is about $15 per unit.

(f) What quantity is demanded at a price of $60 per unit?

The point (0, 60) on the demand graph shows that the demand is 0 at a price of $60 (that is, there is no demand at such a high price). ◀

▶ **EXAMPLE 7**   Suppose the economist of Example 6 concludes that the supply q of siding is related to its price p by the equation

$$p = .85q.$$

(a) Find the supply if the price is $51 per unit.

$$51 = .85q \qquad \text{Let } p = 51.$$
$$60 = q$$

If the price is $51 per unit, then 60 units will be supplied to the marketplace.

(b) Find the price per unit if the supply is 20 units.

$$p = .85(20) = 17 \qquad \text{Let } q = 20.$$

If the supply is 20 units, then the price is $17 per unit.

(c) Graph the supply equation $p = .85q$.

As with demand, each point on the graph has quantity q as its first coordinate and the corresponding price p as its second coordinate. Part (a) shows that the ordered pair (60, 51) is on the graph of the supply equation, and part (b) shows that (20, 17) is on the graph. Using these points we obtain the supply graph in Figure 3.21.

FIGURE 3.21

(d) Use the graph in Figure 3.21 to find the approximate price at which 35 units will be supplied. Then use algebra to find the exact price.

The point on the graph with first coordinate $q = 35$ is approximately (35, 30). Therefore, 35 units will be supplied when the price is approximately $30. To determine the exact price algebraically, substitute $q = 35$ in the supply equation:

$$p = .85q = .85(35) = \$29.75 \blacktriangleleft$$

▶ **EXAMPLE 8** The supply and demand curves of Examples 6 and 7 are shown in Figure 3.22. Graphically determine whether there is a surplus or a shortage of supply at a price of $40 per unit.

FIGURE 3.22

Find 40 on the vertical axis in Figure 3.22 and read across to the point where the horizontal line $p = 40$ crosses the supply graph (that is, the point corresponding to a price of $40). This point lies above the demand graph, so supply is greater than demand at a price of $40 and there is a surplus of supply. ◀

Supply and demand are equal at the point where the supply curve intersects the demand curve. This is the **equilibrium point.** Its second coordinate is the **equilibrium price,** the price at which the same quantity will be supplied as is demanded. Its first coordinate is the quantity that will be demanded and supplied at the equilibrium price; this number is called the **equilibrium quantity.**

▶**EXAMPLE 9**   In the situation described in Examples 6–8, what is the equilibrium quantity? What is the equilibrium price?

The equilibrium point is where the supply and demand curves in Figure 3.22 intersect. To find the quantity q at which the price given by the demand equation $p = 60 - .75q$ (Example 6) is the same as that given by the supply equation $p = .85q$ (Example 7), set these two expressions for p equal and solve the resulting equation.

$$60 - .75q = .85q$$
$$60 = 1.6q$$
$$37.5 = q$$

Therefore, the equilibrium quantity is 37.5 units, the number of units for which supply will equal demand. Substituting $q = 37.5$ in either the demand or supply equation shows that

$$p = 60 - .75(37.5) = 31.875 \quad \text{or} \quad p = .85(37.5) = 31.875.$$

So the equilibrium price is $31.875 (or $31.88 rounded). (To avoid error, it's a good idea to substitute in both equations, as we did here, to be sure the same value of p results; if it doesn't, a mistake has been made.) In this case, the equilibrium point—the point whose coordinates are the equilibrium quantity and price—is (37.5, 31.875). ◀ ⑥

⑥

The demand for a certain commodity is related to the price by $p = 80 - (2/3)q$. The supply is related to the price by $p = (4/3)q$. Find

(a) the equilibrium quantity;
(b) the equilibrium price.

Answers:

(a) 40
(b) $160/3 \approx \$53.33$

TECHNOLOGY TIP   The equilibrium point (37.5, 31.875) can be found on a graphing calculator by graphing the supply and demand curves on the same screen and using the calculator's intersection finder to locate their point of intersection.

3.3 Exercises

Business   *Write a cost function for each of the following. Identify all variables used. (See Example 1.)*

1. A chain-saw rental firm charges $12 plus $1 per hour.

2. A trailer-hauling service charges $45 plus $2 per mile.

3. A parking garage charges 35¢ plus 30¢ per half hour.

4. For a 1-day rental, a car rental firm charges $14 plus 6¢ per mile.

Business   *Assume that each of the following can be expressed as a linear cost function. Find the appropriate cost function in each case. (See Example 1.)*

5. Fixed cost, $100; 50 items cost $1600 to produce.

6. Fixed cost, $1000; 40 items cost $2000 to produce.

7. Marginal cost, $120; 100 items cost $15,800 to produce.

8. Marginal cost, $90; 150 items cost $16,000 to produce.

Business *In each of the following, a cost function is given. Find the average cost per item when the required number of items is produced. (See Example 2.)*

9. $C(x) = 12x + 1800$; 50 items; 500 items; 1000 items

10. $C(x) = 50x + 12,000$; 100 items; 1000 items; 10,000 items

11. $C(x) = 6.5x + 9800$; 200 items; 2000 items; 5000 items

12. $C(x) = 8.75x + 16,500$; 1000 items; 10,000 items; 75,000 items

Business *Work these problems. (See Example 3.)*

13. The total cost (in dollars) to produce x college algebra books is $C(x) = 42.5x + 80,000$.
   (a) What are the fixed costs?
   (b) What is the marginal cost per book?
   (c) What is the total cost of producing 1000 books? Of 32,000 books?
   (d) What is the average cost per book when 1000 are produced? When 32,000 are produced?

14. The total cost (in dollars) to produce x compact disks is $C(x) = 6.80x + 450,000$.
   (a) What are the fixed costs?
   (b) What is the marginal cost per disk?
   (c) What is the total cost of producing 50,000 disks? Of 600,000 disks?
   (d) What is the average cost per disk when 50,000 are produced? When 500,000 are produced?

15. The manager of a restaurant found that the cost to produce 100 cups of coffee is $11.02, while the cost to produce 400 cups is $40.12. Assume the cost $C(x)$ is a linear function of x, the number of cups produced.
   (a) Find a formula for $C(x)$.
   (b) Find the total cost of producing 1000 cups.
   (c) Find the total cost of producing 1001 cups.
   (d) Find the marginal cost of the 1001st cup.
   (e) What is the marginal cost of *any* cup?

16. In deciding whether to set up a new manufacturing plant, company analysts have decided that a linear function is a reasonable estimation for the total cost $C(x)$ in dollars to produce x items. They estimate the cost to produce 10,000 items as $547,500 and the cost to produce 50,000 items as $737,500.
   (a) Find formula for $C(x)$.
   (b) Find the total cost to produce 100,000 items.
   (c) Find the marginal cost of the items to be produced in this plant.

Business *Assume that each of the following has a linear cost function. Find (a) the cost function; (b) the revenue function; (c) the profit function; (d) the profit on 100 items. (See Example 4.)*

	Fixed Cost	Marginal Cost per Item	Item Sells For
17.	$500	$10	$35
18.	$180	$11	$20
19.	$250	$18	$28
20.	$15,000	$30	$80
21.	$18,000	$12.50	$25

22. **Business** In the profit-volume chart below, EF and GH represent the profit-volume graphs of a single-product company for 1989 and 1990, respectively.*

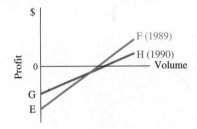

If 1989 and 1990 unit sales prices are identical, how did total fixed costs and unit variable costs of 1990 change compared to 1989? Choose one:

	1990 Total Fixed Costs	1990 Unit Variable Costs
(a)	Decreased	Increased
(b)	Decreased	Decreased
(c)	Increased	Increased
(d)	Increased	Decreased

Use algebra to find the intersection points of the graphs of the given equations. (See Examples 5 and 9.)

23. $2x - y = 7$ and $y = 8 - 3x$

24. $6x - y = 2$ and $y = 4x + 7$

25. $y = 3x - 7$ and $y = 7x + 4$

26. $y = 3x + 5$ and $y = 12 - 2x$

*Uniform CPA Examintion, May 1991. American Institute of Certified Public Accountants.

Business   *Work the following problems. (See Example 5.)*

27. For x thousand policies, an insurance company claims that their monthly revenue in dollars is given by $R = 125x$ and their monthly cost in dollars is given by $C = 100x + 5000$.
    (a) Find the break-even point.
    (b) Graph the revenue and cost equations on the same axes.
    (c) From the graph, estimate the revenue and cost when $x = 100$ (100 thousand policies).

28. The owners of a parking lot have determined that their weekly revenue and cost in dollars are given by $R = 80x$ and $C = 50x + 2400$, where x is the number of long-term parkers.
    (a) Find the break-even point.
    (b) Graph R and C on the same axes.
    (c) From the graph, estimate the revenue and cost when there are 60 long-term parkers.

29. The revenue (in millions of dollars) from the sale of x units at a home supply outlet is given by $r = .21x$. The profit (in millions of dollars) from the sale of x units is given by $p = .084x - 1.5$.
    (a) Find the cost equation.
    (b) What is the cost of producing 7 units?
    (c) What is the break-even point?

30. The profit (in millions of dollars) from the sale of x million units of Blue Glue is given by $p = .7x - 25.5$. The cost is given by $c = .9x + 25.5$.
    (a) Find the revenue equation.
    (b) What is the revenue from selling 10 million units?
    (c) What is the break-even point?

Business   *Suppose you are the manager of a firm. The accounting department has provided cost estimates and the sales department sales estimates on a new product. You must analyze the data they give you, determine what it will take to break even, and decide whether or not to go ahead with production of the new product. (See Example 5.)*

31. Cost estimate is given by $c = 80x + 7000$ and revenue estimate by $r = 95x$; no more than 400 units can be sold.

32. Cost is $c = 140x + 3000$ and revenue is $r = 125x$.

33. Cost is $c = 125x + 42,000$ and revenue is $r = 165.5x$; no more than 2000 units can be sold.

34. Cost is $c = 1750x + 95,000$ and revenue is $r = 1975x$; no more than 600 units can be sold.

35. **Business**   The graphs show the hourly compensation (wages and benefits) in U.S. dollars of production workers in Austria and the United States since 1975.* Estimate the break-even point (the point at which workers in both countries had the same compensation).

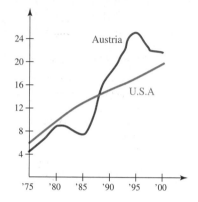

36. **Business**   The graphs show U.S. exports and imports of goods and services (in billions of dollars) since 1965.† Estimate the break-even point (where the values of exports and imports were the same).

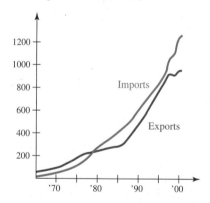

37. **Business**   Direct U.S. investments in Ireland and Italy (in billions of dollars) in 1990 and 1999 are shown in the chart.‡

	1990	1999
Ireland	6.88	19.82
Italy	13.12	17.60

*U.S. Bureau of Labor Statistics.

†U.S. Bureau of Economic Analysis.

‡U.S. Department of Commerce.

(a). Let $x = 0$ correspond to 1990. Assume the change in investment in each case is linear and find linear functions that give the investment in year x for each country. Round all coefficients to two decimal places.

(b) Graph the functions from part (a) on the same coordinate axes.

(c) Find the intersection point of the graphs and interpret your answer.

38. Social Science   The approximate median family income (in thousands of dollars) for white families in the United States is given by $f(x) = .15x + 37$, where x is the number of years since 1980. The approximate median family income for black families is given by $g(x) = .2x + 21$.

(a) Graph both functions on the same coordinate axes for $0 \le x \le 30$.

(b) Do the graphs intersect? What conclusions do you infer from this?

Business   *Use the supply and demand curves graphed below to answer Exercises 39–42. (See Examples 6–9.)*

39. At what price are 20 items supplied?

40. At what price are 20 items demanded?

41. Find the equilibrium quantity.

42. Find the equilibrium price.

Business   *Work the following exercises. (See Examples 6–9.)*

43. Suppose that the demand and price for a certain brand of shampoo are related by

$$p = 16 - \frac{5}{4}q,$$

where p is price, in dollars, and q is demand. Find the price for a demand of

(a) 0 units;   **(b)** 4 units;   **(c)** 8 units.

Find the demand for the shampoo at a price of
(d) \$6;   **(e)** \$11;   **(f)** \$16.

(g) Graph $p = 16 - (5/4)q$. Suppose the price and supply of the shampoo are related by

$$p = \frac{3}{4}q,$$

where q represents the supply and p the price. Find the supply when the price is

(h) \$0;   **(i)** \$10;   **(j)** \$20.

(k) Graph $p = (3/4)q$ on the same axes used for part (g).

(l) Find the equilibrium quantity.

(m) Find the equilibrium price.

44. Let the supply and demand for radial tires in dollars be given by

$$\text{supply: } p = \frac{3}{2}q \quad \text{and} \quad \text{demand: } p = 81 - \frac{3}{4}q.$$

(a) Graph these on the same axes.

(b) Find the equilibrium quantity.

(c) Find the equilibrium price.

45. Let the supply and demand for bananas in cents per pound be given by

$$\text{supply: } p = \frac{2}{5}q \quad \text{and} \quad \text{demand: } p = 100 - \frac{2}{5}q.$$

(a) Graph these on the same axes.

(b) Find the equilibrium quantity.

(c) Find the equilibrium price.

(d) On what interval does demand exceed supply?

46. Let the supply and demand for sugar be given by

$$\text{supply: } p = 1.4q - .6$$

and

$$\text{demand: } p = -2q + 3.2,$$

where p is in dollars.

(a) Graph these on the same axes.

(b) Find the equilibrium quantity.

(c) Find the equilibrium price.

(d) On what interval does supply exceed demand?

47. Explain why the graph of the (total) cost function is always above the x-axis and can never move downward as you go from left to right. Is the same thing true of the graph of the average cost function?

48. Explain why the graph of the profit function may rise or fall (as you go from left to right) and may be below the x-axis.

3.4 QUADRATIC FUNCTIONS

A **quadratic function** is a function whose rule is given by a quadratic polynomial, such as

$$f(x) = x^2, \qquad g(x) = 3x^2 + 30x + 67, \qquad \text{and} \quad h(x) = -x^2 + 4x.$$

Thus, a quadratic function is one whose rule can be written in the form

$$f(x) = ax^2 + bx + c$$

for some constants $a, b, c,$ with $a \neq 0$.

▶ **EXAMPLE 1**   Graph each of these quadratic functions.

(a) $f(x) = x^2$

Choose several negative, zero, and positive values of x, find the values of $f(x)$, and plot the corresponding points. Connecting these points with a smooth curve, we obtain Figure 3.23 below.

① Graph each quadratic function.

(a) $f(x) = x^2 - 4$

(b) $f(x) = -(x - 3)^2$

Answers:

(a)

$f(x) = x^2 - 4$

(b)

$f(x) = -(x - 3)^2$

x	y
2	4
1	1
0	0
−1	1
−2	4

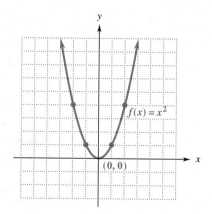

$f(x) = x^2$

(0, 0)

FIGURE 3.23

(b) $h(x) = -(x + 2)^2$

Since $h(x) = -(x + 2)^2 = -(x^2 + 4x + 4) = -x^2 - 4x - 4$, we see that h actually is a quadratic function. When $x = -2$, then

$$h(x) = h(-2) = -(-2 + 2)^2 = 0.$$

Therefore $(-2, 0)$ is on the graph. When $x \neq -2$, then $x + 2 \neq 0$, so that $(x + 2)^2$ is positive, and hence, $h(x) = -(x + 2)^2$ is negative. Thus, the graph lies below the x-axis whenever $x \neq -2$. Using these facts and plotting some points lead to Figure 3.24. ◀ ①

x	y
0	−4
−1	−1
−2	0
−3	−1
−4	−4

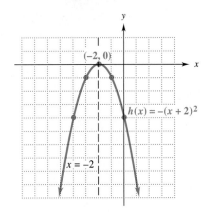

FIGURE 3.24

The curves in Figures 3.23 and 3.24 and in problem 1 in the margin are called **parabolas.** It can be shown that the graph of every quadratic function is a parabola. All parabolas have the same basic "cup" shape, though the cup may be broad or narrow. The parabola opens upward when the coefficient of the x^2-term is positive (as in Figure 3.23) and downward when this coefficient is negative (as in Figure 3.24).

Parabolas have many useful properties. Cross sections of radar dishes and spotlights form parabolas. Disks often visible on the sidelines of televised football games are microphones having reflectors with parabolic cross sections. These microphones are used by the television networks to pick up the shouted signals of the quarterbacks.

When a parabola opens upward (as in Figure 3.23), its lowest point is called the **vertex.** When a parabola opens downward (as in Figure 3.24), its highest point is called the **vertex.** The vertical line through the vertex of a parabola is called the **axis of the parabola.** For example, (0, 0) is the vertex of the parabola in Figure 3.23 and its axis is the y-axis. If you were to fold this graph along its axis, the two halves of the parabola would match exactly. This means that a parabola is *symmetric* about its axis.

The vertex of a parabola can be roughly approximated by using the trace feature or accurately approximated by using the minimum or maximum finder on a graphing calculator. However, there are algebraic techniques for finding the vertex precisely.

▶ **EXAMPLE 2** The function $g(x) = 2(x - 3)^2 + 1$ is quadratic because its rule can be written in the required form:

$$g(x) = 2(x - 3)^2 + 1 = 2(x^2 - 6x + 9) + 1 = 2x^2 - 12x + 19.$$

Its graph will be an upward-opening parabola. Note that

$$g(3) = 2(3 - 3)^2 + 1 = 1,$$

so that (3, 1) is on the graph. We claim that (3, 1) is the vertex—the lowest point on the graph. To see why this is true, note that when $x \neq 3$, the quantity $2(x - 3)^2$ is positive and hence

$$g(x) = 2(x - 3)^2 + 1 = \text{(a positive number)} + 1,$$

so that $g(x) > 1$. Therefore, every point $(x, g(x))$ on the graph with $x \neq 3$ has second coordinate $g(x)$ larger than 1 and hence lies *above* the point (3, 1). In other words, (3, 1) is the lowest point on the graph, the vertex of the parabola. Knowing this fact makes it easy to plot some points on both sides of the vertex and obtain the graph in Figure 3.25. Since the vertex is (3, 1), the vertical line $x = 3$ is the axis of the parabola. ◀

x	y
1	9
2	3
3	1
4	3
5	9

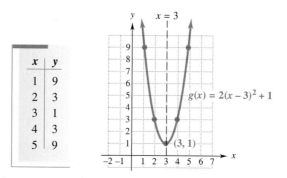

FIGURE 3.25

Example 2 illustrates some of the following facts, whose proofs are omitted.

If f is a quadratic function defined by $y = a(x - h)^2 + k$, then the graph of the function f is a parabola having its vertex at (h, k) and axis of symmetry $x = h$.

If $a > 0$, the parabola opens upward; if $a < 0$, it opens downward.

If $0 < |a| < 1$, the parabola is "broader" than $y = x^2$, while if $|a| > 1$, the parabola is "narrower" than $y = x^2$.

▶ **EXAMPLE 3**   Determine algebraically whether the parabola opens upward or downward and find its vertex.

(a) $f(x) = -3(x - 4)^2 - 7$

The rule of the function is in the form $f(x) = a(x - h)^2 + k$ (with $a = -3$, $h = 4$, and $k = -7$). The parabola opens downward ($a < 0$) and its vertex is $(h, k) = (4, -7)$.

(b) $g(x) = 2(x + 3)^2 + 5$

Be careful here; the vertex is *not* (3, 5). To put the rule of $g(x)$ in the form $a(x - h)^2 + k$, we must rewrite it so that there is a minus sign inside the parentheses:

$$g(x) = 2(x + 3)^2 + 5$$
$$= 2(x - (-3))^2 + 5.$$

②

③

This is the required form with $a = 2$, $h = -3$, and $k = 5$. The parabola opens upward and its vertex is $(-3, 5)$. ◀ ②

▶ **EXAMPLE 4** Find the rule of a quadratic function whose graph has vertex $(3, 4)$ and passes through the point $(6, 22)$.

The graph of $f(x) = a(x - h)^2 + k$ has vertex (h, k). We want $h = 3$ and $k = 4$, so that $f(x) = a(x - 3)^2 + 4$. Since $(6, 22)$ is on the graph, we must have $f(6) = 22$. Therefore

$$f(x) = a(x - 3)^2 + 4$$
$$f(6) = a(6 - 3)^2 + 4$$
$$22 = a(3)^2 + 4$$
$$9a = 18$$
$$a = 2.$$

Thus, the graph of $f(x) = 2(x - 3)^2 + 4$ is a parabola with vertex $(3, 4)$ that passes through $(6, 22)$. ◀

The vertex of each parabola in Examples 2 and 3 was easily determined because the rule of the function had the form

$$f(x) = a(x - h)^2 + k.$$

The rule of *any* quadratic function can be put in this form by using the technique of completing the square, which was discussed in Section 1.7.

▶ **EXAMPLE 5** By completing the square, determine the vertex of the graph of $f(x) = x^2 - 2x + 3$. Then graph the parabola.

In order to get the equation in the form $f(x) = a(x - h)^2 + k$, we first write $f(x) = x^2 - 2x + 3$ as

$$f(x) = (x^2 - 2x \quad) + 3.$$

Now complete the square for the expression in parentheses. Take half the coefficient of x, namely, $(\frac{1}{2})(-2) = -1$, and square the result: $(-1)^2 = 1$. In order to complete the square we must add 1 inside the parentheses, but in order not to change the rule of the function we must also subtract 1:

$$f(x) = (x^2 - 2x + 1 - 1) + 3.$$

By the associative property,

$$f(x) = (x^2 - 2x + 1) + (-1 + 3).$$

Factor $x^2 - 2x + 1$ as $(x - 1)^2$, to get

$$f(x) = (x - 1)^2 + 2.$$

With the rule in this form, we can see that the graph is an upward-opening parabola with vertex $(1, 2)$, as shown in Figure 3.26 on the next page. ◀ ③

④

Complete the square, find the vertex, and graph each function.

(a) $f(x) = 3x^2 - 12x + 14$

(b) $f(x) = -x^2 + 6x - 12$

Answers:

(a) $f(x) = 3(x - 2)^2 + 2$

$f(x) = 3x^2 - 12x + 14$

(b) $f(x) = -(x - 3)^2 - 3$

$f(x) = -x^2 + 6x - 12$

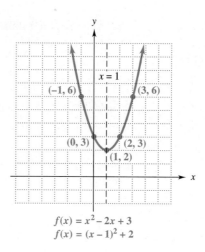

$$f(x) = x^2 - 2x + 3$$
$$f(x) = (x - 1)^2 + 2$$

FIGURE 3.26

▶ **EXAMPLE 6** Find the vertex of the graph of $f(x) = -2x^2 + 12x - 19$ and graph the parabola.

Get $f(x)$ in the form $f(x) = a(x - h)^2 + k$ by first factoring -2 from the expression $-2x^2 + 12x$.

$$f(x) = -2x^2 + 12x - 19 = -2(x^2 - 6x) - 19$$

Working inside the parentheses, take half of -6 (coefficient of x): $(1/2)(-6) = -3$. Square this result: $(-3)^2 = 9$. Add and subtract 9 inside the parentheses.

$$y = -2(x^2 - 6x + 9 - 9) - 19$$
$$y = -2(x^2 - 6x + 9) + (-2)(-9) - 19 \qquad \text{Distributive property}$$

Simplify and factor to get

$$y = -2(x - 3)^2 - 1.$$

Since $a = -2$ is negative, the parabola opens downward. The vertex is at $(3, -1)$, and the parabola is narrower than $y = x^2$. Use these results and plot additional ordered pairs as needed to get the graph in Figure 3.27. ◀ ④

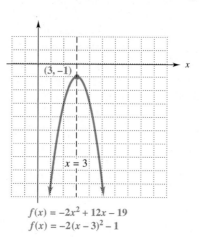

$$f(x) = -2x^2 + 12x - 19$$
$$f(x) = -2(x - 3)^2 - 1$$

FIGURE 3.27

The technique of completing the square can be used to rewrite the general equation $f(x) = ax^2 + bx + c$ in the form $f(x) = a(x - h)^2 + k$, as is shown in Exercise 34. When this is done we obtain a formula for the coordinates of the vertex.

The graph of $f(x) = ax^2 + bx + c$ is a parabola with vertex (h, k), where

$$h = \frac{-b}{2a} \quad \text{and} \quad k = f(h).$$

▶ **EXAMPLE 7** Find the vertex, the axis, and the x- and y-intercepts of the graph of $f(x) = x^2 - x - 6$.

Since $a = 1$ and $b = -1$, the x-value of the vertex is

$$\frac{-b}{2a} = \frac{-(-1)}{2 \cdot 1} = \frac{1}{2}.$$

The y-value of the vertex is

$$f\left(\frac{1}{2}\right) = \left(\frac{1}{2}\right)^2 - \frac{1}{2} - 6 = -\frac{25}{4}.$$

The vertex is $(1/2, -25/4)$ and the axis of the parabola is $x = 1/2$, as shown in Figure 3.28. The intercepts are found by setting each variable equal to 0.

x-intercepts	y-intercept
Set $f(x) = y = 0$:	Set $x = 0$:
$0 - x^2 - x - 6$	$f(x) = y = 0^2 - 0 - 6 = -6$
$0 = (x + 2)(x - 3)$	The y-intercept is -6.
$x + 2 = 0 \quad \text{or} \quad x - 3 = 0$	
$x = -2 \qquad\qquad x = 3$	

The x-intercepts are -2 and 3. ◀ ⑤

⑤

Find the vertex of the graph.

(a) $f(x) = 2x^2 - 7x + 12$

(b) $k(x) = -.4x^2 + 1.6x + 5$

Answers:

(a) $(7/4, 47/8)$

(b) $(2, 6.6)$

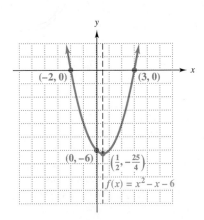

FIGURE 3.28

TECHNOLOGY TIP The maximum or minimum finder on a graphing calculator can approximate the vertex of a parabola with a high degree of accuracy. The max/min finder is in the CALC menu or in the MATH or FCN submenu of the GRAPH menu. Similarly, the calculator's graphical root finder can approximate the x-intercepts of a parabola.

3.4 Exercises

Without graphing, determine the vertex of each of the following parabolas and state whether it opens upward or downward. (See Examples 2, 3, 5–7.)

1. $f(x) = 3(x - 5)^2 + 2$ **2.** $g(x) = -6(x - 2)^2 - 5$

3. $f(x) = x^2 + 12x + 1$ **4.** $g(x) = x^2 - 10x + 3$

5. $f(x) = 2x^2 + 4x + 1$ **6.** $g(x) = -3x^2 + 6x + 5$

Without graphing, determine the x- and y-intercepts of each of the following parabolas. (See Example 7.)

7. $f(x) = 3(x - 2)^2 - 3$ **8.** $f(x) = x^2 - 4x - 1$

9. $g(x) = 2x^2 + 8x + 6$ **10.** $g(x) = x^2 - 10x + 20$

Find the rule of a quadratic function whose graph has the given vertex and passes through the given point. (See Example 4.)

11. Vertex $(1, 2)$; point $(5, 6)$

12. Vertex $(-3, 2)$; point $(2, 1)$

13. Vertex $(-1, -2)$; point $(1, 2)$

14. Vertex $(2, -4)$; point $(5, 2)$

Graph each parabola. Find the vertex and axis of symmetry of each. (See Examples 1–7.)

15. $f(x) = (x + 2)^2$ **16.** $f(x) = -(x + 5)^2$

17. $f(x) = (x - 1)^2 - 3$ **18.** $f(x) = (x - 2)^2 + 1$

19. $f(x) = x^2 - 4x + 6$ **20.** $f(x) = x^2 + 6x + 3$

21. $f(x) = 2x^2 - 4x + 5$ **22.** $f(x) = -3x^2 + 24x - 46$

Use a graphing calculator to work these problems.

23. Transportation According to data from the National Highway Traffic Safety Administration, the fatal accident rate per 1000 licensed drivers every 100 million miles can be approximated by the function $f(x) = .0031x^2 - .291x + 7.1$, where x is the age of the driver ($16 \le x \le 88$). At what age is the rate the lowest?

24. Social Science According to data from the U.S. Census Bureau, the population of Cleveland, Ohio (in thousands), in year x can be approximated by $g(x) = -.172x^2 + 16.82x + 487.62$, where $x = 0$ corresponds to 1900. In what year did Cleveland have its largest population?

25. Social Science Between 1992 and 1998, the percent of college freshmen who planned eventually to get some type of medical degree can be approximated by $f(x) = -.2369x^2 + 1.425x + 6.905$, where $x = 0$ corresponds to 1992. In what year did this percentage reach a maximum?

26. Transportation The money spent per year during the 1990s for research and development of "smart" highways is approximated by the function $g(x) = -19.321x^2 + 3608.7x - 168,310$, where $x = 0$ corresponds to 1900 and $g(x)$ is in millions of dollars. In what year was the maximum amount spent? What was that maximum amount?

In Exercises 27–32, graph the functions in parts (a)–(d) on the same set of axes; then answer part (e).

27. (a) $k(x) = x^2$
 (b) $f(x) = 2x^2$
 (c) $g(x) = 3x^2$
 (d) $h(x) = 3.5x^2$
 (e) Explain how the coefficient a in the function given by $f(x) = ax^2$ affects the shape of the graph, in comparison with the graph of $k(x) = x^2$, when $a > 1$.

28. (a) $k(x) = x^2$
 (b) $f(x) = .8x^2$
 (c) $g(x) = .5x^2$
 (d) $h(x) = .3x^2$
 (e) Explain how the coefficient a in the function given by $f(x) = ax^2$ affects the shape of the graph, in comparison with the graph of $k(x) = x^2$, when $0 < a < 1$.

29. **(a)** $k(x) = -x^2$
    (b) $f(x) = -2x^2$
    (c) $g(x) = -3x^2$
    (d) $h(x) = -3.5x^2$
    (e) Compare these graphs to the ones in Exercise 27 and explain how changing the sign of the coefficient a in the function given by $f(x) = ax^2$ affects the graph.

30. **(a)** $k(x) = x^2$
    (b) $f(x) = x^2 + 2$
    (c) $g(x) = x^2 + 3$
    (d) $h(x) = x^2 + 5$
    (e) Explain how the graph of $f(x) = x^2 + c$ (where c is a positive constant) can be obtained from the graph of $k(x) = x^2$.

31. **(a)** $k(x) = x^2$
    (b) $f(x) = x^2 - 1$
    (c) $g(x) = x^2 - 2$
    (d) $h(x) = x^2 - 4$

(e) Explain how the graph of $f(x) = x^2 - c$ (where c is a positive constant) can be obtained from the graph of $k(x) = x^2$.

32. **(a)** $k(x) = x^2$
    (b) $f(x) = (x + 2)^2$
    (c) $g(x) = (x - 2)^2$
    (d) $h(x) = (x - 4)^2$
    (e) Explain how the graph of $f(x) = (x + c)^2$ and $f(x) = (x - c)^2$ (where c is a positive constant) can be obtained from the graph of $k(x) = x^2$.

33. Find the rule of a quadratic function whose graph is a parabola with vertex $(0, 0)$ that includes the point $(2, 12)$.

34. Verify that the right side of the equation

$$ax^2 + bx + c = a\left(x - \left(\frac{-b}{2a}\right)\right)^2 + \left(c - \frac{b^2}{4a}\right)$$

equals the left side. Since the right side of the equation is in the form $a(x - h)^2 + k$, we conclude that the vertex of the parabola $f(x) = ax^2 + bx + c$ has x-coordinate $h = -b/2a$.

3.5 APPLICATIONS OF QUADRATIC FUNCTIONS

The fact that the vertex of a parabola $y = ax^2 + bx + c$ is the highest or lowest point on the graph can be used in applications to find a maximum or a minimum value.

▶ **EXAMPLE 1**   Anne Kelly owns and operates Aunt Emma's Blueberry Pies. She has hired a consultant to analyze her business operations. The consultant tells her that her profits, $P(x)$, from the sale of x cases of pies, are given by

$$P(x) = 120x - x^2.$$

How many cases of pies should she sell in order to maximize profit? What is the maximum profit?

The profit function can be rewritten as $P(x) = -x^2 + 120x$. Its graph is a downward-opening parabola (see Figure 3.29 on the next page). Its vertex, which may be found as in Section 3.4, is $(60, 3600)$. For each point on the graph,

the x-coordinate is the number of cases;

the y-coordinate is the profit on that number of cases.

Only the portion of the graph in Quadrant I (where both coordinates are positive) is relevant here because she cannot sell a negative number of pies and she is not interested in a negative profit. *Maximum* profit occurs at the point with the largest y-coordinate, namely, the vertex, as shown in Figure 3.29. The maximum profit of $3600 is obtained when 60 cases of pies are sold. ◀

When a company sells x units of a product, its profit is
$$P(x) = -2x^2 + 40x + 280.$$
Find

(a) the number of units that should be sold so that maximum profit is received;

(b) the maximum profit.

Answers:

(a) 10 units

(b) $480

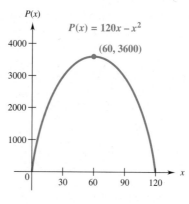

FIGURE 3.29

Supply and demand curves were introduced in Section 3.3. Here is a quadratic example.

▶ **EXAMPLE 2** Suppose that the price and demand for an item are related by

$$p = 150 - 6q^2, \qquad \text{Demand function}$$

where p is the price (in dollars) and q is the number of items demanded (in hundreds). The price and supply are related by

$$p = 10q^2 + 2q, \qquad \text{Supply function}$$

where q is the number of items supplied (in hundreds). Find the equilibrium quantity and the equilibrium price.

The graphs of both of these equations are parabolas (Figure 3.30). Only those portions of the graphs that lie in the first quadrant are included, because neither supply, demand, nor price can be negative.

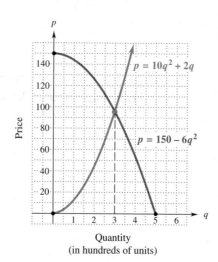

FIGURE 3.30

The point where the demand and supply curves intersect is the equilibrium point. Its first coordinate is the equilibrium quantity, and its second coordinate is the equilibrium price. These coordinates may be found in two ways.

Algebraic Method   At the equilibrium point, the second coordinate of the demand curve must be the same as the second coordinate of the supply curve so that

$$150 - 6q^2 = 10q^2 + 2q.$$

Write this quadratic equation in standard form as follows.

$$0 = 16q^2 + 2q - 150 \qquad \text{Add } -150 \text{ and } 6q^2 \text{ to both sides.}$$

$$0 = 8q^2 + q - 75 \qquad \text{Multiply both sides by } \frac{1}{2}.$$

This equation can be solved by the quadratic formula given in Section 1.7. Here $a = 8$, $b = 1$, and $c = -75$.

$$q = \frac{-1 \pm \sqrt{1 - 4(8)(-75)}}{2(8)}$$

$$= \frac{-1 \pm \sqrt{1 + 2400}}{16} \qquad -4(8)(-75) = 2400$$

$$= \frac{-1 \pm 49}{16} \qquad \sqrt{1 + 2400} = \sqrt{2401} = 49$$

$$q = \frac{-1 + 49}{16} = \frac{48}{16} = 3 \quad \text{or} \quad q = \frac{-1 - 49}{16} = -\frac{50}{16} = -\frac{25}{8}$$

It is not possible to make $-25/8$ units, so discard that answer and use only $q = 3$. Hence the equilibrium quantity is 300. Find the equilibrium price by substituting 3 for q in either the supply or the demand function (and check your answer by using the other one). Using the supply function gives

$$p = 10q^2 + 2q$$
$$p = 10 \cdot 3^2 + 2 \cdot 3 \qquad \text{Let } q = 3.$$
$$= 10 \cdot 9 + 6$$
$$p = 96.$$

Graphical Method   Graph the two functions on a graphing calculator and use the intersection finder to determine that the equilibrium point is (3, 96), as in Figure 3.31. ◀ ②

160

0   7
Intersection
X=3   Y=96
0

FIGURE 3.31

---

②

The price and demand for an item are related by $p = 32 - x^2$, while price and supply are related by $p = x^2$. Find

(a) the equilibrium quantity;

(b) the equilibrium price.

Answers:

(a) 4

(b) 16

▶ EXAMPLE 3   The rental manager of a small apartment complex with 16 units has found from experience that each $40 increase in the monthly rent results in an empty apartment. All 16 apartments will be rented at a monthly rent of $500. How many $40 increases will produce maximum monthly income for the complex?

Let x represent the number of $40 increases. Then the number of apartments rented will be $16 - x$. Also, the monthly rent per apartment will be $500 + 40x$. (There are x increases of $40 for a total increase of $40x$.) The monthly income, $I(x)$, is given by the number of apartments rented times the rent per apartment, so

$$I(x) = (16 - x)(500 + 40x)$$
$$= 8000 + 640x - 500x - 40x^2$$
$$= 8000 + 140x - 40x^2.$$

Since x represents the number of $40 increases and each $40 increase causes one empty apartment, x must be a whole number. Since there are only 16 apartments, $0 \leq x \leq 16$. Because there is a small number of possibilities, the value of x that produces maximum income may be found in several ways.

Brute Force Method   Use a scientific calculator or the table feature of a graphing calculator (as in Figure 3.32) to evaluate $I(x)$ when $x = 1, 2, \ldots, 16$ and find the largest value.

X	Y₁
0	8000
1	8100
2	8120
3	8060
4	7920
5	7700
6	7400
X=0	

X	Y₁
7	7020
8	6560
9	6020
10	5400
11	4700
12	3920
13	3060
X=7	

X	Y₁
14	2120
15	1100
16	0
17	-1180
18	-2440
19	-3780
20	-5200
X=14	

FIGURE 3.32

The tables show that a maximum income of $8120 occurs when $x = 2$. So the manager should charge a rent of $500 + 2(40) = 580, leaving two apartments vacant.

Algebraic Method   The graph of $I(x) = 8000 + 140x - 40x^2$ is a downward-opening parabola (why?) and the value of x that produces maximum income occurs at the vertex. The methods of Section 3.4 show that the vertex is $(1.75, 8122.50)$. Since x must be a whole number, evaluate $I(x)$ at $x = 1$ and $x = 2$ to see which one gives the best result.

$$\text{If } x = 1, \text{ then } I(1) = -40(1)^2 + 140(1) + 8000 = 8100.$$
$$\text{If } x = 2, \text{ then } I(2) = -40(2)^2 + 140(2) + 8000 = 8120.$$

So maximum income occurs when $x = 2$. The manager should charge a rent of $500 + 2(40) = 580 and leave two apartments vacant.   ◀

Quadratic Models   Real-world data can sometimes be used to construct a quadratic function that approximates the data. Such **quadratic models** can then be used (subject to limitations) to predict future behavior.

▶ EXAMPLE 4   The number of women employed full-time in civilian jobs has increased dramatically in the past century. The table shows the number of employed women (in millions) in selected years.*

Year	1910	1920	1930	1940	1950	1960	1970	1980	1990	2000
Working Women	7.4	8.6	10.8	12.8	18.6	23.6	31.4	45.3	56.7	65.6

(a) Display this information graphically.

Let $x = 0$ correspond to 1900. Plot the points given by the table: $(10, 7.4)$, $(20, 8.6)$, and so on, as in Figure 3.33.

FIGURE 3.33   FIGURE 3.34

(b) The shape of the data points in Figure 3.33 resembles the right half of a parabola. Find a quadratic model $f(x) = a(x - h)^2 + k$ for this data.

Recall that when a quadratic function is written in this form, the vertex of its graph is (h, k). Based on Figure 3.33, let $(10, 7.4)$ be the vertex so that

$$f(x) = a(x - 10)^2 + 7.4.$$

To find a, choose another data point, say, $(100, 65.6)$. If this point is on the parabola, then $f(100) = 65.6$, so that

$$f(x) = a(x - 10)^2 + 7.4$$
$$65.5 = a(100 - 10)^2 + 7.4 \qquad \text{Substitute 100 for } x \text{ and 65.6 for } f(x)$$
$$65.6 = 8100a + 7.4$$
$$8100a = 58.2$$
$$a = \frac{58.2}{8100} \approx .007185$$

---

*U.S. Bureau of Labor Statistics.

Find another quadratic model in Example 4(b) by using (10, 7.4) as the vertex and (70, 31.4) as the other point.

Answer:

$$f(x) = \frac{1}{150}(x - 10)^2 + 7.4$$

Therefore $f(x) = .007185(x - 10)^2 + 7.4$ is a quadratic model for this data. The graph of $f(x)$ in Figure 3.34 on the preceding page appears to fit the data reasonably well.

(c) Use the quadratic model in part (b) to estimate the number of women in the work force in 2005.

The year 2005 corresponds to $x = 105$, so the number is approximately

$$f(105) = .007185(105 - 10)^2 + 7.4 \approx 72.2 \text{ million.} \blacktriangleleft \text{③}$$

Quadratic Regression

Linear regression was used in Section 2.3 to construct a linear function that modeled a set of data points. When the data points appear to lie on a parabola rather than on a straight line (as in Example 4), a similar least squares regression procedure is available on most graphing calculators and spreadsheet programs to construct a quadratic model for the data. Simply follow the same steps as in linear regression, with one exception: choose quadratic rather than linear regression (they're on the same menu).

▶ **EXAMPLE 5**   Use a graphing calculator to do the following.

(a) Find a quadratic regression model for the data in Example 4.

Enter the first coordinates of the data points as list L_1 and the second coordinates as list L_2. Performing the quadratic regression, as in Figure 3.35, leads to the model

$$g(x) = .007996x^2 - .21667x + 9.21167.$$

The number R^2 in Figure 3.35 is very close to 1, which indicates that this model fits the data well. The graph of $g(x)$ and the data points in Figure 3.36 visually confirm this.

FIGURE 3.35   **FIGURE 3.36**

(b) Use the regression model to estimate the number of women in the workforce in 2005.

Evaluate $g(x)$ when $x = 105$:

$$g(105) = .007996(105^2) - .21667(105) + 9.21167 \approx 74.6 \text{ million.}$$

This estimate is about 2.4 million higher than the one in Example 4.   ◀

3.5 Exercises

Work these problems. (See Example 1.)

1. **Business** Shannise Cole makes and sells candy. She has found that the cost per box for making x boxes of candy is given by

$$C(x) = x^2 - 40x + 405.$$

   (a) How much does it cost per box to make 15 boxes? 18 boxes? 30 boxes?
   (b) Graph the cost function $C(x)$ and mark the points corresponding to 15, 18, and 30 boxes.
   (c) What point on the graph corresponds to the number of boxes that will make the cost per box as small as possible?
   (d) How many boxes should she make in order to keep the cost per box at a minimum? What is the minimum cost per box?

2. **Business** Greg Tobin sells bottled water. He has found that the average amount of time he spends with each customer is related to his weekly sales volume by the function

$$f(x) = x(60 - x),$$

   where x is the number of minutes per customer and $f(x)$ is the number of cases sold per week.
   (a) How many cases does he sell if he spends 10 minutes with each customer? 20 minutes? 45 minutes?
   (b) Choose an appropriate scale for the axes and sketch the graph of $f(x)$. Mark the points on the graph corresponding to 10, 20, and 45 minutes.
   (c) Explain what the vertex of the graph represents.
   (d) How long should Greg spend with each customer in order to sell as many cases per week as possible? If he does, how many cases will he sell?

3. **Natural Science** A researcher in physiology has decided that a good mathematical model for the number of impulses fired after a nerve has been stimulated is given by $y = -x^2 + 20x - 60$, where y is the number of responses per millisecond and x is the number of milliseconds since the nerve was stimulated.
   (a) When will the maximum firing rate be reached?
   (b) What is the maximum firing rate?

4. **Physical Science** If an object is thrown upward with an initial velocity of 32 feet per second, then its height, in feet, above the ground after t seconds is given by

$$h = 32t - 16t^2.$$

   Find the maximum height attained by the object. Find the number of seconds it takes for the object to hit the ground.

5. **Business** Colleen Davis owns a factory that manufactures souvenir key chains. Her weekly profit (in hundreds of dollars) is given by $P(x) = -2x^2 + 60x - 120$, where x is the number of cases of key chains sold.
   (a) What is the largest number of cases she can sell and still make a profit?
   (b) Explain how it is possible for her to lose money if she sells more cases than your answer in part (a).
   (c) How many cases should she make and sell in order to maximize her profits?

6. **Business** The manager of a bicycle shop has found that, at a price (in dollars) of $p(x) = 150 - \dfrac{x}{4}$ per bicycle, x bicycles will be sold.
   (a) Find an expression for the total revenue from the sale of x bicycles. (*Hint:* revenue = demand $\times$ price.)
   (b) Find the number of bicycle sales that leads to maximum revenue.
   (c) Find the maximum revenue.

Work the following problems. (See Example 2.)

7. **Business** Suppose the supply and demand for a certain textbook are given by

$$\text{supply}: p = \frac{1}{5} q^2; \qquad \text{demand}: p = -\frac{1}{5} q^2 + 40,$$

   where p is price and q is quantity. How many books are demanded at a price of
   (a) 10?   (b) 20?
   (c) 30?   (d) 40?
   How many books are supplied at a price of
   (e) 5?   (f) 10?
   (g) 20?   (h) 30?
   (i) Graph the supply and demand functions on the same axes.

8. **Business** Find the equilibrium quantity and the equilibrium price in Exercise 7.

9. **Business** Suppose the price p of widgets is related to the quantity q that is demanded by

$$p = 640 - 5q^2,$$

   where q is measured in hundreds of widgets. Find the price when the number of widgets demanded is
   (a) 0;   (b) 5;   (c) 10.
   Suppose the supply function for widgets is given by $p = 5q^2$, where q is the number of widgets (in hundreds) that are supplied at price p.
   (d) Graph the demand function $p = 640 - 5q^2$ and the supply function $p = 5q^2$ on the same axes.
   (e) Find the equilibrium quantity.
   (f) Find the equilibrium price.

10. Business The supply function for a commodity is given by $p = q^2 + 200$ and the demand function is given by $p = -10q + 3200$.

(a) Graph the supply and demand functions on the same axes.

(b) Find the equilibrium point.

(c) What is the equilibrium quantity? The equilibrium price?

Business *Find equilibrium quantity and equilibrium price for the commodity whose supply and demand functions are given.*

11. Supply: $p = 45q$; Demand: $p = -q^2 + 10,000$

12. Supply: $p = q^2 + q + 10$; Demand: $p = -10q + 3060$

13. Supply: $p = q^2 + 20q$;

Demand: $p = -2q^2 + 10q + 3000$

14. Supply: $p = .2q + 51$; Demand: $p = \dfrac{3000}{q + 5}$

Business *The revenue function R and the cost function C for a particular product are given. These functions are valid only for the specified range of values. Find the number of units that must be produced to break even.*

15. $R(x) = 200x - x^2$; $C(x) = 70x + 2200$; $0 \le x \le 100$

16. $R(x) = 300x - x^2$; $C(x) = 65x + 7000$; $0 \le x \le 150$

17. $R(x) = 400x - 2x^2$; $C(x) = -x^2 + 200x + 1900$; $0 \le x \le 100$

18. $R(x) = 500x - 2x^2$; $C(x) = -x^2 + 270x + 5125$; $0 \le x \le 125$

Business *Work each problem. (See Example 3.)*

19. A charter flight charges a fare of $200 per person, plus $4 per person for each unsold seat on the plane. If the plane holds 100 passengers and if x represents the number of unsold seats, find the following.

(a) An expression for the total revenue received for the flight. (*Hint:* Multiply the number of people flying, $100 - x$, by the price per ticket.)

(b) The graph for the expression of part (a)

(c) The number of unsold seats that will produce the maximum revenue

(d) The maximum revenue

20. The revenue of a charter bus company depends on the number of unsold seats. If 100 seats are sold, the price is $50. Each unsold seat increases the price per seat by $1. Let x represent the number of unsold seats.

(a) Write an expression for the number of seats that are sold.

(b) Write an expression for the price per seat.

(c) Write an expression for the revenue.

(d) Find the number of unsold seats that will produce the maximum revenue.

(e) Find the maximum revenue.

21. Farmer Linton wants to find the best time to take her hogs to market. The current price is 88 cents per pound and her hogs weigh an average of 90 pounds. The hogs gain 5 pounds per week and the market price for hogs is falling each week by 2 cents per pound. How many weeks should Ms. Linton wait before taking her hogs to market in order to receive as much money as possible? At this time, how much money (per hog) will she get?

22. The manager of a peach orchard is trying to decide when to arrange for picking the peaches. If they are picked now, the average yield per tree will be 100 pounds, which can be sold for 40¢ per pound. Past experience shows that the yield per tree will increase about 5 pounds per week, while the price will decrease about 2¢ per pound per week.

(a) Let x represent the number of weeks that the manager should wait. Find the income per pound.

(b) Find the number of pounds per tree.

(c) Find the total revenue from a tree.

(d) When should the peaches be picked in order to produce maximum revenue?

(e) What is the maximum revenue?

Unless directed otherwise by your instructor, do these problems as follows. If you do not have a graphing calculator or spreadsheet program, omit part (b). If you do have a graphing calculator or spreadsheet program, omit part (a). (See Examples 4 and 5.)

23. Health The table lists the total cumulative number of deaths in the United States to date known to have been caused by AIDS.*

Year	Deaths
1981	159
1985	12,652
1987	41,262
1989	90,218
1991	158,911
1993	245,662
1995	345,331
1997	403,206

(a) Let $x = 0$ correspond to 1980. Use $(1, 159)$ as the vertex and the data for 1997 to find a quadratic function $f(x) = a(x - h)^2 + k$ that models this data.

(b) Let $x = 0$ correspond to 1980 and plot the data points. Use quadratic regression to find a function g that models this data.

(c) Use the quadratic model from part (a) or (b) to estimate the cumulative number of deaths by 1992, 1996, and 2000. For comparison purposes, the actual figures for 1992 and 1996 are 200,391 and 382,261.

*U.S. Department of Health and Human Services, Centers for Disease Control and Prevention, *HIV/AIDS Surveillance Report,* vol. 11, no. 2.

24. Finance   Federal government revenues (in billions of dollars) for selected years are shown in the table.*

Year	Revenue
1980	517.1
1985	734.1
1990	1031.3
1995	1351.5
2000	2025.0

(a) Let $x = 0$ correspond to 1980. Use $(0, 517.1)$ as the vertex and the data for 2000 to find a quadratic function $f(x) = a(x - h)^2 + k$ that models this data.

(b) Let $x = 0$ correspond to 1980 and plot the data points. Use quadratic regression to find a function g that models this data.

(c) Use the quadratic model from part (a) or (b) to estimate government revenues in the years 1983, 1991, and 1996. How well do the estimates compare with the actual revenues of 600.6, 1054.3, and 1452.8?

(d) Estimate government revenue in 2001.

25. Transportation   The table shows fuel consumption (in billions of gallons) by all nonmilitary motor vehicles in selected years.†

Year	Fuel Consumption
1975	109.0
1980	115.0
1985	121.3
1990	130.8
1995	143.8
1998	154.9

(a) Let $x = 0$ correspond to 1970. Use $(5, 109)$ as the vertex and the data for 1998 to find a quadratic function $f(x) = a(x - h)^2 + k$ that models this data.

(b) Let $x = 0$ correspond to 1970 and plot the data points. Use quadratic regression to find a function g that models this data. Based on the value of r^2, how good a model is this?

(c) Use the quadratic model from part (a) or (b) to estimate a fuel consumption in 1993 and 1997. For comparison purposes, the actual figures for 1993 and 1997 are 137.3 and 150.3.

(d) Assuming these models remain accurate, estimate fuel consumption in 2000 and 2003.

26. Business   The table shows the number of personal computers (in thousands) sold in selected years.*

Year	Unit Sales
1984	3,975
1985	3,200
1990	4,000
1995	8,400
1996	9,400
1998	12,800
1999	14,900
2000	16,800

(a) Let $x = 0$ correspond to 1980. Use $(5, 3200)$ as the vertex and the data for 2000 to find a quadratic function $f(x) = a(x - h)^2 + k$ that models this data.

(b) Let $x = 0$ correspond to 1980 and plot the data points. Use quadratic regression to find a function g that models this data. Based on the value of r^2, how good a model is this?

(c) Use the quadratic model from part (a) or (b) to estimate the number of computers sold in 1992.

(d) Assuming these models remained accurate, how many computers were sold in 2002?

Work these problems.

27. Geometry   A field bounded on one side by a river is to be fenced on three sides to form a rectangular enclosure. There are 320 ft of fencing available. What should the dimensions be to have an enclosure with the maximum possible area?

28. Geometry   A rectangular garden bounded on one side by a river is to be fenced on the other three sides. Fencing material for the side parallel to the river costs $30 per foot and material for the other two sides costs $10 per foot. What are the dimensions of the garden of largest possible area, if $1200 is to be spent for fencing material?

Business   *Recall that profit equals revenue minus cost. In Exercises 29 and 30 find the following.*

(a) *The break-even point (to the nearest tenth)*

(b) *The x-value that makes profit a maximum*

(c) *The maximum profit*

(d) *For what x-values will a loss occur?*

(e) *For what x-values will a profit occur?*

29. $R(x) = 400x - 2x^2$ and $C(x) = 200x + 2000$, with $0 \le x \le 100$.

30. $R(x) = 900x - 3x^2$ and $C(x) = 450x + 5000$, with $0 \le x \le 150$.

---

*Financial Management Service of the U.S. Department of the Treasury.

†U.S. Federal Highway Administration.

---

*Consumer Electronics Manufacturers Association.

3.6 POLYNOMIAL FUNCTIONS

A **polynomial function of degree n** is a function whose rule is given by a polynomial of degree n.* For example, linear functions, such as

$$f(x) = 3x - 2 \quad \text{and} \quad g(x) = 7x + 5,$$

are polynomial functions of degree 1 and quadratic functions, such as

$$f(x) = 3x^2 + 4x - 6 \quad \text{and} \quad g(x) = -2x^2 + 9$$

are polynomial functions of degree 2. Similarly, the function given by

$$f(x) = 2x^4 + 5x^3 - 6x^2 + x - 3$$

is a polynomial function of degree 4.

Basic Graphs

The simplest polynomial functions are those whose rules are of the form $f(x) = ax^n$ (where a is a constant).

▶ **EXAMPLE 1** Graph $f(x) = x^3$.

First, find several ordered pairs belonging to the graph. Be sure to choose some negative x-values, $x = 0$, and some positive x-values to get representative ordered pairs. Find as many ordered pairs as you need in order to see the shape of the graph. Then plot the ordered pairs and draw a smooth curve through them, getting the graph in Figure 3.37. ◀ ①

①

Graph $f(x) = x^3 - 5$.

Answer:

$f(x) = x^3 - 5$

x	y
2	8
1	1
0	0
−1	−1
−2	−8

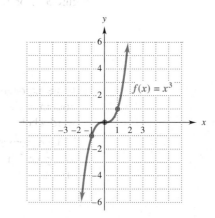

FIGURE 3.37

---

*The degree of a polynomial was defined on page 13.

② Graph these functions.

(a) $f(x) = -.25x^5 - 2$

(b) $f(x) = -x^4 + 3$

Answers:

(a)

(b)

▶ EXAMPLE 2 Graph $f(x) = (3/2)x^4$.

The following table gives some typical ordered pairs.

x	-2	-1	0	1	2
y	24	$3/2$	0	$3/2$	24

The graph is shown in Figure 3.38. ◀ ②

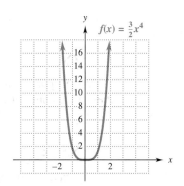

FIGURE 3.38

The graph of $f(x) = ax^n$ has the same general shape as one of the graphs in Figure 3.37 or 3.38.

Graph of $f(x) = ax^n$

If the exponent n is even, then the graph of $f(x) = ax^n$ is cup-shaped with the bottom of the cup at the origin and the y-axis through the middle; it opens upward when $a > 0$ and downward when $a < 0$.

If the exponent n is odd, then the graph of $f(x) = ax^n$ moves upward from left to right when $a > 0$ and downward when $a < 0$, with a single bend at the origin.

Properties of Polynomial Graphs

Unlike the graphs in the preceding figures, the graphs of more complicated polynomial functions may have several "peaks" and "valleys," as illustrated in Figure 3.39 on the next page. The location of the peaks and valleys can be accurately approximated by a maximum or minimum finder on a graphing calculator. Calculus is needed to determine their exact location.

FIGURE 3.39

Figure 3.39 illustrates two important facts about polynomial functions.

1. The total number of peaks and valleys on the graph of a polynomial function of degree n is at most $n - 1$.

2. The number of x-intercepts on the graph of a polynomial function of degree n is at most n.

For instance, the polynomial in Figure 3.39(b) has degree $n = 5$ and the number of peaks and valleys on its graph is $n - 1 = 4$. The graph has 5 x-intercepts. Similarly, the polynomial in Figure 3.39(d) has degree $n = 6$, so that $n - 1 = 5$. The graph has 3 peaks and valleys and 2 x-intercepts.

The domain of every polynomial function is the set of all real numbers, which means that its graph extends forever to the left and right. The shape of the graph far from the center depends on the degree of the polynomial. For instance, compare the graph of $g(x) = 1.5x^4 + x^3 - 4x^2 - 3x + 4$ in Figure 3.39(c) with the graph of $f(x) = 1.5x^4$ in Figure 3.38. At the far left and far right both graphs shoot away from the x-axis in a similar fashion. When both functions are graphed in the same large viewing window on a graphing calculator (Figure 3.40), the ends of the graphs look almost the same. In other words, as you move out from the center, the graph of $g(x) = 1.5x^4 + x^3 - 4x^2 - 3x + 4$ looks very much like the graph of its highest degree term, $1.5x^4$. The same thing is true in the general case.

> The graph of a polynomial function is a smooth, unbroken curve that extends forever to the left and right. When $|x|$ is large, the graph resembles the graph of its highest degree term and moves sharply away from the x-axis.

FIGURE 3.40

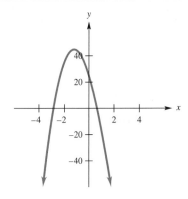

FIGURE 3.41

▶ **EXAMPLE 3**   Let $g(x) = x^3 - 11x^2 - 32x + 24$.

(a) Is Figure 3.41 a complete graph of $g(x)$; that is, does it show all the important features of the graph?

　　The far ends of the graph of $g(x)$ should resemble the graph of its highest degree term x^3. The graph of $f(x) = x^3$ in Figure 3.37 moves upward at the far right, but Figure 3.41 does not. So Figure 3.41 is not a complete graph.

 (b) Use a graphing calculator to find a complete graph of $g(x)$.

　　Since the graph of $g(x)$ must eventually start rising on the right side (as does the graph of x^3), a viewing window that shows a complete graph must extend beyond $x = 4$. By experimenting with various windows, we obtain Figure 3.42. This graph shows a total of 2 peaks and valleys and 3 x-intercepts (the maximum possible for a polynomial of degree 3). At the far ends, the graph of $g(x)$ in Figure 3.42 resembles the graph of $f(x) = x^3$. Therefore, Figure 3.42 is a complete graph of $g(x)$. ◀ ③

③

Find a viewing window on a graphing calculator that shows a complete graph of

$f(x) = -.7x^4 + 119x^2 + 400.$

(*Hint:* The graph has two x-intercepts and the maximum possible number of peaks and valleys.)

Answer:

Many correct answers, including $-15 \le x \le 15$ and $-2000 \le y \le 6000.$

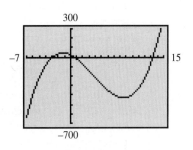

FIGURE 3.42

Graphing Techniques

Accurate graphs of first- and second-degree polynomial functions (lines and parabolas) are easily found algebraically, as we saw in Sections 2.2 and 3.4. All polynomial functions of degree 3, and some of higher degree, can be accurately graphed by hand by using calculus and algebra to locate the peaks and valleys. When a polynomial can be completely factored, the general shape of its graph can be determined algebraically by using the basic properties of polynomial graphs, as illustrated in Example 4 below. Accurate graphs of other polynomial functions generally require technology.

Multiply out the expression for $f(x)$ in Example 4 and determine its degree.

Answer:

$f(x) = 2x^3 + 5x^2 - x - 6$; degree 3.

▶ **EXAMPLE 4** Graph $f(x) = (2x + 3)(x - 1)(x + 2)$.

Note that $f(x)$ is a polynomial of degree 3 (if you don't see why, do problem ④ at the side). Begin by finding any x-intercepts. Set $f(x) = 0$ and solve for x:

$$f(x) = 0$$
$$(2x + 3)(x - 1)(x + 2) = 0$$

Solve this equation by placing each of the three factors equal to 0.

$$2x + 3 = 0 \quad \text{or} \quad x - 1 = 0 \quad \text{or} \quad x + 2 = 0$$
$$x = -\frac{3}{2} \qquad\qquad x = 1 \qquad\qquad x = -2$$

The three numbers, $-3/2$, 1, and -2, divide the x-axis into four intervals:

$$x < -2, \quad -2 < x < -\frac{3}{2}, \quad -\frac{3}{2} < x < 1, \quad \text{and} \quad 1 < x.$$

These intervals are shown in Figure 3.43.

FIGURE 3.43

Since the graph is an unbroken curve, it can only change from above the x-axis to below it by passing through the x-axis. As we have seen, this occurs only at the x-intercepts $x = -2$, $-3/2$, and 1. Consequently, in the interval between two intercepts (or to the left of $x = -2$ or to the right of $x = 1$), the graph of $f(x)$ must lie entirely above or entirely below the x-axis.

We can determine where the graph lies over an interval, by evaluating $f(x) = (2x + 3)(x - 1)(x + 2)$ at a number in that interval. For example, $x = -3$ is in the interval where $x < -2$ and

$$f(-3) = (2(-3) + 3)(-3 - 1)(-3 + 2)$$
$$= -12.$$

Therefore, $(-3, -12)$ is on the graph. Since this point lies below the x-axis, all points in this interval (that is, all points with $x < -2$) must lie below the x-axis. By testing numbers in the other intervals, we obtain this chart.

Interval	$x < -2$	$-2 < x < -3/2$	$-3/2 < x < 1$	$x > 1$
Test Number	-3	$-7/4$	0	2
Value of $f(x)$	-12	$11/32$	-6	28
Sign of $f(x)$	Negative	Positive	Negative	Positive
Graph	Below x-axis	Above x-axis	Below x-axis	Above x-axis

⑤

Graph
$f(x) = .5(x - 1)(x + 2)(x - 3)$.

Answer:

$f(x) = .5(x - 1)(x + 2)(x - 3)$

Since the graph intersects the x-axis at the intercepts $x = -2$ and $x = -3/2$ and is above the x-axis between these intercepts, there must be at least one peak there. Similarly, there must be at least one valley between $x = -3/2$ and $x = 1$ because the graph is below the x-axis there. However, a polynomial function of degree 3 can have a total of at most $3 - 1 = 2$ peaks and valleys. So there must be exactly one peak and exactly one valley on this graph.

Furthermore, when $|x|$ is large, the graph must resemble the graph of $y = 2x^3$ (the highest degree term). The graph of $y = 2x^3$, like the graph of $y = x^3$ in Figure 3.37, moves upward to the right and downward to the left. Using these facts and plotting the x-intercepts shows that the graph must have the general shape shown in Figure 3.44. Plotting additional points leads to the reasonably accurate graph in Figure 3.45. We say "reasonably accurate" because we cannot be sure of the exact locations of the peaks and valleys on the graph without using calculus. ◀ ⑤

FIGURE 3.44

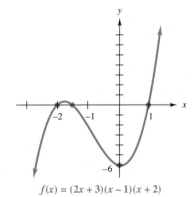

$f(x) = (2x + 3)(x - 1)(x + 2)$

FIGURE 3.45

◥ Polynomial Models

Regression procedures (similar to linear regression in Section 2.3 and quadratic regression in Section 3.5) can be used to find cubic and quartic (degree 4) polynomial models for appropriate data.

▶ **EXAMPLE 5** The table shows the population of San Francisco in selected years.

Year	1900	1950	1970	1980	1990	1998
Population	342,782	775,357	715,674	678,974	723,959	745,774

(a) Plot the data on graphing calculator, with $x = 0$ corresponding to 1900.

The points in Figure 3.46 suggest the general shape of a cubic polynomial graph.

(b) Use cubic regression to obtain a model for this data.

The procedure is the same as for linear regression (just choose "cubic" in place of "linear"). It produces the function

$$f(x) = 2.708x^3 - 493.867x^2 + 26{,}563.818x + 342{,}853.161*$$

Its graph in Figure 3.47 appears to fit the data well.

(c) Use the model to estimate the population of San Francisco in 1985 and 2000.

Now 1985 and 2000 correspond to $x = 85$ and $x = 100$, respectively. Verify that

$$f(85) \approx 695{,}803 \quad \text{and} \quad f(100) \approx 768{,}831.$$

So the population was approximately 695,803 in 1985 and 768,831 in 2000. ◀

FIGURE 3.46

FIGURE 3.47

---

*Coefficients are rounded for easy reading, but the full coefficients are used for graphs and function evaluation.

3.6 Exercises

Graph each of the following polynomial functions. (See Examples 1 and 2.)

1. $f(x) = x^4$

2. $g(x) = -.5x^6$

3. $h(x) = -.2x^5$

4. $f(x) = x^7$

In Exercises 5–8, state whether the graph could possibly be the graph of (a) some polynomial function; (b) a polynomial function of degree 3; (c) a polynomial function of degree 4; (d) a polynomial function of degree 5. (See Example 3.)

5.

6.

7.

8.

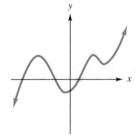

In Exercises 9–14, match the given polynomial function to its graph (A–F), without using a graphing calculator. (See Example 3 and the two boxes preceding it.)

(a)   (b)

(c)   (d)

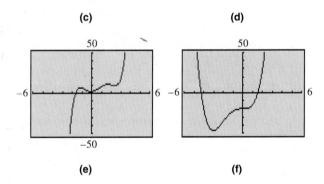

(e)   (f)

9. $f(x) = x^3 - 7x - 9$

10. $f(x) = -x^3 + 4x^2 + 3x - 8$

11. $f(x) = x^4 - 5x^2 + 7$

12. $f(x) = x^4 + 4x^3 - 20$

13. $f(x) = .7x^5 - 2.5x^4 - x^3 + 8x^2 + x + 2$

14. $f(x) = -x^5 + 4x^4 + x^3 - 16x^2 + 12x + 5$

Graph each of the following polynomial functions. (See Example 4.)

15. $f(x) = (x + 2)(x - 3)(x + 4)$

16. $f(x) = (x - 3)(x - 1)(x + 1)$

17. $f(x) = x^2(x - 2)(x + 3)$

18. $f(x) = x^2(x + 1)(x - 1)$

19. $f(x) = x^3 + x^2 - 6x$

20. $f(x) = x^3 - 2x^2 - 8x$

21. $f(x) = x^3 + 3x^2 - 4x$

22. $f(x) = x^4 - 5x^2$

Exercises 23–26 require a graphing calculator. Find a viewing window that shows a complete graph of the polynomial function (that is, a graph that includes all the peaks and valleys and indicates how the curve moves away from the x-axis at the far left and far right). There are many possible correct answers. Consider your answer correct if it shows all the features that appear in the window given in the answers. (See Example 3.)

23. $g(x) = x^3 - 3x^2 - 4x - 5$

24. $f(x) = x^4 - 10x^3 + 35x^2 - 50x + 24$

25. $f(x) = 2x^5 - 3.5x^4 - 10x^3 + 5x^2 + 12x + 6$

26. $g(x) = x^5 + 8x^4 + 20x^3 + 9x^2 - 27x - 7$

In Exercises 27–31, use a calculator to evaluate the functions. Do the graph by plotting points or by using a graphing calculator.

27. Finance   An idealized version of the Laffer curve (originated by economist Arthur Laffer) is shown here. According to this theory, decreasing the tax rate, say, from x_2 to x_1, may actually increase the total revenue to the government. The theory is that people will work harder and earn more if they are taxed at a lower rate, which means higher total tax revenues than would be the case at a higher rate. Suppose that the Laffer curve is given by the function

$$f(x) = \frac{x(x - 100)(x - 160)}{240} \quad (0 \le x \le 100),$$

where $f(x)$ is government revenue (in billions of dollars) from a tax rate of x percent. Find the revenue from the following tax rates.

(a) 20%

(b) 40%

(c) 50%

(d) 70%

(e) Graph $f(x)$

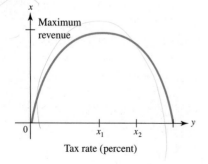

Tax rate (percent)

28. Health   A technique for measuring cardiac output depends on the concentration of a dye after a known amount is injected into a vein near the heart. In a normal heart, the concentration of the dye at time x (in seconds) is given by the function defined by

$$g(x) = -.006x^4 + .140x^3 - .053x^2 + 1.79x.$$

(a) Find the following: $g(0)$; $g(1)$; $g(2)$; $g(3)$.

(b) Graph $g(x)$ for $x \ge 0$.

29. Physical Science   The pressure of the oil in a reservoir tends to drop with time. By taking sample pressure readings for a particular oil reservoir, petroleum engineers have found that the change in pressure is given by

$$P(t) = t^3 - 18t^2 + 81t,$$

where t is time in years from the date of the first reading.

(a) Find the following: $P(0)$; $P(3)$; $P(7)$; $P(10)$.

(b) Graph $P(t)$.

(c) For what time period is the change in pressure (drop) increasing? decreasing?

30. Natural Science   During the early part of the 20th century, the deer population of the Kaibab Plateau in Arizona experienced a rapid increase because hunters had reduced the number of natural predators. The increase in population depleted the food resources and eventually caused the population to decline. For the period from 1905 to 1930, the deer population was approximated by

$$D(x) = -.125x^5 + 3.125x^4 + 4000,$$

where x is time in years from 1905.

(a) Find the following: $D(0)$; $D(5)$; $D(10)$; $D(15)$; $D(20)$; $D(25)$.

(b) Graph $D(x)$.

(c) From the graph, over what period of time (from 1905 to 1930) was the population increasing? relatively stable? decreasing?

31. Transportation   According to the U.S. Department of Transportation, the number of consumer complaints against top U.S. airlines in year x can be approximated by the polynomial function.

$$f(x) = .2196x^4 - 5.454x^3 + 50.258x^2 - 202.68x + 306.134 \quad (5 \le x \le 9),$$

where $x = 5$ corresponds to 1995 and $f(x)$ is in thousands.

(a) How many complaints were made during 1996?

(b) How many complaints were made during 1999?

Assume this model remains valid in later years.

(c) How many complaints were made during 2000?

(d) In what year would the number of complaints reach 113,000? Does this seem plausible?

32. A partial graph of a cubic polynomial function $f(x)$ is shown on the calculator screen below. The coefficient of x^3 in the rule of $f(x)$ is negative. Sketch a graph that has the same shape as the entire graph of $f(x)$.

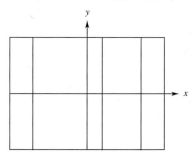

Use a graphing calculator to do the following problems. (See Example 5.)

33. Social Science   The table shows the number of abandoned cars (in thousands) in New York City over a ten-year period.*

Year	Abandoned Cars
1986	82
1987	120
1988	140
1989	148
1990	135
1991	95
1992	75
1993	51
1994	38
1995	27
1996	20

*New York Times, February 15, 1997.

(a) Plot the data points given by the table, with $x = 0$ corresponding to 1986.

(b) Use quartic regression to find a fourth-degree polynomial function $f(x)$ that models this data.

(c) Graph $f(x)$ on the same screen as the data points. Does the graph appear to fit the data well?

(d) Does this model predict an increase or a decrease in the number of abandoned cars after 1996?

34. Physical Science   The table shows U.S. coal consumption (in millions of short tons) in selected years.*

Year	Millions of Short Tons
1950	494.1
1960	398.1
1970	523.2
1980	702.7
1985	818.0
1990	895.5
1995	962.0
1996	1005.6
1997	1027.1

(a) Let $x = 0$ correspond to 1900. Plot the data points in the viewing window with $50 \le x \le 100$ and $350 \le y \le 1100$.

(b) Use cubic regression to find a cubic polynomial function $g(x)$ that models this data.

(c) Graph $g(x)$ on the same screen as the data points. Does the graph appear to fit the data well?

(d) Assuming that this model remains valid, estimate the amount of coal consumed in 2003.

*U.S. Department of Energy, *Annual Energy Review* (1997).

3.7 RATIONAL FUNCTIONS

A **rational function** is a function whose rule is the quotient of two polynomials, such as

$$f(x) = \frac{2}{1+x}, \qquad g(x) = \frac{3x+2}{2x+4}, \qquad h(x) = \frac{x^2 - 2x - 4}{x^3 - 2x^2 + x}.$$

Thus a rational function is one whose rule can be written in the form

$$f(x) = \frac{P(x)}{Q(x)},$$

where $P(x)$ and $Q(x)$ are polynomials, with $Q(x) \ne 0$. The function is undefined for any values of x that make $Q(x) = 0$, so there are breaks in the graph at these numbers.

Linear Rational Functions

We begin with rational functions in which both numerator and denominator are first-degree or constant polynomials. Such functions are sometimes called **linear rational functions.**

▶**EXAMPLE 1**   Graph the rational function defined by $y = \dfrac{2}{1 + x}$.

This function is undefined for $x = -1$, since -1 leads to a 0 denominator. For this reason, the graph of this function will not intersect the vertical line $x = -1$. Since x can take on any value except -1, the values of x can approach -1 as closely as desired from either side of -1, as shown in the following table of values.

x approaches -1
↓

x	-1.5	-1.2	-1.1	-1.01	$-.99$	$-.9$	$-.8$	$-.5$
$1 + x$	$-.5$	$-.2$	$-.1$	$-.01$	$.01$	$.1$	$.2$	$.5$
$\dfrac{2}{1 + x}$	-4	-10	-20	-200	200	20	10	4

↑
$|f(x)|$ gets larger and larger

The table above suggests that as x gets closer and closer to -1 from either side, the denominator $1 + x$ gets closer and closer to 0, and $|2/(1 + x)|$ gets larger and larger. The part of the graph near $x = -1$ in Figure 3.48 on the next page shows this behavior. The vertical line $x = -1$ that is approached by the curve is called a *vertical asymptote*. For convenience, the vertical asymptote is indicated by a dashed line in Figure 3.48, but this line is *not* part of the graph of the function.

As $|x|$ gets larger and larger so does the absolute value of the denominator $1 + x$. Hence $y = 2/(1 + x)$ gets closer and closer to 0, as shown in the table below.

x	-101	-11	-2	0	9	99
$1 + x$	-100	-10	-1	1	10	100
$\dfrac{2}{1 + x}$	$-.02$	$-.2$	-2	2	$.2$	$.02$

The horizontal line $y = 0$ is called a *horizontal asymptote* for this graph. Using the asymptotes and plotting the intercept and other points gives the graph of Figure 3.48. ◀ ①

Graph the following.

(a) $f(x) = \dfrac{3}{5 - x}$

(b) $f(x) = \dfrac{-4}{x + 4}$

Answers:

(a)

$f(x) = \dfrac{3}{5 - x}$

(b)

$f(x) = \dfrac{-4}{x + 4}$

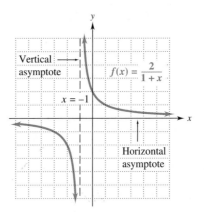

FIGURE 3.48

Example 1 suggests the following conclusion.

If a number c makes the denominator 0 but the numerator nonzero in the expression defining a rational function, then the line $x = c$ is a **vertical asymptote** for the graph of the function.

  Also, wherever the values of y approach but do not equal some number k as $|x|$ gets larger and larger, the line $y = k$ is a **horizontal asymptote** for the graph.

▶ **EXAMPLE 2**   Graph $f(x) = \dfrac{3x + 2}{2x + 4}$.

  Find the vertical asymptote by setting the denominator equal to 0 and then solving for x.

$$2x + 4 = 0$$
$$x = -2$$

  In order to see what the graph looks like when $|x|$ is very large, we rewrite the rule of the function. When $x \neq 0$, dividing both numerator and denominator by x does not change the value of the function:

$$f(x) = \frac{3x + 2}{2x + 4} = \frac{\dfrac{3x + 2}{x}}{\dfrac{2x + 4}{x}}$$

$$= \frac{\dfrac{3x}{x} + \dfrac{2}{x}}{\dfrac{2x}{x} + \dfrac{4}{x}} = \frac{3 + \dfrac{2}{x}}{2 + \dfrac{4}{x}}.$$

Now when $|x|$ is very large, the fractions $2/x$ and $4/x$ are very close to 0 (for instance, when $x = 200$, $4/x = 4/200 = .02$). Therefore the numerator of $f(x)$ is very

②

Graph the following.

(a) $f(x) = \dfrac{3x + 5}{x - 3}$

(b) $f(x) = \dfrac{2 - x}{x + 3}$

Answers:

(a)

$$f(x) = \frac{3x + 5}{x - 3}$$

(b)

$$f(x) = \frac{2 - x}{x + 3}$$

close to $3 + 0 = 3$ and the denominator is very close to $2 + 0 = 2$. Hence $f(x)$ is very close to 3/2 when $|x|$ is large, so the line $y = 3/2$ is the horizontal asymptote of the graph, as shown in Figure 3.49. ◄ ②

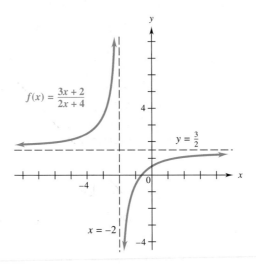

$$f(x) = \frac{3x + 2}{2x + 4}$$

FIGURE 3.49

TECHNOLOGY TIP Depending on the viewing window, a graphing calculator may not accurately represent the graph of a rational function. For example, the graph of $f(x) = \dfrac{3x + 2}{2x + 4}$ in Figure 3.50, which should look like Figure 3.49, has an erroneous vertical line at the place where the graph has a vertical asymptote. This problem can usually be avoided by using a window that has the vertical asymptote at the center of the x-axis, as in Figure 3.51.

FIGURE 3.50

FIGURE 3.51

In Examples 1 and 2, each graph has a single vertical asymptote, determined by the root of the denominator, and a horizontal asymptote determined by the coefficients of x. In Example 2, for instance, $f(x) = \dfrac{3x + 2}{2x + 4}$ has horizontal asymptote $y = \dfrac{3}{2}$, and in Example 1, $f(x) = \dfrac{2}{1 + x} = \dfrac{0x + 2}{1x + 1}$ has horizontal asymptote $y = \dfrac{0}{1} = 0$ (the x-axis). Similar arguments work for any linear rational function.

The graph of $f(x) = \dfrac{ax + b}{cx + d}$ (where $c \neq 0$ and $ad \neq bc$) has a vertical asymptote at the root of the denominator and horizontal asymptote $y = \dfrac{a}{c}$.

Other Rational Functions

When the numerator or denominator of a rational function has degree greater than 1, its graph can be more complicated than those in Examples 1 and 2. The graph may have several vertical asymptotes, as well as peaks and valleys.

▶ **EXAMPLE 3**   Graph $f(x) = \dfrac{2x^2}{x^2 - 4}$.

Find the vertical asymptotes by setting the denominator equal to 0 and solving for x.

$$x^2 - 4 = 0$$
$$(x + 2)(x - 2) = 0$$
$$x + 2 = 0 \quad \text{or} \quad x - 2 = 0$$
$$x = -2 \qquad\qquad x = 2$$

Since neither of these numbers makes the numerator 0, the lines $x = -2$ and $x = 2$ are vertical asymptotes of the graph. The horizontal asymptote can be determined by dividing both the numerator and denominator of $f(x)$ by x^2 (the highest power of x that appears in either one).

$$f(x) = \frac{2x^2}{x^2 - 4}$$

$$= \frac{\dfrac{2x^2}{x^2}}{\dfrac{x^2 - 4}{x^2}}$$

$$= \frac{\dfrac{2x^2}{x^2}}{\dfrac{x^2}{x^2} - \dfrac{4}{x^2}}$$

$$= \frac{2}{1 - \dfrac{4}{x^2}}$$

When $|x|$ is very large, the fraction $4/x^2$ is very close to 0, so that the denominator is very close to 1 and $f(x)$ is very close to 2. Hence the line $y = 2$ is the horizontal asymptote of the graph. Using this information and plotting several points in each of the three regions determined by the vertical asymptotes, we obtain Figure 3.52 on the next page. ◀ ③

③

List the vertical and horizontal asymptotes of the function.

(a) $f(x) = \dfrac{3x + 5}{x + 5}$

(b) $g(x) = \dfrac{2 - x^2}{x^2 - 4}$

Answers:

(a) Vertical $x = -5$; horizontal $y = 3$.

(b) Vertical $x = -2$ and $x = 2$; horizontal $y = -1$.

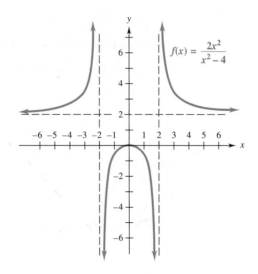

FIGURE 3.52

TECHNOLOGY TIP When a function whose graph has more than one vertical asymptote (as in Example 3) is graphed on a graphing calculator, erroneous vertical lines can sometimes be avoided by using a *decimal window* (with the *y*-range adjusted to show the graph). On TI, Sharp, and HP-39, use (Z)DECIMAL in the ZOOM or VIEWS menu. On Casio, use INIT in the V-WINDOW menu. Figure 3.53 shows the function of Example 3 graphed in a decimal window on a TI-83. (The *x*-range may be different on other calculators.)

FIGURE 3.53

The arguments used to find the horizontal asymptotes in Examples 1–3 work in the general case and lead to this conclusion.

If the numerator of the rational function $f(x)$ has *smaller* degree than the denominator, then the *x*-axis (the line $y = 0$) is the horizontal asymptote of the graph. If the numerator and denominator have the *same* degree, say,
$$f(x) = \frac{ax^n + \cdots}{cx^n + \cdots}, \text{ then the line } y = \frac{a}{c} \text{ is the horizontal asymptote.*}$$

*When the numerator has larger degree than the denominator, the graph has no horizontal asymptote but may have non-horizontal lines or other curves as asymptotes; see Exercises 32 and 33 at the end of this section for examples.

Applications

Rational functions have a variety of applications, some of which are explored here.

▶ **EXAMPLE 4**   In many situations involving environmental pollution, much of the pollutant can be removed from the air or water at a fairly reasonable cost, but the last, small part of the pollutant can be very expensive to remove.

Cost as a function of the percentage of pollutant removed from the environment can be calculated for various percentages of removal, with a curve fitted through the resulting data points. This curve then leads to a function that approximates the situation. Rational functions often are a good choice for these **cost-benefit functions.**

For example, suppose a cost-benefit function is given by

$$f(x) = \frac{18x}{106 - x},$$

where $f(x)$ or y is the cost (in thousands of dollars) of removing x percent of a certain pollutant. The domain of x is the set of all numbers from 0 to 100, inclusive; any amount of pollutant from 0% to 100% can be removed. To remove 100% of the pollutant here would cost

$$y = \frac{18(100)}{106 - 100} = 300,$$

or $300,000. Check that 95% of the pollutant can be removed for about $155,000, 90% for about $101,000, and 80% for about $55,000, as shown in Figure 3.54 (in which the displayed y-coordinates are rounded to the nearest integer).   ◀   ④

④

Using the function of Example 4, find the cost to remove the following percents of pollutants.

(a)   70%

(b)   85%

(c)   98%

Answers:

(a) $35,000

(b)   About $73,000

(c)   About $221,000

FIGURE 3.54

In management, **product-exchange functions** give the relationship between quantities of two items that can be produced by the same machine or factory. For example, an oil refinery can produce gasoline, heating oil, or a combination of the two; a winery can produce red wine, white wine, or a combination of the two. The next example discusses a product-exchange function.

▶ **EXAMPLE 5** The product-exchange function for the Golden Grape Winery for red wine x and white wine y, in tons, is

$$y = \frac{100,000 - 50x}{1000 + x}.$$

Graph the function and find the maximum quantity of each kind of wine that can be produced.

Only nonnegative values of x and y make sense in this situation, so we graph the function in the first quadrant (Figure 3.55). Note that the y-intercept of the graph (found by setting $x = 0$) is 100 and the x-intercept (found by setting $y = 0$ and solving for x) is 2000. Since we are interested only in the Quadrant I portion of the graph, we can find a few more points in Quadrant I and complete the graph as shown in Figure 3.55.

⑤

Rework Example 5 with the product-exchange function

$$y = \frac{60,000 - 10x}{60 + x}$$

to find the maximum amount of each wine that can be produced.

Answer:

6000 tons of red, 1000 tons of white

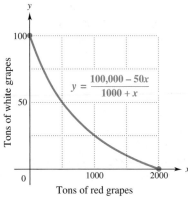

FIGURE 3.55

The maximum value of y occurs when $x = 0$, so the maximum amount of white wine that can be produced is 100 tons, given by the y-intercept. The x-intercept gives the maximum amount of red wine that can be produced, 2000 tons. ◀ ⑤

▶ **EXAMPLE 6** A retailer buys 2500 specialty lightbulbs from a distributor each year. In addition to the cost of each bulb, there is an order fee for each order, so she wants to order as few times as possible. On the other hand, storage costs are higher when there are fewer orders (and hence more bulbs per order to store). Past experience shows that the total annual cost (for the bulbs, ordering fees, and storage costs) is given by the rational function.

$$C(x) = \frac{.98x^2 + 1200x + 22,000}{x},$$

where x is the number of bulbs ordered each time. How many bulbs should be ordered each time in order to have the smallest possible cost?

Graph the cost function $C(x)$ in a window with $0 \leq x \leq 2500$ (because she can't order a negative number and needs only 2500 for the year).

3000   3000

0   2500   0   2500

−1000

FIGURE 3.56   **FIGURE 3.57**

For each point on the graph in Figure 3.56,
   the x-coordinate is the number of bulbs ordered each time;
   the y-coordinate is the annual cost when x bulbs are ordered each time.
Use the minimum finder on a graphing calculator to find the point with smallest y-coordinate, which is approximately (149.83, 1493.67), as shown in Figure 3.57. Since the retailer can't order part of a lightbulb, she should order 150 bulbs each time for an approximate annual cost of $1494. ◀

3.7 Exercises

Graph each function. Give the equations of the vertical and horizontal asymptotes (see Examples 1–3).

1. $f(x) = \dfrac{1}{x + 5}$

2. $g(x) = \dfrac{-7}{x - 6}$

3. $f(x) = \dfrac{-3}{2x + 5}$

4. $h(x) = \dfrac{-4}{2 - x}$

5. $f(x) = \dfrac{3x}{x - 1}$

6. $g(x) = \dfrac{x - 2}{x}$

7. $f(x) = \dfrac{x + 1}{x - 4}$

8. $f(x) = \dfrac{x - 3}{x + 5}$

9. $f(x) = \dfrac{2 - x}{x - 3}$

10. $g(x) = \dfrac{3x - 2}{x + 3}$

11. $f(x) = \dfrac{2x - 1}{4x + 2}$

12. $f(x) = \dfrac{3x - 6}{6x - 1}$

13. $h(x) = \dfrac{x + 1}{x^2 + 3x - 4}$

14. $g(x) = \dfrac{1}{x(x + 1)^2}$

15. $f(x) = \dfrac{x^2 + 1}{x^2 - 1}$

16. $f(x) = \dfrac{x - 1}{x^2 - x - 6}$

Find the equations of the vertical asymptotes of each of the following rational functions.

17. $f(x) = \dfrac{x - 3}{x^2 + x - 2}$

18. $g(x) = \dfrac{x + 2}{x^2 - 1}$

19. $g(x) = \dfrac{x^2 + 2x}{x^2 - 4x - 5}$

20. $f(x) = \dfrac{x^2 - 2x - 4}{x^3 - 2x^2 + x}$

Work these problems. (See Example 4.)

21. Natural Science Suppose a cost-benefit model is given by

$$f(x) = \frac{4.3x}{100 - x},$$

where $f(x)$ is the cost, in thousands of dollars, of removing x percent of a given pollutant. Find the cost of removing each of the following percents of pollutants.

(a) 50%   **(b)** 70%
(c) 80%   **(d)** 90%
(e) 95%   **(f)** 98%
(g) 99%
(h) Is it possible, according to this model, to remove *all* the pollutant?
(i) Graph the function.

22. Natural Science Suppose a cost-benefit model is given by

$$f(x) = \frac{4.5x}{101 - x},$$

where $f(x)$ is the cost, in thousands of dollars, of removing x percent of a certain pollutant. Find the cost of removing the following percents of pollutants.

(a) 0%   **(b)** 50%
(c) 80%   **(d)** 90%
(e) 95%   **(f)** 99%
(g) 100%   **(h)** Graph the function.

23. Natural Science   The function

$$f(x) = \frac{\lambda x}{1 + (ax)^b}$$

is used in population models to give the size of the next generation ($f(x)$) in terms of the current generation (x).[*]
(a) What is a reasonable domain for this function, considering what x represents?
(b) Graph this function for $\lambda = a = b = 1$ and $x \geq 0$.
(c) Graph this function for $\lambda = a = 1$ and $b = 2$ and $x \geq 0$.
(d) What is the effect of making b larger?

24. Natural Science   The function

$$f(x) = \frac{Kx}{A + x}$$

is used in biology to give the growth rate of a population in the presence of a quantity x of food. This is called Michaelis–Menten kinetics.[†]
(a) What is a reasonable domain for this function, considering what x represents?
(b) Graph this function for $K = 5$, $A = 2$ and $x \geq 0$.
(c) Show that $y = K$ is a horizontal asymptote.
(d) What do you think K represents?
(e) Show that A represents the quantity of food for which the growth rate is half of its maximum.

25. Social Science   The average waiting time in a line (or queue) before getting served is given by

$$W = \frac{S(S - A)}{A},$$

where A is the average rate at which people arrive at the line and S is the average service time. At a certain fast food restaurant, the average service time is 3 minutes. Find W for each of the following average arrival times.
(a) 1 minute
(b) 2 minutes
(c) 2.5 minutes
(d) What is the vertical asymptote?
(e) Graph the equation on the interval (0, 3).
(f) What happens to W when $A > 3$? What does this mean?

---

[*]See J. Maynard Smith, *Models in Ecology* (Cambridge University Press, 1974).

[†]See Leah Edelstein-Keshet, *Mathematical Models in Biology* (Random House, 1988).

Business   *Sketch the Quadrant I portion of the graph of each of the functions defined below, and then estimate the maximum quantities of each product that can be produced. (See Example 5.)*

26. The product-exchange function for gasoline, x, and heating oil, y, in hundreds of gallons per day, is

$$y = \frac{125{,}000 - 25x}{125 + 2x}.$$

27. A drug factory found the product-exchange function for a red tranquilizer, x, and a blue tranquilizer, y, is

$$y = \frac{900{,}000{,}000 - 30{,}000x}{x + 90{,}000}.$$

28. Physical Science   The failure of several O-rings in field joints was the cause of the fatal crash of the *Challenger* space shuttle in 1986. NASA data from 24 successful launches prior to *Challenger* suggested that O-ring failure was related to launch temperature by a function similar to

$$N(t) = \frac{600 - 7t}{4t - 100} \qquad (50 \leq t \leq 85),$$

where t is the temperature (in °F) at launch and N is the approximate number of O-rings that fail. Assume that this function accurately models the number of O-ring failures that would occur at lower launch temperatures (an assumption NASA did not make).
(a) Does $N(t)$ have a vertical asymptote? At what value of t does it occur?
(b) Without graphing, what would you conjecture that the graph would look like just to the right of the vertical asymptote? What does this suggest about the number of O-ring failures that might be expected near that temperature? (The temperature at the *Challenger* launching was 31°.)
(c) Confirm your conjecture by graphing $N(t)$ between the vertical asymptote and $t = 85$.

29. Business   A company has fixed costs of \$40,000 and marginal cost of \$2.60 per unit.
(a) Find the linear cost function.
(b) Find the average cost function. (Average cost was defined in Section 3.3.)
(c) Find the horizontal asymptote of the graph of the average cost function. Explain what the asymptote means in this situation (how low can the average cost possibly be?).

Use a graphing calculator to do Exercises 30–33. (See Example 6.)

30. Finance   Another model of a Laffer curve (see Exercise 27 of Section 3.6) is given by

$$f(x) = \frac{300x - 3x^2}{10x + 200},$$

where $f(x)$ is government revenue (in billions of dollars) from a tax rate of x percent. Find the revenue from the following tax rates.

(a) 16%   **(b)** 25%
(c) 40%   **(d)** 55%
(e) Graph $f(x)$.
(f) What tax rate produces maximum revenue? What is the maximum revenue?

31. *Business* When no more than 110 units are produced, the cost of producing x units is given by

$$C(x) = .2x^3 - 25x^2 + 1531x + 25,000.$$

How many units should be produced in order to have the lowest possible average cost?

32. **(a)** Graph $f(x) = \dfrac{x^3 + 3x^2 + x + 1}{x^2 + 2x + 1}$

(b) Does the graph appear to have a horizontal asymptote? Does the graph appear to have some nonhorizontal straight line as an asymptote?

(c) Graph $f(x)$ and the line $y = x + 1$ on the same screen. Does this line appear to be an asymptote of the graph of $f(x)$?

33. **(a)** Graph $g(x) = \dfrac{x^3 - 2}{x - 1}$ in the window with $-4 \le x \le 6$ and $-6 \le y \le 12$.

(b) Graph $g(x)$ and the parabola $y = x^2 + x + 1$ on the same screen. How do the two graphs compare when $|x| \ge 2$?

CHAPTER 3 **SUMMARY**

Key Terms and Symbols

3.1 function
domain
range
functional notation
3.2 graph
linear function
piecewise linear function
absolute value function
greatest integer function
step function
vertical line test

3.3 fixed costs
variable cost
average cost
average rate of change
marginal cost
linear cost function
break-even point
supply and demand curves
equilibrium point
equilibrium price
equilibrium quantity
3.4 quadratic function
parabola

vertex
axis
3.5 quadratic model
3.6 polynomial function
graph of $f(x) = ax^n$
properties of polynomial graphs
polynomial models
3.7 rational function
linear rational function
vertical asymptote
horizontal asymptote

Key Concepts

A **function** consists of a set of inputs called the **domain,** a set of outputs called the **range,** and a rule by which each input determines exactly one output.

If a vertical line intersects a graph in more than one point, the graph is not that of a function.

A **linear cost function** has equation $C(x) = mx + b$ where m is the **marginal cost** (the cost of producing one more item) and b is the **fixed cost.**

If $p = f(q)$ gives the price per unit when x units can be supplied, and $p = g(q)$ gives the price per unit when q units are demanded, then the **equilibrium price** and **equilibrium quantity** occur at the q-value such that $f(q) = g(q)$.

The **quadratic function** defined by $f(x) = a(x - h)^2 + k$ has a graph that is a **parabola** with vertex (h, k) and axis of symmetry $x = h$. The parabola opens upward if $a > 0$, downward if $a < 0$.

If the equation is in the form $f(x) = ax^2 + bx + c$, the vertex is $\left(-\dfrac{b}{2a}, f\left(-\dfrac{b}{2a}\right)\right)$.

When $|x|$ is large, the graph of a **polynomial function** resembles the graph of its highest degree term ax^n. The graph of $f(x) = ax^n$ is described on page 169.

On the graph of a polynomial function of degree n:

The total number of peaks and valleys is at most $n - 1$.

The total number of x-intercepts is at most n.

If a number c makes the denominator of a **rational function** 0, but the numerator nonzero, then the line $x = c$ is a **vertical asymptote** for the graph.

Whenever the values of y approach, but never equal, some number k as $|x|$ gets larger and larger, the line $y = k$ is a **horizontal asymptote** for the graph.

If the numerator of a rational function has *smaller* degree than the denominator, then the x-axis is the horizontal asymptote of the graph.

If the numerator and denominator of a rational function have the *same* degree, say, $f(x) = \dfrac{ax^n + \cdots}{cx^n + \cdots}$, then the line $y = \dfrac{a}{c}$ is the horizontal asymptote of the graph.

Chapter 3 Review Exercises

Which of the following rules defines a function?

1.

x	3	2	1	0	1	2
y	8	5	2	0	-2	-5

2.

x	2	1	0	-1	-2
y	5	3	1	-1	-3

3. $y = \sqrt{x}$

4. $x = |y|$

5. $x = y^2 + 1$

6. $y = 5x - 2$

For each function, find

   (a) $f(6)$ **(b)** $f(-2)$ **(c)** $f(p)$ **(d)** $f(r + 1)$.

7. $f(x) = 4x - 1$

8. $f(x) = 3 - 4x$

9. $f(x) = -x^2 + 2x - 4$

10. $f(x) = 8 - x - x^2$

11. Let $f(x) = 5x - 3$ and $g(x) = -x^2 + 4x$. Find each of the following.

   (a) $f(-2)$ **(b)** $g(3)$ **(c)** $g(-k)$

   (d) $g(3m)$ **(e)** $g(k - 5)$ **(f)** $f(3 - p)$

12. Let $f(x) = x^2 + x + 1$. Find each of the following.

   (a) $f(3)$ **(b)** $f(1)$ **(c)** $f(4)$

   (d) Based on your answers in parts (a)–(c), is it true that $f(a + b) = f(a) + f(b)$ for all real numbers a, b?

Graph each function.

13. $f(x) = |x| - 3$

14. $f(x) = -|x| - 2$

15. $f(x) = -|x + 1| + 3$

16. $f(x) = 2|x - 3| - 4$

17. $f(x) = [x - 3]$

18. $f(x) = \left[\dfrac{1}{2}x - 2\right]$

19. $f(x) = \begin{cases} -4x + 2 & \text{if } x \le 1 \\ 3x - 5 & \text{if } x > 1 \end{cases}$

20. $f(x) = \begin{cases} 3x + 1 & \text{if } x < 2 \\ -x + 4 & \text{if } x \ge 2 \end{cases}$

21. $f(x) = \begin{cases} |x| & \text{if } x < 3 \\ 6 - x & \text{if } x \ge 3 \end{cases}$

22. $f(x) = \sqrt{x^2}$

23. $g(x) = x^2/8 - 3$

24. $h(x) = \sqrt{x} + 2$

25. Business Let f be a function that gives the cost to rent a floor polisher for x hours. The cost is a flat \$3 for renting the polisher plus \$4 per day or fraction of a day for using the polisher.

   (a) Graph f.

   (b) Give the domain and range of f.

   (c) David Fleming wants to rent a polisher, but he can spend no more than \$15. At most how many days can he use it?

26. **Business**   A trailer hauling service charges $45, plus $2 per mile or part of a mile.

    (a) Is $90 enough for a 20-mile haul?

    (b) Graph the ordered pairs (miles, cost).

    (c) Give the domain and range.

27. **Social Science**   The percent of children born to unmarried mothers since 1970 can be approximated by

    $$f(x) = \begin{cases} .9x + 11 & \text{from 1970 to 1994} \\ 32.6 & \text{from 1994 to 2000} \end{cases}$$

    Let $x = 0$ correspond to 1970 and graph the function. What does the graph suggest about births to unmarried mothers?

28. **Social Science**   According to data from the U.S. Department of Agriculture, the annual per capita consumption of red meat in the United States has declined in a linear fashion from 131.7 pounds in 1970 to 107.7 in 2000.

    (a) Let $x = 0$ correspond to 1900 and find the rule of a linear function that gives the approximate annual per capita consumption of red meat in year x.

    (b) If this trend continues, in what year will per capita consumption be 100 pounds?

Business   In Exercises 29–32, find the following.

   (a) the linear cost function

   (b) the marginal cost

   (c) the average cost per unit to produce 100 units

29. Eight units cost $300; fixed cost is $60.

30. Fixed cost is $2000; 36 units cost $8480.

31. Twelve units cost $445; 50 units cost $1585.

32. Thirty units cost $1500; 120 units cost $5640.

33. **Business**   The cost of producing x ink cartridges for a printer is given by $C(x) = 24x + 18,000$. Each cartridge can be sold for $28.

    (a) What are the fixed costs?

    (b) Find the revenue function.

    (c) Find the break-even point.

    (d) If the company sells exactly the number of cartridges needed to break even, what is its revenue?

34. **Business**   The prime time viewing share of network TV stations in year x is approximated by

    $$f(x) = -1.2x + 68 \qquad (9 \le x \le 22),$$

    where $x = 0$ corresponds to 1980. The share of basic cable stations is approximated by $g(x) = 2.3x - 3.6$.* In what year did network and cable stations have the same viewing share?

35. **Business**   Suppose the demand and price for the HBO cable channel are related by $p = -.5q + 30.95$, where p is the monthly price in dollars and q is measured in millions of subscribers. If the price and supply are related by $p = .3q + 2.15$, what are the equilibrium quantity and price?

36. **Business**   Suppose the supply and price for prescription strength Tylenol are related by $p = .0015q + 1$, where p is the price (in dollars) of a 30-day prescription. If the demand is related to price by $p = -.0025q + 64.36$, what are the equilibrium quantity and price?

Without graphing, determine whether each of the following parabolas opens upward or downward and find its vertex.

37. $f(x) = 3(x - 2)^2 + 6$ 38. $f(x) = 2(x + 3)^2 - 5$

39. $g(x) = -4(x + 1)^2 + 8$ 40. $g(x) = -5(x - 4)^2 - 6$

Graph each of the following and label its vertex.

41. $f(x) = x^2 - 4$

42. $f(x) = 6 - x^2$

43. $f(x) = x^2 + 2x - 3$

44. $f(x) = -x^2 + 6x - 3$

45. $f(x) = -x^2 - 4x + 1$

46. $f(x) = 4x^2 - 8x + 3$

47. $f(x) = 2x^2 + 4x - 3$

48. $f(x) = -3x^2 - 12x - 8$

Determine whether each of the following functions has a minimum or a maximum value and find that value.

49. $f(x) = x^2 + 6x - 2$

50. $f(x) = x^2 + 4x + 5$

51. $g(x) = -4x^2 + 8x + 3$

52. $g(x) = -3x^2 - 6x + 3$.

Solve each problem.

53. **Business**   The commodity market is very unstable; money can be made or lost quickly when investing in soybeans, wheat, pork bellies, and the like. Suppose that an investor kept track of her total profit, P, at time t, measured in months, after she began investing and found that $P = -4t^2 + 32t - 20$. At what time is her profit largest? (*Hint:* $t > 0$ in this case.)

54. **Physical Science**   The height h (in feet) of a rocket at time t seconds after liftoff is given by $h = -16t^2 + 800t$.

    (a) How long does it take the rocket to reach 3200 feet?

    (b) What is the maximum height of the rocket?

55. **Business**   The manager of a large apartment complex has found that the profit is given by $P = -x^2 + 250x - 15,000$, where x is the number of units rented. For what value of x does the complex produce the largest profit?

56. **Business**   A rectangular enclosure is to be built with three sides made out of redwood fencing at a cost of $15 per running foot and the fourth side made out of cement blocks at a cost of $30 per running foot. $900 is available for the project. What are the dimensions of the enclosure with maximum possible area and what is this area?

*Based on data from the National Cable Television Association.

57. Finance The table shows the expenditures (in billions of dollars) of the federal government on transportation in selected years.*

Year	1995	1997	1999	2000	2001
Expenditures	39	41	43	47	50

(a) Let $x = 0$ correspond to 1990. Find a quadratic function $f(x) = a(x - h)^2 + k$ that models this data, using $(5, 39)$ as the vertex and the data point for 2001.

(b) Assuming the model remains valid, what are the transportation expenditures (to the nearest billion) in 2002?

58. Health The table shows the number of AIDS patients admitted to hospital in the state of New York in selected years.[†]

Year	1986	1988	1990	1992
Patients	11,875	23,440	40,940	57,500

Year	1994	1996	1998	1999
Patients	61,875	60,625	47,500	45,000

(a) Let $x = 0$ correspond to 1980 and plot the data points on a graphing calculator.

(b) Use quadratic regression to find a quadratic model for the data.

(c) Estimate the number of patients admitted in 1991.

(d) To the nearest year, in what year were the most patients admitted?

Graph each of the following polynomial functions.

59. $f(x) = x^4 - 2$

60. $g(x) = x^3 - x$

61. $f(x) = x(x - 2)(x + 3)$

62. $f(x) = (x - 1)(x + 2)(x - 3)$

63. $f(x) = 3x(3x + 2)(x - 1)$

64. $f(x) = x^3 - 3x^2 - 4x$

65. $f(x) = x^4 - 5x^2 - 6$

66. $f(x) = x^4 - 7x^2 - 8$

Use a graphing calculator to do Problems 67–70.

67. Business The demand equation for automobile oil filters is
$$p = -.000012q^3 - .00498q^2 + .1264q + 1508,$$

where p is in dollars and q is in thousands of items. The supply equation is
$$p = -.000001q^3 + .00097q^2 + 2q.$$

Find the equilibrium quantity and the equilibrium price.

68. Business The average cost (in dollars) per item of manufacturing x thousand cans of spray paint is given by
$$A(x) = -.000006x^4 + .0017x^3 + .03x^2 - 24x + 1110.$$

How many cans should be manufactured if the average cost is to be as low as possible? What is the average cost in that case?

69. Business Plastic racks for holding compact discs sell for $23 each. The cost of manufacturing x racks is given by $C(x) = -.000006x^3 + .07x^2 + 2x + 1200$. At most 600 racks can be produced.

(a) Find the revenue and profit functions.

(b) What is the break-even point?

(c) What is the maximum number of racks that can be made without losing money?

(d) How many racks should be made in order to have as large a profit as possible? What is that profit?

70. Health The table shows the average remaining life expectancy (in years) of a male at selected ages.*

Current Age	Birth	20	40	60	80	90	100
Life Expectancy	73.6	54.7	36.2	19.4	7.5	4	2.4

(a) Let birth $= 0$ and plot the data points.

(b) Use quartic regression to find a fourth-degree polynomial $f(x)$ that models the data.

(c) What is the life expectancy of a man of age 25? Age 35? Age 50?

(d) What is the life expectancy of a man who is exactly your age?

List the vertical and horizontal asymptotes of each function and sketch its graph.

71. $f(x) = \dfrac{1}{x - 3}$

72. $f(x) = \dfrac{-2}{x + 4}$

73. $f(x) = \dfrac{-3}{2x - 4}$

74. $f(x) = \dfrac{5}{3x + 7}$

75. $g(x) = \dfrac{5x - 2}{4x^2 - 4x - 3}$

76. $g(x) = \dfrac{x^2}{x^2 - 1}$

77. Business The average cost per carton of producing x cartons of cocoa is given by
$$C(x) = \dfrac{650}{2x + 40}.$$

---

*U.S. Department of Treasury and the Office of Management and Budget.

†New York State Department of Health.

*National Center for Health Statistics.

Find the average cost per carton to make the given number of cartons.

(a) 10 cartons

(b) 50 cartons

(c) 70 cartons

(d) 100 cartons

(e) Graph $C(x)$.

78. Business   The cost and revenue functions (in dollars) for a frozen yogurt shop are given by

$$C(x) = \frac{400x + 400}{x + 4} \quad \text{and} \quad R(x) = 100x,$$

where x is measured in hundreds of units.

(a) Graph $C(x)$ and $R(x)$ on the same axes.

(b) What is the break-even point for this shop?

(c) If the profit function is given by $P(x)$, does $P(1)$ represent a profit or a loss?

(d) Does $P(4)$ represent a profit or a loss?

79. Business   The supply and demand functions for the yogurt shop in Exercise 78 are as follows:

$$\text{supply: } p = \frac{q^2}{4} + 25; \quad \text{demand: } p = \frac{500}{q},$$

where p is the price in dollars for q hundred units of yogurt.

(a) Graph both functions on the same axes, and from the graph, estimate the equilibrium point.

(b) Give the q-intervals where supply exceeds demand.

(c) Give the q-intervals where demand exceeds supply.

80. Business   A cost-benefit curve for pollution control is given by

$$y = \frac{9.2x}{106 - x},$$

where y is the cost in thousands of dollars of removing x percent of a specific industrial pollutant. Find y for each of the following values of x.

(a) $x = 50$

(b) $x = 98$

(c) What percent of the pollutant can be removed for $22,000?

Chapter 4

Exponential and Logarithmic Functions

Exponential and logarithmic functions play a key role in management, economics, the social and physical sciences, and engineering. They are used to study the growth of money and organizations; learning curves; the growth of human, animal, and bacterial populations; the spread of disease; and radioactive decay.

4.1 EXPONENTIAL FUNCTIONS

In polynomial functions, the variable is raised to various constant exponents; for instance, $f(x) = x^2 + 3x - 5$. In *exponential functions,* such as

$$f(x) = 10^x \quad \text{or} \quad g(x) = 2^{-x^2} \quad \text{or} \quad h(x) = 3^{.6x},$$

the variable is in the exponent and the base is a positive constant. We begin with the simplest type of exponential function.

> If a is a positive constant other than 1 and k is any nonzero constant, then the function defined by
> $$f(x) = a^{kx}$$
> is called an **exponential function with base a.** Its domain is the set of all real numbers.

The function $f(x) = 1^x$ is the constant function $f(x) = 1$. Exponential functions with negative bases are not of interest because when a is negative, a^x may not be defined for some values of x; for instance, $(-4)^{1/2} = \sqrt{-4}$ is not a real number.

▶ **EXAMPLE 1** Graph these exponential functions and discuss the characteristics of their graphs.

(a) $f(x) = 2^x$

Either use a graphing calculator or graph by hand: make a table of values, plot the corresponding points, and join them by a smooth curve, as in Figure 4.1. The graph has the following properties.

The graph rises as x increases (that is, $f(x)$ is an increasing function).

The y-intercept is at 1.

The graph is always above the x-axis (because *every* power of 2 is positive), so there is no x-intercept.

There is no vertical asymptote, but the negative x-axis is a horizontal asymptote (the graph approaches but never touches it).

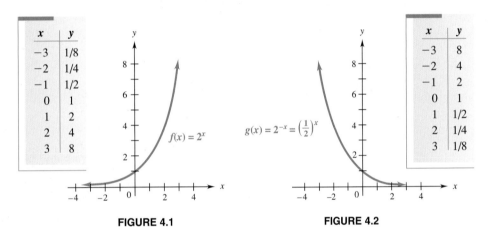

x	y
-3	1/8
-2	1/4
-1	1/2
0	1
1	2
2	4
3	8

$f(x) = 2^x$

$g(x) = 2^{-x} = \left(\frac{1}{2}\right)^x$

x	y
-3	8
-2	4
-1	2
0	1
1	1/2
2	1/4
3	1/8

FIGURE 4.1   **FIGURE 4.2**

①

Use a graphing calculator to graph $f(x) = 4^x$ and

$g(x) = \left(\frac{1}{4}\right)^x$ on the same screen.

Answer:

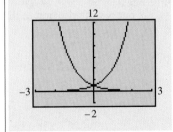

(b) $g(x) = 2^{-x}$

Again, use a graphing calculator or graph by hand, as in Figure 4.2. The graph has the following properties.

The graph falls as x increases (that is, $f(x)$ is a decreasing function).

The y-intercept is at 1.

The graph is always above the x-axis (because *every* power of 2 is positive), so there is no x-intercept.

There is no vertical asymptote, but the positive x-axis is a horizontal asymptote (the graph approaches but never touches it).

Notice that

$$2^{-x} = \frac{1}{2^x} = \left(\frac{1}{2}\right)^x,$$

by properties of exponents. The graph of $g(x) = (1/2)^x$ is the mirror image of the graph of $f(x) = 2^x$, with the y-axis as the mirror. ◀ ①

The graph of $f(x) = 2^x$ in Figure 4.1 illustrates **exponential growth,** which is far more explosive than polynomial growth. For example, if the graph in Figure 4.1 were extended on the same scale to include the point $(50, 2^{50})$, the graph would be more than 2 billion *miles* high! Similarly, the graph of $g(x) = 2^{-x} = (1/2)^x$ illustrates **exponential decay.**

When $a > 1$, the graph of the exponential function $h(x) = a^x$ has the same basic shape as the graph of $f(x) = 2^x$ in Figure 4.1. The larger the base a is, the more steeply the graph rises, as illustrated in Figure 4.3(a).

When $0 < a < 1$, the graph of $h(x) = a^x$ has the same basic shape as the graph of $g(x) = (1/2)^x$ in Figure 4.2. Graphs of some typical exponential functions with $0 < a < 1$ are shown in Figure 4.3(b).

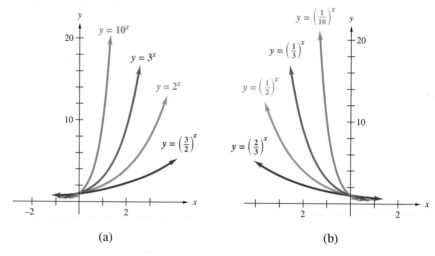

(a)   (b)

FIGURE 4.3

The graphs of exponential functions such as $f(x) = 3^{1-x}$ or $g(x) = 2^{.6x}$ or $h(x) = 3 \cdot 10^{2x+1}$ have the same general shape as the exponential graphs above. The only differences are that the graph may rise or fall at a different rate or the entire graph may be shifted vertically or horizontally.

▶ **EXAMPLE 2**   Graph $f(x)$ and $g(x)$ as defined below and explain how the two graphs are related.

(a) $f(x) = 3^{1-x}$ and $g(x) = 3^{-x}$

Choose values of x that make the exponent positive, zero, and negative, and plot the corresponding points. The graphs are shown in Figure 4.4. The graph of $f(x) = 3^{1-x}$ has the same shape as the graph of $g(x) = 3^{-x}$ but is shifted 1 unit to the right, making the y-intercept $(0, 3)$ rather than $(0, 1)$.

FIGURE 4.4   **FIGURE 4.5**

2

Graph $f(x) = 2^{x+1}$.

Answer:

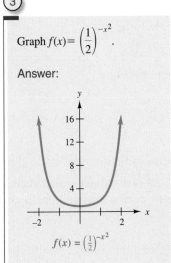

3

Graph $f(x) = \left(\dfrac{1}{2}\right)^{-x^2}$.

Answer:

$f(x) = \left(\frac{1}{2}\right)^{-x^2}$

4

Evaluate the following powers of e.

(a) $e^{.06}$

(b) $e^{-.06}$

(c) $e^{2.30}$

(d) $e^{-2.30}$

Answers:

(a) 1.06184

(b) .94176

(c) 9.97418

(d) .10026

(b) $f(x) = 2^{.6x}$ and $g(x) = 2^x$

Comparing the graphs of $f(x) = 2^{.6x}$ and $g(x) = 2^x$ in Figure 4.5, we see that the graphs are both increasing but the graph of $f(x)$ rises at a slower rate. This happens because of the .6 in the exponent. If the coefficient of x were greater than 1, the graph would rise at a faster rate than the graph of $f(x) = 2^x$. ◀ **2**

When the exponent involves a nonlinear expression in x, an exponential function graph may have a much different shape than the preceding ones.

▶ **EXAMPLE 3**   Graph $f(x) = 2^{-x^2}$.

Either use a graphing calculator or plot points and connect them with a smooth curve, as in Figure 4.6. The graph is symmetric about the y-axis—that is, if the figure were folded on the y-axis, the two halves would match. Like the graphs above, this graph has the x-axis as a horizontal asymptote. The domain is still all real numbers, but here the range is $0 < y \le 1$. Graphs such as this are important in probability, where the normal curve has an equation similar to $f(x)$ in this example. ◀ **3**

x	y
-2	1/16
-1.5	.21
-1	1/2
$-.5$	.84
0	1
.5	.84
1	1/2
1.5	.21
2	1/16

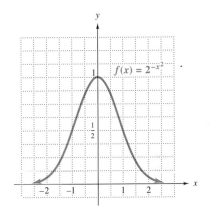

FIGURE 4.6

The Number e

There is an irrational number, denoted e, that arises naturally in a variety of mathematical situations. To 12 decimal places,

$$e \approx 2.718281828459.$$

Perhaps the single most useful exponential function is the function defined by $f(x) = e^x$.

T E C H N O L O G Y   T I P   To evaluate powers of e with a calculator, use the $\boxed{e^x}$ key. On some calculators, you will need to use the two keys $\boxed{\text{INV}}$ $\boxed{\text{LN}}$ or $\boxed{\text{2nd}}$ $\boxed{\text{LN}}$. For example, your calculator should show that $e^{.14} = 1.150273799$ to 8 decimal places. If these keys do not work, consult the instruction manual for your calculator. **4**

NOTE To display the decimal expansion of e on your calculator screen, calculate e^1.

In Figure 4.7 the functions defined by

$$g(x) = 2^x, \qquad f(x) = e^x, \qquad \text{and} \quad h(x) = 3^x$$

are graphed for comparison.

FIGURE 4.7

⑤

The number of organisms present at time t is given by $f(t) = 75e^{.458t}$.

(a) Is this a growth function or a decay function?

Find the number of organisms present at

(b) $t = 0$;

(c) $t = 2$;

(d) $t = 4$.

Answers:

(a) A growth function

(b) 75

(c) About 188

(d) About 469

FIGURE 4.8

▶ **EXAMPLE 4**   If current trends of burning fossil fuel and deforestation continue, then the increase in the amounts of atmospheric carbon dioxide in parts per million (ppm) will be given by

$$f(t) = 375e^{.00609t},$$

where $t = 0$ corresponds to 2000.*

(a) What is the amount of carbon dioxide in 2000?

Since 2000 corresponds to $t = 0$, evaluate the function at $t = 0$:

$$f(0) = 375e^{.00609(0)} = 375e^0 = 375\,(1) = 375 \text{ ppm.}$$

(b) What will the amount of carbon dixode be in 2025?

Since 2025 corresponds to $t = 25$, the amount will be

$$f(25) = 375e^{.00609(25)} = 436.7 \text{ ppm.}$$

(c) Use a graphing calculator to find the year in which the amount of carbon dioxide will reach 500 ppm. In other words, solve the equation $375e^{.00609t} = 500$.

Graph $y = 375e^{.00609x}$ and $y = 500$ on the same screen. Find the intersection point, as in Figure 4.8, which shows that the 500 ppm level will be reached when t is about 47.2, that is, during 2047. ◀ ⑤

---

*Based on projections from the International Panel on Climate Change.

4.1 Exercises

Classify each function as linear, quadratic, or exponential.

1. $f(x) = 2x^2 + x - 3$ **2.** $g(x) = 5^{x-3}$

3. $h(x) = 3 \cdot 6^{2x^2+4}$ **4.** $f(x) = 3x + 1$

Without graphing,
(a) *describe the shape of the graph of each function, and*
(b) *complete the ordered pairs* (0,   ) *and* (1,   ) *for each function.*
(See Examples 1 and 2.)

5. $f(x) = .8^x$ decay   **6.** $g(x) = 6^{-x}$ grow

7. $f(x) = 5^{.4x}$ gr   **8.** $g(x) = -e^x$ decay

Graph each function. (See Examples 1 and 2.)

9. $f(x) = 3^x$ **10.** $g(x) = 3^{.5x}$

11. $f(x) = 2^{x/2}$ **12.** $g(x) = e^{x/4}$

13. $f(x) = (1/5)^x$ **14.** $g(x) = 2^{3x}$

15. Graph these functions on the same axes.
  (a) $f(x) = 2^x$ **(b)** $g(x) = 2^{x+3}$ **(c)** $h(x) = 2^{x-4}$
  (d) If c is a positive constant, explain how the graphs of $y = 2^{x+c}$ and $y = 2^{x-c}$ are related to the graph of $f(x) = 2^x$.

16. Graph these functions on the same axes.
  (a) $f(x) = 3^x$ **(b)** $g(x) = 3^x + 2$
  (c) $h(x) = 3^x - 4$
  (d) If c is a positive constant, explain how the graphs of $y = 3^x + c$ and $y = 3^x - c$ are related to the graph of $f(x) = 3^x$.

The figure shows the graphs of $y = a^x$ for a = 1.8, 2.3, 3.2, .4, .75, and .31. They are identified by letter but not necessarily in the same order as the values of a just given. Use your knowledge of how the exponential function behaves for various powers of a to match each lettered graph with the correct value of a.

17. A   **18.** B   **19.** C   **20.** D   **21.** E   **22.** F

In Exercises 23 and 24, the graph of an exponential function with base a is given. Follow the directions in parts (a)–(f) in each exercise.

23.

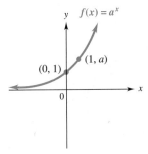

(a) Is $a > 1$ or is $0 < a < 1$?
(b) Give the domain and range of f.
(c) Sketch the graph of $g(x) = -a^x$.
(d) Give the domain and range of g.
(e) Sketch the graph of $h(x) = a^{-x}$.
(f) Give the domain and range of h.

24.

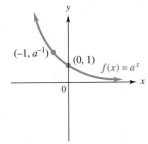

(a) Is $a > 1$ or is $0 < a < 1$?
(b) Give the domain and range of f.
(c) Sketch the graph of $g(x) = a^x + 2$.
(d) Give the domain and range of g.
(e) Sketch the graph of $h(x) = a^{x+2}$.
(f) Give the domain and range of h.

25. If $f(x) = a^x$ and $f(3) = 27$, find the following values of $f(x)$.
  (a) $f(1)$ **(b)** $f(-1)$ **(c)** $f(2)$ **(d)** $f(0)$

26. Give a rule of the form $f(x) = a^x$ to define the exponential function whose graph contains the given point.
  (a) $(3, 8)$ **(b)** $(-3, 64)$

Graph each function. (See Example 3.)

27. $f(x) = 2^{-x^2+2}$ **28.** $g(x) = 2^{x^2-2}$

29. $f(x) = x \cdot 2^x$ **30.** $f(x) = x^2 \cdot 2^x$

Work the following exercises.

31. Finance   If $1 is deposited into an account paying 6% per year compounded annually, then after t years the account will contain

$$y = (1 + .06)^t = (1.06)^t$$

dollars.

(a) Use a calculator to complete the following table.

t	0	1	2	3	4	5	6	7	8	9	10
y	1					1.34					1.79

(b) Graph $y = (1.06)^t$.

32. **Finance** If money loses value at the rate of 3% per year, the value of \$1 in t years is given by

$$y = (1 - .03)^t = (.97)^t.$$

(a) Use a calculator to complete the following table.

t	0	1	2	3	4	5	6	7	8	9	10
y	1				.86						.74

(b) Graph $y = (.97)^t$.

Work these problems. See Example 4.

33. **Finance** If money loses value, then it takes more dollars to buy the same item. Use the results of Exercise 32(a) to answer the following questions.
  (a) Suppose a house costs \$105,000 today. Estimate the cost of a similar house in 10 years. (*Hint:* Solve the equation $.74t = \$105,000$.)
  (b) Estimate the cost of a \$50 textbook in 8 years.

34. **Natural Science** Biologists studying salmon have found that the oxygen consumption of yearling salmon is given by $g(x) = 100e^{.7x}$, where x is the speed in feet per second. Find each of the following.
  (a) The oxygen consumption when the fish are still
  (b) The oxygen consumption at a speed of 2 feet per second

35. **Social Science** Cellular phone use increased exponentially in the 1990s. The number of U.S. cellular phone accounts (in millions) can be approximated by

$$f(x) = 25.49(1.2821)^x \qquad (0 \le x \le 7)$$

where $x = 0$ corresponds to 1994.* Use this function to estimate the number of cell phone accounts in
  (a) 2000; **(b)** 2001.
  (c) Will this model continue to be accurate over the long run?

*Based on data from the CTIA Semi-Annual Wireless Survey.

36. **Business** The monthly payment on a car loan at 12% interest per year on the unpaid balance is given by

$$f(n) = \frac{P}{\dfrac{1 - 1.01^{-n}}{.01}},$$

where P is the amount borrowed and n is the number of months over which the loan is paid back. Find the monthly payment for each of the following loans.
  (a) \$8000 for 48 months
  (b) \$8000 for 24 months
  (c) \$6500 for 36 months
  (d) \$6500 for 60 months

37. **Natural Science** The amount of plutonium remaining from one kilogram after x years is given by the function $W(x) = 2^{-x/24360}$. How much will be left after
  (a) 1000 years?
  (b) 10,000 years?
  (c) 15,000 years?
  (d) Estimate how long it will take for the one kilogram to decay to half its original weight. This may help to explain why nuclear waste disposal is a serious problem.

38. **Health** National health expenditures (in billions of dollars) in year x can be approximated by

$$f(x) = 976.8e^{.06062x},$$

where $x = 0$ corresponds to 1995.* Estimate the health care expenditures in the following years.
  (a) 1999 **(b)** 2001 **(c)** 2005

39. **Social Science** There were fewer than a billion people on earth when Thomas Jefferson died in 1826, and there are now more than six billion. If world population continues to grow as expected, the population (in billions) in year t will be given by the function $P(t) = 4.834(1.01^{(t-1980)})$.† Estimate the world population in the following years.
  (a) 2005 **(b)** 2010 **(c)** 2030
  (d) What will the world population be when you reach 65 years old?

40. **Social Science** Using the function in Exercise 39 and a calculator, estimate the year in which the population will be double what it was in 2005. (*Hint:* If you have a graphing calculator, solve the appropriate equation, as in Example 4(c). If you have a scientific calculator, experiment with various values of t.)

Business *The scrap value of a machine is the value of the machine at the end of its useful life. By one method of calculating scrap value, where it is assumed a constant percentage of value is lost annually, the scrap value S is given by*

$$S = C(1 - r)^n,$$

*Based on projections of the U.S. Healthcare Financing Administration.

†Based on current data and projections by the U.S. Census Bureau.

where C is the original cost, n is the useful life of the machine in years, and r is the constant annual percentage of value lost. Find the scrap value for each of the following machines.

41. Original cost, $54,000; life, 8 years; annual rate of value loss, 12%

42. Original cost, $178,000; life, 11 years; annual rate of value loss, 14%

43. Use the graphs of $f(x) = 2^x$ and $g(x) = 2^{-x}$ (not a calculator) to explain why $2^x + 2^{-x}$ is approximately equal to 2^x when x is very large.

44. Business   For a short period in 2001, Purina Mills' stock price could be approximated by the function

$$f(x) = 8.13(1.3084)^x,$$

where x is measured in months and $x = 0$ corresponds to March 1.

(a) What was the approximate stock price on April 1? On May 15? On June 21?

 (b) Suppose this trend continued. Use a graphing calculator to determine the day when the stock price would have reached $30.

(c) What price does the model predict for today's date? Does this price seem realistic? Check the stock price in today's paper (it's traded on the Nasdaq). What conclusions can you draw?

4.2 APPLICATIONS OF EXPONENTIAL FUNCTIONS

In many situations in biology, economics, and the social sciences, a quantity changes at a rate proportional to the quantity present. In such cases, the amount present at time t is a function of t, called the **exponential growth function.**

Exponential Growth Function

Under normal conditions, growth can be described by a function of the form

$$f(t) = y_0 e^{kt} \quad \text{or} \quad f(t) = y_0 b^t,$$

where $f(t)$ is the amount present at time t, y_0 is the amount present at time $t = 0$, and k and b are constants that depend on the rate of growth.

It is understood that this growth can involve either growing larger or growing smaller. Here is an example of growing smaller.

Suppose the number of bacteria in a culture at time t is

$$y = 500e^{.4t},$$

where t is measured in hours.

(a) How many bacteria are present initially?

(b) How many bacteria are present after 10 hours?

Answers:

(a) 500

(b) About 27,300

▶ **EXAMPLE 1**   Radioactive lead-210 decays to polonium-210. The amount y of radioactive lead-210 at time t is given by $y = y_0 e^{-.032t}$, where t is time in years. How much of an initial 500 grams of lead-210 will remain after 5 years?

Since the initial amount is 500 grams, $y_0 = 500$. Substitute 500 for y_0 and 5 for t in the definition of the function.

$$y = 500e^{-.032(5)} \approx 426$$

About 426 grams will remain. ◀

When a quantity is known to grow exponentially, it is sometimes possible to find a function that models this growth from a small amount of data.

▶ **EXAMPLE 2**   When money is placed in a bank account that pays compound interest, the amount in the account grows exponentially, as we shall see in Chapter 5. Suppose such an account grows from $1000 to $1316 in seven years.

(a) Find a growth function of the form $f(t) = y_0 b^t$ that gives the amount in the account at time t years.

    The value of the account at time $t = 0$ and $t = 7$ are given, that is, $f(0) = 1000$ and $f(7) = 1316$. Solve the first of these equations for y_0:

$$f(0) = 1000$$
$$y_0 b^0 = 1000 \qquad \text{Rule of } f$$
$$y_0 = 1000 \qquad b^0 = 1$$

So the rule of f has the form $f(t) = 1000 b^t$. Now solve the equation $f(7) = 1316$ for b:

$$f(7) = 1316$$
$$1000 b^7 = 1316 \qquad \text{Rule of } f$$
$$b^7 = 1.316 \qquad \text{Divide both sides by 1000.}$$
$$b = (1.316)^{1/7} \approx 1.04 \qquad \text{Take the seventh root of each side.}$$

So the rule of the function is $f(t) = 1000(1.04)^t$.

(b) How much is in the account after 12 years?

$$f(12) = 1000(1.04)^{12} = \$1601.03. \quad \blacktriangleleft \quad ②$$

---

② **Suppose an investment grows exponentially from $500 to $587.12 in three years.**

(a) Find a function of the form $f(t) = y_0 b^t$ that gives the value of the investment after t years.

(b) How much was the investment worth after 10 years?

Answers:

(a) $f(t) = 500(1.055)^t$

(b) $854.07

▶ **EXAMPLE 3**   Infant mortality rates in the United States are shown in the following table.*

Year	Rate	Year	Rate
1920	76.7	1970	20.0
1930	60.4	1980	12.6
1940	47.0	1985	10.6
1950	29.2	1990	9.2
1960	26.0	1995	7.6
		1998	7.2

(a) Let $t = 0$ correspond to 1920. Use the data for 1920 and 1998 to find a function of the form $f(t) = y_0 b^t$ that models this data.

    Since the rate is 76.7 when $t = 0$, we have $y_0 = 76.7$, so that $f(t) = 76.7 b^t$. Since $t = 78$ corresponds to 1998, we have $f(78) = 7.2$, that is,

$$76.7 b^{78} = 72 \qquad \text{Rule of } f$$
$$b^{78} = \frac{7.2}{76.7} \qquad \text{Divide both sides by 76.7.}$$
$$b = \left(\frac{7.2}{76.7}\right)^{\frac{1}{78}} \approx .97 \qquad \text{Take 78th roots on both sides.}$$

Therefore, the function is $f(t) = 76.7(.97)^t$.

---

*Infants under 1 year, deaths per 1000 live births. U.S. National Center for Health Statistics.

(b) Use exponential regression on a graphing calculator to find another model of this data.

The procedure for entering the data and finding the function is the same as for linear regression (just choose "exponential" instead of "linear"), as explained in Section 2.3. Depending on the calculator, one of these functions will be produced:

$$g(t) = 82.14(.9694)^t \quad \text{or} \quad h(t) = 82.14e^{-.03104t}.$$

Both functions give the same values (except for slight differences due to rounding the coefficients displayed above). They fit the data reasonably well, as shown in Figure 4.9.

FIGURE 4.9

(c) Use the preceding models to estimate the mortality rate in 1986 and 2001.

Evaluating each of the models at $t = 66$ and $t = 81$ shows that the models in parts (a) and (b) give similar, but not identical, results.

t	$f(t)$	$g(t)$	$h(t)$
66	10.3	10.6	10.6
81	6.5	6.6	6.6

Other Exponential Models

When a quantity changes exponentially, but does not either grow very large or decrease practically to 0, as in Examples 1–3, different functions are needed.

▶**EXAMPLE 4**   Sales of a new product often grow rapidly at first and then begin to level off with time. Suppose the annual sales of an inexpensive can opener are given by

$$S(x) = 10,000(1 - e^{-.5x}),$$

where $x = 0$ corresponds to the time when the can opener went on the market.

(a) What were the sales in each of the first three years?

At the end of one year ($x = 1$), sales were

$$S(1) = 10,000(1 - e^{-.5(1)}) \approx 3935.$$

Sales in the next two years were

$$S(2) = 10,000(1 - e^{-.5(2)}) \approx 6321 \quad \text{and} \quad S(3) = 10,000(1 - e^{-.5(3)}) \approx 7769.$$

③

Suppose the value of the assets (in thousands of dollars) of a certain company after t years is given by

$$V(t) = 100 - 75e^{-.2t}.$$

(a) What is the initial value of the assets?

(b) What is the limiting value of the assets?

(c) Find the value after 10 years.

(d) Graph $V(t)$.

Answers:

(a) $25,000

(b) $100,000

(c) $89,850

(d)

(b) What were the sales at the end of the tenth year?

$$S(10) = 10,000(1 - e^{-.5(10)}) \approx 9933.$$

(c) Graph the functions S. What does it suggest?

The graph can be obtained by plotting points and connecting them with a smooth curve, or by using a graphing calculator, as in Figure 4.10. The graph indicates that sales will level off after the tenth year to around 10,000 can openers per year. ◀ ③

FIGURE 4.10

A variety of activities can be modeled by **logistic functions,** whose rules are the form

$$f(x) = \frac{c}{1 + ae^{kx}}.$$

The logistic function in the next example is sometimes called a **forgetting curve.**

▶ **EXAMPLE 5** Psychologists have measured people's ability to remember facts that they have memorized. In one such experiment, it was found that the number of facts $N(t)$ remembered after t days was given by

$$N(t) = \frac{10.003}{1 + .0003e^{.8t}}.$$

(a) How many facts were remembered at the beginning of the experiment?
When $t = 0$, then

$$N(0) = \frac{10.003}{1 + .0003e^{.8(0)}} = \frac{10.003}{1.0003} = 10.$$

So 10 facts were remembered at the beginning.

(b) How many facts were remembered after one week? After two weeks?
One and two weeks correspond to $t = 7$ and $t = 14$.

$$N(7) = \frac{10.003}{1 + .003e^{.8(7)}} = \frac{10.003}{1.0811} \approx 9.25$$

$$N(14) = \frac{10.003}{1 + .0003e^{.8(14)}} = \frac{10.003}{22.9391} \approx .44$$

④

In Example 5,

(a) find the number of facts remembered after 10 days;

(b) use the graph to estimate when just 1 fact will be remembered.

Answers:

(a) 5

(b) After about 12 days

So 9 facts were remembered after one week, but effectively none were remembered after two weeks. (.44 is less than "half a fact"). The graph of the function in Figure 4.11 gives a picture of this forgetting process. ◀ ④

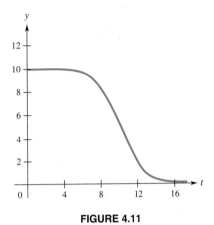

FIGURE 4.11

TECHNOLOGY TIP Many graphing calculators can find a logistic model for appropriate data.

4.2 Exercises

Work these problems. (See Example 1.)

1. **Physical Science** A sealed box contains radium. The number of grams present at time t is given by
$$Q(t) = 100e^{-.00043t},$$
where t is measured in years. Find the amount of radium in the box at the following times.
    (a) $t = 0$
    (b) $t = 800$
    (c) $t = 1600$
    (d) $t = 5000$

2. **Physical Science** The pressure of the atmosphere, $p(h)$, in pounds per square inch, is given by
$$p(h) = p_0e^{-.0000425h},$$
where h is the height above sea level and p_0 is a constant. The pressure at sea level is 15 pounds per square inch.
    (a) Find p_0.
    (b) Find the pressure at an altitude of 6000 feet.

    (c) What would be the pressure encountered by a spaceship at an altitude of 150,000 feet?

3. **Business** The total revenues of the securities industry since 1980 are approximated by the function
$$f(t) = 21.53e^{.13t},$$
where $t = 0$ corresponds to 1980 and $f(t)$ is in billions of dollars. Find the revenues in the following years.
    (a) 1987 7   **(b)** 1994 14   **(c)** 2000 20

4. **Natural Science** The graph on the next page shows how the risk of chromosomal abnormality in a child rises with the age of the mother.*
    (a) Read from the graph the risk of chromosomal abnormality (per 1000) at ages 20, 35, 42, and 49.

_____

*New York Times, February 5, 1994, p. 24.

(b) Verify by substitution that the exponential equation $y = .590e^{.061t}$ "fits" the graph for ages 20 and 35.

(c) Does the equation in part (b) also fit the graph for ages 42 and 49? What does this mean?

Maternity

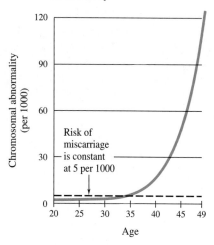

Source: American College of Obstetricians and Gynecologists.

In each of the following problems, find an exponential function of the form $f(t) = y_0 b^t$ to model the data. (See Examples 2 and 3.)

5. Social Sciences   The U.S. Census Bureau predicts that the Hispanic population in the United States will increase from 32.5 million in 2000 to 98.2 million in 2050.*

(a) Find a model for this data, in which $t = 0$ corresponds to 2000.

(b) What is the projected Hispanic population in 2010? In 2035?

(c) By experimenting with different values of t (or by using a graphing calculator to solve an appropriate equation), estimate the year in which the Hispanic population will reach 54 million.

6. Social Sciences   The U.S. Census Bureau predicts that the African-American population will increase from 35.3 million in 2000 to 59.2 million in 2050.*

(a) Find a model for this data, in which $t = 0$ corresponds to 2000.

(b) What is the projected African-American population in 2004? In 2030?

(c) By experimenting with different values of t (or by using a graphing calculator to solve an appropriate equation), estimate the year in which the African-American population will reach 55 million.

7. Finance   Suppose you owe $800 on your credit account and you stop making any payments or new purchases. In

the unlikely event that the credit card company allows this situation to go on for a year, you will then owe $901.46.

(a) Find a model for this data, in which t is measured in *months* and $t = 0$ is the present, when your balance is $800.

(b) If your account does not get turned over to a collection agency (unlikely), how much will you owe after 18 months? After two years?

8. Business   The number of Internet users in the United States was estimated to be 54 million in the fall of 1999 and 85 million in the fall of 2002.*

(a) Find a model for this data, in which t is measured in years with $t = 0$ corresponding to fall 1999 and $f(t)$ is measured in millions.

(b) Use the results of part (a) to determine $f(6)$. Interpret your result.

In the following exercises, find the exponential model as follows. If you do not have suitable technology, use the first and last data points to find a function. (See Examples 2 and 3.) If you have a graphing calculator or other suitable technology, use exponential regression to find a function. (See Example 3.)

9. Business   The table shows the purchasing power of a dollar in various years, with the 1982–1984 average being the base.[†] For example, the entry .600 for 1999 means that a dollar in 1999 bought what 60 cents did in 1982–1984.

Year	1983	1985	1987	1989	1991
Purchasing Power of $1	1.003	.928	.880	.807	.734

Year	1993	1995	1997	1999
Purchasing Power of $1	.692	.656	.623	.600

(a) Find an exponential model for this data, with $t = 0$ corresponding to 1983.

(b) Find the purchasing power of a dollar in 2001, 2003, and 2005.

(c) Use a graphing calculator (or trial and error) to determine the year in which the purchasing power of the 1982–84 dollar will drop to 40 cents.

10. Physical Science   The table at the top of the next column shows the atmospheric pressure (in millibars) at various altitudes (in meters).

Altitude	Pressure
0	1,013
1000	899
2000	795
3000	701
4000	617
5000	541
6000	472
7000	411
8000	357
9000	308
10,000	265

(a) Find an exponential model for this data, in which t is measured in thousands (for instance, $t = 2$ is 2000 meters).

(b) Use the function in part (a) to estimate the atmospheric pressure at 1500 meters and 11,000 meters. Compare your results to the actual values of 846 millibars and 227 millibars, respectively.

11. Health The table shows age-adjusted death rates (per 100,000 population) for heart disease.*

Year	1970	1980	1990	1995	1997
Death Rate	253.6	202.2	152.0	138.3	130.5

(a) Find an exponential model for this data, with $t = 0$ corresponding to 1970.

(b) Estimate the death rate in 1985 and 2000.

(c) Assuming this model remains accurate, use a graphing calculator (or trial and error) to determine the year in which the rate will be 100.

12. Business The table shows outstanding consumer credit (in billions of dollars) in various years.†

Year	1980	1985	1990	1995
Credit	349.4	584.7	789.3	1096.0

Year	1996	1997	1998	1999
Credit	1182.4	1234.1	1300.5	1395.4

(a) Find an exponential model for this data, with $t = 0$ corresponding to 1980.

(b) If this model remains accurate, what will the outstanding consumer credit be in 2003 and 2006?

(c) In what year will consumer credit reach $2500 billion?

Work the following problems. (See Example 4.)

13. Business Assembly line operations tend to have a high turnover of employees, forcing the companies involved to spend much time and effort in training new workers. It has been found that a worker new to the operation of a certain task on the assembly line will produce $P(t)$ items on day t, where

$$P(t) = 25 - 25e^{-.3t}.$$

(a) How many items will be produced on the first day?

(b) How many items will be produced on the eighth day?

(c) What is the maximum number of items, according to the function, the worker can produce?

14. Social Science The number of words per minute that an average typist can type is given by

$$W(t) = 60 - 30e^{-.5t},$$

where t is time in months after the beginning of a typing class. Find each of the following.

(a) $W(0)$

(b) $W(1)$

(c) $W(4)$

(d) $W(6)$

Natural Science *Newton's Law of Cooling says that the rate at which a body cools is proportional to the difference in temperature between the body and an environment into which it is introduced. Using calculus, the temperature $F(t)$ of the body at time t after being introduced into an environment having constant temperature T_0 is*

$$F(t) = T_0 + Cb^t,$$

where C and b are constants. Use this result in Exercises 15 and 16.

15. Boiling water, at 100° Celsius, is placed in a freezer at 0° Celsius. The temperature of the water is 50° Celsius after 24 minutes. Find the temperature of the water after 96 minutes.

16. Paisley refuses to drink coffee cooler than 95°F. She makes coffee with a temperature of 170°F in a room with a temperature of 70°F. The coffee cools to 120°F in 10 minutes. What is the longest time she can let the coffee sit before she drinks the coffee?

---

*U.S. Department of Health and Human Services.

†*Federal Reserve Bulletin.*

17. Social Science A sociologist has shown that the fraction $y(t)$ of people in a group who have heard a rumor after t days is approximated by

$$y(t) = \frac{y_0 e^{kt}}{1 - y_0(1 - e^{kt})},$$

where y_0 is the fraction of people who have heard the rumor at time $t = 0$, and k is a constant. A graph of $y(t)$ for a particular value of k is shown in the figure.
   (a) If $k = .1$ and $y_0 = .05$, find $y(10)$.
   (b) If $k = .2$ and $y_0 = .10$, find $y(5)$.
    (c) Assume the situation in part (b). How many *weeks* will it take for 65% of the people to have heard the rumor?

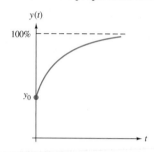

18. Social Science Data from the National Highway Traffic Safety Administration indicate that in year t, the approximate percentage of people in the United States who wear seat belts when driving is given by

$$g(t) = \frac{76.13}{1 + 10.1e^{-.3167t}} \quad (3 \le t \le 18),$$

where $t = 0$ corresponds to 1980. What percentage used seat bealts in
   (a) 1992;   (b) 1994;   (c) 1998?

Assuming this function is accurate after 1998, what percentage of people used seat belts in
   (d) 2000;   (e) 2002;   (f) 2003?
   (g) Graph the function. In California, seat belt usage was 87% in the late 1990s. Does the graph suggest that this will happen nationwide in the near future?

 Use a graphing calculator or other technology to do the following problems. See Example 5.

19. Business The number of people who subscribe to basic cable TV can be approximated by

$$f(x) = \frac{76.7}{1 + 16(.8444)^x},$$

where $x = 0$ corresponds to 1970 and $f(x)$ is in millions.*

(a) Estimate the number of subscribers in 1990, and 2005.
(b) Graph $f(x)$ for the period 1970–2020.
(c) Use the graph to determine the year in which the number of subscribers reached 70 million.
(d) Based on the graph, will the number of subscribers reach 90 million in the foreseeable future?

20. Natural Science The population of fish in a certain lake at time t months is given by the function

$$p(t) = \frac{20,000}{1 + 24(2^{-.36t})}.$$

(a) Graph the population function from $t = 0$ to $t = 48$ (a four-year period).
(b) What was the population at the beginning?
(c) Use the graph to estimate the one-year period in which the population grew most rapidly.
(d) When do you think the population will reach 25,000? What factors in nature might explain your answer?

21. Finance Since 1980, the national debt can be approximated by the logistic model

$$f(x) = \frac{7070.4}{1 + 172.36e^{-.1666x}} \quad (x \ge 20),$$

where $x = 0$ corresponds to 1960 and $f(x)$ is measured in billions of dollars.
(a) Estimate the national debt in 1995, 2000, and 2005.
(b) Graph the function for the period 1980–2030.
(c) Determine when the debt will reach 6.8 trillion dollars (that's 6800 billion).
(d) According to this model, will the national debt level off at any point in the future?

22. Social Science The probability P percent of having an automobile accident is related to the alcohol level of the driver's blood t by the function $P(t) = e^{21.459t}$.
(a) Graph $P(t)$ in a viewing window with $0 \le t \le .2$ and $0 \le P(t) \le 100$.
(b) At what blood alcohol level is the probability of an accident at least 50%? What is the legal blood alcohol level in your state?

*Based on data from *The Cable TV Financial Databook* and *The Pay TV Newsletter*.

4.3 LOGARITHMIC FUNCTIONS

Until the development of computers and calculators, logarithms were the only effective tool for large-scale numerical computations. They are no longer needed for this, but logarithmic functions still play a crucial role in calculus and in many applications.

Common Logarithms

Logarithms are simply *a new language for old ideas*—essentially a special case of exponents.

Definition of Common (Base 10) Logarithms

$$y = \log x \quad \text{means} \quad 10^y = x.$$

"Log x," which is read "the logarithm of *x*," is the answer to the question

To what exponent must 10 be raised to produce *x*?

▶ **EXAMPLE 1**   To find log 10,000 ask yourself, "To what exponent must 10 be raised to produce 10,000?" Since $10^4 = 10,000$, we see that log $10,000 = 4$. Similarly,

$$\log 1 = 0 \quad \text{because} \quad 10^0 = 1;$$

$$\log .01 = -2 \quad \text{because} \quad 10^{-2} = \frac{1}{10^2} = \frac{1}{100} = .01;$$

$$\log \sqrt{10} = 1/2 \quad \text{because} \quad 10^{1/2} = \sqrt{10}. \; ◀ \; ①$$

▶ **EXAMPLE 2**   Log(-25) is the exponent to which 10 must be raised to produce -25. But every power of 10 is positive! So there is no exponent that will produce -25. *Logarithms of negative numbers and 0 are not defined.* ◀

▶ **EXAMPLE 3**   (a) We know that log 359 must be a number between 2 and 3 because $10^2 = 100$ and $10^3 = 1000$. By using the log key on a calculator, we find that log 359 (to four decimal places) is 2.5551. You can verify this by computing $10^{2.5551}$; the result (rounded) is 359. See Figure 4.12.

(b) When 10 is raised to a negative exponent, the result is a number less than 1. Consequently, the logarithms of numbers between 0 and 1 are negative. For instance, log $.026 = -1.5850$, as shown in Figure 4.12. ◀ ②

---

① Find each common logarithm.

(a) log 100

(b) log 1000

(c) log .1

Answers:

(a) 2

(b) 3

(c) −1

② Find each common logarithm.

(a) log 27

(b) log 1089

(c) log .00426

Answers:

(a) 1.4314

(b) 3.0370

(c) −2.3706

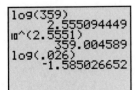

```
log(359)
        2.555094449
10^(2.5551)
        359.004589
log(.026)
        -1.585026652
```

FIGURE 4.12

NOTE On scientific calculators, enter the number followed by the log key. On graphing calculators, press the log key followed by the number, as in Figure 4.12.

Natural Logarithms

Although common logarithms still have some uses (one of which is discussed in Section 4.4), the most widely used logarithms today are defined in terms of the number e (whose decimal expansion begins 2.71828 . . .) rather than 10. They have a special name and notation.

> ### Definition of Natural (Base e) Logarithms
>
> $$y = \ln x \quad \text{means} \quad e^y = x.$$

③

Find the following.

(a) ln 6.1

(b) ln 20

(c) ln .8

(d) ln .1

Answers:

(a) 1.8083

(b) 2.9957

(c) −.2231

(d) −2.3026

Thus the number **ln x** (which is sometimes read "el-en x") is the exponent to which e must be raised to produce the number x. For instance, ln 1 = 0 because $e^0 = 1$. Although logarithms to base e may not seem as "natural" as common logarithms, there are several reasons for using them, some of which are discussed in Section 4.4.

▶ **EXAMPLE 4**   **(a)** To find ln 85 use the [LN] key of your calculator. The result is 4.4427. Thus, 4.4427 is the exponent (to four decimal places) to which e must be raised to produce 85. You can verify this by computing $e^{4.4427}$; the answer (rounded) is 85.

(b) A calculator shows that ln .38 = −.9676 (rounded), which means that $e^{-.9676} \approx .38$.

These result are shown in Figure 4.13.   ◀   ③

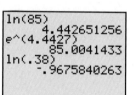

```
ln(85)
         4.442651256
e^(4.4427)
         85.0041433
ln(.38)
         -.9675840263
```

FIGURE 4.13

▶ **EXAMPLE 5**   You don't need a calculator to find ln e^8. Just ask yourself, "To what exponent must e be raised to produce e^8?" The answer, obviously, is 8. So ln $e^8 = 8$.   ◀

Other Logarithms

The procedure used to define common and natural logarithms can be carried out with any positive number $a \neq 1$ as the base in place of 10 or e.

Definition of Logarithms to the Base a

$$y = \log_a x \quad \text{means} \quad a^y = x.$$

Read $y = \log_a x$ as "y is the logarithm of x to the base a." For example, the exponential statement $2^4 = 16$ can be translated into the equivalent logarithmic statement $4 = \log_2 16$. Thus, $\log_a x$ is an *exponent;* it is the answer to the question

To what power must a be raised to produce x?

This key definition should be memorized. It is important to remember the location of the base and exponent in each part of the definition.

Exponent
↓
Logarithmic form: $y = \log_a x$
↑
Base

Exponent
↓
Exponential form: $a^y = x$
↑
Base

Common and natural logarithms are the special cases when $a = 10$ and when $a = e$, respectively. Both $\log u$ and $\log_{10} u$ mean the same thing. Similarly, $\ln u$ and $\log_e u$ mean the same thing.

▶ **EXAMPLE 6** This example shows several statements written in both exponential and logarithmic forms.

Exponential Form	*Logarithmic Form*
(a) $3^2 = 9$	$\log_3 9 = 2$
(b) $(1/5)^{-2} = 25$	$\log_{1/5} 25 = -2$
(c) $10^5 = 100,000$	$\log_{10} 100,000$ (or $\log 100,000$) $= 5$
(d) $4^{-3} = 1/64$	$\log_4(1/64) = -3$
(e) $2^{-4} = 1/16$	$\log_2(1/16) = -4$
(f) $e^0 = 1$	$\log_e 1$ (or $\ln 1$) $= 0$

---

④

Write the logarithmic form of

(a) $5^3 = 125$;

(b) $3^{-4} = 1/81$;

(c) $8^{2/3} = 4$.

Answers:

(a) $\log_5 125 = 3$

(b) $\log_3(1/81) = -4$

(c) $\log_8 4 = 2/3$

---

⑤

Write the exponential form of

(a) $\log_{16} 4 = 1/2$;

(b) $\log_3(1/9) = -2$;

(c) $\log_{16} 8 = 3/4$.

Answers:

(a) $16^{1/2} = 4$

(b) $3^{-2} = 1/9$

(c) $16^{3/4} = 8$

◀ ④ ⑤

Properties of Logarithms

Some of the important properties of logarithms arise directly from their definition.

> Let x and a be any positive real numbers, with $a \neq 1$, and r be any real number. Then
>
> **(a)** $\log_a 1 = 0$ **(b)** $\log_a a = 1$
>
> **(c)** $\log_a a^r = r$ **(d)** $a^{\log_a x} = x$

FIGURE 4.14

Property (a) was discussed in Example 1 (with $a = 10$). Property (c) was illustrated in Example 5 (with $a = e$ and $r = 8$). Property (b) is property (c) with $r = 1$. To understand property (d), recall that $\log_a x$ is the exponent to which a must be raised to produce x. Consequently, when you raise a to this exponent, the result is x, as illustrated for common and natural logarithms in Figure 4.14.

The following properties are part of the reason why logarithms are so useful. They will be used in the next section to solve exponential and logarithmic equations.

> ### The Product, Quotient, and Power Properties
>
> Let x, y, and a be any positive real numbers, with $a \neq 1$. Let r be any real number. Then
>
> **(a)** $\log_a xy = \log_a x + \log_a y$ Product property
>
> **(b)** $\log_a \dfrac{x}{y} = \log_a x - \log_a y$ Quotient property
>
> **(c)** $\log_a x^r = r \log_a x$ Power property

To prove property (a), let

$$m = \log_a x \quad \text{and} \quad n = \log_a y.$$

Then, by the definition of logarithm,

$$a^m = x \quad \text{and} \quad a^n = y.$$

Multiply to get

$$a^m \cdot a^n = x \cdot y,$$

or, by a property of exponents,

$$a^{m+n} = xy.$$

Use the definition of logarithm to rewrite this last statement as

$$\log_a xy = m + n.$$

Replace m with $\log_a x$ and n with $\log_a y$ to get

$$\log_a xy = \log_a x + \log_a y.$$

Properties (b) and (c) can be proved similarly. You can also illustrate these properties on a calculator, as in Figure 4.15.

FIGURE 4.15

▶ **EXAMPLE 7**   Using the properties of logarithms, each of the following can be written as a single logarithm.*

(a) $\log_a x + \log_a(x - 1) = \log_a x(x - 1);$

(b) $\log_a(x^2 + 4) - \log_a(x + 6) = \log_a \dfrac{x^2 + 4}{x + 6};$

(c) $\log_a 9 + 5 \log_a x = \log_a 9 + \log_a x^5 = \log_a 9x^5.$ ◀ ⑥

CAUTION   There is no logarithm property that allows you to simplify the logarithm of a sum, such as $\log_a(x^2 + 4)$. In particular, $\log_a(x^2 + 4)$ is *not* equal to $\log_a x^2 + \log_a 4$. Property (a) of logarithms in Example 7 shows that $\log_a x^2 + \log_a 4 = \log_a 4x^2.$

▶ **EXAMPLE 8**   Assume that $\log_6 7 \approx 1.09$ and $\log_6 5 \approx .90$. Use the properties of logarithms to find each of the following.

(a) $\log_6 35 = \log_6(7 \cdot 5) = \log_6 7 + \log_6 5 \approx 1.09 + .90 = 1.99;$

(b) $\log_6 5/7 = \log_6 5 - \log_6 7 \approx .90 - 1.09 = -.19;$

(c) $\log_6 5^3 = 3 \log_6 5 \approx 3(.90) = 2.70;$

(d) $\log_6 6 = 1;$

(e) $\log_6 1 = 0.$ ◀ ⑦

In Example 8 several logarithms to base 6 were given. However, they could have been found by using a calculator and the following formula.

Change of Base Theorem

For any positive numbers a and x (with $a \neq 1$),

$$\log_a x = \frac{\ln x}{\ln a}.$$

---

*Here and elsewhere, we assume that variable expressions represent positive numbers and that the base a is positive, with $a \neq 1.$

---

⑥

Write as a single logarithm, using the properties of logarithms.

(a) $\log_a 5x + \log_a 3x^4$

(b) $\log_a 3p - \log_a 5q$

(c) $4 \log_a k - 3 \log_a m$

Answers:

(a) $\log_a 15x^5$

(b) $\log_a(3p/5q)$

(c) $\log_a(k^4/m^3)$

---

⑦

Use the properties of logarithms to rewrite and evaluate each of the following, given $\log_3 7 \approx 1.77$ and $\log_3 5 \approx 1.46.$

(a) $\log_3 35$

(b) $\log_3 7/5$

(c) $\log_3 25$

(d) $\log_3 3$

(e) $\log_3 1$

Answers:

(a) 3.23

(b) .31

(c) 2.92

(d) 1

(e) 0

▶ **EXAMPLE 9** To find $\log_7 3$, use the theorem with $a = 7$ and $x = 3$:

$$\log_7 3 = \frac{\ln 3}{\ln 7} \approx \frac{1.0986}{1.9459} \approx .5646.$$

You can check this on your calculator by verifying that $7^{.5646} \approx 3$. ◀

Logarithmic Functions

For a given *positive* value of x, the definition of logarithm leads to exactly one value of y, so that $y = \log_a x$ defines a logarithmic function of base a. (The base a must be positive, with $a \neq 1$.)

If $a > 0$ and a $\neq 1$, the **logarithmic function** with base a is defined as

$$f(x) = \log_a x.$$

The most important logarithmic function is the natural logarithmic function.

▶ **EXAMPLE 10** Graph $f(x) = \ln x$ and $g(x) = e^x$ on the same axes.

For each function, use a calculator to compute some ordered pairs. Then plot the corresponding points and connect them with a curve to obtain the graphs in Figure 4.16.

⑧

Graph $f(x) = \log x$ and $g(x) = 10^x$ on the same axes.

Answer:

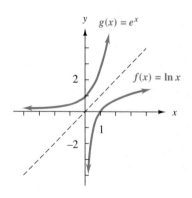

FIGURE 4.16

In Figure 4.16, the dashed straight line is the graph of $y = x$. Observe that the graph of $f(x) = \ln x$ is the mirror image of the graph of $g(x) = e^x$, with the line $y = x$ being the mirror. ◀ ⑧

When the base $a > 1$, the graph of $f(x) = \log_a x$ has the same basic shape as the graph of the natural logarithmic function in Figure 4.16, as summarized below.

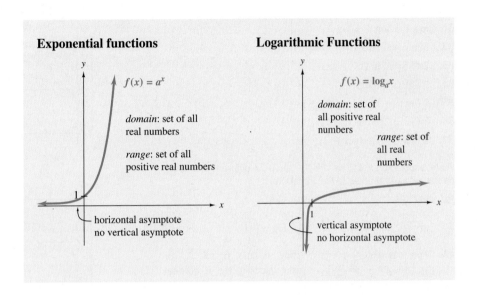

Exponential functions

$f(x) = a^x$

domain: set of all real numbers

range: set of all positive real numbers

horizontal asymptote
no vertical asymptote

Logarithmic Functions

$f(x) = \log_a x$

domain: set of all positive real numbers

range: set of all real numbers

vertical asymptote
no horizontal asymptote

As the information in the box suggests, the graph of $f(x) = \log_a x$ is the mirror image of the graph of $g(x) = a^x$, with the line $y = x$ being the mirror. Functions whose graphs are related in this way are said to be **inverses** of each other. A more complete discussion of inverse functions is given in most standard college algebra books.

4.3 Exercises

Complete each statement in Exercises 1–4.

1. $y = \log_a x$ means $x =$ _____.

2. The statement $\log_5 125 = 3$ tells us that _____ is the power of _____ that equals _____.

3. What is wrong with the expression $y = \log_b$?

4. Logarithms of negative numbers are not defined because _____.

Translate each logarithmic statement into an equivalent exponential statement. (See Examples 1, 5, and 6.)

5. $\log 100,000 = 5$

6. $\log .001 = -3$

7. $\log_3 81 = 4$

8. $\log_2(1/4) = -2$

Translate each exponential statement into an equivalent logarithmic statement. (See Examples 5 and 6.)

9. $10^{1.8751} = 75$

10. $e^{3.2189} = 25$

11. $3^{-2} = 1/9$

12. $16^{1/2} = 4$

Without using a calculator, evaluate each of the following. (See Examples 1, 5, and 6.)

13. $\log 1000$

14. $\log .0001$

15. $\log_5 25$

16. $\log_9 81$

17. $\log_4 64$

18. $\log_6 216$

19. $\log_2 \dfrac{1}{4}$

20. $\log_3 \dfrac{1}{27}$

21. $\ln \sqrt{e}$

22. $\ln(1/e)$

23. $\ln e^{3.78}$

24. $\log 10^{56.9}$

Use a calculator to evaluate each logarithm to three decimal places. (See Examples 3 and 4.)

25. $\log 47$

26. $\log .004$

27. $\ln .351$

28. $\ln 2160$

29. Why does $\log_a 1$ always equal 0 for any valid base a?

Write each expression as the logarithm of a single number or expression. Assume all variables represent positive numbers. (See Example 7.)

30. $\log 15 - \log 3$

31. $\log 4 + \log 8 - \log 2$

32. $3 \ln 2 + 2 \ln 3$

33. $2 \ln 5 - \frac{1}{2} \ln 25$

34. $3 \log x - 2 \log y$

35. $2 \log u + 3 \log w - 6 \log v$

36. $\ln(3x + 2) + \ln(x + 4)$

37. $2 \ln(x + 1) - \ln(x + 2)$

Write each expression as a sum and/or a difference of logarithms with all variables to the first degree.

38. $\log 5x^2 y^3$

39. $\ln \sqrt{6m^4 n^2}$

40. $\ln \dfrac{3x}{5y}$

41. $\log \dfrac{\sqrt{xz}}{z^3}$

42. The calculator-generated table in the figure is for $y_1 = \log(4 - x)$. Why do the values in the y_1 column show ERROR for $x \geq 4$?

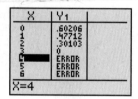

Express each of the following in terms of u and v, where u = ln x and v = ln y. For example, ln x³ = 3(ln x) = 3u.

43. $\ln(x^2 y^5)$

44. $\ln(\sqrt{x} \cdot y^2)$

45. $\ln(x^3/y^2)$

46. $\ln(\sqrt{x}/y)$

Evaluate each expression. (See Example 9.)

47. $\log_6 543$

48. $\log_{20} 97$

49. $\log_{35} 6874$

50. $\log_5 50 - \log_{50} 5$

Find numerical values for b and c for which the given statement is FALSE.

51. $\log(b + c) = \log b + \log c$

52. $\dfrac{\ln b}{\ln c} = \ln\left(\dfrac{b}{c}\right)$

Graph each of the following. (See Example 10.)

53. $y = \ln(x + 2)$

54. $y = \ln x + 2$

55. $y = \log(x - 3)$

56. $y = \log x - 3$

57. Graph $f(x) = \log x$ and $g(x) = \log(x/4)$ for $-2 \leq x \leq 8$. How are these graphs related? How does the quotient rule support your answer?

In Exercises 58 and 59 the coordinates of a point on the graph of the indicated function are displayed at the bottom of the screen. Write the logarithmic and exponential equations associated with the display.

58.

$f(x) = \log x$

59.

$g(x) = \ln x$

60. Match each equation with its graph. Each tick mark represents one unit.

(a) $y = \log x$ (b) $y = 10^x$

(c) $y = \ln x$ (d) $y = e^x$

A.

B.

C.

D.

61. Finance The doubling function

$$D(r) = \frac{\ln 2}{\ln(1 + r)}$$

gives the number of years required to double your money when it is invested at interest rate r (expressed as a decimal), compounded annually. How long does it take to double your money at each of the following rates?

(a) 4%   (b) 8%
(c) 18%   (d) 36%
(e) Round each of your answers in parts (a)–(d) to the nearest year and compare them with these numbers: 72/4, 72/8, 72/18, 72/36. Use this evidence to state a "rule of thumb" for determining approximate doubling time without using the function D. This rule, which has long been used by bankers, is called the *rule of 72*.

62. Health Two people with the flu visited the campus of Big State U. The number of days T that it took for the flu virus to infect n people is given by

$$T = -1.43 \ln\left(\frac{10{,}000 - n}{4998n}\right).$$

How many days will it take for the virus to infect
(a) 500 people?   (b) 5000 people?

63. Business The graph in the next column shows the percent increase in commercial rents in California from 1992 to 1999.
(a) Describe the growth in rents during the period shown in the graph.
(b) The graph can be approximated by the equation $f(x) = -650 + 143 \ln x$, where x is the number of years since 1900 and y is the corresponding percent change in rents. Find $f(92)$ and $f(99)$. Compare your results with the corresponding y-values from the graph.

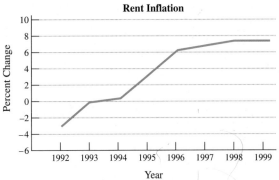

Rent Inflation

Source: CB Commercial/Torto Wheaton Research.

64. Physical Science The barometric pressure p (in inches of mercury) is related to the height h above sea level (in miles) by the equation

$$h = -5 \ln\left(\frac{p}{29.92}\right).$$

The following pressure readings were made by a weather balloon. At what heights were they made?
(a) 29.92 in.   (b) 20.05 in.   (c) 11.92 in.
(d) Use a graphing calculator to determine the pressure at a height of 3 miles.

4.4 LOGARITHMIC AND EXPONENTIAL EQUATIONS

Many applications involve solving logarithmic and exponential equations. So we begin with solution methods for such equations.

Logarithmic Equations

When an equation involves only logarithmic terms, use the logarithm properties to write each side as a single logarithm. Then use this fact.

Let u, v, and a be positive real numbers, with $a \neq 1$.

If $\log_a u = \log_a v$, then $u = v$.

▶ **EXAMPLE 1** Solve $\log x = \log (x + 3) - \log (x - 1)$

First, use the quotient property of logarithms to write the right side as a single logarithm.

$$\log x = \log (x + 3) - \log (x - 1)$$

$$\log x = \log\left(\frac{x + 3}{x - 1}\right)$$

The fact in the preceding box now shows that

$$x = \frac{x + 3}{x - 1}$$

$$x(x - 1) = x + 3$$

$$x^2 - x = x + 3$$

$$x^2 - 2x - 3 = 0$$

$$(x - 3)(x + 1) = 0$$

$$x = 3 \quad \text{or} \quad x = -1$$

Since $\log x$ is not defined when $x = -1$, the only possible solution is $x = 3$. Use a calculator to verify that 3 actually is a solution. ◀ ①

①

Solve each equation.

(a) $\log_2(p + 9) - \log_2 p$
$= \log_2(p + 1)$

(b) $\log_3(m + 1)$
$- \log_3(m - 1) = \log_3 m$

Answers:

(a) 3

(b) $1 + \sqrt{2} \approx 2.414$

When an equation involves constants and logarithmic terms, use algebra and the logarithm properties to write one side as a single logarithm and the other as a constant. Then use the following property of logarithms.

If a and u are positive real numbers, with $a \neq 1$, then

$$a^{\log_a u} = u$$

▶ **EXAMPLE 2** Solve each equation.

(a) $\log_5 (2x - 3) = 2$

Since the logarithm base is 5, raise 5 to the exponents given by the equation.

$$5^{\log_5 (2x - 3)} = 5^2$$

Now use the fact in the preceding box (with $a = 5$ and $u = 2x - 3$) to conclude that

$$2x - 3 = 25$$

$$2x = 28$$

$$x = 14.$$

Verify that 14 is a solution of the original equation.

(b) $\log (x - 16) = 2 - \log (x - 1)$

First rearrange the terms to obtain a single logarithm on the left side.

$$\log (x - 16) + \log (x - 1) = 2$$
$$\log [(x - 16)(x - 1)] = 2 \qquad \text{Product property of logarithms}$$
$$\log (x^2 - 17x + 16) = 2$$

Since the logarithm base is 10, raise 10 to these powers.

$$10^{\log(x^2 - 17x + 16)} = 10^2$$

Apply the logarithm property in the preceding box (with $a = 10$ and $u = x^2 - 17x + 16$).

$$x^2 - 17x + 16 = 100$$
$$x^2 - 17x - 84 = 0$$
$$(x + 4)(x - 21) = 0$$
$$x = -4 \quad \text{or} \quad x = 21$$

When $x = -4$, then in the original equation, $\log (x - 16) = \log (-20)$, which is not defined. So -4 is not a solution. You should verify that 21 is a solution of the original equation.

(c) $\log_2 x - \log_2 (x - 1) = 1$

Proceed as above, with 2 as the logarithm base.

$$\log_2 \frac{x}{x - 1} = 1 \qquad \text{Quotient property of logarithms}$$
$$2^{\log_2 x/(x-1)} = 2^1 \qquad \text{Exponentiate to the base 2.}$$
$$\frac{x}{x - 1} = 2 \qquad \text{Use the fact in the preceding box.}$$
$$x = 2(x - 1) \qquad \text{Multiply both sides by } x - 1.$$
$$x = 2x - 2$$
$$-x = -2$$
$$x = 2$$

Verify that 2 is a solution of the original equation. ②

2

Solve each equation.

(a) $\log_5 x + 2 \log_5 x = 3$

(b) $\log_6 (a + 2)$
$\quad - \log_6 \dfrac{a - 7}{5} = 1$

Answers:

(a) 5

(b) 52

Exponential Equations

An equation in which all the variables are exponents is called an exponential equation. When such an equation can be written as two powers of the same base, it can be solved by using this fact.

Let a be a positive real number, with $a \neq 1$.

If $a^u = a^v$, then $u = v$.

Solve each equation.

(a) $8^{2x} = 4$

(b) $5^{3x} = 25^4$

(c) $36^{-2x} = 6$

Answers:

(a) 1/3

(b) 8/3

(c) −1/4

►EXAMPLE 3   Solve $9^x = 27$.

First, write both sides as powers of the same base. Since $9 = 3^2$ and $27 = 3^3$, we have

$$9^x = 27$$
$$(3^2)^x = 3^3$$
$$3^{2x} = 3^3$$

Apply the fact in the preceding box (with $a = 3$, $u = 2x$, and $v = 3$).

$$2x = 3$$
$$x = \frac{3}{2}$$

Verify that $x = 3/2$ is a solution of the original equation. ◄ ③

Exponential equations involving different bases can often be solved by using the power property of logarithms, which is repeated here for natural logarithms.

If u and r are real numbers, with u positive, then

$$\ln u^r = r \ln u.$$

Although natural logarithms are used in the following examples, logarithms to any base will produce the same solutions.

④

Solve each equation. Round solutions to the nearest thousandth.

(a) $2^x = 7$

(b) $5^m = 50$

(c) $3^y = 17$

Answers:

(a) 2.807

(b) 2.431

(c) 2.579

►EXAMPLE 4   Solve $3^x = 5$.

Take logarithms on both sides.

$$\ln 3^x = \ln 5$$

Apply the power property of logarithms in the preceding box (with $r = x$) to the left side.

$$x \ln 3 = \ln 5$$
$$x = \frac{\ln 5}{\ln 3} \approx 1.465 \qquad \text{Divide both sides by the constant ln 3.}$$

To check, evaluate $3^{1.465}$; the answer should be approximately 5, which verifies that the solution of the given equation is 1.465 to the nearest thousandth. ◄ ④

CAUTION   Be careful; $\dfrac{\ln 5}{\ln 3}$ is *not* equal to $\ln\left(\dfrac{5}{3}\right)$ or $\ln 5 - \ln 3$.

▶ **EXAMPLE 5**   Solve $3^{2x-1} = 4^{x+2}$.

Taking natural logarithms on both sides gives

$$\ln 3^{2x-1} = \ln 4^{x+2}.$$

Now use the power property of logarithms and the fact that $\ln 3$ and $\ln 4$ are constants to rewrite the equation.

$(2x - 1)(\ln 3) = (x + 2)(\ln 4)$	Power property
$2x(\ln 3) - 1(\ln 3) = x(\ln 4) + 2(\ln 4)$	Distributive property
$2x(\ln 3) - x(\ln 4) = 2(\ln 4) + 1(\ln 3)$	Collect terms with x on one side.

Factor out x on the left side to get

$$[2(\ln 3) - \ln 4]x = 2(\ln 4) + \ln 3.$$

Divide both sides by the coefficient of x:

$$x = \frac{2(\ln 4) + \ln 3}{2(\ln 3) - \ln 4}.$$

Using a calculator to evaluate this last expression, we find that

$$x = \frac{2 \ln 4 + \ln 3}{2 \ln 3 - \ln 4} \approx 4.774. \quad \blacktriangleleft \quad ⑤$$

⑤

Solve each equation. Round solutions to the nearest thousandth.

(a) $6^m = 3^{2m-1}$

(b) $5^{6a-3} = 2^{4a+1}$

Answers:

(a) 2.710

(b) .802

Recall that $\ln e = 1$ (because 1 is the exponent to which e must be raised to produce e). This fact simplifies the solution of equations involving powers of e.

▶ **EXAMPLE 6**   Solve $3e^{x^2} = 600$.

First divide each side by 3 to get

$$e^{x^2} = 200.$$

Now take natural logarithms on both sides; then use the power property of logarithms.

$e^{x^2} = 200$	
$\ln e^{x^2} = \ln 200$	
$x^2 \ln e = \ln 200$	Power property
$x^2 = \ln 200$	$\ln e = 1$
$x = \pm\sqrt{\ln 200}$	
$x \approx \pm 2.302$	

The solutions are ± 2.302, rounding to the nearest thousandth. (The symbol $\pm$ is used as a shortcut for writing the two solutions, 2.302 and -2.302.) $\blacktriangleleft$ ⑥

⑥

Solve each equation. Round solutions to the nearest thousandth.

(a) $e^{.1x} = 11$

(b) $e^{3+x} = .893$

(c) $e^{2x^2-3} = 9$

Answers:

(a) 23.979

(b) -3.113

(c) ± 1.612

TECHNOLOGY   TIP Logarithmic and exponential equations can be solved on a graphing calculator. To solve $3^x = 5^{2x-1}$, for example, graph $y = 3^x$ and $y = 5^{2x-1}$ on the same screen. Then use the intersection finder to determine the x-coordinates of their intersection points. Alternatively, graph $y = 3^x - 5^{2x-1}$ and use the root finder to determine the x-intercepts of the graph.

Applications

Some of the most important applications of exponential and logarithmic functions arise in banking and finance. They will be thoroughly discussed in Chapter 5. The applications here are from other fields.

▶ **EXAMPLE 7**   According to projections by the U.S. Census Bureau, the world population (in billions) in year x is approximated by the function $P(x) = 4.834(1.01^{x - 1980})$. When will the population reach 7 billion?

You are asked to find the value of x for which $P(x) = 7$, that is, to solve the equation

$$4.834(1.01^{x - 1980}) = 7$$

$$1.01^{x - 1980} = \frac{7}{4.834} \qquad \text{Divide both sides by 4.834.}$$

$$\ln 1.01^{x - 1980} = \ln\left(\frac{7}{4.834}\right) \qquad \text{Take logarithms on both sides.}$$

$$(x - 1980) \ln 1.01 = \ln\left(\frac{7}{4.834}\right) \qquad \text{Power property of logarithms.}$$

$$x - 1980 = \frac{\ln(7/4.834)}{\ln 1.01} \qquad \text{Divide both sides by } \ln 1.01.$$

$$x = \frac{\ln(7/4.834)}{\ln 1.01} + 1980 \approx 2017.2$$

Hence, the world population will reach 7 billion around 2017. ◀

Use the function in Example 7 to determine when the earth's population will be 10 billion.

Answer:

In 2053

The **half-life** of a radioactive substance is the time it takes for a given quantity to decay to one-half its original mass. The half-life depends only on the substance, not on the size of the sample. It can be shown that the amount of a radioactive substance at time t is given by the function

$$f(t) = y_0\left(\frac{1}{2}\right)^{t/h} = y_0(.5^{t/h}),$$

where y_0 is the initial amount (at time $t = 0$) and h is the half-life of the substance.

Radioactive carbon-14 is found in every living plant and animal. After the plant or animal dies, its carbon-14 decays exponentially, with a half-life of 5730 years. This fact is the basis for a technique called *carbon dating* for determining the age of fossils.

▶ **EXAMPLE 8**   A round wood table hanging in Winchester Castle (England) was alleged to have belonged to King Arthur, who lived in the fifth century. A recent chemical analysis showed that the wood had lost 9% of its carbon-14.* How old is the table?

---

*This is done by measuring the ratio of carbon-14 to nonradioactive carbon-12 in the table (which is approximately constant over long time periods) and comparing it to the ratio in living wood.

The decay function for carbon-14 is

$$f(t) = y_0(.5^{t/5730}),$$

where $t = 0$ corresponds to the time when the wood was cut to make the table (that's when the tree died). Since the wood has lost 9% of its carbon-14, the amount now in the table is 91% of the initial amount y_0, that is, $.91y_0$. We must find the value of t for which $f(t) = .91y_0$. So we must solve the equation

$$y_0(.5^{t/5730}) = .91y_0 \qquad \text{Definition of } f(t)$$
$$.5^{t/5730} = .91 \qquad \text{Divide both sides by } y_0.$$
$$\ln .5^{t/5730} = \ln .91 \qquad \text{Take logarithms on both sides.}$$
$$\left(\frac{t}{5730}\right)\ln .5 = \ln. 91 \qquad \text{Power property of logarithms}$$
$$t \ln .5 = 5730 \ln .91 \qquad \text{Multiply both sides by 5730.}$$
$$t = \frac{5730 \ln .91}{\ln .5} \approx 779.63 \qquad \text{Divide both sides by ln.5.}$$

⑧

> How old is a skeleton that has lost 65% of its carbon-14?
>
> Answer:
>
> About 8679 years

The table is about 780 years old and, therefore, could not have belonged to King Arthur. ◀ ⑧

Earthquakes are often in the news. What you may not have known is that the standard method of measuring their size, the **Richter scale,** is a logarithmic function (base 10).

▶ **EXAMPLE 9**   The intensity $R(i)$ of an earthquake, measured on the Richter scale, is given by

$$R(i) = \log\left(\frac{i}{i_0}\right),$$

where i is the intensity of the ground motion of the earthquake and i_0 is the intensity of the ground motion of the so-called *zero earthquake* (the smallest detectable earthquake, against which others are measured). The 1989 San Francisco earthquake measured 7.1 on the Richter scale.

(a) How did the ground motion of this earthquake compare with that of the zero earthquake?

In this case $R(i) = 7.1$, that is, $\log(i/i_0) = 7.1$. Thus 7.1 is the exponent to which 10 must be raised to produce i/i_0. In other words,

$$10^{7.1} = \frac{i}{i_0}, \quad \text{or equivalently,} \quad i = 10^{7.1}i_0.$$

So this earthquake had $10^{7.1}$ (approximately 12.6 million) times more ground motion than the zero earthquake.

(b) What is the Richter scale intensity of an earthquake with 10 times as much ground motion as the 1989 San Francisco earthquake?

Using the result from (a), the ground motion of such a quake would be

$$i = 10(10^{7.1}i_0) = 10^1 \cdot 10^{7.1}i_0 = 10^{8.1}i_0$$

Find the Richter scale intensity of an earthquake whose ground motion is 100 times greater than the ground motion of the 1989 San Francisco earthquake discussed in Example 9.

Answer:

9.1

so that its Richter scale measure would be

$$R(i) = \log\left(\frac{i}{i_0}\right) = \log\left(\frac{10^{8.1}i_0}{i_0}\right) = \log 10^{8.1} = 8.1.$$

Therefore, a tenfold increase in ground motion increases the Richter scale intensity by just 1.   ◀   ⑨

Earlier, we saw how rational functions describe cost-benefit functions. The final example illustrates another type of cost-benefit function.

▶ **EXAMPLE 10**   One action that government could take to reduce carbon emissions into the atmosphere is to place a tax on fossil fuel. This tax would be based on the amount of carbon dioxide that is emitted into the air when the fuel is burned. The *cost-benefit* equation $\ln(1 - P) = -.0034 - .0053T$ describes the approximate relationship between a tax of T dollars per ton of carbon dioxide and the corresponding percent reduction P (in decimals) of emission of carbon dioxide.*

(a) Write P as a function of T.

We begin by writing the cost-benefit equation in exponential form.

$$\ln(1 - P) = -.0034 - .0053T$$
$$1 - P = e^{-.0034 - .0053T}$$
$$P = P(T) = 1 - e^{-.0034 - .0053T}$$

FIGURE 4.17

A calculator-generated graph of $P(T)$ is shown in Figure 4.17.

(b) Discuss the benefit of continuing to raise taxes on carbon dioxide emissions.

From the graph we see that initially there is a rapid reduction of carbon dioxide emissions. However, after a while there is little benefit in raising taxes further.   ◀

---

*W. Nordhause, "To Slow or Not to Slow: The Economics of the Greenhouse Effect." Yale University, New Haven, Connecticut.

4.4 Exercises

Solve each logarithmic equation. (See Example 1.)

1. $\ln(3x - 1) - \ln(2 + x) = \ln 2$

2. $\ln(8k - 7) - \ln(3 + 4k) = \ln(9/11)$

3. $\ln x + 1 = \ln(x - 4)$

4. $\ln(4x - 2) = \ln 4 - \ln(x - 2)$

5. $2 \ln(x - 3) = \ln(x + 5) + \ln 4$

6. $\ln(k + 5) + \ln(k + 2) = \ln 14k$

Solve each logarithmic equation. (See Example 2.)

7. $\log_3(5x + 1) = 2$

8. $\log_5(9x - 4) = 1$

9. $\log x - \log(x + 3) = -1$

10. $\log m - \log(m - 4) = -2$

11. $\log_3(y + 2) = \log_3(y - 7) + \log_3 4$

12. $\log_8(z - 6) = 2 - \log_8(z + 15)$

13. $\ln(x + 9) - \ln x = 1$

14. $\ln(2x + 1) - 1 = \ln(x - 2)$

15. $\log x + \log(x - 3) = 1$

16. $\log(x - 1) + \log(x + 2) = 1$

Solve each equation for c.

17. $\log(3 + b) = \log(4c - 1)$

18. $\ln(b + 7) = \ln(6c + 8)$

19. $2 - b = \log(6c + 5)$

20. $8b + 6 = \ln(2c) + \ln c$

21. Suppose you overhear the following statement: "I must reject any negative answer when I solve an equation involving logarithms." Is this correct? Write an explanation of why it is or is not correct.

22. What values of x could not possibly be solutions of the following equation?
$$\log_a(4x - 7) + \log_a(x^2 + 4) = 0$$

Solve these exponential equations without using logarithms. (See Example 3.)

23. $3^{x-1} = 9$

24. $16^{-x+1} = 8$

25. $25^{-2x} = 3125$

26. $81^{-2x} = 3^{x-1}$

27. $7^{-x} = 49^{x+3}$

28. $16^x = 64$

29. $\left(\frac{3}{4}\right)^x = \frac{16}{9}$

30. $2^{x^2-4x} = \frac{1}{16}$

Use logarithms to solve these exponential equations. (See Examples 4–6.)

31. $3^x = 5$

32. $5^x = 4$

33. $2^x = 3^{x-1}$

34. $4^{x+2} = 2^{x-1}$

35. $3^{1-2x} = 5^{x+5}$

36. $4^{3x-1} = 3^{x-2}$

37. $e^{2x} = 5$

38. $e^{-3x} = 2$

39. $2e^{5a+2} = 8$

40. $10e^{3z-7} = 5$

Solve each equation for c.

41. $10^{4c-3} = d$

42. $3 \cdot 10^{2c+1} = 4d$

43. $e^{2c-1} = b$

44. $3e^{5c-7} = b$

Solve these equations. (See Examples 1–6.)

45. $\log_5(r + 2) + \log_5(r - 2) = 1$

46. $\log_4(z + 3) + \log_4(z - 3) = 1$

47. $\log_3(a - 3) = 1 + \log_3(a + 1)$

48. $\log w + \log(3w - 13) = 1$

49. $\log_2 \sqrt{2y^2 - 1} = 1/2$

50. $\log_2(\log_2 x) = 1$

51. $\log z = \sqrt{\log z}$

52. $\log x^2 = (\log x)^2$

53. $5^{-2x} = \frac{1}{25}$

54. $5^{x^2+x} = 1$

55. $2^{|x|} = 16$

56. $5^{-|x|} = \frac{1}{25}$

57. $2^{x^2-1} = 12$

58. $3^{2-x^2} = 4$

59. $2(e^x + 1) = 10$

60. $5(e^{2x} - 2) = 15$

61. Explain why the equation $4^{x^2+1} = 2$ has no solutions.

62. Explain why the equation $\log(-x) = -4$ does have a solution and find that solution.

Work these problems. (See Example 7.)

63. Business The number of private cars per one million people in China is approximated by
$$g(x) = 751.635(1.2072^x),$$
where $x = 0$ corresponds to 1990.*
(a) What was the number of cars per million people in 1998 and 2003? (For comparison purposes, there were 768,333 cars per million people in the United States in 1998.)
(b) If this model remains accurate, when will there be 15,000 cars per million people?

64. Business According to data from the World Intellectual Property Organization, the number of patent applications filed worldwide (in millions) during the 1990s is approximated by
$$f(x) = 1.2174e^{.1885x},$$
where $x = 0$ corresponds to 1990.
(a) How many patent applications had been filled by 1995 and by 2000?
(b) If this model remains accurate, when will the number of patent applications reach 18 million?

65. Health A drug's effectiveness decreases over time. If each hour a drug is only 90% as effective as the previous hour, at some point the patient will not be receiving enough medication and must receive another dose. This situation can be modeled by an exponential function with $y = y_0(.90)^{t-1}$. In this equation, y_0 is the amount of the initial dose and y is the percent of medication still available t hours after the drug was administered. Suppose 200 mg of the drug is administered. How long will it take for this initial dose to reach the dangerously low level of 50 mg?

66. Business The U.S. gross domestic product (GDP) in billions of dollars is approximated by
$$f(x) = -6.168 + 5.069 \ln x \qquad (x \geq 10),$$
where $x = 10$ corresponds to 1990.[†] Assuming this model remains accurate, in what year will the GDP be 10 billion dollars?

Work these exercises. (See Example 8.)

67. Natural Science The amount of cobalt-60 (in grams) in a storage facility at time t is given by
$$C(t) = 25e^{-.14t},$$
where time is measured in years.

---

*Chinese Ministry of Communications and the *Economist* magazine.

[†]U.S. Bureau of Economic Analysis.

(a) How much cobalt-60 was present initially?

(b) What is the half-life of cobalt-60? (*Hint:* for what value of t is $C(t) = 12.5$?)

68. Natural Science A Native American mummy was found recently. It had 73.6% of the amount of radiocarbon present in living beings. Approximately how long ago did this person die?

69. Natural Science How old is a piece of ivory that has lost 36% of its radiocarbon?

70. Natural Science A sample from a refuse deposit near the Strait of Magellan had 60% of the carbon 14 of a contemporary living sample. How old was the sample?

Natural Science *Do the following problems. (See Example 9.)*

71. On July 14, 1991, Peshawar, Pakistan, was shaken by an earthquake that measured 6.6 on the Richter scale.

(a) Express this reading in terms of i_0.

(b) In February of the same year, a quake measuring 6.5 on the Richter scale killed about 900 people in the mountains of Pakistan and Afghanistan. Express the intensity of a 6.5 reading in terms of i_0.

(c) How many times more intense was the July earthquake than the February one?

72. Find the Richter scale intensity of earthquakes whose ground motion is

(a) $1000i_0$

(b) $100{,}000i_0$

(c) $10{,}000{,}000i_0$.

(d) Fill in the blank in this statement: Increasing the ground motion by a factor of 10^k increases the Richter intensity by _____ units.

73. The loudness of sound is measured in units called decibels. The decibel rating of a sound is given by

$$D(i) = 10 \cdot \log\left(\frac{i}{i_0}\right),$$

where i is the intensity of the sound and i_0 is the minimum intensity detectable by the human ear (the so-called *threshold sound*). Find the decibel rating of each of the following sounds whose intensities are given. Round answers to the nearest whole number.

(a) Whisper, $115i_0$

(b) Average sound level in the movie *Godzilla*, $10^{10}i_0$

(c) Jackhammer, $31{,}600{,}000{,}000i_0$

(d) Rock music, $895{,}000{,}000{,}000i_0$

(e) Jetliner at takeoff, $109{,}000{,}000{,}000{,}000i_0$

74. (a) How much more intense is a sound that measures 100 decibels than the threshold sound?

(b) How much more intense is a sound that measures 50 decibels than the threshold sound?

(c) How much more intense is a sound measuring 100 decibels than one measuring 50 decibels?

75. Natural Science Refer to Example 10.

(a) Determine the percent reduction in carbon dioxide when the tax is $60.

(b) What tax will cause a 50% reduction in carbon dioxide emissions?

76. Social Science The number of years, $N(r)$, since two independently evolving languages split off from a common ancestral language is approximated by

$$N(r) = -5000 \ln r,$$

where r is the proportion of the words from the ancestral language that is common to both languages now. Find each of the following.

(a) $N(.9)$ (b) $N(.5)$ (c) $N(.3)$

(d) How many years have elapsed since the split if 70% of the words of the ancestral language are common to both languages today?

(e) If two languages split off from a common ancestral language about 1000 years ago, find r.

77. Natural Science In the central Sierra Nevada of California, the percent of moisture that falls as snow rather than rain is approximated reasonably well by

$$p = 86.3 \ln h - 680,$$

where p is the percent of moisture as snow at an altitude of h feet (with $3000 \le h < 8500$).

(a) Graph p.

(b) At what altitude is 50 percent of the moisture snow?

78. Physical Science The table gives some of the planets' average distances D from the sun and their period P of revolution around the sun in years. The distances have been normalized so that Earth is one unit from the sun. Thus, Jupiter's distance of 5.2 means that Jupiter's distance from the sun is 5.2 times farther than Earth's.*

Planet	D	P
Earth	1	1
Jupiter	5.2	11.9
Saturn	9.54	29.5
Uranus	19.2	84.0

(a) Plot the points (D, P) for these planets. Would a straight line or an exponential curve fit these points best?

(b) Plot the points $(\ln D, \ln P)$ for these planets. Do these points appear to lie on a line?

*C. Ronan, *The Natural History of the Universe*, (New York: Macmillan Publishing Co., 1991).

(c) Determine a linear equation that approximates the data points with $x = \ln D$ and $y = \ln P$. Use the first and last data points (rounded to 2 decimal places). Graph your line and the data on the same coordinate axes.

(d) Use the linear equation to predict the period of the planet Pluto if its distance is 39.5. Compare your answer to the true value of 248.5 years.

 Use logarithmic regression for these exercises.

79. Health   The table shows the number of kidney transplants in selected years.*

Year	1985	1990	1995	1996	1997	1998
Transplants	7695	9880	11,901	12,152	12,307	13,139

(a) Find a logarithmic function that models this data with $x = 5$ corresponding to 1985.

(b) Estimate the number of transplants in 2002.

(c) If this model remains accurate, in what year will there be 14,500 transplants?

80. Social Science   The table shows the number of visitors to U.S. national parks (in millions) in selected years.†

Year	1950	1960	1970	1980	1990	1994	1998
Visitors	14	28	46	47	47.7	63	64.5

(a) Find a logarithmic function that models this data with $x = 50$ corresponding to 1950.

(b) Assuming the model remains accurate, estimate the number of visitors in 2000, 2003, and 2005.

(c) In what year will the number of visitors reach 75 million?

*U.S. Department of Health and Human Services and the United Network for Organ Sharing.

†National Park Service.

CHAPTER 4   SUMMARY

Key Terms and Symbols

4.1   exponential function
exponential growth and decay
the number $e \approx 2.71828 \ldots$

4.2   exponential growth function
logistic function

4.3 $\log x$ common (base 10 logarithm) of x
$\ln x$ natural (base e logarithm) of x
$\log_a x$ base a logarithm of x
product, quotient, and power properties of logarithms
change of base theorem

logarithmic function
inverses

4.4   logarithmic equations
exponential equations
half-life
Richter scale

Key Concepts

An important application of exponents is the **exponential growth function,** defined as $f(t) = y_0 e^{kt}$ or $f(t) = y_0 b^t$ where y_0 is the amount of a quantity present at time $t = 0$, $e \approx 2.71828$, and k and b are constants.

The **logarithm** of x to the base a is defined as follows. For $a > 0$ and $a \neq 1$, $y = \log_a x$ means $a^y = x$. Thus, $\log_a x$ is an *exponent*, the power to which a must be raised to produce x.

Properties of Logarithms

Let x, y, and a be positive real numbers, $a \neq 1$, and let r be any real number.

$$\log_a 1 = 0 \qquad\qquad \log_a a = 1$$
$$\log_a a^r = r \qquad\qquad a^{\log_a x} = x$$

Product property: $\log_a xy = \log_a x + \log_a y$

Quotient property: $\log_a \dfrac{x}{y} = \log_a x - \log_a y$

Power property: $\log_a x^r = r \log_a x$

Solving Exponential and Logarithmic Equations

Let $a > 0$, $a \neq 1$.

If $\log_a u = \log_a v$, then $u = v$.

If $a^u = a^v$, $u = v$.

Chapter 4 Review Exercises

Match each equation with the letter of the graph that most closely resembles its graph. Assume that $a > 1$.

1. $y = a^{x+2}$ **2.** $y = a^x + 2$

3. $y = -a^x + 2$ **4.** $y = a^{-x} + 2$

(a)

(b)

(c)

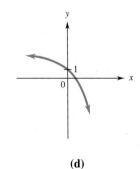

(d)

Consider the exponential function $y = f(x) = a^x$ graphed here. Answer each question based on the graph.

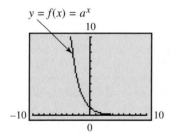

$y = f(x) = a^x$

5. What is true about the value of a in comparison to 1?

6. What is the domain of f?

7. What is the range of f?

8. What is the value of $f(0)$?

Graph each function.

9. $f(x) = 4^x$

10. $g(x) = 4^{-x}$

11. $f(x) = \ln x + 5$

12. $g(x) = \log x - 3$

Work these problems.

13. Business India has become an important exporter of software to the United States. Since 1985 India's exports (in millions of dollars) can be approximated by the function

$$f(x) = 6.2(10^{-12})(1.4^x),$$

where x is the number of years since 1900.
(a) According to this model, what was the value of software exports in 2000?
(b) When will software exports double their value in 2000?

14. Business A person learning certain skills involving repetition tends to learn quickly at first. Then learning tapers off and approaches some upper limit. Suppose the number of symbols per minute a textbook typesetter can produce is given by $p(t) = 250 - 120(2.8)^{-.5t}$, where t is the number of months the typesetter has been in training. Find each of the following:
(a) $p(2)$
(b) $p(4)$
(c) $p(10)$.
(d) Graph $y = p(t)$.

Translate each exponential statement into an equivalent logarithmic one.

15. $10^{1.6721} = 47$

16. $5^4 = 625$

17. $e^{3.6636} = 39$

18. $5^{1/2} = \sqrt{5}$

Translate each logarithmic statement into an equivalent exponential one.

19. $\log 1000 = 3$

20. $\log 16.6 = 1.2201$

21. $\ln 95.4 = 4.5581$

22. $\log_2 64 = 6$

Evaluate each expression without using a calculator.

23. $\ln e^3$

24. $\log \sqrt{10}$

25. $10^{\log 7.4}$

26. $\ln e^{4k}$

27. $\log_8 16$

28. $\log_{25} 5$

Write each expression as a single logarithm. Assume all variables represent positive quantities.

29. $\log 4k + \log 5k^3$

30. $4 \log x - 2 \log x^3$

31. $2 \log b - 3 \log c$

32. $4 \ln x - 2(\ln x^3 + 4 \ln x)$

Solve each equation. Round to the nearest thousandth.

33. $\ln(m + 3) - \ln m = \ln 2$

34. $2 \ln(y + 1) = \ln(y^2 - 1) + \ln 5$

35. $\log(m + 2) = 1$

36. $\log x^2 = 2$

37. $\log_2(3k - 2) = 4$

38. $\log_5\left(\dfrac{5z}{z - 2}\right) = 2$

39. $\log x + \log(x + 3) = 1$

40. $\log_2 r + \log_2(r - 2) = 3$

41. $2^{3x} = \dfrac{1}{8}$

42. $\left(\dfrac{9}{16}\right)^x = \dfrac{3}{4}$

43. $9^{2y-1} = 27^y$

44. $\dfrac{1}{2} = \left(\dfrac{b}{4}\right)^{1/4}$

45. $8^p = 19$

46. $3^z = 11$

47. $5 \cdot 2^{-m} = 35$

48. $2 \cdot 15^{-k} = 18$

49. $e^{-5-2x} = 5$

50. $e^{3x-1} = 12$

51. $6^{2-m} = 2^{3m+1}$

52. $5^{3r-1} = 6^{2r+5}$

53. $(1 + .003)^k = 1.089$

54. $(1 + .094)^z = 2.387$

Work these problems.

55. Business Suppose the gross national product (GNP) of a small country (in millions of dollars) is approximated by the function, $G(t) = 15 + 2 \log t$, where t is time in years, for $1 \le t \le 6$. Find the GNP at the following times.
(a) 1 year (b) 2 years (c) 5 years

56. Natural Science A population is increasing according to the growth law $y = 2e^{.02t}$, where y is in millions and t is in years. Match each of the questions (a), (b), (c), and (d) with one of the solutions (A), (B), (C), or (D).
(a) How long will it take for the population to triple?
   (A) Evaluate $2e^{.02(1/3)}$.
(b) When will the population reach 3 million?
   (B) Solve $2e^{.02t} = 3 \cdot 2$ for t.
(c) How large will the population be in 3 years?
   (C) Evaluate $2e^{.02(3)}$.
(d) How large will the population be in 4 months?
   (D) Solve $2e^{.02t} = 3$ for t.

57. Natural Science The amount of polonium (in grams) present after t days is given by

$$A(t) = 10e^{-.00495t}.$$

(a) How much polonium was present initially?

(b) What is the half-life of polonium?

(c) How long will it take for the polonium to decay to 3 grams?

58. Natural Science   One earthquake measures 4.6 on the Richter scale. A second earthquake has ground motion 1000 times greater than the first. What does the second one measure on the Richter scale?

Natural Science   *Here is another form of Newton's Law of Cooling (see Section 4.2, Exercises 15 and 16):* $F(t) = T_0 + Ce^{-kt}$, *where C and k are constants.*

59. A piece of metal is heated to 300° Celsius and then placed in a cooling liquid at 50° Celsius. After 4 minutes the metal has cooled to 175° Celsius. Find its temperature after 12 minutes.

60. A frozen pizza has a temperature of 3.4° Celsius when it is taken from the freezer and left out in a room at 18° Celsius. After half an hour its temperature is 7.2° Celsius. How long will it take for the pizza to thaw to 10° Celsius?

In Exercises 61 and 62, do part (a) and skip part (b) if you do not have a graphing calculator. If you have a graphing calculator, then do part (b) and skip part (a).

61. Business   The number of music cassette tapes sold each year (in millions) has been steadily dropping, as shown in the table.

Year	1994	1995	1996	1997	1998	1999
Cassettes Sold	345.4	272.6	225.3	172.6	158.5	123.6

(a) Let $x = 0$ correspond to 1994. Use the data points from 1994 and 1999 to find a function of the form $f(x) = a(b^x)$ that models this data.

(b) Use exponential regression to find a function $g(x)$ that models this data, with $x = 0$ corresponding to 1994.

(c) Estimate the number of cassetes sold in 2002.

(d) According to this model, when will cassette sales drop below 50 million?

62. Physical Science   The atmospheric pressure (in millibars) at a given altitude (in thousands of meters) is listed in the table.

Altitude	0	2000	4000	6000	8000	10,000
Pressure	1013	795	617	472	357	265

(a) Use the data points for altitudes 0 and 10,000 to find a function of the form $f(x) = a(b^x)$ that models this data, where altitude x is measured in thousands of meters (for example, $x = 4$ means 4000 meters).

(b) Use exponential regression to find a function $g(x)$ that models this data, with x measured in thousands of meters.

(c) Estimate the pressure at 1500 m and 11,000 m and compare the results to the actual values of 846 and 227 millibars.

(d) At what height is the pressure 500 millibars?

Use a graphing calculator to do the following exercises.

63. Business   The table shows the number of Southwest Airlines passengers (in millions) in selected years.

Year	1988	1990	1992	1994	1996	2000
Passengers	14.9	19.8	27.8	42.7	49.6	63.7

(a) Let $x = 0$ correspond to 1980. If your calculator has the capability, find a logistic model for this data.

(b) If this model remains accurate, when will the number of passengers reach 70 million?

64. Physical Science   The power of personal computers has increased dramatically as a result of the ability to place an increasing number of transistors on a single processor chip. The table lists the number of transistors on some popular computer chips by Intel.*

Year	Chip	Transistors
1971	4004	2,300
1986	386DX	275,000
1989	486DX	1,200,000
1993	Pentium	3,100,000
1995	P6	5,500,000
2000	Pentium 4	42,000,000

(a) Let x be the year, where $x = 0$ corresponds to 1971, and y be the number of transistors. Plot the data.

(b) Would a linear, exponential, or logarithmic function fit this data best?

(c) Use regression to find a model for this data.

(d) Graph the function from part (c) with the data points. Does the function fit the data?

(e) If this model remains accurate, when will the number of transistors be 70 million?

*Data provided by Intel.

Case 3

Characteristics of the Monkeyface Prickleback*

The monkeyface prickleback (*Cebidichthys violaceus*), known to anglers as the monkeyface "eel," is found in rocky intertidal and subtidal habitats ranging from San Quintin Bay, Baja California, to Brookings, Oregon. Pricklebacks are prime targets of the few sports anglers who "poke pole" in the rocky intertidal zone at low tide. Little is known about the life history of this species. The results of a study of the length, weight, and age of this species are discussed in this case.

Data on standard length (*SL*) and total length (*TL*) were collected. Early in the study only *TL* was measured, so a conversion to *SL* was necessary. The equation relating the two lengths, calculated from 177 observations for which both lengths had been measured, is

$$SL = TL(.931) + 1.416.$$

Ages (determined by standard aging techniques) were used to estimate parameters of the von Bertanfany growth model

$$L_t = L_x(1 - e^{-kt}) \qquad (1)$$

where
  L_t = length at age t,
  L_x = asymptotic age of the species,
  k = growth completion rate, and
  t_0 = theoretical age at zero length.

The constants a and b in the model

$$W = aL^b \qquad (2)$$

where
  W = weight in g,
  L = standard length in cm,

were determined using 139 fish ranging from 27 cm and 145 g to 60 cm and 195 g.

Growth curves giving length as a function of age are shown in the accompanying figure. For the data marked opercle, the lengths were computed from the ages using equation (1).

*From William H. Marshall and Tina Wyllie Echeverria, "Characteristics of the Monkeyface Prickleback," *California Fish & Game*, Vol. 78, No. 2, Spring 1992. Copyright 1992. American Association for the Advancement of Science. Reprinted with permission.

Estimated length from equation (1) at a given age was larger for males than females after age eight. See the table below.

Structure/ Sex	Age (yr)	Length (cm)	L_x	k	t_0	n
Otolith						
Est.	2–18	23–67	72	.10	−1.89	91
S.D.			8	.03	1.08	
Opercle						
Est.	2–18	23–67	71	.10	−2.63	91
S.D.			8	.04	1.31	
Opercle– Females						
Est.	0–18	15–62	62	.14	−1.95	115
S.D.			2	.02	.28	
Opercle– Males						
Est.	0–18	13–67	70	.12	−1.91	74
S.D.			5	.02	.29	

Weight/length relationships found with equation (2) are shown in the accompanying figure, along with data from other studies.

Length (cm standard length)

Exercises

1. Use equation (1) to estimate the lengths at ages 4, 11, and 17. Let $L_x = 71.5$ and $k = .1$. Compare your answers with the results in the first figure. What do you find?

2. Use equation (2) with $a = .01289$ and $b = 2.9$ to estimate the weights for lengths of 25 cm, 40 cm, and 60 cm. Compare with the results in the second figure. Are your answers reasonable compared to the curve?

Chapter 5

Mathematics of Finance

It is important for both businesspersons and consumers to understand the mathematics of finance in order to make sound financial decisions. Eventually, almost everyone takes out a loan to buy a car, a household appliance, or a home. We must all plan for eventual retirement, which involves investing in various types of investments, including Individual Retirement Accounts (IRAs), stocks, bonds, and annuities. The interest formulas for borrowing or investing money are introduced in this chapter.

5.1 SIMPLE INTEREST AND DISCOUNT

Interest is the fee paid to use someone else's money. Interest on loans of a year or less is frequently calculated as **simple interest,** which is paid only on the amount borrowed or invested and not on past interest. The amount borrowed or deposited is called the **principal.** The **rate** of interest is given as a percent per year, expressed as a decimal. For example, $6\% = .06$ and $11\frac{1}{2}\% = .115$. The time the money is earning interest is calculated in years. Simple interest is the product of the principal, rate, and time.

Simple Interest

The simple interest, I, on P dollars at a rate of interest r per year for t years is

$$I = Prt.$$

It is customary in financial problems to round interest to the nearest cent.

Find the simple interest for each loan.

(a) $2000 at 8.5% for 10 months

(b) $3500 at 10.5% for $1\frac{1}{2}$ years

Answers:

(a) $141.67

(b) $551.25

▶**EXAMPLE 1**   To furnish her new apartment, Eleanor Chin borrowed $6000 at 9.2% interest for 11 months. How much interest will she pay?

Use the formula, $I = Prt$, with $P = 6000$, $r = .092$, and $t = 11/12$ years.

$$I = 6000(.092)(11/12) = 506,$$

or $506.00 ◀

A deposit of P dollars today at a rate of interest r for t years produces interest of $I = Prt$. The interest, added to the original principal P, gives

$$P + Prt = P(1 + rt).$$

This amount is called the **future value** of P dollars at an interest rate r for time t in years. When loans are involved, the future value is often called the **maturity value** of the loan. This idea is summarized as follows.

Future or Maturity Value for Simple Interest

The **future value** or **maturity value**, A, of P dollars for t years at a rate of interest r per year is

$$A = P(1 + rt).$$

▶**EXAMPLE 2**   Find each maturity value and the amount of interest paid.

(a) Rick borrows $20,000 from his parents at 5.25% to add a room on his house. He plans to repay the loan in 9 months with a bonus he expects to receive at that time.

The loan is for 9 months, or 9/12 of a year, so $t = .75$, $P = 20,000$, and $r = .0525$. Use the formula.

$$\begin{aligned} A &= P(1 + rt) \\ &= 20,000[1 + .0525(.75)] \\ &\approx 20,787.5, \qquad \text{Use a calculator.} \end{aligned}$$

Find each future value.

(a) $1000 at 4.6% for 6 months

(b) $8970 at 11% for 9 months

(c) $95,106 at 9.8% for 76 days

Answers:

(a) $1023

(b) $9710.03

(c) $97,073.64

or $20,787.50. The maturity value is the sum of the principal and interest, so the interest on the loan is

$$20,787.50 - 20,000 = 787.50.$$

(b) A loan of $11,280 for 85 days at 11% interest

It is common to assume 360 days in a year when working with simple interest. We shall usually make such an assumption in this book. The maturity value in this example is

$$A = 11,280\left[1 + .11\left(\frac{85}{360}\right)\right] \approx 11,280[1.0259722] \approx 11,572.97,$$

or $11,572.97. The interest is

$$11,572.97 - 11,280 = 292.97. \quad ◀ \text{②}$$

TECHNOLOGY TIP The table feature of a graphing calculator makes it easy to compare the effect on the future value of a loan of a small difference in the interest rate. The screen in Figure 5.1 shows the portion of a year in the X column and the future value of a $2500 loan with interest rates of 12.1% in the Y_1 column and 11.6% in the Y_2 column. From the table we see that when X = .75 the future value is $9.50 less at 11.6% than at 12.1%.

X	Y₁	Y₂
.7	2711.8	2703
.71	2714.8	2705.9
.72	2717.8	2708.8
.73	2720.8	2711.7
.74	2723.9	2714.6
.75	2726.9	2717.5
.76	2729.9	2720.4

X=.75

FIGURE 5.1

Present Value

A sum of money that can be deposited today to yield some larger amount in the future is called the **present value** of that future amount. Present value refers to the principal to be invested or loaned, so we use the same variable P as we did for principal. In interest problems, P always represents the amount at the beginning of the time period, and A always represents the amount at the end of the time period. To find a formula for P, we begin with the future value formula

$$A = P(1 + rt).$$

Dividing each side by $1 + rt$ gives the following formula for present value.

Present Value for Simple Interest

The **present value** P of a future amount of A dollars at a simple interest rate r for t years is

$$P = \frac{A}{1 + rt}.$$

③

Find the present value of the following future amounts. Assume 6% interest.

(a) $7500 in 1 year

(b) $89,000 in 5 months

(c) $164,200 in 125 days

Answers:

(a) $7075.47

(b) $86.829.27

(c) $160,848.98

▶ **EXAMPLE 3** Find the present value of $32,000 in 4 months at 9% interest.

$$P = \frac{32,000}{1 + (.09)\left(\frac{4}{12}\right)} = \frac{32,000}{1.03} = 31,067.96$$

A deposit of $31,067.96 today, at 9% interest, would produce $32,000 in 4 months. These two sums, $31,067.96 today and $32,000.00 in 4 months, are equivalent (at 9%) because the first amount becomes the second amount in 4 months. ◀ ③

Jerrell Davis is owed $19,500 by Christine O'Brien. The money will be paid in 11 months, with no interest. If the current interest rate is 10%, how much should Davis be willing to accept today in settlement of the debt?

Answer:

$17,862.60

▶**EXAMPLE 4**   Because of a court settlement, Carter Fenton owes $5000 to Jay Beckenstein. The money must be paid in 10 months, with no interest. Suppose Fenton wishes to pay the money today and Beckenstein can invest it at an annual rate of 5%. What amount should he be willing to accept?

The amount that Beckenstein should be willing to accept is given by the present value:

$$P = \frac{5000}{1 + (.05)\left(\frac{10}{12}\right)} = 4800.00.$$

Beckenstein should be willing to accept $4800 today in settlement of the obligation.   ◀ ④

▶**EXAMPLE 5**   Suppose you borrow $40,000 today and are required to pay $41,400 in 4 months to pay off the loan and interest. What is the simple interest rate?

We can use the future value formula, with $P = 40,000$, $A = 41,400$, and $t = 4/12 = 1/3$, and solve for r.

$$A = P(1 + rt)$$

$$41,400 = 40,000\left(1 + r \cdot \frac{1}{3}\right)$$

$$41,400 = 40,000 + \frac{40,000r}{3}$$

$$1400 = \frac{40,000r}{3}$$

$$40,000r = 3 \cdot 1400 = 4200$$

$$r = \frac{4200}{40,000} = .105$$

Therefore, the interest rate is 10.5%.   ◀ ⑤

⑤

You lend a friend $500. She agrees to pay you $520 in 6 months. What is the interest rate?

Answer:

8%

Simple Discount Notes

The loans discussed up to this point are called **simple interest notes,** where interest on the face value of the loan is added to the loan and paid at maturity. Another common type of note, called a **simple discount note,** has the interest deducted in advance from the amount of a loan before giving the *balance* to the borrower. The *full* value of the note must be paid at maturity. The money that is deducted is called the **bank discount** or just the **discount,** and the money actually received by the borrower is called the **proceeds.**

For example, consider a loan of $3000 at 6% interest for 9 months. We can compare the two types of loan arrangements as follows.

	Simple Interest Note	*Bank Discount Note*
Interest on the note	3000(.06)(9/12) = $135	3000(.06)(9/12) = $135
Borrower receives	$3000	$2865
Borrower pays back	$3135	$3000

Kelly Bell signs an agreement at her bank to pay the bank $25,000 in 5 months. The bank charges a 13% discount rate. Find the amount of the discount and the amount Bell actually receives.

Answer:

$1354.17; $23,645.83

▶ **EXAMPLE 6**   Theresa Vasquez needs a loan from her bank and agrees to pay $8500 to her banker in 9 months. The banker subtracts a discount of 12% and gives the balance to Vasquez. Find the amount of the discount and the proceeds.

As shown above, the discount is found in the same way that simple interest is found, except that it is based on the amount to be repaid.

$$\text{Discount} = 8500(.12)\left(\frac{9}{12}\right) = 765.00$$

The proceeds are found by subtracting the discount from the original amount.

$$\text{Proceeds} = \$8500 - \$765.00 = \mathbf{\$7735.00} \quad ◀ \quad ⑥$$

In Example 6, the borrower was charged a discount of 12%. However, 12% is *not* the interest rate paid, since 12% applies to the $8500, while the borrower actually received only $7735. In the next example, we find the rate of interest actually paid by the borrower.

⑦

Refer to problem 6 and find the actual rate of interest paid by Bell.

Answer:

13.7% (to the nearest tenth)

▶ **EXAMPLE 7**   Find the actual rate of interest paid by Vasquez in Example 6.

Use the formula for simple interest, $I = Prt$, with r the unknown. Since the borrower received only $7735 and must repay $8500, $I = 8500 - 7735 = 765$. Here, $P = 7735$ and $t = 9/12 = .75$. Substitute these values into $I = Prt$.

$$I = Prt$$
$$765 = 7735(r)(.75)$$
$$\frac{765}{7735(.75)} = r$$
$$.132 \approx r$$

The actual interest rate paid by the borrower is about 13.2%. ◀ ⑦

Let D represent the amount of discount on a loan. Then $D = Art$, where A is the maturity value of the loan (the amount borrowed plus interest) and r is the stated rate of interest. The amount actually received, the proceeds, can be written as $P = A - D$, or $P = A - Art = A(1 - rt)$.

The formulas for discount are summarized below.

Discount

If D is the discount on a loan having a maturity value A at a rate of interest r for t years, and if P represents the proceeds, then

$$P = A - D \quad \text{or} \quad P = A(1 - rt).$$

▶ **EXAMPLE 8**   John Cross owes $4250 to Jane Fleming. The loan is payable in 1 year at 10% interest. Fleming needs cash to buy a new car, so 3 months before the loan is payable she goes to the bank to have the loan discounted. That is, she sells the loan (note) to the bank. The bank charges an 11% discount fee. Find the amount of cash she will receive from the bank.

First find the maturity value of the loan, the amount (with interest) Cross must pay Fleming. By the formula for maturity value,

$$A = P(1 + rt)$$
$$= 4250[1 + (.10)(1)]$$
$$= 4250(1.10) = 4675,$$

or $4675.00.

The bank applies its discount rate to this total:

Amount of discount $= 4675(.11)(3/12) \approx$ **128.56.**

(Remember that the loan was discounted 3 months before it was due.) Fleming actually receives

$$\$4675 - \$128.56 = \$4546.44$$

in cash from the bank. Three months later, the bank will get $4675 from Cross. ◀ ⑧

---

⑧

A firm accepts a $21,000 note due in 7 months with interest of 10.5%. Suppose the firm discounts the note at a bank 75 days before it is due. Find the amount the firm would receive if the bank charges a 12.4% discount rate. (Use 360 days in a year.)

Answer:

$21,710.52

5.1 Exercises

1. What factors determine the amount of interest earned on a fixed principal?

Find the simple interest. (See Example 1.)

2. $25,000 at 7% for 9 months

3. $3850 at 9% for 8 months

4. $1974 at 6.3% for 7 months

5. $3724 at 8.4% for 11 months

Find the simple interest. Assume a 360-day year and a 30-day month. (See Example 2.)

6. $5147.18 at 10.1% for 58 days

7. $2930.42 at 11.9% for 123 days

8. $7980 at 10%; loan made on May 7 and due September 19

9. $5408 at 12%; loan made on August 16 and due December 30

Find the simple interest. Assume 365 days in a year and use the exact number of days in a month. (Assume 28 days in February.)

10. $7800 at 11%; made on July 7 and due October 25

11. $11,000 at 10%; made on February 19 and due May 31

12. $2579 at 9.6%; made on October 4 and due March 15

13. $37,098 at 11.2%; made on September 12 and due July 30

 14. In your own words, describe the *maturity value* of a loan.

15. What is meant by the *present value* of money?

Find the present value of each future amount. Assume 360 days in a year. (See Examples 3 and 4.)

16. $15,000 for 8 months; money earns 6%

17. $48,000 for 9 months; money earns 5%

18. $15,402 for 125 days; money earns 6.3%

19. $29,764 for 310 days; money earns 7.2%

Find the proceeds for each loan. Assume 360 days in a year. (See Example 6.)

20. $7150; discount rate 12%; length of loan 11 months

21. $9450; discount rate 10%; length of loan 7 months

22. $35,800; discount rate 9.1%; length of loan 183 days

23. $50,900; discount rate 8.2%; length of loan 238 days

24. Why is the discount rate charged on a simple discount note different from the actual interest rate paid on the proceeds?

Find the interest rate to the nearest tenth on the proceeds for the following simple discount notes. (See Example 7.)

25. $6200; discount rate 10%; length of loan 8 months

26. $5000; discount rate 8.1%; length of loan 6 months

27. $58,000; discount rate 10.8%; length of loan 9 months

28. $43,000; discount rate 9%; length of loan 4 months

Finance   *Work the following applied problems.*

29. Anne Kelly borrowed $25,900 from her father to start a flower shop. She repaid him after 11 months, with interest of 8.4%. Find the total amount she repaid.

30. An accountant for a corporation forgot to pay the firm's income tax of $725,896.15 on time. The government charged a penalty of 12.7% interest for the 34 days the money was late. Find the total amount (tax and penalty) that was paid. (Use a 365-day year.)

31. Mike Branson invested his summer earnings of $3000 in a savings account for college. The account pays 2.5% interest. How much will this amount to in 9 months?

32. To pay for textbooks, a student borrows $450 dollars from a credit union at 6.5% simple interest. He will repay the loan in 38 days, when he expects to be paid for tutoring. How much interest will he pay? (Assume a 360-day year.)

33. An account invested in a money market fund grew from $67,081.20 to $67,359.39 in a month. What was the interest rate to the nearest tenth?

34. A $100,000 certificate of deposit held for 60 days is worth $101,133.33. To the nearest tenth of a percent, what interest rate was earned?

35. A stock that sold for $22 at the beginning of the year was selling for $24 at the end of the year. If the stock paid a dividend of $.50 per share, what is the simple interest rate on an investment in this stock? (*Hint:* Consider the interest to be the increase in value plus the dividend.)

36. A bond with a face value of $10,000 in 10 years can be purchased now for $5988.02. What is the simple interest rate?

37. Tuition of $1769 will be due when the spring term begins in 4 months. What amount should a student deposit today, at 6.25%, to have enough to pay the tuition?

38. A firm of accountants has ordered 7 new computers at a cost of $5104 each. The machines will not be delivered for 7 months. What amount could the firm deposit in an account paying 6.42% to have enough to pay for the machines?

39. John Sun Yee needs $6000 to pay for remodeling work on his house. A contractor agrees to do the work in 10 months. How much should Yee deposit at 4.6% to accumulate the $6000 at that time?

40. Lorie Reilly decides to go back to college. For transportation, she borrows money from her parents to buy a small car for $7200. She plans to repay the loan in 7 months. What amount can she deposit today at 5.25% to have enough to pay off the loan?

41. John Matthews signs a $4200 note at the bank. The bank charges a 12.2% discount rate. Find the net proceeds if the note is for 10 months. Find the actual interest rate (to the nearest hundredth) charged by the bank.

42. A building contractor gives a $13,500 note to a plumber. (The plumber loans $13,500 to the contractor.) The note is due in 9 months, with interest of 9%. Three months after the note is signed, the plumber discounts it at the bank. The bank charges a 10.1% discount rate. How much will the plumber receive? Will it be enough to pay a bill for $13,582?

43. Shalia Johnson owes $7000 to the Eastside Music Shop. She has agreed to pay the amount in 7 months at an interest rate of 10%. Two months before the loan is due, the store needs $7350 to pay a wholesaler's bill. The bank will discount the note at a rate of 10.5%. How much will the store receive? Is it enough to pay the bill?

44. Fay, Inc., received a $30,000, six-month, 12% interest-bearing note from a customer.* The note was discounted the same day by Carr National Bank at 15%. The amount of cash received by Fay from the bank was

  (a) $30,000   **(b)** $29,550

  (c) $29,415   **(d)** $27,750.

45. Many credit card companies target entering college freshmen with attractive interest rates. What they do not emphasize, however, is that the low rate will be in effect for only a short time, changing automatically to a much higher rate, often in just a few months. Freshman Michael Branson accepts a credit card offer at 8% annual interest on any unpaid charges. The rate will increase to 15% annually after 6 months. To furnish his dorm room, he charges a small refrigerator, some bedding, and personal supplies amounting to $457.80. When he receives his first month's bill, he finds that he can pay just a minimum amount of $87.50.

  (a) If Michael pays the minimum amount and makes no additional charges, how much will he owe the following month?

  (b) Suppose Michael continues to pay the same minimum amount for 4 more months. What will he owe at the end of the next month?

  (c) The minimum amount normally is reduced as the total charge is reduced. Suppose each month, the minimum is reduced by $15 and Michael continues to pay only the minimum. Can he pay off the charges in 6 months? If not, what will he owe at the end of 6 months?

---

*Uniform CPA Examination, May 1989, American Institute of Certified Public Accountants.

5.2 COMPOUND INTEREST

Use the formula

$$A = P(1 + rt)$$

to find the amount in the account after 2 years at 5% simple interest.

Answer:

$1100

Simple interest is normally used for loans or investments of a year or less. For longer periods *compound interest* is used. With **compound interest,** interest is charged (or paid) on interest as well as on principal. For example, if $1000 is deposited at 5% compounded annually, then the interest for the first year is $1000(.05) = $50, just as with simple interest, so that the account balance is $1050 at the end of the year. During the second year, interest is paid on the entire $1050 (not just on the original $1000 as with simple interest), so the amount in the account at the end of the second year is $1050 + $1050(.05) = $1102.50. This is more than simple interest would produce.

To find a formula for compound interest, suppose that P dollars are deposited at interest rate r per year. The amount A on deposit after 1 year is found by the simple interest formula.

$$A = P[1 + r(1)] = P(1 + r)$$

If the deposit earns compound interest, the interest for the second year is paid on the total amount on deposit at the end of the first year, $P(1 + r)$. Using the formula $A = P(1 + rt)$ again, with $P(1 + r)$ (in place of P) and $t = 1$ gives the total amount on deposit at the end of the second year.

$$A = [P(1 + r)](1 + r \cdot 1) = P(1 + r)^2$$

In the same way, the total amount on deposit at the end of the third year is

$$P(1 + r)^3.$$

Continuing in this way, the total amount on deposit after t years is

$$A = P(1 + r)^t,$$

which is called the **compound amount.**

NOTE Compare this formula for compound interest with the formula for simple interest from the previous section.

Compound interest $A = P(1 + r)^t$

Simple interest $A = P(1 + rt)$

The important distinction between the two formulas is that in the compound interest formula, the number of years t is an *exponent,* so that money grows much more rapidly when interest is compounded.

Spreadsheets are ideal for performing financial calculations. Figure 5.2 shows a Microsoft Excel spreadsheet with the formulas for compound and simple interest used to create columns B and C, respectively, when $1000 is invested at an annual rate of 10%. Notice how rapidly the compound amount increases compared to the maturity value with simple interest. For more details on the use of spreadsheets in the mathematics of finance, see *The Spreadsheet Manual* that is available with this book.

	A	B	C
1	period	compound	simple
2	1	1100	1100
3	2	1210	1200
4	3	1331	1300
5	4	1464.1	1400
6	5	1610.51	1500
7	6	1771.561	1600
8	7	1948.7171	1700
9	8	2143.58881	1800
10	9	2357.947691	1900
11	10	2593.74246	2000
12	11	2853.116706	2100
13	12	3138.428377	2200
14	13	3452.271214	2300
15	14	3797.498336	2400
16	15	4177.248169	2500
17	16	4594.972986	2600
18	17	5054.470285	2700
19	18	5559.917313	2800
20	19	6115.909045	2900
21	20	6727.499949	3000

FIGURE 5.2

Interest can be compounded more than once a year. Common **compounding periods** include

semiannually (two periods per year),
quarterly (four periods per year),
monthly (twelve periods), and
daily (usually 365 periods per year).

To find the *interest rate per period, i,* we *divide* the annual interest rate r by the number of compounding periods, m, per year. The total number of compounding periods, n, is found by *multiplying* the number of years t by the number of compounding periods per year, m. Then the general formula for compound interest can be derived in much the same way as the formula given earlier.

Compound Interest

If P dollars are deposited for n compounding periods at a rate of interest i per period, the compound amount (or future value) A is

$$A = P(1 + i)^n.$$

In particular, if the annual interest rate r is compounded m times per year and the number of years is t, then $i = r/m$ and $n = mt$.

Suppose $17,000 is deposited at 4% compounded annually for 11 years.

(a) Find the compound amount.

(b) Find the amount of interest earned.

Answers:

(a) $26,170.72

(b) $9170.72

Find the compound amount.

(a) $10,000 at 8% compounded quarterly for 7 years

(b) $36,000 at 6% compounded monthly for 2 years

Answers:

(a) $17,410.24

(b) $40,577.75

▶**EXAMPLE 1**   Suppose $1000 is deposited for 6 years in an account paying 8.31% per year compounded annually.

(a) Find the compound amount.

In the formula above, $P = 1000$, $i = .0831$, and $n = 6$. The compound amount is

$$A = P(1 + i)^n$$
$$A = 1000(1.0831)^6$$
$$A = \$1614.40.$$

(b) Find the amount of interest earned.

Subtract the initial deposit from the compound amount.

Amount of interest $= \$1614.40 - \$1000 = \$614.40$ ◀

▶**EXAMPLE 2**   Find the amount of interest earned by a deposit of $2450 for 6.5 years at 5.25% compounded quarterly.

Interest compounded quarterly is compounded 4 times a year. In 6.5 years, there are $6.5(4) = 26$ periods. Thus, $n = 26$. Interest of 5.25% per year is 5.25/4 per quarter, so $i = .0525/4$. Now use the formula for compound amount.

$$A = P(1 + i)^n$$
$$A = 2450(1 + .0525/4)^{26}$$

$$= 3438.784190$$

Rounded to the nearest cent, the compound amount is $3438.78, so the interest is $3438.78 - \$2450 = \$988.78$.   ◀ ③

CAUTION   As shown in Example 2, compound interest problems involve two rates—the nominal or stated annual rate r and the rate i per compounding period. Be sure you understand the distinction between them. When interest is compounded annually, these rates are the same. In all other cases, $i \neq r$.

The more often interest is compounded within a given time period, the more interest will be earned. Surprisingly, however, there is a limit on the amount of interest, no matter how often it is compounded. To see this, suppose that $1 is invested at 100% interest per year, compounded n times per year. Then the interest rate (in decimal form) is 1.00 and the interest rate per period is $1/n$. According to the formula (with $P = 1$), the compound amount at the end of 1 year will be

$$A = \left(1 + \frac{1}{n}\right)^n.$$

A computer gives the following results for various values of n.

Interest Is Compounded	n	$\left(1 + \dfrac{1}{n}\right)^n$
Annually	1	$\left(1 + \dfrac{1}{1}\right)^1 = 2$
Semiannually	2	$\left(1 + \dfrac{1}{2}\right)^2 = 2.25$
Quarterly	4	$\left(1 + \dfrac{1}{4}\right)^4 \approx 2.4414$
Monthly	12	$\left(1 + \dfrac{1}{12}\right)^{12} \approx 2.6130$
Daily	365	$\left(1 + \dfrac{1}{365}\right)^{365} \approx 2.71457$
Hourly	8760	$\left(1 + \dfrac{1}{8760}\right)^{8760} \approx 2.718127$
Every minute	525,600	$\left(1 + \dfrac{1}{525,600}\right)^{525,600} \approx 2.7182792$
Every second	31,536,000	$\left(1 + \dfrac{1}{31,536,000}\right)^{31,536,000} \approx 2.7182818$

Because interest is rounded to the nearest penny, the compound amount never exceeds $2.72, no matter how big n is. (If you try computing the values in the table with your calculator, you may notice that your answers do not agree exactly. This is because of round-off error.)

The table above suggests that as n takes larger and larger values, then the corresponding values of $\left(1 + \dfrac{1}{n}\right)^n$ get closer and closer to a specific real number, whose decimal expansion begins 2.71828. . . . This is indeed the case, as is shown in calculus, and the number 2.71828 . . . is denoted e.

The preceding example is typical of what happens when interest is compounded n times per year, with larger and larger values of n. It can be shown that no matter what interest rate or principal is used, there is always an upper limit on the compound amount, which is called the compound amount from **continuous compounding.**

Continuous Compounding

The compound amount A for a deposit of P dollars at interest rate r per year compounded continuously for t years is given by

$$A = Pe^{rt}.*$$

Most calculators have an e^x key for computing powers of e. See Chapter 4 for details on using a calculator to evaluate e^x.

---

*Other applications of the exponential function $f(x) = e^x$ are discussed in Chapter 4.

Find the compound amount, assuming continuous compounding.

(a) $12,000 at 10% for 5 years

(b) $22,867 at 7.2% for 9 years

Answers:

(a) $19,784.66

(b) $43,715.15

▶ **EXAMPLE 3** Suppose $5000 is invested at an annual rate of 4% compounded continuously for 5 years. Find the compound amount.

In the formula for continuous compounding, let $P = 5000$, $r = .04$, and $t = 5$. Then a calculator with an e^x key shows that

$$A = 5000e^{(.04)5} = 5000e^{.2} \approx \$6107.01.$$

You can readily verify that daily compounding would have produced a compound amount about 6¢ less. ◀ ④

Effective Rate

If $1 is deposited at 4% compounded quarterly, a calculator can be used to find that at the end of one year, the compound amount is $1.0406, an increase of 4.06% over the original $1. The actual increase of 4.06% in the money is somewhat higher than the stated increase of 4%. To differentiate between these two numbers, 4% is called the **nominal** or **stated rate** of interest, while 4.06% is called the **effective rate.** To avoid confusion between stated rates and effective rates, we shall continue to use r for the stated rate and we will use r_E for the effective rate.*

Find the effective rate corresponding to a nominal rate of

(a) 12% compounded monthly;

(b) 8% compounded quarterly.

Answers:

(a) 12.68%

(b) 8.24%

▶ **EXAMPLE 4** Find the effective rate corresponding to a stated rate of 6% compounded semiannually.

A calculator shows that $100 at 6% compounded semiannually will grow to

$$A = 100\left(1 + \frac{.06}{2}\right)^2 = 100(1.03)^2 = \$106.09.$$

Thus, the actual amount of compound interest is $106.09 − $100 = $6.09. Now if you earn $6.09 interest on $100 in 1 year with annual compounding, your rate is $6.09/100 = .0609 = 6.09\%$. Thus, the effective rate is $r_E = 6.09\%$. ◀ ⑤

If you examine Example 4, you see that the effective rate is $\dfrac{106.09 - 100}{100}$, that is,

$$\frac{100\left(1 + \frac{.06}{2}\right)^2 - 100}{100} = \frac{100\left[\left(1 + \frac{.06}{2}\right)^2 - 1\right]}{100} = \left(1 + \frac{.06}{2}\right)^2 - 1$$

The same argument works with any values of P, r, and m in place of 100, .06 and 2 and leads to this formula.

Effective Rate

The effective rate corresponding to a stated rate of interest r compounded m times per year is

$$r_E = \left(1 + \frac{r}{m}\right)^m - 1.$$

---

*When applied to consumer finance, the effective rate is called the annual percentage rate, APR, or annual percentage yield, APY.

Using the effective rate formula, a graphing calculator generated the table in Figure 5.3, in which the interest rate r is 10%, Y_1 denotes the effective rate, and X is the number of compoundings. The table shows how the effective rate increases with more frequent compounding.

X	Y1
1	.1
2	.1025
4	.10381
6	.10426
8	.10449
10	.10462
12	.10471

X=4

FIGURE 5.3

6

Find the effective rate corresponding to a nominal rate of

(a) 8% compounded monthly;

(b) 10% compounded quarterly.

Answers:

(a) 8.30%

(b) 10.38%

▶ **EXAMPLE 5** A bank pays interest of 4.9% compounded monthly. Find the effective rate.

Use the formula given above with $r = .049$ and $m = 12$. The effective rate is

$$r_E = \left(1 + \frac{.049}{12}\right)^{12} - 1$$
$$= 1.050115575 - 1 \approx .0501,$$

or 5.01%. ◀ **6**

▶ **EXAMPLE 6** Bank A is now lending money at 10% interest compounded annually. The rate at Bank B is 9.6% compounded monthly, and the rate at Bank C is 9.7% compounded quarterly. If you need to borrow money, at which bank will you pay the least interest?

Compare the effective rates.

Bank A: $\left(1 + \frac{.10}{1}\right)^1 - 1 = .10 = 10\%$

Bank B: $\left(1 + \frac{.096}{12}\right)^{12} - 1 \approx .10034 = 10.034\%$

Bank C: $\left(1 + \frac{.097}{4}\right)^4 - 1 \approx .10057 = 10.057\%$

The lowest effective interest rate is at Bank A, which has the highest nominal rate. ◀

Present Value for Compound Interest

The formula for compound interest, $A = P(1 + i)^n$, has four variables, A, P, i, and n. Given the values of any three of these variables, the value of the fourth can be found. In particular, if A (the future amount), i, and n are known, then P can be found. Here P is the amount that should be deposited today to produce A dollars in n periods.

▶**EXAMPLE 7** Keisha Jones must pay a lump sum of $6000 in 5 years. What amount deposited today at 6.2% compounded annually will amount to $6000 in 5 years?

Here $A = 6000$, $i = .062$, $n = 5$, and P is unknown. Substituting these values into the formula for the compound amount gives

$$6000 = P(1.062)^5$$

$$P = \frac{6000}{(1.062)^5} = 4441.49,$$

or $4441.49. If Jones leaves $4441.49 for 5 years in an account paying 6.2% compounded annually, she will have $6000 when she needs it. To check your work, use the compound interest formula with $P = \$4441.49$, $i = .062$, and $n = 5$. You should get $A = \$6000.00$. ◀ ⑦

As Example 7 shows, $6000 in 5 years is (approximately) the same as $4441.49 today (if money can be deposited at 6.2% annual interest). An amount that can be deposited today to yield a given amount in the future is called the *present value* of the future amount. By solving $A = P(1 + i)^n$ for P, we get the following general formula for present value.

Present Value for Compound Interest

The **present value** of A dollars compounded at an interest rate i per period for n periods is

$$P = \frac{A}{(1 + i)^n} \quad \text{or} \quad P = A(1 + i)^{-n}.$$

NOTE Since this is just the compound interest formula solved for P, it is not necessary to remember a new formula for present value. You can use the compound interest formula, if you understand what each of the variables represents.

▶**EXAMPLE 8** Find the present value of $16,000 in 9 years if money can be deposited at 6% compounded semiannually.

In 9 years there are $2 \cdot 9 = 18$ semiannual periods. A rate of 6% per year is 3% in each semiannual period. Apply the formula with $A = 16,000$, $i = .03$, and $n = 18$.

$$P = \frac{A}{(1 + i)^n} = \frac{16,000}{(1 + .03)^{18}} \approx \frac{16,000}{1.702433} \approx 9398.31$$

A deposit of $9398.31 today, at 6% compounded semiannually, will produce a total of $16,000 in 9 years. ◀ ⑧

The formula for compound amount can also be solved for n.

---

⑦

Find P in Example 7 if the interest rate is

(a) 6%;

(b) 10%.

Answers:

(a) $4483.55

(b) $3725.53

Find the present value in Example 8 if money is deposited at 10% compounded semiannually.

Answer:

$6648.33

▶ **EXAMPLE 9**   Suppose the general level of inflation in the economy averages 8% per year. Find the number of years it would take for the overall level of prices to double.

To find the number of years it will take for $1 worth of goods or services to cost $2, we must solve for n in the equation

$$2 = 1(1 + .08)^n,$$

where $A = 2$, $P = 1$, and $i = .08$. This equation simplifies to

$$2 = (1.08)^n.$$

To solve for n, we will use base 10 logarithms as in Chapter 4.

$$\log 2 = \log (1.08)^n \qquad \text{Take the logarithm of each side.}$$
$$\log 2 = n \log 1.08 \qquad \text{Property (c) of logarithms}$$
$$n = \frac{\log 2}{\log 1.08} \qquad \text{Divide both sides by log 1.08.}$$
$$n \approx 9.01$$

To check, use a calculator to get $1.08^{9.01} = 2.00$ to the nearest hundredth. ◀

When interest is compounded continuously, the present value can be found by solving the continuous compounding formula $A = Pe^{rt}$ for P.

▶ **EXAMPLE 10**   Assuming a continuous inflation rate of 3.5%, how much did a $100 item today cost 4 years ago?

Here the future value is $A = 100$, $r = .035$, and $t = 4$. We want to find the present value P.

$$A = Pe^{rt}$$
$$100 = Pe^{(.035)4}$$
$$100 = Pe^{.14}$$
$$P = \frac{100}{e^{.14}} \approx \$86.94$$

The item cost $86.94 four years ago. ◀ ⑩

Using a calculator, find the number of years it will take for $500 to increase to $750 in an account paying 6% interest compounded semiannually.

Answer:

About 7 years

⑩

Find the value of a $100 item in 3 years at a continuous inflation rate of 4%.

Answer:

$112.75

Summary

At this point, it seems helpful to summarize the notation and the most important formulas for simple and compound interest. We use the following variables.

P = principal or present value

A = future or maturity value

r = annual (stated or nominal) interest rate

t = number of years

m = number of compounding periods per year

i = interest rate per period ($i = r/m$)

n = total number of compounding periods ($n = tm$)

r_E = effective rate

Simple Interest	*Compound Interest*	*Continuous Compounding*
$A = P(1 + rt)$	$A = P(1 + i)^n$	$A = Pe^{rt}$
$P = \dfrac{A}{1 + rt}$	$P = \dfrac{A}{(1 + i)^n} = A(1 + i)^{-n}$	$P = \dfrac{A}{e^{rt}}$
	$r_E = \left(1 + \dfrac{r}{m}\right)^m - 1$	

5.2 Exercises

1. In the compound interest formula in the box on page 239, what is the difference between r and i? between t and n?

2. Explain the difference between simple interest and compound interest.

3. What factors determine the amount of interest earned on a fixed principal?

4. In your own words, describe the *maturity value* of a loan.

5. What is meant by the *present value* of money?

6. If interest is compounded more than once per year, which rate is higher, the stated rate or the effective rate?

Find the compound amount for each of the following deposits. (See Examples 1 and 2.)

7. $1000 at 6% compounded annually for 8 years

8. $1000 at 7% compounded annually for 10 years

9. $470 at 10% compounded semiannually for 12 years

10. $15,000 at 6% compounded semiannually for 11 years

11. $6500 at 12% compounded quarterly for 6 years

12. $9100 at 8% compounded quarterly for 4 years

Find the amount of interest earned by each of the following deposits. (See Examples 1 and 2.)

13. $26,000 at 7% compounded annually for 5 years

14. $32,000 at 5% compounded annually for 10 years

15. $8000 at 4% compounded semiannually for 6.4 years

16. $2500 at 4.5% compounded semiannually for 8 years

17. $5124.98 at 6.3% compounded quarterly for 5.2 years

18. $27,630.35 at 7.1% compounded quarterly for 3.7 years

Find the compound amount if $25,000 is invested at 6% compounded continuously for the following number of years. (See Example 3.)

19. 1

20. 5

21. 10

22. 15

23. 20

24. How do the nominal or stated interest rate and the effective interest rate differ?

Find the effective rate corresponding to the following nominal rates. (See Examples 4–6.)

25. 4% compounded semiannually

26. 8% compounded quarterly

27. 8% compounded semiannually

28. 10% compounded semiannually

29. 12% compounded semiannually

30. 12% compounded quarterly

Find the present value of the following future amounts. (See Examples 7 and 8.)

31. $12,000 at 5% compounded annually for 6 years

32. $8500 at 6% compounded annually for 9 years

33. $4253.91 at 6.8% compounded semiannually for 4 years

34. $27,692.53 at 4.6% compounded semiannually for 5 years

35. $17,230 at 4% compounded quarterly for 10 years

36. $5240 at 8% compounded quarterly for 8 years

37. If money can be invested at 8% compounded quarterly, which is larger: $1000 now or $1210 in 5 years? Use present value to decide.

38. If money can be invested at 6% compounded annually, which is larger: $10,000 now or $15,000 in 10 years? Use present value to decide.

Finance *Work the following applied problems.*

39. A small business borrows $50,000 for expansion at 12% compounded monthly. The loan is due in 4 years. How much interest will the business pay?

40. A developer needs $80,000 to buy land. He is able to borrow the money at 10% per year compounded quarterly. How much will the interest amount to if he pays off the loan in 5 years?

41. A company has agreed to pay $2.9 million in 5 years to settle a lawsuit. How much must they invest now in an account paying 8% compounded monthly to have that amount when it is due?

42. Bill Poole wants to have $20,000 available in 5 years for a down payment on a house. He has inherited $15,000. How much of the inheritance should he invest now to accumulate the $20,000, if he can get an interest rate of 8% compounded quarterly?

43. An advertisement for WingspanBank.com boasted the "highest CD yield in the universe," with an APY (or effective yield) of 7.35%.* The actual rate was not stated. Given that interest was compounded monthly, find the actual rate.

44. According to a financial Web site, on October 16, 2000, DeepGreen Bank of Cleveland, Ohio, paid 6.96% interest, compounded daily, on a one-year CD, while Flagstar Bank of Bloomfield Hills, Michigan, paid 7.01% compounded quarterly.† What are the effective rates for the two CDs, and which bank pays a higher effective rate?

45. A New York bank offered the following special on CD (certificate of deposit) rates. The rates are annual percentage yield, or effective rates, which are higher than the corresponding nominal rates. Assume quarterly compounding. Solve for r to approximate the corresponding nominal rates to the nearest hundredth.

Term	6 mo	1 yr	18 mo	2 yr	3 yr
APY (%)	5.00	5.30	5.45	5.68	5.75

46. The pie graph shows the percent of baby boomers ages 46 to 49 who said they had investments with a total value as shown in each category.‡ Note that 30% say they have saved less than $10,000 and 28% don't know or gave no

answer. Assume the money is invested at an average rate of 8% compounded quarterly for 20 years, when this age group will be ready for retirement. Find the range of amounts each group (except the "don't know or no answer" group) in the graph will have saved for retirement if no more is added.

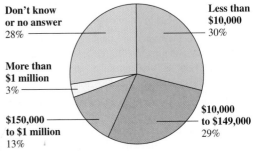

Note: Figures add to more than 100% because of rounding.

Sources: Census Bureau (age distribution); Merrill Lynch Baby Boom Retirement Index (investments); William M. Mercer Inc. (life expectancy).

47. Scott Silva borrowed $5200 from his friend Joe Vetere to buy computer equipment. He repaid the loan 10 months later with simple interest at 7%. Vetere then invested the proceeds in a 5-year certificate of deposit paying 6.3% compounded quarterly. How much will he have at the end of 5 years? (*Hint:* You need to use both simple and compound interest.)

48. Lora Reilly has inherited $10,000 from her uncle's estate. She will invest the money for 2 years. She is considering two investments: a money market fund that pays a guaranteed 5.25% interest compounded monthly or a 2-year Treasury note at 6% annual interest. Which investment pays the most interest over the 2-year period?

49. You decide to invest a $16,000 bonus in a money market fund that guarantees a 5.5% annual interest rate compounded monthly for 7 years. A one-time fee of $30 is charged to set up the account. In addition there is an annual administrative charge of 1.25% of the balance in the account at the end of each year.
 (a) How much is in the account at the end of the first year?
 (b) How much is in the account at the end of the seventh year?

50. Two partners agree to invest equal amounts in their business. One will contribute $10,000 immediately. The other plans to contribute an equivalent amount in 3 years, when she expects to acquire a large sum of money. How much should she contribute at that time to match her partner's investment now, assuming an interest rate of 6% compounded semi-annually?

51. As the prize in a contest, you are offered $1000 now or $1210 in 5 years. If money can be invested at 6% compounded annually, which is larger?

New York Times, October 15, 2000, p. 39.

†www.bankrate.com.

‡*New York Times,* December 31, 1995, sec. 3, p.5.

Use the approach in Example 9 to find the time it would take for the general level of prices in the economy to double at the average annual inflation rates in Exercises 52 and 53.

52. 4%   **53.** 5%

54. The consumption of electricity has increased historically at 6% per year. If it continues to increase at this rate indefinitely, find the number of years before the electric utilities will need to double their generating capacity.

55. Suppose a conservation campaign coupled with higher rates causes the demand for electricity to increase at only 2% per year, as it has recently. Find the number of years before the utilities will need to double generating capacity.

56. An article in the *New York Times* discussed how long it will take for Bill Gates, the world's second richest person (behind the Sultan of Brunei), to become the world's first trillionaire.* His birthday is October 28, 1955, and on July 16, 1997, he was worth $42 billion. (*Note:* A trillion dollars is 1000 billion dollars.)

(**a**) Assume that Bill Gates' fortune grows at an annual rate of 58%, the historical growth rate through 1997 of Microsoft stock, which made up most of his wealth in 1997. Find the age at which he becomes a trillionaire. (*Hint:* Use the formula for interest compounded annually, $A = P(1 + i)^n$, with $P = 42$. Graph the future value as a function of n on a graphing calculator and find where the graph crosses the line $y = 1000$.)

(**b**) Repeat part (a) using 10.9% growth, the average return on all stocks since 1926.

(**c**) What rate of growth would be necessary for Bill Gates to become a trillionaire by the time he is eligible for Social Security on January 1, 2022, after he has turned 66?

(**d**) An article on September 19, 1999, gave Bill Gates's wealth as roughly $90 billion.[†] What was the rate of growth of his wealth between the 1997 and 1999 articles?

57. In 1995, O. G. McClain of Houston, Texas, mailed a $100 check to a descendant of Texas independence hero Sam Houston to repay a $100 debt of McClain's great-great-grandfather, who died in 1835, to Sam Houston.[‡] A bank estimated the interest on the loan to be $420 million for

the 160 years it was due. Find the interest rate the bank was using, assuming interest is compounded annually.

58. In the New Testament, Jesus commends a widow who contributed 2 mites to the temple treasury (Mark 12:42–44). A mite was worth roughly 1/8 of a cent. Suppose the temple had invested those 2 mites at 4% interest compounded quarterly. How much would the money be worth 2000 years later?

59. Suppose $10,000 is invested at an annual rate of 5% for 10 yr. Find the future value if interest is compounded as follows.

(**a**) annually   (**b**) quarterly
(**c**) monthly   (**d**) daily (365 days)

60. In Exercise 59, notice that as the money is compounded more often, the compound amount becomes larger and larger. Is it possible to compound often enough so that the compound amount is $17,000 after 10 years? Explain.

The following exercises are from professional examinations.

61. On January 1, 1980, Jack deposited $1000 into Bank X to earn interest at the rate of j per annum compounded semiannually. On January 1, 1985, he transferred his account to Bank Y to earn interest at the rate of k per annum compounded quarterly. On January 1, 1988, the balance at Bank Y was $1990.76. If Jack could have earned interest at the rate of k per annum compounded quarterly from January 1, 1980, through January 1, 1988, his balance would have been $2203.76. Which of the following represents the ratio k/j?*

(**a**) 1.25   (**b**) 1.30   (**c**) 1.35   (**d**) 1.40
(**e**) 1.45

62. On January 1, 1987, Tone Company exchanged equipment for a $200,000 noninterest bearing note due on January 1, 1990. The prevailing rate of interest for a note of this type at January 1, 1987, was 10%. The present value of $1 at 10% for three periods is 0.75. What amount of interest revenue should be included in Tone's 1988 income statement?[†]

(**a**) $7500   (**b**) $15,000   (**c**) $16,500
(**d**) $20,000

New York Times, July 20, 1997, sec. 4, p. 2.

†*New York Times,* September 19, 1999, WK rev., p. 2.

‡*New York Times,* March 30, 1995.

---

*Problem from "Course 140 Examination, Mathematics of Compound Interest" of the Education and Examination Committee of The Society of Actuaries. Reprinted by permission of The Society of Actuaries.

†Uniform CPA Examination, May 1989, American Institute of Certified Public Accountants.

5.3 FUTURE VALUE OF AN ANNUITY AND SINKING FUNDS

So far in this chapter, only lump sum deposits and payments have been discussed. Many financial situations, however, involve a sequence of equal payments at regular intervals, such as weekly deposits in a savings account or monthly payments on a mortgage or car loan. To develop formulas to deal with periodic payments like these, we first discuss *sequences*.

Geometric Sequences

If a and r are fixed nonzero numbers, then the infinite list of numbers, $a, ar, ar^2, ar^3, ar^4, \ldots$ is called a **geometric sequence.** For instance, if $a = 5$ and $r = 2$, we have the sequence

$$5, 5 \cdot 2, 5 \cdot 2^2, 5 \cdot 2^3, 5 \cdot 2^4, \ldots,$$

or

$$5, 10, 20, 40, 80, \ldots.$$

In the sequence $a, ar, ar^2, ar^3, ar^4, \ldots$, the number a is the *first term* of the sequence, ar the *second term*, ar^2 the *third term*, ar^3 the *fourth term*, and so on. Thus, for any $n \geq 1$,

$$ar^{n-1} \text{ is the } n\text{th term of the sequence.}$$

Each term in the sequence is r times the preceding term. The number r is called the **common ratio** of the sequence.

▶ **EXAMPLE 1** Find the seventh term of the geometric sequence with $a = 6$ and $r = 4$. Using ar^{n-1}, with $n = 7$, the seventh term is

$$ar^6 = 6(4)^6 = 6(4096) = 24{,}576. \quad ◀$$

TECHNOLOGY TIP Some graphing calculators have the ability to produce a list of the terms of a sequence, given the expression for the nth term. With other calculators you can use the TABLE feature by entering the expression for the nth term as a function. The first four terms of the sequence discussed in Example 1 are shown in Figure 5.4.

FIGURE 5.4

Write the first four terms of the geometric sequence with $a = 5$ and $r = -2$. Then find the seventh term.

Answer:

$5, -10, 20, -40; 320$

▶EXAMPLE 2 $100 is deposited in an account that pays interest of 10% compounded annually. How much is in the account at the beginning of each of the first five years?

According to the compound interest formula (with $P = 100$ and $i = .1$), the value of the account at the beginning of each of the first five years is given by the first five terms of the geometric sequence with $a = 100$ and $r = 1.1$. The first five terms are

$$100, \quad 100(1.1), \quad 100(1.1)^2, \quad 100(1.1)^3, \quad 100(1.1)^4,$$

so the value of the account at the beginning of each of the five years is

$$\$100, \quad \$110, \quad \$121, \quad \$133.10, \quad \$146.41. \quad ◀ \ ①$$

The sum S_n of the first n terms of a geometric sequence can be written as

$$S_n = a + ar + ar^2 + ar^3 + \cdots + ar^{n-1}. \tag{1}$$

If $r = 1$, this sum is

$$S_n = \underbrace{a + a + a + \cdots + a}_{n \text{ terms}} = na.$$

To derive a formula for S_n when $r \neq 1$, multiply both sides of equation (1) by r, obtaining

$$rS_n = ar + ar^2 + ar^3 + ar^4 + \cdots + ar^n. \tag{2}$$

Now subtract corresponding sides of equation (1) from equation (2), and solve the equation for S_n.

$$rS_n = \qquad ar + ar^2 + ar^3 + \cdots + ar^{n-1} + ar^n$$
$$\underline{-S_n = -(a + ar + ar^2 + ar^3 + \cdots + ar^{n-1})}$$
$$rS_n - S_n = ar^n - a$$
$$S_n(r - 1) = a(r^n - 1)$$
$$S_n = \frac{a(r^n - 1)}{r - 1}$$

Sum of Terms

If a geometric sequence has first term a and common ratio r, then the **sum of the first n terms** is given by

$$S_n = \frac{a(r^n - 1)}{r - 1}, \qquad r \neq 1.$$

②

(a) Find S_4 and S_7 for the geometric sequence 5, 15, 45, 135, ...

(b) Find S_5 for the geometric sequence having $a = -3$ and $r = -5$.

Answers:

(a) $S_4 = 200$, $S_7 = 5465$

(b) $S_5 = -1563$

▶ **EXAMPLE 3** Find the sum of the first six terms of the geometric sequence 3, 12, 48, ...

Here $a = 3$ and $r = 4$. Find S_6 by the result above.

$$S_6 = \frac{3(4^6 - 1)}{4 - 1} \qquad \text{Let } n = 6, a = 3, r = 4.$$

$$= \frac{3(4096 - 1)}{3}$$

$$= 4095 \quad ◀ \quad ②$$

TECHNOLOGY TIP Graphing calculators with sequence capability can also find the sum of the first n terms of a sequence, given the expression for the nth term. In Example 3, for instance, the sequence is given by $ar^{n-1} = 3(4^{n-1})$. The sum of the first six terms is shown in Figure 5.5.

FIGURE 5.5

Ordinary Annuities

A sequence of equal payments made at equal periods of time is called an **annuity.** If the payments are made at the end of the time period, and if the frequency of payments is the same as the frequency of compounding, the annuity is called an **ordinary annuity.** The time between payments is the **payment period,** and the time from the beginning of the first payment period to the end of the last period is called the **term of the annuity.** The **future value of the annuity,** the final sum on deposit, is defined as the sum of the compound amounts of all the payments, compounded to the end of the term.

Two common uses of annuities are to accumulate funds for some goal or to withdraw funds from an account. For example, an annuity may be used to save money for a large purchase, such as an automobile, an expensive trip, or a down payment on a home. An annuity also may be used to provide monthly payments for retirement. We explore these options in this and the next section.

For example, suppose $1500 is deposited at the end of the year for the next 6 years in an account paying 8% per year compounded annually. Figure 5.6 on the next page shows this annuity schematically.

Term of annuity

End of year

| | 1 | 2 | 3 | 4 | 5 | 6 |

Period 1   Period 2   Period 3   Period 4   Period 5   Period 6

$1500 $1500 $1500 $1500 $1500 $1500

↑—— The $1500 is deposited at the end of the year.

FIGURE 5.6

To find the future value of this annuity, look separately at each of the $1500 payments. The first of these payments will produce a compound amount of

$$1500(1 + .08)^5 = 1500(1.08)^5.$$

Use 5 as the exponent instead of 6 since the money is deposited at the *end* of the first year and earns interest for only 5 years. The second payment of $1500 will produce a compound amount of $1500(1.08)^4$.

The computation of the compound amounts for the remaining payments is shown in Figure 5.7. (Note that the last payment earns no interest.)

③

Complete these steps for an annuity of $2000 at the end of each year for 3 years. Assume interest of 6% compounded annually.

(a) The first deposit of $2000 produces a total of _____.

(b) The second deposit becomes _____.

(c) No interest is earned on the third deposit, so the total in the account is _____.

Answers:

(a) $2247.20

(b) $2120.00

(c) $6367.20

FIGURE 5.7

Figure 5.7 shows that the total amount after six years is the sum

$$1500 + 1500 \cdot 1.08^1 + 1500 \cdot 1.08^2 + 1500 \cdot 1.08^3$$
$$+ 1500 \cdot 1.08^4 + 1500 \cdot 1.08^5,$$

which is the sum of the first six terms of a geometric sequence with $a = 1500$ and $r = 1.08$. Applying the sum formula with $n = 6$, we see that the sum is

$$\frac{a(r^n - 1)}{r - 1} = \frac{1500[(1.08)^6 - 1]}{1.08 - 1} = \$11,003.89. \quad ③$$

To generalize this result, suppose that payments of R dollars each are deposited into an account at the *end of each period* for *n periods,* at a rate of interest *i per period.* The first payment of R dollars will produce a compound amount of $R(1 + i)^{n-1}$ dollars, the second payment will produce $R(1 + i)^{n-2}$ dollars, and so on; the final payment earns no interest and contributes just R dollars to the total. If S represents the future value of the annuity, then (as shown in Figure 5.8),

$$S = R + R(1 + i)^1 + R(1 + i)^2 + \cdots + R(1 + i)^{n-1}.$$

FIGURE 5.8

This result is the sum of the first n terms of the geometric sequence having first term R and common ratio $1 + i$. Using the formula for the sum of the first n terms of a geometric sequence,

$$S = \frac{R[(1 + i)^n - 1]}{(1 + i) - 1} = \frac{R[(1 + i)^n - 1]}{i} = R\left[\frac{(1 + i)^n - 1}{i}\right].$$

The quantity in brackets is sometimes written $s_{\overline{n}|i}$ (read "s-angle-n at i"), so that

$$S = R \cdot s_{\overline{n}|i}.$$

We summarize the preceding work as follows.*

Future Value of an Ordinary Annuity

The future value S of an ordinary annuity is given by

$$S = R\left[\frac{(1 + i)^n - 1}{i}\right] \quad \text{or} \quad S = R \cdot s_{\overline{n}|i},$$

where
S is future value,
R is the payment at the end of each period,
i is the interest rate per period,
n is the number of periods.

T E C H N O L O G Y   T I P   A calculator will be very helpful in computations with annuities. The TI-83 graphing calculator has a special finance menu that is designed to give any desired result after the basic information is entered. A similar finance package for TI-86 can be downloaded from the TI Web site. If your graphing calculator does not have this feature, it probably can be programmed to evaluate the formulas introduced in this section and the next. These programs are in the Program Appendix, which is available at www.aw.com/MWA8.

---

*We use S for the future value here, instead of A as in the compound interest formula, to help avoid confusing the two formulas.

Johnson Building Materials deposits $2500 at the end of each year into an account paying 8% per year compounded annually. Find the total amount on deposit after

(a) 6 years;

(b) 10 years.

Answers:

(a) $18,339.82

(b) $36,216.41

▶ **EXAMPLE 4** Chris Webber is an athlete who feels that his playing career will last 7 years. To prepare for his future, he deposits $22,000 at the end of each year for 7 years in an account paying 6% compounded annually. How much will he have on deposit after 7 years?

His payments form an ordinary annuity with $R = 22,000$, $n = 7$, and $i = .06$. The future value of this annuity (by the formula above) is

$$S = 22,000\left[\frac{(1.06)^7 - 1}{.06}\right] = \$184,664.43. \quad ◀ \ (4)$$

Sinking Funds

A fund set up to receive periodic payments is called a **sinking fund.** The periodic payments, together with the interest earned by the payments, are designed to produce a given sum at some time in the future. For example, a sinking fund might be set up to receive money that will be needed to pay off the principal on a loan at some future time. If the payments are all the same amount and are made at the end of a regular time period, they form an ordinary annuity.

▶ **EXAMPLE 5** Experts say that the baby boom generation (born between 1946 and 1960) cannot count on a company pension or Social Security to provide a comfortable retirement, as their parents did. It is recommended that they start to save early and regularly. Michael Boezi, a baby boomer, has decided to deposit $200 at the end of each month in a sinking fund that pays interest of 7.2% compounded monthly for retirement in 20 years.

(a) How much will be in the account at that time?

This savings plan is an annuity with $R = 200$, $i = .072/12$, and $n = 12(20)$. The future value is

$$S = 200\left[\frac{1 + (.072/12)^{12(20)} - 1}{.072/12}\right] = 106,752.47,$$

or $106,752.47.

(b) Michael believes he needs to accumulate $130,000 in the 20-year period to have enough for retirement. What interest rate would provide that amount?

Once again we have an annuity with $R = 200$ and $n = 12(20)$. Now the future value $S = 130,000$ is given and we must find the interest rate. If x is the annual interest rate in decimal form, then the interest rate per month is $i = x/12$ and we have

$$R\left[\frac{(1 + i)^n - 1}{i}\right] = S$$

$$200\left[\frac{(1 + x/12)^{12(20)} - 1}{x/12}\right] = 130,000. \qquad (1)$$

This last equation is difficult to solve algebraically. We can get a rough approximation by using a calculator to compute the left side of the equation for various values of x, to see which one comes closest to producing $130,000. For example, $x = .08$ produces 117,804 and $x = .09$ produces 133,577, so the solution is between .08 and .09. Further experimentation of this type leads to the approximate solution

(5)

Pete's Pizza deposits $5800 at the end of each quarter for 4 years.

(a) Find the final amount on deposit if the money earns 6.4% compounded quarterly.

(b) Pete wants to accumulate $110,000 in the four-year period. What interest rate (to the nearest tenth) will be required?

Answers:

(a) $104,812.44
(b) 8.9%

$x = .0879$, that is, 8.79%. A more exact approximation can be obtained with a graphing calculator (see the following Technology Tip). ◀ (5)

TECHNOLOGY TIP To solve equation (1) in Example 5 with a graphing calculator, graph the equations

$$y = 200\left[\frac{[(1 + x/12)^{12(20)} - 1]}{x/12}\right] \quad \text{and} \quad y = 130,000$$

on the same screen and find the intersection of their graphs, as in Figure 5.9(a).

The same result can be obtained on the TI-83 without graphing, by using the TVM solver in the FINANCE menu. Enter 240 for N (the number of periods), 0 for PV (present value), −200 for PMT (negative because the money is being paid out), 130,000 for FV (future value), and 12 for P/Y (payments per year). Put the cursor next to I% (interest rate) and press SOLVE. The result, shown in Figure 5.9(b) indicates that an interest rate of 8.79% is needed.

(a)   **(b)**

FIGURE 5.9

▶ **EXAMPLE 6** The Chinns are close to retirement. They agree to sell an antique urn to the local museum for $17,000. Their tax adviser suggests that they defer receipt of this money until they retire, 5 years in the future. (At that time, they might well be in a lower tax bracket.) Find the amount of each payment the museum must make into a sinking fund so that it will have the necessary $17,000 in 5 years. Assume that the museum can earn 6% compounded annually on its money. Also, assume that the payments are made annually.

These payments are the periodic payments into an ordinary annuity. The annuity will amount to $17,000 in 5 years at 6% compounded annually. Using the formula and a calculator,

$$17,000 = R\left[\frac{(1.06)^5 - 1}{.06}\right]$$

$$17,000 \approx R(5.637093)$$

$$R \approx \frac{17,000}{5.637093} = 3015.74,$$

or $3015.74. If the museum management deposits $3015.74 at the end of each year for 5 years in an account paying 6% compounded annually, it will have the total amount needed. This result is shown in the following sinking fund table. In these tables the last payment may differ slightly from the others because of rounding in the preceding lines.

Francisco Arce needs $8000 in 6 years so that he can go on an archaeological dig. He wants to deposit equal payments at the end of each quarter so that he will have enough to go on the dig. Find the amount of each payment if the bank pays

(a) 12% compounded quarterly;

(b) 8% compounded quarterly.

Answers:

(a) $232.38

(b) $262.97

Payment Number	Amount of Deposit	Interest Earned	Total in Account
1	$3015.74	$0	$3,015.74
2	3015.74	180.94	6,212.42
3	3015.74	372.75	9,600.91
4	3015.74	576.05	13,192.70
5	3015.74	791.56	17,000.00

To construct the table, notice that the first payment does not earn interest until the second payment is made. Line 2 of the table shows the second payment, 6% interest of $180.94 on the first payment, and the sum of these amounts added to the total in line 1. Line 3 shows the third payment, 6% interest of $372.75 on the total from line 2, and the new total found by adding these amounts to the total in line 2. This procedure is continued to complete the table. ◀ ⑥

Annuities Due

The formula developed above is for *ordinary annuities*—those with payments at the *end* of each time period. These results can be modified slightly to apply to **annuities due**—annuities where payments are made at the *beginning* of each time period.

To find the future value of an annuity due, treat each payment as if it were made at the end of the *preceding* period; that is, find $s_{\overline{n}|i}$ for *one additional period*. No payment is made at the end of the last period, so to compensate for this, subtract the amount of one payment. Thus, the **future value of an annuity due** of n payments of R dollars each at the beginning of consecutive interest periods with interest compounded at the rate of i per period is

$$S = R\left[\frac{(1+i)^{n+1} - 1}{i}\right] - R \quad \text{or} \quad S = R \cdot s_{\overline{n+1}|i} - R.$$

⑦

(a) Ms. Black deposits $800 at the beginning of each 6-month period for 5 years. Find the final amount if the account pays 6% compounded semiannually.

(b) Find the final amount if this account were an ordinary annuity.

Answers:

(a) $9446.24

(b) $9171.10

▶ **EXAMPLE 7** Payments of $500 are made at the beginning of each quarter for 7 years in an account paying 12% compounded quarterly.

(a) Find the future value of this annuity due.

In 7 years there are 28 quarterly periods. Add one period to get $n = 29$ and use the formula with $i = 12\%/4 = 3\%$.

$$S = 500\left[\frac{(1.03)^{29} - 1}{.03}\right] - 500$$

$$S \approx 500(45.21885) - 500 = \$22,109.43$$

The account will contain a total of $22,109.43 after 7 years.

(b) Compare the result from part (a) with the future value of an ordinary annuity. In this case the future value is given by

$$S = 500\left[\frac{1.03^{28} - 1}{.03}\right] = \$21,465.46.$$

The future amount is a little less because the payments are each made one month later than the payments for an annuity due. ◀ ⑦

5.3 Exercises

1. In your own words, describe a geometric sequence.

2. What is the nth term of a geometric sequence with first term a and common ratio r?

3. According to T. Rowe Price Associates, a person with a moderate investment strategy and n years to retirement should have accumulated savings of a_n percent of his or her annual salary. The geometric sequence with

$$a_n = 1276(.916)^n$$

gives the appropriate percent for each year n.
(a) Find a_1 and r.
(b) Find and interpret the terms a_{10} and a_{20}.

4. Refer to Exercise 3. For someone who has a conservative investment strategy with n years to retirement, the geometric sequence is

$$a_n = 1278(.935)^n.*$$

(a) Repeat part (a) of Exercise 3.
(b) Repeat part (b) of Exercise 3.
(c) Why are the answers in parts (a) and (b) larger than in Exercise 3?

Find the fourth term of each of the following geometric sequences. (See Example 1.)

5. $a = 5, r = 3$ **6.** $a = 20, r = 2$

7. $a = 48, r = .5$ **8.** $a = 80, r = .1$

9. $a = 2000, r = 1.05$ **10.** $a = 10,000, r = 1.01$

Find the sum of the first four terms for each of the following geometric sequences. (See Example 3.)

11. $a = 1, r = 2$ **12.** $a = 3, r = 3$

13. $a = 5, r = .2$ **14.** $a = 6, r = .5$

15. $a = 128, r = 1.1$ **16.** $a = 100, r = 1.05$

_____
*T. Rowe Price Associates.

Find each of the following values.

17. $s_{\overline{12}|.05}$ **18.** $s_{\overline{20}|.06}$

19. $s_{\overline{16}|.04}$ **20.** $s_{\overline{40}|.02}$

21. $s_{\overline{20}|.01}$ **22.** $s_{\overline{18}|.015}$

Find the future value of the following ordinary annuities. Payments are made and interest is compounded as given. (See Examples 4 and 5(a).)

23. $R = \$12,000$, 6.2% interest compounded annually for 10 years

24. $R = \$20,000$, 5.5% interest compounded annually for 12 years

25. $R = \$865$, 6% interest compounded semiannually for 8 years

26. $R = \$7300$, 9% interest compounded semiannually for 6 years

27. $R = \$1200$, 8% interest compounded quarterly for 10 years

28. $R = \$20,000$, 6% interest compounded quarterly for 12 years

Find the periodic payment that will amount to the given sums under the given conditions if payments are made at the end of each period. (See Example 6.)

29. $S = \$14,500$; interest is 5% compounded semiannually for 8 years

30. $S = \$43,000$; interest is 6% compounded semiannually for 5 years

31. $S = \$62,000$; interest is 8% compounded quarterly for 6 years

32. $S = \$12,800$; interest is 6% compounded monthly for 4 years

33. What is meant by a sinking fund? List some reasons for establishing a sinking fund.

Find the amount of each payment to be made into a sinking fund to accumulate the following amounts. Payments are made at the end of each period. (See Example 6.)

34. $11,000; money earns 6% compounded semiannually, for 6 years

35. $75,000; money earns 6% compounded semiannually, for $4\frac{1}{2}$ years

36. $50,000; money earns 10% compounded quarterly, for $2\frac{1}{2}$ years

37. $25,000; money earns 12% compounded quarterly, for $3\frac{1}{2}$ years

38. $6000; money earns 8% compounded monthly, for 3 years

39. $9000; money earns 12% compounded monthly, for $2\frac{1}{2}$ years

40. Explain the difference between an ordinary annuity and an annuity due.

Find the future value of each annuity due. (See Example 7.)

41. Payments of $500 for 10 years at 5% compounded annually

42. Payments of $1050 for 6 years at 3.5% compounded annually

43. Payments of $16,000 for 8 years at 4.7% compounded annually

44. Payments of $25,000 for 12 years at 6% compounded annually

45. Payments of $1000 for 9 years at 8% compounded semi-annually

46. Payments of $750 for 15 years at 6% compounded semi-annually

47. Payments of $100 for 7 years at 12% compounded quarterly

48. Payments of $1500 for 11 years at 12% compounded quarterly

Finance *Work the following applied problems.*

49. A typical pack-a-day smoker spends about $100 per month on cigarettes. Suppose the smoker invests that amount at the end of each month in a savings account at 4.8% compounded monthly. What would the account be worth after 40 years?

50. Ron Hampton is saving for a computer. At the end of each month he puts $60 in a savings account that pays 8% interest compounded monthly. How much is in the account after 3 years?

51. Hassi is paid on the first day of the month, and $80 is automatically deducted from his pay and deposited in a savings account. If the account pays 7.5% interest com-

pounded monthly, how much will be in the account after 3 years and 9 months?

52. A father opened a savings account for his daughter on the day she was born, depositing $1000. Each year on her birthday he deposits another $1000, making the last deposit on her twenty-first birthday. If the account pays 9.5% interest compounded annually, how much is in the account at the end of the day on the daughter's twenty-first birthday?

53. Becky Anderson deposits $12,000 at the end of each year for 9 years in an account paying 8% interest compounded annually.
  (a) Find the final amount she will have on deposit.
  (b) Becky's brother-in-law works in a bank that pays 6% compounded annually. If she deposits money in this bank instead of the one above, how much will she have in her account?
  (c) How much would Becky lose over 9 years by using her brother-in-law's bank?

54. Raul Vasquez, a 25-year-old professional, puts $750 in a retirement fund at the end of each quarter until he reaches age 60. The account pays 8% interest compounded quarterly.
  (a) How much will be in the account when he is 60?
  (b) If Raul makes no further deposits after age 60, how much will he have for retirement at age 65?

55. At the end of each quarter a 50-year-old woman puts $1200 in a retirement account that pays 7% interest compounded quarterly. When she reaches age 60, she withdraws the entire amount and places it in a mutual fund that pays 9% interest compounded monthly. From then on she deposits $300 in the mutual fund at the end of each month. How much is in the account when she reaches age 65?

56. Jasspreet Kaur deposits $2435 at the beginning of each semiannual period for 8 years in an account paying 6% compounded semiannually. She then leaves that money alone, with no further deposits, for an additional 5 years. Find the final amount on deposit after the entire 13-year period.

57. Chuck Hickman deposits $10,000 at the beginning of each year for 12 years in an account paying 5% compounded annually. He then puts the total amount on deposit in another account paying 6% compounded semiannually for another 9 years. Find the final amount on deposit after the entire 21-year period.

58. Suppose Michael, in Example 5, cannot get the higher interest rate to produce $130,000 in 20 years. To meet that goal at an interest rate of 7.2%, he must increase his monthly payment. What payment should he make each month?

59. David Horwitz needs $10,000 in 8 years.
  (a) What amount should he deposit at the end of each quarter at 8% compounded quarterly so that he will have his $10,000?
  (b) Find Horwitz's quarterly deposit if the money is deposited at 6% compounded quarterly.

60. Harv's Meats knows that it must buy a new deboner machine in 4 years. The machine costs $12,000. In order to accumulate enough money to pay for the machine, Harv decides to deposit a sum of money at the end of each 6 months in an account paying 6% compounded semiannually. How much should each payment be?

61. Karin Sandberg wants to buy an $18,000 car in 6 years. How much money must she deposit at the end of each quarter in an account paying 12% compounded quarterly so that she will have enough to pay for her car?

In Exercises 62 and 63, use a graphing calculator to find the value of i that produces the given value of S. (See the Technology Tip following Example 5.)

62. To save for retirement, Karla Harby put $300 each month into an ordinary annuity for 20 years. Interest was compounded monthly. At the end of the 20 years, the annuity was worth $147,126. What annual interest rate did she receive?

63. Jennifer Wall made payments of $250 per month at the end of each month to purchase a piece of property. At the end of 30 years, she completely owned the property, which she sold for $330,000. What annual interest rate would she need to earn on an ordinary annuity for a comparable rate of return?

64. In a 1992 Virginia lottery, the jackpot was $27 million. An Australian investment firm tried to buy all possible combinations of numbers, which would have cost $7 million. In fact, the firm ran out of time and was unable to buy all combinations, but ended up with the only winning ticket anyway. The firm received the jackpot in 20 equal annual payments of $1.35 million.* Assume these payments meet the conditions of an ordinary annuity.
  (a) Suppose the firm can invest money at 8% interest compounded annually. How many years would it take until the investors would be further ahead than if they had simply invested the $7 million at the same rate? (*Hint:* Experiment with different values of *n*, the number of years, or use a graphing calculator to plot the value of both investments as a function of the number of years.)
  (b) How many years would it take in part (a) at an interest rate of 12%?

65. Diane Gray sells some land in Nevada. She will be paid a lump sum of $60,000 in 7 years. Until then, the buyer pays 8% simple interest quarterly.
  (a) Find the amount of each quarterly interest payment.
  (b) The buyer sets up a sinking fund so that enough money will be present to pay off the $60,000. The buyer wants to make semiannual payments into the sinking fund; the account pays 6% compounded semiannually. Find the amount of each payment into the fund.
  (c) Prepare a table showing the amount in the sinking fund after the first 6 deposits.

66. Joe Seniw bought a rare stamp for his collection. He agreed to pay a lump sum of $4000 after 5 years. Until then, he pays 6% simple interest semiannually.
  (a) Find the amount of each semiannual interest payment.
  (b) Seniw sets up a sinking fund so that enough money will be present to pay off the $4000. He wants to make annual payments into the fund. The account pays 8% compounded annually. Find the amount of each payment.
  (c) Prepare a table showing the amount in the sinking fund after each deposit.

---

*Washington Post, March 10, 1992, p. A1.

5.4 PRESENT VALUE OF AN ANNUITY AND AMORTIZATION

Suppose that at the end of each year, for the next 10 years, $500 is deposited in a savings account paying 7% interest compounded annually. This is an example of an ordinary annuity. The **present value** of this annuity is the amount that would have to be deposited in one lump sum today (at the same compound interest rate) in order to produce exactly the same balance at the end of 10 years. We can find a formula for the present value of an annuity as follows.

Suppose deposits of R dollars are made at the end of each period for n periods at interest rate i per period. Then the amount in the account after n periods is the future value of this annuity:

$$S = R\left[\frac{(1 + i)^n - 1}{i}\right].$$

On the other hand, if P dollars are deposited today at the same compound interest rate i, then at the end of n periods, the amount in the account is $P(1 + i)^n$. This amount must be the same as the amount S in the formula above, that is,

$$P(1 + i)^n = R\left[\frac{(1 + i)^n - 1}{i}\right].$$

To solve this equation for P, multiply both sides by $(1 + i)^{-n}$.

$$P = R(1 + i)^{-n}\left[\frac{(1 + i)^n - 1}{i}\right]$$

Use the distributive property; also recall that $(1 + i)^{-n}(1 + i)^n = (1 + i)^0 = 1$.

$$P = R\left[\frac{(1 + i)^{-n}(1 + i)^n - (1 + i)^{-n}}{i}\right]$$

$$P = R\left[\frac{1 - (1 + i)^{-n}}{i}\right]$$

The amount P is called the *present value of the annuity.* The quantity in brackets is abbreviated as $a_{\overline{n}|i}$, so

$$a_{\overline{n}|i} = \frac{1 - (1 + i)^{-n}}{i}.$$

(The symbol $a_{\overline{n}|i}$ is read "a-angle-n at i." Compare this quantity with $s_{\overline{n}|i}$ in the previous section.) The formula for the present value of an annuity is summarized below.

Present Value of an Annuity

The present value P of an annuity of n payments of R dollars each at the end of consecutive interest periods with interest compounded at a rate of interest i per period is

$$P = R\left[\frac{1 - (1 + i)^{-n}}{i}\right] \quad \text{or} \quad P = R \cdot a_{\overline{n}|i}.$$

CAUTION Don't confuse the formula for the present value of an annuity with the one for the future value of an annuity. Notice the difference: the numerator of the fraction in the present value formula is $1 - (1 + i)^{-n}$, but in the future value formula, it is $(1 + i)^n - 1$.

▶ **EXAMPLE 1** Surinder Sinah and Maria Gonzalez are both graduates of the Forestvire Institute of Technology. They both agree to contribute to the endowment fund of FIT. Sinah says that he will give $500 at the end of each year for 9 years.

What lump sum deposited today would be equivalent to equal payments of

(a) $650 at the end of each year for 9 years at 4% compounded annually?

(b) $1000 at the end of each quarter for 4 years at 4% compounded quarterly?

Answers:

(a) $4832.97

(b) $14,717.87

Gonzalez prefers to give a lump sum today. What lump sum can she give that will equal the present value of Sinah's annual gifts, if the endowment fund earns 7.5% compounded annually?

Here $R = 500$, $n = 9$, and $i = .075$ and we have

$$P = 500\left[\frac{1 - (1.075)^{-9}}{.075}\right]$$

$$= 3189.44.$$

Therefore, Gonzalez must donate a lump sum of $3189.44 today. ◄

TECHNOLOGY TIP The financial feature of the TI-83 calculator can be used to find the present value of an annuity by choosing that option from the menu and entering the required information. If your graphing calculator does not have this built-in feature, it will be useful to store a program to calculate present value of an annuity. See the Program Appendix at www.aw.com/MWA8.

Kelly Erin buys a small business for $174,000. She agrees to pay off the cost in payments at the end of each semiannual period for 7 years, with interest of 8% compounded semiannually on the unpaid balance. Find the amount of each payment.

Answer:

$16,472.40

▶**EXAMPLE 2**   A car costs $22,000. After a down payment of $4000, the balance will be paid off in 48 equal monthly payments with interest of 12% per year on the unpaid balance. Find the amount of each payment.

A single lump sum payment of $18,000 today would pay off the loan. So, $18,000 is the present value of an annuity of 48 monthly payments with interest of 12%/12 = 1% per month. Thus, $P = 18,000$, $n = 48$, and $i = .01$, and we must find the monthly payment R in the formula

$$P = R\left[\frac{1 - (1 + i)^{-n}}{i}\right].$$

$$18,000 = R\left[\frac{1 - (1.01)^{-48}}{.01}\right]$$

$$R \approx 474.009378$$

A monthly payment of $474.01 will be needed. ◄

Amortization

A loan is **amortized** if both the principal and interest are paid by a sequence of equal periodic payments. In Example 2 above, a loan of $18,000 at 12% interest compounded monthly could be amortized by paying $474.01 per month for 48 months.

The periodic payment needed to amortize a loan may be found, as in Example 2, by solving the present value equation for R.

Amortization Payments

A loan of P dollars at interest rate i per period may be amortized in n equal periodic payments of R dollars made at the end of each period, where

$$R = \frac{P}{a_{\overline{n}|i}} = \frac{P}{\left[\dfrac{1 - (1 + i)^{-n}}{i}\right]} = \frac{Pi}{1 - (1 + i)^{-n}}.$$

▶ **EXAMPLE 3** The Beckenstein family buys a house for $114,000 with a down-payment of $36,000. They take out a 30-year mortgage for $78,000 at an annual interest rate of 9.6%.

(a) Find the amount of the monthly payment needed to amortize this loan.

Here $P = 78,000$ and the monthly interest rate is $9.6\%/12 = .096/12 = .008.$* The number of monthly payments is $12 \cdot 30 = 360$. Therefore,

$$R = \frac{78,000}{a\,\overline{_{360|}}\,.008} = \frac{78,000}{\left[\dfrac{1 - (1.008)^{-360}}{.008}\right]} \approx \frac{78,000}{117.90229} \approx 661.56.$$

Monthly payments of $661.56 are required to amortize the loan.

(b) Find the total amount of interest paid when the loan is amortized over 30 years.

The Beckenstein family makes 360 payments of $661.56 each, for a total of $238,161.60. Since the amount of the loan was $78,000, the total interest paid is

$$\$238,161.60 - \$78,000 = \$160,161.60.$$

This large amount of interest is typical of what happens with a long mortgage. A 15-year mortgage would have higher payments but would involve significantly less interest. ③

(c) Find the part of the first payment that is interest and the part that is applied to reducing the debt.

As we saw in part (a), the monthly interest rate is .008. During the first month, the entire $78,000 is owed. Interest on this amount for 1 month is found by the formula for simple interest.

$$I = Prt = 78,000(.008)(1) = 624$$

At the end of the month, a payment of $661.56 is made; since $624 of this is interest, a total of

$$\$661.56 - \$624 = \$37.56$$

is applied to the reduction of the original debt. Thus, the remaining balance after one payment is $78,000 - \$37.56 = \$77,962.44.$ ◀

Amortization Schedules

In Example 3, 360 payments are made to amortize a $78,000 loan. The loan balance after the first payment is reduced by only $37.56, which is much less than $(1/360)(78,000) \approx \$216.67$. Therefore, even though equal *payments* are made to amortize a loan, the loan *balance* does not decrease in equal steps. This fact is very important if a loan is paid off early.

▶ **EXAMPLE 4** Jill Stuart borrows $1000 for 1 year at 12% annual interest compounded monthly. Verify that her monthly loan payment is $88.85. After making three payments, she decides to pay off the remaining balance all at once. How much must she pay?

If the mortgage in Example 3 runs for 15 years, find

(a) the monthly payment;

(b) the total amount of interest paid.

Answers:

(a) $819.21

(b) $69,457.80

---

*Mortgage rates are quoted in terms of annual interest, but it is always understood that the monthly rate is 1/12 of the annual rate and that interest is compounded monthly.

Find the following for a car loan of $10,000 for 48 months at 9% interest compounded monthly.

(a) The monthly payment

(b) How much is still due after 12 payments have been made?

Answers:

(a) $248.85

(b) $7825.54

Since nine payments remain to be paid, they can be thought of as an annuity consisting of nine payments of $88.85 at 1% interest per period. The present value of this annuity is

$$88.85 \left[\frac{1 - (1.01)^{-9}}{.01} \right] = 761.09.$$

So Jill's remaining balance, computed by this method, is $761.09. ◄

The method used to compute the remaining balance in Example 4 applies in the general case. After x payments, there are $n - x$ payments remaining. So the balance is the present value of an annuity of $n - x$ payments of the same amount at the same interest rate. As in Example 4 (where $n = 12$ and $x = 3$), this can be found by using $n - x$ in place of n in the present value formula.

Remaining Balance

If a loan can be amortized by n payments of R dollars each at interest rate i per period, then the *approximate* remaining balance after x payments is

$$y = R \left[\frac{1 - (1 + i)^{-(n-x)}}{i} \right].$$

Because payments are rounded to the nearest penny, the formula in the preceding box gives an approximate rather than exact remaining balance, as the next example illustrates.

▶ **EXAMPLE 5** In Example 3, we saw that the monthly payment on a $78,000 mortgage for 30 years, with an annual interest rate of 9.6%, was $661.56.

(a) What was the approximate balance after the first payment?

Apply the formula with $R = 661.56$, $n = 30$, $x = 1$, and $I = .096/12 = .008$ (the monthly rate of interest). The approximate remaining balance is

$$y = 661.56 \left[\frac{1 - (1 + .008)^{-(360-1)}}{.008} \right] = 661.56 \left[\frac{1 - 1.008^{-359}}{.008} \right] \approx \$77,961.87.$$

This differs by only 57 cents from the actual remaining balance of $77,962.44 found in Example 3.

(b) Use a graphing calculator to estimate how long it will take to repay half the loan.

Half the loan is $39,000. Graph $y = 39,000$ and the remaining balance function

$$y = 661.56 \left[\frac{1 - 1.008^{-(360-x)}}{.008} \right]$$ on the same screen and find their intersection

Figure 5.10 shows that the intersection occurs when $x = 279.94$, that is, after approximately 280 monthly payments. This means that it takes more than 23 years to pay off half of the 30-year mortgage. ◄

FIGURE 5.10

The remaining balance formula is a quick and convenient way to get a reasonable estimate of the remaining balance on a loan, but it is not accurate enough for a bank or business, which must keep its books exactly. To determine the exact remaining balance after each loan payment, financial institutions normally use an **amortization schedule,** which lists how much of each payment is interest, how much goes to reduce the balance, and how much is still owed after each payment.

▶**EXAMPLE 6**   Determine the exact amount Jill Stuart in Example 4 owes after three monthly payments.

An amortization table for the loan is shown below. It is obtained as follows. The annual interest rate is 12% compounded monthly, so the interest rate per month is 12%/12 = 1% = .01. When the first payment is made, 1 month's interest, namely, .01(1000) = $10, is owed. Subtracting this from the $88.85 payment leaves $78.85 to be applied to repayment. Hence, the principal at the end of the first payment period is 1000 − 78.85 = $921.15, as shown in "payment 1" line of the table.

When payment 2 is made, 1 month's interest on $921.15 is owed, namely, .01(921.15) = $9.21. Subtracting this from the $88.85 payment leaves $79.64 to reduce the principal. So the principal at the end of payment 2 is 921.15 − 79.64 = $841.51. The interest portion of payment 3 is based on this amount, and the remaining lines of the table are found in a similar fashion.

Payment Number	Amount of Payment	Interest for Period	Portion to Principal	Principal at End of Period
0	—	—	—	$1000.00
1	$88.85	$10.00	$78.85	921.15
2	88.85	9.21	79.64	841.51
3	88.85	8.42	80.43	761.08
4	88.85	7.61	81.24	679.84
5	88.85	6.80	82.05	597.79
6	88.85	5.98	82.87	514.92
7	88.85	5.15	83.70	431.22
8	88.85	4.31	84.54	346.68
9	88.85	3.47	85.38	261.30
10	88.85	2.61	86.24	175.06
11	88.85	1.75	87.10	87.96
12	88.84	.88	87.96	0

The schedule shows that after three payments, she still owes $761.08, an amount that differs slightly from that obtained in Example 4.   ◀

The amortization schedule in Example 6 is typical. In particular, note that all payments are the same except that last one. It is often necessary to adjust the amount of the final payment to account for rounding off earlier and to ensure that the final balance is exactly 0.

An amortization schedule also shows how the periodic payments are applied to interest and principal. The amount going to interest decreases with each payment, while the amount going to reduce the principal owed increases with each payment.

TECHNOLOGY TIP Graphing calculator programs to produce amortization schedules are in the Program Appendix (at www.aw.com/MWA8). The TI-83 includes a built-in program to find the amortization payment. Spreadsheets are another useful tool for creating amortization tables. Microsoft Excel has a built-in feature for calculating monthly payments. Figure 5.11 shows an Excel amortization table for Example 6. For more details, see *The Spreadsheet Manual*, also available with this book.

	A	B	C	D	E	F
1	Prnt#	Payment	Interest	Principal	End Principal	
2	0				1000	
3	1	88.85	10.00	78.85	921.15	
4	2	88.85	9.21	79.64	841.51	
5	3	88.85	8.42	80.43	761.08	
6	4	88.85	7.61	81.24	679.84	
7	5	88.85	6.80	82.05	597.79	
8	6	88.85	5.98	82.87	514.92	
9	7	88.85	5.15	83.70	431.22	
10	8	88.85	4.31	84.54	346.68	
11	9	88.85	3.47	85.38	261.30	
12	10	88.85	2.61	86.24	175.06	
13	11	88.85	1.75	87.10	87.96	
14	12	88.85	0.88	87.97	-0.01	

FIGURE 5.11

5.4 Exercises

1. Which of the following is represented by $a_{\overline{n}|i}$?

(a) $\dfrac{(1 + i)^{-n} - 1}{i}$ (b) $\dfrac{(1 + i)^{n} - 1}{i}$

(c) $\dfrac{1 - (1 + i)^{-n}}{i}$ (d) $\dfrac{1 - (1 + i)^{n}}{i}$

2. Which of the choices in Exercise 1 represents $s_{\overline{n}|i}$?

Find each of the following.

3. $a_{\overline{15}|.06}$ 4. $a_{\overline{10}|.03}$

5. $a_{\overline{18}|.04}$ 6. $a_{\overline{30}|.01}$

7. $a_{\overline{16}|.01}$ 8. $a_{\overline{32}|.02}$

9. Explain the difference between the present value of an annuity and the future value of an annuity. For a given annuity, which is larger? Why?

Find the present value of each ordinary annuity. (See Example 1.)

10. Payments of $890 each year for 16 years at 8% compounded annually

11. Payments of $1400 each year for 8 years at 8% compounded annually

12. Payments of $10,000 semiannually for 15 years at 10% compounded semiannually

13. Payments of $50,000 quarterly for 10 years at 8% compounded quarterly

14. Payments of $15,806 quarterly for 3 years at 10.8% compounded quarterly

15. Payments of $18,579 every 6 months for 8 years at 9.4% compounded semiannually

Find the lump sum deposited today that will yield the same total amount as payments of $10,000 at the end of each year for 15 years, at each of the given interest rates. (See Example 1.)

16. 4% compounded annually

17. 5% compounded annually

18. 6% compounded annually

19. What does it mean to amortize a loan?

Find the payment necessary to amortize each of the following loans. (See Example 2.)

20. $2500; 8% compounded quarterly; 6 quarterly payments

21. $41,000; 10% compounded semiannually; 10 semiannual payments

22. $90,000; 8% compounded annually; 12 annual payments

23. $140,000; 12% compounded quarterly; 15 quarterly payments

24. $7400; 8.2% compounded semiannually; 18 semiannual payments

25. $5500; 12.5% compounded monthly; 24 monthly payments

Find the monthly house payment necessary to amortize the following loans. (See Example 3.)

26. $149,560 at 7.75% for 25 years

27. $170,892 at 8.11% for 30 years

28. $153,762 at 8.45% for 30 years

29. $96,511 at 9.57% for 25 years

Use the amortization table in Example 6 to answer the questions in Exercises 30–33.

30. How much of the fifth payment is interest?

31. How much of the tenth payment is used to reduce the debt?

32. How much interest is paid in the first 5 months of the loan?

33. How much interest is paid in the last 5 months of the loan?

34. What sum deposited today at 5% compounded annually for 8 years will provide the same amount as $1000 deposited at the end of each year for 8 years at 6% compounded annually?

35. What lump sum deposited today at 8% compounded quarterly for 10 years will yield the same final amount as deposits of $4000 at the end of each six-month period for 10 years at 6% compounded semiannually?

Finance *Work the following applied problems.*

36. An auto stereo dealer sells a stereo system for $600 down and monthly payments of $30 for the next 3 years. If the interest rate is 1.25% per month on the unpaid balance, find
(a) the cost of the stereo system;
(b) the total amount of interest paid.

37. John Kushida buys a car costing $6000. He agrees to make payments at the end of each monthly period for 4 years. He pays 12% interest, compounded monthly.
(a) What is the amount of each payment?
(b) Find the total amount of interest Kushida will pay.

38. A speculator agrees to pay $15,000 for a parcel of land; this amount, with interest, will be paid over 4 years, with semi-annual payments, at an interest rate of 10% compounded semiannually. Find the amount of each payment.

39. In many states, the winnings of million-dollar lottery jackpots are divided into equal payments given annually for 20 years. (In Colorado, the results are distributed over 25 years.)* This means that the present value of the jackpot is worth less than the stated prize, with the actual value determined by the interest rate at which the money could be invested.
(a) Find the present value of a $1 million lottery jackpot distributed in equal annual payments over 20 years, using an interest rate of 5%.
(b) Find the present value of a $1 million lottery jackpot distributed in equal annual payments over 20 years, using an interest rate of 9%.
(c) Calculate the answer for part (a) using the 25-year distribution time in Colorado.
(d) Calculate the answer for part (b) using the 25-year distribution time in Colorado.

Finance *Student borrowers now have more options to choose from when selecting repayment plans.† The standard plan repays the loan in 10 years with equal monthly payments. The extended plan allows from 12 to 30 years to repay the loan. A student borrows $35,000 at 7.43% compounded monthly.*

40. Find the monthly payment and total interest paid under the standard plan.

*New York Times, November 12, 1996, pp. A1, A22.

†New York Times, April 2, 1995, "Money and College," Saul Hansell, p. 28.

41. Find the monthly payment and total interest paid under the extended plan with 20 years to pay off the loan.

Finance *Use the formula for the approximate remaining balance to work each problem. (See Example 5.)*

42. When Teresa Flores opened her law office, she bought $14,000 worth of law books and $7200 worth of office furniture. She paid $1200 down and agreed to amortize the balance with semiannual payments for 5 years, at 12% compounded semiannually.
  (a) Find the amount of each payment.
  (b) When her loan had been reduced below $5000, Flores received a large tax refund and decided to pay off the loan. How many payments were left at this time?

43. Kareem Adams buys a house for $285,000. He pays $60,000 down and takes out a mortgage at 9.5% on the balance. Find his monthly payment and the total amount of interest he will pay if the length of the mortgage is
  (a) 15 years;   **(b)** 20 years;   **(c)** 25 years.
  (d) When will half the 20-year loan be paid off?

44. The Beyes plan to purchase a home for $212,000. They will pay 20% down and finance the remainder for 30 years at 8.9% interest, compounded monthly.*
  (a) How large are their monthly payments?
  (b) What will be their loan balance right after they have made their 96th payment?
  (c) How much interest will they pay during the 7th year of the loan?
  (d) If they were to increase their monthly payments by $150, how long would it take to pay off the loan?

Finance *Work each problem.*

45. Sandi Goldstein has inherited $25,000 from her grandfather's estate. She deposits the money in an account offering 6% interest compounded annually. She wants to make equal annual withdrawals from the account so that the money (principal and interest) lasts exactly 8 years.
  (a) Find the amount of each withdrawal.
  (b) Find the amount of each withdrawal if the money must last 12 years.

---

*Exercises 44–48 were supplied by Norman Lindquist at Western Washington University.

46. Jeni Ramirez plans to retire in 20 years. She will make 240 equal monthly contributions to her retirement account. One month after her last contribution, she will begin the first of 120 monthly withdrawals from the account. She expects to withdraw $3500 per month. How large must her monthly contributions be in order to accomplish her goal if her account is assumed to earn interest of 10.5%, compounded monthly throughout the life of this problem?

47. Ron Okimura also plans to retire in 20 years. Ron will make 120 equal monthly contributions to his account. Ten years after his last contribution (120 months) he will begin the first of 120 monthly withdrawals from his account. Ron also expects to withdraw $3500 per month. His account also earns interest of 10.5%, compounded monthly throughout the life of this problem. How large must Ron's monthly contributions be in order to accomplish his goals?

48. Madeline and Nick Swenson took out a 30-year mortgage for $160,000 at 9.8% interest, compounded monthly. After they had made 12 years of payments (144 payments) they decided to refinance the remaining loan balance for 25 years at 7.2% interest, compounded monthly. What will be the balance on their loan 5 years after they refinance?

Finance *In Exercises 49–52, prepare an amortization schedule showing the first 4 payments for each loan. (See Example 5.)*

49. An insurance firm pays $4000 for a new printer for its computer. It amortizes the loan for the printer in 4 annual payments at 8% compounded annually.

50. Large semitrailer trucks cost $72,000 each. Ace Trucking buys such a truck and agrees to pay for it by a loan that will be amortized with 9 semiannual payments at 6% compounded semiannually.

51. One retailer charges $1048 for a certain computer. A firm of tax accountants buys 8 of these computers. They make a down payment of $1200 and agree to amortize the balance with monthly payments at 12% compounded monthly for 4 years.

52. Joan Varozza plans to borrow $20,000 to stock her small boutique. She will repay the loan with semiannual payments for 5 years at 7% compounded semiannually.

CHAPTER 5 SUMMARY

Key Terms and Symbols

5.1 simple interest
principal
rate
time
future value (maturity value)
present value
discount (bank discount)
proceeds

5.2 compound interest
compound amount
compounding period
continuous compounding
nominal rate (stated rate)
effective rate

5.3 geometric sequence
common ratio
annuity
ordinary annuity
payment period

term of an annuity
future value of an annuity
sinking fund
annuity due

5.4 present value of an annuity
amortize a loan
amortization schedule

Key Concepts

A Strategy for Solving Finance Problems

We have presented a lot of new formulas in this chapter. By answering the following questions, you can decide which formula to use for a particular problem.

1. Is simple or compound interest involved?
   Simple interest is normally used for investments or loans of a year or less; compound interest is normally used in all other cases.
2. If simple interest is being used, what is being sought: interest amount, future value, present value, or discount?
3. If compound interest is being used, does it involve a lump sum (single payment) or an annuity (sequence of payments)?
   (a) For a lump sum,
     (i) Is ordinary compound interest or continuous interest involved?
     (ii) What is being sought: present value, future value, number of periods, or effective rate?
   (b) For an annuity,
     (i) Is it an ordinary annuity (payment at the end of each period) or an annuity due (payment at the beginning of each period)?
     (ii) What is being sought: present value, future value, or payment amount?

Once you have answered these questions, choose the appropriate formula and work the problem. As a final step, consider whether the answer you get makes sense. For instance, present value should always be less than future value. The amount of interest or the payments in an annuity should be fairly small compared to the total future value.

Key Formulas

LIST OF VARIABLES

r is the annual interest rate.

m is the number of periods per year.

i is the interest rate per period. $i = \dfrac{r}{m}$

t is the number of years.

n is the number of periods. $\qquad n = tm$

P is the principal or present value.

A is the future value of a lump sum.

S is the future value of an annuity.

R is the periodic payment in an annuity.

INTEREST

	Simple Interest	**Compound Interest**
Interest	$I = Prt$	$I = A - P$
Future value	$A = P(1 + rt)$	$A = P(1 + i)^n$
Present value	$P = \dfrac{A}{1 + rt}$	$P = \dfrac{A}{(1 + i)^n} = A(1 + i)^{-n}$

Effective rate $\quad r_E = \left(1 + \dfrac{r}{m}\right)^m - 1$

DISCOUNT

If D is the **discount** on a loan having maturity value A at simple interest rate r for t years, then $\boldsymbol{D = Art}$. If D is the discount and P the **proceeds** of a loan having maturity value A at simple interest rate r for t years, then $\boldsymbol{P = A - D}$ or $\boldsymbol{P = A(1 - rt)}$.

CONTINUOUS INTEREST

If P dollars are deposited for t years at interest rate r per year, compounded continuously, the **compound amount (future value)** A is $\boldsymbol{A = Pe^{rt}}$.

The **present value** P of A dollars at interest rate r per year compounded continuously for t years is

$$P = \frac{A}{e^{rt}}.$$

ANNUITIES

Ordinary annuity	Future value	$S = R\left[\dfrac{(1 + i)^n - 1}{i}\right] = R \cdot s_{\overline{n}|i}$
	Present value	$P = R\left[\dfrac{1 - (1 + i)^{-n}}{i}\right] = R \cdot a_{\overline{n}|i}$
Annuity due	Future value	$S = R\left[\dfrac{(1 + i)^{n+1} - 1}{i}\right] - R$

Chapter 5 Review Exercises

Find the simple interest for the following loans.

1. $4902 at 9.5% for 11 months

2. $42,368 at 15.22% for 5 months

3. $3478 at 7.4% for 88 days (assume a 360-day year)

4. $2390 at 18.7% from May 3 to July 28 (assume 365 days in a year)

5. What is meant by the present value of an amount A?

Find the present value of the following future amounts. Assume 360 days in a year; use simple interest.

6. $459.57 in 7 months; money earns 8.5%

7. $80,612 in 128 days; money earns 6.77%

8. Explain what happens when a borrower is charged a discount. What are the proceeds?

Find the proceeds in Exercises 9 and 10. Assume 360 days in a year.

9. $802.34; discount rate 8.6%; length of loan 11 months

10. $12,000; discount rate 7.09%; length of loan 145 days

11. For a given amount of money at a given interest rate for a given time period greater than 1 year, does simple interest or compound interest produce more interest? Explain.

Find the compound amount and the amount of interest earned in each of the following.

12. $2800 at 6% compounded annually for 10 years

13. $57,809.34 at 4% compounded quarterly for 5 years

14. $12,903.45 at 6.37% compounded quarterly for 29 quarters

15. $4677.23 at 4.57% compounded monthly for 32 months

Find the present value of the following amounts.

16. $42,000 in 7 years; 12% compounded monthly

17. $17,650 in 4 years; 8% compounded quarterly

18. $1347.89 in 3.5 years; 6.77% compounded semiannually

19. $2388.90 in 44 months; 5.93% compounded monthly

20. Write the first five terms of the geometric sequence with $a = 2$ and $r = 3$.

21. Write the first four terms of the geometric sequence with $a = 4$ and $r = 1/2$.

22. Find the sixth term of the geometric sequence with $a = -3$ and $r = 2$.

23. Find the fifth term of the geometric sequence with $a = -2$ and $r = -2$.

24. Find the sum of the first four terms of the geometric sequence with $a = -3$ and $r = 3$.

25. Find the sum of the first five terms of the geometric sequence with $a = 8000$ and $r = -1/2$.

26. Find $s_{\overline{30}|.01}$.

27. What is meant by the future value of an annuity?

Find the future value of each annuity.

28. $1288 deposited at the end of each year for 14 years; money earns 8% compounded annually

29. $4000 deposited at the end of each quarter for 7 years; money earns 6% compounded quarterly

30. $233 deposited at the end of each month for 4 years; money earns 12% compounded monthly

31. $672 deposited at the beginning of each quarter for 7 years; money earns 8% compounded quarterly

32. $11,900 deposited at the beginning of each month for 13 months; money earns 12% compounded monthly

33. What is the purpose of a sinking fund?

Find the amount of each payment that must be made into a sinking fund to accumulate the following amounts. Assume payments are made at the end of each period.

34. $6500; money earns 5% compounded annually; 6 annual payments

35. $57,000; money earns 6% compounded semiannually for $8\frac{1}{2}$ years

36. $233,188; money earns 5.7% compounded quarterly for $7\frac{3}{4}$ years

37. $56,788; money earns 6.12% compounded monthly for $4\frac{1}{2}$ years

Find the present value of each ordinary annuity.

38. Payments of $850 annually for 4 years at 5% compounded annually

39. Payments of $1500 quarterly for 7 years at 8% compounded quarterly

40. Payments of $4210 semiannually for 8 years at 8.6% compounded semiannually

41. Payments of $877.34 monthly for 17 months at 6.4% compounded monthly

42. Give two examples of the types of loans that are commonly amortized.

Find the amount of the payment necessary to amortize each of the following loans.

43. $32,000 at 9.4% compounded quarterly; 10 quarterly payments

44. $5607 at 7.6% compounded monthly; 32 monthly payments

Find the monthly house payments for the following mortgages.

45. $56,890 at 10.74% for 25 years

46. $77,110 at 8.45% for 30 years

A portion of an amortization table is given below for a $127,000 loan at 8.5% interest compounded monthly for 25 years.

Payment Number	Amount of Payment	Interest for Period	Portion to Principal	Principal at End of Period
0	—	—	—	$127,000.00
1	$1022.64	$899.58	$123.06	126,876.94
2	1022.64	898.71	123.93	126,753.02
3	1022.64	897.83	124.80	126,628.21
4	1022.64	896.95	125.69	126,502.53
5	1022.64	896.06	126.58	126,375.95
6	1022.64	895.16	127.48	126,248.47

Use the table to answer the following questions.

47. How much of the fifth payment is interest?

48. How much of the sixth payment is used to reduce the debt?

49. How much interest is paid in the first 3 months of the loan?

50. How much has the debt been reduced at the end of the first 6 months?

Finance *Work the following applied problems.*

51. Larry DiCenso needs $9812 to buy new equipment for his business. The bank charges a discount rate of 12%. Find the amount of DiCenso's loan if he borrows the money for 7 months.

52. Tom Wilson owes $5800 to his mother. He has agreed to pay back the money in 10 months, at an interest rate of 6%. Three months before the loan is due, Tom's mother discounts the loan at the bank to get cash for $6000 worth of new furniture. The bank charges a 10.45% discount rate. How much money does Tom's mother receive? Is it enough to buy the furniture?

53. A firm of attorneys deposits $15,000 of profit-sharing money in an account at 6% compounded semiannually for $7\frac{1}{2}$ years. Find the amount of interest earned.

54. Tom, a graduate student, is considering investing $500 now, when he is 23, or waiting until he is 40 to invest $500. How much more money will he have at the age of 65 if he invests now, given that he can earn 5% interest compounded quarterly?

55. The graph shows the growth of a $10,000 investment in Seligman Communications and Information Fund Class A shares at the end of each year from 1990 through 2000.*

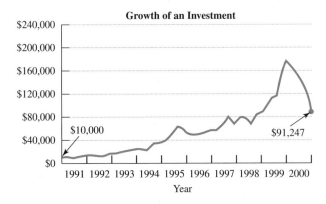

Growth of an Investment

(a) The maximum value of $175,000 during this time period occurred at the beginning of 2000. What was the annual rate of return (interest rate) from the beginning of 1991 to that date?

(b) What was the rate of return from the beginning of 1991 to the end of 2000?

56. According to a financial Web site, on October 16, 2000, Guarantee Bank Milwaukee, Wisconsin, paid 6.90% interest, compounded quarterly, on a 1-year CD, while Capital Crossing Bank of Boston, Massachusetts, paid 6.88% compounded monthly.* What are the effective rates for the two CDs, and which bank pays a higher effective rate?

57. The Consumer Price Index for 1999 was approximately double that for 1980.† What was the inflation rate for that period. (*Hint:* Use continuous compounding.)

58. In 1999 an item that cost $100 in 1983 sold for $166.60.† What inflation rate does this represent? (*Hint:* Use continous compounding.)

59. Chalon Bridges deposits semiannual payments of $3200, received in payment of a debt, in an ordinary annuity at 6.8% compounded semiannually. Find the final amount in the account and the interest earned at the end of 3.5 years.

60. Each year a firm must set aside enough funds to provide employee retirement benefits of $52,000 in 20 years. If the firm can invest money at 7.5% compounded monthly, what amount must be invested at the end of each month for this purpose?

61. In 1995, Oseola McCarty donated $150,000 to the University of Southern Mississippi to establish a scholarship fund.‡ What is unusual about her is that the entire amount came from what she was able to save each month from her work as a washerwoman, a job she began in 1916 at the age of 8, when she dropped out of school. How much would Ms. McCarthy have to put into her savings account at the end of every 3 months to accumulate $150,000 over 79 years? Assume she received an interest rate of 5.25% compounded quarterly.

62. Pension experts recommend that you start drawing at least 40% of your full pension as early as possible.§ Suppose you have built up a pension with $12,000 annual payments by working 10 years for a company, when you leave to accept

*www.bankrate.com

†U.S. Bureau of Labor Statistics.

‡*New York Times*, November 12, 1996, p. A1, A22.

§*Smart Money*, October 1994, "Pocket That Pension," p. 33.

*Seligman Communications and Information Fund Annual Report, December 31, 2000.

a better job. The company gives you the option of collecting half the full pension when you reach age 55 or the full pension at age 65. Assume an interest rate of 8% compounded annually. By age 75, how much will each plan produce? Which plan would produce the larger amount?

63. In 3 years Ms. Nguyen must pay a pledge of $7500 to her favorite charity. What lump sum can she deposit today, at 10% compounded semiannually, so that she will have enough to pay the pledge?

64. To finance the $15,000 cost of their kitchen remodeling, the Chews will make equal payments at the end of each month for 36 months. They will pay interest at the rate of 7.2% compounded monthly. Find the amount of each payment.

65. To expand her business, the owner of a small restaurant borrows $40,000. She will repay the money in equal payments at the end of each semiannual period for 8 years at 9% interest compounded semiannually. What payments must she make?

66. The Taggart family bought a house for $91,000. They paid $20,000 down and took out a 30-year mortgage for the balance at 9%.
(a) Find their monthly payment.
(b) How much of the first payment is interest?

After 180 payments, the family sells their house for $136,000. They must pay closing costs of $3700 plus 2.5% of the sale price.
(c) Estimate the current mortgage balance at the time of the sale using one of the methods from Examples 4 and 5 in Section 5.4.
(d) Find the total closing costs.
(e) Find the amount of money they receive from the sale after paying off the mortgage.

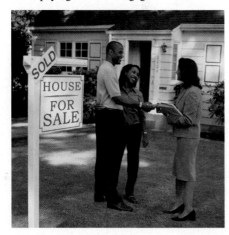

67. The proceeds of a $10,000 death benefit are left on deposit with an insurance company for 7 years at an annual effective interest rate of 5%.* The balance at the end of

7 years is paid to the beneficiary in 120 equal monthly payments of X, with the first payment made immediately. During the payout period, interest is credited at an annual effective interest rate of 3%. Calculate X.
(a) 117   (b) 118   (c) 129   (d) 135   (e) 158

68. Gene deposits $500 per quarter to his nest egg account. The account earns interest of 9%, compounded quarterly. Right after Gene makes his 16th deposit, he loses his job and cannot make any deposits for the next 5 years (20 quarters). Eventually Gene gets another job and again begins making deposits to his account. Since he missed so many deposits while he was out of work, Gene now deposits $750 per quarter. His first $750 deposit comes exactly 20 quarters after his last $500 deposit. What will be Gene's account balance right after he has made his 32nd $750 deposit?*

69. Cathy wants to retire on $55,000 per year for her life expectancy of 20 years. She estimates that she will be able to earn interest of 9%, compounded annually, throughout her lifetime. To reach her retirement goal, Cathy will make annual contributions to her account for the next 25 years. One year after making her last contribution, she will take her first retirement check. How large must her yearly contributions be?

70. The *New York Times* posed a scenario with two individuals, Sue and Joe, who each have $1200 a month to spend on housing and investing. Each takes out a mortgage for $140,000. Sue gets a 30-year mortgage at a rate of 6.625%. Joe gets a 15-year mortgage at a rate of 6.25%. Whatever money is left after the mortgage payment is invested in a mutual fund with a return of 10% annually.[†]
(a) What annual interest rate, when compounded monthly, gives an effective annual rate of 10%?
(b) What is Sue's monthly payment?
(c) If Sue invests the remainder of her $1200 each month, after the payment in part (b), in a mutual fund with the interest rate found in part (a), how much money will she have in the fund at the end of 30 years?
(d) What is Joe's monthly payment?
(e) You found in part (d) that Joe has nothing left to invest until his mortgage is paid off. If he then invests the entire $1200 monthly in a mutual fund with the interest rate in part (a), how much money will he have at the end of 30 years (i.e., after 15 years of paying the mortgage and 15 years of investing)?
(f) Who is ahead at the end of the 30 years and by how much?
(g) Discuss to what extent the difference found in part (f) is due to the different interest rates or to the different amounts of time.

*Problem from "Course 140 Examination, Mathematics of Compound Interest" of the Education and Examination Committee of The Society of Actuaries. Reprinted by permission of The Society of Actuaries.

*Exercises 68 and 69 supplied by Norman Lindquist, Western Washington University.

†*New York Times,* September 27, 1998, p. BU 10.

Case 4

Time, Money, and Polynomials*

A *time line* is often helpful for evaluating complex investments. For example, suppose you buy a $1000 CD at time t_0. After one year $2500 is added to the CD at t_1. By time t_2, after another year, your money has grown to $3851 with interest. What rate of interest, called *yield to maturity* (*YTM*), did your money earn? A time line for this situation is shown in the next figure.

Assuming interest is compounded annually at a rate i and using the compound interest formula gives the following description of the YTM.

$$1000(1 + i)^2 + 2500(1 + i) = 3851$$

To determine the yield to maturity, we must solve this equation for i. Since the quantity $1 + i$ is repeated, let $x = 1 + i$ and first solve the second-degree (quadratic) polynomial equation for x.

$$1000x^2 + 2500x - 3851 = 0$$

We can use the quadratic formula with $a = 1000$, $b = 2500$, and $c = -3851$.

$$x = \frac{-2500 \pm \sqrt{2500^2 - 4(1000)(-3851)}}{2(1000)}$$

We get $x = 1.0767$ and $x = -3.5767$. Since $x = 1 + i$, the two values for i are $.0767 = 7.67\%$ and $-4.5767 = -457.67\%$. We reject the negative value because the final accumulation is greater than the sum of the deposits. In some applications, however, negative rates may be meaningful. By checking in the first equation, we see that the yield to maturity for the CD is 7.67%.

Now let us consider a more complex but realistic problem. Suppose Curt Reynolds has contributed for four years to a retirement fund. He contributed $6000 at the beginning of the first year. At the beginning of the next three years, he con-

tributed $5840, $4000, and $5200, respectively. At the end of the fourth year, he had $29,912.38 in his fund. The interest rate earned by the fund varied between 21% and -3%, so Reynolds would like to know the YTM $= i$ for his hard-earned retirement dollars. Look at the time line in the next figure.

We can set up the following in $1 + i$ for Reynold's savings program.

$$6000(1 + i)^4 + 5840(1 + i)^3 + 4000(1 + i)^2$$
$$+ 5200(1 + i) = 29,912.38$$

Let $x = 1 + i$. We need to solve the fourth-degree polynomial equation

$$f(x) = 6000x^4 + 5840x^3 + 4000x^2 + 5200x$$
$$- 29,912.38 = 0.$$

If you have a graphing calculator or computer algebra system, you can solve this equation graphically, or by using the equation solver. If the only technology you have is an ordinary calculator, you can solve the equation by the guess and check method, as follows.

We expect that $0 < i < 1$, so that $1 < x < 2$. Let us calculate $f(1)$ and $f(2)$. If there is a change of sign, we will know that there is a solution to $f(x) = 0$ between 1 and 2. We find that

$$f(1) = -8872.38 \quad \text{and} \quad f(2) = 139,207.62.$$

There is a change in sign as expected. Now we find $f(1.1)$, $f(1.2), f(1.3)$, and so on, and look for another change in sign. Right away we find

$$f(1.1) = -2794.74 \quad \text{and} \quad f(1.2) = 4620.74.$$

*From *Time, Money, and Polynomials*, COMAP.

This process can be repeated for values of x between 1.1 and 1.2, to get $f(1.11) = -2116.59, f(1.12) = -1424.88, f(1.13) = -719.42$, and $f(1.14) = 0$. We were lucky; the solution for $f(x) = 0$ is exactly 1.14, so $i = \text{YTM} = .14 = 14\%$.

Exercises

1. Brenda Bravener received $50 on her 16th birthday, and $70 on her 17th birthday, both of which she immediately invested in the bank with interest compounded annually. On her 18th birthday, she had $127.40 in her account. Draw a time line, set up a polynomial equation, and calculate the YTM.

2. At the beginning of the year, Jay Beckenstein invested $10,000 at 5% for the first year. At the beginning of the second year, he added $12,000 to the account. The total account earned 4.5% for the second year.
   (a) Draw a time line for this investment.
   (b) How much was in the fund at the end of the second year?
   (c) Set up and solve a polynomial equation and determine the YTM. What do you notice about the YTM?

3. On January 2 each year for three years, Greg Tobin deposited bonuses of $1025, $2200, and $1850, respectively, in an account. He received no bonus the following year, so he made no deposit. At the end of the fourth year, there was $5864.17 in the account.
   (a) Draw a time line for these investments.

   (b) Write a polynomial equation in x ($x = 1 + i$) and use technology or the guess and check method to find the YTM for these investments.

4. Carter Fenton invested yearly in a fund for his children's college education. At the beginning of the first year, he invested $1000; at the beginning of the second year, $2000; at the third through the sixth, $2500 each year; and at the beginning of the seventh, he invested $5000. At the beginning of the eighth year, there was $21,259 in the fund.
   (a) Draw a time line for this investment program.
   (b) Write a seventh-degree polynomial equation in $1 + i$ that gives the YTM for this investment program.
   (c) Use a graphing calculator to show that the YTM is less than 5.07% and greater than 5.05%.
   (d) Use a graphing calculator to calculate the solution for $1 + i$ and find the YTM.

5. People often lose money on investments. Christine O'Brien invested $50 at the beginning of each of two years in a mutual fund, and at the end of two years her investment was worth $90.
   (a) Draw a time line and set up a polynomial equation in $1 + i$. Solve for i.
   (b) Examine each negative solution (rate of return on the investment) to see if it has a reasonable interpretation in the context of the problem. To do this, use the compound interest formula on each value of i to trace each $50 payment to maturity.

Chapter 6

Systems of Linear Equations and Matrices

Many applications of mathematics require finding the solution of a *system* of first-degree equations. This chapter presents methods for solving such systems, including matrix methods. Matrix algebra and other applications of matrices are also discussed.

6.1 SYSTEMS OF LINEAR EQUATIONS

This section deals with **linear** (or **first-degree**) **equations** such as

$$2x + 3y = 14 \qquad \text{linear equation in two variables,}$$
$$4x - 2y + 5z = 8 \qquad \text{linear equation in three variables,}$$

and so on. A **solution** of such an equation is an ordered list of numbers that, when substituted for the variables in the order they appear, produces a true statement. For instance, $(1, 4)$ is a solution of the equation $2x + 3y = 14$ because substituting $x = 1$ and $y = 4$ produces the true statement $2(1) + 3(4) = 14$. Similarly, $(0, -4, 0)$ is a solution of $4x - 2y + 5z = 8$ because $4(0) - 2(-4) + 5(0) = 8$.

Many applications involve **systems of linear equations,** such as these two:

Two equations in two variables

$$5x - 3y = 7$$
$$2x + 4y = 8$$

Three equations in four variables

$$2x + y + z \qquad\; = 3$$
$$x + y + z + w = 5$$
$$-4x + \quad\; z + w = 0$$

275

Which of $(-8, 3, 0)$, $(8, -2, -2)$, $(-6, 5, -2)$ are solutions of the system

$$x + 2y + 3z = -2$$
$$2x + 6y + z = 2$$
$$3x + 3y + 10z = -2?$$

Answer:

Only $(8, -2, -2)$

A **solution of a system** is a solution that satisfies *all* the equations in the system. For instance, in the right-hand system of equations at the bottom of the previous page, $(1, 0, 1, 3)$ is a solution of all three equations (check it) and hence is a solution of the system. On the other hand, $(1, 1, 0, 3)$ is a solution of the first two equations but not the third. Hence $(1, 1, 0, 3)$ is not a solution of the system. ①

Systems of Two Equations in Two Variables

The graph of a linear equation in two variables is a straight line. Earlier we saw that the coordinates of each point on the graph represent a solution of the equation. Thus, the solution of a system of two such equations is represented by the point or points where the two lines intersect. There are exactly three geometric possibilities for two lines: they intersect at a single point, or they coincide, or they are distinct and parallel. As illustrated in Figure 6.1, each of these geometric possibilities corresponds to an algebraic outcome for which special terminology is used.

Lines intersect at one point: exactly one solution.

Lines coincide: infinitely many solutions.

Lines are parallel: no solutions.

Independent system
(a)

Dependent system
(b)

Inconsistent system
(c)

FIGURE 6.1

In theory, every system of two equations in two variables can be solved by graphing the lines and finding their intersection points (if any). In practice, however, algebraic techniques are usually needed to determine the solutions precisely. Such systems can be solved by the **elimination method,** as illustrated in the following examples.

▶ **EXAMPLE 1**   Solve the system

$$3x - 4y = 1 \tag{1}$$
$$2x + 3y = 12. \tag{2}$$

Multiply the first equation by 2 and the second equation by -3 to get

$$6x - 8y = 2$$
$$-6x - 9y = -36$$

②

Solve the system of equations

$$3x + 2y = -1$$
$$5x - 3y = 11.$$

Draw the graph of each equation on the same axes.

Answer:

$(1, -2)$

The multipliers 2 and -3 were chosen so that the coefficients of x in the two equations would be negatives of each other. Any solution of both these equations must also be a solution of their sum:

$$
\begin{array}{r}
6x - 8y = 2 \\
-6x - 9y = -36 \\
\hline
-17y = -34 \\
y = 2
\end{array}
$$

Variable x is eliminated.

To find the corresponding value of x, substitute 2 for y in either of the original equations. We choose the first equation.

$$3x - 4(2) = 1$$
$$3x - 8 = 1$$
$$x = 3$$

Therefore, the solution of the system is $(3, 2)$. The graphs of both equations of the system are shown in Figure 6.1(a). They intersect at the point $(3, 2)$, the solution of the system.   ◀ ②

TECHNOLOGY TIP   To find approximate solutions of a system of two equations in two variables with a graphing calculator, solve each equation for y and graph both of them on the same screen. In Example 1, for instance, we would graph

$$y_1 = \frac{3x - 1}{4} \quad \text{and} \quad y_2 = \frac{12 - 2x}{3}.$$

The point where the graphs intersect (the solution of the system) can then be found by using the calculator's intersection finder, as shown in Figure 6.2.

FIGURE 6.2

▶ **EXAMPLE 2**   Solve the system

$$-4x + y = 2$$
$$8x - 2y = -4.$$

Solve the following system.
$$3x - 4y = 13$$
$$12x - 16y = 52$$

Answer:

All ordered pairs that satisfy the equation $3x - 4y = 13$ (or $12x - 16y = 52$)

Solve the system
$$x - y = 4$$
$$2x - 2y = 3.$$
Draw the graph of each equation on the same axes.

Answer:

No solution

Eliminate x by multiplying both sides of the first equation by 2 and adding the results to the second equation.

$$
\begin{array}{rcr}
-8x + 2y = & 4 \\
8x - 2y = & -4 \\
\hline
0 = & 0
\end{array}
\qquad \text{Both variables eliminated.}
$$

Although both variables have been eliminated, the resulting statement "$0 = 0$" is true, which is the algebraic indication that the two equations have the same graph, as shown in Figure 6.1(b). Therefore the system is dependent and has an infinite number of solutions: every ordered pair that is a solution of the equation $-4x + y = 2$ is a solution of the system. ◀ ③

▶ **EXAMPLE 3**   Solve the system

$$3x - 2y = 4$$
$$-6x + 4y = 7.$$

The graphs of these equations are parallel lines (each has slope 3/2), as shown in Figure 6.1(c). Therefore the system has no solution. However, you do not need the graphs to discover this fact. If you try to solve the system algebraically by multiplying both sides of the first equation by 2 and adding the results to the second equation, you obtain

$$
\begin{array}{rcr}
6x - 4y = & 8 \\
-6x + 4y = & 7 \\
\hline
0 = & 15.
\end{array}
$$

The false statement "$0 = 15$" is the algebraic signal that the system is inconsistent and has no solution. ◀ ④

Larger Systems of Linear Equations

Two systems of equations are said to be **equivalent** if they have the same solutions. The basic procedure for solving a large system of equations is to use properties of algebra to transform the system into a simpler, equivalent system and then solve this simpler system.

> Performing any one of the following **elementary operations** on a system of linear equations produces an equivalent system.
>
> 1. Interchange any two equations.
> 2. Multiply both sides of an equation by a nonzero constant.
> 3. Replace an equation by the sum of itself and a constant multiple of another equation in the system.

Performing either of the first two elementary operations produces an equivalent system because rearranging the order of the equations or multiplying both sides of an equation by a constant does not affect the solutions of the individual equations, and

Verify that $x = 2$, $y = 1$ is the solution of the system

$$x - 3y = -1$$
$$3x + 2y = 8.$$

(a) Replace the second equation by the sum of itself and -3 times the first equation.

(b) What is the solution of the system in part (a)?

Answers:

(a) The system becomes

$$x - 3y = -1$$
$$11y = 11.$$

(b) $x = 2$, $y = 1$

hence it does not affect the solutions of the system. No formal proof will be given here that performing the third elementary operation produces an equivalent system, but side problem 5 illustrates this fact. ⑤

Example 4 shows how to use elementary operations on a system to eliminate certain variables and produce an equivalent system that is easily solved. The italicized statements provide an outline of the procedure.

▶ **EXAMPLE 4** Use the elimination method to solve the system

$$2x + y - z = 2$$
$$x + 3y + 2z = 1$$
$$x + y + z = 2.$$

First, *use elementary operations to produce an equivalent system in which 1 is the coefficient of x in the first equation.* One way to do this is to interchange the first two equations (another would be to multiply both sides of the first equation by 1/2).

$$x + 3y + 2z = 1 \qquad \text{Interchange } R_1, R_2.$$
$$2x + y - z = 2$$
$$x + y + z = 2$$

Here and below we use R_1 to denote the first equation in a system, R_2 the second equation, and so on.

Next, *use elementary operations to produce an equivalent system in which the x-term has been eliminated from the second and third equations.* To eliminate the x-term from the second equation above, replace the second equation by the sum of itself and -2 times the first equation.

$$
\begin{array}{r|r}
-2R_1 & -2x - 6y - 4z = -2 \\
R_2 & 2x + y - z = 2 \\
\hline
-2R_1 + R_2 & -5y - 5z = 0
\end{array}
$$

$$x + 3y + 2z = 1$$
$$-5y - 5z = 0 \qquad -2R_1 + R_2 \leftarrow$$
$$x + y + z = 2$$

To eliminate the x-term from the third equation of this last system, replace the third equation by the sum of itself and -1 times the first equation.

$$
\begin{array}{r|r}
-1R_1 & -x - 3y - 2z = -1 \\
R_3 & x + y + z = 2 \\
\hline
-1R_1 + R_3 & -2y - z = 1
\end{array}
$$

$$x + 3y + 2z = 1$$
$$-5y - 5z = 0$$
$$-2y - z = 1 \qquad -1R_1 + R_3 \leftarrow$$

Now that x has been eliminated from all but the first equation, we ignore the first equation and work on the remaining ones. *Use elementary operations to produce an equivalent system in which 1 is the coefficient of y in the second equation.*

This can be done by multiplying the second equation in the system above by $-1/5$.

$$x + 3y + 2z = 1$$
$$y + z = 0 \qquad -\frac{1}{5}R_2$$
$$-2y - z = 1$$

Then *use elementary operations to obtain an equivalent system in which y has been eliminated from the third equation.* Replace the third equation by the sum of itself and 2 times the second equation:

$$x + 3y + 2z = 1$$
$$y + z = 0$$
$$z = 1. \qquad 2R_2 + R_3$$

The solution of the third equation is obvious: $z = 1$. Now work backward in the system. Substitute 1 for z in the second equation and solve for y, obtaining $y = -1$. Finally, substitute 1 for z and -1 for y in the first equation and solve for x, obtaining $x = 2$. This process is known as **back substitution.** When it is finished we have the solution of the original system, namely, $(2, -1, 1)$. It is always wise to check the solution by substituting the values for x, y, and z in all equations of the original system. ◀

The procedure used in Example 4 to eliminate variables and produce a system in which back substitution works can be carried out with any system, as summarized below. In this summary, the first variable that appears in an equation with nonzero coefficient is called the **leading variable** of that equation, and its nonzero coefficient is called the **leading coefficient.**

> ## The Elimination Method for Solving Large Systems of Linear Equations
>
> Use elementary operations to transform the given system into an equivalent one as follows:
> 1. Make the leading coefficient of the first equation 1 either by interchanging equations or by multiplying the first equation by a suitable constant.
> 2. Eliminate the leading variable of the first equation from each later equation by replacing the later equation by the sum of itself and a suitable multiple of the first equation.
> 3. Repeat Steps 1 and 2 for the second equation: make its leading coefficient 1 and eliminate its leading variable from each later equation by replacing the later equation by the sum of itself and a suitable multiple of the second equation.
> 4. Repeat Steps 1 and 2 for the third equation, fourth equation, and so on, until it is not possible to go further.
>
> Then solve the resulting system by back substitution.

6

Use the elimination method to solve each system.

(a) $\begin{aligned} 2x + y &= -1 \\ x + 3y &= 2 \end{aligned}$

(b) $\begin{aligned} 2x - y + 3z &= 2 \\ x + 2y - z &= 6 \\ -x - y + z &= -5 \end{aligned}$

Answers:

(a) $(-1, 1)$

(b) $(3, 1, -1)$

At various stages in the elimination process, you may have a choice of elementary operations that can be used. As long as the final result is a system in which back

7

a) Write the augmented matrix of this system.

$4x - 2y + 3z = 4$

$3x + 5y + z = -7$

$5x - y + 4z = 6$

(b) Write the system of equations associated with this augmented matrix.

$$\begin{bmatrix} 2 & -2 & \vline & -2 \\ 1 & 1 & \vline & 4 \\ 3 & 5 & \vline & 8 \end{bmatrix}$$

Answers:

(a) $\begin{bmatrix} 4 & -2 & 3 & \vline & 4 \\ 3 & 5 & 1 & \vline & -7 \\ 5 & -1 & 4 & \vline & 6 \end{bmatrix}$

(b) $2x - 2y = -2$

$x + y = 4$

$3x + 5y = 8$

substitution can be used, the choice does not matter. To avoid unnecessary errors, choose elementary operations that minimize the amount of computation and, as far as possible, avoid complicated fractions. **6**

Matrix Methods

You may have noticed that the variables in a system of equations remain unchanged during the solution process. So we only need to keep track of the coefficients and the constants. For instance, consider the system in Example 4:

$$2x + y - z = 2$$
$$x + 3y + 2z = 1$$
$$x + y + z = 2.$$

This system can be written in an abbreviated form as

$$\begin{bmatrix} 2 & 1 & -1 & 2 \\ 1 & 3 & 2 & 1 \\ 1 & 1 & 1 & 2 \end{bmatrix}.$$

Such a rectangular array of numbers, consisting of horizontal **rows** and vertical **columns**, is called a **matrix** (plural **matrices**). Each number in the array is an **element** or **entry**. To separate the constants in the last column of the matrix from the coefficients of the variables, we sometimes use a vertical line, producing the following **augmented matrix**.

$$\begin{bmatrix} 2 & 1 & -1 & \vline & 2 \\ 1 & 3 & 2 & \vline & 1 \\ 1 & 1 & 1 & \vline & 2 \end{bmatrix}$$ **7**

The rows of the augmented matrix can be transformed in the same way as the equations of the system, since the matrix is just a shortened form of the system. The following **row operations** on the augmented matrix correspond to the elementary operations used on systems of equations.

8

Perform the following row operations on the matrix

$$\begin{bmatrix} -1 & 5 \\ 3 & -2 \end{bmatrix}.$$

(a) Interchange R_1 and R_2.

(b) $2R_1$

(c) Replace R_2 by $-3R_1 + R_2$.

(d) Replace R_1 by $2R_2 + R_1$.

Answers:

(a) $\begin{bmatrix} 3 & -2 \\ -1 & 5 \end{bmatrix}$

(b) $\begin{bmatrix} -2 & 10 \\ 3 & -2 \end{bmatrix}$

(c) $\begin{bmatrix} -1 & 5 \\ 6 & -17 \end{bmatrix}$

(d) $\begin{bmatrix} 5 & 1 \\ 3 & -2 \end{bmatrix}$

Performing any one of the following **row operations** on the augmented matrix of a system of linear equations produces the augmented matrix of an equivalent system.

1. Interchange any two rows.
2. Multiply each element of a row by a nonzero constant.
3. Replace a row by the sum of itself and a constant multiple of another row of the matrix.

Row operations on a matrix are indicated by the same notation we used for elementary operations on a system of equations. For example, $2R_3 + R_1$ indicates the sum of 2 times row 3 and row 1. **8**

▶ **EXAMPLE 5** Use matrices to solve the system

$$x - 2y = 6 - 4z$$
$$x + 13z = 6 - y$$
$$-2x + 6y - z = -10.$$

First, put the system in the required form, with the constants on the right side of the equal sign and the terms with variables *in the same order* in each equation on the left side of the equal sign. Then write the augmented matrix of the system.

$$
\begin{array}{l}
x - 2y + 4z = 6 \\
x + y + 13z = 6 \\
-2x + 6y - z = -10
\end{array}
\qquad
\left[\begin{array}{ccc|c}
1 & -2 & 4 & 6 \\
1 & 1 & 13 & 6 \\
-2 & 6 & -1 & -10
\end{array}\right]
$$

The matrix method is the same as the elimination method, except that row operations are used on the augmented matrix instead of elementary operations on the corresponding system of equations, as shown in this side-by-side comparison.

Equation Method	*Matrix Method*
Replace the second equation by the sum of itself and -1 times the first equation.	Replace the second row by the sum of itself and -1 times the first row.

$$
\begin{array}{l}
x - 2y + 4z = 6 \\
3y + 9z = 0 \\
-2x + 6y - z = -10
\end{array}
\quad \leftarrow -1R_1 + R_2 \rightarrow
\left[\begin{array}{ccc|c}
1 & -2 & 4 & 6 \\
0 & 3 & 9 & 0 \\
-2 & 6 & -1 & -10
\end{array}\right]
$$

Replace the third equation by the sum of itself and 2 times the first equation.

Replace the third row by the sum of itself and 2 times the first row.

$$
\begin{array}{l}
x - 2y + 4z = 6 \\
3y + 9z = 0 \\
2y + 7z = 2
\end{array}
\quad \leftarrow 2R_1 + R_3 \rightarrow
\left[\begin{array}{ccc|c}
1 & -2 & 4 & 6 \\
0 & 3 & 9 & 0 \\
0 & 2 & 7 & 2
\end{array}\right]
$$

Multiply both sides of the second equation by 1/3.

Multiply each element of row 2 by 1/3.

$$
\begin{array}{l}
x - 2y + 4z = 6 \\
y + 3z = 0 \\
2y + 7z = 2
\end{array}
\quad \leftarrow \frac{1}{3}R_2 \rightarrow
\left[\begin{array}{ccc|c}
1 & -2 & 4 & 6 \\
0 & 1 & 3 & 0 \\
0 & 2 & 7 & 2
\end{array}\right]
$$

Replace the third equation by the sum of itself and -2 times the second equation.

Replace the third row by the sum of itself and -2 times the second row.

$$
\begin{array}{l}
x - 2y + 4z = 6 \\
y + 3z = 0 \\
z = 2
\end{array}
\quad \leftarrow -2R_2 + R_3 \rightarrow
\left[\begin{array}{ccc|c}
1 & -2 & 4 & 6 \\
0 & 1 & 3 & 0 \\
0 & 0 & 1 & 2
\end{array}\right]
$$

Complete the matrix solution of the system with this augmented matrix.

$$\begin{bmatrix} 1 & 1 & 1 & 2 \\ 1 & -2 & 1 & -1 \\ 0 & 3 & 1 & 5 \end{bmatrix}$$

Answer:

$(-1, 1, 2)$

Now use back substitution.

$$z = 2 \qquad\qquad y + 3(2) = 0 \qquad\qquad x - 2(-6) + 4(2) = 6$$
$$y = -6 \qquad\qquad x + 20 = 6$$
$$x = -14$$

The solution of the system is $(-14, -6, 2)$. ◀

A matrix, such as the last one in Example 5, is said to be in **row echelon form** when

All rows consisting entirely of zeros (if any) are at the bottom.
The first nonzero entry in each row is 1 (called a *leading* 1).
Each leading 1 appears to the right of the leading 1's in any preceding rows.

When a row echelon form matrix is the augmented matrix of a system of equations, as in Example 5, the system can readily be solved by back substitution. So the matrix solution method amounts to transforming the augmented matrix of a system of equations into a row echelon form matrix.

Virtually all graphing calculators have matrix capabilities that make it easy to solve systems of linear equations.

EXAMPLE 6 Use a graphing calculator to solve the system in Example 5.

Enter the augmented matrix of the system in the calculator (Figure 6.3) Then use the REF key (in the MATH or OPS submenu of the MATRIX menu) to put this matrix into row echelon form. Figure 6.4 shows the resulting row echelon form matrix and the corresponding system of equations.

$$x - 3y + \quad .5z = 5$$
$$y + 3.125z = .25$$
$$z = 2$$

FIGURE 6.3   **FIGURE 6.4**

Because the calculator used a different sequence of row operations than was used in Example 5, it produced a different row echelon form matrix (and corresponding system). However, back substitution shows that the solutions of this system are the same ones found in Example 5.

$$z = 2 \qquad y + 3.125(2) = .25 \qquad x - 3(-6) + .5(2) = 5$$
$$y + 6.25 = .25 \qquad x + 18 + 1 = 5$$
$$y = -6 \qquad x = -14 \ ◀$$

Dependent and Inconsistent Systems

Recall that a system of equations in two variables may have exactly one solution, or an infinite number of solutions, or no solutions at all. This fact was illustrated geometrically in Figure 6.1. The same thing is true for systems with three or more variables (and the same terminology is used): a system has exactly one solution (an **independent system**), or an infinite number of solutions (a **dependent system**), or no solutions at all (an **inconsistent system**). Both the equation method and the matrix method always produce the unique solution of an independent system. The matrix method also provides a useful way of describing the infinitely many solutions of a dependent system, as we now see.

▶ **EXAMPLE 7** Solve the system

$$2x - 3y + 4z = 6$$
$$x - 2y + z = 9.$$

Use the steps of the matrix method as far as possible. The augmented matrix is

$$\begin{bmatrix} 2 & -3 & 4 & | & 6 \\ 1 & -2 & 1 & | & 9 \end{bmatrix}.$$

$$\begin{bmatrix} 1 & -2 & 1 & | & 9 \\ 2 & -3 & 4 & | & 6 \end{bmatrix} \qquad \text{Interchange } R_1 \text{ and } R_2.$$

$$\begin{bmatrix} 1 & -2 & 1 & | & 9 \\ 0 & 1 & 2 & | & -12 \end{bmatrix} \quad -2R_1 + R_2 \qquad\qquad \begin{aligned} x - 2y + z &= 9 \\ y + 2z &= -12 \end{aligned}$$

The last augmented matrix above represents the system shown to its right. Since there are only two rows in the matrix, it is not possible to continue the process. The fact that the corresponding system has one variable (namely, z) that is not the leading variable of an equation indicates a dependent system. Its solutions can be found as follows. Solve the second equation for y.

$$y = -2z - 12$$

Now substitute the result for y in the first equation and solve for x.

$$x - 2y + z = 9$$
$$x - 2(-2z - 12) + z = 9$$
$$x + 4z + 24 + z = 9$$
$$x + 5z = -15$$
$$x = -5z - 15$$

Use the following values of z to find additional solutions for the system of Example 7.

(a) $z = 7$

(b) $z = -14$

(c) $z = 5$

Answers:

(a) $(-50, -26, 7)$

(b) $(55, 16, -14)$

(c) $(-40, -22, 5)$

Each choice of a value for z leads to values for x and y. For example,

$$\text{if } z = 1, \quad \text{then } x = -20 \quad \text{and} \quad y = -14;$$
$$\text{if } z = -6, \quad \text{then } x = 15 \quad \text{and} \quad y = 0;$$
$$\text{if } z = 0, \quad \text{then } x = -15 \quad \text{and} \quad y = -12.$$

There are infinitely many solutions for the original system, since z can take on infinitely many values. The solutions are all ordered triples in the form

$$(-5z - 15, -2z - 12, z),$$

where z is any real number. ◀ ⑩

Since both x and y in Example 7 were expressed in terms of z, the variable z is called a **parameter.** If we solved the system in a different way, x or y could be the parameter. The system in Example 7 had one more variable than equations. If there are two more variables than equations, there usually will be two parameters, and so on.

FIGURE 6.5

▶ **EXAMPLE 8**   Row operations were used to reduce the augmented matrix of a system of three equations in four variables $(x, y, z,$ and $w)$ to the row echelon form matrix in Figure 6.5. Solve the system.

First write out the system represented by the matrix.

$$x + 2y + 4z - w = 0$$
$$y - 2z + 3w = 2$$
$$z - 2w = 1$$

Let w be the parameter and use back substitution.

$$z = 2w + 1 \quad y - 2(2w + 1) + 3w = 2 \quad x + 2(w + 4) + 4(2w + 1) - w = 0$$
$$y - 4w - 2 + 3w = 2 \quad x + 2w + 8 + 8w + 4 - w = 0$$
$$y = w + 4 \quad x = -9w - 12$$

The solutions are given by $(-9w - 12, w + 4, 2w + 1, w)$, where w is any real number. ◀

When a system is inconsistent, the matrix method will indicate this fact, too, as in the next example.

▶ **EXAMPLE 9**   Solve the system

$$4x + 12y + 8z = -4$$
$$2x + 8y + 5z = 0$$
$$3x + 9y + 6z = 2.$$

Write the augmented matrix and go through the steps of the matrix method.

$$\left[\begin{array}{ccc|c} 4 & 12 & 8 & -4 \\ 2 & 8 & 5 & 0 \\ 3 & 9 & 6 & 2 \end{array}\right]$$

$$\left[\begin{array}{ccc|c} 1 & 3 & 2 & -1 \\ 2 & 8 & 5 & 0 \\ 3 & 9 & 6 & 2 \end{array}\right] \quad (1/4)R_1$$

$$\left[\begin{array}{ccc|c} 1 & 3 & 2 & -1 \\ 0 & 2 & 1 & 2 \\ 3 & 9 & 6 & 2 \end{array}\right] \quad -2R_1 + R_2$$

$$\left[\begin{array}{ccc|c} 1 & 3 & 2 & -1 \\ 0 & 2 & 1 & 2 \\ 0 & 0 & 0 & 5 \end{array}\right] \quad -3R_1 + R_3$$

(11)

Complete the matrix solution of the system.

$$\left[\begin{array}{ccc|c} -1 & 3 & -2 & -1 \\ 1 & -2 & 3 & 1 \\ 2 & -4 & 6 & 5 \end{array}\right]$$

Answer:

No solution

Stop! The last row has all zeros to the left of the vertical bar, so the corresponding equation is $0x + 0y + 0z = 5$. This equation has no solution because its left side is 0 and its right side is 5, resulting in the false statement "$0 = 5$." Therefore the entire system cannot have any solutions and is inconsistent. ◀ **(11)**

CAUTION It is possible for the elimination method to result in an equation that is 0 on *both* sides, such as $0x + 0y + 0z = 0$. Unlike the situation in Example 9, such an equation has infinitely many solutions, so a system that contains it may also have solutions.

Applications

To apply the mathematical techniques in this text to practical problems, begin by reading the problem carefully. Next, identify what must be found. Let each unknown quantity be represented by a variable. (It is a good idea to *write down* exactly what each variable represents.) Now reread the problem, looking for all necessary data. Write that down, too. Finally, look for one or more sentences that lead to equations or inequalities.

▶**EXAMPLE 10** Kelly Karpet Kleaners sells rug cleaning machines. The EZ model weighs 10 pounds and comes in a 10 cubic foot box. The compact model weighs 20 pounds and comes in an 8 cubic foot box. The commercial model weighs 60 pounds and comes in a 28 cubic foot box. Each of their delivery vans has 248 cubic feet of space and can hold a maximum of 440 pounds. In order for a van to be fully loaded, how many of each model should it carry?

Let x be the number of EZ, y the number of compact, and z the number of commercial models carried by a van. Then we can summarize the information in this chart.

Model	*Number*	*Weight*	*Volume*
EZ	x	10	10
Compact	y	20	8
Commercial	z	60	28
Total for a load		440	248

Since a fully loaded van can carry 440 pounds and 248 cubic feet, we must solve this system of equations.

$$10x + 20y + 60z = 440 \qquad \text{Weight equation}$$
$$10x + 8y + 28z = 248 \qquad \text{Volume equation}$$

The augmented matrix of this system is

$$\begin{bmatrix} 10 & 20 & 60 & | & 440 \\ 10 & 8 & 28 & | & 248 \end{bmatrix}.$$

Transform it into row echelon form, either by using a graphing calculator or by performing row operations by hand, as in problem 12 at the side. ⑫

A graphing calculator or problem 12 at the side shows that the original system is equivalent to the following one.

$$x + 2y + 6z = 44$$
$$y + \frac{8}{3}z = 16$$

---

⑫

List a sequence of row operations that will transform the augmented matrix in Example 10 into row echelon form.

Answer:

Many sequences are possible, including this one:

replace R_1 by $\frac{1}{10}R_1$;

replace R_2 by $\frac{1}{2}R_2$.

replace R_2 by $-5R_1 + R_2$;

replace R_2 by $-\frac{1}{6}R_2$.

$$\begin{bmatrix} 1 & 2 & 6 & | & 44 \\ 0 & 1 & \frac{8}{3} & | & 16 \end{bmatrix}$$

Solving this dependent system by back substitution, we have

$$y = 16 - \frac{8}{3}z$$

$$x = 44 - 2y - 6z = 44 - 2\left(16 - \frac{8}{3}z\right) - 6z = 12 - \frac{2}{3}z,$$

so that all solutions of the system are given by $\left(12 - \frac{2}{3}z, 16 - \frac{8}{3}z, z\right)$. The only solutions that apply in this situation, however, are those given by $z = 0, 3,$ or 6 because all other values of z lead to fractions or negative numbers (you can't deliver a negative number of boxes or part of a box). Hence, there are three ways to have a fully loaded van.

Solution	Van Load
$(12, 16, 0)$	12 EZ, 16 compact, 0 commercial
$(10, 8, 3)$	10 EZ, 8 compact, 3 commercial
$(8, 0, 6)$	8 EZ, 0 compact, 6 commercial ◀

6.1 Exercises

Determine whether the given ordered list of numbers is a solution of the system of equations.

1. $(-1, 3)$
$$2x + y = 1$$
$$-3x + 2y = 9$$

2. $(2, 1.5, -.5)$
$$3x + 4y - 2z = -.5$$
$$.5x + 8z = -3$$
$$x - 3y + 5z = -5$$

Solve each of the following systems of two equations in two variables. (See Examples 1–3.) $1-18$

3. $x - 2y = 5$
$2x + y = 3$

4. $3x - y = 1$
$-x + 2y = 4$ old fashion

5. $2x - 2y = 12$
$-2x + 3y = 10$

6. $3x + 2y = -4$
$4x - 2y = -10$ elimination

7. $x + 3y = -1$
$2x - y = 5$

8. $4x - 3y = -1$
$x + 2y = 19$

9. $2x + 3y = 15$
$8x + 12y = 40$

10. $2x + 5y = 8$
$6x + 15y = 18$

11. $2x - 8y = 2$
$3x - 12y = 3$

12. $3x - 2y = 4$
$6x - 4y = 8$

13. $3x + 2y = 5$
$6x + 4y = 8$

14. $9x - 5y = 1$
$-18x + 10y = 1$

15. Only one of the three screens at the right gives the correct graphs for the system in Exercise 8. Which is it? (*Hint:* Solve for y first in each equation and use the slope-intercept form to help you answer the question.)

(a)

(b)

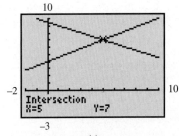

(c)

In Exercises 16 and 17, multiply both sides of each equation by a common denominator to eliminate the fractions. Then solve the system.

16. $\dfrac{x}{2} + \dfrac{y}{3} = 8$

$\dfrac{2x}{3} + \dfrac{3y}{2} = 17$

17. $\dfrac{x}{5} + 3y = 31$

$2x - \dfrac{y}{5} = 8$

18. **Business** When Neil Simon opens a new play, he has to decide whether to open the show on Broadway or off Broadway. For example, he decided to open his play *London Suite* off Broadway. From information provided by Emanuel Azenberg, his producer, the following equations were developed:

$$43,500x - y = 1,295,000$$
$$27,000x - y = 440,000,$$

where x represents the number of weeks that the show has run and y represents the profit or loss from the show (first equation is Broadway and second equation is off Broadway).*

(a) Solve this system of equations to determine when the profit/loss from the show will be equal for each venue. What is the profit?

(b) Discuss which venue is favorable for the show.

19. **Social Sciences** One of the factors that contribute to the success or failure of a particular army during war is its ability to get new troops ready for service. It is possible to analyze the rate of change in the number of troops of two hypothetical armies with the following simplified model,

Rate of increase (Red Army) $= 200,000 - .5r - .3b$

Rate of increase (Blue Army) $= 350,000 - .5r - .7b$,

where r is the number of soldiers in the Red Army at a given time and b is the number of soldiers in the Blue Army at a given time. The factors .5 and .7 represent each army's efficiency at bringing new soldiers to the fight.†

---

*Albert Goetz, "Basic Economics: Calculating against Theatrical Disaster," *Mathematics Teacher* 89, no. 1 (January 1996): 30–32. Reprinted with permission, ©1996 by the National Council of Teachers of Mathematics. All rights reserved.

†Ian Bellamy, "Modelling War," *Journal of Peace Research* 36, no. 6 (1999): 729–739. ©1999 Sage Publications, Ltd.

(a) Solve this system of equations to determine the number of soldiers in each army when the rate of increase for each is zero.

(b) Describe what might be going on in a war when the rate of increase is zero.

Write the augmented matrix of each of the following systems. Do not solve the systems.

20. $2x + y + z = 3$
$3x - 4y + 2z = -5$
$x + y + z = 2$

21. $3x + 4y - 2z - 3w = 0$
$x - 3y + 7z + 4w = 9$
$2x + 5z - 6w = 0$

Write the system of equations associated with the following augmented matrices. Do not solve the systems.

22. $\begin{bmatrix} 2 & 3 & 8 & | & 20 \\ 1 & 4 & 6 & | & 12 \\ 0 & 3 & 5 & | & 10 \end{bmatrix}$

23. $\begin{bmatrix} 3 & 2 & 6 & | & 18 \\ 2 & -2 & 5 & | & 7 \\ 1 & 0 & 5 & | & 20 \end{bmatrix}$

Use the indicated row operation to transform each matrix.

24. Interchange R_2 and R_3.

$\begin{bmatrix} 1 & 2 & 3 & | & -1 \\ 6 & 5 & 4 & | & 6 \\ 2 & 0 & 7 & | & -4 \end{bmatrix}$

25. Replace R_3 by $-3R_1 + R_3$.

$\begin{bmatrix} 1 & 5 & 2 & 0 & | & -1 \\ 8 & 5 & 4 & 6 & | & 6 \\ 3 & 0 & 7 & 1 & | & -4 \end{bmatrix}$

26. Replace R_2 by $2R_1 + R_2$.

$\begin{bmatrix} -4 & -3 & 1 & -1 & | & 2 \\ 8 & 2 & 5 & 0 & | & 6 \\ 0 & -2 & 9 & 4 & | & 5 \end{bmatrix}$

27. Replace R_3 by $\dfrac{1}{4}R_3$.

$\begin{bmatrix} 2 & 5 & 1 & | & -1 \\ -4 & 0 & 4 & | & 6 \\ 6 & 0 & 8 & | & -4 \end{bmatrix}$

In Exercises 28–33, perform row operations on the augmented matrix as far as necessary to determine whether the system is independent, dependent, or inconsistent. (See Examples 7–9.)

28. $x + 2y = 0$
$y - z = 2$
$x + y + z = -2$

29. $x + 2y + z = 0$
$y + 2z = 0$
$x + y - z = 0$

30. $x + 2y + 4z = 6$
$y + z = 1$
$x + 3y + 5z = 10$

31. $x + y + 2z + 3w = 1$
$2x + y + 3z + 4w = 1$
$3x + y + 4z + 5w = 2$

32. $a - 3b - 2c = -3$
$3a + 2b - c = 12$
$-a - b + 4c = 3$

33. $2x + 2y + 2z = 6$
$3x - 3y - 4z = -1$
$x + y + 3z = 11$

Write the augmented matrix of the system and use the matrix method to solve the system. If the system has an infinite number of solutions, express them in terms of the parameter z. (See Examples 4–9.)

34. $x + y + z = 2$
$2x + y - z = 5$
$x - y + z = -2$

35. $2x + y + z = 9$
$-x - y + z = 1$
$3x - y + z = 9$

36. $x + 3y + 4z = 14$
$2x - 3y + 2z = 10$
$3x - y + z = 9$
$4x + 2y + 5z = 23$

37. $4x - y + 3z = -2$
$3x + 5y - z = 15$
$-2x + y + 4z = 14$
$x + 6y + 3z = 29$

38.
$$x + 2y + 3z = 8$$
$$3x - y + 2z = 5$$
$$-2x - 4y - 6z = 5$$

39.
$$3x - 2y - 8z = 1$$
$$9x - 6y - 24z = -2$$
$$x - y + z = 1$$

40.
$$2x - 4y + z = -4$$
$$x + 2y - z = 0$$
$$-x + y + z = 6$$
$$2x - y + z = 2$$

41.
$$4x - 3y + z = 9$$
$$3x + 2y - 2z = 4$$
$$x - y + 3z = 5$$
$$2x + 3y - 5z = -1$$

42.
$$5x + 3y + 4z = 19$$
$$3x - y + z = -4$$

43.
$$3x + y - z = 0$$
$$2x - y + 3z = -7$$

44.
$$11x + 10y + 9z = 5$$
$$x + 2y + 3z = 1$$
$$3x + 2y + z = 1$$

45.
$$x + y = 3$$
$$5x - y = 3$$
$$9x - 4y = 1$$

46. Find constants a, b, c such that the points $(2, 3)$, $(-1, 0)$, and $(-2, 2)$ lie on the graph of the equation $y = ax^2 + bx + c$. (*Hint:* Since $(2, 3)$ is on the graph, we must have $3 = a(2^2) + b(2) + c$, that is, $4a + 2b + c = 3$. Similarly, the other two points lead to two more equations. Solve the resulting system for a, b, c.)

47. **(a)** Find the equation of the straight line through $(1, 2)$ and $(3, 4)$.
(b) Find the equation of the line through $(-1, 1)$ with slope 3.
(c) Find a point that lies on both the lines in (a) and (b).

48. Graph the equations in the following system. Then explain why the graphs show that the system is inconsistent.
$$2x + 3y = 8$$
$$x - y = 4$$
$$5x + y = 7$$

49. Explain why a system with more variables than equations cannot have a unique solution (that is, be an independent system). (*Hint:* When you apply the elimination method to such a system, what must necessarily happen?)

Work the following problems by writing and solving a system of equations. (See Example 10.)

50. **Physical Science**   Linear systems occur in the design of roof trusses for new homes and buildings. The simplest type of roof truss is a triangle. The truss shown in the figure at the top of the next column is used to frame roofs of small buildings. If a 100-pound force is applied at the peak of the truss, then the forces or weights W_1 and W_2 exerted parallel to each rafter of the truss are determined by the following system of linear equations.
$$\frac{\sqrt{3}}{2}(W_1 + W_2) = 100$$
$$W_1 - W_2 = 0$$

Solve the system to find W_1 and W_2.*

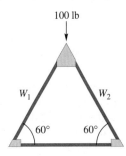

51. **Physical Science**   (Refer to Exercise 50.) Use the following system of equations to determine the forces or weights W_1 and W_2 exerted on each rafter for the truss shown in the figure.
$$W_1 + \sqrt{2}W_2 = 300$$
$$\sqrt{3}W_1 - \sqrt{2}W_2 = 0$$

52. **Finance**   Shirley Cicero has \$16,000 invested in Boeing and GE stock. The Boeing stock currently sells for \$30 a share and the GE stock for \$70 a share. If GE stock triples in value and Boeing stock goes up 50%, her stock will be worth \$34,500. How many shares of each stock does she own?

53. **Business**   An apparel shop sells skirts for \$45 and blouses for \$35. Its entire stock is worth \$51,750. But sales are slow and only half the skirts and two-thirds of the blouses are sold, for a total of \$30,600. How many skirts and blouses are left in the store?

54. **Business**   A company produces two models of bicycles, model 201 and model 301. Model 201 requires 2 hours of assembly time, and model 301 requires 3 hours of assembly time. The parts for model 201 cost \$25 per bike, and the parts for model 301 cost \$30 per bike. If the company has a total of 34 hours of assembly time and \$365 available per day for these two models, how many of each can be made in a day?

55. **Social Science**   The relationship between a professional basketball player's height H (in inches) and weight W (in pounds) was modeled using two different samples of players. The resulting equations that modeled each sample were $W = 7.46H - 374$ and $W = 7.93H - 405$.

---

*R. Hibbeler, *Structural Analysis* (Englewood Cliffs, N.J.: Prentice-Hall, 1995).

(a) Use both equations to predict the weight of a 6'11" professional basketball player.

(b) According to each model, what change in weight is associated with a 1-inch increase in height?

(c) Determine the weight and height where the two models agree.

56. Finance An investor wants to invest $30,000 in corporate bonds that are rated AAA, A, and B. The lower rated ones pay higher interest but are higher risk as well. The average yield is 5% on AAA bonds, 6% on A bonds and 10% on B bonds. Being conservative, the investor wants to have twice as much in AAA bonds as in B bonds. How much should she invest in each type of bond to have an interest income of $2000?

57. Business A candy store sells cashews for $4.40 per pound and peanuts for $1.20 per pound. If you want to buy exactly 3 pounds of nuts for $6, how many pounds of each kind should you buy? (*Hint:* If you buy x pounds of cashews and y pounds of peanuts, then $x + y = 3$. Find a second equation by considering cost; then solve the resulting system.)

58. Business A winemaker has two large casks of wine. One wine is 8% alcohol and the other is 18% alcohol. How many gallons of each wine should be mixed to produce 30 gallons of wine that is 12% alcohol? (*Hint:* If x is the number of gallons of the 8% wine and y the number of gallons of the 18% wine, then $x + y = ?$ Use the fact that the amount of alcohol in the mixture is "8% of x + 18% of y" to find a second equation.)

59. Health Computer Aided Tomography (CAT) scanners take X-rays of a part of the body from different directions and put the information together to create a picture of a cross section of the body.* The amount by which the energy of the X-ray decreases, measured in linear-attenuation units, tells whether the X-ray has passed through healthy tissue, tumorous tissue, or bone, based on the following table.

Type of Tissue	Linear-Attenuation Values
Healthy tissue	.1625–.2977
Tumorous tissue	.2679–.3930
Bone	.3857–.5108

The part of the body to be scanned is divided into cells. If an X-ray passes through more than one cell, the total linear-attenuation value is the sum of the values for the cells. For example, in the figure at the top of the next column, let a, b, and c be the values for cells A, B, and C. The attenuation value for beam 1 is $a + b$ and for beam 2 is $a + c$.

(a) Find the attenuation value for beam 3.

(b) Suppose that the attenuation values are .8, .55, and .65 for beams 1, 2, and 3, respectively. Set up and solve the system of three equations for a, b, and c. What can you conclude about cells A, B, and C?

60. Health (Refer to Exercise 59.) Four X-ray beams are aimed at four cells, as shown in the following figure.

(a) Suppose the attenuation values for beams 1, 2, 3, and 4 are .60, .75, .65, and .70, respectively. Do we have enough information to determine the values of a, b, c, and d? Explain.

(b) Suppose we have the data from part (a), as well as the following values for d. Find the values for a, b, and c and make conclusions about cells A, B, C, and D in each case.

(i) .33   (ii) .43

(c) Two X-ray beams are added, as shown in the figure. In addition to the data in part (a), we now have attenuation values for beams 5 and 6 of .85 and .50. Find the values for a, b, c, and d and make conclusions about cells A, B, C, and D.

*Exercises 59 and 60 are based on the article "Medical Applications of Linear Equations" by David Jabon, Gail Nord, Bryce W. Wilson, and Penny Coffman, *Mathematics Teacher* 89, no. 5 (May 1996) p. 98: 398.

61. *Transportation* At rush hours, substantial traffic congestion is encountered at the traffic intersections shown in the figure. (The streets are one-way, as shown by the arrows.)

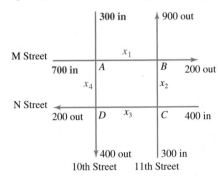

The city wishes to improve the signals at these corners to speed the flow of traffic. The traffic engineers first gather data. As the figure shows, 700 cars per hour come down M Street to intersection A, and 300 cars per hour come down 10th Street to intersection A. A total of x_1 of these cars leave A on M Street, and x_4 cars leave A on 10th Street. The number of cars entering A must equal the number leaving, so that

$$x_1 + x_4 = 700 + 300$$

or

$$x_1 + x_4 = 1000.$$

For intersection B, x_1 cars enter on M street and x_2 on 11th Street. The figure shows that 900 cars leave B on 11th and 200 on M. We have

$$x_1 + x_2 = 900 + 200$$
$$x_1 + x_2 = 1100.$$

(a) Write two equations representing the traffic entering and leaving intersections C and D.

(b) Solve the system of four equations using x_4 as the parameter.

(c) Based on your solution to part (b) what are the largest and smallest possible values for the number of cars leaving intersection A on 10th Street?

(d) Answer the question in part (c) for the other three variables.

(e) Verify that you could have discarded any one of the four original equations without changing the solution. What does this tell you about the original problem?

62. *Transportation* The diagram shows the traffic flow at four intersections during rush hour, as in Exercise 61.

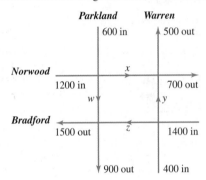

(a) What are the possible values x, y, z, w in order to avoid any congestion? (Use w as the parameter.)

(b) What are the possible values of w?

6.2 THE GAUSS-JORDAN METHOD

In Example 5 of the previous section, we used matrix methods to rewrite the system

$$x - 2y = 6 - 4z$$
$$x + 13z = 6 - y$$
$$-2x + 6y - z = -10$$

as an augmented matrix. We carried out the steps of the matrix method until the final matrix was

$$\begin{bmatrix} 1 & -2 & 4 & | & 6 \\ 0 & 1 & 3 & | & 0 \\ 0 & 0 & 1 & | & 2 \end{bmatrix}.$$

Use the Gauss-Jordan method to solve the system
$$x + 2y = 11$$
$$-4x + y = -8,$$
as follows. Give the shorthand notation and the new matrix in (b)–(d).

(a) Set up the augmented matrix.

(b) Get 0 in row two, column one.

(c) Get 1 in row two, column two.

(d) Finally, get 0 in row one, column two.

(e) The solution for the system is _____.

Answers:

(a) $\begin{bmatrix} 1 & 2 & | & 11 \\ -4 & 1 & | & -8 \end{bmatrix}$

(b) $4R_1 + R_2$; $\begin{bmatrix} 1 & 2 & | & 11 \\ 0 & 9 & | & 36 \end{bmatrix}$

(c) $\frac{1}{9}R_2$; $\begin{bmatrix} 1 & 2 & | & 11 \\ 0 & 1 & | & 4 \end{bmatrix}$

(d) $-2R_2 + R_1$; $\begin{bmatrix} 1 & 0 & | & 3 \\ 0 & 1 & | & 4 \end{bmatrix}$

(e) $(3, 4)$

We then used back substitution to solve it. In the **Gauss-Jordan method,** additional elimination of variables replaces back substitution, as follows.

$$\begin{bmatrix} 1 & -2 & 4 & | & 6 \\ 0 & 1 & 0 & | & -6 \\ 0 & 0 & 1 & | & 2 \end{bmatrix} \quad -3R_3 + R_2$$

$$\begin{bmatrix} 1 & -2 & 0 & | & -2 \\ 0 & 1 & 0 & | & -6 \\ 0 & 0 & 1 & | & 2 \end{bmatrix} \quad -4R_3 + R_1$$

$$\begin{bmatrix} 1 & 0 & 0 & | & -14 \\ 0 & 1 & 0 & | & -6 \\ 0 & 0 & 1 & | & 2 \end{bmatrix} \quad 2R_2 + R_1$$

The solution of the system is now obvious; it is $(-14, -6, 2)$. Note that this solution is the last column of the final augmented matrix. ①

A matrix, such as the last one above, is said to be in **reduced row echelon form** if it is in echelon form *and* every column containing a leading 1 has zeros in all its entries. In the Gauss-Jordan method, row operations may be performed in any order, but it is best to transform the matrix systematically. Either follow the procedure in the preceding example (which first puts the system in a form where back substitution could be used and then eliminates additional variable terms) or work column by column from left to right, as in the next example.

▶ **EXAMPLE 1** Use the Gauss-Jordan method to solve the system
$$x \qquad + 5z = -6 + y$$
$$3x + 3y \qquad = 10 + z$$
$$x + 3y + 2z = 5.$$

The system must first be rewritten in proper form as follows.
$$x - y + 5z = -6$$
$$3x + 3y - z = 10$$
$$x + 3y + 2z = 5$$

Begin the solution by writing the augmented matrix of the linear system.
$$\begin{bmatrix} 1 & -1 & 5 & | & -6 \\ 3 & 3 & -1 & | & 10 \\ 1 & 3 & 2 & | & 5 \end{bmatrix}$$

The first element in column one is already 1. Get 0 for the second element in column one by multiplying each element in the first row by -3 and adding the results to the corresponding elements in row two.
$$\begin{bmatrix} 1 & -1 & 5 & | & -6 \\ 0 & 6 & -16 & | & 28 \\ 1 & 3 & 2 & | & 5 \end{bmatrix} \quad -3R_1 + R_2$$

②

Continue the solution of the system in Example 1 as follows. Give the shorthand notation and the matrix for each step.

(a) Get 1 in row two, column two.

(b) Get 0 in row one, column two.

(c) Now get 0 in row three, column two.

Answers:

(a) $\frac{1}{6}R_2$;

$$\begin{bmatrix} 1 & -1 & 5 & | & -6 \\ 0 & 1 & -\frac{8}{3} & | & \frac{14}{3} \\ 0 & 4 & -3 & | & 11 \end{bmatrix}$$

(b) $R_2 + R_1$;

$$\begin{bmatrix} 1 & 0 & \frac{7}{3} & | & -\frac{4}{3} \\ 0 & 1 & -\frac{8}{3} & | & \frac{14}{3} \\ 0 & 4 & -3 & | & 11 \end{bmatrix}$$

(c) $-4R_2 + R_3$;

$$\begin{bmatrix} 1 & 0 & \frac{7}{3} & | & -\frac{4}{3} \\ 0 & 1 & -\frac{8}{3} & | & \frac{14}{3} \\ 0 & 0 & \frac{23}{3} & | & -\frac{23}{3} \end{bmatrix}$$

(Solution continued in the text.)

③

Use the Gauss-Jordan method to solve

$$x + y - z = 6$$
$$2x - y + z = 3$$
$$-x + y + z = -4.$$

Answer:

$(3, 1, -2)$

Now, change the first element in row three to 0 by multiplying each element of the first row by -1 and adding the results to the corresponding elements of the third row.

$$\begin{bmatrix} 1 & -1 & 5 & | & -6 \\ 0 & 6 & -16 & | & 28 \\ 0 & 4 & -3 & | & 11 \end{bmatrix} \quad -1R_1 + R_3$$

This transforms the first column. Transform the second column in a similar manner, as directed in side problem 2. ②

Complete the solution by transforming the third column of the matrix in part (c) of side problem 2.

$$\begin{bmatrix} 1 & 0 & \frac{7}{3} & | & -\frac{4}{3} \\ 0 & 1 & -\frac{8}{3} & | & \frac{14}{3} \\ 0 & 0 & 1 & | & -1 \end{bmatrix} \quad \frac{3}{23}R_3$$

$$\begin{bmatrix} 1 & 0 & 0 & | & 1 \\ 0 & 1 & -\frac{8}{3} & | & \frac{14}{3} \\ 0 & 0 & 1 & | & -1 \end{bmatrix} \quad -\frac{7}{3}R_3 + R_1$$

$$\begin{bmatrix} 1 & 0 & 0 & | & 1 \\ 0 & 1 & 0 & | & 2 \\ 0 & 0 & 1 & | & -1 \end{bmatrix} \quad \frac{8}{3}R_3 + R_2$$

The linear system associated with this last augmented matrix is

$$x \qquad\qquad = 1$$
$$\quad y \qquad = 2$$
$$\qquad z = -1,$$

and the solution is $(1, 2, -1)$. ◀ ③

Most graphing calculators can transform a matrix into reduced row echelon form in a single keystroke, as illustrated in the next example.

▶ **EXAMPLE 2** Use a graphing calculator to solve the system

$$x - y + 5z = -6$$
$$3x + 3y - z = 10$$
$$x - 5y + 8z = -17$$
$$x + 3y + 2z = 5.$$

Enter the augmented matrix in the calculator (Figure 6.6 on the next page). Then use the RREF key (in the MATH or OPS submenu of the MATRIX menu) or a REFF program (in the Graphing Calculator Appendix that is available at www.aw.com/MWA8) to put this matrix in reduced row echelon form (Figure 6.7 on the next page).

(4)

Use a graphing calculator to solve the system.

$$x + 3y = 4$$
$$4x + 8y = 4$$
$$6x + 12y = 6$$

Answer:

$(-5, 3)$

```
[A]
[[1  -1  5  -6 ]
 [3   3 -1  10 ]
 [1  -5  8 -17 ]
 [1   3  2   5 ]]
```

FIGURE 6.6

```
rref([A])
[[1  0  0  1 ]
 [0  1  0  2 ]
 [0  0  1 -1 ]
 [0  0  0  0 ]]
```

FIGURE 6.7

The solutions can now be read from the last column of the matrix in Figure 6.7: $x = 1, y = 2, z = -1.$ ◀ (4)

▶ **EXAMPLE 3** Use the Gauss-Jordan method to solve the system

$$2x + 4y = 4$$
$$3x + 6y = 8$$
$$2x + y = 7.$$

There are several possible ways to proceed.

Manual Method Write the augmented matrix and perform row operations to obtain a first column whose entries (from top to bottom) are 1, 0, 0.

$$\begin{bmatrix} 2 & 4 & | & 4 \\ 3 & 6 & | & 8 \\ 2 & 1 & | & 7 \end{bmatrix}$$

$$\begin{bmatrix} 1 & 2 & | & 2 \\ 3 & 6 & | & 8 \\ 2 & 1 & | & 7 \end{bmatrix} \quad \frac{1}{2}R_1$$

$$\begin{bmatrix} 1 & 2 & | & 2 \\ 0 & 0 & | & 2 \\ 2 & 1 & | & 7 \end{bmatrix} \quad -3R_1 + R_2$$

Stop! The second row of the matrix denotes the equation $0x + 0y = 2$. Since the left side of this equation is always 0 and the right side is 2, it has no solution. Therefore, the original system has no solution.

(5)

Solve each system.

(a) $x - y = 4$
$-2x + 2y = 1$

(b) $3x - 4y = 0$
$2x + y = 0$

Answers:

(a) No solution

(b) $(0, 0)$

Calculator Method Enter the augmented matrix into a graphing calculator and use the RREF key to put it in reduced row echelon form, as in Figure 6.8. The last row of that matrix corresponds to $0x + 0y = 1$, which has no solution. Hence the original system has no solution. ◀ (5)

```
rref([A])
[[1  0  0]
 [0  1  0]
 [0  0  1]]
```

FIGURE 6.8

NOTE Whenever the Gauss-Jordan method produces a row whose elements are all 0 except the last one, as in Example 3, the system is inconsistent and has no solutions. On the other hand, if a row with *every* element 0 is produced, the system may have solutions. In that case, continue carrying out the Gauss-Jordan method.

▶ **EXAMPLE 4**   Use the Gauss-Jordan method to solve the system

$$x + 2y - z = 0$$
$$3x - y + z = 6.$$

When working manually, start with the augmented matrix and use row operations to obtain a first column whose entries (from top to bottom) are $1, 0$.

$$\begin{bmatrix} 1 & 2 & -1 & | & 0 \\ 3 & -1 & 1 & | & 6 \end{bmatrix}$$

$$\begin{bmatrix} 1 & 2 & -1 & | & 0 \\ 0 & -7 & 4 & | & 6 \end{bmatrix} \quad -3R_1 + R_2$$

Now use row operations to obtain a second column whose entries (from top to bottom) are $0, 1$.

$$\begin{bmatrix} 1 & 2 & -1 & | & 0 \\ 0 & 1 & -\frac{4}{7} & | & -\frac{6}{7} \end{bmatrix} \quad -\frac{1}{7}R_2$$

$$\begin{bmatrix} 1 & 0 & \frac{1}{7} & | & \frac{12}{7} \\ 0 & 1 & -\frac{4}{7} & | & -\frac{6}{7} \end{bmatrix} \quad -2R_2 + R_1$$

The RREF key on a graphing calculator produces the same matrix (but see the Technology Tip below). In any case, this last matrix is the augmented matrix of the system

$$x + \frac{1}{7}z = \frac{12}{7}$$

$$y - \frac{4}{7}z = -\frac{6}{7}.$$

Solving the first equation for x and the second for y gives the solution

$$z \text{ arbitrary}$$

$$y = -\frac{6}{7} + \frac{4}{7}z$$

$$x = \frac{12}{7} - \frac{1}{7}z,$$

or $(12/7 - z/7, -6/7 + 4z/7, z)$.   ◀   ⑥

⑥

Use the Gauss-Jordan method to solve the following.

(a) $3x + 9y = -6$
$-x - 3y = 2$

(b) $2x + 9y = 12$
$4x + 18y = 5$

Answers:

(a) y arbitrary,
$x = -3y - 2$
or $(-3y - 2, y)$

(b)   No solution

TECHNOLOGY TIP   When the RREF key is used, a calculator may produce a matrix with some long decimal entries. Such entries can often be displayed as fractions by using the FRAC key on TI calculators (or changing the number format mode to "fraction" on HP-39). Figure 6.9 illustrates this for the matrix in Example 4.

FIGURE 6.9

The techniques used in Example 1–4 can be summarized as follows:

The Gauss-Jordan Method for Solving a System of Linear Equations

1. Arrange the equations with the variable terms in the same order on the left of the equal sign and the constants on the right.
2. Write the augmented matrix of the system.
3. Use row operations or a graphing calculator to transform the augmented matrix into reduced row echelon form.
4. Stop the process in Step 3 if you obtain a row whose elements are all zero except the last one. In that case, the system is inconsistent and has no solutions. Otherwise, finish Step 3 and read the solutions of the system from the final matrix.

Applications

The key to solving applied problems is to determine the system of equations that must be solved.

▶ **EXAMPLE 5**   An animal feed is to be made from corn, soybeans, and cottonseed. Determine how many units of each ingredient are needed to make a feed that supplies 1800 units of fiber, 2800 units of fat, and 2200 units of protein, given that one unit of each ingredient provides the numbers of units shown in the table below. The table states, for example, that a unit of corn provides 10 units of fiber, 30 units of fat, and 20 units of protein.

	Corn	*Soybeans*	*Cottonseed*	*Totals*
Units of Fiber	10	20	30	1800
Units of Fat	30	20	40	2800
Units of Protein	20	40	25	2200

Let x represent the required number of units of corn, y the number of units of soybeans, and z the number of units of cottonseed. Since the total amount of fiber is to be 1800,

$$10x + 20y + 30z = 1800.$$

The feed must supply 2800 units of fat, so

$$30x + 20y + 40z = 2800.$$

Finally, since 2200 units of protein are required,

$$20x + 40y + 25z = 2200.$$

List a sequence of row operations that will transform the augmented matrix of system (1) in Example 5 into the augmented matrix of system (2).

Answer:

Many sequences are possible, including this one:

replace R_1 by $\frac{1}{10}R_1$;

replace R_2 by $\frac{1}{10}R_2$;

replace R_3 by $\frac{1}{5}R_3$;

replace R_2 by $-3R_1 + R_2$;
replace R_3 by $-4R_1 + R_3$;

replace R_2 by $-\frac{1}{4}R_2$;

replace R_3 by $-\frac{1}{7}R_3$.

Thus we must solve this system of equations.

$$\begin{aligned} 10x + 20y + 30z &= 1800 \\ 30x + 20y + 40z &= 2800 \\ 20x + 40y + 25z &= 2200 \end{aligned} \qquad (1)$$

Now solve this system, either manually or by using technology.

Manual Method Write the augmented matrix and use row operations to transform it into row echelon form, as in problem 7 at the side. The resulting matrix represents the following system.

$$\begin{aligned} x + 2y + 3z &= 180 \\ y + \frac{5}{4}z &= 65 \\ z &= 40 \end{aligned} \qquad (2)$$

Back substitution now shows that

$$z = 40, \quad y = 65 - \frac{5}{4}(40) = 15, \quad \text{and} \quad x = 180 - 2(15) - 3(40) = 30.$$

Thus, the feed should contain 30 units of corn, 15 units of soybeans, and 40 units of cottonseed.

Calculator Method Enter the augmented matrix of the system into the calculator (top of Figure 6.10). Use the RREF key or a REFF program to transform it into reduced row echelon form (bottom of Figure 6.10), which shows that $x = 30$, $y = 15$, and $z = 40$. ◄

FIGURE 6.10

▶ **EXAMPLE 6** The U-Drive Rent-a-Truck Company plans to spend 3 million dollars on 200 new vehicles. Each van will cost \$10,000; each small truck, \$15,000; and each large truck, \$25,000. Past experience shows that they need twice as many vans as small trucks. How many of each kind of vehicle can they buy?

Let x be the number of vans, y the number of small trucks, and z the number of large trucks. Then

$$\left(\begin{array}{c}\text{Number of} \\ \text{vans}\end{array}\right) + \left(\begin{array}{c}\text{Number of} \\ \text{small trucks}\end{array}\right) + \left(\begin{array}{c}\text{Number of} \\ \text{large trucks}\end{array}\right) = \text{Total number of vehicles}$$

$$x + y + z = 200. \qquad (1)$$

Similarly,

$$\left(\begin{array}{c}\text{Cost of } x \\ \text{vans}\end{array}\right) + \left(\begin{array}{c}\text{Cost of } y \\ \text{small trucks}\end{array}\right) + \left(\begin{array}{c}\text{Cost of } z \\ \text{large trucks}\end{array}\right) = \text{Total cost}$$

$$10{,}000x + 15{,}000y + 25{,}000z = 3{,}000{,}000$$

Dividing both sides by 5000 produces the equivalent equation

$$2x + 3y + 5z = 600 \tag{2}$$

Finally, the number of vans is twice the number of small trucks, that is, $x = 2y$, or equivalently,

$$x - 2y = 0. \tag{3}$$

We must solve the system given by equations (1)–(3).

$$\begin{aligned}x + y + z &= 200 \\ 2x + 3y + 5z &= 600 \\ x - 2y \quad\;\; &= 0\end{aligned}$$

Form the augmented matrix and transform it into reduced row echelon form (either manually, as shown below, or by using technology).

$$\left[\begin{array}{ccc|c} 1 & 1 & 1 & 200 \\ 2 & 3 & 5 & 600 \\ 1 & -2 & 0 & 0 \end{array}\right]$$

$$\left[\begin{array}{ccc|c} 1 & 1 & 1 & 200 \\ 0 & 1 & 3 & 200 \\ 0 & -3 & -1 & -200 \end{array}\right] \begin{array}{l} -2R_1 + R_2 \\ -R_1 + R_3 \end{array}$$

$$\left[\begin{array}{ccc|c} 1 & 0 & -2 & 0 \\ 0 & 1 & 3 & 200 \\ 0 & 0 & 8 & 400 \end{array}\right] \begin{array}{l} -R_2 + R_1 \\[6pt] 3R_2 + R_3 \end{array}$$

$$\left[\begin{array}{ccc|c} 1 & 0 & -2 & 0 \\ 0 & 1 & 3 & 200 \\ 0 & 0 & 1 & 50 \end{array}\right] \begin{array}{l} \\ \tfrac{1}{8}R_3 \end{array}$$

$$\left[\begin{array}{ccc|c} 1 & 0 & 0 & 100 \\ 0 & 1 & 0 & 50 \\ 0 & 0 & 1 & 50 \end{array}\right] \begin{array}{l} 2R_3 + R_1 \\ -3R_3 + R_2 \end{array}$$

The final matrix corresponds to the system

$$\begin{aligned} x &= 100 \\ y &= 50 \\ z &= 50. \end{aligned}$$

Therefore, U-Drive should buy 100 vans, 50 small trucks, and 50 large trucks. ◀

In Example 6, suppose that the U-Drive Company can spend only 2 million dollars on 150 new vehicles and that they need three times as many vans as small trucks. Write a system of equations to express these conditions.

Answer:

$$\begin{aligned} x + y + z &= 150 \\ 2x + 3y + 5z &= 400 \\ x - 3y \quad\;\; &= 0 \end{aligned}$$

6.2 Exercises

In Exercises 1–4, the reduced row echelon form of the augmented matrix of a system of equations is given. Find the solutions of the system.

1.
$$\begin{bmatrix} 1 & 0 & 0 & 0 & 3/2 \\ 0 & 1 & 0 & 0 & 5 \\ 0 & 0 & 1 & 0 & -2 \\ 0 & 0 & 0 & 1 & 0 \end{bmatrix}$$

2.
$$\begin{bmatrix} 1 & 0 & 0 & 0 & 0 & 5 \\ 0 & 1 & 0 & 0 & 0 & 4 \\ 0 & 0 & 1 & 0 & 0 & 3 \\ 0 & 0 & 0 & 0 & 1 & 2 \\ 0 & 0 & 0 & 0 & 0 & 1 \end{bmatrix}$$

3.
$$\begin{bmatrix} 1 & 0 & 0 & 1 & 2 \\ 0 & 1 & 0 & 2 & -3 \\ 0 & 0 & 1 & 0 & 4 \\ 0 & 0 & 0 & 0 & 0 \end{bmatrix}$$

4.
$$\begin{bmatrix} 1 & 0 & 0 & 0 & 7 \\ 0 & 1 & 0 & 0 & 1 \\ 0 & 0 & 1 & 0 & -5 \\ 0 & 0 & 0 & 1 & 4 \\ 0 & 0 & 0 & 0 & 0 \\ 0 & 0 & 0 & 0 & 0 \end{bmatrix}$$

Use the Gauss-Jordan method to solve each of the following systems of equations. (See Examples 1–4.)

5.
$$\begin{aligned} x + 2y + z &= 5 \\ 2x + y - 3z &= -2 \\ 3x + y + 4z &= -5 \end{aligned}$$

6.
$$\begin{aligned} 3x - 2y + z &= 6 \\ 3x + y - z &= -4 \\ -x + 2y - 2z &= -8 \end{aligned}$$

7.
$$\begin{aligned} x + 3y - 6z &= 7 \\ 2x - y + 2z &= 0 \\ x + y + 2z &= -1 \end{aligned}$$

8.
$$\begin{aligned} x &= 1 - y \\ 2x &= z \\ 2z &= -2 - y \end{aligned}$$

9.
$$\begin{aligned} 3x + 5y - z &= 0 \\ 4x - y + 2z &= 1 \\ -6x - 10y + 2z &= 0 \end{aligned}$$

10.
$$\begin{aligned} x + y &= -1 \\ y + z &= 4 \\ x + z &= 1 \end{aligned}$$

11.
$$\begin{aligned} x + y - z &= 6 \\ 2x - y + z &= -9 \\ x - 2y + 3z &= 1 \end{aligned}$$

12.
$$\begin{aligned} y &= x - 1 \\ y &= 6 + z \\ z &= -1 - x \end{aligned}$$

13.
$$\begin{aligned} x - 2y + z &= 5 \\ 2x + y - z &= 2 \\ -2x + 4y - 2z &= 2 \end{aligned}$$

14.
$$\begin{aligned} 2x + 3y + z &= 9 \\ 4x + y - 3z &= -7 \\ 6x + 2y - 4z &= -8 \end{aligned}$$

15.
$$\begin{aligned} -8x - 9y &= 11 \\ 24x + 34y &= 2 \\ 16x + 11y &= -57 \end{aligned}$$

16.
$$\begin{aligned} 2x + y &= 7 \\ x - y &= 3 \\ x + 3y &= 4 \end{aligned}$$

17.
$$\begin{aligned} x + y - z &= -20 \\ 2x - y + z &= 11 \end{aligned}$$

18.
$$\begin{aligned} 4x + 3y + z &= 1 \\ -2x - y + 2z &= 0 \end{aligned}$$

19.
$$\begin{aligned} 2x + y + 3z - 2w &= -6 \\ 4x + 3y + z - w &= -2 \\ x + y + z + w &= -5 \\ -2x - 2y + 2z + 2w &= -10 \end{aligned}$$

20.
$$\begin{aligned} x + y + z + w &= -1 \\ -x + 4y + z - w &= 0 \\ x - 2y + z - 2w &= 11 \\ -x - 2y + z + 2w &= -3 \end{aligned}$$

21.
$$\begin{aligned} x + 2y - z &= 3 \\ 3x + y + w &= 4 \\ 2x - y + z + w &= 2 \end{aligned}$$

22.
$$\begin{aligned} x - 2y - z - 3w &= -3 \\ -x + y + z &= 2 \\ 4y + 3z - 6w &= -2 \end{aligned}$$

Use systems of equations to answer these questions. (See Examples 5 and 6.)

23. Business McFrugal Snack Shops plan to hire two public relations firms to survey 500 customers by phone, 750 by mail, and 250 by in-person interviews. The Garcia firm has personnel to do 10 phone surveys, 30 mail surveys, and 5 interviews per hour. The Wong firm can handle 20 phone surveys, 10 mail surveys, and 10 interviews per hour. For how many hours should each firm be hired to produce the exact number of surveys needed?

24. Business A knitting shop ordered yarn from three suppliers, I, II, and III. One month the shop ordered a total of 100 units of yarn from these suppliers. The delivery costs were $80, $50, and $65 per unit for the orders from suppliers I, II, and III, respectively, with total delivery costs of $5990. The shop ordered the same amount from suppliers I and III. How many units were ordered from each supplier?

25. Business Pretzels cost $3 per pound, dried fruit $4 per pound, and nuts $8 per pound. How many pounds of each should be used to produce 140 pounds of trail mix costing $6 per pound in which there are twice as many pretzels (by weight) as dried fruit?

26. Business An auto manufacturer sends cars from two plants, I and II, to dealerships A and B located in a midwestern city. Plant I has a total of 28 cars to send, and plant II has 8. Dealer A needs 20 cars, and dealer B needs 16. Transportation costs based on the distance of each dealership from each plant are $220 from I to A, $300 from I to B, $400 from II to A, and $180 from II to B. The manufacturer wants to limit transportation costs to $10,640. How many cars should be sent from each plant to each of the two dealerships?

27. Physical Sciences The stopping distance of a car traveling 25 mph is 61.7 feet, and for a car traveling 35 mph it is 106 feet.* The stopping distance in feet can be described by the equation $y = ax^2 + bx$, where x is the speed in mph.
(a) Find the values of a and b.
(b) Use your answers from part (a) to find the stopping distance for a car traveling 55 mph.

---

*National Traffic Safety Institute Student Workbook, 1993, p. 7.

28. Natural Science An animal breeder can buy four types of tiger food. Each case of Brand A contains 25 units of fiber, 30 units of protein, and 30 units of fat. Each case of Brand B contains 50 units of fiber, 30 units of protein, and 20 units of fat. Each case of Brand C contains 75 units of fiber, 30 units of protein, and 20 units of fat. Each case of Brand D contains 100 units of fiber, 60 units of protein, and 30 units of fat. How many cases of each brand should the breeder mix together to obtain a food that provides 1200 units of fiber, 600 units of protein, and 400 units of fat?

29. Finance An investor plans to invest $70,000 in a mutual fund, corporate bonds, and a fast food franchise. She plans to put twice as much in bonds as in the mutual fund. Based on past performance, she expects the mutual fund to pay a 2% dividend, the bonds 10%, and the franchise 6%. She would like a dividend income of $4800. How much should she put in each of three investments?

30. Business According to data from a Texas agricultural report, the amount of nitrogen (in lbs/acre), phosphate (in lbs/acre), and labor (in hr/acre) needed to grow honeydews, yellow onions, and lettuce is given by the following table.*

	Honeydews	Yellow Onions	Lettuce
Nitrogen	120	150	180
Phosphate	180	80	80
Labor	4.97	4.45	4.65

(a) If the farmer has 220 acres, 29,100 pounds of nitrogen, 32,600 pounds of phosphate, and 480 hours of labor, is it possible to use all resources completely? If so, how many acres should he allot for each crop?

(b) Suppose everything is the same as in part (a), except that 1061 hours of labor are available. Is it possible to use all resources completely? If so, how many acres should he allot for each crop?

⟍ *A graphing calculator or other technology is recommended for the following exercises.*

31. Health The table shows the calories, sodium, and protein in one cup of various kinds of soup.

	Progresso Roasted Chicken Rotini	Healthy Choice Hearty Chicken	Campbell's Chunky Chicken Noodle
Calories	100	130	130
Sodium (mg)	970	480	880
Protein (g)	6	8	8

*Miguel Paredes, Mohammad Fatehi, and Richard Hinthorn, "The Transformation of an Inconsistent Linear System into a Consistent System," *AMATYC Review*, 13, no. 2 (spring 1992).

How many cups of each kind of soup should be mixed together to produce ten servings of soup, each of which provides 203 calories, 1190 milligrams of sodium, and 12.4 grams of protein? What is the serving size (in cups)? (*Hint:* In ten servings, there must be 2030 calories, 11,900 milligrams of sodium, and 124 grams of protein.)

32. Health The table shows the calories, sodium, and fat in one ounce of various snack foods (all produced by Planters).

	Cashews	Dry Roasted Honey Peanuts	Cajun Crunch Trail Mix
Calories	170	150	130
Sodium (mg)	120	95	10
Fat (g)	6	7	3

How many ounces of each kind of snack should be combined to produce ten servings, each of which provides 288 calories, 115 milligrams of sodium, and 9.6 grams of fat? What is the serving size?

33. Finance An investment firm recommends that a client invest in AAA, A, and B rated bonds. The average yield on AAA bonds is 6%, on A bonds 7%, and on B bonds 10%. The client wants to invest twice as much in AAA bonds as in B bonds. How much should be invested in each type of bond under the following conditions?

(a) The total investment is $25,000, and the investor wants an annual return of $1810 on the three investments.

(b) The values in part (a) are changed to $30,000 and $2150, respectively.

(c) The values in part (a) are changed to $40,000 and $2900, respectively.

34. Business An electronics company produces transistors, resistors, and computer chips. Each transistor requires 3 units of copper, 1 unit of zinc, and 2 units of glass. Each resistor requires 3, 2, and 1 units of the three materials, and each computer chip requires 2, 1, and 2 units of these materials, respectively. How many of each product can be made with the following amounts of materials?

(a) 810 units of copper, 410 units of zinc, and 490 units of glass

(b) 765 units of copper, 385 units of zinc, and 470 units of glass

(c) 1010 units of copper, 500 units of zinc, and 610 units of glass

35. Business At a pottery factory, fuel consumption for heating the kilns varies with the size of the order being fired. In the past, the company recorded the following figures.

$x =$ Number of Platters	$y =$ Fuel Cost per Platter
6	$2.80
8	2.48
10	2.24

(a) Find an equation of the form $y = ax^2 + bx + c$ whose graph contains the three points corresponding to the data in the table.

(b) How many platters should be fired at one time in order to minimize the fuel cost per platter? What is the minimum fuel cost per platter?

36. Business The business analyst for Melcher Manufacturing wants to find an equation that can be used to project sales of a relatively new product. For the years 2000, 2001, and 2002, sales were $15,000, $32,000, and $123,000, respectively.

(a) Graph the sales for the years 2000, 2001, and 2002, letting the year 2000 equal 0 on the x-axis. Let the values on the vertical axis be in thousands. (For example, the point (2001, 32,000) will be graphed as (1, 32).)

(b) Find the equation of the straight line $ax + by = c$ through the points for 2000 and 2002.

(c) Find the equation of the parabola $y = ax^2 + bx + c$ through the three given points.

(d) Find the projected sales for 2005 first by using the equation from part (b) and then by using the equation from part (c). If you were to estimate sales of the product in 2005, which result would you choose? Why?

37. Natural Science Determining the amount of carbon dioxide in the atmosphere is important because carbon dioxide is known to be a greenhouse gas. Carbon dioxide concentrations (in parts per million) have been measured at Mauna Loa, Hawaii, over the past 45 years. The concentrations have increased quadratically.* The table lists readings for three years.

Year	CO_2
1958	315
1978	335
1998	367

(a) If the relationship between the carbon dioxide concentration C and the year t is expressed as $C = at^2 + bt + c$, where $t = 0$ corresponds to 1958, use a linear system of equations to determine the constants a, b, and c.

(b) Predict the year when the amount of carbon dioxide in the atmosphere will double from its 1958 level.

38. Physical Science For certain aircraft there exists a quadratic relationship between an airplane's maximum speed S (in knots) and its ceiling C, its highest altitude possible (in thousands of feet).* The table lists three airplanes that conform to this relationship.

Airplane	Maximum Speed	Ceiling
Hawkeye	320	33
Corsair	600	40
Tomcat	1283	50

(a) If the relationship between C and S is written as $C = aS^2 + bS + c$, use a linear system of equations to determine the constants a, b, and c.

(b) A new aircraft of this type has a ceiling of 45,000 feet. Predict its top speed.

39. Physical Science Estimates of the total amount of lead emissions in the United States are made by the Environmental Protection Agency's Office of Air Quality Planning and Standards each year. The table lists lead estimates for four years.[†]

Year	Lead Emissions (in short tons)
1988	7053
1992	3808
1994	4043
1997	3915

(a) If the relationship between the lead emissions L and the year t is expressed as $L = at^2 + bt + c$, where $t = 0$ corresponds to 1988, use data from 1988, 1992, and 1997 and a linear system of equations to determine the constants a, b, and c.

(b) Use the equation from part (a) to predict the emissions in 1994 and compare the result with the actual data.

(c) If the relationship between the lead emissions L and the year t is expressed as $L = at^3 + bt^2 + ct + d$, where $t = 0$ corresponds to 1988, use all four data points and a linear system of equations to determine the constants a, b, c, and d.

(d) Discuss the appropriateness of the functions used in parts (a) and (b) to model this data.

---

*C. D. Keeling and T. P. Whorf, Scripps Institution of Oceanography.

*D. Sanders, *Statistics: A First Course*, Fifth Edition (McGraw Hill, 1995).

†U.S. Environmental Protection Agency.

6.3 BASIC MATRIX OPERATIONS

Until now we have used matrices only as a convenient shorthand to solve systems of equations. However, matrices are also important in the fields of management, natural science, engineering, and social science as a way to organize data, as Example 1 demonstrates.

▶ **EXAMPLE 1** The EZ Life Company manufactures sofas and armchairs in three models, A, B, and C. The company has regional warehouses in New York, Chicago, and San Francisco. In its August shipment, the company sends 10 model A sofas, 12 model B sofas, 5 model C sofas, 15 model A chairs, 20 model B chairs, and 8 model C chairs to each warehouse.

This data might be organized by first listing it as follows.

Sofas   10 model A   12 model B   5 model C
Chairs   15 model A   20 model B   8 model C

Alternatively, we might tabulate the data in a chart.

		MODEL		
		A	*B*	*C*
FURNITURE	Sofa	10	12	5
	Chair	15	20	8

With the understanding that the numbers in each row refer to the furniture type (sofa, chair) and the numbers in each column refer to the model (A, B, C), the same information can be given by a matrix, as follows.

$$M = \begin{bmatrix} 10 & 12 & 5 \\ 15 & 20 & 8 \end{bmatrix}$$ ①

A matrix with m rows and n columns has dimensions, or size, $m \times n$. The number of rows is always given first.

▶ **EXAMPLE 2** (a) The matrix $\begin{bmatrix} 6 & 5 \\ 3 & 4 \\ 5 & -1 \end{bmatrix}$ is a 3×2 matrix.

(b) $\begin{bmatrix} 5 & 8 & 9 \\ 0 & 5 & -3 \\ -4 & 0 & 5 \end{bmatrix}$ is a 3×3 matrix. (c) $[1 \quad 6 \quad 5 \quad -2 \quad 5]$ is a 1×5 matrix.

(d) A graphing calculator displays a 4×1 matrix like this.

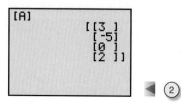

◀ ②

① Rewrite this information in a matrix with three rows and two columns.

Answer:
$\begin{bmatrix} 10 & 15 \\ 12 & 20 \\ 5 & 8 \end{bmatrix}$

② Give the size of each of the following matrices.
(a) $\begin{bmatrix} 2 & 1 & -5 & 6 \\ 3 & 0 & 7 & -4 \end{bmatrix}$
(b) $\begin{bmatrix} 1 & 2 & 3 \\ 4 & 5 & 6 \\ 9 & 8 & 7 \end{bmatrix}$

Answers:
(a) 2×4
(b) 3×3

A matrix with only one row, as in Example 2(c), is called a **row matrix** or **row vector.** A matrix with only one column, as in Example 2(d), is called a **column matrix** or **column vector.** A matrix with the same number of rows as columns is called a **square matrix.** The matrix in Example 2(b) above is a square matrix, as are

3

Use the numbers $2, 5, -8, 4$ to write

(a) a row matrix;

(b) a column matrix;

(c) a square matrix.

Answers:

(a) $[2 \quad 5 \quad -8 \quad 4]$

(b) $\begin{bmatrix} 2 \\ 5 \\ -8 \\ 4 \end{bmatrix}$

(c) $\begin{bmatrix} 2 & 5 \\ -8 & 4 \end{bmatrix}$ or $\begin{bmatrix} 2 & -8 \\ 5 & 4 \end{bmatrix}$

(Other answers are possible.)

$$A = \begin{bmatrix} -5 & 6 \\ 8 & 3 \end{bmatrix} \quad \text{and} \quad B = \begin{bmatrix} 0 & 0 & 0 & 0 \\ -2 & 4 & 1 & 3 \\ 0 & 0 & 0 & 0 \\ -5 & -4 & 1 & 8 \end{bmatrix}.$$

When a matrix is denoted by a single letter, such as the matrix A above, then the element in row i and column j is denoted a_{ij}. For example, $a_{21} = 8$ (the element in row 2, column 1). Similarly, in matrix B above, $b_{42} = -4$ (the element in row 4, column 2).

Addition

The matrix given in Example 1,

$$M = \begin{bmatrix} 10 & 12 & 5 \\ 15 & 20 & 8 \end{bmatrix},$$

shows the August shipment from the EZ Life plant to each of its warehouses. If matrix N below gives the September shipment to the New York warehouse, what is the total shipment for each item of furniture to the New York warehouse for these two months?

$$N = \begin{bmatrix} 45 & 35 & 20 \\ 65 & 40 & 35 \end{bmatrix}$$

If 10 model A sofas were shipped in August and 45 in September, then altogether 55 model A sofas were shipped in the 2 months. Adding the other corresponding entries gives a new matrix, Q, that represents the total shipment to the New York warehouse for the 2 months.

$$Q = \begin{bmatrix} 55 & 47 & 25 \\ 80 & 60 & 43 \end{bmatrix}$$

It is convenient to refer to Q as the "sum" of M and N.

The way these two matrices were added illustrates the following definition of addition of matrices.

The **sum** of two $m \times n$ matrices X and Y is the $m \times n$ matrix $X + Y$ in which each element is the sum of the corresponding elements of X and Y.

It is important to remember that only matrices that are the same size can be added.

Find each sum when possible.

(a) $\begin{bmatrix} 2 & 5 & 7 \\ 3 & -1 & 4 \end{bmatrix}$

$+ \begin{bmatrix} -1 & 2 & 0 \\ 10 & -4 & 5 \end{bmatrix}$

(b) $\begin{bmatrix} 1 \\ 2 \\ 3 \end{bmatrix} + \begin{bmatrix} 2 & -1 \\ 4 & 5 \\ 6 & 0 \end{bmatrix}$

(c) $[5 \quad 4 \quad -1] + [-5 \quad 2 \quad 3]$

Answers:

(a) $\begin{bmatrix} 1 & 7 & 7 \\ 13 & -5 & 9 \end{bmatrix}$

(b) Not possible

(c) $[0 \quad 6 \quad 2]$

▶ **EXAMPLE 3**   Find each sum if possible.

(a) $\begin{bmatrix} 5 & -6 \\ 8 & 9 \end{bmatrix} + \begin{bmatrix} -4 & 6 \\ 8 & -3 \end{bmatrix} = \begin{bmatrix} 5 + (-4) & -6 + 6 \\ 8 + 8 & 9 + (-3) \end{bmatrix} = \begin{bmatrix} 1 & 0 \\ 16 & 6 \end{bmatrix}$

(b) The matrices

$$A = \begin{bmatrix} 5 & 8 \\ 6 & 2 \end{bmatrix} \quad \text{and} \quad B = \begin{bmatrix} 3 & 9 & 1 \\ 4 & 2 & 5 \end{bmatrix}$$

are different sizes, so it is not possible to find the sum $A + B$. ◀ ④

TECHNOLOGY TIP   Spreadsheets and graphing calculators can find matrix sums, as illustrated in Figure 6.11.

[A]
   [[-5 4 6]
    [2 0 4]]
[B]
   [[-8 3 2]
    [6 2 -4]]

[A]+[B]
   [[-13 7 8]
    [8 2 0]]

FIGURE 6.11

▶ **EXAMPLE 4**   The September shipments of the three models of sofas and chairs from the EZ Life Company to the New York, San Francisco, and Chicago warehouses are given in matrices N, S, and C below.

$$N = \begin{bmatrix} 45 & 35 & 20 \\ 65 & 40 & 35 \end{bmatrix}, \quad S = \begin{bmatrix} 30 & 32 & 28 \\ 43 & 47 & 30 \end{bmatrix}, \quad C = \begin{bmatrix} 22 & 25 & 38 \\ 31 & 34 & 35 \end{bmatrix}$$

What was the total amount shipped to the three warehouses in September?

The total of the September shipments is represented by the sum of the three matrices N, S, and C.

$$N + S + C = \begin{bmatrix} 45 & 35 & 20 \\ 65 & 40 & 35 \end{bmatrix} + \begin{bmatrix} 30 & 32 & 28 \\ 43 & 47 & 30 \end{bmatrix} + \begin{bmatrix} 22 & 25 & 38 \\ 31 & 34 & 35 \end{bmatrix}$$

$$= \begin{bmatrix} 97 & 92 & 86 \\ 139 & 121 & 100 \end{bmatrix}$$

For example, from this sum the total number of model C sofas shipped to the three warehouses in September was 86. ◀ ⑤

⑤

From the result of Example 4, find the total number of the following shipped to the three warehouses.

(a)   Model A chairs

(b)   Model B sofas

(c)   Model C chairs

Answers:

(a)   139

(b)   92

(c)   100

As mentioned in Section 1.1, the additive inverse of the real number a is $-a$; a similar definition is given for the additive inverse of a matrix.

The **additive inverse** (or *negative*) of a matrix X is the matrix $-X$ in which each element is the additive inverse of the corresponding element of X.

If

$$A = \begin{bmatrix} 1 & 2 & 3 \\ 0 & -1 & 5 \end{bmatrix} \quad \text{and} \quad B = \begin{bmatrix} -2 & 3 & 0 \\ 1 & -7 & 2 \end{bmatrix},$$

then by the definition of the additive inverse of a matrix,

$$-A = \begin{bmatrix} -1 & -2 & -3 \\ 0 & 1 & -5 \end{bmatrix} \quad \text{and} \quad -B = \begin{bmatrix} 2 & -3 & 0 \\ -1 & 7 & -2 \end{bmatrix}.$$

TECHNOLOGY TIP A graphing calculator gives the additive inverse of a matrix by preceding the matrix with a negative sign. See Figure 6.12.

```
[A]
        [[2 3 ]
         [1 1.5]]
-[A]
        [[-2 -3 ]
         [-1 -1.5]]
```

FIGURE 6.12

By the definition of matrix addition, for each matrix X, the sum $X + (-X)$ is a **zero matrix,** O, whose elements are all zeros. There is an $m \times n$ zero matrix for each pair of values of m and n.

$$\begin{bmatrix} 0 & 0 \\ 0 & 0 \end{bmatrix} \qquad \begin{bmatrix} 0 & 0 & 0 & 0 \\ 0 & 0 & 0 & 0 \end{bmatrix}$$
2×2 zero matrix 2×4 zero matrix

Zero matrices have the following *identity property.*

If O is the $m \times n$ zero matrix, and A is any $m \times n$ matrix, then

$$A + O = O + A = A.$$

Compare this with the identity property for real numbers: for any real number a, $a + 0 = 0 + a = a$.

Subtraction

The **subtraction** of matrices can be defined in a manner comparable to subtraction for real numbers.

For two $m \times n$ matrices X and Y, the **difference** of X and Y is the $m \times n$ matrix $X - Y$ in which each element is the difference of the corresponding elements of X and Y, or, equivalently,

$$X - Y = X + (-Y).$$

According to this definition, matrix subtraction can be performed by subtracting corresponding elements. For example,

$$\begin{bmatrix} 1 & 2 & 3 \\ 0 & -1 & 5 \end{bmatrix} - \begin{bmatrix} -2 & 3 & 0 \\ 1 & -7 & 2 \end{bmatrix} = \begin{bmatrix} 1-(-2) & 2-3 & 3-0 \\ 0-1 & -1-(-7) & 5-2 \end{bmatrix}$$

$$= \begin{bmatrix} 3 & -1 & 3 \\ -1 & 6 & 3 \end{bmatrix}.$$

⑥

Find each of the following differences when possible.

(a) $\begin{bmatrix} 2 & 5 \\ -1 & 0 \end{bmatrix} - \begin{bmatrix} 6 & 4 \\ 3 & -2 \end{bmatrix}$

(b) $\begin{bmatrix} 1 & 5 & 6 \\ 2 & 4 & 8 \end{bmatrix} - \begin{bmatrix} 2 & 1 \\ 10 & 3 \end{bmatrix}$

(c) $[5 \quad -4 \quad 1] - [6 \quad 0 \quad -3]$

Answers:

(a) $\begin{bmatrix} -4 & 1 \\ -4 & 2 \end{bmatrix}$

(b) Not possible

(c) $[-1 \quad -4 \quad 4]$

▶**EXAMPLE 5** **(a)** $[8 \quad 6 \quad -4] - [3 \quad 5 \quad -8] = [5 \quad 1 \quad 4]$

(b) The matrices

$$\begin{bmatrix} -2 & 5 \\ 0 & 1 \end{bmatrix} \quad \text{and} \quad \begin{bmatrix} 3 \\ 5 \end{bmatrix}$$

are different sizes and cannot be subtracted.

(c) Spreadsheets and graphing calculators can do matrix subtraction, as shown in Figure 6.13. ◀ ⑥

```
[B]
      [[-4  5  0   2]
       [6   3  -7  1]]
[C]
      [[4 12 8 -5]
       [2  8 5  0]]
```

```
[B]-[C]
      [[-8 -7 -8   7]
       [4  -5 -12  1]]
```

FIGURE 6.13

▶**EXAMPLE 6** During September the Chicago warehouse of the EZ Life Company shipped out the following numbers of each model.

$$K = \begin{bmatrix} 5 & 10 & 8 \\ 11 & 14 & 15 \end{bmatrix}$$

What was the Chicago warehouse inventory on October 1, taking into account only the number of items received and sent out during the month?

The number of each kind of item received during September is given by matrix C from Example 4; the number of each model sent out during September is given by matrix K above. The October 1 inventory will be represented by the matrix $C - K$ as shown below.

$$C - K = \begin{bmatrix} 22 & 25 & 38 \\ 31 & 34 & 35 \end{bmatrix} - \begin{bmatrix} 5 & 10 & 8 \\ 11 & 14 & 15 \end{bmatrix} = \begin{bmatrix} 17 & 15 & 30 \\ 20 & 20 & 20 \end{bmatrix}$$ ◀

▶ **EXAMPLE 7** A drug company is testing 200 patients to see if Painoff (a new headache medicine) is effective. Half the patients receive Painoff and half receive a placebo. The data on the first 50 patients is summarized in this matrix.

Pain Relief Obtained

	Yes	No
Patient took Painoff	22	3
Patient took placebo	8	17

For example, row 2 shows that of the people who took the placebo, 8 got relief but 17 did not. The test was repeated on three more groups of 50 patients each, with the results summarized by these matrices.

$$\begin{bmatrix} 21 & 4 \\ 6 & 19 \end{bmatrix} \quad \begin{bmatrix} 19 & 6 \\ 10 & 15 \end{bmatrix} \quad \begin{bmatrix} 23 & 2 \\ 3 & 22 \end{bmatrix}$$

The total results of the test can be obtained by adding these four matrices.

$$\begin{bmatrix} 22 & 3 \\ 8 & 17 \end{bmatrix} + \begin{bmatrix} 21 & 4 \\ 6 & 19 \end{bmatrix} + \begin{bmatrix} 19 & 6 \\ 10 & 15 \end{bmatrix} + \begin{bmatrix} 23 & 2 \\ 3 & 22 \end{bmatrix} = \begin{bmatrix} 85 & 15 \\ 27 & 73 \end{bmatrix}$$

Because 85 of 100 patients got relief with Painoff and only 27 of 100 with the placebo, it appears that Painoff is effective. ◀ ⑦

⑦

Later it was discovered that the data in the last group of 50 patients in Example 7 was invalid. Use a matrix to represent the total test results after those data were eliminated.

Answer:

$$\begin{bmatrix} 62 & 13 \\ 24 & 51 \end{bmatrix}$$

Suppose one of the EZ Life Company warehouses receives the following order, written in matrix form, where the entries have the same meaning as given earlier.

$$\begin{bmatrix} 5 & 4 & 1 \\ 3 & 2 & 3 \end{bmatrix}$$

Later, the store that sent the order asks the warehouse to send six more of the same order. The six new orders can be written as one matrix by multiplying each element in the matrix by 6, giving the product

$$6 \begin{bmatrix} 5 & 4 & 1 \\ 3 & 2 & 3 \end{bmatrix} = \begin{bmatrix} 30 & 24 & 6 \\ 18 & 12 & 18 \end{bmatrix}.$$

In work with matrices, a real number, like the 6 in the multiplication above, is called a **scalar.**

The **product** of a scalar k and a matrix X is the matrix kX, each of whose elements is k times the corresponding element of X.

⑧

Here are two examples, one done by hand and the other on a graphing calculator (Figure 6.14).

Find each product.

(a)

$$-3 \begin{bmatrix} 4 & -2 \\ 1 & 5 \end{bmatrix}$$

(b)

$$4 \begin{bmatrix} 2 & 4 & 7 \\ 8 & 2 & 1 \\ 5 & 7 & 3 \end{bmatrix}$$

Answers:

(a)

$$\begin{bmatrix} -12 & 6 \\ -3 & -15 \end{bmatrix}$$

(b)

$$\begin{bmatrix} 8 & 16 & 28 \\ 32 & 8 & 4 \\ 20 & 28 & 12 \end{bmatrix}$$

$$(-3) \begin{bmatrix} 2 & -5 \\ 1 & 7 \\ 4 & -6 \end{bmatrix} = \begin{bmatrix} -6 & 15 \\ -3 & -21 \\ -12 & 18 \end{bmatrix}$$

```
[A]
            [[3 -4]
             [5 2 ]]
2*[A]
            [[6  -8]
             [10 4 ]]
■
```

FIGURE 6.14   ⑧

6.3 Exercises

Find the size of each of the following. Identify any square, column, or row matrices. (See Example 2.) Give the additive inverse of each matrix.

1. $\begin{bmatrix} 7 & -8 & 4 \\ 0 & 13 & 9 \end{bmatrix}$ 2×3

2. $\begin{bmatrix} -7 & 23 \\ 5 & -6 \end{bmatrix}$

3. $\begin{bmatrix} -3 & 0 & 11 \\ 1 & \frac{1}{4} & -7 \\ 5 & -3 & 9 \end{bmatrix}$

4. $[6 \quad -4 \quad \frac{2}{3} \quad 12 \quad 2]$

5. $\begin{bmatrix} 7 \\ 11 \end{bmatrix}$

6. $[-5]$

7. If A is a 5×3 matrix and $A + B = A$, what do you know about B?

8. If C is a 3×3 matrix and D is a 3×4 matrix, then $C + D$ is _____.

Perform the indicated operations where possible. (See Examples 3–6.)

9. $\begin{bmatrix} 1 & 2 & 5 & -1 \\ 3 & 0 & 2 & -4 \end{bmatrix} + \begin{bmatrix} 8 & 10 & -5 & 3 \\ -2 & -1 & 0 & 0 \end{bmatrix}$

10. $\begin{bmatrix} 1 & 5 \\ 2 & -3 \\ 3 & 7 \end{bmatrix} + \begin{bmatrix} 2 & 3 \\ 8 & 5 \\ -1 & 9 \end{bmatrix}$

11. $\begin{bmatrix} 1 & 5 & 7 \\ 2 & 2 & 3 \end{bmatrix} + \begin{bmatrix} 4 & 8 & -7 \\ 1 & -1 & 5 \end{bmatrix}$

12. $\begin{bmatrix} 2 & 4 \\ -8 & 1 \end{bmatrix} + \begin{bmatrix} 9 & -3 \\ 8 & 5 \end{bmatrix}$

13. $\begin{bmatrix} 4 & -2 & 5 \\ 3 & 7 & 0 \end{bmatrix} - \begin{bmatrix} 1 & 5 & -2 \\ -1 & 3 & 8 \end{bmatrix}$

14. $\begin{bmatrix} 9 & 1 \\ 0 & -3 \\ 4 & 10 \end{bmatrix} - \begin{bmatrix} 1 & 9 & -4 \\ -1 & 1 & 0 \end{bmatrix}$ *undefined*

Let $A = \begin{bmatrix} -2 & 4 \\ 0 & 3 \end{bmatrix}$ *and* $B = \begin{bmatrix} -6 & 2 \\ 4 & 0 \end{bmatrix}$. *Find each of the following.*

15. $2A$ **16.** $-3B$ **17.** $-4B$

18. $5A$ **19.** $-4A + 5B$ **20.** $3A - 10B$

Let $A = \begin{bmatrix} 1 & -2 \\ 4 & 3 \end{bmatrix}$ *and* $B = \begin{bmatrix} 2 & -1 \\ 0 & 5 \end{bmatrix}$. *Find a matrix X satisfying the given equation.*

21. $2X = 2A + 3B$ **22.** $3X = A - 3B$

Using matrices

$$O = \begin{bmatrix} 0 & 0 \\ 0 & 0 \end{bmatrix}, P = \begin{bmatrix} m & n \\ p & q \end{bmatrix}, T = \begin{bmatrix} r & s \\ t & u \end{bmatrix}, \text{ and }$$

$$X = \begin{bmatrix} x & y \\ z & w \end{bmatrix},$$

verify that the statements in Exercises 23–28 are true.

23. $X + T$ is a 2×2 matrix.

24. $X + T = T + X$ (Commutative property of addition of matrices)

25. $X + (T + P) = (X + T) + P$ (Associative property of addition of matrices)

26. $X + (-X) = O$ (Inverse property of addition of matrices)

27. $P + O = P$ (Identity property of addition of matrices)

28. Which of the above properties are valid for matrices that are not square?

Work the following exercises. (See Examples 1 and 7.)

29. Business   When ticket holders fail to attend, major league sports teams lose the money these fans would have spent on refreshments, souvenirs, etc. The percentage of fans who don't show up is 16% in basketball and hockey, 20% in football, and 18% in baseball. The lost revenue per fan is $18.20 in basketball, $18.25 in hockey, $19 in football, and $15.40 in baseball. The total annual lost revenue is $22.7 million in basketball, $35.8 million in hockey, $51.9 million in football, and $96.3 million in baseball.* Express this information in matrix form; specify what the rows and columns represent.

30. Finance   67% of undergraduate students had credit cards in 1998 and 78% in 2000. The average credit card debt was $1879 in 1998 and $2748 in 2000. The percentage of students with balances over $7000 was 10% in 1998 and 9% in 2000. The average number of credit cards per stu-

dent was 3.5 in 1998 and 3 in 2000.* Write this information first as a 4×2 matrix and then as a 2×4 matrix.

31. Health   The shortage of organs for transplants is a continuing problem in the United States. In 1990 there were 1788 people waiting for a heart transplant, 308 for a lung transplant, 1237 for a liver transplant, and 17,883 for a kidney transplant. Corresponding figures for 1998 were 4121, 3171, 12,070, and 38,270. In mid-2000, the figures were 4143, 3614, 15,359, and 45,273.[†] Express this information as a matrix, labeling rows and columns.

Work these exercises. (See Examples 1, 4, 6, and 7.)

32. Management   There are three convenience stores in Gambier. This week, Store I sold 88 loaves of bread, 48 quarts of milk, 16 jars of peanut butter, and 112 pounds of cold cuts. Store II sold 105 loaves of bread, 72 quarts of milk, 21 jars of peanut butter, and 147 pounds of cold cuts. Store III sold 60 loaves of bread, 40 quarts of milk, no peanut butter, and 50 pounds of cold cuts.

   (a) Use a 3×4 matrix to express the sales information for the three stores.

   (b) During the following week, sales on these products at Store I increased by 25%; sales at Store II increased by 1/3; and sales at Store III increased by 10%. Write the sales matrix for that week.

   (c) Write a matrix that represents total sales over the two-week period.

33. Social Science   Some of the fastest growing college sports for both men and women are soccer, rowing, and equestrian sports. In 1981–82, there were 80 women's college soccer teams, 43 rowing teams, and 7 equestrian teams. The numbers of men's college teams in these sports were 744, 48, and 2. In 1998–99, there were 926 women's soccer teams, 122 rowing teams, and 41 equestrian teams. The numbers of men's teams in these sports were 879, 70, and 33.[‡]

   (a) Express the information for women's teams as a 3×2 matrix A.

   (b) Express the information for men's teams as a 3×2 matrix B.

   (c) Find $A + B$ and explain what it means.

   (d) Suppose the number of women's teams in each sport has increased by 10% since 1998–99 (rounded to the nearest whole number). Write a matrix C that expresses the new numbers.

*American Demographics.

*Nellie Mae.

†United Network of Organ Sharing and National Organ Procurement and Transplantation Network.

‡General Accounting Office.

34. *Social Sciences* The following table gives the educational attainment of the U.S. population 25 years and older in various years.*

	MALE		FEMALE	
	Four Years of High School or More	*Four Years of College or More*	*Four Years of High School or More*	*Four Years of College or More*
1940	22.7%	5.5%	26.3%	3.8%
1950	32.6	7.3	36.0	5.2
1960	39.5	9.7	42.5	5.8
1970	51.9	13.5	52.8	8.1
1980	67.3	20.1	65.8	12.8
1990	77.7	24.4	77.5	18.4
1999	83.4	27.5	83.4	23.1

(a) Write a matrix for the educational attainment of males.

(b) Write a matrix for the educational attainment of females.

(c) Use the matrices from parts (a) and (b) to write a matrix showing how much more (or less) education males have attained than females.

35. *Transportation* The tables in the next column give the death rates (per million person trips) for male and female drivers for various ages and number of passengers.[†]

---

*U.S. Census Bureau.

†Li-Hui Chen, Susan Baker, Elisa Braver, and Guohua Li, "Carrying Passengers as a Risk Factor for Crashes Fatal to 16- and 17-Year-Old Drivers," JAMA 283, no. 12 (March 22/29, 2000): 1578–1582.

MALE DRIVERS

Age	Number of Passengers			
	0	1	2	≥3
16	2.61	4.39	6.29	9.08
17	1.63	2.77	4.61	6.92
30–59	.92	.75	.62	.54

FEMALE DRIVERS

Age	Number of Passengers			
	0	1	2	≥3
16	1.38	1.72	1.94	3.31
17	1.26	1.48	2.82	2.28
30–59	.41	.33	.27	.40

(a) Write a matrix A for the death rate of male drivers.

(b) Write a matrix B for the death rate of female drivers.

(c) Use the matrices from parts (a) and (b) to write a matrix showing the difference between the death rate of males and females.

36. *Transportation* Use matrix operations on the matrices found in Exercise 35(a)(b) to find one matrix that gives the combined death rates for males and females (per million person trips) of drivers of various ages, with varying numbers of passengers. (*Hint:* Consider $\frac{1}{2}(A + B)$.)

6.4 MATRIX PRODUCTS AND INVERSES

In the previous section we showed how to multiply a matrix by a scalar. Finding the product of two matrices is more involved but is important in solving practical problems. To understand the reasoning behind the definition of matrix multiplication, look again at the EZ Life Company. Suppose sofas and chairs of the same model are often sold as sets with matrix W showing the number of each model set in each warehouse.

$$\begin{array}{c} \\ \text{New York} \\ \text{Chicago} \\ \text{San Francisco} \end{array} \begin{array}{ccc} A & B & C \end{array} \\ \begin{bmatrix} 10 & 7 & 3 \\ 5 & 9 & 6 \\ 4 & 8 & 2 \end{bmatrix} = W$$

If the selling price of a model A set is $800, of a model B set $1000, and of a model C set $1200, find the total value of the sets in the New York warehouse as follows.

In this example of the EZ Life Company, find the total value of the New York sets if model A sets sell for $1200, model B for $1600, and model C for $1300.

Answer:

$27,100

Type	Number of Sets		Price of Set		Total
A	10	×	$ 800	=	$ 8,000
B	7	×	1000	=	7,000
C	3	×	1200	=	3,600
			Total for New York		$18,600

The total value of the three kinds of sets in New York is $18,600. (1)

The work done in the table above is summarized as follows:

$$10(\$800) + 7(\$1000) + 3(\$1200) = \$18,600.$$

In the same way, the Chicago sets have a total value of

$$5(\$800) + 9(\$1000) + 6(\$1200) = \$20,200,$$

and in San Francisco, the total value of the sets is

$$4(\$800) + 8(\$1000) + 2(\$1200) = \$13,600.$$

The selling prices can be written as a column matrix, P, and the total value in each location as a column matrix, V.

$$P = \begin{bmatrix} 800 \\ 1000 \\ 1200 \end{bmatrix} \quad \text{and} \quad V = \begin{bmatrix} 18,600 \\ 20,200 \\ 13,600 \end{bmatrix}$$

Consider how the first row of the matrix W and the single column P lead to the first entry of V.

$$\begin{matrix} W & P & & V \end{matrix}$$

$$\begin{bmatrix} \mathbf{10} & \mathbf{7} & \mathbf{3} \\ 5 & 9 & 6 \\ 4 & 8 & 2 \end{bmatrix} \begin{bmatrix} \mathbf{800} \\ \mathbf{1000} \\ \mathbf{1200} \end{bmatrix} \quad 10(800) + 7(1000) + 3(1200) = \begin{bmatrix} 18,600 \\ 20,200 \\ 13,600 \end{bmatrix}$$

Product of first entries + Product of second entries + Product of third entries

Similarly, adding the products of corresponding entries in the second row of W and the column P produces the second entry in V. The third entry in V is obtained in the same way by using the third row of W and column P. This suggests that it is reasonable to *define* the product WP to be V.

$$WP = \begin{bmatrix} 10 & 7 & 3 \\ 5 & 9 & 6 \\ 4 & 8 & 2 \end{bmatrix} \begin{bmatrix} 800 \\ 1000 \\ 1200 \end{bmatrix} = \begin{bmatrix} 18,600 \\ 20,200 \\ 13,600 \end{bmatrix} = V$$

Note the sizes of the matrices here: the product of a 3×3 matrix and a 3×1 matrix is a 3×1 matrix.

Multiplying Matrices

We first define the **product of a row of a matrix and a column of a matrix** (with the same number of entries in each) to be the *number* obtained by multiplying the corresponding entries (first by first, second by second, etc.) and adding the results. For instance,

$$[3 \quad -2 \quad 1] \cdot \begin{bmatrix} 4 \\ 5 \\ 0 \end{bmatrix} = 3 \cdot 4 + (-2) \cdot 5 + 1 \cdot 0 = 12 - 10 + 0 = 2.$$

Now **matrix multiplication** is defined as follows.

> Let A be an $m \times n$ matrix and let B be an $n \times k$ matrix. The **product matrix** AB is the $m \times k$ matrix whose entry in the ith row and jth column is
>
> the product of the ith row of A and the jth column of B.

CAUTION Be careful when multiplying matrices. Remember that the number of *columns* of A must equal the number of *rows* of B in order to get the product matrix AB. The final product will have as many rows as A and as many columns as B.

▶ **EXAMPLE 1** Suppose matrix A is 2×2 and matrix B is 2×4. Can the product AB be calculated? What is the size of the product?

The following diagram helps decide the answers to these questions.

②

Matrix A is 4×6 and matrix B is 2×4.

(a) Can AB be found? If so, give its size.

(b) Can BA be found? If so, give its size.

Answers:

(a) No

(b) Yes; 2×6

matrix A
size
2×2

matrix B
size
2×4

must match

size of AB
2×4

The product AB can be calculated because A has two columns and B has two rows. The product will be a 2×4 matrix. ◀ ②

▶ **EXAMPLE 2** Find the product CD when

$$C = \begin{bmatrix} -3 & 4 & 2 \\ 5 & 0 & 4 \end{bmatrix} \quad \text{and} \quad D = \begin{bmatrix} -6 & 4 \\ 2 & 3 \\ 3 & -2 \end{bmatrix}.$$

Here matrix C is 2×3 and matrix D is 3×2, so matrix CD can be found and will be 2×2.

Step 1: row 1, column 1

$$\begin{bmatrix} -3 & 4 & 2 \\ 5 & 0 & 4 \end{bmatrix} \begin{bmatrix} -6 & 4 \\ 2 & 3 \\ 3 & -2 \end{bmatrix} \qquad (-3) \cdot (-6) + 4 \cdot 2 + 2 \cdot 3 = 32$$

Hence, 32 is the entry in row 1 column 1 of CD, as shown in Step 5 below.

Step 2: row 1, column 2

$$\begin{bmatrix} -3 & 4 & 2 \\ 5 & 0 & 4 \end{bmatrix} \begin{bmatrix} -6 & 4 \\ 2 & 3 \\ 3 & -2 \end{bmatrix} \qquad (-3) \cdot 4 + 4 \cdot 3 + 2 \cdot (-2) = -4$$

So -4 is the entry in row 1, column 2 of CD, as shown in Step 5. Continue in this manner to find the remaining entries of CD.

Step 3: row 2, column 1

$$\begin{bmatrix} -3 & 4 & 2 \\ 5 & 0 & 4 \end{bmatrix} \begin{bmatrix} -6 & 4 \\ 2 & 3 \\ 3 & -2 \end{bmatrix} \qquad 5 \cdot (-6) + 0 \cdot 2 + 4 \cdot 3 = -18$$

Step 4: row 2, column 2

$$\begin{bmatrix} -3 & 4 & 2 \\ 5 & 0 & 4 \end{bmatrix} \begin{bmatrix} -6 & 4 \\ 2 & 3 \\ 3 & -2 \end{bmatrix} \qquad 5 \cdot 4 + 0 \cdot 3 + 4 \cdot (-2) = 12$$

Step 5:   The product is

$$CD = \begin{bmatrix} -3 & 4 & 2 \\ 5 & 0 & 4 \end{bmatrix} \begin{bmatrix} -6 & 4 \\ 2 & 3 \\ 3 & -2 \end{bmatrix} = \begin{bmatrix} 32 & -4 \\ -18 & 12 \end{bmatrix}. \quad \blacktriangleleft \ \text{③}$$

▶ **EXAMPLE 3**   Find BA given

$$A = \begin{bmatrix} 1 & -3 \\ 7 & 2 \end{bmatrix} \quad \text{and} \quad B = \begin{bmatrix} 1 & 0 & -1 \\ 3 & 1 & 4 \end{bmatrix}.$$

Since B is a 2×3 matrix and A is a 2×2 matrix, the product BA cannot be found. $\blacktriangleleft$ ④

T E C H N O L O G Y   T I P   Graphing calculators can find matrix products. However, if you use a graphing calculator to try to find the product in Example 3, the calculator will display an error message.

Matrix multiplication has some similarities with the multiplication of numbers.

---

③

Find the product CD given

$$C = \begin{bmatrix} 1 & 3 & 5 \\ 2 & -4 & -1 \end{bmatrix}$$

and

$$D = \begin{bmatrix} 2 & -1 \\ 4 & 3 \\ 1 & -2 \end{bmatrix}.$$

Answer:

$$CD = \begin{bmatrix} 19 & -2 \\ -13 & -12 \end{bmatrix}$$

Give the size of each of the following products that can be found.

(a) $\begin{bmatrix} 2 & 4 \\ 6 & 8 \end{bmatrix} \begin{bmatrix} 1 & 2 & 3 \\ 0 & -1 & 2 \end{bmatrix}$

(b) $\begin{bmatrix} 1 & 2 \\ 5 & 10 \\ 12 & 7 \end{bmatrix} \begin{bmatrix} 2 & 4 \\ 3 & 6 \\ 9 & 1 \end{bmatrix}$

(c) $\begin{bmatrix} 5 \\ 2 \\ 4 \end{bmatrix} [1 \ \ 0 \ \ 6]$

Answers:

(a) 2×3

(b) Not possible

(c) 3×3

For any matrices A, B, C, such that all the indicated sums and products exist, matrix multiplication is associative and distributive.

$$A(BC) = (AB)C \qquad A(B + C) = AB + AC \qquad (B + C)A = BA + CA$$

However, there are important differences between matrix multiplication and multiplication of numbers. (See Exercises 19–22 at the end of this section.) In particular, matrix multiplication is *not* commutative.

If A and B are matrices such that the products AB and BA exist,

$$AB \text{ may not equal } BA.$$

Figure 6.15 shows an example of this situation.

```
[A]
          [[1  -4]
           [3  2 ]]
[B]
          [[-2  4]
           [5   3]]
```

```
[A] [B]
          [[-22  -8]
           [4    18]]
[B] [A]
          [[10  16 ]
           [14  -14]]
```

FIGURE 6.15

▶ **EXAMPLE 4**   A contractor builds three kinds of houses, models A, B, and C, with a choice of two styles, Spanish or contemporary. Matrix P shows the number of each kind of house planned for a new 100-home subdivision.

$$
\begin{array}{c}
\phantom{\text{Model A}} \\
\text{Model A} \\
\text{Model B} \\
\text{Model C}
\end{array}
\begin{array}{c}
\text{Spanish} \quad \text{Contemporary} \\
\left[
\begin{array}{cc}
0 & 30 \\
10 & 20 \\
20 & 20
\end{array}
\right] = P
\end{array}
$$

The amounts for each of the exterior materials used depend primarily on the style of the house. These amounts are shown in matrix Q. (Concrete is in cubic yards, lumber in units of 1000 board feet, brick in 1000s, and shingles in units of 100 square feet.)

$$
\begin{array}{c}
\phantom{\text{Contemporary}} \\
\text{Spanish} \\
\text{Contemporary}
\end{array}
\begin{array}{c}
\text{Concrete} \quad \text{Lumber} \quad \text{Brick} \quad \text{Shingles} \\
\left[
\begin{array}{cccc}
10 & 2 & 0 & 2 \\
50 & 1 & 20 & 2
\end{array}
\right] = Q
\end{array}
$$

Matrix R gives the cost for each kind of material.

$$
\begin{array}{c} \text{Cost per Unit} \\[4pt]
\begin{array}{r} \text{Concrete} \\ \text{Lumber} \\ \text{Brick} \\ \text{Shingles} \end{array}
\begin{bmatrix} 20 \\ 180 \\ 60 \\ 25 \end{bmatrix} = R
\end{array}
$$

(a) What is the total cost for each model house?

First find PQ. The product PQ shows the amount of each material needed for each model house.

$$
PQ = \begin{bmatrix} 0 & 30 \\ 10 & 20 \\ 20 & 20 \end{bmatrix} \begin{bmatrix} 10 & 2 & 0 & 2 \\ 50 & 1 & 20 & 2 \end{bmatrix}
$$

$$
PQ = \begin{array}{c}
\begin{array}{cccc} \text{Concrete} & \text{Lumber} & \text{Brick} & \text{Shingles} \end{array} \\
\begin{bmatrix} 1500 & 30 & 600 & 60 \\ 1100 & 40 & 400 & 60 \\ 1200 & 60 & 400 & 80 \end{bmatrix}
\begin{array}{l} \text{Model A} \\ \text{Model B} \\ \text{Model C} \end{array}
\end{array}
$$

Now multiply PQ and R, the cost matrix, to get the total cost for each model house.

$$
\begin{bmatrix} 1500 & 30 & 600 & 60 \\ 1100 & 40 & 400 & 60 \\ 1200 & 60 & 400 & 80 \end{bmatrix} \begin{bmatrix} 20 \\ 180 \\ 60 \\ 25 \end{bmatrix} = \begin{array}{c} \text{Cost} \\ \begin{bmatrix} 72,900 \\ 54,700 \\ 60,800 \end{bmatrix} \begin{array}{l} \text{Model A} \\ \text{Model B} \\ \text{Model C} \end{array} \end{array}
$$

(b) How much of each of the four kinds of material must be ordered?

The totals of the columns of matrix PQ will give a matrix whose elements represent the total amounts of each material needed for the subdivision. Call this matrix T and write it as a row matrix.

$$
T = [3800 \quad 130 \quad 1400 \quad 200]
$$

(c) What is the total cost for material?

Find the total cost of all the materials by taking the product of matrix T, the matrix showing the total amounts of each material, and matrix R, the cost matrix. (To multiply these and get a 1×1 matrix, representing total cost, we must multiply a 1×4 matrix by a 4×1 matrix. This is why T was written as a row matrix in (b) above.)

$$
TR = [3800 \quad 130 \quad 1400 \quad 200] \begin{bmatrix} 20 \\ 180 \\ 60 \\ 25 \end{bmatrix} = [188,400]
$$

(d) Suppose the contractor builds the same number of homes in five subdivisions. Calculate the total amount of each material for each model for all five subdivisions. Multiply PQ by the scalar 5, as follows.

$$
5 \begin{bmatrix} 1500 & 30 & 600 & 60 \\ 1100 & 40 & 400 & 60 \\ 1200 & 60 & 400 & 80 \end{bmatrix} = \begin{bmatrix} 7500 & 150 & 3000 & 300 \\ 5500 & 200 & 2000 & 300 \\ 6000 & 300 & 2000 & 400 \end{bmatrix}
$$

Let matrix A be

$$\begin{array}{c} & \textit{Vitamin} \\ & \begin{array}{ccc} C & E & K \end{array} \\ \textit{Brand} \begin{array}{c} X \\ Y \end{array} & \begin{bmatrix} 2 & 7 & 5 \\ 4 & 6 & 9 \end{bmatrix} \end{array}$$

and matrix B be

$$\begin{array}{c} & \textit{Cost} \\ & \begin{array}{cc} X & Y \end{array} \\ \textit{Vitamin} \begin{array}{c} C \\ E \\ K \end{array} & \begin{bmatrix} 12 & 14 \\ 18 & 15 \\ 9 & 10 \end{bmatrix}. \end{array}$$

(a) What quantities do matrices A and B represent?

(b) What quantities does the product AB represent?

(c) What quantities does the product BA represent?

Answers:

(a) A = brand/vitamin, B = vitamin/cost

(b) AB = brand/cost

(c) Not meaningful, although the product BA can be found

We can introduce a notation to help keep track of the quantities a matrix represents. For example, we can say that matrix P, from Example 4, represents models/styles, matrix Q represents styles/materials, and matrix R represents materials/cost. In each case, the meaning of the rows is written first and the columns second. When we found the product PQ in Example 4, the rows of the matrix represented models and the columns represented materials. Therefore, we can say the matrix product PQ represents models/materials. The common quantity, styles, in both P and Q was eliminated in the product PQ. Do you see that the product $(PQ)R$ represents models/cost?

In practical problems this notation helps decide in what order to multiply two matrices so that the results are meaningful. In Example 4(c) we could have found either product RT or product TR. However, since T represents subdivisions/materials and R represents materials/cost, the product TR gives subdivisions/cost.

Identity and Inverse Matrices

Recall from Section 1.1 that the real number 1 is the identity element for multiplication of real numbers: for any real number a, $a \cdot 1 = 1 \cdot a = a$. In this section, an **identity matrix I** is defined that has properties similar to those of the number 1.

If I is to be the identity matrix, the products AI and IA must both equal A. The 2×2 identity matrix that satisfies these conditions is

$$I = \begin{bmatrix} 1 & 0 \\ 0 & 1 \end{bmatrix}. \quad \text{⑥}$$

To check that I is really the 2×2 identity matrix, let

$$A = \begin{bmatrix} a & b \\ c & d \end{bmatrix}.$$

Then AI and IA should both equal A.

$$AI = \begin{bmatrix} a & b \\ c & d \end{bmatrix}\begin{bmatrix} 1 & 0 \\ 0 & 1 \end{bmatrix} = \begin{bmatrix} a(1) + b(0) & a(0) + b(1) \\ c(1) + d(0) & c(0) + d(1) \end{bmatrix} = \begin{bmatrix} a & b \\ c & d \end{bmatrix} = A$$

$$IA = \begin{bmatrix} 1 & 0 \\ 0 & 1 \end{bmatrix}\begin{bmatrix} a & b \\ c & d \end{bmatrix} = \begin{bmatrix} 1(a) + 0(c) & 1(b) + 0(d) \\ 0(a) + 1(c) & 0(b) + 1(d) \end{bmatrix} = \begin{bmatrix} a & b \\ c & d \end{bmatrix} = A$$

This verifies that I has been defined correctly. (It can also be shown that I is the only 2×2 identity matrix.)

The identity matrices for 3×3 matrices and 4×4 matrices, respectively, are

Let $A = \begin{bmatrix} 3 & -2 \\ 4 & -1 \end{bmatrix}$

and $I = \begin{bmatrix} 1 & 0 \\ 0 & 1 \end{bmatrix}.$

Find IA and AI.

Answer:

$IA = \begin{bmatrix} 3 & -2 \\ 4 & -1 \end{bmatrix} = A$ and

$AI = \begin{bmatrix} 3 & -2 \\ 4 & -1 \end{bmatrix} = A$

$$I = \begin{bmatrix} 1 & 0 & 0 \\ 0 & 1 & 0 \\ 0 & 0 & 1 \end{bmatrix} \quad \text{and} \quad I = \begin{bmatrix} 1 & 0 & 0 & 0 \\ 0 & 1 & 0 & 0 \\ 0 & 0 & 1 & 0 \\ 0 & 0 & 0 & 1 \end{bmatrix}.$$

By generalizing, an identity matrix can be found for any n by n matrix: this identity matrix will have 1's on the main diagonal from upper left to lower right, with all other entries equal to 0.

TECHNOLOGY TIP An $n \times n$ identity matrix can be displayed on most graphing calculators by using IDENTITY n or IDENT n or IDENMAT(n). Look in the MATH or OPS submenu of the TI MATRIX menu, or the OPTN MAT menu of Casio, or the MATRIX submenu of the HP-39 MATH menu.

Recall that for every nonzero real number a, the equation $ax = 1$ has a solution, namely, $x = 1/a = a^{-1}$. Similarly, for a square matrix A, we consider the matrix equation $AX = I$. This equation does not always have a solution, but when it does, we use special terminology. If there is a matrix A^{-1} satisfying

$$AA^{-1} = I,$$

(that is, A^{-1} is a solution of $AX = I$), then A^{-1} is called the **inverse matrix** of A. In this case it can be proved that $A^{-1}A = I$ and that A^{-1} is unique (that is, a square matrix has no more than one inverse). When a matrix has an inverse, it can be found by using the row operations given in Section 6.2, as we shall see below.

CAUTION Only square matrices have inverses, but not every square matrix has one. A matrix that does not have an inverse is called a **singular matrix.** Note that the symbol A^{-1} (read A-inverse) does *not* mean $1/A$ or I/A; the symbol A^{-1} is just the notation for the inverse of matrix A. There is no such thing as matrix division.

▶ **EXAMPLE 5** Given matrices A and B below, determine whether B is the inverse of A.

$$A = \begin{bmatrix} 1 & 2 \\ 4 & 6 \end{bmatrix} \qquad B = \begin{bmatrix} -3 & 1 \\ 2 & -\frac{1}{2} \end{bmatrix}$$

B is the inverse of A if $AB = I$ and $BA = I$, so we find those products.

$$AB = \begin{bmatrix} 1 & 2 \\ 4 & 6 \end{bmatrix}\begin{bmatrix} -3 & 1 \\ 2 & -\frac{1}{2} \end{bmatrix} = \begin{bmatrix} 1 & 0 \\ 0 & 1 \end{bmatrix} = I$$

$$BA = \begin{bmatrix} -3 & 1 \\ 2 & -\frac{1}{2} \end{bmatrix}\begin{bmatrix} 1 & 2 \\ 4 & 6 \end{bmatrix} = \begin{bmatrix} 1 & 0 \\ 0 & 1 \end{bmatrix} = I$$

Therefore, B is inverse of A, that is, $A^{-1} = B$. (It is also true that A is the inverse of B, that is, $B^{-1} = A$.) ◀ ⑦

To see how to find the multiplicative inverse of a matrix, let us look for the inverse of

$$A = \begin{bmatrix} 2 & 4 \\ 1 & -1 \end{bmatrix}.$$

Let the unknown inverse matrix be

$$A^{-1} = \begin{bmatrix} x & y \\ z & w \end{bmatrix}.$$

⑦

Given $A = \begin{bmatrix} 2 & 3 \\ 1 & 8 \end{bmatrix}$ and

$B = \begin{bmatrix} -1 & 3 \\ 1 & -2 \end{bmatrix}$, determine if they are inverses.

Answer:

No, because $AB \neq I$.

By the definition of matrix inverse, $AA^{-1} = I$, or

$$AA^{-1} = \begin{bmatrix} 2 & 4 \\ 1 & -1 \end{bmatrix}\begin{bmatrix} x & y \\ z & w \end{bmatrix} = \begin{bmatrix} 1 & 0 \\ 0 & 1 \end{bmatrix}.$$

Use matrix multiplication to get

$$\begin{bmatrix} 2x + 4z & 2y + 4w \\ x - z & y - w \end{bmatrix} = \begin{bmatrix} 1 & 0 \\ 0 & 1 \end{bmatrix}.$$

Setting corresponding elements equal gives the system of equations

$$2x + 4z = 1 \tag{1}$$
$$2y + 4w = 0 \tag{2}$$
$$x - z = 0 \tag{3}$$
$$y - w = 1. \tag{4}$$

Since equations (1) and (3) involve only x and z, while equations (2) and (4) involve only y and w, these four equations lead to two systems of equations,

$$\begin{array}{cc} 2x + 4z = 1 \\ x - z = 0 \end{array} \quad \text{and} \quad \begin{array}{cc} 2y + 4w = 0 \\ y - w = 1. \end{array}$$

Writing the two systems as augmented matrices gives

$$\begin{bmatrix} 2 & 4 & | & 1 \\ 1 & -1 & | & 0 \end{bmatrix} \quad \text{and} \quad \begin{bmatrix} 2 & 4 & | & 0 \\ 1 & -1 & | & 1 \end{bmatrix}.$$

Note that the row operations needed to transform both matrices are the same because the first two columns of both matrices are identical. Consequently, we can save time by combining these matrices into the single matrix

$$\begin{bmatrix} 2 & 4 & | & 1 & 0 \\ 1 & -1 & | & 0 & 1 \end{bmatrix} \tag{5}$$

Columns 1–3 represent the first system and columns 1, 2, and 4 represent the second system. Now use row operations as follows.

$$\begin{bmatrix} 1 & -1 & | & 0 & 1 \\ 2 & 4 & | & 1 & 0 \end{bmatrix} \qquad \text{Interchange } R_1, R_2$$

$$\begin{bmatrix} 1 & -1 & | & 0 & 1 \\ 0 & 6 & | & 1 & -2 \end{bmatrix} \qquad -2R_1 + R_2$$

$$\begin{bmatrix} 1 & -1 & | & 0 & 1 \\ 0 & 1 & | & \frac{1}{6} & -\frac{1}{3} \end{bmatrix} \qquad \frac{1}{6}R_2$$

$$\begin{bmatrix} 1 & 0 & | & \frac{1}{6} & \frac{2}{3} \\ 0 & 1 & | & \frac{1}{6} & -\frac{1}{3} \end{bmatrix} \qquad R_2 + R_1 \tag{6}$$

The left half of the augmented matrix (6) is the identity matrix, so the Gauss-Jordan process is finished and the solutions can be read from the right half of the augmented matrix. The numbers in the first column to the right of the vertical bar give the values of x and z. The second column to the right of the bar gives the values of y and w. That is,

(8)

(a) Find A^{-1} if
$$A = \begin{bmatrix} 2 & 2 \\ 4 & 1 \end{bmatrix}.$$

(b) Check your answer by finding AA^{-1}.

Answers:

(a) $\begin{bmatrix} -\frac{1}{6} & \frac{1}{3} \\ \frac{2}{3} & -\frac{1}{3} \end{bmatrix}$

(b) $\begin{bmatrix} 1 & 0 \\ 0 & 1 \end{bmatrix}$

$$\begin{bmatrix} 1 & 0 & | & x & y \\ 0 & 1 & | & z & w \end{bmatrix} = \begin{bmatrix} 1 & 0 & | & \frac{1}{6} & \frac{2}{3} \\ 0 & 1 & | & \frac{1}{6} & -\frac{1}{3} \end{bmatrix}$$

so that

$$A^{-1} = \begin{bmatrix} x & y \\ z & w \end{bmatrix} = \begin{bmatrix} \frac{1}{6} & \frac{2}{3} \\ \frac{1}{6} & -\frac{1}{3} \end{bmatrix}.$$

Check by multiplying A and A^{-1}. The result should be I.

$$AA^{-1} = \begin{bmatrix} 2 & 4 \\ 1 & -1 \end{bmatrix} \begin{bmatrix} \frac{1}{6} & \frac{2}{3} \\ \frac{1}{6} & -\frac{1}{3} \end{bmatrix} = \begin{bmatrix} \frac{1}{3} + \frac{2}{3} & \frac{4}{3} - \frac{4}{3} \\ \frac{1}{6} - \frac{1}{6} & \frac{2}{3} + \frac{1}{3} \end{bmatrix} = \begin{bmatrix} 1 & 0 \\ 0 & 1 \end{bmatrix} = I \quad (8)$$

Thus the original augmented matrix (5) has A as its left half and the identity matrix as its right half, while the final augmented matrix (6), at the end of the Gauss-Jordan process, has the identity matrix as its left half and the inverse matrix A^{-1} as its right half.

$$[A \mid I] \rightarrow [I \mid A^{-1}]$$

This procedure for finding the inverse of a matrix can be generalized as follows.

> To obtain an **inverse matrix** A^{-1} for any $n \times n$ matrix A for which A^{-1} exists, follow these steps.
>
> **1.** Form the augmented matrix $[A \mid I]$ where I is the $n \times n$ identity matrix.
> **2.** Perform row operations on $[A \mid I]$ to get a matrix of the form $[I \mid B]$.
> **3.** Matrix B is A^{-1}.

▶ **EXAMPLE 6** Find A^{-1} if $A = \begin{bmatrix} 1 & 0 & 1 \\ 2 & -2 & -1 \\ 3 & 0 & 0 \end{bmatrix}.$

First write the augmented matrix $[A \mid I]$.

(9)

(a) Complete this step.

(b) Write this row transformation as _____ .

Answers:

(a)
$$\begin{bmatrix} 1 & 0 & 1 & | & 1 & 0 & 0 \\ 0 & -2 & -3 & | & -2 & 1 & 0 \\ 0 & 0 & -3 & | & -3 & 0 & 1 \end{bmatrix}$$

(b) $-3R_1 + R_3$

$$[A \mid I] = \begin{bmatrix} 1 & 0 & 1 & | & 1 & 0 & 0 \\ 2 & -2 & -1 & | & 0 & 1 & 0 \\ 3 & 0 & 0 & | & 0 & 0 & 1 \end{bmatrix}$$

The augmented matrix already has 1 in the upper left-hand corner as needed, so begin by selecting the row operation that will result in a 0 for the first element in row two. Multiply row one by -2 and add the result to row two. This gives

$$\begin{bmatrix} 1 & 0 & 1 & | & 1 & 0 & 0 \\ 0 & -2 & -3 & | & -2 & 1 & 0 \\ 3 & 0 & 0 & | & 0 & 0 & 1 \end{bmatrix}. \quad -2R_1 + R_2$$

Get 0 for the first element in row three by multiplying row one by -3 and adding to row three as directed in Side Problem 9. (9)

(a) Complete these steps.

(b) Write these row transformations as _____.

Answers:

(a)
$$\left[\begin{array}{ccc|ccc} 1 & 0 & 0 & 0 & 0 & \frac{1}{3} \\ 0 & 1 & 0 & -\frac{1}{2} & -\frac{1}{2} & \frac{1}{2} \\ 0 & 0 & 1 & 1 & 0 & -\frac{1}{3} \end{array}\right]$$

(b) $-1R_3 + R_1; -\frac{3}{2}R_3 + R_2$

Get 1 for the second element in row two by multiplying row two of the matrix found in problem 9 at the side by $-1/2$, obtaining the new matrix

$$\left[\begin{array}{ccc|ccc} 1 & 0 & 1 & 1 & 0 & 0 \\ 0 & 1 & \frac{3}{2} & 1 & -\frac{1}{2} & 0 \\ 0 & 0 & -3 & -3 & 0 & 1 \end{array}\right]. \qquad -\frac{1}{2}R_2$$

Get 1 for the third element in row three by multiplying row three by $-1/3$, with the result

$$\left[\begin{array}{ccc|ccc} 1 & 0 & 1 & 1 & 0 & 0 \\ 0 & 1 & \frac{3}{2} & 1 & -\frac{1}{2} & 0 \\ 0 & 0 & 1 & 1 & 0 & -\frac{1}{3} \end{array}\right]. \qquad -\frac{1}{3}R_3$$

Now do side problem 10 to get 0's for the third elements in rows one and two. ⑩

The answer for side problem 10(a) gives the desired inverse:

$$A^{-1} = \left[\begin{array}{ccc} 0 & 0 & \frac{1}{3} \\ -\frac{1}{2} & -\frac{1}{2} & \frac{1}{2} \\ 1 & 0 & -\frac{1}{3} \end{array}\right].$$

Verify that AA^{-1} is I.

A graphing calculator provides a fast way to find matrix inverses.

EXAMPLE 7 Use a graphing calculator to find the inverse of the following matrices (if they have inverses).

$$A = \left[\begin{array}{ccc} 2 & 1 & -1 \\ 1 & 3 & 2 \\ 1 & 1 & 1 \end{array}\right] \qquad \text{and} \qquad B = \left[\begin{array}{cc} 2 & 4 \\ 3 & 6 \end{array}\right]$$

Enter matrix A in the calculator (Figure 6.16(a)). Then use the x^{-1} key to find the inverse matrix, as in Figure 6.16(b). (Using $\wedge$ and -1 for the inverse results in an error message on most calculators.)

(a)   (b)

FIGURE 6.16

(11)

Use a graphing calculator to find the inverses of these matrices (if they exist).

(a) $A = \begin{bmatrix} 1 & 2 & 3 \\ 4 & -1 & 0 \\ 5 & 1 & 3 \end{bmatrix}$

(b) $B = \begin{bmatrix} 2 & 3 \\ -1 & 4 \end{bmatrix}$

Answers:

(a) No inverse

(b) Use the FRAC key (if you have one) to simplify the answer, which is

$$B^{-1} = \begin{bmatrix} 4/11 & -3/11 \\ 1/11 & 2/11 \end{bmatrix}.$$

Now enter matrix B in the calculator and use the x^{-1} key. The result is an error message (Figure 6.17), which indicates that the matrix is singular; it does not have an inverse. (If you were working by hand, you would have found that the appropriate system of equations has no solution.) ◀ **(11)**

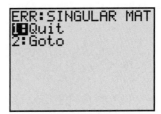

FIGURE 6.17

TECHNOLOGY TIP Because of round-off error, a graphing calculator may sometimes display an "inverse" for a matrix that doesn't actually have one. So always verify your results by multiplying A and A^{-1}. If the product is not the identity matrix, then A does not have an inverse.

6.4 Exercises

In Exercises 1–6, the sizes of two matrices A and B are given. Find the sizes of the product AB and the product BA whenever these products exist. (See Example 1.)

1. A is 2×2 and B is 2×2. 2×2

2. A is 3×3 and B is 3×3. 3×3

3. A is 3×5 and B is 5×2. 3×2

4. A is 4×3 and B is 3×6. 4×6

5. A is 4×2 and B is 3×4. 4×4

6. A is 7×3 and B is 2×7. 7×7

7. To find the product matrix AB, the number of _____ of A must be the same as the number of _____ of B.

8. The product matrix AB has the same number of _____ as A and the same number of _____ as B.

Find each of the following matrix products. (See Examples 2–4.)

9. $\begin{bmatrix} 1 & 2 \\ 3 & 4 \end{bmatrix} \begin{bmatrix} -1 \\ 7 \end{bmatrix}$

10. $\begin{bmatrix} -1 & 5 \\ 7 & 0 \end{bmatrix} \begin{bmatrix} 6 \\ 2 \end{bmatrix}$

11. $\begin{bmatrix} 2 & 2 & -1 \\ 3 & 0 & 1 \end{bmatrix} \begin{bmatrix} 0 & 2 \\ -1 & 4 \\ 0 & 2 \end{bmatrix}$

12. $\begin{bmatrix} -9 & 2 & 1 \\ 3 & 0 & 0 \end{bmatrix} \begin{bmatrix} 2 \\ -1 \\ 4 \end{bmatrix}$

13. $\begin{bmatrix} -4 & 1 \\ 2 & -3 \end{bmatrix} \begin{bmatrix} 1 & 0 \\ 0 & 1 \end{bmatrix}$

14. $\begin{bmatrix} 1 & 0 \\ 0 & 1 \end{bmatrix} \begin{bmatrix} 3 & -2 \\ 1 & -5 \end{bmatrix}$

15. $\begin{bmatrix} 1 & 0 & 0 \\ 0 & 1 & 0 \\ 0 & 0 & 1 \end{bmatrix} \begin{bmatrix} 3 & -5 & 7 \\ -2 & 1 & 6 \\ 0 & -3 & 4 \end{bmatrix}$

16. $\begin{bmatrix} -8 & 9 \\ 3 & -4 \\ -1 & 6 \end{bmatrix} \begin{bmatrix} 1 & 0 & 0 \\ 0 & 1 & 0 \end{bmatrix}$

17. $\begin{bmatrix} 1 & 2 & 3 \\ 4 & 5 & 6 \\ 7 & 8 & 9 \end{bmatrix} \begin{bmatrix} -1 & 5 \\ 7 & 0 \\ 1 & 2 \end{bmatrix}$

18. $\begin{bmatrix} -2 & 0 & 3 \\ 5 & -3 & -1 \end{bmatrix} \begin{bmatrix} 2 & 0 & -1 & 3 \\ 0 & 1 & 0 & -1 \\ 4 & 2 & 5 & -4 \end{bmatrix}$

In Exercises 19–21, use the matrices

$$A = \begin{bmatrix} -3 & -9 \\ 2 & 6 \end{bmatrix} \quad \text{and} \quad B = \begin{bmatrix} 4 & 6 \\ 2 & 3 \end{bmatrix}.$$

19. Show that $AB \neq BA$. Hence, matrix multiplication is not commutative.

20. Show that $(A + B)^2 \neq A^2 + 2AB + B^2$.

21. Show that $(A + B)(A - B) \neq A^2 - B^2$.

22. Show that $D^2 = D$, where

$$D = \begin{bmatrix} 1 & 0 & 0 \\ \frac{1}{2} & 0 & \frac{1}{2} \\ 0 & 0 & 1 \end{bmatrix}.$$

Given matrices

$$P = \begin{bmatrix} m & n \\ p & q \end{bmatrix}, \quad X = \begin{bmatrix} x & y \\ z & w \end{bmatrix}, \quad T = \begin{bmatrix} r & s \\ t & u \end{bmatrix},$$

verify that the statements in Exercises 23–26 are true.

23. $(PX)T = P(XT)$ (Associative property)

24. $P(X + T) = PX + PT$ (Distributive property)

25. $k(X + T) = kX + kT$ for any real number k

26. $(k + h)P = kP + hP$ for any real numbers k and h

Determine whether the given matrices are inverses of each other by computing their product. (See Example 5.)

27. $\begin{bmatrix} 5 & 2 \\ 3 & -1 \end{bmatrix}$ and $\begin{bmatrix} -1 & 2 \\ 3 & -4 \end{bmatrix}$

28. $\begin{bmatrix} 0 & 1 \\ 1 & 0 \end{bmatrix}$ and $\begin{bmatrix} 3 & 5 \\ 7 & 9 \end{bmatrix}$

29. $\begin{bmatrix} 3 & -1 \\ -4 & 2 \end{bmatrix}$ and $\begin{bmatrix} 1 & \frac{1}{2} \\ 2 & \frac{3}{2} \end{bmatrix}$

30. $\begin{bmatrix} 1 & 1 \\ .1 & .2 \end{bmatrix}$ and $\begin{bmatrix} 2 & -10 \\ -1 & 10 \end{bmatrix}$

31. $\begin{bmatrix} 1 & 1 & 1 \\ 2 & 3 & 0 \\ 1 & 2 & 1 \end{bmatrix}$ and $\begin{bmatrix} 1.5 & .5 & -1.5 \\ -1 & 0 & 1 \\ .5 & -2 & 2 \end{bmatrix}$

32. $\begin{bmatrix} 2 & 5 & 4 \\ 1 & 4 & 3 \\ 1 & 3 & 2 \end{bmatrix}$ and $\begin{bmatrix} 1 & 2 & 1 \\ -5 & 8 & 2 \\ 7 & -11 & -3 \end{bmatrix}$

Find the inverse, if it exists, for each of the following matrices. (See Example 6.)

33. $\begin{bmatrix} 2 & 3 \\ 1 & 2 \end{bmatrix}$

34. $\begin{bmatrix} -1 & 2 \\ 1 & -1 \end{bmatrix}$

35. $\begin{bmatrix} 2 & 4 \\ 3 & 6 \end{bmatrix}$

36. $\begin{bmatrix} -3 & -5 \\ 6 & 10 \end{bmatrix}$

37. $\begin{bmatrix} 2 & 6 \\ 1 & 4 \end{bmatrix}$

38. $\begin{bmatrix} 1 & 2 \\ 3 & 4 \end{bmatrix}$

39. $\begin{bmatrix} 1 & -1 & 1 \\ 0 & 2 & -1 \\ 2 & 3 & 0 \end{bmatrix}$

40. $\begin{bmatrix} 1 & 2 & 3 \\ 1 & 1 & 2 \\ 0 & 1 & 2 \end{bmatrix}$

41. $\begin{bmatrix} 1 & 4 & 3 \\ 1 & -3 & -2 \\ 2 & 5 & 4 \end{bmatrix}$

42. $\begin{bmatrix} 1 & 2 & 0 \\ 3 & -1 & 2 \\ -2 & 3 & -2 \end{bmatrix}$

43. $\begin{bmatrix} 1 & -3 & 4 \\ 2 & -5 & 7 \\ 0 & -1 & 1 \end{bmatrix}$

44. $\begin{bmatrix} 5 & 0 & 2 \\ 2 & 2 & 1 \\ -3 & 1 & -1 \end{bmatrix}$

Use a graphing calculator to find the inverse of each matrix.

45. $\begin{bmatrix} 2 & 4 & 6 \\ -1 & -4 & -3 \\ 0 & 1 & -1 \end{bmatrix}$

46. $\begin{bmatrix} 2 & 2 & -4 \\ 2 & 6 & 0 \\ -3 & -3 & 5 \end{bmatrix}$

47. $\begin{bmatrix} 1 & -2 & 3 & 0 \\ 0 & 1 & -1 & 1 \\ -2 & 2 & -2 & 4 \\ 0 & 2 & -3 & 1 \end{bmatrix}$

48. $\begin{bmatrix} 1 & 1 & 0 & 2 \\ 2 & -1 & 1 & -1 \\ 3 & 3 & 2 & -2 \\ 1 & 2 & 1 & 0 \end{bmatrix}$

A graphing calculator or other technology is recommended for part (c) of Exercises 49–51.

49. Social Sciences The average birth and death rates per million for several regions of the world and their populations (in millions) are given below.*

Region	Births	Deaths
Asia	.024	.008
Latin America	.025	.007
North America	.015	.009
Europe	.011	.011

Year	Asia	Latin America	North America	Europe
1970	1996	286	226	460
1980	2440	365	252	484
1990	2906	455	277	499
2000	3683	519	310	729
2025 (projected)	4723	697	364	702

(a) Write the information in the first table as a 4×2 matrix R.

(b) Write the information in the second table as a 5×4 matrix P.

(c) Find the product PR.

(d) Explain what PR represents.

(e) From matrix PR, what was the total number of births in 2000? What total number of deaths is projected for 2025?

*U.S. Bureau of Census and the United Nations Population Fund.

50. Business The first table shows the number of telecommunications devices sold (in millions) over a four-year period. The second gives the average price of each such device (in dollars) over the same period.*

Year	1998	1999	2000	2001
Cordless Phone	31.6	32.6	34.0	35.0
Answering Machine	18.3	18.9	19.5	20.1
Home Fax Machine	2.5	2.6	2.7	2.8

Year	1998	1999	2000	2001
Cordless Phone	71.20	68.71	66.47	64.51
Answering Machine	67.76	64.29	62.05	60.20
Home Fax Machine	262	240.38	229.63	219.64

(a) Write the information in the first table as a 3 × 4 matrix A.

(b) Write the information in the second table as a 4 × 3 matrix B.

(c) Find the product AB.

(d) Explain what the following entries in AB represent: row 1, column 1; row 2, column 2; row 3, column 3. What do the other entries represent?

(e) From matrix AB, what were the total sales (in dollars) of answering machines over this four-year period?

51. Health The first table shows the number of live births (in thousands). The second shows the infant mortality rates (deaths per 1000 live births) in those years.[†]

Year	1995	1996	1997	1998
Black	394	595	601	610
White	2753	3093	3085	3119

Year	1995	1996	1997	1998
Black	15.1	14.2	13.7	14.1
White	6.3	6.0	6.0	6.0

(a) Write the information in the first table as a 4 × 2 matrix C.

(b) Write the information in the second table as a 2 × 4 matrix D.

(c) Find the product DC.

---

*Multimedia Telecommunications Association.

†U.S. National Center for Health Statistics.

(d) Which entry in DC gives the total number of black infant deaths in 1995–98?

(e) Which entry in DC gives the total number of white infant deaths in 1995–98?

52. Business Burger Barn's three locations sell hamburgers, fries, and soft drinks. Barn I sells 900 burgers, 600 orders of fries, and 750 soft drinks each day. Barn II sells 1500 burgers a day and Barn III sells 1150. Soft drink sales number 900 a day at Barn II and 825 a day at Barn III. Barn II sells 950 and Barn III sells 800 orders of fries per day.

(a) Write a 3 × 3 matrix S that displays daily sales figures for all locations.

(b) Burgers cost $1.50 each, fries $.90 an order, and soft drinks $.60 each. Write a 1 × 3 matrix P that displays the prices.

(c) What matrix product displays the daily revenue at each of the three locations?

(d) What is the total daily revenue from all locations?

53. Business The four departments of Stagg Enterprises need to order the following amounts of the same products.

	Paper	Tape	Printer Ribbon	Memo Pads	Pens
Department 1	10	4	3	5	6
Department 2	7	2	2	3	8
Department 3	4	5	1	0	10
Department 4	0	3	4	5	5

The unit price (in dollars) of each product is given below for two suppliers.

Product	Supplier A	Supplier B
Paper	2	3
Tape	1	1
Printer Ribbon	4	3
Memo Pads	3	3
Pens	1	2

(a) Use matrix multiplication to get a matrix showing the comparative costs for each department for the products from the two suppliers.

(b) Find the total cost to buy products from each supplier. From which supplier should the company make the purchase?

54. Business The Perulli Candy Company makes three types of chocolate candy: Cheery Cherry, Mucho Mocha, and Almond Delight. The company produces its products in San Diego, Mexico City, and Managua using two main ingredients: chocolate and sugar.

(a) Each kilogram of Cheery Cherry requires .5 kg of sugar and .2 kg of chocolate; each kilogram of Mucho Mocha requires .4 kg of sugar and .3 kg of chocolate; and each kilogram of Almond Delight requires .3 kg of sugar and .3 kg of chocolate. Put this information into a 2 × 3 matrix, labeling the rows and columns.

(b) The cost of 1 kg of sugar is $3 in San Diego, $2 in Mexico City, and $1 in Managua. The cost of 1 kg of chocolate is $3 in San Diego, $3 in Mexico City, and $4 in Managua. Put this information into a matrix in such a way that when you multiply it with your matrix from part (a), you get a matrix representing the ingredient cost of producing each type of candy in each city.

(c) Multiply the matrices in parts (a) and (b), labeling the product matrix.

(d) From part (c) what is the combined sugar-and-chocolate cost to produce 1 kg of Mucho Mocha in Managua?

(e) Perulli Candy needs to quickly produce a special shipment of 100 kg of Cheery Cherry, 200 kg of Mucho Mocha, and 500 kg of Almond Delight, and it decides to select one factory to fill the entire order. Use matrix multiplication to determine in which city the total sugar-and-chocolate cost to produce the order is the smallest.

6.5 APPLICATIONS OF MATRICES

This section gives a variety of applications of matrices.

Solving Systems with Matrices

Consider this system of linear equations.

$$2x - 3y = 4$$
$$x + 5y = 2$$

Let

$$A = \begin{bmatrix} 2 & -3 \\ 1 & 5 \end{bmatrix}, \qquad X = \begin{bmatrix} x \\ y \end{bmatrix}, \qquad B = \begin{bmatrix} 4 \\ 2 \end{bmatrix}.$$

Since

$$AX = \begin{bmatrix} 2 & -3 \\ 1 & 5 \end{bmatrix}\begin{bmatrix} x \\ y \end{bmatrix} = \begin{bmatrix} 2x - 3y \\ x + 5y \end{bmatrix} \quad \text{and} \quad B = \begin{bmatrix} 4 \\ 2 \end{bmatrix},$$

the original system is equivalent to the single matrix equation $AX = B$. Similarly, any system of linear equations can be written as a matrix equation $AX = B$. The matrix A is called the **coefficient matrix.** ①

①

Write the matrix of coefficients, the matrix of variables, and the matrix of constants for the system

$$2x + 6y = -14$$
$$-x - 2y = 3.$$

Answers:

$$A = \begin{bmatrix} 2 & 6 \\ -1 & -2 \end{bmatrix},$$

$$X = \begin{bmatrix} x \\ y \end{bmatrix},$$

$$B = \begin{bmatrix} -14 \\ 3 \end{bmatrix}.$$

A matrix equation $AX = B$ can be solved if A^{-1} exists. Assuming A^{-1} exists and using the facts that $A^{-1}A = I$ and $IX = X$ along with the associative property of multiplication of matrices gives

$$AX = B$$
$$A^{-1}(AX) = A^{-1}B \qquad \text{Multiply both sides by } A^{-1}.$$
$$(A^{-1}A)X = A^{-1}B \qquad \text{Associative property}$$
$$IX = A^{-1}B \qquad \text{Inverse property}$$
$$X = A^{-1}B. \qquad \text{Identity property}$$

When multiplying by matrices on both sides of a matrix equation, be careful to multiply in the same order on both sides of the equation, since multiplication of matrices is not commutative (unlike multiplication of real numbers). This discussion is summarized below.

> A system of equations $AX = B$, where A is the matrix of coefficients, X is the matrix of variables, and B is the matrix of constants, is solved by first finding A^{-1}. Then, if A^{-1} exists, $X = A^{-1}B$.

This method is very efficient when technology is available. The elimination method (the matrix version) or the Gauss-Jordan method is the best choice when solving a system by hand.

▶ **EXAMPLE 1**   Use the inverse of the coefficient matrix to solve the system

$$-x - 2y + 2z = 9$$
$$2x + y - z = -3$$
$$3x - 2y + z = -6.$$

We use the fact that $X = A^{-1}B$, with

$$A = \begin{bmatrix} -1 & -2 & 2 \\ 2 & 1 & -1 \\ 3 & -2 & 1 \end{bmatrix} \quad \text{and} \quad B = \begin{bmatrix} 9 \\ -3 \\ -6 \end{bmatrix}.$$

Using either the method explained in Section 6.4 or a graphing calculator,

$$A^{-1} = \begin{bmatrix} \frac{1}{3} & \frac{2}{3} & 0 \\ \frac{5}{3} & \frac{7}{3} & -1 \\ \frac{7}{3} & \frac{8}{3} & -1 \end{bmatrix}$$

and

$$X = A^{-1}B = \begin{bmatrix} \frac{1}{3} & \frac{2}{3} & 0 \\ \frac{5}{3} & \frac{7}{3} & -1 \\ \frac{7}{3} & \frac{8}{3} & -1 \end{bmatrix} \begin{bmatrix} 9 \\ -3 \\ -6 \end{bmatrix} = \begin{bmatrix} 1 \\ 14 \\ 19 \end{bmatrix}.$$

Thus, the solution is $(1, 14, 19)$. ◀ ②

②

Use the inverse matrix to solve the system in Example 1, if the constants for the three equations are 12, 0, and 8, respectively.

Answer:

$(4, 12, 20)$

▶ **EXAMPLE 2** Use the inverse of the coefficient matrix to solve the system

$$x + 1.5y = 8$$
$$2x + 3y = 10.$$

The coefficient matrix is $A = \begin{bmatrix} 1 & 1.5 \\ 2 & 3 \end{bmatrix}$. A graphing calculator will indicate that A^{-1} does not exist. If we try to carry out the row operations, we see why.

$$\begin{bmatrix} 1 & 1.5 & | & 1 & 0 \\ 2 & 3 & | & 0 & 1 \end{bmatrix}$$

$$\begin{bmatrix} 1 & 1.5 & | & 1 & 0 \\ 0 & 0 & | & -2 & 1 \end{bmatrix}$$

The next step cannot be performed because of the zero in the second row, second column. Verify that the original system has no solution. ◀ ③

③

Solve the system in Example 2 if the constants are, respectively, 3 and 6.

Answer:

$(3 - 1.5y, y)$ for all real numbers y

Input-Output Analysis

An interesting application of matrix theory to economics was developed by Nobel Prize winner Wassily Leontief. His application of matrices to the interdependencies in an economy is called **input-output** analysis. In practice, input-output analysis is very complicated with many variables. We shall discuss only simple examples with a few variables.

Input-output models are concerned with the production and flow of goods (and perhaps services). In an economy with n basic commodities (or sectors), the production of each commodity uses some (perhaps all) of the commodities in the economy as inputs. The amounts of each commodity used in the production of 1 unit of each commodity can be written as an $n \times n$ matrix A, called the **technological** or **input-output matrix** of the economy.

▶ **EXAMPLE 3** Suppose a simplified economy involves just three commodity categories: agriculture, manufacturing, and transportation, all in appropriate units. Production of 1 unit of agriculture requires 1/2 unit of manufacturing and 1/4 unit of transportation. Production of 1 unit of manufacturing requires 1/4 unit of agriculture and 1/4 unit of transportation; while production of 1 unit of transportation requires 1/3 unit of agriculture and 1/4 unit of manufacturing. Write the input-output matrix of this economy.

The matrix is shown below.

$$\text{Input} \begin{array}{c} \text{Agriculture} \\ \text{Manufacturing} \\ \text{Transportation} \end{array} \overset{\displaystyle \overset{\textit{Output}}{\begin{array}{ccc} \text{Agricul-} & \text{Manufac-} & \text{Trans-} \\ \text{ture} & \text{turing} & \text{portation} \end{array}}}{\begin{bmatrix} 0 & \frac{1}{4} & \frac{1}{3} \\ \frac{1}{2} & 0 & \frac{1}{4} \\ \frac{1}{4} & \frac{1}{4} & 0 \end{bmatrix}} = A$$

The first column of the input-output matrix represents the amount of each of the three commodities consumed in the production of 1 unit of agriculture. The second column gives the corresponding amounts required to produce 1 unit of manufacturing, and the last column gives the amounts needed to produce 1 unit of transportation. (Although it is unrealistic, perhaps, that production of 1 unit of a commodity requires none of that commodity, the simpler matrix involved is useful for our purposes.) ◄ ④

---

④

Write a 2 × 2 technological matrix in which 1 unit of electricity requires 1/2 unit of water and 1/3 unit of electricity, while 1 unit of water requires no water but 1/4 unit of electricity.

Answer:

$$\begin{array}{c} \\ \text{Elec.} \\ \text{Water} \end{array} \begin{array}{cc} \text{Elec.} & \text{Water} \\ \begin{bmatrix} \frac{1}{3} & \frac{1}{4} \\ \frac{1}{2} & 0 \end{bmatrix} \end{array}$$

Another matrix used with the input-output matrix is a matrix giving the amount of each commodity produced, called the **production matrix,** or the **vector of gross output.** In an economy producing n commodities, the production matrix can be represented by a column matrix x with entries $x_1, x_2, x_3, \ldots , x_n$.

► **EXAMPLE 4**   In Example 3, suppose the production matrix is

$$X = \begin{bmatrix} 60 \\ 52 \\ 48 \end{bmatrix}.$$

Then 60 units of agriculture, 52 units of manufacturing, and 48 units of transportation are produced. As 1/4 unit of agriculture is used for each unit of manufacturing produced, $1/4 \times 52 = 13$ units of agriculture must be used up in the "production" of manufacturing. Similarly, $1/3 \times 48 = 16$ units of agriculture will be used up in the "production" of transportation. Thus $13 + 16 = 29$ units of agriculture are used for production in the economy. Look again at the matrices A and X. Since X gives the number of units of each commodity produced and A gives the amount (in units) of each commodity used to produce 1 unit of the various commodities, the matrix product AX gives the amount of each commodity used up in production.

$$AX = \begin{bmatrix} 0 & \frac{1}{4} & \frac{1}{3} \\ \frac{1}{2} & 0 & \frac{1}{4} \\ \frac{1}{4} & \frac{1}{4} & 0 \end{bmatrix} \begin{bmatrix} 60 \\ 52 \\ 48 \end{bmatrix} = \begin{bmatrix} 29 \\ 42 \\ 28 \end{bmatrix}$$

This product shows that 29 units of agriculture, 42 units of manufacturing, and 28 units of transportation are used to produce 60 units of agriculture, 52 units of manufacturing, and 48 units of transportation. ◄

We have seen that the matrix product AX represents the amount of each commodity used in the production process. The remainder (if any) must be enough to satisfy the demand for the various commodities from outside the production system. In an n-commodity economy, this demand can be represented by a **demand matrix** D with entries $d_1, d_2, \ldots , d_n$. The difference between the production matrix, X, and the amount, AX, used in the production process must equal the demand, D, or

$$D = X - AX.$$

In Example 4,

$$D = \begin{bmatrix} 60 \\ 52 \\ 48 \end{bmatrix} - \begin{bmatrix} 29 \\ 42 \\ 28 \end{bmatrix} = \begin{bmatrix} 31 \\ 10 \\ 20 \end{bmatrix}.$$

This result shows that production of 60 units of agriculture, 52 units of manufacturing, and 48 units of transportation would satisfy a demand of 31, 10, and 20 units of each, respectively. ⑤

---

⑤

(a) Write a 2 × 1 matrix X to represent gross production of 9000 units of electricity and 12,000 units of water.

(b) Find AX using A from the last side problem.

(c) Find D using $D = X - AX$.

Answers:

(a) $\begin{bmatrix} 9000 \\ 12,000 \end{bmatrix}$

(b) $\begin{bmatrix} 6000 \\ 4500 \end{bmatrix}$

(c) $\begin{bmatrix} 3000 \\ 7500 \end{bmatrix}$

In practice, A and D usually are known and X must be found. That is, we need to decide what amounts of production are necessary to satisfy the required demands. Matrix algebra can be used to solve the equation $D = X - AX$ for X.

$$D = X - AX$$
$$D = IX - AX \qquad \text{Identity property}$$
$$D = (I - A)X \qquad \text{Distributive property}$$

If the matrix $I - A$ has an inverse, then

$$X = (I - A)^{-1}D.$$

▶ **EXAMPLE 5** Suppose, in the 3-commodity economy of Examples 3 and 4, there is a demand for 516 units of agriculture, 258 units of manufacturing, and 129 units of transportation. What should production of each commodity be?

The demand matrix is

$$D = \begin{bmatrix} 516 \\ 258 \\ 129 \end{bmatrix}.$$

Find the production matrix by first calculating $I - A$.

$$I - A = \begin{bmatrix} 1 & 0 & 0 \\ 0 & 1 & 0 \\ 0 & 0 & 1 \end{bmatrix} - \begin{bmatrix} 0 & \frac{1}{4} & \frac{1}{3} \\ \frac{1}{2} & 0 & \frac{1}{4} \\ \frac{1}{4} & \frac{1}{4} & 0 \end{bmatrix} = \begin{bmatrix} 1 & -\frac{1}{4} & -\frac{1}{3} \\ -\frac{1}{2} & 1 & -\frac{1}{4} \\ -\frac{1}{4} & -\frac{1}{4} & 1 \end{bmatrix}$$

Using a calculator with matrix capability or row operations, find the inverse of $I - A$.

$$(I - A)^{-1} = \begin{bmatrix} 1.3953 & .4961 & .5891 \\ .8372 & 1.3643 & .6202 \\ .5581 & .4651 & 1.3023 \end{bmatrix}$$

(The entries are rounded to four decimal places.)* Since $X = (I - A)^{-1}D$,

$$X = \begin{bmatrix} 1.3953 & .4961 & .5891 \\ .8372 & 1.3643 & .6202 \\ .5581 & .4651 & 1.3023 \end{bmatrix} \begin{bmatrix} 516 \\ 258 \\ 129 \end{bmatrix} = \begin{bmatrix} 924 \\ 864 \\ 576 \end{bmatrix}$$

(rounded to the nearest whole numbers).

From the last result, we see that production of 924 units of agriculture, 864 units of manufacturing, and 576 units of transportation is required to satisfy demands of 516, 258, and 129 units, respectively. ◀

▶ **EXAMPLE 6** An economy depends on two basic products, wheat and oil. To produce 1 metric ton of wheat requires .25 metric ton of wheat and .33 metric

---

*Although we show the matrix $(I - A)^{-1}$ with entries rounded to four decimal places, we did not round off in calculating $(I - A)^{-1}D$. If the rounded figures are used, the numbers in the product may vary slightly in the last digit.

A simple economy depends on just two products, beer and pretzels.

(a) Suppose 1/2 unit of beer and 1/2 unit of pretzels are needed to make 1 unit of beer, and 3/4 unit of beer is needed to make 1 unit of pretzels. Write the technological matrix A for the economy.

(b) Find $I - A$.

(c) Find $(I - A)^{-1}$.

(d) Find the gross production X that will be needed to get a net production of

$$D = \begin{bmatrix} 100 \\ 1000 \end{bmatrix}.$$

Answers:

(a) $\begin{bmatrix} \frac{1}{2} & \frac{3}{4} \\ \frac{1}{2} & 0 \end{bmatrix}$

(b) $\begin{bmatrix} \frac{1}{2} & -\frac{3}{4} \\ -\frac{1}{2} & 1 \end{bmatrix}$

(c) $\begin{bmatrix} 8 & 6 \\ 4 & 4 \end{bmatrix}$

(d) $\begin{bmatrix} 6800 \\ 4400 \end{bmatrix}$

ton of oil. Production of 1 metric ton of oil consumes .08 metric ton of wheat and .11 metric ton of oil. Find the production that will satisfy a demand of 500 metric tons of wheat and 1000 metric tons of oil.

The input-output matrix, A, and the matrix $I - A$ are

$$A = \begin{bmatrix} .25 & .08 \\ .33 & .11 \end{bmatrix} \quad \text{and} \quad I - A = \begin{bmatrix} .75 & -.08 \\ -.33 & .89 \end{bmatrix}.$$

Next, calculate $(I - A)^{-1}$.

$$(I - A)^{-1} = \begin{bmatrix} 1.3882 & .1248 \\ .5147 & 1.1699 \end{bmatrix} \quad \text{(rounded)}$$

To find the production matrix X, use the equation $X = (I - A)^{-1}D$, with

$$D = \begin{bmatrix} 500 \\ 1000 \end{bmatrix}.$$

The production matrix is

$$X = \begin{bmatrix} 1.3882 & .1248 \\ .5147 & 1.1699 \end{bmatrix} \begin{bmatrix} 500 \\ 1000 \end{bmatrix} = \begin{bmatrix} 819 \\ 1427 \end{bmatrix}.$$

The production numbers were rounded to the nearest whole number. Production of 819 metric tons of wheat and 1427 metric tons of oil is required to satisfy the indicated demand. ◀ ⑥

TECHNOLOGY TIP If you are using a graphing calculator to determine X, you can calculate $(I - A)^{-1}D$ in one step without finding the intermediate matrices $I - A$ and $(I - A)^{-1}$.

Code Theory

Governments need sophisticated methods of coding and decoding messages. One example of such an advanced code uses matrix theory. Such a code takes the letters in the words and divides them into groups. (Each space between words is treated as a letter; punctuation is disregarded.) Then, numbers are assigned to the letters of the alphabet. For our purposes, let the letter a correspond to 1, b to 2, and so on. Let the number 27 correspond to a space between words.

For example, the message

mathematics is for the birds

can be divided into groups of three letters each.

mat   hem   ati   cs–   is–   for   –th   e–b   ird   s––

(We used – to represent a space between words.) We now write a column matrix for each group of three symbols using the corresponding numbers, as determined above, instead of letters. For example, the letters *mat* can be encoded as

$$\begin{bmatrix} 13 \\ 1 \\ 20 \end{bmatrix}.$$

Write the message *"when"* using 2 × 1 matrices.

Answer:

$$\begin{bmatrix} 23 \\ 8 \end{bmatrix}, \begin{bmatrix} 5 \\ 14 \end{bmatrix}$$

Use the matrix given below to find the 2 × 1 matrices to be transmitted for the message you encoded in side problem 7.

$$\begin{bmatrix} 2 & 1 \\ 5 & 0 \end{bmatrix}$$

Answer:

$$\begin{bmatrix} 54 \\ 115 \end{bmatrix}, \begin{bmatrix} 24 \\ 25 \end{bmatrix}$$

The coded message then consists of the 3 × 1 column matrices:

$$\begin{bmatrix} 13 \\ 1 \\ 20 \end{bmatrix}, \begin{bmatrix} 8 \\ 5 \\ 13 \end{bmatrix}, \begin{bmatrix} 1 \\ 20 \\ 9 \end{bmatrix}, \begin{bmatrix} 3 \\ 19 \\ 27 \end{bmatrix}, \begin{bmatrix} 9 \\ 19 \\ 27 \end{bmatrix}, \begin{bmatrix} 6 \\ 15 \\ 18 \end{bmatrix}, \begin{bmatrix} 27 \\ 20 \\ 8 \end{bmatrix}, \begin{bmatrix} 5 \\ 27 \\ 2 \end{bmatrix}, \begin{bmatrix} 9 \\ 18 \\ 4 \end{bmatrix}, \begin{bmatrix} 19 \\ 27 \\ 27 \end{bmatrix}.$$ ⑦

We can further complicate the code by choosing a matrix that has an inverse (in this case a 3 × 3 matrix, call it *M*) and finding the products of this matrix and each of the above column matrices. The size of each group, the assignment of numbers to letters, and the choice of matrix *M* must all be predetermined.

Suppose we choose

$$M = \begin{bmatrix} 1 & 3 & 3 \\ 1 & 4 & 3 \\ 1 & 3 & 4 \end{bmatrix}.$$

If we find the products of *M* and the column matrices above, we have a new set of column matrices,

$$\begin{bmatrix} 1 & 3 & 3 \\ 1 & 4 & 3 \\ 1 & 3 & 4 \end{bmatrix}\begin{bmatrix} 13 \\ 1 \\ 20 \end{bmatrix} = \begin{bmatrix} 76 \\ 77 \\ 96 \end{bmatrix}, \quad \text{and so on.}$$

The entries of these matrices can then be transmitted to an agent as the message 76, 77, 96, and so on. ⑧

When the agent receives the message, it is divided into groups of three numbers with each group formed into a column matrix. After multiplying each column matrix by the matrix M^{-1}, the message can be read. For example,

$$M^{-1} \cdot \begin{bmatrix} 76 \\ 77 \\ 96 \end{bmatrix} = \begin{bmatrix} 13 \\ 1 \\ 20 \end{bmatrix}.$$

Although this type of code is relatively simple, it is actually difficult to break. Many complications are possible. For example, a long message might be placed in groups of 20, thus requiring a 20 × 20 matrix for coding and decoding. Finding the inverse of such a matrix would require an impractical amount of time if calculated by hand. For this reason some of the largest computers are used by government agencies involved in coding.

Routing

The diagram in Figure 6.18 shows the roads connecting four cities. Another way of representing this information is shown in matrix *A*, where the entries represent the number of roads connecting two cities without passing through another city.* For example, from the diagram we see that there are two roads connecting city 1 to city

---

*From *Matrices with Applications*, section 3.2, example 5, by Hugh G. Campbell. Copyright © 1968, pp. 50–51. Adapted by permission of Prentice-Hall, Englewood Cliffs, New Jersey.

4 without passing through either city 2 or 3. This information is entered in row one, column four and again in row four, column one, of matrix A.

$$A = \begin{bmatrix} 0 & 1 & 2 & 2 \\ 1 & 0 & 1 & 0 \\ 2 & 1 & 0 & 1 \\ 2 & 0 & 1 & 0 \end{bmatrix}$$

Note that there are zero roads connecting each city to itself. Also, there is one road connecting cities 3 and 2.

FIGURE 6.18

Use a graphing calculator to find the following

(a) A^3

(b) $A + A^2$

Answers:

(a)

```
[A]^3
 [[12 12 22 21]
  [12  4  9  6 ]
  [22  9 12 12]
  [21  6 12  8 ]]
■
```

(b)

```
[A]+[A]²
 [[9 3 5 4]
  [3 2 3 3]
  [5 3 6 5]
  [4 3 5 5]]
```

How many ways are there to go from city 1 to city 2, for example, by going through exactly one other city? Because we must go through one other city, we must go through either city 3 or city 4. On the diagram in Figure 6.18, we see that we can go from city 1 to city 2 through city 3 in two ways. We can go from city 1 to city 3 in two ways and then from city 3 to city 2 in one way, so there are $2 \cdot 1 = 2$ ways to get from city 1 to city 2 through city 3. It is not possible to go from city 1 to city 2 through city 4, because there is no direct route between cities 4 and 2.

The matrix A^2 gives the number of ways to travel between any two cities by passing through exactly one other city. Multiply matrix A by itself, to get A^2. Let the first row, second column, entry of A^2 be b_{12}. (We use a_{ij} to denote the entry in the ith row and jth column of matrix A.) The entry b_{12} is found as follows.

$$b_{12} = a_{11}a_{12} + a_{12}a_{22} + a_{13}a_{32} + a_{14}a_{42}$$
$$= 0 \cdot 1 + 1 \cdot 0 + 2 \cdot 1 + 2 \cdot 0$$
$$= 2$$

The first product $0 \cdot 1$ in the calculations above represents the number of ways to go from city 1 to city 1 (that is, 0) and then from city 1 to city 2 (that is, 1). The 0 result indicates that such a trip does not involve a third city. The only nonzero product $(2 \cdot 1)$ represents the two routes from city 1 to city 3 and the one route from city 3 to city 2, which result in the $2 \cdot 1$ or 2 routes from city 1 to city 2 by going through city 3.

Similarly, A^3 gives the number of ways to travel between any two cities by passing through exactly two cities. Also, $A + A^2$ represents the total number of ways to travel between two cities with at most one intermediate city. ⑨

The diagram can be given many other interpretations. For example, the lines could represent lines of mutual influence between people or nations, or they could represent communication lines such as telephone lines.

6.5 Exercises

Solve the matrix equation AX = B for X. (See Example 1.)

1. $A = \begin{bmatrix} 1 & -1 \\ 5 & -6 \end{bmatrix}, B = \begin{bmatrix} 2 \\ 4 \end{bmatrix}$

2. $A = \begin{bmatrix} 3 & -2 \\ -1 & 1 \end{bmatrix}, B = \begin{bmatrix} -3 \\ 5 \end{bmatrix}$

3. $A = \begin{bmatrix} 3 & 1 \\ 4 & 2 \end{bmatrix}, B = \begin{bmatrix} 3 & 4 \\ 5 & 6 \end{bmatrix}$

4. $A = \begin{bmatrix} 7 & -3 \\ -2 & 1 \end{bmatrix}, B = \begin{bmatrix} 0 & 8 \\ 4 & 1 \end{bmatrix}$

5. $A = \begin{bmatrix} 1 & -2 & -3 \\ -1 & 4 & 6 \\ 1 & -1 & -2 \end{bmatrix}, B = \begin{bmatrix} 2 \\ 7 \\ 4 \end{bmatrix}$

6. $A = \begin{bmatrix} 3 & -1 & 0 \\ 0 & 1 & 2 \\ 6 & 0 & 5 \end{bmatrix}, B = \begin{bmatrix} -6 \\ 12 \\ 15 \end{bmatrix}$

Use the inverse of the coefficient matrix to solve each system of equations. (The inverses for Exercises 9–14 were found in Exercises 41 and 44–48 of Section 6.4.) (See Example 1.)

7. $x + 2y + 3z = 5$
   $2x + 3y + 2z = 2$
   $-x - 2y - 4z = -1$

8. $x + y - 3z = 4$
   $2x + 4y - 4z = 8$
   $-x + y + 4z = -3$

9. $x + 4y + 3z = -12$
   $x - 3y - 2z = 0$
   $2x + 5y + 4z = 7$

10. $5x \qquad + 2z = 3$
   $2x + 2y + z = 4$
   $-3x + y - z = 5$

11. $2x + 4y + 6z = 4$
   $-x - 4y - 3z = 8$
   $y - z = -4$

12. $2x + 2y - 4z = 12$
   $2x + 6y \qquad = 16$
   $-3x - 3y + 5z = -20$

13. $x - 2y + 3z \qquad = 4$
   $y - z + w = -8$
   $-2x + 2y - 2z + 4w = 12$
   $2y - 3z + w = -4$

14. $x + y \qquad + 2w = 3$
   $2x - y + z - w = 3$
   $3x + 3y + 2z - 2w = 5$
   $x + 2y + z \qquad = 3$

Use matrix algebra to solve the following matrix equations for X. Then use the given matrices to find X and check your work.

15. $N = X - MX, N = \begin{bmatrix} 8 \\ -12 \end{bmatrix}, M = \begin{bmatrix} 0 & 1 \\ -2 & 1 \end{bmatrix}$

16. $A = BX + X, A = \begin{bmatrix} 4 & 6 \\ -2 & 2 \end{bmatrix}, B = \begin{bmatrix} -2 & -2 \\ 3 & 3 \end{bmatrix}$

Find the production matrix for the given input-output and demand matrices. (See Examples 3–6.)

17. $A = \begin{bmatrix} \frac{1}{2} & \frac{2}{5} \\ \frac{1}{4} & \frac{1}{5} \end{bmatrix}, D = \begin{bmatrix} 2 \\ 4 \end{bmatrix}$

18. $A = \begin{bmatrix} .1 & .03 \\ .07 & .6 \end{bmatrix}, D = \begin{bmatrix} 5 \\ 10 \end{bmatrix}$

 19. **Business** In his work *Input-Output Economics,* Leontief provides an example of a simplified economy with just three sectors: agriculture, manufacturing, and households (that is, the sector of the economy that produces labor.)* It has the following input-output matrix.

	Agriculture	Manufacturing	Households
Agriculture	.25	.40	.133
Manufacturing	.14	.12	.100
Households	.80	3.60	.133

He also gives the demand matrix

$$D = \begin{bmatrix} 35 \\ 38 \\ 40 \end{bmatrix}$$

Find the amount of each commodity that should be produced.

20. **Business** A much simplified version of Leontief's 42-sector analysis of the 1947 American economy has the following input-output matrix:†

	Agriculture	Manufacturing	Households
Agriculture	.245	.102	.051
Manufacturing	.099	.291	.279
Households	.433	.372	.011

The demand matrix (in billions of dollars) is

$$D = \begin{bmatrix} 2.88 \\ 31.45 \\ 30.91 \end{bmatrix}$$

Find the amount of each commodity that should be produced.

---

*Wassily Leontief, *Input-Output Economics,* Second Edition (Oxford University Press, 1966), pp. 20–27.

†Ibid, pp. 6–9.

21. **Business**   An analysis of the 1958 Israeli economy is simplified here by grouping the economy into three sectors, with the following input-output matrix.*

	Agriculture	Manufacturing	Energy
Agriculture	.293	0	0
Manufacturing	.014	.207	.017
Energy	.044	.010	.216

The demand (in thousands of Israeli pounds) as measured by exports is

$$D = \begin{bmatrix} 138,213 \\ 17,597 \\ 1786 \end{bmatrix}$$

Find the amount of each commodity that should be produced.

22. **Business**   The 1981 Chinese economy can be simplified to three sectors: agriculture, industry and construction, and transportation and commerce.[†] The input-output matrix is given below.

	Agriculture	Industry/Constr.	Trans./Commerce
Agriculture	.158	.156	.009
Industry/Constr.	.136	.432	.071
Trans./Commerce	.013	.041	.011

The demand (in 100,000 RMB, the unit of money in China) is

$$D = \begin{bmatrix} 106,674 \\ 144,739 \\ 26,725 \end{bmatrix}$$

(a) Find the amount of each commodity that should be produced.

(b) Interpret the economic value of an increase in demand of 1 RMB in agriculture exports.

Write a system of equations and use the inverse of the coefficient matrix to solve the system.

23. **Business**   Felsted Furniture makes dining room furniture. A buffet requires 30 hours for construction and 10 hours for finishing. A chair requires 10 hours for construction and 10 hours for finishing. A table requires 10 hours for construction and 30 hours for finishing. The construction department has 350 hours of labor and the finishing department 150 hours of labor available each week. How many pieces of each type of furniture should be produced each week if the factory is to run at full capacity?

24. **Natural Science**   **(a)** A hospital dietician is planning a special diet for a certain patient. The total amount per meal of food groups A, B, and C must equal 400 grams. The diet should include one-third as much of group A as of group B, and the sum of the amounts of group A and group C should equal twice the amount of group B. How many grams of each food group should be included?

(b) Suppose we drop the requirement that the diet include one-third as much of group A as of group B. Describe the set of all possible solutions.

(c) Suppose that, in addition to the conditions given in part (a), foods A and B cost 2 cents per gram and food C costs 3 cents per gram and that a meal must cost $8. Is a solution possible?

25. **Natural Science**   Three species of bacteria are fed three foods, I, II, and III. A bacterium of the first species consumes 1.3 units each of foods I and II and 2.3 units of food III each day. A bacterium of the second species consumes 1.1 units of food I, 2.4 units of food II, and 3.7 units of food III each day. A bacterium of the third species consumes 8.1 units of I, 2.9 units of II, and 5.1 units of III each day. If 16,000 units of I, 28,000 units of II, and 44,000 units of III are supplied each day, how many of each species can be maintained in this environment?

26. **Business**   A company produces three combinations of mixed vegetables which sell in 1-kilogram packages. Italian style combines .3 kilogram of zucchini, .3 of broccoli, and .4 of carrots. French style combines .6 kilogram of broccoli and .4 of carrots. Oriental style combines .2 kilogram of zucchini, .5 of broccoli, and .3 of carrots. The company has a stock of 16,200 kilograms of zucchini, 41,400 kilograms of broccoli, and 29,400 kilograms of carrots. How many packages of each style should they prepare to use up their supplies?

27. **Business**   A national chain of casual clothing stores recently sent shipments of jeans, jackets, sweaters, and shirts to its stores in various cities. The number of items shipped to each city and their total wholesale cost are shown in the table below. Find the wholesale price of one pair of jeans, one jacket, one sweater, and one shirt.

City	Jeans	Jackets	Sweaters	Shirts	Total Cost
Cleveland	3000	3000	2200	4200	$507,650
St. Louis	2700	2500	2100	4300	459,075
Seattle	5000	2000	1400	7500	541,225
Phoenix	7000	1800	600	8000	571,500

---

*Ibid, pp. 174–175.

[†]*Input-Output Tables of China, 1981*, China Statistical Information and Consultancy Service Centre, 1987, pp. 17–19.

28. **Health** A 100-bed nursing home provides two levels of long-term care: regular and maximum. Patients at each level have a choice of a private room or a less expensive semiprivate room. The tables below show the number of patients in each category at various times last year. The total daily costs for all patients were $18,824 in January, $18,738 in April, $18,606 in July, and $18,824 in October. Find the daily cost of each of the following: a private room (regular care), a private room (maximum care), a semiprivate room (regular care), and a semiprivate room (maximum care).

REGULAR CARE PATIENTS		
Month	*Semiprivate*	*Private*
January	22	8
April	26	8
July	24	14
October	20	10

MAXIMUM CARE PATIENTS		
Month	*Semiprivate*	*Private*
January	60	10
April	54	12
July	56	6
October	62	8

Exercises 29 and 30 refer to Example 6.

29. **Business** If the demand is changed to 690 metric tons of wheat and 920 metric tons of oil, how many units of each commodity should be produced?

30. **Business** Change the technological matrix so that production of 1 metric ton of wheat requires 1/5 metric ton of oil (and no wheat) and the production of 1 metric ton of oil requires 1/3 metric ton of wheat (and no oil). To satisfy the same demand matrix, how many units of each commodity should be produced?

Work these problems. (See Examples 3–6.)

31. **Business** A simplified economy has only two industries, the electric company and the gas company. Each dollar's worth of the electric company's output requires $.40 of its own output and $.50 of the gas company's output. Each dollar's worth of the gas company's output requires $.25 of its own output and $.60 of the electric company's output. What should the production of electricity and gas be (in dollars) if there is a $12 million demand for gas and a $15 million demand for electricity?

32. **Business** A two-segment economy consists of manufacturing and agriculture. To produce one unit of manufacturing output requires .40 unit of its own output and .20 unit of agricultural output. To produce one unit of agricultural output requires .30 unit of its own output and .40 unit of manufacturing output. If there is a demand of 240 units of manufacturing and 90 units of agriculture, what should be the output of each segment?

33. **Business** A primitive economy depends on two basic goods, yams and pork. Production of 1 bushel of yams requires 1/4 bushel of yams and 1/2 of a pig. To produce 1 pig requires 1/6 bushel of yams. Find the amount of each commodity that should be produced to get
    (a) 1 bushel of yams and 1 pig;
    (b) 100 bushels of yams and 70 pigs.

34. **Business** A simplified economy is based on agriculture, manufacturing, and transportation. Each unit of agricultural output requires .4 unit of its own output, .3 unit of manufacturing, and .2 unit of transportation output. One unit of manufacturing output requires .4 unit of its own output, .2 unit of agricultural, and .3 unit of transportation output. One unit of transportation output requires .4 unit of its own output, .1 unit of agricultural, and .2 unit of manufacturing output. There is demand for 35 units of agricultural, 90 units of manufacturing, and 20 units of transportation output. How many units should each segment of the economy produce?

35. **Business** The use of input-output analysis can also be used to model how changes in one city can affect cities that are connected with it in some way.* For example, if a large manufacturing company shuts down in one city, it is very likely that the economic welfare of all of the cities around it will suffer. Consider three Pennsylvania communities: Sharon, Farrell, and Hermitage. Due to their proximity to each other, residents of these three communities regularly spend time and money in the other communities. Suppose that we have gathered information in the form of an input-output matrix

$$\begin{bmatrix} .2 & .1 & .1 \\ .1 & .1 & 0 \\ .5 & .6 & .7 \end{bmatrix}.$$

This matrix can be thought of as the likelihood that a person from a particular community will spend money in each of the communities.
    (a) Treat this matrix like an input-output matrix and calculate $(I - A)^{-1}$, where A is the input-output matrix above.
    (b) Interpret the entries of this inverse matrix.

---

*The idea for this problem came from an example created by Thayer Watkins, Department of Economics, San Jose State University.

Work these coding problems.

36. Social Science   Use the method discussed in the text to encode the message

Anne is home.

Break the message into groups of two letters and use the matrix

$$M = \begin{bmatrix} 1 & 3 \\ 2 & 7 \end{bmatrix}$$

37. Social Science   Use the matrix of Exercise 36 to encode the message

Head for the hills!

38. Social Science   Decode the following message, which was encoded by using the matrix M of Exercise 36.

$$\begin{bmatrix} 90 \\ 207 \end{bmatrix}, \begin{bmatrix} 39 \\ 87 \end{bmatrix}, \begin{bmatrix} 26 \\ 57 \end{bmatrix}, \begin{bmatrix} 66 \\ 145 \end{bmatrix}, \begin{bmatrix} 61 \\ 142 \end{bmatrix}, \begin{bmatrix} 89 \\ 205 \end{bmatrix},$$

Work these routing problems.

39. Social Science   Use matrix A in the discussion on routing in the text to find A^2. Then answer the following questions. How many ways are there to travel from
   (a) City 1 to city 3 by passing through exactly one city?
   (b) City 2 to city 4 by passing through exactly one city?
   (c) City 1 to city 3 by passing through at most one city?
   (d) City 2 to city 4 by passing through at most one city?

40. Social Sciences   The matrix A^3 (See Exercise 39) was found in side problem 9 of the text. Use it to answer the following questions.
   (a) How many ways are there to travel between cities 1 and 4 by passing through exactly two cities?
   (b) How many ways are there to travel between cities 1 and 4 by passing through at most two cities?

41. Business   A small telephone system connects three cities. There are four lines between cities 3 and 2, three lines connecting city 3 with city 1, and two lines between cities 1 and 2.
   (a) Write a matrix B to represent this information.
   (b) Find B^2.

(c) How many lines that connect cities 1 and 2 go through exactly one other city (city 3)?
(d) How many lines that connect cities 1 and 2 go through at most one other city?

42. Transportation   The figure shows four southern cities served by Supersouth Airlines.

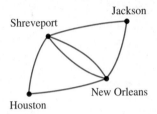

(a) Write a matrix to represent the number of nonstop routes between cities.
(b) Find the number of one-stop flights between Houston and Jackson.
(c) Find the number of flights between Houston and Shreveport that require at most one stop.
(d) Find the number of one-stop flights between New Orleans and Houston.

43. Natural Science   The figure shows a food web. The arrows indicate the food sources of each population. For example, cats feed on rats and on mice.

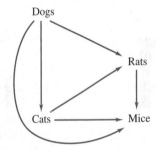

(a) Write a matrix C in which each row and corresponding column represent a population in the food chain. Enter a 1 when the population in a given row feeds on the population in the given column.
(b) Calculate and interpret C^2.

CHAPTER 6 SUMMARY

Key Terms and Symbols

6.1 linear equation
system of linear equations
solution of a system
independent system
dependent system
inconsistent system
elimination method
equivalent systems
elementary operations
row
column
matrix (matrices)

element (entry)
augmented matrix
row operations
row echelon form
parameter
6.2 Gauss-Jordan method
reduced row echelon form
6.3 row matrix (row vector)
column matrix (column vector)
square matrix
additive inverse of a matrix
zero matrix
scalar
product of a scalar and a matrix

6.4 product matrix
identity matrix
inverse matrix
singular matrix
6.5 coefficient matrix
input-output model
technological matrix (input-output matrix)
production matrix (vector of gross output)
demand matrix
code theory
routing theory

Key Concepts

Solving Systems of Equations

The following **elementary operations** are used to transform a system of equations into a simpler equivalent system.

1. Interchange any two equations.
2. Multiply both sides of an equation by a nonzero constant.
3. Replace an equation by the sum of itself and a constant multiple of another equation in the system.

The **elimination method** is a systematic way of using elementary operations to transform a system into an equivalent system that can be solved by **back substitution.** See Section 6.1 for details.

The matrix version of the elimination method uses the following **matrix row operations** that correspond to using elementary row operations with back substitution on a system of equations.

1. Interchange any two rows.
2. Multiply each element of a row by a nonzero constant.
3. Replace a row by the sum of itself and a constant multiple of another row in the matrix.

The **Gauss-Jordan method** is an extension of the elimination method for solving a system of linear equations. It uses row operations on the augmented matrix of the system. See Section 6.2 for details.

Operations on Matrices

The **sum** of two $m \times n$ matrices X and Y is the $m \times n$ matrix $X + Y$ in which each element is the sum of the corresponding elements of X and Y. The **difference** of two $m \times n$ matrices X and Y is the $m \times n$ matrix $X - Y$ in which each element is the difference of the corresponding elements of X and Y.

The **product** of a scalar k and a matrix X is the matrix kX, with each element k times the corresponding element of X.

The **product matrix** AB of an $m \times n$ matrix A and an $n \times k$ matrix B is the $m \times k$ matrix whose entry in the ith row and jth column is the product of the ith row of A and the jth column of B.

The **inverse matrix** A^{-1} for any $n \times n$ matrix A for which A^{-1} exists is found as follows. Form the augmented matrix $[A \mid I]$; perform row operations on $[A \mid I]$ to get the matrix $[I \mid A^{-1}]$.

Chapter 6 Review Exercises

Use the elimination or matrix method to solve each of the following systems. Identify any dependent or inconsistent systems.

1. $-5x - 3y = 4$
$\quad 2x + y = -3$

2. $3x - y = 6$
$\quad 2x + 3y = 7$

3. $3x - 5y = 10$
$\quad 4x - 3y = 6$

4. $\dfrac{1}{4}x - \dfrac{1}{3}y = -\dfrac{1}{4}$
$\quad \dfrac{1}{10}x + \dfrac{2}{5}y = \dfrac{2}{5}$

5. $\quad x - 2y = 1$
$\quad 4x + 4y = 2$
$\quad 10x + 8y = 4$

6. $\quad x + y - 4z = 0$
$\quad 2x + y - 3z = 2$

7. $\quad 3x + y - z = 13$
$\quad x \quad\ + 2z = 9$
$\quad -3x - y + 2z = 9$

8. $4x - y - 2z = 4$
$\quad x - y - \dfrac{1}{2}z = 1$
$\quad 2x - y - z = 8$

9. **Business** An office supply manufacturer makes two kinds of paper clips, standard and extra large. To make 1000 standard paper clips requires 1/4 hour on a cutting machine and 1/2 hour on a machine that shapes the clips. One thousand extra large paper clips require 1/3 hour on each machine. The manager of paper clip production has 4 hours per day available on the cutting machine and 6 hours per day on the shaping machine. How many of each kind of clip can he make?

10. **Business** Gretchen Schmidt plans to buy shares of two stocks. One costs $32 per share and pays dividends of $1.20 per share. The other costs $23 per share and pays dividends of $1.40 per share. She has $10,100 to spend and wants to earn dividends of $540. How many shares of each stock should she buy?

11. **Business** Joyce Pluth has money in two investment funds. Last year the first fund paid a dividend of 8% and the second a dividend of 2% and Joyce received a total of $780. This year the first fund paid a 10% dividend and the second only 1% and Joyce received $810. How much does she have invested in each fund?

12. **Finance** You are given $144 in one, five, and ten dollar bills. There are 35 bills. There are two more ten dollar bills than five dollar bills. How many bills of each type are there?

13. **Social Science** A social service agency provides counseling, meals, and shelter to clients referred by sources I, II, and III. Clients from source I require an average of $100 for food, $250 for shelter, and no counseling. Source II clients require an average of $100 for counseling, $200 for food, and nothing for shelter. Source III clients require an average of $100 for counseling, $150 for food, and $200 for shelter. The agency has funding of $25,000 for counseling, $50,000 for food, and $32,500 for shelter. How many clients from each source can be served?

14. **Business** The Waputi Indians make woven blankets, rugs, and skirts. Each blanket requires 24 hours for spinning the yarn, 4 hours for dying the yarn, and 15 hours for weaving. Rugs require 30, 5, and 18 hours and skirts 12, 3, and 9 hours, respectively. If there are 306, 59, and 201 hours available for spinning, dying, and weaving, respectively, how many of each item can be made? (*Hint:* Simplify the equations you write, if possible, before solving the system.)

Use the Gauss-Jordan method to solve the following systems.

15. $\quad x - z = -3$
$\quad y + z = 6$
$\quad 2x - 3z = -9$

16. $\quad 2x - y + 4z = -1$
$\quad -3x + 5y - z = 5$
$\quad 2x + 3y + 2z = 3$

17. $\quad 5x - 8y + z = 1$
$\quad 3x - 2y + 4z = 3$
$\quad 10x - 16y + 2z = 3$

18. $\quad x - 2y + 3z = 4$
$\quad 2x + y - 4z = 3$
$\quad -3x + 4y - z = -2$

19. $\quad 3x + 2y - 6z = 9$
$\quad x + y + 2z = 4$
$\quad 2x + 2y + 5z = 0$

20. **Business** Each week at a furniture factory, there are 2000 work hours available in the construction department, 1400 work hours in the painting department, and 1300 work hours in the packing department. Producing a chair requires 2 hours of construction, 1 hour of painting, and 2 hours for packing. Producing a table requires 4 hours of construction, 3 hours of painting, and 3 hours for packing. Producing a chest requires 8 hours of construction, 6 hours of painting, and 4 hours for packing. If all available time is used in every department, how many of each item are produced each week?

For each of the following, find the dimensions of the matrix and identify any square, row, or column matrices.

21. $\begin{bmatrix} 2 & 3 \\ 5 & 9 \end{bmatrix}$

22. $\begin{bmatrix} 2 & -1 \\ 4 & 6 \\ 5 & 7 \end{bmatrix}$

23. $[12 \quad 4 \quad -8 \quad -1]$

24. $\begin{bmatrix} -7 & 5 & 6 \\ 3 & 2 & -1 \\ -1 & 12 & 8 \end{bmatrix}$

25. $\begin{bmatrix} 6 & 8 & 10 \\ 5 & 3 & -2 \end{bmatrix}$

26. $\begin{bmatrix} -9 \\ 15 \\ 4 \end{bmatrix}$

27. Natural Science   The activities of a grazing animal can be classified roughly into three categories: grazing, moving, and resting. Suppose horses spend 8 hours grazing, 8 moving, and 8 resting; cattle spend 10 grazing, 5 moving, and 9 resting; sheep spend 7 grazing, 10 moving, and 7 resting; and goats spend 8 grazing, 9 moving, and 7 resting. Write this information as a 4 × 3 matrix.

28. Business   The New York Stock Exchange reports in the daily newspapers give the dividend, price-to-earnings ratio, sales (in hundreds of shares), last price, and change in price for each company. Write the following stock reports as a 4 × 5 matrix. American Telephone & Telegraph: 5, 7, 2532, $52\frac{3}{8}$, $-\frac{1}{4}$. General Electric: 3, 9, 1464, 56, $+\frac{1}{8}$. Gulf Oil: 2.50, 5, 4974, 41, $-1\frac{1}{2}$. Sears: 1.36, 10, 1754, 18, $+\frac{1}{2}$.

Given the matrices

$A = \begin{bmatrix} 4 & 10 \\ -2 & -3 \\ 6 & 9 \end{bmatrix}$, $B = \begin{bmatrix} 2 & 3 & -2 \\ 2 & 4 & 0 \\ 0 & 1 & 2 \end{bmatrix}$, $C = \begin{bmatrix} 5 & 0 \\ -1 & 3 \\ 4 & 7 \end{bmatrix}$,

$D = \begin{bmatrix} 6 \\ 1 \\ 0 \end{bmatrix}$, $E = [1 \quad 3 \quad -4]$, $F = \begin{bmatrix} -1 & 4 \\ 3 & 7 \end{bmatrix}$,

$G = \begin{bmatrix} 2 & 5 \\ 1 & 6 \end{bmatrix}$,

find each of the following (if possible).

29. $-B$

30. $-D$

31. $3A - 2C$

32. $F + 3G$

33. $2B - 5C$

34. $G - 2F$

35. Business   Refer to Exercise 28. Write a 4 × 2 matrix using the sales and price changes for the four companies. The next day's sales and price changes for the same four companies were 2310, 1258, 5061, 1812 and $-1/4$, $-1/4$, $+1/2$, $+1/2$, respectively. Write a 4 × 2 matrix using these new sales and price change figures. Use matrix addition to find the total sales and price changes for the two days.

36. Business   An oil refinery in Tulsa sent 110,000 gallons of oil to a Chicago distributor, 73,000 to a Dallas distributor, and 95,000 to an Atlanta distributor. Another refinery in New Orleans sent the following amounts to the same three distributors: 85,000, 108,000, 69,000. The next month the two refineries sent the same distributors new shipments of oil as follows: from Tulsa, 58,000 to Chicago, 33,000 to Dallas, and 80,000 to Atlanta; from New Orleans, 40,000, 52,000, and 30,000, respectively.
(a) Write the monthly shipments from the two distributors to the three refineries as 3 × 2 matrices.
(b) Use matrix addition to find the total amounts sent to the refineries from each distributor.

Use the matrices given above Exercise 29 to find each of the following (if possible).

37. AG

38. EB

39. GF

40. CA

41. AGF

42. Health   In a recent study, the numbers of head and neck injuries among hockey players wearing full face shields and half face shields were compared. The following table provides the rates per 1000 athlete-exposures for specific injuries that caused a player wearing either shield to miss one or more events.*

	Half Shield	Full Shield
Head and Face Injuries excluding Concussions	3.54	1.41
Concussions	1.53	1.57
Neck Injuries	.34	.29
Other	7.53	6.21

If an equal number of players in a large league wear each type of shield and the total number of athlete-exposures for the league in a season is 8000, use matrix operations to estimate the total number of injuries of each type.

43. Business   An office supply manufacturer makes two kinds of paper clips, standard and extra large. To make a unit of standard paper clips requires 1/4 hour on a cutting machine and 1/2 hour on a machine that shapes the clips. A unit of extra large paper clips requires 1/3 hour on each machine.
(a) Write this information as a 2 × 2 matrix (size/machine).
(b) If 48 units of standard and 66 units of extra large clips are to be produced, use matrix multiplication to find out how many hours each machine will operate. (*Hint:* Write the units as a 1 × 2 matrix.)

---

*Brian Benson, Nicholas Nohtadi, Sarah Rose, and Willem Meeuwisse, "Head and Neck Injuries among Ice Hockey Players Wearing Full Face Shields vs. Half Face Shields," *JAMA*, 282, no. 24, (December 22/29, 1999): 2328–2332.

44. Business Theresa DePalo buys shares of three stocks. Their cost per share and dividend earnings per share are $32, $23, and $54, and $1.20, $1.49, and $2.10, respectively. She buys 50 shares of the first stock, 20 shares of the second, and 15 shares of the third.

(a) Write the cost per share and earnings per share of the stocks as a 3×2 matrix.

(b) Write the number of shares of each stock as a 1×3 matrix.

(c) Use matrix multiplication to find the total cost and total dividend earnings of these stocks.

45. If $A = \begin{bmatrix} 3 & 0 \\ 2 & 1 \end{bmatrix}$, find a matrix B such that both AB and BA are defined and $AB \neq BA$.

46. Is it possible to do Exercise 45 if $A = \begin{bmatrix} 4 & 0 \\ 0 & 4 \end{bmatrix}$? Explain.

Find the inverse of each of the following matrices that has an inverse.

47. $\begin{bmatrix} -4 & 2 \\ 0 & 3 \end{bmatrix}$

48. $\begin{bmatrix} 2 & 1 \\ 5 & 3 \end{bmatrix}$

49. $\begin{bmatrix} 6 & 4 \\ 3 & 2 \end{bmatrix}$

50. $\begin{bmatrix} 2 & 0 \\ -1 & 5 \end{bmatrix}$

51. $\begin{bmatrix} 2 & 0 & 4 \\ 1 & -1 & 0 \\ 0 & 1 & -2 \end{bmatrix}$

52. $\begin{bmatrix} 2 & -1 & 0 \\ 1 & 0 & 1 \\ 1 & -2 & 0 \end{bmatrix}$

53. $\begin{bmatrix} 2 & 3 & 5 \\ -2 & -3 & -5 \\ 1 & 4 & 2 \end{bmatrix}$

54. $\begin{bmatrix} 1 & 3 & 6 \\ 4 & 0 & 9 \\ 5 & 15 & 30 \end{bmatrix}$

55. $\begin{bmatrix} 1 & 3 & -2 & -1 \\ 0 & 1 & 1 & 2 \\ -1 & -1 & 1 & -1 \\ 1 & -1 & -3 & -2 \end{bmatrix}$

56. $\begin{bmatrix} 3 & 2 & 0 & -1 \\ 2 & 0 & 1 & 2 \\ 1 & 2 & -1 & 0 \\ 2 & -1 & 1 & 1 \end{bmatrix}$

Refer again to the matrices given above Exercise 29 to find each of the following (if possible).

57. F^{-1}

58. G^{-1}

59. $(G - F)^{-1}$

60. $(F + G)^{-1}$

61. B^{-1}

62. Explain why the matrix $\begin{bmatrix} a & 0 \\ c & 0 \end{bmatrix}$, where a and c are nonzero constants, cannot possibly have an inverse.

Solve each of the following matrix equations AX = B for X.

63. $A = \begin{bmatrix} 2 & 4 \\ -1 & -3 \end{bmatrix}$, $B = \begin{bmatrix} 8 \\ 3 \end{bmatrix}$

64. $A = \begin{bmatrix} 1 & 3 \\ -2 & 4 \end{bmatrix}$, $B = \begin{bmatrix} 15 \\ 10 \end{bmatrix}$

65. $A = \begin{bmatrix} 1 & 0 & 2 \\ -1 & 1 & 0 \\ 3 & 0 & 4 \end{bmatrix}$, $B = \begin{bmatrix} 8 \\ 4 \\ -6 \end{bmatrix}$

66. $A = \begin{bmatrix} 2 & 4 & 0 \\ 1 & -2 & 0 \\ 0 & 0 & 3 \end{bmatrix}$, $B = \begin{bmatrix} 72 \\ -24 \\ 48 \end{bmatrix}$

Use the method of matrix inverses to solve each of the following systems.

67. $x + y = 4$
$2x + 3y = 10$

68. $5x - 3y = -2$
$2x + 7y = -9$

69. $2x + y = 5$
$3x - 2y = 4$

70. $x - 2y = 7$
$3x + y = 7$

71. $x + y + z = 1$
$2x - y = -2$
$3y + z = 2$

72. $x = -3$
$y + z = 6$
$2x - 3z = -9$

73. $3x - 2y + 4z = 4$
$4x + y - 5z = 2$
$-6x + 4y - 8z = -2$

74. $x + 2y = -1$
$3y - z = -5$
$x + 2y - z = -3$

Solve each of the following problems by any method.

75. Business A wine maker has two large casks of wine. One is 9% alcohol and the other is 14% alcohol. How many liters of each wine should be mixed to produce 40 liters of wine that is 12% alcohol?

76. Business A gold merchant has some 12 carat gold (12/24 pure gold), and some 22 carat gold (22/24 pure). How many grams of each could be mixed to get 25 grams of 15 carat gold?

77. Natural Science A chemist has some 40% acid solution and some 60% solution. How many liters of each should be used to get 40 liters of a 45% solution?

78. Business How many pounds of tea worth $4.60 a pound should be mixed with tea worth $6.50 a pound to get 10 pounds of a mixture worth $5.74 a pound?

79. Business A machine in a pottery factory takes 3 minutes to form a bowl and 2 minutes to form a plate. The material for a bowl costs $.25 and the material for a plate costs $.20. If the machine runs for 8 hours and exactly $44 is spent for material, how many bowls and plates can be produced?

80. Transportation A boat travels at a constant speed a distance of 57 kilometers downstream in 3 hours, then turns around and travels 55 kilometers upstream in 5 hours. What is the speed of the boat and of the current?

81. Business Ms. Tham invests $50,000 three ways—at 8%, $8\frac{1}{2}$%, and 11%. In total, she receives $4436.25 per year in interest. The interest from the 11% investment is

$80 more than the interest on the 8% investment. Find the amount she has invested at each rate.

82. Business Tickets to a band concert cost $2 for children, $3 for teenagers, and $5 for adults. 570 people attended the concert and total ticket receipts were $1950. Three-fourths as many teenagers as children attended. How many children, teenagers, and adults were at the concert?

83. Given the input-output matrix $A = \begin{bmatrix} 0 & \frac{1}{4} \\ \frac{1}{2} & 0 \end{bmatrix}$ and the

demand matrix $D = \begin{bmatrix} 2100 \\ 1400 \end{bmatrix}$, find each of the following.

(a) $I - A$ (b) $(I - A)^{-1}$
(c) the production matrix X

84. Business The 1987 economy of the state of Washington has been simplified to four sectors: natural resource, manufacturing, trade and services, and personal consumption. The input-output matrix is given below.*

	natural resources	manu-facturing	trade and services	personal consumption
natural resources	.1045	.0428	.0029	.0031
manufacturing	.0826	.1087	.0584	.0321
trade and services	.0867	.1019	.2032	.3555
personal consumption	.6253	.3448	.6106	.0798

Suppose the demand (in millions of dollars) is

$$D = \begin{bmatrix} 450 \\ 300 \\ 125 \\ 100 \end{bmatrix}.$$

Find the amount of each commodity that should be produced.

85. Business The 1963 economy of the state of Nebraska has been condensed to six sectors; livestock, crops, food products, mining and manufacturing, households, and other. The input-output matrix is in the next column.†

$$\begin{bmatrix} .178 & .018 & .411 & 0 & .005 & 0 \\ .143 & .018 & .088 & 0 & .001 & 0 \\ .089 & 0 & .035 & 0 & .060 & .003 \\ .001 & .010 & .012 & .063 & .007 & .014 \\ .141 & .252 & .088 & .089 & .402 & .124 \\ .188 & .156 & .103 & .255 & .008 & .474 \end{bmatrix}$$

(a) Find the matrix $(I - A)^{-1}$ and interpret the value in row two, column one, of this matrix.
(b) Suppose the demand (in millions of dollars) is

$$D = \begin{bmatrix} 1980 \\ 650 \\ 1750 \\ 1000 \\ 2500 \\ 3750 \end{bmatrix}.$$

Find the dollar amount of each commodity that should be produced.

86. Business An economy depends on two commodities, goats and cheese. It takes 2/3 of a unit of goats to produce 1 unit of cheese and 1/2 unit of cheese to produce 1 unit of goats.
(a) Write the input-output matrix for this economy.
(a) Find the production required to satisfy a demand of 400 units of cheese and 800 units of goats.

87. Business In a simple economic model, a country has two industries: agriculture and manufacturing. To produce $1 of agricultural output requires $.10 of agricultural output and $.40 of manufacturing output. To produce $1 of manufacturing output requires $.70 of agricultural output and $.20 of manufacturing output. If agricultural demand is $60,000 and manufacturing demand is $20,000, what must each industry produce? (Round answers to the nearest whole number.)

88. Business The matrix below represents the number of direct flights between four cities.

$$\begin{array}{c} \\ A \\ B \\ C \\ D \end{array} \begin{array}{cccc} A & B & C & D \\ \begin{bmatrix} 0 & 1 & 0 & 1 \\ 1 & 0 & 0 & 1 \\ 0 & 0 & 0 & 1 \\ 1 & 1 & 1 & 0 \end{bmatrix} \end{array}$$

(a) Find the number of one-stop flights between cities A and C.
(b) Find the total number of flights between cities B and C that are either direct or one-stop.
(c) Find the matrix that gives the number of two-stop flights between these cities.

89. Social Science (a) Use the matrix $M = \begin{bmatrix} 2 & 6 \\ 1 & 4 \end{bmatrix}$ to encode the message "leave now."
(b) What matrix should be used to decode this message?

*Robert Chase, Philip Bourque, and Richard Conway, Jr., "The 1987 Washington State Input-Output Study," Report to the Graduate School of Business Administration, University of Washington, September 1993.

†F. Charles Lamphear and Theodore Roesler, "1970 Nebraska Input-Output Tables," *Nebraska Economic and Business Report No. 10*, Bureau of Business Research, University of Nebraska-Lincoln.

Case 5

Matrix Operations and Airline Route Maps

Airline route maps are usually published on airline web sites, as well as in in-flight magazines. The purpose of these maps is to show what cities are connected to each other by non-stop flights provided by the airline. We can think of these maps as another type of **graph,** and we can use matrix operations to answer questions of interest about the graph. In order to study these graphs, a bit of terminology will be helpful. A **graph** is a set of points (called **vertices,** or **nodes**) and a set of lines called **edges** connecting some pairs of vertices. Two vertices connected by an edge are said to be **adjacent.** Consider for example the northern route map for Cape Air from May 2001, which is Figure 1 below.* Here the vertices are the cities to which Cape Air flies, and two vertices are connected if there is a non-stop flight between them.

FIGURE 1 Cape Air Northern Route Map—May 2001

Some natural questions arise about graphs. It might be important to know if two vertices are connected by a sequence of two edges, even if they are not connected by a single edge. In the route map, Provincetown and Hyannis are connected by a two-edge sequence, meaning that a passenger would have to

stop in Boston while flying between those cities on Cape Air. It might be important to know if it is possible to get from a vertex to another vertex in a given number of flights. In the example, a passenger on Cape Air can get from any city in their network to any other city given enough flights. But how many flights are enough? This is another issue of interest: what is the minimum number of steps to get from one vertex to another? What is the minimum number of steps to get from **any** vertex on the graph to any other? While these questions are relatively easy to answer for a small graph, as the number of vertices and edges grows, it becomes harder to keep track of all the different ways the vertices are connected. Matrix notation and computation can help to answer these questions.

The **adjacency matrix** for a graph with n vertices is an $n \times n$ matrix whose (i, j) entry is 1 if the ith vertex and jth are connected and 0 if they are not. If the vertices in the Cape Air graph respectively correspond to Boston (B), Hyannis (H), Martha's Vineyard (M), Nantucket (N), New Bedford (NB), Providence (P), and Provincetown (PT), then the adjacency matrix for Cape Air is as follows

$$
A = \begin{array}{c} \\ \\ \\ \\ \\ \\ \\ \end{array}
\begin{array}{c}
 \begin{array}{ccccccc} \text{B} & \text{H} & \text{M} & \text{N} & \text{NB} & \text{P} & \text{PT} \end{array} \\
\left[\begin{array}{ccccccc}
0 & 1 & 1 & 1 & 0 & 0 & 1 \\
1 & 0 & 1 & 1 & 0 & 0 & 0 \\
1 & 1 & 0 & 1 & 1 & 1 & 0 \\
1 & 1 & 1 & 0 & 1 & 1 & 0 \\
0 & 0 & 1 & 1 & 0 & 0 & 0 \\
0 & 0 & 1 & 1 & 0 & 0 & 0 \\
1 & 0 & 0 & 0 & 0 & 0 & 0
\end{array} \right]
\begin{array}{c} \text{B} \\ \text{H} \\ \text{M} \\ \text{N} \\ \text{NB} \\ \text{P} \\ \text{PT} \end{array}
\end{array}
$$

Adjacency matrices can be used to address the questions about graphs raised earlier. Which vertices are connected by a two-edge sequence? How many different two-edge sequences connect each pair of vertices? Consider the matrix A^2, which is A multiplied by itself. For example, let the $(6, 2)$ entry in the matrix A^2 be named b_{62}. This entry in A^2 is the product of the 6th row of A and the 2nd column of A, or

$$b_{62} = a_{61}a_{12} + a_{62}a_{22} + a_{63}a_{32} + a_{64}a_{42} + a_{65}a_{52} + a_{66}a_{62} + a_{67}a_{72}$$

$$= 0 \cdot 1 + 0 \cdot 0 + 1 \cdot 1 + 1 \cdot 1 + 0 \cdot 0 + 0 \cdot 0 + 0 \cdot 0$$

$$= 2,$$

*Compliments of Cape Air. See www.flycapeair.com/rout_cc_frameset.htm

which happens to be the number of two-flight sequences which connect city 6 (Providence) and city 2 (Hyannis). A careful look at Figure 1 confirms this fact. This calculation works because in order for a two-flight sequence to occur between Providence and Hyannis, Providence and Hyannis must both connect to an intermediate city. Since Providence connects to Martha's Vineyard (city 3) and Martha's Vineyard connects to Hyannis, $a_{63}a_{32} = 1 \cdot 1 = 1$. Thus there is 1 two-flight sequence from Providence to Hyannis that passes through Martha's Vineyard. Since Providence does not connect to Boston (city 1) but Boston does connect with Hyannis, $a_{61}a_{12} = 0 \cdot 1 = 0$. Thus there is no two-flight sequence from Providence to Hyannis that passes through Boston. To find the total number of two-flight sequences between Providence and Hyannis, simply sum over all intermediate points. Notice that this sum, which is

$$a_{61}a_{12} + a_{62}a_{22} + a_{63}a_{32} + a_{64}a_{42}$$
$$+ a_{65}a_{52} + a_{66}a_{62} + a_{67}a_{72},$$

is just the b_{62}, the (6, 2) entry in the matrix A^2. And so it is seen that the number of two-step sequences between vertex i and vertex j in a graph with adjacency matrix A is the (i, j) entry in A^2. A more general result is the following:

The number of k-step sequences between vertex i and vertex j in a graph with adjacency matrix A is the (i, j) entry in A^k.

If A is the adjacency matrix for Figure 1,

$$A^2 = \begin{array}{c} \\ \\ \\ \\ \\ \\ \\ \end{array}\begin{bmatrix} 4 & 2 & 2 & 2 & 2 & 2 & 0 \\ 2 & 3 & 2 & 2 & 2 & 2 & 1 \\ 2 & 2 & 5 & 4 & 1 & 1 & 1 \\ 2 & 2 & 4 & 5 & 1 & 1 & 1 \\ 2 & 2 & 1 & 1 & 2 & 2 & 0 \\ 2 & 2 & 1 & 1 & 2 & 2 & 0 \\ 0 & 1 & 1 & 1 & 0 & 0 & 1 \end{bmatrix}\begin{array}{c} B \\ H \\ M \\ N \\ NB \\ P \\ PT \end{array}$$

$$\begin{array}{c} B \;\; H \;\; M \;\; N \;\; NB \;\; P \;\; PT \end{array}$$

and

$$A^3 = \begin{bmatrix} 6 & 8 & 12 & 12 & 4 & 4 & 4 \\ 8 & 6 & 11 & 11 & 4 & 4 & 2 \\ 12 & 11 & 10 & 11 & 9 & 9 & 2 \\ 12 & 11 & 11 & 10 & 9 & 9 & 2 \\ 4 & 4 & 9 & 9 & 2 & 2 & 2 \\ 4 & 4 & 9 & 9 & 2 & 2 & 2 \\ 4 & 2 & 2 & 2 & 2 & 2 & 0 \end{bmatrix}\begin{array}{c} B \\ H \\ M \\ N \\ NB \\ P \\ PT \end{array}$$

$$\begin{array}{c} B \;\; H \;\; M \;\; N \;\; NB \;\; P \;\; PT \end{array}$$

Since the (6,3) entry in A^2 is 1, there is 1 two-step sequence from Providence to Martha's Vineyard. Likewise, there are 4 three-step sequences between Hyannis and New Bedford since the (2,5) entry in A^3 is 4.

In observing the figures, note that some two-step or three step sequences may not be meaningful. On the Cape Air route map, Nantucket is reachable in two steps from Boston (via Hyannis or Martha's Vineyard), but in reality this does not matter, since there is a non-stop flight between the two cities. A better question to ask of a graph might be, "What is the least number of edges that must be traversed to go from vertex i to vertex j?"

To answer this question, consider the matrix $S_k = A + A^2 + \cdots + A^k$. The (i, j) entry in this matrix tallies the number of ways to get from vertex i to vertex j in k steps or less. If such a trip is impossible, this entry will be zero. Thus to find the shortest number of steps between the vertices, continue to compute S_k as k increases; the first k for which the (i, j) entry in S_k is nonzero is the shortest number of steps between i and j. Note that although the shortest number of steps may be computed, the method does not determine what those steps are.

If you are interested in other airlines' route maps, visit the web site dir.yahoo.com/Business_and_Economy/Shopping_and_Services/Travel_and_Transportation/Airlines/. This includes links to many obscure and smaller airlines, as well as the more well known carriers.

Exercises

1. Which Cape Air cities may be reached by a two-flight sequence from New Bedford? Which may be reached by a three-flight sequence?

2. It was shown above that there are 4 three-step sequences between Hyannis and New Bedford. Describe these three-step sequences.

3. Which trips in the Cape Air network take the greatest number of flights?

4. The route map for the northern routes of Big Sky Airlines for May 2001 is given in Figure 2.* Produce an adjacency matrix for this map. The current route map is available at hosts3 .in-tch.com/www.bigskyair.com/northern_routes. html.

FIGURE 2 Big Sky Airlines Northern Route Map—May 2001

5. Which Big Sky cities may be reached by a three-flight sequence from Helena? For which cities does it take *at least* three flights to get from them to Helena?

6. Which trips in the Big Sky network take the largest number of flights? How many flights do these trips take?

*Airline route map courtesy of Big Sky Airlines, Billings, MT, www.bigskyair.com.

Chapter 7

Linear Programming

Many realistic problems involve inequalities. For example, a factory may have no more than 200 workers on a shift and must manufacture at least 3000 units at a cost of no more than $35 each. How many workers should it have per shift in order to produce the required units at minimal cost? *Linear programming* is a method for finding the optimal (best possible) solution for such problems, if there is one.

In this chapter we shall study two methods of solving linear programming problems: the graphical method and the simplex method. The graphical method requires a knowledge of **linear inequalities,** those involving only first-degree polynomials in x and y. So we begin with a study of such inequalities.

7.1 GRAPHING LINEAR INEQUALITIES IN TWO VARIABLES

Examples of linear inequalities in two variables include

$$x + 2y < 4, \qquad 3x + 2y > 6, \qquad \text{and} \quad 2x - 5y \geq 10.$$

A solution of a linear inequality is an ordered pair that satisfies the inequality. For example (4, 4) is a solution of

$$3x - 2y \leq 6.$$

(Check by substituting 4 for x and 4 for y.) A linear inequality has an infinite number of solutions, one for every choice of a value for x. The best way to show these solutions is with a graph that consists of all the points in the plane whose coordinates satisfy the inequality.

▶EXAMPLE 1   Graph the inequality $3x - 2y \leq 6$.

First, solve the inequality for y.

$$3x - 2y \leq 6$$
$$-2y \leq -3x + 6$$
$$y \geq \frac{3}{2}x - 3 \qquad \text{Multiply by } -1/2;$$
$$\text{reverse the inequality.}$$
$$y \leq 1.5x - 3$$

①

Graph.

(a) $2x + 5y \leq 10$

(b) $x - y \geq 4$

Answers:

(a)

$2x + 5y \leq 10$

(b)

$x - y \geq 4$

This inequality has the same solutions as the original one. To solve it, note that the points on the line $y = 1.5x - 3$ certainly satisfy $y \geq 1.5x - 3$. Plot some points and graph this line, as in Figure 7.1.

FIGURE 7.1

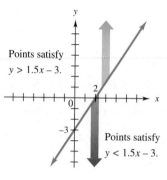

FIGURE 7.2

The points on the line satisfy "y *equals* $1.5x - 3$." The points satisfying "y is *greater than* $1.5x - 3$" are the points *above* the line (because they have larger second coordinates than the points on the line). Similarly, the points satisfying

$$y < 1.5x - 3$$

lie below the line (because they have smaller second coordinates), as shown in Figure 7.2. The line $y = 1.5x - 3$ is the **boundary line.**

In summary, the solutions of $y \geq 1.5x - 3$ are all points *on or above* the line $y = 1.5x - 3$. The line and the shaded region of Figure 7.3 make up the graph of the inequality $y \geq 1.5x - 3$. ◀ ①

FIGURE 7.3

Graph.

(a) $2x + 3y > 12$

(b) $3x - 2y < 6$

Answers:

(a)

(b)

▶ **EXAMPLE 2**   Graph $x + 4y < 4$.

First obtain an equivalent inequality by solving for y.

$$4y < -x + 4$$

$$y < -\frac{1}{4}x + 1$$

$$y < -.25x + 1$$

The boundary line is $y = -.25x + 1$, but *it is not* part of the solution since points on the line do not satisfy $y < -.25x + 1$. To indicate this, the line is drawn dashed in Figure 7.4. The points *below* the boundary line are the solutions of $y < -.25x + 1$ because they have smaller second coordinates than the points on the line $y = -.25x + 1$. The shaded region in Figure 7.4 (excluding the dashed line) is the graph of the inequality $y < -.25x + 1$. ◀ ②

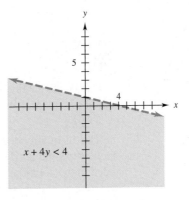

FIGURE 7.4

Examples 1 and 2 show that the solutions of a linear inequality form a **half-plane** consisting of all points on one side of the boundary line (and possibly the line itself). When an inequality is solved for y, the inequality symbol immediately tells whether the points above ($>$), on ($=$), or below ($<$) the boundary line satisfy the inequality, as summarized here.

Inequality	*Solution Consists of All Points*
$y \geq mx + b$	*on or above* the line $y = mx + b$
$y > mx + b$	*above* the line $y = mx + b$
$y \leq mx + b$	*on or below* the line $y = mx + b$
$y < mx + b$	*below* the line $y = mx + b$

When graphing by hand, draw the boundary line $y = mx + b$ solid when it is included in the solution ($\geq$ or $\leq$ inequalities) and dashed when it is not part of the solution ($>$ or $<$ inequalities).

FIGURE 7.5

③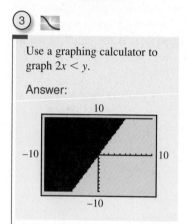

Use a graphing calculator to graph $2x < y$.

Answer:

▶ EXAMPLE 3 Graph $5y - 2x \leq 10$.

Solve the inequality for y.

$$5y \leq 2x + 10$$

$$y \leq \frac{2}{5}x + 2$$

$$y \leq .4x + 2$$

The graph consists of all points on or below the boundary line $y \leq .4x + 2$, as shown in Figure 7.5. (Check your instruction manual for directions on shading calculator generated graphs.) ◀ ③

CAUTION You cannot tell from a calculator-produced graph whether or not the boundary line is included. It is included in Figure 7.5 but not in the answer to problem 3 at the side.

▶ EXAMPLE 4 Graph each of the following.

(a) $y \geq 2$

The boundary line is the horizontal line $y = 2$. The graph consists of all points on or above this line (Figure 7.6).

(b) $x \leq -1$

This inequality does not fit the pattern discussed above, but it can be solved by a similar technique. Here the boundary line is the vertical line $x = -1$ and it is included in the solution. The points satisfying $x < -1$ are all points to the left of this line (because they have x-coordinates smaller than -1). So the graph consists of the points that are *on or to the left of* the vertical line $x = -1$, as shown in Figure 7.7.

◀ ④

④

Graph each of the following.

(a) $x \geq 3$

(b) $y - 3 \leq 0$

Answers:

(a)

(b)

FIGURE 7.6   FIGURE 7.7

An alternative technique for solving inequalities, that does not require solving for y is illustrated in the next example. Feel free to use it, if you find it easter than the technique presented in Examples 1–4.

▶ **EXAMPLE 5** Graph $4y - 2x > 6$.

The boundary line is $4y - 2x = 6$, which can be graphed by finding its intercepts.

x-intercept: let $y = 0$	y-intercept: let $x = 0$
$4(0) - 2x = 6$	$4y - 2(0) = 6$
$x = -3$	$y = \dfrac{6}{4} = 1.5$

The graph is the half-plane above or below this line. To determine which, choose a test point—any point not on the boundary line, say, $(0, 0)$. Letting $x = 0$ and $y = 0$ in the inequality produces

$$4(0) - 2(0) > 6, \quad a\ \textit{false}\ \text{statement.}$$

Therefore, $(0, 0)$ is not in the solution. So the solution is the half-plane that does *not* include $(0, 0)$, as shown in Figure 7.8.

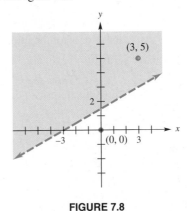

FIGURE 7.8

If a different test point is used, say $(3, 5)$, then substituting $x = 3$ and $y = 5$ in the inequality produces

$$4(5) - 2(3) > 6, \quad a\ \textit{true}\ \text{statement.}$$

Therefore, the solution of the inequality is the half-plane containing $(3, 5)$, as shown in Figure 7.8. ◀

Systems of Inequalities

Realistic problems often involve many inequalities. For example, a manufacturing problem might produce inequalities resulting from production requirements, as well as inequalities about cost requirements. A set of at least two inequalities is called a **system of inequalities.** The **graph** of a system of inequalities is made up of all those points that satisfy all the inequalities of the system at the same time.

▶ **EXAMPLE 6** Graph the system

$$y < -3x + 12$$
$$x < 2y.$$

First, graph the solution of $y < -3x + 12$. The boundary, the line with equation $y = -3x + 12$, is dashed in Figure 7.9 on the next page. The points below the

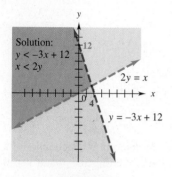

Solution:
$y < -3x + 12$
$x < 2y$

$2y = x$

$y = -3x + 12$

FIGURE 7.9

boundary satisfy $y < -3x + 12$. Now graph the solution of $x < 2y$ on the same axes. Again, the boundary line $x = 2y$ is dashed. Use a test point to see that the region above the boundary satisfies $x < 2y$. The purple region in Figure 7.8 shows all the points that satisfy both inequalities of the system. ◀ ⑤

The purple region of Figure 7.9 is sometimes called the **region of feasible solutions**, or just the **feasible region**, since it is made up of all the points that satisfy (are feasible for) each inequality of the system.

▶ **EXAMPLE 7**   Graph the feasible region for the system

$$2x - 5y \leq 10$$
$$x + 2y \leq 8$$
$$x \geq 0, y \geq 0.$$

The feasible region may be found by using either of the methods illustrated above or a combination of them.

⑤

Graph the system

$x + y \leq 6, 2x + y \geq 4.$

Answer:

⑥

Graph the feasible region of the system

$$x + 4y \leq 8$$
$$x - y \geq 3$$
$$x \geq 0, y \geq 0.$$

Answer:

"Solve for y" Method
Solve the first two inequalities for y:

$$2x - 5y \leq 10$$
$$-5y \leq -2x + 10$$
$$y \geq .4x - 2$$

and

$$x + 2y \leq 8$$
$$2y \leq -x + 8$$
$$y \leq -.5x + 4.$$

The inequalities $x \geq 0$ and $y \geq 0$ restrict the graph to the first quadrant. Therefore, the feasible region consists of all points in the first quadrant that are on or above the line $y = .4x - 2$ and on or below the line $y = -.5x + 4$. A calculator-generated graph, with the feasible region shaded by hand, is shown in Figure 7.10.

Test Point Method
Graph $2x - 5y = 10$ by finding its intercepts, which give the points $(5, 0)$ and $(0, -2)$. Choose a test point and determine that the solutions of $2x - 5y \leq 10$ are the points on or above the boundary line.

Similarly, determine the intercepts of $x + 2y = 8$, which show that the line goes through $(8, 0)$ and $(0, 4)$. Choose a test point and conclude that the solutions of $x + 2y \leq 8$ are the points that lie on or below the boundary line. The inequalities $x \geq 0$ and $y \geq 0$ restrict the graph to the first quadrant. Hence, the feasible region consists of all points in the first quadrant that are on or above $2x - 5y = 10$ and on or below $x + 2y = 8$, as shown in Figure 7.11. ◀ ⑥

FIGURE 7.10

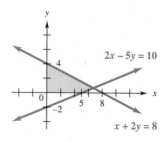

$2x - 5y = 10$

$x + 2y = 8$

FIGURE 7.11

Applications

As we shall see in the rest of this chapter, many realistic problems lead to systems of linear inequalities. The next example is typical of such problems.

▶**EXAMPLE 8** Midtown Manufacturing Company makes plastic plates and cups, both of which require time on two machines. A unit of plates requires 1 hour on machine A and 2 on machine B, while a unit of cups requires 3 hours on machine A and 1 on machine B. Each machine is operated for at most 15 hours per day. Write a system of inequalities expressing these conditions and graph the feasible region.

Let x represent the number of units of plates to be made and y represent the number of units of cups. Then make a chart that summarizes the given information.

	Number of Units	TIME ON MACHINE A	TIME ON MACHINE B
Plates	x	1	2
Cups	y	3	1
Maximum Time Available		15	15

On machine A, x units of plates require a total of $1 \cdot x = x$ hours while y units of cups require $3 \cdot y = 3y$ hours. Since machine A is available no more than 15 hours a day,

$$x + 3y \leq 15. \qquad \text{Machine A}$$

The requirement that machine B be used no more than 15 hours a day gives

$$2x + y \leq 15. \qquad \text{Machine B}$$

It is not possible to produce a negative number of cups or plates, so that

$$x \geq 0 \quad \text{and} \quad y \geq 0.$$

The feasible region for this system of inequalities is shown in Figure 7.12. ◀

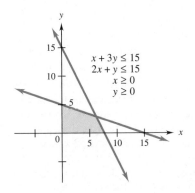

$$x + 3y \leq 15$$
$$2x + y \leq 15$$
$$x \geq 0$$
$$y \geq 0$$

FIGURE 7.12

7.1 Exercises

Graph each of the following linear inequalities. (See Examples 1–5.)

1. $y < 5 - 2x$

2. $y > x + 3$

3. $3x - 2y \geq 18$

4. $2x + 5y \leq 10$

5. $2x - y \leq 4$

6. $4x - 3y \geq 24$

7. $y \leq -4$

8. $x \geq -2$

9. $x + 4y \leq 2$

10. $3x + 2y \geq 6$

11. $4x + 3y > -3$

12. $5x + 3y > 15$

13. $2x - 4y < 3$

14. $4x - 3y < 12$

15. $x \leq 5y$

16. $2x \geq y$

17. $-3x < y$

18. $-x > 6y$

19. $y < x$

20. $y > -2x$

21. In your own words, explain how to determine whether the boundary of an inequality is solid or dashed.

22. When graphing $y \leq 3x - 6$, would you shade above or below the line $y = 3x - 6$? Explain your answer.

Graph the feasible region for the following systems of inequalities. (See Examples 6 and 7.)

23. $x - y \geq 1$
    $x \leq 3$

24. $2x + y \leq 5$
    $x + 2y \leq 5$

25. $4x + y \geq 9$
    $2x + 3y \leq 7$

26. $2x + y > 8$
    $4x - y < 3$

27. $x + y > 5$
    $x - 2y < 2$

28. $3x - 4y < 6$
    $2x + 5y > 15$

29. $2x - y < 1$
    $3x + y < 6$

30. $x + 3y \leq 6$
    $2x + 4y \geq 7$

31. $-x - y < 5$
    $2x - y < 4$

32. $6x - 4y > 8$
    $3x + 2y > 4$

33. $3x + y \geq 6$
    $x + 2y \geq 7$
    $x \geq 0$
    $y \geq 0$

34. $2x + 3y \geq 12$
    $x + y \geq 4$
    $x \geq 0$
    $y \geq 0$

35. $-2 < x < 3$
    $-1 \leq y \leq 5$
    $2x + y < 6$

36. $-2 < x < 2$
    $y > 1$
    $x - y > 0$

37. $2y - x \geq -5$
    $y \leq 3 + x$
    $x \geq 0$
    $y \geq 0$

38. $2x + 3y \leq 12$
    $2x + 3y > -6$
    $3x + y < 4$
    $x \geq 0$
    $y \geq 0$

39. $3x + 4y > 12$
    $2x - 3y < 6$
    $0 \leq y \leq 2$
    $x \geq 0$

40. $0 \leq x \leq 9$
    $x - 2y \geq 4$
    $3x + 5y \leq 30$
    $y \geq 0$

In Exercises 41 and 42, find a system of inequalities whose feasible region is the interior of the given polygon.

41. Rectangle with vertices $(2, 3)$, $(2, -1)$, $(7, 3)$, $(7, -1)$

42. Triangle with vertices $(2, 4)$, $(-4, 0)$, $(2, -1)$

43. **Business** Cindi Herring and Kent Merrill produce handmade shawls and afghans. They spin the yarn, dye it, and then weave it. A shawl requires 1 hour of spinning, 1 hour of dyeing, and 1 hour of weaving. An afghan needs 2 hours of spinning, 1 of dyeing, and 4 of weaving. Together, they spend at most 8 hours spinning, 6 hours dyeing, and 14 hours weaving.

    (a) Complete the following chart.

	Number	Hours Spinning	Hours Dyeing	Hours Weaving
Shawls	x	1	1	1
Afghans	y	2	1	4
Maximum Number of Hours Available		8	6	14

(b) Use the chart to write a system of inequalities that describes the situation.

(c) Graph the feasible region of this system of inequalities.

44. **Business** An electric shaver manufacturer makes two models, the regular and the flex. Because of demand, the number of regular shavers made is never more than half the number of flex shavers. The factory's production cannot exceed 1200 shavers per week.

    (a) Write a system of inequalities that describes the possibilities for making x regular and y flex shavers per week.

    (b) Graph the feasible region of this system of inequalities.

In each of the following, write a system of inequalities that describes all the conditions and graph the feasible region of the system. (See Example 8.)

45. **Business** Southwestern Oil supplies two distributors located in the Northwest. One distributor needs at least 3000 barrels of oil, and the other needs at least 5000 barrels. Southwestern can send out at most 10,000 barrels. Let x = the number of barrels of oil sent to distributor 1 and y = the number sent to distributor 2.

46. **Business** The California Almond Growers have 2400 boxes of almonds to be shipped from their plant in Sacramento to Des Moines and San Antonio. The Des Moines market needs at least 1000 boxes, while the San Antonio market must have at least 800 boxes. Let x = the number of boxes to be shipped to Des Moines and y = the number of boxes to be shipped to San Antonio.

47. **Business** A cement manufacturer produces at least 3.2 million barrels of cement annually. He is told by the Environmental Protection Agency that his operation emits 2.5 pounds of dust for each barrel produced. The EPA has ruled that annual emissions must be reduced to 1.8 million pounds. To do this, the manufacturer plans to replace the present dust collectors with two types of electronic precipitators. One type would reduce emissions to .5 pound per barrel and would cost 16¢ per barrel. The other would reduce the dust to .3 pound per barrel and would cost 20¢ per barrel. The manufacturer does not want to spend more than .8 million dollars on the precipitators. He needs to know how many barrels he should produce with each type. Let x = the number of barrels in millions produced with the first type and y = the number of barrels in millions produced with the second type.

48. **Health** A dietician is planning a snack package of fruit and nuts. Each ounce of fruit will supply 1 unit of protein, 2 units of carbohydrates, and 1 unit of fat. Each ounce of nuts will supply 1 unit of protein, 1 unit of carbohydrates, and 1 unit of fat. Every package must provide at least 7 units of protein, at least 10 units of carbohydrates, and no more than 9 units of fat. Let x be the ounces of fruit and y the ounces of nuts to be used in each package.

7.2 LINEAR PROGRAMMING: THE GRAPHICAL METHOD

Many problems in business, science, and economics involve finding the optimal value of a function (for instance, the maximum value of the profit function or the minimum value of the cost function) subject to various **constraints** (such as transportation costs, environmental protection laws, availability of parts, interest rates, etc.). **Linear programming** deals with such situations in which the function to be optimized, called the **objective function,** is linear and the constraints are given by linear inequalities. Linear programming problems that involve only two variables can be solved by the graphical method, explained in Example 1.

▶**EXAMPLE 1**   Find the maximum and minimum values of the objective function $z = 2x + 5y$, subject to the following constraints.

$$3x + 2y \leq 6$$
$$-2x + 4y \leq 8$$
$$x + y \geq 1$$
$$x \geq 0, y \geq 0$$

First, graph the feasible region of the system of inequalities (Figure 7.13). The points in this region or on its boundaries are the only ones that satisfy all the constraints. However, each such point may produce a different value of the objective function. For instance, the points (.5, 1) and (1, 0) in the feasible region lead to the values

$$z = 2(.5) + 5(1) = 6 \quad \text{and} \quad z = 2(1) + 5(0) = 2.$$

We must find the points that produce the maximum and minimum values of z.

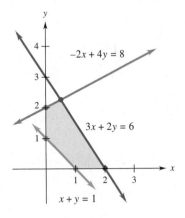

FIGURE 7.13

To find the maximum value, consider various possible values for z. For instance, when $z = 0$, then the objective function is $0 = 2x + 5y$, whose graph is a straight line. Similarly, when z is 5, 10, and 15, the objective function becomes (in turn)

$$5 = 2x + 5y, \qquad 10 = 2x + 5y, \qquad 15 = 2x + 5y.$$

These four lines are graphed in Figure 7.14 on the next page. (All the lines are parallel because they have the same slope.) The figure shows that z cannot take on the value 15 because the graph for $z = 15$ is entirely outside the feasible region. The maximum possible value of z will be obtained from a line parallel to the others and between the lines representing the objective function when $z = 10$ and $z = 15$. The value of z will be as large as possible and all constraints will be satisfied if this line just touches the feasible region. This occurs with the green line through point A.

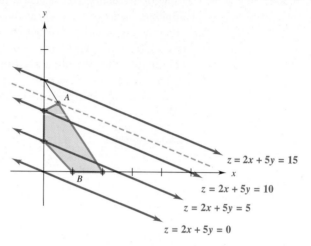

FIGURE 7.14

The point A is the intersection of the graphs of $3x + 2y = 6$ and $-2x + 4y = 8$. Its coordinates can be found either algebraically or graphically (using a graphing calculator).

Algebraic Method	*Graphical Method*
Solve the system	Solve the two equations for y.
$$3x + 2y = 6$$ $$-2x + 4y = 8,$$	$$y = -1.5x + 3$$ $$y = .5x + 2$$
as in Section 6.1, to get $x = 1/2$ and $y = 9/4$. Hence A has coordinates $(1/2, 9/4) = (.5, 2.25)$.	Graph both equations on the same screen and use the intersection finder to find that the coordinates of the intersection point A are $(.5, 2.25)$.

The value of z at the point A is

$$z = 2x + 5y = 2(.5) + 5(2.25) = 12.25.$$

Thus, the maximum possible value of z is 12.25. Similarly, the minimum value of z occurs at point B, which has coordinates $(1, 0)$. The minimum value of z is $2(1) + 5(0) = 2$. ◀ ①

Points such as A and B in Example 1 are called corner points. A **corner point** is a point in the feasible region where the boundary lines of two constraints cross. The feasible region in Figure 7.13 is **bounded,** because the region is enclosed by boundary lines on all sides. Linear programming problems with bounded regions always have solutions. However, if Example 1 did not include the constraint $3x + 2y \le 6$, the feasible region would be **unbounded,** and there would be no way to *maximize* the value of the objective function.

Some general conclusions can be drawn from the method of solution used in Example 1. Figure 7.15 shows various feasible regions and the lines that result from

---

① Suppose the objective function in Example 1 is changed to $z = 5x + 2y$.

(a) Sketch the graphs of the objective function when $z = 0$, $z = 5$, and $z = 10$ on the region of feasible solutions given in Figure 7.13.

(b) From the graph, decide what values of x and y will maximize the objective function.

Answers:

(a)

$z = 0$ $z = 5$ $z = 10$

(b) $(2, 0)$

various values of z. (Figure 7.15 shows the situation in which the lines are in order from left to right as z increases.) In part (a) of the figure, the objective function takes on its minimum value at corner point Q and its maximum value at P. The minimum is again at Q in part (b), but the maximum occurs at P_1 or P_2, or any point on the line segment connecting them. Finally, in part (c), the minimum value occurs at Q, but the objective function has no maximum value because the feasible region is unbounded.

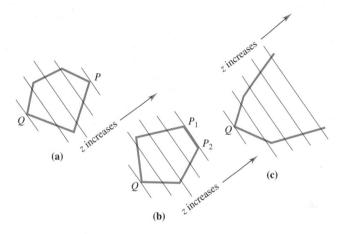

FIGURE 7.15

The preceding discussion suggests the truth of the **corner point theorem.**

Corner Point Theorem

If the feasible region is bounded, then the objective function has both a maximum and a minimum value and each occurs at one or more corner points.

If the feasible region is unbounded, the objective function may not have a maximum or minimum. But if a maximum or minimum value exists, it will occur at one or more corner points.

This theorem simplifies the job of finding an optimum value: First, graph the feasible region and find all corner points. Then test each point in the objective function. Finally, identify the corner point producing the optimum solution.

With the theorem, the problem in Example 1 could have been solved by identifying the five corner points of Figure 7.13: $(0, 1)$, $(0, 2)$, $(.5, 2.25)$, $(2, 0)$, and $(1, 0)$. Then, substituting each of these points into the objective function $z = 2x + 5y$ would identify the corner points that produce the maximum and the minimum values of z.

Corner Point	Value of $z = 2x + 5y$
$(0, 1)$	$2(0) + 5(1) = 5$
$(0, 2)$	$2(0) + 5(2) = 10$
$(.5, 2.25)$	$2(.5) + 5(2.25) = 12.25$ (maximum)
$(2, 0)$	$2(2) + 5(0) = 4$
$(1, 0)$	$2(1) + 5(0) = 2$ (minimum)

(a) Identify the corner points in the graph.

(b) Which corner point would minimize $z = 2x + 3y$?

Answers:

(a) $(0, 4), (1, 1), (4, 0)$

(b) $(1, 1)$

From these results, the corner point $(.5, 2.25)$ yields the maximum value of 12.25 and the corner point $(1, 0)$ gives the minimum value of 2. These are the same values found earlier. ②

A summary of the steps in solving a linear programming problem by the graphical method is given here.

Solving a Linear Programming Problem Graphically

1. Write the objective function and all necessary constraints.
2. Graph the feasible region.
3. Determine the coordinates of each of the corner points.
4. Find the value of the objective function at each corner point.
5. If the feasible region is bounded, the solution is given by the corner point producing the optimum value of the objective function.
6. If the feasible region is an unbounded region in the first quadrant and both coefficients of the objective function are positive,* then the minimum value of the objective function occurs at a corner point and there is no maximum value.

▶ **EXAMPLE 2**   Sketch the feasible region for the following set of constraints:

$$3y - 2x \geq 0$$
$$y + 8x \leq 52$$
$$y - 2x \leq 2$$
$$x \geq 3.$$

Then find the maximum and minimum values of the objective function $z = 5x + 2y$.

The graph in Figure 7.16(a) shows that the feasible region is bounded. The corner points are found by solving systems of two equations either algebraically, using the methods of Chapter 6, or graphically, using the intersection finder on a graphing calculator. Figure 7.16(b) shows the calculator graphs of the first three inequalities. With the graphical method, corner points on the line $x = 3$ are found by observation.

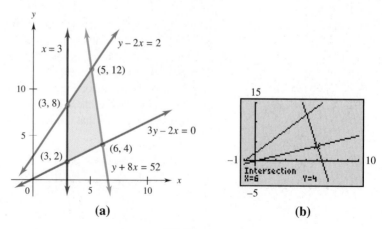

(a)   **(b)**

FIGURE 7.16

*This is the only case of an unbounded region that occurs in the applications considered here.

Use the region of feasible solutions in the sketch to find the following.

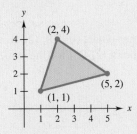

(a) The values of x and y that maximize $z = 2x - y$.

(b) The maximum value of $z = 2x - y$.

(c) The values of x and y that minimize $z = 4x + 3y$.

(d) The minimum value of $z = 4x + 3y$.

Answers:

(a) $(5, 2)$

(b) 8

(c) $(1, 1)$

(d) 7

Use the corner points from the graph to find the maximum and minimum values of the objective function.

Corner Point	Value of $z = 5x + 2y$
$(3, 2)$	$5(3) + 2(2) = 19$ (minimum)
$(6, 4)$	$5(6) + 2(4) = 38$
$(5, 12)$	$5(5) + 2(12) = 49$ (maximum)
$(3, 8)$	$5(3) + 2(8) = 31$

The minimum value of $z = 5x + 2y$ is 19 at the corner point $(3, 2)$. The maximum value is 49 at $(5, 12)$. ◀ ③

▶ **EXAMPLE 3** Solve the following linear programming problem.

$$\text{Minimize} \quad z = x + 2y$$
$$\text{subject to:} \quad x + y \le 10$$
$$3x + 2y \ge 6$$
$$x \ge 0, y \ge 0.$$

The feasible region is shown in Figure 7.17. From the figure, the corner points are $(0, 3)$, $(0, 10)$, $(10, 0)$, and $(2, 0)$. These corner points give the following values of z.

Corner Point	Value of $z = x + 2y$
$(0, 3)$	$0 + 2(3) = 6$
$(0, 10)$	$0 + 2(10) = 20$
$(10, 0)$	$10 + 2(0) = 10$
$(2, 0)$	$2 + 2(0) = 2$ (minimum)

The minimum value of z is 2; it occurs at $(2, 0)$. ◀ ④

④

The sketch shows a feasible region. Let $z = x + 3y$. Use the sketch to find the values of x and y that

(a) minimize z;

(b) maximize z.

Answers:

(a) $(3, 1)$

(b) $(5, 4)$

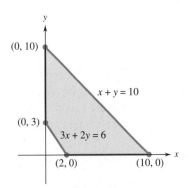

FIGURE 7.17

▶ **EXAMPLE 4** Solve the following linear programming problem.

Minimize $z = 2x + 4y$

subject to: $x + 2y \geq 10$

$3x + y \geq 10$

$x \geq 0, y \geq 0.$

Figure 7.18 shows the hand-drawn graph with corner points (0, 10), (2, 4), and (10, 0), as well as the calculator graph with the corner point (2, 4). Find the value of z for each point.

FIGURE 7.18

⑤

The sketch below shows a region of feasible solutions. From the sketch decide what ordered pair would minimize $z = 2x + 4y$.

Answer:

(1, 1)

Corner Point	Value of $z = 2x + 4y$
(0, 10)	$2(0) + 4(10) = 40$
(2, 4)	$2(2) + 4(4) = 20$ (minimum)
(10, 0)	$2(10) + 4(0) = 20$ (minimum)

In this case, both (2, 4) and (10, 0), as well as all the points on the boundary line between them, give the same optimum value of z. There is an infinite number of equally "good" values of x and y which give the same minimum value of the objective function $z = 2x + 4y$. The minimum value is 20. ◀ ⑤

7.2 Exercises

Exercises 1–6 show regions of feasible solutions. Use these regions to find maximum and minimum values of each given objective function. (See Examples 1 and 2.)

1. $z = 3x + 5y$

2. $z = 6x + y$

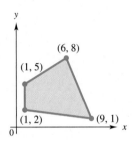

3. $z = .40x + .75y$

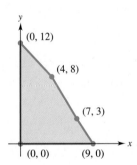

4. $z = .35x + 1.25y$

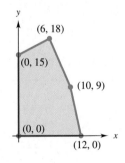

5. **(a)** $z = 4x + 2y$
(b) $z = 2x + 3y$
(c) $z = 2x + 4y$
(d) $z = x + 4y$

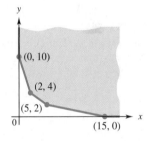

6. **(a)** $z = 4x + y$
(b) $z = 5x + 6y$
(c) $z = x + 2y$
(d) $z = x + 6y$

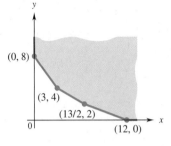

Use graphical methods to solve Exercises 7–12. (See Examples 2–4.)

7. Maximize $z = 5x + 2y$
subject to: $2x + 3y \le 6$
$4x + y \le 6$
$x \ge 0, y \ge 0.$

8. Minimize $z = x + 3y$
subject to: $2x + y \le 10$
$5x + 2y \ge 20$
$-x + 2y \ge 0$
$x \ge 0, y \ge 0.$

9. Minimize $z = 2x + y$
subject to: $3x - y \ge 12$
$x + y \le 15$
$x \ge 2, y \ge 3.$

10. Maximize $z = x + 3y$
subject to: $2x + 3y \le 100$
$5x + 4y \le 200$
$x \ge 10, y \ge 20.$

11. Maximize $z = 4x + 2y$
subject to: $x - y \le 10$
$5x + 3y \le 75$
$x \ge 0, y \ge 0.$

12. Maximize $z = 4x + 5y$
   subject to: $10x - 5y \leq 100$
             $20x + 10y \geq 150$
             $x \geq 0, y \geq 0.$

Find the minimum and maximum values of $z = 3x + 4y$ (if possible) for each of the following sets of constraints. (See Examples 2–4.)

13. $3x + 2y \geq 6$
      $x + 2y \geq 4$
      $x \geq 0, y \geq 0$

14. $2x + y \leq 20$
      $10x + y \geq 36$
      $2x + 5y \geq 36$

15. $x + y \leq 6$
      $-x + y \leq 2$
      $2x - y \leq 8$

16. $-x + 2y \leq 6$
      $3x + y \geq 3$
      $x \geq 0, y \geq 0$

17.   Find values of $x \geq 0$ and $y \geq 0$ that maximize $z = 10x + 12y$ subject to each of the following sets of constraints.
   (a) $x + y \leq 20$
         $x + 3y \leq 24$
   (b) $3x + y \leq 15$
         $x + 2y \leq 18$
   (c) $x + 2y \geq 10$
         $2x + y \geq 12$
         $x - y \leq 8$

18.   Find values of $x \geq 0$ and $y \geq 0$ that minimize $z = 3x + 2y$ subject to each of the following sets of constraints.
   (a) $10x + 7y \leq 42$
         $4x + 10y \geq 35$
   (b) $6x + 5y \geq 25$
         $2x + 6y \geq 15$
   (c) $2x + 5y \leq 22$
         $4x + 3y \leq 28$
         $2x + 2y \leq 17$

19.   Explain why it is impossible to maximize the function $z = 3x + 4y$ subject to the constraints:
      $x + y \geq 8$
      $2x + y \leq 10$
      $x + 2y \leq 8$
      $x \geq 0, \; y \geq 0$

20.   You are given the following linear programming problem:*
         Maximize $z = c_1x_1 + c_2x_2$
         subject to: $2x_1 + x_2 \leq 11$
                $-x_1 + 2x_2 \leq 2$
                $x_1 \geq 0, x_2 \geq 0.$

   If $c_2 > 0$, determine the range of c_1/c_2 for which $(x_1, x_2) = (4, 3)$ is an optimal solution.

   (a) $[-2, 1/2]$

   (b) $[-1/2, 2]$

   (c) $[-11, -1]$

   (d) $[1, 11]$

   (e) $[-11, 11]$

---

*Problem from "Course 130 Examination Operations Research" of the *Education and Examination Committee of The Society of Actuaries*. Reprinted by permission of The Society of Actuaries.

7.3 APPLICATIONS OF LINEAR PROGRAMMING

In this section we show several applications of linear programming with two variables.

▶ **EXAMPLE 1**   A 4-H Club member raises geese and pigs. She wants to raise no more than 16 animals including no more than 10 geese. She spends $15 to raise a goose and $45 to raise a pig, and she has $540 available for this project. Find the 0maximum profit she can make if each goose produces a profit of $7 and each pig a profit of $20.

The total profit is determined by the number of geese and pigs. So let x be the number of geese to be produced and let y be the number of pigs. Then summarize the information of the problem in a table.

	Number	Cost to Raise	Profit Each
Geese	x	$ 15	$ 7
Pigs	y	$ 45	$20
Maximum Available	16	$540	

Use this table to write the necessary constraints. Since the total number of animals cannot exceed 16, the first constraint is

$$x + y \le 16.$$

"No more than 10 geese" leads to

$$x \le 10.$$

The cost to raise x geese at $15 per goose is $15x$ dollars, while the cost for y pigs at $45 each is $45y$ dollars. Only $540 is available, so

$$15x + 45y \le 540.$$

Dividing both sides by 15 gives the equivalent inequality

$$x + 3y \le 36.$$

The number of geese and pigs cannot be negative, so

$$x \ge 0, \quad y \ge 0.$$

The 4-H Club member wants to know the number of geese and the number of pigs that should be raised for maximum profit. Each goose produces a profit of $7 and each pig, $20. If z represents total profit, then

$$z = 7x + 20y$$

is the objective function, which is to be maximized.

We must solve the following linear programming problem.

Maximize $z = 7x + 20y$ Objective function
subject to: $x + y \le 16$
$x \le 10$
$x + 3y \le 36$ Constraints
$x \ge 0, y \ge 0.$

Using the methods of the previous section, graph the feasible region for the system of inequalities given by the constraints, as in Figure 7.19.

FIGURE 7.19

Find the corner points P and Q in Figure 7.19.

Answer:

$P = (6, 10)$
$Q = (10, 6)$

The corner points $(0, 12)$, $(0, 0)$, and $(10, 0)$ can be read directly from the graph. Find the coordinates of the other corner points by solving a system of equations or with a graphing calculator. ①

Test each corner point in the objective function to find the maximum profit.

Corner Point	$z = 7x + 20y$
$(0, 12)$	$7(0) + 20(12) = 240$
$(6, 10)$	$7(6) + 20(10) = 242$ (maximum)
$(10, 6)$	$7(10) + 20(6) = 190$
$(10, 0)$	$7(10) + 20(0) = 70$
$(0, 0)$	$7(0) + 20(0) = 0$

The maximum value for z of 242 occurs at $(6, 10)$. Thus 6 geese and 10 pigs will produce a maximum profit of $242. ◄

▶ **EXAMPLE 2** An office manager needs to purchase new filing cabinets. He knows that Ace cabinets cost \$40 each, require 6 square feet of floor space, and hold 8 cubic feet of files. On the other hand, each Excello cabinet costs \$80, requires 8 square feet of floor space, and holds 12 cubic feet. His budget permits him to spend no more than \$560 on files, while the office has room for no more than 72 square feet of cabinets. The manager desires the greatest storage capacity within the limitations imposed by funds and space. How many of each type of cabinet should he buy?

Let x represent the number of Ace cabinets to be bought and let y represent the number of Excello cabinets. The information given in the problem can be summarized as follows.

	Number	Cost of Each	Space Required	Storage Capacity
Ace	x	\$ 40	6 sq ft	8 cu ft
Excello	y	\$ 80	8 sq ft	12 cu ft
Maximum Available		\$560	72 sq ft	

The constraints imposed by cost and space are

$$40x + 80y \leq 560 \qquad \text{Cost}$$
$$6x + 8y \leq 72. \qquad \text{Floor space}$$

The number of cabinets cannot be negative, so $x \geq 0$ and $y \geq 0$. The objective function to be maximized gives the amount of storage capacity provided by some combination of Ace and Excello cabinets. From the information in the chart, the objective function is

$$\text{Storage space} = z = 8x + 12y.$$

Find the corner point labeled P on the region of feasible solutions given below.

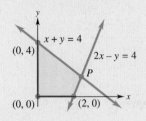

Answer:

$(8/3, 4/3)$

A popular cereal combines oats and corn. At least 27 tons of the cereal is to be made. For the best flavor, the amount of corn should be no more than twice the amount of oats. Oats cost \$300 per ton and corn costs \$200 per ton. How much of each grain should be used to minimize the cost?

(a) Make a chart to organize the information given in the problem.

(b) Write an equation for the objective function.

(c) Write four inequalities for the constraints.

Answers:

(a)

	Number of Tons	Cost/ Ton
Oats	x	\$300
Corn	y	\$200
	27	

(b) $z = 300x + 200y$

(c) $x + y \geq 27$

$y \leq 2x$

$x \geq 0$

$y \geq 0$

In summary, the given problem has produced the following linear programming problem.

$$\text{Maximize} \quad z = 8x + 12y$$

$$\text{subject to:} \quad 40x + 80y \leq 560$$

$$6x + 8y \leq 72$$

$$x \geq 0, y \geq 0.$$

A graph of the feasible region is shown in Figure 7.20. Three of the corner points can be identified from the graph as $(0, 0)$, $(0, 7)$, and $(12, 0)$. The fourth corner point, labeled Q in the figure, can be found algebraically or with a graphing calculator to be $(8, 3)$. ②

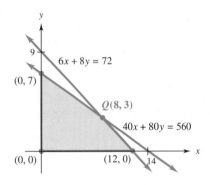

FIGURE 7.20

Use the corner point theorem to find the maximum value of z.

Corner Point	Value of $z = 8x + 12y$
$(0, 0)$	0
$(0, 7)$	84
$(12, 0)$	96
$(8, 3)$	100 (maximum)

The objective function, which represents storage space, is maximized when $x = 8$ and $y = 3$. The manager should buy 8 Ace cabinets and 3 Excello cabinets. ◀ ③

▶ **EXAMPLE 3** Certain laboratory animals must have at least 30 grams of protein and at least 20 grams of fat per feeding period. These nutrients come from food A, which costs 18¢ per unit and supplies 2 grams of protein and 4 of fat, and food B, with 6 grams of protein and 2 of fat, costing 12¢ per unit. Food B is bought under a long-term contract requiring that at least 2 units of B be used per serving. How much of each food must be bought to produce minimum cost per serving?

Let x represent the amount of food A needed and y the amount of food B. Use the given information to produce the following table.

Food	Number of Units	Grams of Protein	Grams of Fat	Cost
A	x	2	4	18¢
B	y	6	2	12¢
Minimum Required		30	20	

The linear programming problem can be stated as follows.

$$\text{Minimize} \quad z = .18x + .12y$$
$$\text{subject to:} \quad 2x + 6y \geq 30$$
$$4x + 2y \geq 20$$
$$y \geq 2$$
$$x \geq 0, y \geq 0.$$

(The constraint $y \geq 0$ is redundant because of the constraint $y \geq 2$.) A graph of the feasible region with the corner points identified is shown in Figure 7.21.

Use the information in side problem 3 to do the following.

(a) Graph the feasible region and find the corner points.

(b) Determine the minimum value of the objective function and the point where it occurs.

(c) Is there a maximum cost?

Answers:

.**(a)**

Corner points: (27, 0), (9, 18)

(b) $6300 at (9, 18)

(c) No

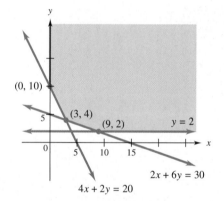

FIGURE 7.21

Use the corner point theorem to find the minimum value of z as shown in the chart below.

Corner Points	$z = .18x + .12y$
(0, 10)	$.18(0) + .12(10) = 1.20$
(3, 4)	$.18(3) + .12(4) = 1.02$ (minimum)
(9, 2)	$.18(9) + .12(2) = 1.86$

The minimum value of 1.02 occurs at (3, 4). Thus, 3 units of A and 4 units of B will produce a minimum cost of $1.02 per serving. ◀ ④

The feasible region in Figure 7.21 is an unbounded feasible region—the region extends indefinitely to the upper right. With this region it would not be possible to *maximize* the objective function, because the total cost of the food could always be increased by encouraging the animals to eat more.

7.3 Exercises

Write the constraints in Exercises 1–5 as linear inequalities and identify all variables used. In some instances, not all of the information is needed to write the constraints. (See Examples 1–3.)

1. A canoe requires 6 hours of fabrication and a rowboat 4 hours. The fabrication department has at most 90 hours of labor available each week.

2. Doug Gilbert needs at least 2400 mg of Vitamin C per day. Each Supervite pill provides 250 mg and each Vitahealth pill provides 350 mg.

3. A candidate can afford to spend no more than $8500 on radio and TV advertising. Each radio spot costs $150 and each TV ad costs $750.

4. A hospital dietician has two meal choices, one for patients on solid food that costs $2.25 and one for patients on liquids that costs $3.75. There are a maximum of 400 patients in the hospital.

Solve these linear programming problems, which are somewhat simpler than the examples in the text.

5. **Business** A chain saw requires 4 hours of assembly and a wood chipper 6 hours. A maximum of 48 hours of assembly time are available. The profit is $150 on a chain saw and $220 on a chipper. How many of each should be assembled for maximum profit?

6. **Health** Don Garcia needs at least 24 units of vitamin A per day. Each green pill provides 4 units and each red pill provides 1 unit. Green pills cost 3.5¢ each and red pills 1¢ each. How many of each kind of pill should he take to minimize his cost?

7. **Business** Deluxe coffee is to be mixed with regular coffee to make at least 50 pounds of a blended coffee. The mixture must contain at least 10 pounds of deluxe coffee. Deluxe coffee costs $6 per pound and regular coffee $5 per pound. How many pounds of each kind of coffee should be used to minimize costs?

8. **Business** Pauline Wong spends 4 hours selling a used car and 6 hours selling a new car. She works no more than 40 hours per week. In order to receive a bonus, she must sell at least one used car and four new cars each week. In that case, she receives a bonus of $180 for each used car and $290 for each new car. How many used and how many new cars should she try to sell to maximize her bonus?

Solve the following linear programming problems. (See Examples 1–3.)

9. **Business** A company is considering two insurance plans with the types of coverage and premiums shown in the table below.

	Policy A	*Policy B*
Fire/Theft	$10,000	$15,000
Liability	$180,000	$120,000
Premium	$50	$40

(For example, this means that $50 buys one unit of plan A, consisting of $10,000 fire and theft insurance and $180,000 of liability insurance.)

(a) The company wants at least $300,000 fire/theft insurance and at least $3,000,000 liability insurance from these plans. How many units should be purchased from each plan to minimize the cost of the premiums? What is the minimum premium?

(b) Suppose the premium for policy A is reduced to $25. Now how many units should be purchased from each plan to minimize the cost of the premiums? What is the minimum premium?

10. **Business** A manufacturer of refrigerators must ship at least 100 refrigerators to its two West Coast warehouses. Each warehouse holds a maximum of 100 refrigerators. Warehouse A holds 25 refrigerators already, and warehouse B has 20 on hand. It costs $12 to ship a refrigerator to warehouse A and $10 to ship one to warehouse B. Union rules require that at least 300 workers be hired. Shipping a refrigerator to Warehouse A requires 4 workers, while shipping a refrigerator to Warehouse B requires 2 workers. How many refrigerators should be shipped to each warehouse to minimize costs? What is the minimum cost?

11. **Health** Sing, who is dieting, requires two food supplements, I and II. He can get these supplements from two different products, A and B, as shown in the table below.

		GRAMS OF SUPPLEMENT PER SERVING	
		I	II
PRODUCT	A	3	2
	B	2	4

Sing's physician has recommended that he include at least 15 g of each supplement in his daily diet. If product A costs 25¢ per serving and product B costs 40¢ per serving, how can he satisfy his dietetic requirements most economically? Find the minimum cost.

12. **Business** The manufacturing process requires that oil refineries must manufacture at least 2 gallons of gasoline for every gallon of fuel oil. To meet the winter demand for fuel oil, at least 3 million gallons a day must be produced. The demand for gasoline is no more than 6.4 million gallons per day. It takes .25 hour to ship each million gallons of gasoline and 1 hour to ship each million gallons of fuel oil out of the warehouse. No more than 4.65 hours are available for shipping. If the refinery sells gasoline for $1.25 per gallon and fuel oil for $1 per gallon, how much of each should be produced to maximize revenue? Find the maximum revenue.

13. **Business** A machine shop manufactures two types of bolts. The bolts require time on each of three groups of machines, but the time required on each group differs, as shown in the table below.

		MACHINE GROUP	
	I	*II*	*III*
BOSS Type 1	.1 min	.1 min	.1 min
Type 2	.1 min	.4 min	.02 min

Production schedules are made up one day at a time. In a day there are 240, 720, and 160 minutes available, respectively, on these machines. Type 1 bolts sell for 10¢ and type 2 bolts for 12¢. How many of each type of bolt should be manufactured per day to maximize revenue? What is the maximum revenue?

14. **Health** Kim Walrath has a nutritional deficiency and is told to take at least 2400 mg of iron, 2100 mg of Vitamin B-1, and 1500 mg of Vitamin B-2. One Maxivite pill contains 40 mg of iron, 10 mg of B-1, and 5 mg of B-2 and costs 6¢. One Healthovite pill provides 10 mg of iron, 15 mg of B-1, and 15 mg of B-2 and costs 8¢. What combination of Maxivite and Healthovite pills will meet the requirement at lowest cost? What is the minimum cost?

15. **Social Sciences** An anthropology article describes a hypothetical situation that could be described by a linear programming model.* Suppose a population gathers plants and animals for survival They need at least 360 units of energy, 300 units of protein, and 8 hides during some time period. One unit of plants provides 30 units of

energy, 10 units of protein, and no hides. One animal provides 20 units of energy, 25 units of protein, and 1 hide. Only 25 units of plants and 25 animals are available. It costs the population 30 hours of labor to gather one unit of a plant and 15 hours for an animal. Find how many units of plants and how many animals should be gathered to meet the requirements with a minimum number of hours of labor.

16. **Business** The Miers Company produces small engines for several manufacturers. The company receives orders from two assembly plants for their Topflight engine. Plant I needs at least 50 engines, and plant II needs at least 27 engines. The company can send at most 85 engines to these two assembly plants. It costs $20 per engine to ship to plant I and $35 per engine to ship to plant II. Plant I gives Miers $15 in rebates toward its products for each engine they buy, while plant II gives similar $10 rebates. Miers estimates that they need at least $1110 in rebates to cover products they plan to buy from the two plants. How many engines should be shipped to each plant to minimize shipping costs? What is the minimum cost?

17. **Business** A greeting card manufacturer has 370 boxes of a particular card in warehouse I and 290 boxes of the same card in warehouse II. A greeting card shop in San Jose orders 350 boxes of the card, and another shop in Memphis orders 300 boxes. The shipping costs per box to these shops from the two warehouses are shown in the following table.

		DESTINATION	
		San Jose	*Memphis*
WAREHOUSE	I	$.25	$.22
	II	$.23	$.21

How many boxes should be shipped to each city from each warehouse to minimize shipping costs? What is the minimum cost? (*Hint:* Use x, $350 - x$, y, and $300 - y$ as the variables.)

18. **Business** Hotnews Magazine publishes a U.S. and a Canadian edition each week. There are 30,000 subscribers in the United States and 20,000 in Canada. Other copies are sold at newsstands. Postage and shipping costs average $80 per thousand copies for the U.S. and $60 per thousand copies for Canada. Surveys show that no more than 120,000 copies of each issue can be sold (including subscriptions) and that the number of copies of the Canadian edition should not exceed twice the number of copies of the U.S. edition. The publisher can spend at most $8400 a month on postage and shipping. If the profit is $200 for each thousand copies of the U.S. edition and $150 for each thousand copies of the Canadian edition,

---

*Van A. Reidhead, "Linear Programming Models in Archaeology," *Annual Review of Anthropology* 8 (1979): 543–578.

how many copies of each version should be printed to earn as large a profit as possible? What is that profit?

19. **Finance** A pension fund manager decides to invest at most $40 million in U.S. Treasury Bonds paying 8% annual interest and in mutual funds paying 12% annual interest. He plans to invest at least $20 million in bonds and at least $15 million in mutual funds. Bonds have an initial fee of $300 per million dollars, while the fee for mutual funds is $100 per million. The fund manager is allowed to spend no more than $8400 on fees. How much should be invested in each to maximize annual interest? What is the maximum annual interest?

20. **Natural Science** A certain predator requires at least 10 units of protein and 8 units of fat per day. One prey of Species I provides 5 units of protein and 2 units of fat; one prey of Species II provides 3 units of protein and 4 units of fat. Capturing and digesting each Species II prey requires 3 units of energy, and capturing and digesting each Species I prey requires 2 units of energy. How many of each prey would meet the predator's daily food requirements with the least expenditure of energy? Are the answers reasonable? How could they be interpreted?

21. **Social Science** Students at Upscale U are required to take at least 3 humanities and 4 science courses. The maximum allowable number of science courses is 12. Each humanities course carries 4 credits and each science course 5 credits. The total number of credits in science and humanities cannot exceed 80. Quality points for each course are assigned in the usual way: the number of credit hours times 4 for an A grade; times 3 for a B grade; times 2 for a C grade. Susan Katz expects to get B's in all her science courses. She expects to get C's in half her humanities courses, B's in one-fourth of them, and A's in the rest. Under these assumptions, how many courses of each kind should she take in order to earn the maximum possible number of quality points?

22. **Social Science** In Exercise 21, find Susan's grade point average (the total number of quality points divided by the total number of credit hours) at each corner point of the feasible region. Does the distribution of courses that produces the highest number of quality points also yield the highest grade point average? Is this a contradiction?

*The importance of linear programming is shown by the inclusion of linear programming problems on most qualification examinations for Certified Public Accountants. Exercises 23–25 are reprinted from one such examination.**

---

*Material from *Uniform CPA Examinations and Unofficial Answers,* copyright © 1973, 1974, 1975 by the American Institute of Certified Public Accountants, Inc., is reprinted with permission.

The Random Company manufactures two products, Zeta and Beta. Each product must pass through two processing operations. All materials are introduced at the start of Process No. 1. There are no work-in-process inventories. Random may produce either one product exclusively or various combinations of both products subject to the following constraints.

	Process No. 1	Process No. 2	Contribution Margin per Unit
Hours required to produce 1 unit of:			
Zeta	1 hour	1 hour	$4.00
Beta	2 hours	3 hours	$5.25
Total capacity in hours per day	1000 hours	1275 hours	

A shortage of technical labor has limited Beta production to 400 units per day. There are no constraints on the production of Zeta other than the hour constraints in the above schedule. Assume that all the relationships between capacity and production are linear.

23. Given the objective to maximize total contribution margin, what is the production constraint for Process No. 1?
    (a) Zeta + Beta ≤ 1000
    (b) Zeta + 2 Beta ≤ 1000
    (c) Zeta + Beta ≥ 1000
    (d) Zeta + 2 Beta ≥ 1000

24. Given the objective to maximize total contribution margin, what is the labor constraint for production of Beta?
    (a) Beta ≤ 400   (b) Beta ≥ 400
    (c) Beta ≤ 425   (d) Beta ≥ 425

25. What is the objective function of the data presented?
    (a) Zeta + 2 Beta = $9.25
    (b) $4.00 Zeta + 3($5.25) Beta = total contribution margin
    (c) $4.00 Zeta + $5.25 Beta = total contribution margin
    (d) 2($4.00) Zeta + 3($5.25) Beta = total contribution margin

7.4 THE SIMPLEX METHOD: MAXIMIZATION

For linear programming problems with more than two variables or with two variables and many constraints, the graphical method is usually inefficient or impossible, so the **simplex method** is used. The simplex method, which is introduced here, was developed for the U.S. Air Force by George B. Danzig in 1947. It was used successfully during the Berlin airlift in 1948–49 to maximize the amount of cargo delivered under very severe constraints and is widely used today in a variety of industries.

Because the simplex method is used for problems with many variables, it usually is not convenient to use letters such as x, y, z, or w as variable names. Instead, the symbols x_1 (read "x-sub-one"), x_2, x_3, and so on, are used. In the simplex method, all constraints must be expressed in the linear form

$$a_1 x_1 + a_2 x_2 + a_3 x_3 + \cdots \le b,$$

where x_1, x_2, x_3, . . . are variables, a_1, a_2, a_3, . . . are coefficients, and b is a constant.

We first discuss the simplex method for linear programming problems in *standard maximum form*.

Standard Maximum Form

A linear programming problem is in **standard maximum form** if

1. the objective function is to be maximized;
2. all variables are nonnegative ($x_i \ge 0$, $i = 1, 2, 3, . . .$);
3. all constraints involve $\le$;
4. the constants on the right side in the constraints are all nonnegative ($b \ge 0$).

Problems that do not meet all of these conditions are considered in Sections 7.6 and 7.7.

The "mechanics" of the simplex method are demonstrated in Examples 1–5. Although the procedures to be followed will be made clear, as will the fact that they result in an optimal solution, the reasons why these procedures are used may not be immediately apparent. Examples 6 and 7 will supply these reasons and explain the connection between the simplex method and the graphical method used in Section 7.3.

Setting Up the Problem

The first step is to convert each constraint, a linear inequality, into a linear equation. This is done by adding a nonnegative variable, called a **slack variable,** to each constraint. For example, convert the inequality $x_1 + x_2 \le 10$ into an equation by adding the slack variable x_3, to get

$$x_1 + x_2 + x_3 = 10, \quad \text{where } x_3 \ge 0.$$

The inequality $x_1 + x_2 \leq 10$ says that the sum $x_1 + x_2$ is less than or perhaps equal to 10. The variable x_3 "takes up any slack" and represents the amount by which $x_1 + x_2$ fails to equal 10. For example, if $x_1 + x_2$ equals 8, then x_3 is 2. If $x_1 + x_2 = 10$, the value of x_3 is 0.

CAUTION A different slack variable must be used for each constraint.

Rewrite the following set of constraints as equations by adding nonnegative slack variables.

$$x_1 + x_2 + x_3 \leq 12$$
$$2x_1 + 4x_2 \qquad \leq 15$$
$$x_2 + 3x_3 \leq 10$$

Answer:

$$x_1 + x_2 + x_3 + x_4 = 12$$
$$2x_1 + 4x_2 \qquad + x_5 = 15$$
$$x_2 + 3x_3 + x_6 = 10$$

▶ **EXAMPLE 1**   Restate the following linear programming problem by introducing slack variables.

$$\text{Maximize} \quad z = 2x_1 + 3x_2 + x_3$$
$$\text{subject to:} \quad x_1 + x_2 + 4x_3 \leq 100$$
$$x_1 + 2x_2 + x_3 \leq 150$$
$$3x_1 + 2x_2 + x_3 \leq 320$$
$$\text{with} \quad x_1 \geq 0, x_2 \geq 0, x_3 \geq 0.$$

Rewrite the three constraints as equations by introducing nonnegative slack variables x_4, x_5, and x_6, one for each constraint. Then the problem can be restated as

$$\text{Maximize} \quad z = 2x_1 + 3x_2 + x_3$$
$$\text{subject to:} \quad x_1 + x_2 + 4x_3 + x_4 \qquad\qquad = 100$$
$$x_1 + 2x_2 + x_3 \qquad + x_5 \qquad = 150$$
$$3x_1 + 2x_2 + x_3 \qquad\qquad + x_6 = 320$$
$$\text{with} \quad x_1 \geq 0, x_2 \geq 0, x_3 \geq 0, x_4 \geq 0, x_5 \geq 0, x_6 \geq 0. \quad ◀ \;① $$

Adding slack variables to the constraints converts a linear programming problem into a system of linear equations. These equations should have all variables on the left of the equals sign and all constants on the right. All the equations of Example 1 satisfy this condition except for the objective function, $z = 2x_1 + 3x_2 + x_3$, which may be written with all variables on the left as

$$-2x_1 - 3x_2 - x_3 + z = 0.$$

Now the equations of Example 1 (with the constraints listed first) can be written as the following augmented matrix.

$$
\begin{array}{ccccccc}
x_1 & x_2 & x_3 & x_4 & x_5 & x_6 & z \\
\end{array}
$$

$$
\left[
\begin{array}{ccccccc|c}
1 & 1 & 4 & 1 & 0 & 0 & 0 & 100 \\
1 & 2 & 1 & 0 & 1 & 0 & 0 & 150 \\
3 & 2 & 1 & 0 & 0 & 1 & 0 & 320 \\
\underbrace{-2 & -3 & -1 & 0 & 0 & 0}_{} & 1 & 0 \\
\end{array}
\right]
$$

Indicators

This matrix is the initial **simplex tableau.** Except for the last entries, the 1 and 0 on the right end, the numbers in the bottom row of a simplex tableau are called **indicators.** ②

Set up the initial simplex tableau for the following linear programming problem:

$$\text{Maximize} \quad z = 2x_1 + 3x_2$$
$$\text{subject to:} \quad x_1 + 2x_2 \leq 85$$
$$2x_1 + x_2 \leq 92$$
$$x_1 + 4x_2 \leq 104$$
$$\text{with} \quad x_1 \geq 0, x_2 \geq 0.$$

Locate and label the indicators.

Answer:

$$
\begin{array}{cccccc}
x_1 & x_2 & x_3 & x_4 & x_5 & z \\
\end{array}
$$

$$
\left[
\begin{array}{cccccc|c}
1 & 2 & 1 & 0 & 0 & 0 & 85 \\
2 & 1 & 0 & 1 & 0 & 0 & 92 \\
1 & 4 & 0 & 0 & 1 & 0 & 104 \\
-2 & -3 & 0 & 0 & 0 & 1 & 0 \\
\end{array}
\right]
$$

Indicators

This simplex tableau represents a system of 4 linear equations in 7 variables. Since there are more variables than equations, the system is dependent and has infinitely many solutions. Our goal is to find a solution in which all the variables are nonnegative and z is as large as possible. This will be done by using row operations to replace the given system by an equivalent one in which certain variables are eliminated from some of the equations. The process will be repeated until the optimum solution can be read from the matrix, as explained below.

Selecting the Pivot

Recall how row operations are used to eliminate variables in the Gauss-Jordan method. A particular nonzero entry in the matrix is chosen and changed to a 1; then all other entries in that column are changed to zeros. A similar process is used in the simplex method. The chosen entry is called the **pivot.** If we were only interested in solving the system, we could choose the various pivots in many different ways, as in Chapter 6. Here, however, it's not enough just to find a solution. We must find one that is nonnegative (so that it satisfies all the constraints) *and* makes z as a large as possible. Consequently, the pivot must be chosen carefully, as explained in the next example. The reasons why this procedure is used and why it works are discussed in Example 7.

▶ **EXAMPLE 2** Determine the pivot in the simplex tableau for the problem in Example 1.

Look at the indicators (the last row of the tableau) and choose the most negative one.

$$
\begin{array}{ccccccc}
x_1 & x_2 & x_3 & x_4 & x_5 & x_6 & z \\
\left[\begin{array}{ccccccc|c}
1 & 1 & 4 & 1 & 0 & 0 & 0 & 100 \\
1 & 2 & 1 & 0 & 1 & 0 & 0 & 150 \\
3 & 2 & 1 & 0 & 0 & 1 & 0 & 320 \\
\hline
-2 & \boxed{-3} & -1 & 0 & 0 & 0 & 1 & 0
\end{array}\right]
\end{array}
$$

Most negative indicator

The most negative indicator identifies the variable that is to be eliminated from all but one of the equations (rows), in this case x_2. The column containing the most negative indicator is called the **pivot column.** Now for each *positive* entry in the pivot column, divide the number in the far right column of the same row by the positive number in the pivot column.

$$
\begin{array}{ccccccc}
x_1 & x_2 & x_3 & x_4 & x_5 & x_6 & z \\
\left[\begin{array}{ccccccc|c}
1 & \mathbf{1} & 4 & 1 & 0 & 0 & 0 & \mathbf{100} \\
1 & \mathbf{2} & 1 & 0 & 1 & 0 & 0 & \mathbf{150} \\
3 & \mathbf{2} & 1 & 0 & 0 & 1 & 0 & \mathbf{320} \\
\hline
-2 & -3 & -1 & 0 & 0 & 0 & 1 & 0
\end{array}\right]
\end{array}
$$

Quotients

$100/1 = 100$

$150/2 = 75 \leftarrow$Smallest

$320/2 = 160$

The row with the smallest quotient (in this case, the second row) is called the **pivot row.** The entry in the pivot row and pivot column is the pivot.

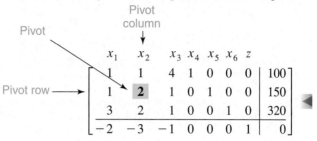

Find the pivot for the following tableau.

$$\begin{array}{cccccc} x_1 & x_2 & x_3 & x_4 & x_5 & z \\ \left[\begin{array}{cccccc|c} 0 & 1 & 1 & 0 & 0 & 0 & 50 \\ -2 & 3 & 0 & 1 & 0 & 0 & 78 \\ 2 & 4 & 0 & 0 & 1 & 0 & 65 \\ \hline -5 & -3 & 0 & 0 & 0 & 1 & 0 \end{array}\right] \end{array}$$

Answer:

2 (in first column)

CAUTION In some simplex tableaus the pivot column may contain zeros or negative entries. Only the positive entries in the pivot column should be used to form the quotients and determine the pivot row. If there are no positive entries in the pivot column (so that a pivot row cannot be chosen), then no maximum solution exists. ③

Pivoting

Once the pivot has been selected, row operations are used to replace the initial simplex tableau by another simplex tableau in which the pivot column variable is eliminated from all but one of the equations. Since this new tableau is obtained by row operations, it represents an equivalent system of equations (that is, a system with the same solutions as the original system). This process, which is called **pivoting,** is explained in the next example.

▶ **EXAMPLE 3** Use the indicated pivot, 2, to perform the pivoting on the simplex tableau of Example 2:

$$\begin{array}{ccccccc} x_1 & x_2 & x_3 & x_4 & x_5 & x_6 & z \\ \left[\begin{array}{ccccccc|c} 1 & 1 & 4 & 1 & 0 & 0 & 0 & 100 \\ 1 & \mathbf{2} & 1 & 0 & 1 & 0 & 0 & 150 \\ 3 & 2 & 1 & 0 & 0 & 1 & 0 & 320 \\ \hline -2 & -3 & -1 & 0 & 0 & 0 & 1 & 0 \end{array}\right] \end{array}.$$

Start by multiplying each entry of row two by 1/2 in order to change the pivot to 1.

$$\begin{array}{ccccccc} x_1 & x_2 & x_3 & x_4 & x_5 & x_6 & z \\ \left[\begin{array}{ccccccc|c} 1 & 1 & 4 & 1 & 0 & 0 & 0 & 100 \\ \frac{1}{2} & \mathbf{1} & \frac{1}{2} & 0 & \frac{1}{2} & 0 & 0 & 75 \\ 3 & 2 & 1 & 0 & 0 & 1 & 0 & 320 \\ \hline -2 & -3 & -1 & 0 & 0 & 0 & 1 & 0 \end{array}\right] \end{array} \quad \frac{1}{2} R_2$$

Now use row operations to make the entry in row one, column two, a 0.

$$
\begin{array}{ccccccc}
x_1 & x_2 & x_3 & x_4 & x_5 & x_6 & z \\
\end{array}
$$

$$
\left[\begin{array}{ccccccc|c}
\frac{1}{2} & 0 & \frac{7}{2} & 1 & -\frac{1}{2} & 0 & 0 & 25 \\
\frac{1}{2} & 1 & \frac{1}{2} & 0 & \frac{1}{2} & 0 & 0 & 75 \\
3 & 2 & 1 & 0 & 0 & 1 & 0 & 320 \\
-2 & -3 & -1 & 0 & 0 & 0 & 1 & 0
\end{array}\right] \quad -R_2 + R_1
$$

Change the 2 in row three, column two, to a 0 by a similar process.

$$
\begin{array}{ccccccc}
x_1 & x_2 & x_3 & x_4 & x_5 & x_6 & z \\
\end{array}
$$

$$
\left[\begin{array}{ccccccc|c}
\frac{1}{2} & 0 & \frac{7}{2} & 1 & -\frac{1}{2} & 0 & 0 & 25 \\
\frac{1}{2} & 1 & \frac{1}{2} & 0 & \frac{1}{2} & 0 & 0 & 75 \\
2 & 0 & 0 & 0 & -1 & 1 & 0 & 170 \\
-2 & -3 & -1 & 0 & 0 & 0 & 1 & 0
\end{array}\right] \quad -2R_2 + R_3
$$

Finally, add 3 times row 2 to the last row in order to change the indicator -3 to 0.

$$
\begin{array}{ccccccc}
x_1 & x_2 & x_3 & x_4 & x_5 & x_6 & z \\
\end{array}
$$

$$
\left[\begin{array}{ccccccc|c}
\frac{1}{2} & 0 & \frac{7}{2} & 1 & -\frac{1}{2} & 0 & 0 & 25 \\
\frac{1}{2} & 1 & \frac{1}{2} & 0 & \frac{1}{2} & 0 & 0 & 75 \\
2 & 0 & 0 & 0 & -1 & 1 & 0 & 170 \\
-\frac{1}{2} & 0 & \frac{1}{2} & 0 & \frac{3}{2} & 0 & 1 & 225
\end{array}\right] \quad -3R_2 + R_4
$$

The pivoting is now complete because the pivot column variable x_2 has been eliminated from all equations except the one represented by the pivot row. The initial simplex tableau has been replaced by a new simplex tableau, which represents an equivalent system of equations. ◀

CAUTION During pivoting, do not interchange rows of the matrix. Make the pivot entry 1 by multiplying the pivot row by an appropriate constant, as in Example 3. ④

When at least one of the indicators in the last row of a simplex tableau is negative (as is the case with the tableau obtained in Example 3), the simplex method requires that a new pivot be selected and the pivoting be performed again. This procedure is repeated until a simplex tableau with no negative indicators in the last row is obtained or a tableau is reached in which no pivot row can be chosen.

▶ **EXAMPLE 4** In the simplex tableau obtained in Example 3, select a new pivot and perform the pivoting.

First, locate the pivot column by finding the most negative indicator in the last row. Then locate the pivot row by computing the necessary quotients and finding the smallest one, as shown here.

---

④

For the simplex tableau below,

(a) find the pivot.

(b) Perform the pivoting and write the new tableau.

$$
\begin{array}{cccccc}
x_1 & x_2 & x_3 & x_4 & x_5 & z \\
\end{array}
$$

$$
\left[\begin{array}{ccccc|c}
1 & 2 & 6 & 1 & 0 & 0 & 16 \\
1 & 3 & 0 & 0 & 1 & 0 & 25 \\
-1 & -4 & -3 & 0 & 0 & 1 & 0
\end{array}\right]
$$

Answers:

(a) 2

(b)

$$
\begin{array}{cccccc}
x_1 & x_2 & x_3 & x_4 & x_5 & z \\
\end{array}
$$

$$
\left[\begin{array}{ccccc|c}
\frac{1}{2} & 1 & 3 & \frac{1}{2} & 0 & 0 & 8 \\
-\frac{1}{2} & 0 & -9 & -\frac{3}{2} & 1 & 0 & 1 \\
1 & 0 & 9 & 2 & 0 & 1 & 32
\end{array}\right]
$$

$$
\begin{array}{c}
\quad\quad x_1\ \ x_2\ \ x_3\ \ x_4\ \ \ x_5\ \ x_6\ \ z \\
\text{Pivot row} \longrightarrow
\left[
\begin{array}{ccccccc|c}
\tfrac{1}{2} & 0 & \tfrac{7}{2} & 1 & -\tfrac{1}{2} & 0 & 0 & 25 \\[4pt]
\tfrac{1}{2} & 1 & \tfrac{1}{2} & 0 & \tfrac{1}{2} & 0 & 0 & 75 \\[4pt]
2 & 0 & 0 & 0 & -1 & 1 & 0 & 170 \\[4pt]
\hline
-\tfrac{1}{2} & 0 & \tfrac{1}{2} & 0 & \tfrac{3}{2} & 0 & 1 & 225
\end{array}
\right]
\end{array}
$$

Quotients

$\dfrac{25}{1/2} = 50$ Smallest

$\dfrac{75}{1/2} = 150$

$170/2 = 85$

Pivot column

So the pivot is the number $1/2$ in row one, column one. Begin the pivoting by multiplying every entry in row one by 2. Then continue as indicated below, to obtain the following simplex tableau.

$$
\begin{array}{c}
x_1\ \ x_2\ \ \ \ x_3\ \ \ x_4\ \ \ \ x_5\ \ x_6\ \ z \\
\left[
\begin{array}{cccccccc|c}
1 & 0 & 7 & 2 & -1 & 0 & 0 & 50 \\
0 & 1 & -3 & -1 & 1 & 0 & 0 & 50 \\
0 & 0 & -14 & -4 & 1 & 1 & 0 & 70 \\
\hline
0 & 0 & 4 & 1 & 1 & 0 & 1 & 250
\end{array}
\right]
\end{array}
\quad
\begin{array}{l}
2R_1 \\
-\tfrac{1}{2}R_1 + R_2 \\
-2R_1 + R_3 \\
\tfrac{1}{2}R_1 + R_4
\end{array}
$$

Since there are no negative indicators in the last row, no further pivoting is necessary and we call this the **final simplex tableau.** ◀

Reading the Solution

The next example shows how to read an optimal solution of the original linear programming problem from the final simplex tableau.

▶ **EXAMPLE 5** Solve the linear programming problem introduced in Example 1. Look at the final simplex tableau for this problem, which was obtained in Example 4.

$$
\begin{array}{c}
x_1\ \ x_2\ \ \ \ x_3\ \ \ x_4\ \ \ \ x_5\ \ x_6\ \ z \\
\left[
\begin{array}{cccccccc|c}
1 & 0 & 7 & 2 & -1 & 0 & 0 & 50 \\
0 & 1 & -3 & -1 & 1 & 0 & 0 & 50 \\
0 & 0 & -14 & -4 & 1 & 1 & 0 & 70 \\
\hline
0 & 0 & 4 & 1 & 1 & 0 & 1 & 250
\end{array}
\right]
\end{array}
$$

The last row of this matrix represents the equation

$$4x_3 + x_4 + x_5 + z = 250, \quad \text{or equivalently,} \quad z = 250 - 4x_3 - x_4 - x_5.$$

If $x_3, x_4,$ and x_5 are all 0, then the value of z is 250. If any one of $x_3, x_4,$ or x_5 is positive, then z will have a smaller value than 250 (why?). Consequently, since we want a solution for this system in which all the variables are nonnegative and z is as large as possible, we must have

$$x_3 = 0, \quad x_4 = 0, \quad x_5 = 0.$$

When these values are substituted in the first equation (represented by the first row of the final simplex tableau), the result is

$$x_1 + 7 \cdot 0 + 2 \cdot 0 - 1 \cdot 0 = 50, \quad \text{that is,} \quad x_1 = 50.$$

Similarly, substituting 0 for x_3, x_4, and x_5 in the last three equations represented by the final simplex tableau shows that

$$x_2 = 50, \qquad x_6 = 70, \qquad z = 250.$$

Therefore, the maximum value of $z = 2x_1 + 3x_2 + x_3$ occurs when

$$x_1 = 50, \qquad x_2 = 50, \qquad x_3 = 0,$$

in which case $z = 2 \cdot 50 + 3 \cdot 50 + 0 = 250$. (The values of the slack variables are irrelevant in stating the solution of the original problem.) ◀

In any simplex tableau, some columns look like columns of an identity matrix (one entry is 1, the rest are 0). The variables corresponding to these columns are called **basic variables** and the variables corresponding to the other columns **nonbasic variables**. In the tableau of Example 5, for instance, the basic variables are x_1, x_2, x_6, and z (shown in blue below), and the nonbasic variables are x_3, x_4, and x_5.

$$
\begin{array}{ccccccc}
x_1 & x_2 & x_3 & x_4 & x_5 & x_6 & z \\
\end{array}
$$

$$
\left[
\begin{array}{ccccccc|c}
1 & 0 & 7 & 2 & -1 & 0 & 0 & 50 \\
0 & 1 & -3 & -1 & 1 & 0 & 0 & 50 \\
0 & 0 & -14 & -4 & 1 & 1 & 0 & 70 \\
0 & 0 & 4 & 1 & 1 & 0 & 1 & 250 \\
\end{array}
\right]
$$

The optimal solution in Example 5 was obtained from the final simplex tableau by setting the nonbasic variables equal to 0 and solving for the basic variables. Furthermore, the values of the basic variables are easy to read off the matrix: find the 1 in the column representing a basic variable; the last entry in that row is the value of that basic variable in the optimal solution. In particular, *the entry in the lower right-hand corner of the final simplex tableau is the maximum value of z.* ⑤

CAUTION If there are two identical columns in a tableau, each of which is a column in an identity matrix, only one of the variables corresponding to these columns can be a basic variable. The other is treated as a nonbasic variable. You may choose either one to be the basic variable, unless one of them is z, in which case z must be the basic variable.

The steps involved in solving a standard maximum linear programming problem by the simplex method have been illustrated in Examples 1–5 and are summarized here.

Simplex Method

1. Determine the objective function.

2. Write all necessary constraints.

3. Convert each constraint into an equation by adding slack variables.

continued

A linear programming problem with slack variables x_4 and x_5 has the final simplex tableau shown below. What is the optimal solution?

$$
\begin{array}{cccccc}
x_1 & x_2 & x_3 & x_4 & x_5 & z \\
\end{array}
$$

$$
\left[
\begin{array}{cccccc|c}
0 & 3 & 1 & 5 & 2 & 0 & 9 \\
1 & -2 & 0 & 4 & 1 & 0 & 6 \\
0 & 5 & 0 & 1 & 0 & 1 & 21 \\
\end{array}
\right]
$$

Answer:

$z = 21$ when $x_1 = 6$, $x_2 = 0$, $x_3 = 9$.

4. Set up the initial simplex tableau.

5. Locate the most negative indicator. If there are two such indicators, choose one. This indicator determines the pivot column.

6. Use the positive entries in the pivot column to form the quotients necessary for determining the pivot. If there are no positive entries in the pivot column, no maximum solution exists. If two quotients are equally the smallest, let either determine the pivot.*

7. Multiply every entry in the pivot row by the reciprocal of the pivot to change the pivot to 1. Then use row operations to change all other entries in the pivot column to 0 by adding suitable multiples of the pivot row to the other rows.

8. If the indicators are all positive or 0, this is the final tableau. If not, go back to Step 5 above and repeat the process until a tableau with no negative indicators is obtained.†

9. In the final tableau, the *basic* variables correspond to the columns that have one entry 1 and the rest 0. The *nonbasic* variables correspond to the other columns. Set each nonbasic variable equal to 0 and solve the system for the basic variables. The maximum value of the objective function is the number in the lower right-hand corner of the final tableau.

⑥

A linear programming problem has the initial tableau given below. Use the simplex method to solve the problem.

$$\begin{array}{c} \quad x_1 \quad\quad x_2\ x_3\ x_4\ z \\ \left[\begin{array}{ccccc|c} 1 & 1 & 1 & 0 & 0 & 40 \\ 2 & 1 & 0 & 1 & 0 & 24 \\ \hline -300 & -200 & 0 & 0 & 1 & 0 \end{array}\right] \end{array}$$

Answer:

$x_1 = 0, x_2 = 24, x_3 = 16,$
$x_4 = 0, z = 4800$

The solution found by the simplex method may not be unique, especially when choices are possible in steps 5, 6, or 9. There may be other solutions that produce the same maximum value of the objective function. (See Exercises 37 and 38 at the end of this section) ⑥

The Simplex Method with Technology

In the preceding discussion and in most of the examples that follow, the simplex method is carried out by hand. This allows the reader to become familiar with the procedure and to understand how it works.

Unfortunately, it's easy to make an error in manual computations, and even a small error can make a big difference in the final result. Furthermore, most real-life applications involve a large number of variables and constraints, which makes the manual approach impractical. Consequently,

we recommend that you use some kind of technology
for doing the simplex method.

Unless marked otherwise, the exercises can be done manually, but using technology will make your life much easier and your answers more accurate. Readily available technology includes the following.

---

*It may be that the first choice of a pivot does not produce a solution. In that case try the other choice.

†Some special circumstances are noted at the end of Section 7.7.

GRAPHING CALCULATORS Any graphing calculator that has matrix capabilities can do row operations. Doing the simplex method this way is an improvement on manual computation but still rather cumbersome. It is much more effective to use the simplex method programs in the Graphing Calculator Appendix. You can install these programs manually in your calculator or use a computer to download them from www.aw.com/MWA8 if you have the appropriate hardware for transferring the programs to the calculator.

SPREADSHEETS Most spreadsheets have a built-in simplex method program. Figure 7.22 shows the Solver of Microsoft Excel. Spreadsheets also provide a "sensitivity analysis," which allows you to see how much the constraints could be varied without changing the maximal solution.

FIGURE 7.22

Geometric Interpretation of the Simplex Method

Although it may not be immediately apparent, the simplex method is based on the same geometrical considerations as the graphical method. This can be seen by looking at a problem that can be readily solved by both methods.

▶ **EXAMPLE 6** In Example 2 of Section 7.3 the following problem was solved graphically (using x, y instead of x_1, x_2):

$$\text{Maximize} \quad z = 8x_1 + 12x_2$$
$$\text{subject to:} \quad 40x_1 + 80x_2 \le 560$$
$$6x_1 + 8x_2 \le 72$$
$$x_1 \ge 0, x_2 \ge 0.$$

Graphing the feasible region (Figure 7.23) and evaluating z at each corner point shows that the maximum value of z occurs at $(8, 3)$.

Corner Point	Value of $z = 8x_1 + 12x_2$
$(0, 0)$	0
$(0, 7)$	84
$(12, 0)$	96
$(8, 3)$	100 (maximum)

FIGURE 7.23

To solve the same problem by the simplex method, add a slack variable to each constraint.

$$40x_1 + 80x_2 + x_3 \qquad\;\; = 560$$
$$6x_1 + \;\; 8x_2 \qquad\quad + x_4 = 72$$

Then write the initial simplex tableau.

$$
\begin{array}{c c c c c}
x_1 & x_2 & x_3 & x_4 & z
\end{array}
$$
$$
\left[
\begin{array}{ccccc|c}
40 & 80 & 1 & 0 & 0 & 560 \\
\hline
6 & 8 & 0 & 1 & 0 & 72 \\
\hline
-8 & -12 & 0 & 0 & 1 & 0
\end{array}
\right]
$$

In this tableau the basic variables are x_3, x_4, and z (why?). By setting the nonbasic variables (namely, x_1 and x_2) equal to 0 and solving for the basic variables, we obtain the following solution (which will be called a **basic feasible solution**):

$$x_1 = 0, \qquad x_2 = 0, \qquad x_3 = 560, \qquad x_4 = 72, \qquad z = 0.$$

Since $x_1 = 0$ and $x_2 = 0$, this solution corresponds to the corner point at the origin in the graphical solution (Figure 7.23). The value $z = 0$ at the origin is obviously not maximal and pivoting in the simplex method is designed to improve it.

The most negative indicator in the initial tableau is -12 and the necessary quotients are

$$\frac{560}{80} = 7 \quad \text{and} \quad \frac{72}{8} = 9.$$

The smaller quotient is 7, giving 80 as the pivot. Performing the pivoting leads to this tableau.

$$
\begin{array}{c c c c c}
x_1 & x_2 & x_3 & x_4 & z
\end{array}
$$
$$
\left[
\begin{array}{ccccc|c}
\frac{1}{2} & 1 & \frac{1}{80} & 0 & 0 & 7 \\
2 & 0 & -\frac{1}{10} & 1 & 0 & 16 \\
\hline
-2 & 0 & \frac{3}{20} & 0 & 1 & 84
\end{array}
\right]
\begin{array}{l}
\frac{1}{80}R_1 \\
-8R_1 + R_2 \\
12R_1 + R_3
\end{array}
$$

The basic variables here are x_2, x_4, and z and the basic feasible solution (found by setting the nonbasic variables equal to 0 and solving for the basic variables) is

$$x_1 = 0, \qquad x_2 = 7, \qquad x_3 = 0, \qquad x_4 = 16, \qquad z = 84,$$

which corresponds to the corner point $(0, 7)$ in Figure 7.23. Note that the new value of the pivot variable x_2 is precisely the smallest quotient, 7, that was used to select the pivot row. Although this value of z is better, further improvement is possible.

Now the most negative indicator is -2 and the quotients are

$$\frac{7}{1/2} = 14 \quad \text{and} \quad \frac{16}{2} = 8.$$

Since 8 is smaller, the pivot is the number 2 in row two, column one. Pivoting produces the final tableau.

$$
\begin{array}{ccccc}
x_1 & x_2 & x_3 & x_4 & z
\end{array}
$$

$$
\left[
\begin{array}{ccccc|c}
0 & 1 & \frac{3}{80} & -\frac{1}{4} & 0 & 3 \\
1 & 0 & -\frac{1}{20} & \frac{1}{2} & 0 & 8 \\
\hline
0 & 0 & \frac{1}{20} & 1 & 1 & 100
\end{array}
\right]
\begin{array}{l}
-\frac{1}{2}R_2 + R_1 \\
\frac{1}{2}R_2 \\
2R_2 + R_3
\end{array}
$$

Here the basic feasible solution is

$$x_1 = 8, \qquad x_2 = 3, \qquad x_3 = 0, \qquad x_4 = 0, \qquad z = 100,$$

which corresponds to the corner point $(8, 3)$ in Figure 7.23. Once again, the new value of the pivot variable x_1 is the smallest quotient, 8, that was used to select the pivot. From the graphical method we know that this solution provides the maximum value of the objective function. This fact can also be seen algebraically by using an algebraic argument similar to the one used in Example 5. Thus, there is no need to move to another corner point and the simplex method ends. ◀

As illustrated in Example 6, the basic feasible solution obtained from a simplex tableau corresponds to a corner point of the feasible region. Pivoting, which replaces one tableau with another, is a systematic way of moving from one corner point to another, each time improving the value of the objective function. The simplex method ends when a corner point that produces the maximum value of the objective function is reached (or when it becomes clear that the problem has no maximum solution).

When there are three or more variables in a linear programming problem, it may be difficult or impossible to draw a picture, but it can be proved that the optimal value of the objective function occurs at a basic feasible solution (corresponding to a corner point in the two variable case). The simplex method provides a means of moving from one basic feasible solution to another until one that produces the optimal value of the objective function is reached.

Explanation of Pivoting

The rules for selecting the pivot in the simplex method can be understood by examining how the first pivot was chosen in Example 6.

▶ EXAMPLE 7   The initial simplex tableau of Example 6 provides a basic feasible solution with $x_1 = 0$, $x_2 = 0$.

$$
\begin{array}{ccccc}
x_1 & x_2 & x_3 & x_4 & z \\
\end{array}
$$

$$
\left[\begin{array}{ccccc|c}
40 & 80 & 1 & 0 & 0 & 560 \\
6 & 8 & 0 & 1 & 0 & 72 \\
\hline
-8 & -12 & 0 & 0 & 1 & 0
\end{array}\right]
$$

This solution certainly does not give a maximum value for the objective function $z = 8x_1 + 12x_2$. Since x_2 has the largest coefficient, z will be increased most if x_2 is increased. In other words, the most negative indicator in the tableau (which corresponds to the largest coefficient in the objective function) identifies the variable that will provide the greatest change in the value of z.

To determine how much x_2 can be increased without leaving the feasible region, look at the first two equations

$$
\begin{aligned}
40x_1 + 80x_2 + x_3 \quad\quad &= 560 \\
6x_1 + 8x_2 \quad\quad + x_4 &= 72
\end{aligned}
$$

and solve for the basic variables x_3 and x_4.

$$
\begin{aligned}
x_3 &= 560 - 40x_1 - 80x_2 \\
x_4 &= 72 - 6x_1 - 8x_2
\end{aligned}
$$

Now x_2 is to be increased while x_1 is to keep the value 0. Hence

$$
\begin{aligned}
x_3 &= 560 - 80x_2 \\
x_4 &= 72 - 8x_2.
\end{aligned}
$$

Since $x_3 \geq 0$ and $x_4 \geq 0$, we must have:

$$
\begin{aligned}
0 &\leq 560 - 80x_2 \quad \text{and} \quad & 0 &\leq 72 - 8x_2 \\
80x_2 &\leq 560 & 8x_2 &\leq 72 \\
x_2 &\leq \frac{560}{80} = 7 & x_2 &\leq \frac{72}{8} = 9.
\end{aligned}
$$

The right sides of these last inequalities are the quotients used to select the pivot row. Since x_2 must satisfy both inequalities, x_2 can be at most 7. In other words, the smallest quotient formed from positive entries in the pivot column identifies the value of x_2 that produces the largest change in z while remaining in the feasible region. By pivoting with the pivot determined in this way, we obtain the second tableau and a basic feasible solution in which $x_2 = 7$, as was shown in Example 6. ◀

An analysis similar to that in Example 7 applies to each occurrence of pivoting in the simplex method. The idea is to improve the value of the objective function by adjusting one variable at a time. The most negative indicator identifies the variable that will account for the largest increase in z. The smallest quotient determines the largest value of that variable that will produce a feasible solution. Pivoting leads to a solution in which the selected variable has this largest value.

The simplex method is easily implemented on a computer and some graphing calculators. Technology is essential for the simplex method in any situation where there are a large number of variables and constraints (and hence an enormous number of corner points to check).

7.4 Exercises

In Exercises 1–4, (a) determine the number of slack variables needed; (b) name them; (c) use the slack variables to convert each constraint into a linear equation. (See Example 1.)

1. Maximize $z = 32x_1 + 9x_2$
   subject to: $4x_1 + 2x_2 \le 20$
   $5x_1 + x_2 \le 50$
   $2x_1 + 3x_2 \le 25$
   $x_1 \ge 0, x_2 \ge 0.$

2. Maximize $z = 3.7x_1 + 4.3x_2$
   subject to: $2.4x_1 + 1.5x_2 \le 10$
   $1.7x_1 + 1.9x_2 \le 15$
   $x_1 \ge 0, x_2 \ge 0.$

3. Maximize $z = 8x_1 + 3x_2 + x_3$
   subject to: $3x_1 - x_2 + 4x_3 \le 95$
   $7x_1 + 6x_2 + 8x_3 \le 118$
   $4x_1 + 5x_2 + 10x_3 \le 220$
   $x_1 \ge 0, x_2 \ge 0, x_3 \ge 0.$

4. Maximize $z = 12x_1 + 15x_2 + 10x_3$
   subject to: $2x_1 + 2x_2 + x_3 \le 8$
   $x_1 + 4x_2 + 3x_3 \le 12$
   $x_1 \ge 0, x_2 \ge 0, x_3 \ge 0.$

Introduce slack variables as necessary and then write the initial simplex tableau for each of these linear programming problems.

5. Maximize $z = 5x_1 + x_2$
   subject to: $2x_1 + 3x_2 \le 6$
   $4x_1 + x_2 \le 6$
   $5x_1 + 2x_2 \le 15$
   $x_1 \ge 0, x_2 \ge 0.$

6. Maximize $z = 5x_1 + 3x_2 + 4x_3$
   subject to: $4x_1 + 3x_2 + 2x_3 \le 60$
   $3x_1 + 4x_2 + x_3 \le 24$
   $x_1 \ge 0, x_2 \ge 0, x_3 \ge 0.$

7. Maximize $z = x_1 + 5x_2 + 10x_3$
   subject to: $x_1 + 2x_2 + 3x_3 \le 10$
   $2x_1 + x_2 + x_3 \le 8$
   $3x_1 + 2x_3 \le 6$
   $x_1 \ge 0, x_2 \ge 0, x_3 \ge 0.$

8. Maximize $z = 5x_1 - x_2 + 3x_3$
   subject to: $3x_1 + 2x_2 + x_3 \le 36$
   $x_1 + 4x_2 + x_3 \le 24$
   $x_1 - x_2 - x_3 \le 32$
   $x_1 \ge 0, x_2 \ge 0, x_3 \ge 0.$

Find the pivot in each of the following simplex tableaus. (See Example 2.)

9.

x_1	x_2	x_3	x_4	x_5	z	
2	2	0	3	1	0	15
3	4	1	6	0	0	20
−2	−1	0	1	0	1	10

10.

x_1	x_2	x_3	x_4	x_5	z	
0	2	1	1	3	0	5
1	−5	0	1	2	0	8
0	−2	0	−1	1	1	10

11.

x_1	x_2	x_3	x_4	x_5	x_6	z	
6	2	1	3	0	0	0	8
0	2	0	1	0	1	0	7
2	1	0	3	1	0	0	6
−3	−2	0	2	0	0	1	12

12.

x_1	x_2	x_3	x_4	x_5	x_6	z	
0	2	0	1	2	2	0	3
0	3	1	0	1	2	0	2
1	4	0	0	3	5	0	5
0	−4	0	0	4	−3	1	20

In Exercises 13–16, use the indicated entry as the pivot and perform the pivoting. (See Examples 3 and 4.)

13.

x_1	x_2	x_3	x_4	x_5	z	
1	2	4	1	0	0	56
2	**2**	1	0	1	0	40
−1	−3	−2	0	0	1	0

14.

x_1	x_2	x_3	x_4	x_5	x_6	z	
2	2	**1**	1	0	0	0	12
1	2	3	0	1	0	0	45
3	1	1	0	0	1	0	20
−2	−1	−3	0	0	0	1	0

15.

x_1	x_2	x_3	x_4	x_5	x_6	z	
1	1	1	1	0	0	0	60
3	1	**2**	0	1	0	0	100
1	2	3	0	0	1	0	200
−1	−1	−2	0	0	0	1	0

16.

x_1	x_2	x_3	x_4	x_5	x_6	z	
4	2	3	1	0	0	0	22
2	2	**5**	0	1	0	0	28
1	3	2	0	0	1	0	45
−3	−2	−4	0	0	0	1	0

For each simplex tableau in Exercises 17–20, (a) list the basic and the nonbasic variables; (b) find the basic feasible solution determined by setting the nonbasic variables equal to 0; (c) decide whether this is a maximum solution. (See Examples 5 and 6.)

17.

x_1	x_2	x_3	x_4	x_5	z	
3	2	0	−3	1	0	29
4	0	1	−2	0	0	16
−5	0	0	−1	0	1	11

18.

$$
\begin{array}{ccccccc}
x_1 & x_2 & x_3 & x_4 & x_5 & x_6 & z
\end{array}
$$

$$
\left[\begin{array}{ccccccc|c}
-3 & 0 & \frac{1}{2} & 1 & -2 & 0 & 0 & 22 \\
2 & 0 & -3 & 0 & 1 & 1 & 0 & 10 \\
4 & 1 & 4 & 0 & \frac{3}{4} & 0 & 0 & 17 \\
\hline
-1 & 0 & 0 & 0 & 1 & 0 & 1 & 120
\end{array}\right]
$$

19.

$$
\begin{array}{ccccccc}
x_1 & x_2 & x_3 & x_4 & x_5 & x_6 & z
\end{array}
$$

$$
\left[\begin{array}{ccccccc|c}
1 & 0 & 2 & \frac{1}{2} & 0 & \frac{1}{3} & 0 & 6 \\
0 & 1 & -1 & 5 & 0 & -1 & 0 & 13 \\
0 & 0 & 1 & \frac{3}{2} & 1 & -\frac{1}{3} & 0 & 21 \\
\hline
0 & 0 & 2 & \frac{1}{2} & 0 & 3 & 1 & 18
\end{array}\right]
$$

20.

$$
\begin{array}{cccccccc}
x_1 & x_2 & x_3 & x_4 & x_5 & x_6 & x_7 & z
\end{array}
$$

$$
\left[\begin{array}{cccccccc|c}
-1 & 0 & 0 & 1 & 0 & 3 & -2 & 0 & 47 \\
2 & 0 & 1 & 0 & 0 & 2 & -\frac{1}{2} & 0 & 37 \\
3 & 0 & 0 & 0 & 1 & -1 & 6 & 0 & 43 \\
\hline
4 & 1 & 0 & 0 & 0 & 6 & 0 & 1 & 86
\end{array}\right]
$$

Use the simplex method to solve Exercises 21–36.

21. Maximize $z = x_1 + 3x_2$
subject to: $x_1 + x_2 \le 10$
$5x_1 + 2x_2 \le 20$
$x_1 + 2x_2 \le 36$
$x_1 \ge 0, x_2 \ge 0.$

22. Maximize $z = 5x_1 + x_2$
subject to: $2x_1 + 3x_2 \le 8$
$4x_1 + 8x_2 \le 12$
$5x_1 + 2x_2 \le 30$
$x_1 \ge 0, x_2 \ge 0.$

23. Maximize $z = 2x_1 + x_2$
subject to: $x_1 + 3x_2 \le 12$
$2x_1 + x_2 \le 10$
$x_1 + x_2 \le 4$
$x_1 \ge 0, x_2 \ge 0.$

24. Maximize $z = 4x_1 + 2x_2$
subject to: $-x_1 - x_2 \le 12$
$3x_1 - x_2 \le 15$
$x_1 \ge 0, x_2 \ge 0.$

25. Maximize $z = 5x_1 + 4x_2 + x_3$
subject to: $-2x_1 + x_2 + 2x_3 \le 3$
$x_1 - x_2 + x_3 \le 1$
$x_1 \ge 0, x_2 \ge 0, x_3 \ge 0.$

26. Maximize $z = 3x_1 + 2x_2 + x_3$
subject to: $2x_1 + 2x_2 + x_3 \le 10$
$x_1 + 2x_2 + 3x_3 \le 15$
$x_1 \ge 0, x_2 \ge 0, x_3 \ge 0.$

27. Maximize $z = 2x_1 + x_2 + x_3$
subject to: $x_1 - 3x_2 + x_3 \le 3$
$x_1 - 2x_2 + 2x_3 \le 12$
$x_1 \ge 0, x_2 \ge 0, x_3 \ge 0.$

28. Maximize $z = 4x_1 + 5x_2 + x_3$
subject to: $x_1 + 2x_2 + 4x_3 \le 10$
$2x_1 + 2x_2 + x_3 \le 10$
$x_1 \ge 0, x_2 \ge 0, x_3 \ge 0.$

29. Maximize $z = 2x_1 + 2x_2 - 4x_3$
subject to: $3x_1 + 3x_2 - 6x_3 \le 51$
$5x_1 + 5x_2 + 10x_3 \le 99$
$x_1 \ge 0, x_2 \ge 0, x_3 \ge 0.$

30. Maximize $z = 4x_1 + x_2 + 3x_3$
subject to: $x_1 + 3x_3 \le 6$
$6x_1 + 3x_2 + 12x_3 \le 40$
$x_1 \ge 0, x_2 \ge 0, x_3 \ge 0.$

31. Maximize $z = 300x_1 + 200x_2 + 100x_3$
subject to: $x_1 + x_2 + x_3 \le 100$
$2x_1 + 3x_2 + 4x_3 \le 320$
$2x_1 + x_2 + x_3 \le 160$
$x_1 \ge 0, x_2 \ge 0, x_3 \ge 0.$

32. Maximize $z = x_1 + 5x_2 - 10x_3$
subject to: $8x_1 + 4x_2 + 12x_3 \le 18$
$x_1 + 6x_2 + 2x_3 \le 45$
$5x_1 + 7x_2 + 3x_3 \le 60$
$x_1 \ge 0, x_2 \ge 0, x_3 \ge 0.$

33. Maximize $z = 4x_1 - 3x_2 + 2x_3$
subject to: $2x_1 - x_2 + 8x_3 \le 40$
$4x_1 - 5x_2 + 6x_3 \le 60$
$2x_1 - 2x_2 + 6x_3 \le 24$
$x_1 \ge 0, x_2 \ge 0, x_3 \ge 0.$

34. Maximize $z = 3x_1 + 2x_2 - 4x_3$
subject to: $x_1 - x_2 + x_3 \le 10$
$2x_1 - x_2 + 2x_3 \le 30$
$-3x_1 + x_2 + 3x_3 \le 40$
$x_1 \ge 0, x_2 \ge 0, x_3 \ge 0.$

35. Maximize $z = x_1 + 2x_2 + x_3 + 5x_4$
subject to: $x_1 + 2x_2 + x_3 + x_4 \le 50$
$3x_1 + x_2 + 2x_3 + x_4 \le 100$
$x_1 \ge 0, x_2 \ge 0, x_3 \ge 0, x_4 \ge 0.$

36. Maximize $z = x_1 + x_2 + 4x_3 + 5x_4$
subject to: $x_1 + 2x_2 + 3x_3 + x_4 \le 115$
$2x_1 + x_2 + 8x_3 + 5x_4 \le 200$
$x_1 + x_3 \le 50$
$x_1 \ge 0, x_2 \ge 0, x_3 \ge 0, x_4 \ge 0.$

37. The initial simplex tableau of a linear programming problem is given below.

$$
\begin{array}{cccccc}
x_1 & x_2 & x_3 & x_4 & x_5 & z
\end{array}
$$

$$
\left[\begin{array}{cccccc|c}
1 & 1 & 1 & 1 & 0 & 0 & 12 \\
2 & 1 & 2 & 0 & 1 & 0 & 30 \\
\hline
-2 & -2 & -1 & 0 & 0 & 1 & 0
\end{array}\right]
$$

(a) Use the simplex method to solve the problem, with column one as the first pivot column.

(b) Now use the simplex method to solve the problem, with column two as the first pivot column.

(c) Does this problem have a unique maximum solution? Why?

38. The final simplex tableau of a linear programming problem is given here.

$$
\begin{array}{ccccc|c}
x_1 & x_2 & x_3 & x_4 & z & \\
\hline
1 & 1 & 2 & 0 & 0 & 24 \\
2 & 0 & 2 & 1 & 0 & 8 \\
4 & 0 & 0 & 0 & 1 & 40
\end{array}
$$

(a) What is the solution given by this tableau?

(b) Even though all the indicators are nonnegative, perform one more round of pivoting on this tableau, using column three as the pivot column and choosing the pivot row by forming quotients in the usual way.

(c) Show that there is more than one solution to the linear programming problem by comparing your answer in part (a) to the basic feasible solution given by the tableau found in part (b). Does it give the same value of z as the solution in part (a)?

7.5 MAXIMIZATION APPLICATIONS

Applications of the simplex method are considered in this section. First, however, we make a slight change in notation. You have noticed that the column representing the variable z in a simplex tableau never changes during pivoting. (Since all the entries except the last one in this column are 0, performing row operations has no effect on these entries—they remain 0.) Consequently, this column is unnecessary and can be omitted without causing any difficulty.

> Hereafter in this text, the column corresponding to the variable z (representing the objective function) will be omitted from all simplex tableaus.

▶ **EXAMPLE 1** A farmer has 110 acres of available land he wishes to plant with a mixture of potatoes, corn, and cabbage. It costs him $400 to produce an acre of potatoes, $160 to produce an acre of corn, and $280 to produce an acre of cabbage. He has a maximum of $20,000 to spend. He makes a profit of $120 per acre of potatoes, $40 per acre of corn, and $60 per acre of cabbage.

(a) How many acres of each crop should he plant to maximize his profit?
  Begin by summarizing the given information as follows.

Crop	Number of Acres	Cost per Acre	Profit per Acre
Potatoes	x_1	$400	$120
Corn	x_2	$160	$ 40
Cabbage	x_3	$280	$ 60
Maximum Available	110	$20,000	

If the number of acres allotted to each of the three crops is represented by x_1, x_2, and x_3, respectively, then the constraints can be expressed as

$$x_1 + x_2 + x_3 \le 110 \qquad \text{Number of acres}$$
$$400x_1 + 160x_2 + 280x_3 \le 20{,}000 \qquad \text{Production costs}$$

where x_1, x_2, and x_3 are all nonnegative. The first of these constraints says that $x_1 + x_2 + x_3$ is less than or perhaps equal to 110. Use x_4 as the slack variable, giving the equation

$$x_1 + x_2 + x_3 + x_4 = 110.$$

Here x_4 represents the amount of the farmer's 110 acres that will not be used. (x_4 may be 0 or any value up to 110.)

In the same way, the constraint $400x_1 + 160x_2 + 280x_3 \le 20{,}000$ can be converted into an equation by adding a slack variable, x_5:

$$400x_1 + 160x_2 + 280x_3 + x_5 = 20{,}000.$$

The slack variable x_5 represents any unused portion of the farmer's \$20,000 capital. (Again, x_5 may have any value from 0 to 20,000.)

The farmer's profit on potatoes is the product of the profit per acre (\$120) and the number x_1 of acres, that is, $120x_1$. His profit on corn and cabbage is computed similarly. Hence his total profit is given by

$$z = \text{profit on potatoes} + \text{profit on corn} + \text{profit on cabbage}$$
$$z = 120x_1 + 40x_2 + 60x_3.$$

The linear programming problem can now be stated as follows:

$$\text{Maximize} \quad z = 120x_1 + 40x_2 + 60x_3$$
$$\text{subject to:} \qquad x_1 + x_2 + x_3 + x_4 \qquad = 110$$
$$400x_1 + 160x_2 + 280x_3 + \qquad x_5 = 20{,}000$$
$$\text{with} \quad x_1 \ge 0, x_2 \ge 0, x_3 \ge 0, x_4 \ge 0, x_5 \ge 0.$$

The initial simplex tableau (without the z column) is

x_1	x_2	x_3	x_4	x_5	
1	1	1	1	0	110
400	160	280	0	1	20,000
-120	-40	-60	0	0	0

The most negative indicator is -120; column one is the pivot column. The quotients needed to determine the pivot row are $110/1 = 110$ and $20{,}000/400 = 50$. So the pivot is 400 in row two, column one. Multiplying row two by $1/400$ and completing the pivoting leads to the final simplex tableau.

x_1	x_2	x_3	x_4	x_5		
0	.6	.3	1	$-.0025$	60	$-1R_2 + R_1$
1	.4	.7	0	.0025	50	$\dfrac{1}{400}R_2$
0	8	24	0	.3	6000	$120R_2 + R_3$

Setting the nonbasic variables x_2, x_3, and x_5 equal to 0, solving for the basic variables x_1 and x_4, and remembering that the value of z is in the lower right-hand corner leads to this maximum solution:

$$x_1 = 50, \qquad x_2 = 0, \qquad x_3 = 0, \qquad x_4 = 60, \qquad x_5 = 0, \qquad z = 6000.$$

Therefore, the farmer will make a maximum profit of $6000 by planting 50 acres of potatoes, and no corn or cabbage.

(b) If the farmer maximizes his profit, how much land will remain unplanted? What is the explanation for this?

Since 50 of 110 acres are planted, 60 acres will remain unplanted. Alternatively, note that the unplanted acres of land are represented by x_4, the slack variable in the "number of acres" constraint. In the maximal solution found in part (a), $x_4 = 60$, which means that 60 acres are left unplanted.

The amount of unused cash is represented by x_5, the slack variable in the "production costs" constraint. Since $x_5 = 0$, all the available money has been used. By using the maximal solution in part (a), the farmer has used his $20,000 most effectively. If he had more cash, he would plant more crops and make a larger profit. ◀

▶**EXAMPLE 2** Ana Pott, who is a candidate for the state legislature, has $96,000 to buy TV advertising time. Ads cost $400 per minute on a local cable channel, $4000 per minute on a regional independent channel, and $12,000 per minute on a national network channel. Because of existing contracts the TV stations can provide at most 30 minutes of advertising time, with a maximum of 6 minutes on the national network channel. At any given time during the evening, approximately 100,000 people watch the cable channel, 200,000 the independent channel, and 600,000 the network channel. To get maximum exposure, how much time should Ana buy from each station?

(a) Set up the initial simplex tableau for this problem.

Let x_1 be the number of minutes of ads on the cable channel, x_2 the number of minutes on the independent channel, and x_3 the number of minutes on the network channel. Exposure is measured in viewer-minutes. For instance, 100,000 people watching x_1 minutes of ads on the cable channel produces $100,000x_1$ viewer-minutes. The amount of exposure is given by the total number of viewer-minutes for all three channels, namely,

$$100,000x_1 + 200,000x_2 + 600,000x_3.$$

Since 30 minutes are available,

$$x_1 + x_2 + x_3 \leq 30.$$

The fact that only 6 minutes can be used on the network channel means that

$$x_3 \leq 6.$$

Expenditures are limited to $96,000, so

$$\text{Cable cost} + \text{independent cost} + \text{network cost} \leq 96,000$$
$$400x_1 + 4000x_2 + 12,000x_3 \leq 96,000.$$

Therefore, Ana must solve the following linear programming problem:

$$\text{Maximize} \quad z = 100,000x_1 + 200,000x_2 + 600,000x_3$$
$$\text{subject to:} \quad x_1 + x_2 + x_3 \leq 30$$
$$x_3 \leq 6$$
$$400x_1 + 4000x_2 + 12,000x_3 \leq 96,000$$
$$\text{with} \quad x_1 \geq 0, x_2 \geq 0, x_3 \geq 0.$$

Introducing slack variables x_4, x_5, and x_6 (one for each constraint), rewriting the constraints as equations, and expressing the objective function as

$$-100{,}000x_1 - 200{,}000x_2 - 600{,}000x_3 + z = 0$$

leads to the initial simplex tableau:

$$
\begin{array}{cccccc}
x_1 & x_2 & x_3 & x_4 & x_5 & x_6 \\
\end{array}
$$

$$
\left[
\begin{array}{cccccc|c}
1 & 1 & 1 & 1 & 0 & 0 & 30 \\
0 & 0 & 1 & 0 & 1 & 0 & 6 \\
400 & 4000 & 12{,}000 & 0 & 0 & 1 & 96{,}000 \\
\hline
-100{,}000 & -200{,}000 & -600{,}000 & 0 & 0 & 0 & 0
\end{array}
\right].
$$

(b) Use the simplex method to find the final simplex tableau.

Work by hand, or use a graphing calculator simplex program or a spreadsheet, to obtain this final tableau.

$$
\begin{array}{cccccc}
x_1 & x_2 & x_3 & x_4 & x_5 & x_6 \\
\end{array}
$$

$$
\left[
\begin{array}{cccccc|c}
1 & 0 & 0 & \frac{10}{9} & \frac{20}{9} & -\frac{25}{90{,}000} & 20 \\
0 & 0 & 1 & 0 & 1 & 0 & 6 \\
0 & 1 & 0 & -\frac{1}{9} & -\frac{29}{9} & \frac{25}{90{,}000} & 4 \\
\hline
0 & 0 & 0 & \frac{800{,}000}{9} & \frac{1{,}600{,}000}{9} & \frac{250}{9} & 6{,}400{.}000
\end{array}
\right].
$$

Therefore, the optimal solution is

$$x_1 = 20, \qquad x_2 = 4, \qquad x_3 = 6, \qquad x_4 = 0, \qquad x_5 = 0, \qquad x_6 = 0.$$

Ana should buy 20 minutes of time on the cable channel, 4 minutes on the independent channel, and 6 minutes on the network channel.

(c) What do the values of the slack variables in the optimal solution tell you?

All three slack variables are 0. This means that all the available minutes have been used ($x_4 = 0$ in the first constraint), the maximum possible six minutes on the national network have been used ($x_5 = 0$ in the second constraint), and all of the $96,000 has been spent ($x_6 = 0$ in the third constraint). ◀ ①

In Example 2, what is the number of viewer-minutes in the optimal solution?

Answer:

$z = 6{,}400{,}000$

▶ **EXAMPLE 3**   A chemical plant makes three products: glaze, solvent, and clay, each of which brings in different revenue per truckload. Production is limited, first by the number of air pollution units the plant is allowed to produce each day and second by the time available in the evaporation tank. The plant manager wants to maximize the daily revenue. Using information not given here, he sets up an initial simplex tableau and uses the simplex method to produce the following final simplex tableau.

$$
\left[
\begin{array}{ccccc|c}
-10 & -25 & 0 & 1 & -1 & 60 \\
3 & 4 & 1 & 0 & .1 & 24 \\
\hline
7 & 13 & 0 & 0 & .4 & 96
\end{array}
\right]
$$

The three variables represent the number of truckloads of glaze, solvent, and clay, respectively. The first slack variable comes from the air pollution constraint and the second slack variable from the time constraint on the evaporation tank. The revenue function is given in hundreds of dollars.

(a) What is the optimal solution?

$$x_1 = 0, \quad x_2 = 0, \quad x_3 = 24, \quad x_4 = 60, \quad x_5 = 0, \quad \text{and} \quad z = 96.$$

(b) Interpret this solution. What do the variables represent and what does it mean?
    The variable x_1 is the number of truckloads of glaze, x_2 the number of truckloads of solvent, x_3 the number of truckloads of clay to be produced and z is the revenue produced (in hundreds of dollars). The plant should produce 24 truckloads of clay and no glaze or solvent for a maximum revenue of $9600. The first slack variable x_4 represents the number of air pollution units below the maximum number allowed. Since $x_4 = 60$, air pollution will be 60 units less than the allowable maximum. The second slack variable x_5 represents the unused time in the evaporation tank. Since $x_5 = 0$, the evaporation tank is fully used. ◀

7.5 Exercises

Set up the initial simplex tableau for each of the following problems.

1. Business A cat breeder has the following amounts of cat food: 90 units of tuna, 80 units of liver, and 50 units of chicken. To raise a Siamese cat, the breeder must use 2 units of tuna, 1 of liver, and 1 of chicken per day, while raising a Persian cat requires 1, 2, and 1 units, respectively, per day. If a Siamese cat sells for $12 while a Persian cat sells for $10, how many of each should be raised in order to obtain maximum gross income? What is the maximum gross income?

2. Business Banal, Inc., produces art for motel rooms. Its painters can turn out mountain scenes, seascapes, and pictures of clowns. Each painting is worked on by three different artists, T, D, and H. Artist T works only 25 hours per week, while D and H work 45 and 40 hours per week, respectively. Artist T spends 1 hour on a mountain scene, 2 hours on a seascape, and 1 hour on a clown. Corresponding times for D and H are 3, 2, and 2 hours, and 2, 1, and 4 hours, respectively. Banal makes $20 on a mountain scene, $18 on a seascape, and $22 on a clown. The head painting packer can't stand clowns, so that no more than 4 clown paintings may be done in a week. Find the number of each type of painting that should be made weekly in order to maximize profit. Find the maximum possible profit.

3. Health A biologist has 500 kg of nutrient A, 600 kg of nutrient B, and 300 kg of nutrient C. These nutrients will be used to make 4 types of food, whose contents (in percent of nutrient per kilogram of food) and whose "growth values" are as shown in the table.

	P	Q	R	S
A	0	0	37.5	62.5
B	0	75	50	37.5
C	100	25	12.5	0
Growth Value	90	70	60	50

How many kilograms of each food should be produced in order to maximize total growth value? Find the maximum growth value.

4. Natural Science A lake is stocked each spring with three species of fish, A, B, and C. The average weights of the fish are 1.62, 2.12, and 3.01 kilograms for species A, B, and C, respectively. Three foods, I, II, and III, are available in the lake. Each fish of species A requires 1.32 units of food I, 2.9 units of food II, and 1.75 units of food III on the average each day. Species B fish each require 2.1 units of food I, .95 units of food II, and .6 units of food III daily. Species C fish require .86, 1.52, and 2.01 units of I, II, and III per day, respectively. If 490 units of food I, 897 units of food II, and 653 units of food III are available daily, how should the lake be stocked to maximize the weight of the fish supported by the lake?

In each of the following exercises: (a) *Use the simplex method to solve the problem;* (b) *explain what the values of the slack variables in the optimal solution mean in the context of the problem.* (See Examples 1–3).

5. Business   A manufacturer of bicycles builds one-, three-, and ten-speed models. The bicycles need both aluminum and steel. The company has available 91,800 units of steel and 42,000 units of aluminum. The one-, three-, and ten-speed models need, respectively, 20, 30, and 40 units of steel and 12, 21, and 16 units of aluminum. How many of each type of bicycle should be made in order to maximize profit if the company makes $8 per one-speed bike, $12 per three-speed, and $24 per ten-speed? What is the maximum possible profit?

6. Social Science   Jayanta is working to raise money for the homeless by sending information letters and making follow-up calls to local labor organizations and church groups. She discovered that each church group requires 2 hours of letter writing and 1 hour of follow-up, while for each labor union she needs 2 hours of letter writing and 3 hours of follow-up. Jayanta can raise $100 from each church group and $200 from each union local, and she has a maximum of 16 hours of letter-writing time and a maximum of 12 hours of follow-up time available per month. Determine the most profitable mixture of groups she should contact and the most money she can raise in a month.

7. Social Science   A political party is planning a half-hour television show. The show will have 3 minutes of direct requests for money from viewers. Three of the party's politicians will be on the show—a senator, a congresswoman, and a governor. The senator, a party "elder statesman," demands that he be on screen at least twice as long as the governor. The total time taken by the senator and the governor must be at least twice the time taken by the congresswoman. Based on a pre-show survey, it is believed that 40, 60, and 50 (in thousands) viewers will watch the program for each minute the senator, congresswoman, and governor, respectively, are on the air. Find the time that should be allotted to each politician in order to get the maximum number of viewers. Find the maximum number of viewers.

8. Business   The Cut-Right Company sells sets of kitchen knives. The Basic Set consists of 2 utility knives and 1 chef's knife. The Regular Set consists of 2 utility knives, 1 chef's knife, and 1 slicer. The Deluxe Set consists of 3 utility knives, 1 chef's knife, and 1 slicer. Their profit is $30 on a Basic Set, $40 on a Regular Set, and $60 on a Deluxe Set. The factory has on hand 800 utility knives, 400 chef's knives, and 200 slicers. Assuming that all sets will be sold, how many of each type should be made up in order to maximize profit? What is the maximum profit?

Use the simplex method to solve the following problems. (See Examples 1–3.)

9. Business   The Fancy Fashions Store has $8000 available each month for advertising. Newspaper ads cost $400 each and no more than 20 can be run per month. Radio ads cost $200 each and no more than 30 can run per month. TV ads cost $1200 each, with a maximum of 6 available each month. Approximately 2000 women will see each newspaper ad, 1200 will hear each radio commercial, and 10,000 will see each TV ad. How much of each type of advertising should be used if the store wants to maximize its ad exposure?

10. Business   Caroline's Quality Candy Confectionery is famous for fudge, chocolate cremes, and pralines. Its candy-making equipment is set up to make 100-pound batches at a time. Currently there is a chocolate shortage and the company can get only 120 pounds of chocolate in the next shipment. On a week's run, the confectionery's cooking and processing equipment is available for a total of 42 machine hours. During the same period the employees have a total of 56 work hours available for packaging. A batch of fudge requires 20 pounds of chocolate while a batch of cremes uses 25 pounds of chocolate. The cooking and processing take 120 minutes for fudge, 150 minutes for chocolate cremes, and 200 minutes for pralines. The packaging times measured in minutes per 1-pound box are 1, 2, 3, respectively, for fudge, cremes, and pralines. Determine how many batches of each type of candy the confectionery should make, assuming that the profit per pound box is 50¢ on fudge, 40¢ on chocolate cremes, and 45¢ on pralines. Also, find the maximum profit for the week.

11. Finance   A political party is planning its fund-raising activities for a coming election. They plan to raise money through large fund-raising parties, letters requesting funds, and dinner parties where people can meet the candidate personally. Each large fund-raising party costs $3000, each mailing costs $1000, and each dinner party costs $12,000. They can spend up to $102,000 for these activities. From experience they know that each large party will raise $200,000, each letter campaign will raise $100,000. and each dinner party will raise $600,000. They are able to carry out as many as 25 of these activities.

   (a) How many of each should they plan in order to raise the maximum amount of money? What is the maximum amount?

   (b) Dinner parties are more expensive than letter campaigns, yet the optimum solution found in part (a) includes dinner parties but no letter campaigns. Explain how this is possible.

12. Business   A baker has 60 units of flour, 132 units of sugar, and 102 units of raisins. A loaf of raisin bread requires 1 unit of flour, 1 unit of sugar, and 2 units of

raisins, while a raisin cake needs 2, 4, and 1 units, respectively. If raisin bread sells for $3 a loaf and a raisin cake for $4, how many of each should be baked so that the gross income is maximized? What is the maximum gross income?

Business *The next two problems come from past CPA examinations.* Select the appropriate answer for each question.*

13. The Ball Company manufactures three types of lamps, labeled A, B, and C. Each lamp is processed in two departments, I and II. Total available man-hours per day for departments I and II are 400 and 600, respectively. No additional labor is available. Time requirements and profit per unit for each lamp type are as follows.

	A	B	C
Man-Hours in I	2	3	1
Man-Hours in II	4	2	3
Profit per Unit	$5	$4	$3

The company has assigned you as the accounting member of its profit planning committee to determine the numbers of types of A, B, and C lamps that it should produce in order to maximize its total profit from the sale of lamps. The following questions relate to a linear programming model that your group has developed.

(a) The coefficients of the objective function would be
   (1) 4, 2, 3;
   (2) 2, 3, 1;
   (3) 5, 4, 3;
   (4) 400, 600.

(b) The constraints in the model would be
   (1) 2, 3, 1;
   (2) 5, 4, 3;
   (3) 4, 2, 3;
   (4) 400, 600.

(c) The constraint imposed by the available man-hours in department I could be expressed as
   (1) $4X_1 + 2X_2 + 3X_3 \leq 400$;
   (2) $4X_1 + 2X_2 + 3X_3 \geq 400$;
   (3) $2X_1 + 3X_2 + 1X_3 \leq 400$;
   (4) $2X_1 + 3X_2 + 1X_3 \geq 400$.

---

*Material from *Uniform CPA Examination Questions and Unofficial Answers,* copyright © 1973, 1974, 1975 by the American Institute of Certified Public Accountants, Inc., is reprinted with permission.

14. The Golden Hawk Manufacturing Company wants to maximize the profits on products A, B, and C. The contribution margin for each product follows.

Product	Contribution Margin
A	$2
B	5
C	4

The production requirements and departmental capacities, by departments, are as follows.

	PRODUCTION REQUIREMENTS BY PRODUCT (HOURS)			DEPARTMENTAL CAPACITY (TOTAL HOURS)
Department	A	B	C	
Assembling	2	3	2	30,000
Painting	1	2	2	38,000
Finishing	2	3	1	28,000

(a) What is the profit-maximization formula for the Golden Hawk Company?
   (1) $2A + $5B + $4C = X   (where X = profit)
   (2) 5A + 8B + 5C ≤ 96,000
   (3) $2A + $5B + $4C ≤ X
   (4) $2A + $5B + $4C = 96,000

(b) What is the constraint for the Painting Department of the Golden Hawk Company?
   (1) 1A + 2B + 2C ≥ 38,000
   (2) $2A + $5B + $4C ≥ 38,000
   (3) 1A + 2B + 2C ≤ 38,000
   (4) 2A + 3B + 2C ≤ 30,000

15. Solve the problem in Exercise 1.

 Use a graphing calculator or a computer program for the simplex method to solve the following linear programming problems.

16. Exercise 2. Your final answer should consist of whole numbers (Banal can't sell half a painting).

17. Exercise 3

18. Exercise 4

7.6 THE SIMPLEX METHOD: DUALITY AND MINIMIZATION

In this section the simplex method is extended to linear programming problems satisfying the following conditions.

1. The objective function is to be *minimized*.
2. All the coefficients of the objective function are nonnegative.
3. All constraints involve $\geq$.
4. All variables are nonnegative.

The method of solving minimization problems presented here is based on an interesting connection between maximizing and minimizing problems: any solution of a maximizing problem produces the solution of an associated minimizing problem, or vice-versa. Each of the associated problems is called the **dual** of the other. Thus, duals enable us to solve minimization problems of the type described above by the simplex method introduced in Section 7.4. (An alternative approach for solving minimization problems is given in the next section.)

When dealing with minimization problems, we use y_1, y_2, y_3, etc., as variables and denote the objective function by w. An example will explain the idea of a dual.

▶**EXAMPLE 1**   Minimize $w = 8y_1 + 16y_2$

subject to: $y_1 + 5y_2 \geq 9$

$2y_1 + 2y_2 \geq 10$

$y_1 \geq 0, y_2 \geq 0.$

Before considering slack variables, write the augmented matrix of the system of inequalities and include the coefficients of the objective function (not their negatives) as the last row in the matrix.

$$\begin{array}{r} \\ \\ \text{Objective function} \longrightarrow \end{array} \overset{\overset{\displaystyle \text{Constants}}{\big\downarrow}}{\left[\begin{array}{cc|c} 1 & 5 & 9 \\ 2 & 2 & 10 \\ \hline 8 & 16 & 0 \end{array}\right]}$$

Look now at the following new matrix, obtained from the one above by interchanging rows and columns.

$$\begin{array}{r} \\ \\ \text{Objective function} \longrightarrow \end{array} \overset{\overset{\displaystyle \text{Constants}}{\big\downarrow}}{\left[\begin{array}{cc|c} 1 & 2 & 8 \\ 5 & 2 & 16 \\ \hline 9 & 10 & 0 \end{array}\right]}$$

The *rows* of the first matrix (for the minimizing problem) are the *columns* of the second matrix.

The entries in this second matrix could be used to write the following maximizing problem in standard form (again ignoring the fact that the numbers in the last row are not negative).

Use the corner points in Figure 7.24(a) to find the minimum value of $w = 8y_1 + 16y_2$ and where it occurs.

Answer:

48 when $y_1 = 4$, $y_2 = 1$

Maximize $z = 9x_1 + 10x_2$
subject to: $x_1 + 2x_2 \leq 8$
$5x_1 + 2x_2 \leq 16$
$x_1 \geq 0, x_2 \geq 0.$

Figure 7.24(a) shows the region of feasible solutions for the minimization problem given above, while Figure 7.24(b) shows the region of feasible solutions for the maximization problem produced by exchanging rows and columns. ◀ ① ②

Use Figure 7.24(b) to find the maximum value of $z = 9x_1 + 10x_2$ and where it occurs.

Answer:

48 when $x_1 = 2$, $x_2 = 3$

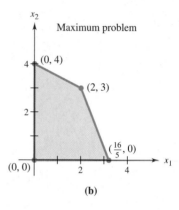

FIGURE 7.24

The two feasible regions in Figure 7.24 are different and the corner points are different, but the values of the objective functions found in Problems 1 and 2 at the side are equal—both are 48. An even closer connection between the two problems is shown by using the simplex method to solve the maximization problem given above.

Maximization problem

$$\begin{array}{cccc} x_1 & x_2 & x_3 & x_4 \end{array}$$
$$\left[\begin{array}{cccc|c} 1 & \mathbf{2} & 1 & 0 & 8 \\ 5 & 2 & 0 & 1 & 16 \\ \hline -9 & -10 & 0 & 0 & 0 \end{array}\right]$$

$$\begin{array}{cccc} x_1 & x_2 & x_3 & x_4 \end{array}$$
$$\left[\begin{array}{cccc|c} \frac{1}{2} & 1 & \frac{1}{2} & 0 & 4 \\ \mathbf{4} & 0 & -1 & 1 & 8 \\ \hline -4 & 0 & 5 & 0 & 40 \end{array}\right] \begin{array}{l} \frac{1}{2}R_1 \\ \\ -2R_1 + R_2 \\ \\ 10R_1 + R_3 \end{array}$$

$$\begin{array}{cccc} x_1 & x_2 & x_3 & x_4 \end{array}$$
$$\left[\begin{array}{cccc|c} 0 & 1 & \frac{5}{8} & -\frac{1}{8} & 3 \\ 1 & 0 & -\frac{1}{4} & \frac{1}{4} & 2 \\ \hline 0 & 0 & 4 & 1 & 48 \end{array}\right] \begin{array}{l} -\frac{1}{2}R_2 + R_1 \\ \\ \frac{1}{4}R_2 \\ \\ 4R_2 + R_3 \end{array}$$

The maximum is 48 when $x_1 = 2$, $x_2 = 3$.

Notice that the solution to the *minimization problem* (namely, $y_1 = 4$, $y_2 = 1$) is found in the bottom row and slack variable columns of the final simplex tableau for the maximization problem. This result suggests that a minimization problem can be solved by forming the dual maximization problem, solving it by the simplex method, and then reading the solution for the minimization problem from the bottom row of the final simplex tableau.

Before using this method to solve a minimization problem, let us find the duals of some typical linear programming problems. The process of exchanging the rows and columns of a matrix, which is used to find the dual, is called **transposing** the matrix, and each of the two matrices is the **transpose** of the other.

▶ **EXAMPLE 2**   Find the transpose of each matrix.

(a) $A = \begin{bmatrix} 2 & -1 & 5 \\ 6 & 8 & 0 \\ -3 & 7 & -1 \end{bmatrix}$

Write the rows of matrix A as the columns of the transpose.

$$\text{Transpose of } A = \begin{bmatrix} 2 & 6 & -3 \\ -1 & 8 & 7 \\ 5 & 0 & -1 \end{bmatrix}$$

(b) The transpose of $\begin{bmatrix} 1 & 2 & 4 & 0 \\ 2 & 1 & 7 & 6 \end{bmatrix}$ is $\begin{bmatrix} 1 & 2 \\ 2 & 1 \\ 4 & 7 \\ 0 & 6 \end{bmatrix}$. ◀ ③

TECHNOLOGY TIP Most graphing calculators can find the transpose of a matrix. Look for this feature in the MATRIX MATH menu (TI) or the OPTN MAT menu (Casio) or the MATH MATRIX menu (HP-39). The transpose of matrix A from Example 2(a) is shown in Figure 7.25.

FIGURE 7.25

▶ **EXAMPLE 3**   Write the duals of the following minimization linear programming problems.

(a)   Minimize $w = 10y_1 + 8y_2$

   subject to: $y_1 + 2y_2 \geq 2$

   $y_1 + y_2 \geq 5$

   $y_1 \geq 0,\ y_2 \geq 0.$

---

③

Give the transpose of each matrix.

(a) $\begin{bmatrix} 2 & 4 \\ 6 & 3 \\ 1 & 5 \end{bmatrix}$

(b) $\begin{bmatrix} 4 & 7 & 10 \\ 3 & 2 & 6 \\ 5 & 8 & 12 \end{bmatrix}$

Answers:

(a) $\begin{bmatrix} 2 & 6 & 1 \\ 4 & 3 & 5 \end{bmatrix}$

(b) $\begin{bmatrix} 4 & 3 & 5 \\ 7 & 2 & 8 \\ 10 & 6 & 12 \end{bmatrix}$

Begin by writing the augmented matrix for the given problem.

$$\left[\begin{array}{cc|c} 1 & 2 & 2 \\ 1 & 1 & 5 \\ \hline 10 & 8 & 0 \end{array}\right]$$

Form the transpose of this matrix to get

$$\left[\begin{array}{cc|c} 1 & 1 & 10 \\ 2 & 1 & 8 \\ \hline 2 & 5 & 0 \end{array}\right].$$

④

The dual problem is stated from this second matrix as follows (using x instead of y).

$$\text{Maximize} \quad z = 2x_1 + 5x_2$$
$$\text{subject to:} \quad x_1 + x_2 \le 10$$
$$2x_1 + x_2 \le 8$$
$$x_1 \ge 0, x_2 \ge 0.$$

(b) Minimize $w = 7y_1 + 5y_2 + 8y_3$
subject to: $\quad 3y_1 + 2y_2 + y_3 \ge 10$
$$y_1 + y_2 + y_3 \ge 8$$
$$4y_1 + 5y_2 \quad \ge 25$$
$$y_1 \ge 0, y_2 \ge 0, y_3 \ge 0.$$

The dual problem is stated as follows.

$$\text{Maximize} \quad z = 10x_1 + 8x_2 + 25x_3$$
$$\text{subject to:} \quad 3x_1 + x_2 + 4x_3 \le 7$$
$$2x_1 + x_2 + 5x_3 \le 5$$
$$x_1 + x_2 \quad \le 8$$
$$x_1 \ge 0, x_2 \ge 0, x_3 \ge 0. \quad ◄ \; ④$$

In Example 3, all the constraints of the minimization problems were $\ge$ inequalities, while all those in the dual maximization problems were $\le$ inequalities. This is generally the case; inequalities are reversed when the dual problem is stated.

The following table shows the close connection between a problem and its dual.

> **4**
>
> Write the dual of the following linear programming problem.
> Minimize
> $w = 2y_1 + 5y_2 + 6y_3$
> subject to:
> $$2y_1 + 3y_2 + y_3 \ge 15$$
> $$y_1 + y_2 + 2y_3 \ge 12$$
> $$5y_1 + 3y_2 \quad \ge 10$$
> $$y_1 \ge 0, y_2 \ge 0, y_3 \ge 0.$$
>
> **Answer:**
>
> Maximize
> $$z = 15x_1 + 12x_2 + 10x_3$$
> subject to:
> $$2x_1 + x_2 + 5x_3 \le 2$$
> $$3x_1 + x_2 + 3x_3 \le 5$$
> $$x_1 + 2x_2 \quad \le 6$$
> $$x_1 \ge 0, x_2 \ge 0, x_3 \ge 0.$$

Given Problem	Dual Problem
m variables	n variables
n constraints	m constraints (m slack variables)
Coefficients from objective function	Constraint constants
Constraint constants	Coefficients from objective function

The next theorem, whose proof requires advanced methods, guarantees that a minimization problem can be solved by forming a dual maximization problem.

Theorem of Duality

The objective function w of a minimizing linear programming problem takes on a minimum value if and only if the objective function z of the corresponding dual maximizing problem takes on a maximum value. The maximum value of z equals the minimum value of w.

This method is illustrated in the following example.

▶ **EXAMPLE 4**   Minimize $w = 3y_1 + 2y_2$

subject to: $y_1 + 3y_2 \geq 6$

$$2y_1 + y_2 \geq 3$$

$$y_1 \geq 0, \; y_2 \geq 0.$$

Use the given information to write the matrix.

$$\begin{bmatrix} 1 & 3 & | & 6 \\ 2 & 1 & | & 3 \\ \hline 3 & 2 & | & 0 \end{bmatrix}$$

Transpose to get the following matrix for the dual problem.

$$\begin{bmatrix} 1 & 2 & | & 3 \\ 3 & 1 & | & 2 \\ \hline 6 & 3 & | & 0 \end{bmatrix}$$

Write the dual problem from this matrix, as follows.

Maximize $z = 6x_1 + 3x_2$

subject to: $x_1 + 2x_2 \leq 3$

$$3x_1 + x_2 \leq 2$$

$$x_1 \geq 0, x_2 \geq 0.$$

Solve this standard maximization problem using the simplex method. Start by introducing slack variables to give the system

$$x_1 + 2x_2 + x_3 \qquad\qquad = 3$$

$$3x_1 + x_2 \qquad + x_4 \qquad = 2$$

$$-6x_1 - 3x_2 - 0x_3 - 0x_4 + z = 0$$

with $x_1 \geq 0, x_2 \geq 0, x_3 \geq 0, x_4 \geq 0$.

The initial tableau for this system is given below with the pivot as indicated.

$$
\begin{array}{cccc}
x_1 & x_2 & x_3 & x_4 \\
\end{array}
$$

					Quotients
1	2	1	0	3	3/1 = 3
3	1	0	1	2	2/3
−6	−3	0	0	0	

The simplex method gives the following final tableau.

$$
\begin{array}{cccc}
x_1 & x_2 & x_3 & x_4 \\
\end{array}
$$

$$
\left[
\begin{array}{cccc|c}
0 & 1 & \frac{3}{5} & -\frac{1}{5} & \frac{7}{5} \\
1 & 0 & -\frac{1}{5} & \frac{2}{5} & \frac{1}{5} \\
\hline
0 & 0 & \frac{3}{5} & \frac{9}{5} & \frac{27}{5}
\end{array}
\right]
$$

$$
 y_1 y_2 w
$$

This tableau shows that the solution of the dual maximization problem is $x_1 = 1/5$, $x_2 = 7/5$ and the maximum value of z is $27/5$. But that is *not* the answer to the original *minimization* problem. To find the minimization answer, we must read the final tableau differently (as suggested in the discussion after Example 1).

As indicated in blue below the final tableau, the last entries in the columns corresponding to the slack variables (x_3 and x_4) give the values of the original variables y_1 and y_2 that produce the minimal value of w. This minimal value of w appears in the lower right-hand corner (and is the same as the maximal value of z in the dual problem). So the solution of the given minimization problems is as follows.

The minimum value of $w = 3y_1 + 2y_2$, subject to the given constraints, is $27/5$ and occurs when $y_1 = 3/5$ and $y_2 = 9/5$. ◀ ⑤

A minimizing problem that meets the conditions listed at the beginning of the section can be solved by the method of duals, as illustrated in Examples 1–4 and summarized here.

Minimize $w = 10y_1 + 8y_2$ subject to:

$$y_1 + 2y_2 \geq 2$$
$$y_1 + y_2 \geq 5$$
$$y_1 \geq 0, y_2 \geq 0.$$

Answer:

$y_1 = 0, y_2 = 5$, for a minimum of 40

Solving Minimization Problems with Duals

1. Find the dual standard maximum problem.*

2. Use the simplex method to solve the dual maximum problem.

3. Read the optimal solution of the original minimization problem from the final simplex tableau:

y_1 is the last entry in the column corresponding to the first slack variable;

y_2 is the last entry in the column corresponding to the second slack variable; and so on.

These values of y_1, y_2, y_3, etc., produce the minimum value of w, which is the entry in the lower right-hand corner of the tableau.

▶ **EXAMPLE 5** A minimization problem in three variables was solved by using duals. The final simplex tableau for the dual maximization problem is shown here.

$$
\begin{array}{cccccc}
x_1 & x_2 & x_3 & x_4 & x_5 & x_6 \\
\end{array}
$$

$$
\left[
\begin{array}{cccccc|c}
3 & 1 & 1 & 0 & 9 & 0 & 1 \\
13 & -1 & 0 & 1 & -2 & 0 & 10 \\
9 & 10 & 0 & 0 & 7 & 1 & 7 \\
\hline
5 & 1 & 0 & 4 & 1 & 7 & 28
\end{array}
\right]
$$

---

*The coefficients of the objective function in the minimization problem are the constants on the right side of the constraints in the dual maximization problem. So when all these coefficients are nonnegative (condition 2), the dual problem is in standard maximum form.

(a) What is the optimal solution of the minimization problem?

Since there are three variables in the minimization problem, there must be three slack variables in the dual maximization problem, namely, x_4, x_5, and x_6. Looking at the bottom of the columns corresponding to these slack variables, we see that the solution of the minimization problem is

$$y_1 = 4, \qquad y_2 = 1, \qquad y_3 = 7, \quad \text{with a minimal value of } w = 28.$$

(b) What is the optimal solution of the dual maximization problem?

Since x_4, x_5, and x_6 are slack variables by part (a), the variables in the dual problem are x_1, x_2, and x_3. Read the solution from the final tableau, as in Sections 7.4 and 7.5:

$$x_1 = 0, \qquad x_2 = 0 \qquad x_3 = 1, \quad \text{with a maximal value of } z = 28. \quad \blacktriangleleft$$

Further Uses of the Dual

The dual is useful not only in solving minimization problems but also in seeing how small changes in one variable will affect the value of the objective function. For example, suppose an animal breeder needs at least 6 units per day of nutrient A and at least 3 units of nutrient B and that the breeder can choose between two different feeds, feed 1 and feed 2. Find the minimum cost for the breeder if each bag of feed 1 costs \$3 and provides 1 unit of nutrient A and 2 units of B, while each bag of feed 2 costs \$2 and provides 3 units of nutrient A and 1 of B.

If y_1 represents the number of bags of feed 1 and y_2 represents the number of bags of feed 2, the given information leads to

$$
\begin{aligned}
\text{Minimize} \quad & w = 3y_1 + 2y_2 \\
\text{subject to:} \quad & y_1 + 3y_2 \geq 6 \\
& 2y_1 + y_2 \geq 3 \\
& y_1 \geq 0, y_2 \geq 0.
\end{aligned}
$$

This minimization linear programming problem is the one we solved in Example 4 of this section. In that example, we formed the dual and reached the following final tableau.

$$
\begin{array}{cccc}
x_1 & x_2 & x_3 & x_4
\end{array}
$$
$$
\left[
\begin{array}{cccc|c}
0 & 1 & \frac{3}{5} & -\frac{1}{5} & \frac{7}{5} \\
1 & 0 & -\frac{1}{5} & \frac{2}{5} & \frac{1}{5} \\
\hline
0 & 0 & \frac{3}{5} & \frac{9}{5} & \frac{27}{5}
\end{array}
\right]
$$

This final tableau shows that the breeder will obtain minimum feed costs by using 3/5 bag of feed 1 and 9/5 bags of feed 2 per day, for a daily cost of $27/5 = 5.40$ dollars.

Now look at the data from the feed problem shown in the table below.

	UNITS OF NUTRIENT (PER BAG)		Cost per Bag
	A	B	
Feed 1	1	2	\$3
Feed 2	3	1	\$2
Minimum Nutrient Needed	6	3	

If x_1 and x_2 are the cost per unit of nutrients A and B, the constraints of the dual problem can be stated as follows.

$$\text{Cost of feed 1:} \quad x_1 + 2x_2 \leq 3$$
$$\text{Cost of feed 2:} \quad 3x_1 + x_2 \leq 2$$

The solution of the dual problem, which maximizes nutrients, also can be read from the final tableau:

$$x_1 = \frac{1}{5} = .20 \quad \text{and} \quad x_2 = \frac{7}{5} = 1.40,$$

which means that a unit of nutrient A costs 1/5 of a dollar = \$.20, while a unit of nutrient B costs 7/5 dollars = \$1.40. The minimum daily cost, \$5.40, is found by the following procedure.

$$(\$.20 \text{ per unit of A}) \times (6 \text{ units of A}) = \$1.20$$
$$\underline{+\ (\$1.40 \text{ per unit of B}) \times (3 \text{ units of B}) = \$4.20}$$
$$\text{Minimum daily cost} = \$5.40$$

The numbers .20 and 1.40 are called the **shadow costs** of the nutrients. These two numbers from the dual, \$.20 and \$1.40, also allow the breeder to estimate feed costs for "small" changes in nutrient requirements. For example, an increase of 1 unit in the requirement for each nutrient would produce total cost as follows.

\$5.40	6 units of A, 3 of B
.20	1 extra unit of A
1.40	1 extra unit of B
\$7.00	Total cost per day

The final tableau of the dual of the problem about filing cabinets in Example 2, Section 7.3, and Example 6, Section 7.4 is given below.

$$
\begin{array}{cccc|c}
x_1 & x_2 & x_3 & x_4 & \\
\hline
0 & 1 & \frac{1}{2} & -\frac{1}{4} & 1 \\
1 & 0 & -\frac{1}{20} & -\frac{1}{80} & \frac{1}{20} \\
0 & 0 & 8 & 3 & 100
\end{array}
$$

(a) What are the imputed amounts of storage for each unit of cost and floor space?

(b) What are the shadow values of the cost and the floor space?

Answers:

(a) Cost: 28 sq ft
floor space: 72 sq ft

(b) $\frac{1}{20}, 1$

7.6 Exercises

Find the transpose of each matrix. (See Example 2.)

1. $\begin{bmatrix} 3 & -4 & 5 \\ 1 & 10 & 7 \\ 0 & 3 & 6 \end{bmatrix}$

2. $\begin{bmatrix} 3 & -5 & 9 & 4 \\ 1 & 6 & -7 & 0 \\ 4 & 18 & 11 & 9 \end{bmatrix}$

3. $\begin{bmatrix} 3 & 0 & 14 & -5 & 3 \\ 4 & 17 & 8 & -6 & 1 \end{bmatrix}$

4. $\begin{bmatrix} 15 & -6 & -2 \\ 13 & -1 & 11 \\ 10 & 12 & -3 \\ 24 & 1 & 0 \end{bmatrix}$

State the dual problem for each of the following, but do not solve it. (See Example 3.)

5. Minimize $w = 3y_1 + 5y_2$
subject to: $3y_1 + y_2 \geq 4$
$-y_1 + 2y_2 \geq 6$
$y_1 \geq 0, y_2 \geq 0.$

6. Minimize $w = 4y_1 + 7y_2$
subject to: $y_1 + y_2 \geq 17$
$3y_1 + 6y_2 \geq 21$
$2y_1 + 4y_2 \geq 19$
$y_1 \geq 0, y_2 \geq 0.$

7. Minimize $w = 2y_1 + 8y_2$
subject to: $y_1 + 7y_2 \geq 18$
$4y_1 + y_2 \geq 15$
$5y_1 + 3y_2 \geq 20$
$y_1 \geq 0, y_2 \geq 0.$

8. Minimize $w = 5y_1 + y_2 + 3y_3$
subject to: $7y_1 + 6y_2 + 8y_3 \geq 18$
$4y_1 + 5y_2 + 10y_3 \geq 20$
$y_1 \geq 0, y_2 \geq 0, y_3 \geq 0.$

9. Minimize $w = y_1 + 2y_2 + 6y_3$
   subject to: $3y_1 + 4y_2 + 6y_3 \geq 8$
   $y_1 + 5y_2 + 2y_3 \geq 12$
   $y_1 \geq 0, y_2 \geq 0, y_3 \geq 0.$

10. Minimize $w = 4y_1 + 3y_2 + y_3$
    subject to: $y_1 + 2y_2 + 3y_3 \geq 115$
    $2y_1 + y_2 + 8y_3 \geq 200$
    $y_1 \qquad - y_3 \geq 50$
    $y_1 \geq 0, y_2 \geq 0, y_3 \geq 0.$

11. Minimize $w = 8y_1 + 9y_2 + 3y_3$
    subject to: $y_1 + y_2 + y_3 \geq 5$
    $y_1 + y_2 \qquad \geq 4$
    $2y_1 + y_2 + 3y_3 \geq 15$
    $y_1 \geq 0, y_2 \geq 0, y_3 \geq 0.$

12. Minimize $w = y_1 + 2y_2 + y_3 + 5y_4$
    subject to: $y_1 + y_2 + y_3 + y_4 \geq 50$
    $3y_1 + y_2 + 2y_3 + y_4 \geq 100$
    $y_1 \geq 0, y_2 \geq 0, y_3 \geq 0, y_4 \geq 0.$

Use duality to solve the following problems. (See Example 4.)

13. Minimize $w = 2y_1 + y_2 + 3y_3$
    subject to: $y_1 + y_2 + y_3 \geq 100$
    $2y_1 + y_2 \qquad \geq 50$
    $y_1 \geq 0, y_2 \geq 0, y_3 \geq 0.$

14. Minimize $w = 2y_1 + 4y_2$
    subject to: $4y_1 + 2y_2 \geq 10$
    $4y_1 + y_2 \geq 8$
    $2y_1 + y_2 \geq 12$
    $y_1 \geq 0, y_2 \geq 0.$

15. Minimize $w = 3y_1 + y_2 + 4y_3$
    subject to: $2y_1 + y_2 + y_3 \geq 6$
    $y_1 + 2y_2 + y_3 \geq 8$
    $2y_1 + y_2 + 2y_3 \geq 12$
    $y_1 \geq 0, y_2 \geq 0, y_3 \geq 0.$

16. Minimize $w = y_1 + y_2 + 3y_3$
    subject to: $2y_1 + 6y_2 + y_3 \geq 8$
    $y_1 + 2y_2 + 4y_3 \geq 12$
    $y_1 \geq 0, y_2 \geq 0, y_3 \geq 0.$

17. Minimize $w = 6y_1 + 4y_2 + 2y_3$
    subject to: $2y_1 + 2y_2 + y_3 \geq 2$
    $y_1 + 3y_2 + 2y_3 \geq 3$
    $y_1 + y_2 + 2y_3 \geq 4$
    $y_1 \geq 0, y_2 \geq 0, y_3 \geq 0.$

18. Minimize $w = 12y_1 + 10y_2 + 7y_3$
    subject to: $2y_1 + y_2 + y_3 \geq 7$
    $y_1 + 2y_2 + y_3 \geq 4$
    $y_1 \geq 0, y_2 \geq 0, y_3 \geq 0.$

19. Minimize $w = 20y_1 + 12y_2 + 40y_3$
    subject to: $y_1 + y_2 + 5y_3 \geq 20$
    $2y_1 + y_2 + y_3 \geq 30$
    $y_1 \geq 0, y_2 \geq 0, y_3 \geq 0.$

20. Minimize $w = 4y_1 + 5y_2$
    subject to: $10y_1 + 5y_2 \geq 100$
    $20y_1 + 10y_2 \geq 150$
    $y_1 \geq 0, y_2 \geq 0.$

21. Minimize $w = 4y_1 + 2y_2 + y_3$
    subject to: $y_1 + y_2 + y_3 \geq 4$
    $3y_1 + y_2 + 3y_3 \geq 6$
    $y_1 + y_2 + 3y_3 \geq 5$
    $y_1 \geq 0, y_2 \geq 0, y_3 \geq 0.$

22. Minimize $w = 3y_1 + 2y_2$
    subject to: $2y_1 + 3y_2 \geq 60$
    $y_1 + 4y_2 \geq 40$
    $y_1 \geq 0, y_2 \geq 0.$

23. **Health** Glenn Russell, who is dieting, requires two food supplements, I and II. He can get these supplements from two different products, A and B, as shown in the following table.

		SUPPLEMENT (GRAMS PER SERVING)	
		I	II
PRODUCT	A	4	2
	B	2	5

Glenn's physician has recommended that he include at least 20 grams of supplement I and 18 grams of supplement II in his diet. If product A costs 24¢ per serving and product B costs 40¢ per serving, how can he satisfy these requirements most economically?

24. **Business** An animal food must provide at least 54 units of vitamins and 60 calories per serving. One gram of soybean meal provides at least 2.5 units of vitamins and 5 calories. One gram of meat by-products provides at least 4.5 units of vitamins and 3 calories. One gram of grain provides at least 5 units of vitamins and 10 calories. If a gram of soybean meal costs 8¢, a gram of meat by-products 9¢, and a gram of grain 10¢, what mixture of these three ingredients will provide the required vitamins and calories at minimum cost?

25. **Business** A brewery produces regular beer and a lower-carbohydrate "light" beer. Steady customers of the brewery buy 12 units of regular beer and 10 units of light beer monthly. While setting up the brewery to produce the beers, the management decides to produce extra beer, beyond the need to satisfy the customers. The cost per unit of regular beer is $36,000 and the cost per unit of light beer is $48,000. Every unit of regular beer brings in $100,000 in revenue, while every unit of light beer brings in $300,000

in revenue. The brewery wants at least $7,000,000 in revenue. At least 20 additional units of beer can be sold. How much of each type of beer should be made so as to minimize total production costs?

26. **Business** Joan McKee has a part-time job conducting public opinion interviews. She has found that a political interview takes 45 minutes and a market interview takes 55 minutes. She needs to minimize the time she spends doing interviews to allow more time for her full-time job. Unfortunately, to keep her part-time job, she must complete at least 8 interviews each week. Also, she must earn at least $60 per week at this job: she earns $8 for each political interview and $10 for each market interview. Finally, to stay in good standing with her supervisor, she must earn at least 40 bonus points per week; she receives 6 bonus points for each political interview and 5 points for each market interview. How many of each interview should she do each week to minimize the time spent?

27. You are given the following linear programming problem (P):*

$$\text{Minimize} \quad z = x_1 + 2x_2$$
$$\text{subject to:} \quad -2x_1 + x_2 \geq 1$$
$$x_1 - 2x_2 \geq 1$$
$$x_1 \geq 0, x_2 \geq 0.$$

The dual of (P) is (D). Which of the statements below is true?

(a) (P) has no feasible solution and the objective function of (D) is unbounded.

(b) (D) has no feasible solution and the objective function of (P) is unbounded.

(c) The objective functions of both (P) and (D) are unbounded.

(d) Both (P) and (D) have optimal solutions.

(e) Neither (P) nor (D) has feasible solutions.

28. **Business** Refer to the end of this section, to the text on minimizing the daily cost of feeds.

---

*Problem 2 from "November 1989 Course 130 Examination Operations Research" of the *Education and Examination Committee of The Society of Actuaries*. Reprinted by permission of The Society of Actuaries.

(a) Find a combination of feeds that will cost $7.00 and give 7 units of A and 4 units of B.

(b) Use the dual variables to predict the daily cost of feed if the requirements change to 5 units of A and 4 units of B. Find a combination of feeds to meet these requirements at the predicted price.

29. **Business** A small toy manufacturing firm has 200 squares of felt, 600 ounces of stuffing, and 90 feet of trim available to make two types of toys, a small bear and a monkey. The bear requires 1 square of felt and 4 ounces of stuffing. The monkey requires 2 squares of felt, 3 ounces of stuffing, and 1 foot of trim. The firm makes $1 profit on each bear and $1.50 profit on each monkey. The linear program to maximize profit is

$$\text{Maximize} \quad z = x_1 + 1.5x_2$$
$$\text{subject to:} \quad x_1 + 2x_2 \leq 200$$
$$4x_1 + 3x_2 \leq 600$$
$$x_2 \leq 90$$
$$x_1 \geq 0, x_2 \geq 0.$$

The final simplex tableau is

$$\begin{bmatrix} 1 & 0 & -.6 & .4 & 0 & | & 120 \\ 0 & 0 & -.8 & .2 & 1 & | & 50 \\ 0 & 1 & .8 & -.2 & 0 & | & 40 \\ 0 & 0 & .6 & .1 & 0 & | & 180 \end{bmatrix}.$$

(a) What is the corresponding dual problem?

(b) What is the optimal solution to the dual problem?

(c) Use the shadow values to estimate the profit the firm will make if their supply of felt increases to 210 squares.

(d) How much profit will the firm make if their supply of stuffing is cut to 590 ounces and their supply of trim is cut to 80 feet?

30. Refer to Example 1 in Section 7.5.

(a) Give the dual problem.

(b) Use the shadow values to estimate the farmer's profit if land is cut to 90 acres but capital increases to $21,000.

(c) Suppose the farmer has 110 acres but only $19,000. Find the optimum profit and the planting strategy that will produce this profit.

7.7 THE SIMPLEX METHOD: NONSTANDARD PROBLEMS

So far the simplex method has been used to solve problems in which the variables are nonnegative and all the other constraints are of one type (either all $\leq$ or all $\geq$). Now we extend the simplex method to linear programming problems with nonnegative variables and mixed constraints ($\leq$, $=$, and $\geq$).

The solution method to be used here requires that all inequality constraints be written so that the constant on the right side is nonnegative. For instance, the inequality

$$4x_1 + 5x_2 - 12x_3 \leq -30$$

can be replaced by the equivalent one obtained by multiplying both sides by -1 and reversing the direction of the inequality sign:

$$-4x_1 - 5x_2 + 12x_3 \geq 30.$$

Maximization with $\leq$ and $\geq$ Constraints

As is always the case when the simplex method is involved, each inequality constraint must be written as an equation. Constraints involving $\leq$ are converted to equations by adding a nonnegative slack variable, as in Section 7.4. Similarly, constraints involving $\geq$ are converted to equations by *subtracting* a nonnegative **surplus variable**. For example, the inequality $2x_1 - x_2 + 5x_3 \geq 12$ is written as

$$2x_1 - x_2 + 5x_3 - x_4 = 12,$$

where $x_4 \geq 0$. The surplus variable x_4 represents the amount by which $2x_1 - x_2 + 5x_3$ exceeds 12.

▶ **EXAMPLE 1** Restate the following problem in terms of equations and write its initial simplex tableau.

$$\text{Maximize} \quad z = 4x_1 + 10x_2 + 6x_3$$
$$\text{subject to:} \quad x_1 + 4x_2 + 4x_3 \geq 8$$
$$x_1 + 3x_2 + 2x_3 \leq 6$$
$$3x_1 + 4x_2 + 8x_3 \leq 22$$
$$x_1 \geq 0, x_2 \geq 0, x_3 \geq 0.$$

In order to write the constraints as equations, subtract a surplus variable from the $\geq$ constraint and add a slack variable to each $\leq$ constraint. So the problem becomes

$$\text{Maximize} \quad z = 4x_1 + 10x_2 + 6x_3$$
$$\text{subject to:} \quad x_1 + 4x_2 + 4x_3 - x_4 \qquad\qquad = 8$$
$$x_1 + 3x_2 + 2x_3 \qquad + x_5 \qquad = 6$$
$$3x_1 + 4x_2 + 8x_3 \qquad\qquad + x_6 = 22$$
$$x_1 \geq 0, x_2 \geq 0, x_3 \geq 0, x_4 \geq 0, x_5 \geq 0, x_6 \geq 0.$$

Write the objective function as $z - 4x_1 - 10x_2 - 6x_3 = 0$ and use the coefficients of the four equations to write the initial simplex tableau (omitting the z column):

$$
\begin{array}{cccccc}
x_1 & x_2 & x_3 & x_4 & x_5 & x_6 \\
\end{array}
$$

$$
\left[
\begin{array}{cccccc|c}
1 & 4 & 4 & -1 & 0 & 0 & 8 \\
1 & 3 & 2 & 0 & 1 & 0 & 6 \\
3 & 4 & 8 & 0 & 0 & 1 & 22 \\
\hline
-4 & -10 & -6 & 0 & 0 & 0 & 0
\end{array}
\right]
$$
◀ ①

The tableau in Example 1 resembles those that have appeared previously, and similar terminology is used. The variables whose columns have one entry ± 1 and the rest 0 will be called **basic variables;** the other variables are nonbasic. A solution obtained by setting the nonbasic variables equal to 0 and solving for the basic variables (by looking at the constants in the right-hand column) will be called a **basic solution.** A basic solution that is feasible is called a **basic feasible solution.** In the tableau of Example 1, for instance, the basic variables are x_4, x_5, and x_6, and the basic solution is:

$$x_1 = 0, \qquad x_2 = 0, \qquad x_3 = 0, \qquad x_4 = -8, \qquad x_5 = 6, \qquad x_6 = 22.$$

However, because one variable is negative, this solution is not feasible. ②

The solution method for problems such as the one in Example 1 consists of two stages. **Stage I** consists of finding a basic *feasible* solution that can be used as the starting point for the simplex method. (This stage is unnecessary in a standard maximization problem because the solution given by the initial tableau is always feasible.) There are many systematic ways of finding a feasible solution, all of which depend on the fact that row operations (such as pivoting) produce a tableau that represents a system with the same solutions as the original one. One such technique is explained in the next example. Since the immediate goal is to find a feasible solution, not necessarily an optimal one, the procedures for choosing pivots differ from those in the ordinary simplex method.

▶ **EXAMPLE 2** Find a basic feasible solution for the problem in Example 1, whose initial tableau is

$$\begin{array}{cccccc} x_1 & x_2 & x_3 & x_4 & x_5 & x_6 \\ \left[\begin{array}{cccccc|c} 1 & 4 & 4 & -1 & 0 & 0 & 8 \\ 1 & 3 & 2 & 0 & 1 & 0 & 6 \\ 3 & 4 & 8 & 0 & 0 & 1 & 22 \\ \hline -4 & -10 & -6 & 0 & 0 & 0 & 0 \end{array}\right] \end{array}.$$

In the basic solution given by this tableau, x_4 has a negative value. The only nonzero entry in its column is the -1 in row one. Choose any *positive* entry in row one except the entry on the far right. The column that the chosen entry is in will be the pivot column. We choose the first positive entry in row one, the 1 in column one. The pivot row is determined in the usual way by considering quotients (constant at the right end of the row divided by the positive entry in the pivot column) in each row except the objective row:

$$8/1 = 8, \qquad 6/1 = 6, \qquad 22/3 = 7\frac{1}{3}.$$

The smallest quotient is 6, so the pivot is the 1 in row two, column one. Pivoting in the usual way leads to this tableau

$$\begin{array}{cccccc} x_1 & x_2 & x_3 & x_4 & x_5 & x_6 \\ \left[\begin{array}{cccccc|c} 0 & 1 & 2 & -1 & -1 & 0 & 2 \\ 1 & 3 & 2 & 0 & 1 & 0 & 6 \\ 0 & -5 & 2 & 0 & -3 & 1 & 4 \\ \hline 0 & 2 & 2 & 0 & 4 & 0 & 24 \end{array}\right] \end{array} \quad \begin{array}{l} -R_2 + R_1 \\ \\ -3R_2 + R_3 \\ 4R_2 + R_4 \end{array}$$

and the basic solution

$$x_1 = 6, \qquad x_2 = 0, \qquad x_3 = 0, \qquad x_4 = -2, \qquad x_5 = 0, \qquad x_6 = 4.$$

Since the basic variable x_4 is negative, this solution is not feasible. So we repeat the pivoting process described above. The x_4 column has a -1 in row one, so we choose a positive entry in that row, namely, the 1 in row one, column two. This choice makes column two the pivot column. The pivot row is determined by the quotients $2/1 = 2$ and $6/3 = 2$ (negative entries in the pivot column and the entry in the objective row are not used). Since there is a tie, we can choose either row one or row two. We choose row one and use the 1 in row one, column 2, as the pivot. Pivoting produces this tableau

$$
\begin{array}{cccccc}
x_1 & x_2 & x_3 & x_4 & x_5 & x_6 \\
\end{array}
$$

$$
\left[
\begin{array}{cccccc|c}
0 & 1 & 2 & -1 & -1 & 0 & 2 \\
1 & 0 & -4 & 3 & 4 & 0 & 0 \\
0 & 0 & 12 & -5 & -8 & 1 & 14 \\
0 & 0 & -2 & 2 & 6 & 0 & 20
\end{array}
\right]
\begin{array}{l}
\\
-3R_1 + R_2 \\
5R_1 + R_3 \\
-2R_1 + R_4
\end{array}
$$

and the basic *feasible* solution

$$x_1 = 0, \qquad x_2 = 2, \qquad x_3 = 0, \qquad x_4 = 0, \qquad x_5 = 0, \qquad x_6 = 14. \blacktriangleleft$$

Once a basic feasible solution has been found, Stage I is ended. The procedures used in Stage I are summarized below.*

---

(3)

The initial tableau of a maximization problem is given below. Use column one as the pivot column for carrying out Stage I and state the basic feasible solution that results.

$$
\begin{array}{cccc}
x_1 & x_2 & x_3 & x_4 \\
\end{array}
$$

$$
\left[
\begin{array}{cccc|c}
1 & 3 & 1 & 0 & 70 \\
2 & 4 & 0 & -1 & 50 \\
-8 & -10 & 0 & 0 & 0
\end{array}
\right]
$$

Answer:

$$
\begin{array}{cccc}
x_1 & x_2 & x_3 & x_4 \\
\end{array}
$$

$$
\left[
\begin{array}{cccc|c}
0 & 1 & 1 & \frac{1}{2} & 45 \\
1 & 2 & 0 & -\frac{1}{2} & 25 \\
0 & 6 & 0 & -4 & 200
\end{array}
\right]
$$

$x_1 = 25, x_2 = 0, x_3 = 45,$
$x_4 = 0.$

Finding a Basic Feasible Solution

1. If any basic variable has a negative value, locate the -1 in that variable's column and note the row it is in.

2. In the row determined in Step 1, choose a positive entry (other than the one at the far right) and note the column it is in. This is the pivot column.

3. Use the positive entries in the pivot column (except in the objective row) to form quotients and select the pivot.

4. Pivot as usual, which results in the pivot column's having one entry 1 and the rest 0's.

5. Repeat Steps 1–4 until every basic variable is nonnegative, so that the basic solution given by the tableau is feasible. If it ever becomes impossible to continue, then the problem has no feasible solution.

One way to make the required choices systematically is to choose the first possibility in each case (going from the top for rows or from the left for columns). However, any choice meeting the required conditions may be used. For maximum efficiency, it is usually best to choose the pivot column in Step 2 so that the pivot is in the same row chosen in Step 1, if this is possible. **(3)**

---

*Except in rare cases that do not occur in this book, this method eventually produces a basic feasible solution or shows that one does not exist. The *two-phase method* using artificial variables, which is discussed in more advanced texts, works in all cases and often is more efficient.

In **Stage II,** the simplex method is applied as usual to the tableau that produced the basic feasible solution in Stage I. Just as in Section 7.4, each round of pivoting replaces the basic feasible solution of one tableau with the basic feasible solution of a new tableau in such a way that the value of the objective function is increased, until an optimal value is obtained (or it becomes clear that no optimal solution exists).

▶ **EXAMPLE 3**   Solve the linear programming problem in Example 1.

A basic feasible solution for this problem was found in Example 2 by using the tableau shown below. However, this solution is not maximal because there is a negative indicator in the objective row. So we use the simplex method: the most negative indicator determines the pivot column and the usual quotients determine that the number 2 in row one, column three, is the pivot.

$$
\begin{array}{cccccc}
x_1 & x_2 & x_3 & x_4 & x_5 & x_6
\end{array}
$$

$$
\left[\begin{array}{cccccc|c}
0 & 1 & \mathbf{2} & -1 & -1 & 0 & 2 \\
1 & 0 & -4 & 3 & 4 & 0 & 0 \\
0 & 0 & 12 & -5 & -8 & 1 & 14 \\
0 & 0 & -2 & 2 & 6 & 0 & 20
\end{array}\right]
$$

Quotients

$2/2 \leftarrow$ Smallest

$14/12$

Most negative indicator

Pivoting leads to the final tableau.

$$
\begin{array}{cccccc}
x_1 & x_2 & x_3 & x_4 & x_5 & x_6
\end{array}
$$

$$
\left[\begin{array}{cccccc|c}
0 & \frac{1}{2} & 1 & -\frac{1}{2} & -\frac{1}{2} & 0 & 1 \\
1 & 0 & -4 & 3 & 4 & 0 & 0 \\
0 & 0 & 12 & -5 & -8 & 1 & 14 \\
0 & 0 & -2 & 2 & 6 & 0 & 20
\end{array}\right]
$$

$\frac{1}{2}R_1$

$$
\begin{array}{cccccc}
x_1 & x_2 & x_3 & x_4 & x_5 & x_6
\end{array}
$$

$$
\left[\begin{array}{cccccc|c}
0 & \frac{1}{2} & 1 & -\frac{1}{2} & -\frac{1}{2} & 0 & 1 \\
1 & 2 & 0 & 1 & 2 & 0 & 4 \\
0 & -6 & 0 & 1 & -2 & 1 & 2 \\
0 & 1 & 0 & 1 & 5 & 0 & 22
\end{array}\right]
$$

$4R_1 + R_2$

$-12R_1 + R_3$

$2R_1 + R_4$

Therefore, the maximum value of z occurs when $x_1 = 4$, $x_2 = 0$, and $x_3 = 1$, in which case $z = 22$. ◀ ④

Complete Stage II and find an optimal solution for side problem 3 on page 399. What is the optimal value of the objective function z?

Answer:

The optimal value $z = 560$ occurs when $x_1 = 70, x_2 = 0,$ $x_3 = 0, x_4 = 90.$

Minimization Problems

The two-stage method for maximization problems illustrated in Examples 1–3 also provides a means of solving minimization problems. To see why, consider this simple fact: when a number t gets smaller, then $-t$ gets larger, and vice versa. For instance, if t goes from 6 down to -8, then $-t$ goes from -6 up to 8. Thus, if w is the objective function of a linear programming problem, the feasible solution that produces the minimum value of w also produces the maximum value of $-w$, and vice versa. Therefore, to solve a minimization problem with objective function w, we need only solve the maximization problem with the same constraints and objective function $z = -w$.

▶ **EXAMPLE 4**   Minimize $w = 2y_1 + y_2 - y_3$
subject to: $-y_1 - y_2 + y_3 \le -4$
$y_1 + 3y_2 + 3y_3 \ge 6$
$y_1 \ge 0, y_2 \ge 0, y_3 \ge 0.$

Make the constant in the first constraint positive by multiplying both sides by -1. Then solve this maximization problem:

Maximize $z = -w = -2y_1 - y_2 + y_3$
subject to: $y_1 + y_2 - y_3 \ge 4$
$y_1 + 3y_2 + 3y_3 \ge 6$
$y_1 \ge 0, y_2 \ge 0, y_3 \ge 0.$

Convert the constraints to equations by subtracting surplus variables and set up the first tableau.

$$
\begin{array}{ccccc}
y_1 & y_2 & y_3 & y_4 & y_5 \\
\end{array}
$$
$$
\left[\begin{array}{ccccc|c}
1 & 1 & -1 & -1 & 0 & 4 \\
1 & 3 & 3 & 0 & -1 & 6 \\
2 & 1 & -1 & 0 & 0 & 0 \\
\end{array}\right]
$$

The basic solution given by this tableau, $y_1 = 0, y_2 = 0, y_3 = 0, y_4 = -4, y_5 = -6$, is not feasible, so the procedures of Stage I must be used to find a basic feasible solution. In the column of the negative basic variable y_4, there is a -1 in row one; we choose the first positive entry in that row, so that column one will be the pivot column. The quotients $4/1 = 4$ and $6/1 = 6$ show that the pivot is the 1 in row one, column one. Pivoting produces this tableau:

$$
\begin{array}{ccccc}
y_1 & y_2 & y_3 & y_4 & y_5 \\
\end{array}
$$
$$
\left[\begin{array}{ccccc|c}
1 & 1 & 1 & 1 & 0 & 4 \\
0 & 2 & 4 & 1 & -1 & 2 \\
0 & -1 & 1 & 2 & 0 & -8 \\
\end{array}\right]. \quad \begin{array}{l} -R_1 + R_2 \\ \\ -2R_1 + R_3 \end{array}
$$

The basic solution $y_1 = 4, y_2 = 0, y_3 = 0, y_4 = 0, y_5 = -2$ is not feasible because y_5 is negative, so we repeat the process. We choose the first positive entry in row two (the row containing the -1 in the y_5 column), which is in column two, so that column two is the pivot column. The relevant quotients are $4/1 = 4$ and $2/2 = 1$, so the pivot is the 2 in row two, column two. Pivoting produces a new tableau.

$$
\begin{array}{ccccc}
y_1 & y_2 & y_3 & y_4 & y_5 \\
\end{array}
$$
$$
\left[\begin{array}{ccccc|c}
1 & 1 & -1 & -1 & 0 & 4 \\
0 & 1 & 2 & \frac{1}{2} & -\frac{1}{2} & 1 \\
0 & -1 & 1 & 2 & 0 & -8 \\
\end{array}\right] \quad \frac{1}{2}R_2
$$

$$
\begin{array}{ccccc}
y_1 & y_2 & y_3 & y_4 & y_5 \\
\end{array}
$$
$$
\left[\begin{array}{ccccc|c}
1 & 0 & -3 & -\frac{3}{2} & \frac{1}{2} & 3 \\
0 & 1 & 2 & \frac{1}{2} & -\frac{1}{2} & 1 \\
0 & 0 & 3 & \frac{5}{2} & -\frac{1}{2} & -7 \\
\end{array}\right] \quad \begin{array}{l} -R_2 + R_1 \\ \\ R_2 + R_3 \end{array}
$$

The basic solution $y_1 = 3, y_2 = 1, y_3 = 0, y_4 = 0, y_5 = 0$ is feasible, so Stage I is complete. However, this solution is not optimal because the objective row contains the negative indicator $-1/2$ in column five. According to the simplex method, column five is the next pivot column. The only positive ratio $3/\frac{1}{2} = 6$ is in row one, so the pivot is $1/2$ in row one, column five. Pivoting produces the final tableau.

$$
\begin{array}{ccccc}
y_1 & y_2 & y_3 & y_4 & y_5 \\
\end{array}
$$

$$
\left[\begin{array}{ccccc|c}
2 & 0 & -6 & -3 & 1 & 6 \\
0 & 1 & 2 & \frac{1}{2} & -\frac{1}{2} & 1 \\
\hline
0 & 0 & 3 & \frac{5}{2} & -\frac{1}{2} & -7 \\
\end{array}\right] \quad 2R_1
$$

$$
\begin{array}{ccccc}
y_1 & y_2 & y_3 & y_4 & y_5 \\
\end{array}
$$

$$
\left[\begin{array}{ccccc|c}
2 & 0 & -6 & -3 & 1 & 6 \\
1 & 1 & -1 & -1 & 0 & 4 \\
\hline
1 & 0 & 0 & 1 & 0 & -4 \\
\end{array}\right] \quad \begin{array}{c} \frac{1}{2}R_1 + R_2 \\ \frac{1}{2}R_1 + R_3 \end{array}
$$

Since there are no negative indicators, the solution given by this tableau ($y_1 = 0$, $y_2 = 4$, $y_3 = 0$, $y_4 = 0$, $y_5 = 6$) is optimal. The maximum value of $z = -w$ is -4. Therefore, the minimum value of the original objective function w is $-(-4) = 4$, which occurs when $y_1 = 0$, $y_2 = 4$, $y_3 = 0$. ◀ ⑤

Minimize $w = 2y_1 + 3y_2$
subject to:

$$y_1 + y_2 \geq 10$$
$$2y_1 + y_2 \geq 16$$
$$y_1 \geq 0, y_2 \geq 0.$$

Answer:

$$y_1 = 10, y_2 = 0; w = 20$$

Equation Constraints

Recall that for any real numbers a and b,

$$a = b \quad \text{exactly when } a \geq b \text{ and simultaneously } a \leq b.$$

Thus, an equation such as $y_1 + 3y_2 + 3y_3 = 6$ is equivalent to this pair of inequalities:

$$y_1 + 3y_2 + 3y_3 \geq 6$$
$$y_1 + 3y_2 + 3y_3 \leq 6.$$

In a linear programming problem, each equation constraint should be replaced in this way by a pair of inequality constraints. Then the problem can be solved by the two-stage method.

▶ **EXAMPLE 5**   Minimize $w = 2y_1 + y_2 - y_3$
subject to: $-y_1 - y_2 + y_3 \leq -4$
$y_1 + 3y_2 + 3y_3 = 6$
$y_1 \geq 0, y_2 \geq 0, y_3 \geq 0.$

Multiply the first inequality by -1 and replace the equation by an equivalent pair of inequalities, as explained above, to obtain this problem:

Maximize $z = -w = -2y_1 - y_2 + y_3$
subject to: $y_1 + y_2 - y_3 \geq 4$
$y_1 + 3y_2 + 3y_3 \geq 6$
$y_1 + 3y_2 + 3y_3 \leq 6$
$y_1 \geq 0, y_2 \geq 0, y_3 \geq 0.$

Convert the constraints to equations by subtracting surplus variables y_4 and y_5 from the first two and adding a slack variable y_6 to the third. Then the first tableau is

$$
\begin{array}{cccccc}
y_1 & y_2 & y_3 & y_4 & y_5 & y_6 \\
\end{array}
$$
$$
\left[\begin{array}{cccccc|c}
1 & 1 & -1 & -1 & 0 & 0 & 4 \\
1 & 3 & 3 & 0 & -1 & 0 & 6 \\
1 & 3 & 3 & 0 & 0 & 1 & 6 \\
\hline
2 & 1 & -1 & 0 & 0 & 0 & 0
\end{array}\right]
$$

The basic solution given by this tableau is $y_1 = 0$, $y_2 = 0$, $y_3 = 0$, $y_4 = -4$, $y_5 = -6$, $y_6 = 6$, which is not feasible. So we carry out the procedures of Stage I.

$$
\begin{array}{cccccc}
y_1 & y_2 & y_3 & y_4 & y_5 & y_6 \\
\end{array}
$$
$$
\left[\begin{array}{cccccc|c}
1 & 1 & -1 & -1 & 0 & 0 & 4 \\
0 & 2 & 4 & 1 & -1 & 0 & 2 \\
0 & 2 & 4 & 1 & 0 & 1 & 2 \\
\hline
0 & -1 & 1 & 2 & 0 & 0 & -8
\end{array}\right]
$$

$$
\begin{array}{cccccc}
y_1 & y_2 & y_3 & y_4 & y_5 & y_6 \\
\end{array}
$$
$$
\left[\begin{array}{cccccc|c}
1 & 0 & -3 & -1.5 & .5 & 0 & 3 \\
0 & 1 & 2 & .5 & -.5 & 0 & 1 \\
0 & 0 & 0 & 0 & 1 & 1 & 0 \\
\hline
0 & 0 & 3 & 2.5 & -.5 & 0 & -7
\end{array}\right]
$$

This tableau gives the basic feasible solution $y_1 = 3$, $y_2 = 1$, $y_3 = 0$, $y_4 = 0$, $y_5 = 0$, $y_6 = 0$, so Stage I is complete. Now apply the simplex method. One round of pivoting produces the final tableau.

$$
\begin{array}{cccccc}
y_1 & y_2 & y_3 & y_4 & y_5 & y_6 \\
\end{array}
$$
$$
\left[\begin{array}{cccccc|c}
1 & 0 & -3 & -1.5 & 0 & -.5 & 3 \\
0 & 1 & 2 & .5 & 0 & .5 & 1 \\
0 & 0 & 0 & 0 & 1 & 1 & 0 \\
\hline
0 & 0 & 3 & 2.5 & 0 & .5 & -7
\end{array}\right]
$$

Therefore, the minimum value of $w = -z$ is $w = -(-7) = 7$, which occurs when $y_1 = 3$, $y_2 = 1$, and $y_3 = 0$. ◀

You may have noticed that Example 5 is just Example 4, with the last inequality constraint replaced by an equation constraint. Note, however, that the optimal solutions are different in the two examples. The minimal value of w found in Example 4 is smaller than the one found in Example 5, but does not satisfy the equation constraint in Example 5.

The two-stage method used in Examples 1–5 is summarized here.

Solving Nonstandard Problems

1. Replace each equation constraint by an equivalent pair of inequality constraints.

2. If necessary, write each constraint with a positive constant.

continued

3. Convert a minimization problem to maximization problem by letting $z = -w$.
4. Add slack variables and subtract surplus variables as needed to convert the constraints into equations.
5. Write the initial simplex tableau.
6. Find a basic feasible solution for the problem, if one exists (Stage I).
7. When a basic feasible solution is found, use the simplex method to solve the problem (Stage II).

NOTE It may happen that the tableau that gives the basic feasible solution in Stage I has no negative indicators in its last row. In this case, the solution found is already optimal and Stage II is not necessary.

Applications

Many real-world applications of linear programming involve mixed constraints. Since they typically involve a large number of variables and constraints, technology is normally required to solve such problems.

▶**EXAMPLE 6**   A college textbook publisher has received orders from two colleges, C_1 and C_2. C_1 needs 500 books, and C_2 needs 1000. The publisher can supply the books from either of two warehouses. Warehouse W_1 has 900 books available and warehouse W_2 has 700. The costs to ship a book from each warehouse to each college are given below.

		TO	
		C_1	C_2
FROM	W_1	\$1.20	\$1.80
	W_2	\$2.10	\$1.50

How many books should be sent from each warehouse to each college to minimize the shipping costs?
To begin, let

$$y_1 = \text{the number of books shipped from } W_1 \text{ to } C_1;$$
$$y_2 = \text{the number of books shipped from } W_2 \text{ to } C_1;$$
$$y_3 = \text{the number of books shipped from } W_1 \text{ to } C_2;$$
$$y_4 = \text{the number of books shipped from } W_2 \text{ to } C_2.$$

C_1 needs 500 books, so $y_1 + y_2 = 500$, which is equivalent to this pair of inequalities:

$$y_1 + y_2 \geq 500$$
$$y_1 + y_2 \leq 500.$$

Similarly, $y_3 + y_4 = 1000$, which is equivalent to

$$y_3 + y_4 \geq 1000$$
$$y_3 + y_4 \leq 1000.$$

Since W_1 has 900 books availble and W_2 has 700 vailable,

$$y_1 + y_3 \leq 900 \quad \text{and} \quad y_2 + y_4 \leq 700.$$

The company wants to minimize shipping costs, so the objective function is

$$w = 1.20y_1 + 2.10y_2 + 1.80y_3 + 1.50y_4.$$

Now write the problem as a system of linear equations, adding slack or surplus variables as needed, and let $z = -w$.

$$
\begin{aligned}
y_1 + \ y_2 \qquad\qquad\qquad - y_5 \qquad\qquad\qquad\quad &= 500 \\
y_1 + \ y_2 + \qquad\qquad\quad y_6 \qquad\qquad\quad &= 500 \\
y_3 + \ y_4 \quad - y_7 \qquad\qquad\quad &= 1000 \\
y_3 + \ y_4 + \quad y_8 \qquad\quad &= 1000 \\
y_1 + \qquad\quad y_3 + \qquad\qquad\quad y_9 \qquad &= 900 \\
y_2 + \qquad\quad y_4 + \qquad\qquad\qquad y_{10} &= 700 \\
1.20y_1 + 2.10y_2 + 1.80y_3 + 1.50y_4 \qquad\qquad\qquad + z &= 0
\end{aligned}
$$

Set up the initial simplex tableau.

y_1	y_2	y_3	y_4	y_5	y_6	y_7	y_8	y_9	y_{10}	
1	1	0	0	-1	0	0	0	0	0	500
1	1	0	0	0	1	0	0	0	0	500
0	0	1	1	0	0	-1	0	0	0	1000
0	0	1	1	0	0	0	1	0	0	1000
1	0	1	0	0	0	0	0	1	0	900
0	1	0	1	0	0	0	0	0	1	700
1.20	2.10	1.80	1.50	0	0	0	0	0	0	0

The basic solution here is not feasible because $y_5 = -500$ and $y_7 = -1000$. Stages I and II could be done by hand here, but because of the large size of the matrix, it's more efficient to use technology such, as the program in the Graphing Calculator Appendix. Stage I takes four rounds of pivoting and produces the feasible solution in Figure 7.26. Because of the small size of a calculator screen, you must scroll to the right to see the entire matrix.

FIGURE 7.26

Now Stage II begins. Two rounds of pivoting produce the final tableau (Figure 7.27).

FIGURE 7.27

The optimal solution is $y_1 = 500$, $y_2 = 0$, $y_3 = 300$, $y_4 = 700$, which results in a minimum shipping cost of \$2190 (remember that the optimal value for the original minimization problem is the negative of the optimal value of the associated maximization problem). ◀

7.7 Exercises

In Exercises 1–4, (a) restate the problem in terms of equations by introducing slack and surplus variables; (b) write the initial simplex tableau. (See Example 1.)

1. Maximize $z = 5x_1 + 2x_2 - x_3$
subject to: $2x_1 + 3x_2 + 5x_3 \geq 8$
$4x_1 - x_2 + 3x_3 \leq 7$
$x_1 \geq 0, x_2 \geq 0, x_3 \geq 0.$

2. Maximize $z = x_1 + 4x_2 + 6x_3$
subject to: $5x_1 + 8x_2 - 5x_3 \leq 10$
$6x_1 + 2x_2 + 3x_3 \geq 7$
$x_1 \geq 0, x_2 \geq 0, x_3 \geq 0.$

3. Maximize $z = 2x_1 - 3x_2 + 4x_3$
subject to: $x_1 + x_2 + x_3 \leq 100$
$x_1 + x_2 + x_3 \geq 75$
$x_1 + x_2 \geq 27$
$x_1 \geq 0, x_2 \geq 0, x_3 \geq 0.$

4. Maximize $z = -x_1 + 5x_2 + x_3$
subject to: $2x_1 + x_3 \leq 40$
$x_1 + x_2 \geq 18$
$x_1 + x_3 = 20$
$x_1 \geq 0, x_2 \geq 0, x_3 \geq 0.$

Convert Exercises 5–8 into maximization problems with positive constants on the right side of each constraint and write the initial simplex tableau. (See Examples 4 and 5.)

5. Minimize $w = 2y_1 + 5y_2 - 3y_3$
subject to: $y_1 + 2y_2 + 3y_3 \geq 115$
$2y_1 + y_2 + y_3 \leq 200$
$y_1 + y_3 \geq 50$
$y_1 \geq 0, y_2 \geq 0, y_3 \geq 0.$

6. Minimize $w = 7y_1 + 6y_2 + y_3$
subject to: $y_1 + y_2 + y_3 \geq 5$
$-y_1 + y_2 \leq -4$
$2y_1 + y_2 + 3y_3 \geq 15$
$y_1 \geq 0, y_2 \geq 0, y_3 \geq 0.$

7. Minimize $w = y_1 - 4y_2 + 2y_3$
subject to: $-7y_1 + 6y_2 - 8y_3 \leq -18$
$4y_1 + 5y_2 + 10y_3 \geq 20$
$y_1 \geq 0, y_2 \geq 0, y_3 \geq 0.$

8. Minimize $w = y_1 + 2y_2 + y_3 + 5y_4$
subject to: $-y_1 + y_2 + y_3 + y_4 \leq -50$
$3y_1 + y_2 + 2y_3 + y_4 = 100$
$y_1 \geq 0, y_2 \geq 0, y_3 \geq 0, y_4 \geq 0.$

Use the two-stage method to solve Exercises 9–20. (See Examples 1–5.)

9. Maximize $z = 12x_1 + 10x_2$
subject to: $x_1 + 2x_2 \geq 24$
$x_1 + x_2 \leq 40$
$x_1 \geq 0, x_2 \geq 0.$

10. Find $x_1 \geq 0$, $x_2 \geq 0$, and $x_3 \geq 0$ such that
$x_1 + x_2 + x_3 \leq 150$
$x_1 + x_2 + x_3 \geq 100$
and $z = 2x_1 + 5x_2 + 3x_3$ is maximized.

11. Find $x_1 \geq 0$, $x_2 \geq 0$, and $x_3 \geq 0$ such that
$x_1 + x_2 + 2x_3 \leq 38$
$2x_1 + x_2 + x_3 \geq 24$
and $z = 3x_1 + 2x_2 + 2x_3$ is maximized.

12. Maximize $z = 6x_1 + 8x_2$
subject to: $3x_1 + 12x_2 \geq 48$
$2x_1 + 4x_2 \leq 60$
$x_1 \geq 0, x_2 \geq 0.$

13. Find $x_1 \geq 0$ and $x_2 \geq 0$ such that
$$x_1 + 2x_2 \leq 18$$
$$x_1 + 3x_2 \geq 12$$
$$2x_1 + 2x_2 \leq 30$$
and $z = 5x_1 + 10x_2$ is maximized.

14. Find $y_1 \geq 0, y_2 \geq 0$ such that
$$10y_1 + 5y_2 \geq 100$$
$$20y_1 + 10y_2 \geq 160$$
and $w = 4y_1 + 5y_2$ is minimized.

15. Minimize $w = 3y_1 + 2y_2$
subject to: $2y_1 + 3y_2 \geq 60$
$y_1 + 4y_2 \geq 40$
$y_1 \geq 0, y_2 \geq 0.$

16. Minimize $w = 3y_1 + 4y_2$
subject to: $y_1 + 2y_2 \geq 10$
$y_1 + y_2 \geq 8$
$2y_1 + y_2 \leq 22$
$y_1 \geq 0, y_2 \geq 0.$

17. Maximize $z = 3x_1 + 2x_2$
subject to: $x_1 + x_2 = 50$
$4x_1 + 2x_2 \geq 120$
$5x_1 + 2x_2 \leq 200$
with $x_1 \geq 0, x_2 \geq 0.$

18. Maximize $z = 10x_1 + 9x_2$
subject to: $x_1 + x_2 = 30$
$x_1 + x_2 \geq 25$
$2x_1 + x_2 \leq 40$
with $x_1 \geq 0, x_2 \geq 0.$

19. Minimize $w = 32y_1 + 40y_2$
subject to: $20y_1 + 10y_2 = 200$
$25y_1 + 40y_2 \leq 500$
$18y_1 + 24y_2 \geq 300$
with $y_1 \geq 0, y_2 \geq 0.$

20. Minimize $w = 15y_1 + 12y_2$
subject to: $y_1 + 2y_2 \leq 12$
$3y_1 + y_2 \geq 18$
$y_1 + y_2 = 10$
with $y_1 \geq 0, \quad y_2 \geq 0.$

In Exercises 21–24, set up the initial simplex tableau, but do not solve the problem. (See Example 6.)

21. **Business**   A company is developing a new additive for gasoline. The additive is a mixture of three liquid ingredients, I, II, and III. For proper performance, the total amount of additive must be at least 10 ounces per barrel of gasoline. However, for safety reasons, the amount of additive should not exceed 15 ounces per barrel of gasoline. At least 1/4 ounce of ingredient I must be used for

every ounce of ingredient II and at least 1 ounce of ingredient III must be used for every ounce of ingredient I. If the costs of I, II, and III are \$.30, \$.09, and \$.27 per ounce, respectively, find the mixture of the three ingredients that produces the minimum cost of the additive. What is the minimum cost?

22. **Business**   A popular soft drink called Sugarlo, which is advertised as having a sugar content of no more than 10%, is blended from five ingredients, each of which has some sugar content. Water may also be added to dilute the mixture. The sugar content of the ingredients and their costs per gallon are given below.

	INGREDIENT					
	1	2	3	4	5	*Water*
Sugar content (%)	.28	.19	.43	.57	.22	0
Cost (\$/gal.)	.48	.32	.53	.28	.43	.04

At least .01 of the content of Sugarlo must come from ingredients 3 or 4, .01 must come from ingredients 2 or 5, and .01 from ingredients 1 or 4. How much of each ingredient should be used in preparing at least 15,000 gallons of Sugarlo to minimize the cost? What is the minimum cost?

23. **Business**   The manufacturer of a popular personal computer has orders from two dealers. Dealer D_1 wants 32 computers, and dealer D_2 wants 20 computers. The manufacturer can fill the orders from either of two warehouses, W_1 or W_2. W_1 has 25 of the computers on hand, and W_2 has 30. The costs (in dollars) to ship one computer to each dealer from each warehouse are given below

		TO	
		D_1	D_2
FROM	W_1	\$14	\$22
	W_2	\$12	\$10

How should the orders be filled to minimize shipping costs? What is the minimum cost?

24. **Natural Science**   Mark, who is ill, takes vitamin pills. Each day he must have at least 16 units of vitamin A, 5 units of vitamin B_1, and 20 units of vitamin C. He can choose between pill #1 which costs 10¢ and contains 8 units of A, 1 of B_1, and 2 of C, and pill #2 which costs 20¢ and contains 2 units of A, 1 of B_1, and 7 of C. How many of each pill should he buy in order to minimize his cost?

Use the two-stage method to solve Exercises 23–30. (See Examples 5 and 6.)

25. **Transportation**   Southwestern Oil supplies two distributors in the Northwest from two outlets, S_1 and S_2.

Distributor D_1 needs at least 3000 barrels of oil, and distributor D_2 needs at least 5000 barrels. The two outlets can each furnish up to 5000 barrels of oil. The costs per barrel to ship the oil are given in the table.

OUTLETS		DISTRIBUTORS	
		D_1	D_2
	S_1	$30	$20
	S_2	$25	$22

There is also a shipping tax per barrel as given in the table below. Southwestern Oil is determined to spend no more than $40,000 on shipping tax.

	D_1	D_2
S_1	$2	$6
S_2	$5	$4

How should the oil be supplied to minimize shipping costs?

26. **Transportation** Change Exercise 25 so that the two outlets each furnish exactly 5000 barrels of oil, with everything else the same. Solve the problem as in Example 5.

27. **Business** Topgrade Turf lawn seed mixture contains three types of seeds: bluegrass, rye, and Bermuda. The costs per pound of the three types of seed are 12¢, 15¢, and 5¢, respectively. In each batch there must be at least 20% bluegrass seed, and the amount of Bermuda must be no more than 2/3 the amount of rye. To fill current orders, the company must make at least 5000 pounds of the mixture. How much of each kind of seed should be used to minimize cost?

28. **Business** Change Exercise 27 so that the company must make exactly 5000 pounds of the mixture. Solve the problem as in Example 5.

29. **Finance** A bank has set aside a maximum of $25 million for commercial and home loans. Every million dollars in commercial loans requires 2 lengthy application forms, while every million dollars in home loans requires 3 lengthy application forms. The bank cannot process more than 72 application forms at this time. The bank's policy is to loan at least four times as much for

home loans as for commercial loans. Because of prior commitments, at least $10 million will be used for these two types of loans. The bank earns 12% on home loans and 10% on commercial loans. What amount of money should be allotted for each type of loan to maximize the interest income?

30. **Finance** Virginia Keleske has decided to invest a $100,000 inheritance in government securities that earn 7% per year, municipal bonds that earn 6% per year, and mutual funds that earn an average of 10% per year. She will spend at least $40,000 on government securities, and she wants at least half the inheritance to go to bonds and mutual funds. Government securities have an initial fee of 2%, municipal bonds have an initial fee of 1%, and mutual funds have an initial fee of 3%. Virginia has $2400 available to pay initial fees. How much should be invested in each way to maximize the interest yet meet the constraints? What is the maximum interest she can earn?

31. **Business** A brewery produces regular beer and a lower-carbohydrate "light" beer. Steady customers of the brewery buy 12 units of regular beer and 10 units of light beer. While setting up the brewery to produce the beers, the management decides to produce extra beer, beyond that needed to satisfy the steady customers. The cost per unit of regular beer is $36,000 and the cost per unit of light beer is $48,000. The number of units of light beer should not exceed twice the number of units of regular beer. At least 20 additional units of beer can be sold. How much of each type beer should be made so as to minimize total production costs?

32. **Business** The chemistry department at a local college decides to stock at least 800 small test tubes and 500 large test tubes. It wants to buy at least 1500 test tubes to take advantage of a special price. Since the small tubes are broken twice as often as the larger, the department will order at least twice as many small tubes as large. If the small test tubes cost 15¢ each and the large ones, made of a cheaper glass, cost 12¢ each, how many of each size should they order to minimize cost?

Business *Use technology to solve the following exercises, whose initial tableaus were set up in Exercises 21–23.*

33. Exercise 21   34. Exercise 22

35. Exercise 23

CHAPTER 7 **SUMMARY**

Key Terms and Symbols

7.1 linear inequality
boundary
half-plane
system of inequalities
region of feasible solutions
(feasible region)

7.2 linear programming
objective function
constraints
corner point

bounded feasible region
unbounded feasible region
corner point theorem

7.4 standard maximum form
slack variable
simplex tableau
indicator
pivot and pivoting
final simplex tableau
basic variables
nonbasic variables
basic feasible solution

7.6 dual
transpose of a matrix
theorem of duality
shadow costs

7.7 surplus variable
basic variables
basic solution
basic feasible solution
Stage I
Stage II

Key Concepts

Graphing a Linear Inequality

Graph the boundary line as a solid line if the inequality includes "or equal," a dashed line otherwise. Shade the half-plane that includes a test point that makes the inequality true. The graph of a system of inequalities, called the **region of feasible solutions,** includes all points that satisfy all the inequalities of the system at the same time.

Solving Linear Programming Problems

Graphically: Determine the objective function and all necessary constraints. Graph the region of feasible solutions. The maximum or minimum value will occur at one or more of the corner points of this region.

Simplex Method: Determine the objective function and all necessary constraints. Convert each constraint into an equation by adding slack variables. Set up the initial simplex tableau. Locate the most negative indicator. Form the quotients to determine the pivot. Use row operations to change the pivot to 1 and all other numbers in that column to 0. If the indicators are all positive or 0, this is the final tableau. If not, choose a new pivot and repeat the process until no indicators are negative. Read the solution from the final tableau. The optimum value of the objective function is the number in the lower right corner of the final tableau. For problems with **mixed constraints,** replace each equation constraint by a pair of inequality constraints. Then add slack variables and subtract surplus variables as needed to convert each constraint into an equation. In Stage I, use row operations to transform the matrix until the solution is feasible. In Stage II, use the simplex method as described above. For **minimum** problems, let the objective function be w and set $-w = z$. Then proceed as with mixed constraints.

Solving Minimum Problems with Duals

Find the dual maximum problem. Solve the dual using the simplex method. The minimum value of the objective function w is the maximum value of the dual objective function z. The optimal solution is found in the entries in the bottom row of the columns corresponding to the slack variables.

Chapter 7 Review Exercises

Graph each of the following linear inequalities.

1. $y \le 3x + 2$

2. $2x - y \ge 6$

3. $3x + 4y \ge 12$

4. $y \le 4$

Graph the solution of each of the following systems of inequalities.

5. $x + y \le 6$
$2x - y \ge 3$

6. $4x + y \ge 8$
$2x - 3y \le 6$

7. $2 \le x \le 5$
$1 \le y \le 7$
$x - y \le 3$

8. $x + 2y \le 4$
$2x - 3y \le 6$
$x \ge 0$
$y \ge 0$

Set up a system of inequalities for each of the following problems; then graph the region of feasible solutions.

9. Business   A bakery makes both cakes and cookies. Each batch of cakes requires 2 hours in the oven and 3 hours in the decorating room. Each batch of cookies needs $1\frac{1}{2}$ hours in the oven and 2/3 of an hour in the decorating room. The oven is available no more than 15 hours a day, while the decorating room can be used no more than 13 hours a day.

10. Business   A company makes two kinds of pizza, special and basic. The special has toppings of cheese, tomatoes, and vegetables. Basic has just cheese and tomatoes. The company sells at least 6 units a day of the special pizza and 4 units a day of the basic. The cost of the vegetables (including tomatoes) is $2 per unit for special and $1 per unit for basic. No more than $32 per day can be spent on vegetables (including tomatoes). The cheese used for the special is $5 per unit, and the cheese for the basic is $4 per unit. The company can spend no more than $100 per day on cheese.

Use the given regions to find the maximum and minimum values of the objective function $z = 2x + 4y$.

11.

12.

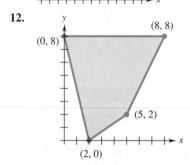

Use the graphical method to solve Exercises 13–16.

13. Maximize $z = 3x + 2y$
subject to: $2x + 7y \le 14$
$2x + 3y \le 10$
$x \ge 0, y \ge 0.$

14. Find $x \ge 0$ and $y \ge 0$ such that
$8x + 9y \ge 72$
$6x + 8y \ge 72$
and $w = 4x + 12y$ is minimized.

15. Find $x \ge 0$ and $y \ge 0$ such that
$x + y \le 50$
$2x + y \ge 20$
$x + 2y \ge 30$
and $w = 8x + 3y$ is minimized.

16. Maximize $z = 2x - 5y$
subject to: $3x + 2y \le 12$
$5x + y \ge 5$
$x \ge 0, y \ge 0.$

17. Business   How many batches of cakes and cookies should the bakery of Exercise 9 make in order to maximize profits if cookies produce a profit of $20 per batch and cakes produce a profit of $30 per batch?

18. Business How many units of each kind of pizza should the company of Exercise 10 make in order to maximize revenue if special sells for $20 per unit and basic for $15 per unit?

For Exercises 19–22, (a) select appropriate variables, (b) write the objective function, (c) write the constraints as inequalities.

19. Business Roberta Hernandez sells three items, A, B, and C, in her gift shop. Each unit of A costs her $2 to buy, $1 to sell, and $2 to deliver. For each unit of B, the costs are $3, $2, and $2, respectively, and for each unit of C the costs are $6, $2, and $4, respectively. The profit on A is $4, on B it is $3, and on C, $3. How many of each should she order to maximize her profit if she can spend $1200 to buy, $800 on selling costs, and $500 on delivery costs?

20. Business An investor is considering three types of investment: a high-risk venture into oil leases with a potential return of 15%, a medium-risk investment in bonds with a 9% return, and a relatively safe stock investment with a 5% return. He has $50,000 to invest. Because of the risk, he will limit his investment in oil leases and bonds to 30% and his investment in oil leases and stock to 50%. How much should he invest in each to maximize his return, assuming investment returns are as expected?

21. Business The Aged Wood Winery makes two white wines, Fruity and Crystal, from two kinds of grapes and sugar. The wines require the following amounts of each ingredient per gallon and produce a profit per gallon as shown below.

	Grape A (bushels)	Grape B (bushels)	Sugar (pounds)	Profit (dollars)
Fruity	2	2	2	12
Crystal	1	3	1	15

The winery has available 110 bushels of grape A, 125 bushels of grape B, and 90 pounds of sugar. How much of each wine should be made to maximize profit?

22. Business A company makes three sizes of plastic bags: 5 gallon, 10 gallon, and 20 gallon. The production time in hours for cutting, sealing, and packaging a unit of each size is shown below.

Size	Cutting	Sealing	Packaging
5 gallon	1	1	2
10 gallon	1.1	1.2	3
20 gallon	1.5	1.3	4

There are at most 8 hours available each day for each of the three operations. If the profit per unit is $1 for 5-gallon bags, $.90 for 10-gallon bags, and $.95 for 20-gallon bags, how many of each size should be made per day to maximize the profit?

23. When is it necessary to use the simplex method rather than the graphical method?

24. What types of problems can be solved using slack variables and surplus variables?

25. What kind of problem can be solved using the method of duals?

26. In solving a linear programming problem, you are given the following initial tableau.

$$\left[\begin{array}{ccccccc|c} 4 & 2 & 3 & 1 & 0 & 0 & 9 \\ 5 & 4 & 1 & 0 & 1 & 0 & 10 \\ \hline -6 & -7 & -5 & 0 & 0 & 1 & 0 \end{array}\right]$$

(a) What is the problem being solved?

(b) If the 1 in row 1, column 4, were a -1 rather than a 1, how would it change your answer to part (a)?

(c) After several steps of the simplex algorithm, the following tableau results.

$$\left[\begin{array}{cccccc|c} 3 & 0 & 5 & 2 & -1 & 0 & 8 \\ 11 & 10 & 0 & -1 & 3 & 0 & 21 \\ \hline 47 & 0 & 0 & 13 & 11 & 10 & 227 \end{array}\right]$$

What is the solution? (List only the values of the original variables and the objective function. Do not include slack or surplus variables.)

(d) What is the dual of the problem you found in part (a)?

(e) What is the solution of the dual you found in part (d)? (Do not perform any steps of the simplex algorithm; just examine the tableau given in part (c).)

For each of the following problems, (a) add slack variables and (b) set up the initial simplex tableau.

27. Maximize $z = 2x_1 + 7x_2$
subject to: $3x_1 + 5x_2 \le 47$
$x_1 + x_2 \le 25$
$5x_1 + 2x_2 \le 35$
$2x_1 + x_2 \le 30$
$x_1 \ge 0, x_2 \ge 0.$

28. Maximize $z = 15x_1 + 10x_2$
subject to: $2x_1 + 5x_2 \le 50$
$x_1 + 3x_2 \le 25$
$4x_1 + x_2 \le 18$
$x_1 + x_2 \le 12$
$x_1 \ge 0, x_2 \ge 0.$

29. Maximize $z = 4x_1 + 6x_2 + 3x_3$
subject to: $x_1 + x_2 + x_3 \le 100$
$2x_1 + 3x_2 \le 500$
$x_1 + 2x_3 \le 350$
$x_1 \ge 0, x_2 \ge 0, x_3 \ge 0.$

30. Maximize $z = x_1 + 4x_2 + 2x_3$
subject to: $x_1 + x_2 + x_3 \le 90$
$2x_1 + 5x_2 + x_3 \le 120$
$x_1 + 3x_2 \le 80$
$x_1 \ge 0, x_2 \ge 0, x_3 \ge 0.$

For each of the following, use the simplex method to solve the maximization linear programming problems with initial tableaus as given.

31.

$$
\begin{array}{ccccc}
x_1 & x_2 & x_3 & x_4 & x_5
\end{array}
$$
$$
\left[\begin{array}{ccccc|c}
1 & 2 & 3 & 1 & 0 & 28 \\
2 & 4 & 8 & 0 & 1 & 32 \\
\hline
-5 & -2 & -3 & 0 & 0 & 0
\end{array}\right]
$$

32.

$$
\begin{array}{cccc}
x_1 & x_2 & x_3 & x_4
\end{array}
$$
$$
\left[\begin{array}{cccc|c}
2 & 1 & 1 & 0 & 10 \\
9 & 3 & 0 & 1 & 15 \\
\hline
-2 & -3 & 0 & 0 & 0
\end{array}\right]
$$

33.

$$
\begin{array}{cccccc}
x_1 & x_2 & x_3 & x_4 & x_5 & x_6
\end{array}
$$
$$
\left[\begin{array}{cccccc|c}
1 & 2 & 2 & 1 & 0 & 0 & 50 \\
4 & 24 & 0 & 0 & 1 & 0 & 20 \\
1 & 0 & 2 & 0 & 0 & 1 & 15 \\
\hline
-5 & -3 & -2 & 0 & 0 & 0 & 0
\end{array}\right]
$$

34.

$$
\begin{array}{ccccc}
x_1 & x_2 & x_3 & x_4 & x_5
\end{array}
$$
$$
\left[\begin{array}{ccccc|c}
1 & -2 & 1 & 0 & 0 & 38 \\
1 & -1 & 0 & 1 & 0 & 12 \\
2 & 1 & 0 & 0 & 1 & 30 \\
\hline
-1 & -2 & 0 & 0 & 0 & 0
\end{array}\right]
$$

Convert the following problems into maximization problems without using duals.

35. Minimize $w = 18y_1 + 10y_2$
subject to: $y_1 + y_2 \geq 17$
$5y_1 + 8y_2 \geq 42$
$y_1 \geq 0, y_2 \geq 0.$

36. Minimize $w = 12y_1 + 20y_2 - 8y_3$
subject to: $y_1 + y_2 + 2y_3 \geq 48$
$y_1 + y_2 \geq 12$
$y_3 \geq 10$
$3y_1 + y_3 \geq 30$
$y_1 \geq 0, y_2 \geq 0, y_3 \geq 0.$

37. Minimize $w = 6y_1 - 3y_2 + 4y_3$
subject to: $2y_1 + y_2 + y_3 \geq 112$
$y_1 + y_2 + y_3 \geq 80$
$y_1 + y_2 \geq 45$
$y_1 \geq 0, y_2 \geq 0, y_3 \geq 0.$

Use the simplex method to solve the following mixed constraint problems.

38. Maximize $z = 2x_1 + 4x_2$
subject to: $3x_1 + 2x_2 \leq 12$
$5x_1 + x_2 \geq 5$
$x_1 \geq 0, x_2 \geq 0.$

39. Minimize $w = 4y_1 - 8y_2$
subject to: $y_1 + y_2 \leq 50$
$2y_1 - 4y_2 \geq 20$
$y_1 - y_2 \leq 22$
$y_1 \geq 0, y_2 \geq 0.$

The following tableaus are the final tableaus of minimizing problems solved by letting $w = -z$. Give the solution and the minimum value of the objective function for each problem.

40.
$$
\left[\begin{array}{cccccc|c}
0 & 1 & 0 & 2 & 5 & 0 & 17 \\
0 & 0 & 1 & 3 & 1 & 1 & 25 \\
1 & 0 & 0 & 4 & 2 & \frac{1}{2} & 8 \\
\hline
0 & 0 & 0 & 2 & 5 & 0 & -427
\end{array}\right]
$$

41.
$$
\left[\begin{array}{ccccccc|c}
0 & 0 & 2 & 1 & 0 & 6 & 6 & 92 \\
1 & 0 & 3 & 0 & 0 & 0 & 2 & 47 \\
0 & 1 & 0 & 0 & 0 & 1 & 0 & 68 \\
0 & 0 & 4 & 0 & 1 & 0 & 3 & 35 \\
\hline
0 & 0 & 5 & 0 & 0 & 2 & 9 & -1957
\end{array}\right]
$$

The tableaus in Exercises 42–44 are the final tableaus of minimizing problems solved by the method of duals. State the solution and the minimum value of the objective function for each problem.

42.
$$
\left[\begin{array}{ccccccc|c}
1 & 0 & 0 & 3 & 1 & 2 & 12 \\
0 & 0 & 1 & 4 & 5 & 3 & 5 \\
0 & 1 & 0 & -2 & 7 & -6 & 8 \\
\hline
0 & 0 & 0 & 5 & 7 & 3 & 172
\end{array}\right]
$$

43.
$$
\left[\begin{array}{cccccc|c}
0 & 0 & 1 & 6 & 3 & 1 & 2 \\
1 & 0 & 0 & 4 & -2 & 2 & 8 \\
0 & 1 & 0 & 10 & 7 & 0 & 12 \\
\hline
0 & 0 & 0 & 9 & 5 & 8 & 62
\end{array}\right]
$$

44.
$$
\left[\begin{array}{cccc|c}
1 & 0 & 7 & -1 & 100 \\
0 & 1 & 1 & 3 & 27 \\
\hline
0 & 0 & 7 & 2 & 640
\end{array}\right]
$$

Business *Solve the following maximization problems, which were begun in Exercises 19–22.*

45. Exercise 19

46. Exercise 20

47. Exercise 21

48. Exercise 22

Business *Solve the following minimization problems.*

49. Cauchy Canners produces canned corn, beans, and carrots. Demand for vegetables requires it to produce at least 1000 cases per month. Based on past sales, it should produce at least twice as many cases of corn as of beans and at least 340 cases of carrots. It costs $10 to produce a case of corn, $15 to produce a case of beans, and $25 to produce a case of carrots. How many cases of each vegetable should be produced to minimize costs? What is the minimum cost?

50. A contractor builds boathouses in two basic models, the atlantic and pacific. Each atlantic model requires 1000 feet of framing lumber, 3000 cubic feet of concrete, and $2000 for advertising. Each pacific model requires 2000 feet of framing lumber, 3000 cubic feet of concrete, and

$3000 for advertising. Contracts call for using at least 8000 feet of framing lumber, 18,000 cubic feet of concrete, and $15,000 worth of advertising. If the total spent on each atlantic model is $3000 and the total spent on each pacific model is $4000, how many of each model should be built to minimize costs?

Business   *Solve these mixed constraint problems.*

51. Brand X Cannery produces canned whole tomatoes and tomato sauce. This season, they have available 3,000,000 kilograms of tomatoes for these two products. To meet the demands of regular customers, they must produce at least 80,000 kilograms of sauce and 800,000 kilograms of whole tomatoes. The cost per kilogram is $4 to produce canned whole tomatoes and $3.25 to produce tomato sauce. Labor agreements require that at least 110,000 person hours be used. Each one-kilogram can of sauce requires 3 minutes for one worker, and each one-

kilogram can of whole tomatoes requires 6 minutes for one worker. How many kilograms of tomatoes should Brand X use for each product to minimize cost? (For simplicity, assume production of y_1 kilograms of canned whole tomatoes and y_2 kilograms of tomato sauce requires $y_1 + y_2$ kilograms of tomatoes.)

52. A steel company produces two types of alloys. A run of type I requires 3000 pounds of molybdenum and 2000 tons of iron ore pellets as well as $2000 in advertising. A run of type II requires 3000 pounds of molybdenum and 1000 tons of iron ore pellets as well as $3000 in advertising. Total costs are $15,000 on a run of type I and $6000 on a run of type II. The company has on hand 18,000 pounds of molybdenum and 7000 tons of iron ore pellets and wants to use all of it. They plan to spend at least $14,000 on advertising. How much of each type should be produced to minimize cost? What is the minimum cost?

Case 6

Cooking with Linear Programming

Constructing a nutritious recipe can be a difficult task. The recipe must produce food that tastes good, and it must also balance the nutrients that each ingredient brings to the dish. This balancing of nutrients is very important in several diet plans that are currently popular. Many of these plans restrict the intake of certain nutrients (usually fat) while allowing for large amounts of other nutrients (protein and carbohydrates are popular choices). The number of calories in the dish is also often minimized. Linear programming can be used to help create recipes that balance nutrients.

In order to develop solutions to this type of problem we will need to know nutritional data for the ingredients in our recipes. This data is available from the United States Department of Agriculture in the USDA Nutrient Database for Standard Reference. This database, which is available at www.na1.udsa .gov/fnic/foodcomp/data/sr13.html, contains the nutrient levels for hundreds of basic foodstuffs. The nutrient levels are given per 100 grams of foodstuff. Unfortunately, grams are not often used in recipes; instead, kitchen measures like cups, tablespoons, and fractions of vegetables are used. Table 1 shows the conversion

factors from grams to more familiar kitchen units and also gives serving sizes for various foodstuffs.

Table 1   Serving Sizes of Various Foodstuffs

Foodstuff	Serving Size
Beef	6 oz = 170 g
Egg	1 egg = 61 g
Feta Cheese	1/4 cup = 38 g
Lettuce	1/2 cup = 28 g
Milk	1 cup = 244 g
Oil	1 Tbsp = 13.5 g
Onion	1 onion = 110 g
Salad Dressing	1 cup = 250 g
Soy Sauce	1 Tbsp = 18 g
Spinach	1 cup = 180 g
Tomato	1 tomato = 123 g

Consider creating a recipe for a spinach omelet from eggs, milk, vegetable oil, and spinach. The nutrients of interest will be protein, fat, and carbohydrates. Calories will also be monitored. The amounts of the nutrients and calories for these ingredients are given in Table 2:

Table 2 **Nutritional Values per 100 g of Foodstuff**

Nutrient (units)	Eggs	Milk	Oil	Spinach
Calories (kcal)	152	61.44	884	23
Protein (g)	10.33	3.29	0	2.9
Fat (g)	11.44	3.34	100	.26
Carbohydrates (g)	1.04	4.66	0	3.75

Let x_1 be the number of 100 gram units of eggs to use in the recipe, x_2 be the number of 100 gram units of milk, x_3 be the number of 100 gram units of oil, and x_4 be the number of 100 gram units of spinach. We will want to minimize the amount of calories in the dish, while providing at least 4 grams of protein, 15 grams of carbohydrates, and 20 grams of fat. The cooking technique specifies that at least 1/8 of a cup of milk (30.5 grams) must be used in the recipe. We should thus minimize the objective function (using 100 gram units of foodstuff):

$$z = 152x_1 + 61.44x_2 + 884x_3 + 23x_4$$

subject to

$$10.33x_1 + 3.29x_2 + \quad 0x_3 + 2.90x_4 \geq 4$$
$$11.44x_1 + 3.34x_2 + 100x_3 + \quad .26x_4 \geq 20$$
$$1.04x_1 + 4.66x_2 + \quad 0x_3 + 3.75x_4 \geq 15$$
$$0x_1 + \quad 1x_2 + \quad 0x_3 + \quad 0x_4 \geq .305.$$

Of course, all variables are subject to nonnegativity constraints:

$$x_1 \geq 0, \qquad x_2 \geq 0, \qquad x_3 \geq 0, \qquad x_4 \geq 0.$$

Using a graphing calculator or a computer with linear programming software, we get the following solution:

$$x_1 = 1.2600, \qquad x_2 = .3050, \qquad x_3 = .0448, \qquad x_4 = .338.$$

This recipe produces an omelet with 257.63 calories. In grams, the amounts of each ingredient are 126 grams of eggs, 30.5 grams of milk, 4.48 grams of oil, and 33.8 grams of spinach. Converting to kitchen units using Table 1, we find the recipe to be approximately 2 eggs, 1/8 cup milk, 1 teaspoon oil, and 1/4 cup spinach.

Exercises

1. Consider preparing a high-carbohydrate Greek salad using feta cheese, lettuce, salad dressing, and tomato. The amount of carbohydrates in the salad should be maximized. In addition the salad should have less than 260 calories, over 210 milligrams of calcium, and over 6 grams of protein. The salad should also weigh in total less than 400 grams and be dressed with at least 2 tablespoons (1/8 cup) of salad dressing. The amounts of the nutrients and calories for these ingredients are given in Table 3:

Table 3 **Nutritional Values per 100 g of Foodstuff**

Nutrient (units)	Feta Cheese	Lettuce	Salad Dressing	Tomato
Calories (kcal)	263	14	448.8	21
Calcium (mg)	492.5	36	0	5
Protein (g)	10.33	1.62	0	.85
Carbohydrates (g)	4.09	2.37	2.5	4.64

Use linear programming to find the number of 100 gram units of each ingredient to be used in such a Greek salad and convert to kitchen units using Table 1.

2. Consider preparing a stir fry using beef, oil, onion, and soy sauce. A low-calorie stir fry is desired, which also contains less than 10 grams of carbohydrates, more than 50 grams of protein, and more than 3.5 grams of vitamin C. In order for the wok to function correctly, at least one teaspoon (or 4.5 grams) of oil must be used in the recipe. The amounts of the nutrients and calories for these ingredients are given in Table 4:

Table 4 **Nutritional Values per 100 g of Foodstuff**

Nutrient (units)	Beef	Oil	Onion	Soy Sauce
Calories (kcal)	215	884	38	60
Protein (g)	26	0	1.16	10.51
Carbohydrates (g)	0	1	8.63	5.57
Vitamin C (g)	0	0	6.4	0

Use linear programming to find the number of 100 gram units of each ingredient to be used in the stir fry and convert to kitchen units using Table 1.

Chapter 8

Sets and Probability

Federal officials cannot predict exactly how traffic deaths are affected by the trends toward fewer drunken drivers and increased use of seat belts. Economists cannot tell exactly how stricter federal regulations on bank loans affect the U.S. economy. The number of traffic deaths and the growth of the economy are subject to many factors that cannot be predicted precisely.

Probability theory enables us to deal with uncertainty. The basic concepts of probability are discussed in this chapter, and applications of probability are discussed in the next chapter. Sets and set operations are the basic tools for the study of probability, so we begin with them.

8.1 SETS

Think of a set as a well-defined collection of objects. A set of coins might include one of each type of coin now put out by the U.S. government. Another set might be made up of all the students in your English class. On the other hand, a collection of "young adults" does not constitute a set unless the designation young adult is clearly defined. For example, this set might be defined as those aged 18–29.

In mathematics sets are often made up of numbers. The set consisting of the numbers 3, 4, and 5 is written

$$\{3, 4, 5\},$$

where **set braces**, { }, are used to enclose the numbers belonging to the set. The numbers, 3, 4, and 5 are called the **elements** or **members** of this set. To show that 4 is an element of the set $\{3, 4, 5\}$, we use the symbol $\in$ and write

$$4 \notin \{3, 4, 5\},$$

read "4 is an element of the set containing 3, 4, and 5."

415

Write *true* or *false*.

(a) $9 \in \{8, 4, -3, -9, 6\}$

(b) $4 \notin \{3, 9, 7\}$

(c) If $M = \{0, 1, 2, 3, 4\}$, then $0 \in M$.

Answers:

(a) False

(b) True

(c) True

Also, $5 \in \{3, 4, 5\}$. Place a slash through the symbol $\in$ to show that 8 is *not* an element of this set.

$$8 \notin \{3, 4, 5\}$$

This is read "8 is not an element of the set $\{3, 4, 5\}$."

Sets are often named with capital letters, so that if

$$B = \{5, 6, 7\},$$

then, for example, $6 \in B$ and $10 \notin B$.

Sometimes a set has no elements. Some examples are the set of female presidents of the United States in the period 1788–2004, the set of counting numbers less than 1, and the set of men more than 10 feet tall. A set with no elements is called the **empty set**. The symbol $\emptyset$ is used to represent the empty set.

CAUTION   Be careful to distinguish between the symbols 0, $\emptyset$ and $\{0\}$. The symbol 0 represents a *number;* $\emptyset$ represents a *set* with no elements; and $\{0\}$ represents a *set* with one element, the number 0. Do not confuse the empty set symbol $\emptyset$ with the zero $\emptyset$ on a computer screen or printout.

Two sets are **equal** if they contain exactly the same elements. The sets $\{5, 6, 7\}$, $\{7, 6, 5\}$, and $\{6, 5, 7\}$ all contain exactly the same elements and are equal. In symbols,

$$\{5, 6, 7\} = \{7, 6, 5\} = \{6, 5, 7\}.$$

This means that the ordering of the elements in a set is unimportant. Sets that do not contain exactly the same elements are *not equal*. For example, the sets $\{5, 6, 7\}$ and $\{5, 6, 7, 8\}$ do not contain exactly the same elements and are not equal. We show this by writing

$$\{5, 6, 7\} \neq \{5, 6, 7, 8\}.$$

Sometimes we describe a set by a common property of its elements rather than by a list of elements. This common property can be expressed with **set-builder notation;** for example,

$$\{x \mid x \text{ has property } P\}$$

List the elements in the following sets.

(a) $\{x \mid x$ is a counting number more than 5 and less than 8$\}$

(b) $\{x \mid x$ is an integer, $-3 < x \le 1\}$

Answers:

(a) $\{6, 7\}$

(b) $\{-2, -1, 0, 1\}$

(read "the set of all elements x such that x has property P") represents the set of all elements x having some property P.

▶ **EXAMPLE 1**   List the elements belonging to each of the following sets.

(a) $\{x \mid x$ is a natural number less than 5$\}$
The natural numbers less than 5 make up the set $\{1, 2, 3, 4\}$.

(b) $\{x \mid x$ is a state that borders Florida$\}$ = $\{$Alabama, Georgia$\}$ ◀

The **universal set** in a particular discussion is a set that contains all of the objects being discussed. In grade-school arithmetic, for example, the set of whole numbers might be the universal set, whereas in a college calculus class the universal set might be the set of all real numbers. When it is necessary to consider the universal set being used, it will be clearly specified or easily understood from the context of the problem.

Sometimes every element of one set also belongs to another set. For example, if

$$A = \{3, 4, 5, 6\}$$

and

$$B = \{2, 3, 4, 5, 6, 7, 8\},$$

then every element of A is also an element of B. This is an example of the following definition.

> A set A is a **subset** of a set B (written $A \subseteq B$) provided that every element of A is also an element of B.

③

Write *true* or *false*.

(a) $\{3, 4, 5\} \subseteq \{2, 3, 4, 6\}$

(b) $\{x \mid x$ is an automobile$\}$ $\subseteq \{x \mid x$ is a motor vehicle$\}$

(c) $\{3, 6, 9, 10\}$ $\subseteq \{3, 9, 11, 13\}$

Answers:

(a) False

(b) True

(c) False

▶ **EXAMPLE 2**   Decide whether or not $M \subseteq N$.

(a) M is the set of all small businesses with less than 20 employees. N is the set of all businesses.

Each business with less than 20 employees is also a business, so $M \subseteq N$.

(b) M is the set of all fourth-grade students in a school at the end of the school year, and N is the set of all nine-year-old students in the school at the end of the school year.

By the end of the school year, some fourth-grade students are ten years old, so there are elements in M that are not in N. Thus, M is not a subset of N, written $M \nsubseteq N.$ ◀ ③

Every set A is a subset of itself because the statement "every element of A is also an element of A" is always true. It is also true that the empty set is a subset of every set.*

> For any set A,
>
> $$\emptyset \subseteq A \quad \text{and} \quad A \subseteq A.$$

▶ **EXAMPLE 3**   List all possible subsets for each of the following sets.

(a) $\{7, 8\}$

There are 4 subsets of $\{7, 8\}$:

$$\emptyset, \quad \{7\}, \quad \{8\}, \quad \{7, 8\}.$$

---

*This fact is not intuitively obvious to most people. If you wish, you can think of it as a convention that we agree to adopt in order to simplify the statements of several results later.

List all subsets of $\{w, x, y, z\}$.

Answer:

$\emptyset$, $\{w\}$, $\{x\}$, $\{y\}$, $\{z\}$, $\{w, x\}$, $\{w, y\}$, $\{w, z\}$, $\{x, y\}$, $\{x, z\}$, $\{y, z\}$, $\{w, x, y\}$, $\{w, x, z\}$, $\{w, y, z\}$, $\{x, y, z\}$, $\{w, x, y, z\}$

(b) $\{a, b, c\}$

There are 8 subsets of $\{a, b, c\}$:

$\emptyset$, $\{a\}$, $\{b\}$, $\{c\}$, $\{a, b\}$, $\{a, c\}$, $\{b, c\}$, $\{a, b, c\}$. ◀

A good way to find the subsets of $\{7, 8\}$ and the subsets of $\{a, b, c\}$ in Example 3 is to use a **tree diagram,** a systematic way of listing all the subsets of a given set. Figures 8.1(a) and (b) show tree diagrams for finding the subsets of $\{7, 8\}$ and $\{a, b, c\}$. ④

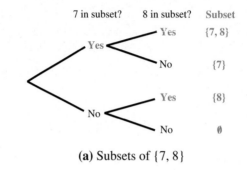

(a) Subsets of $\{7, 8\}$

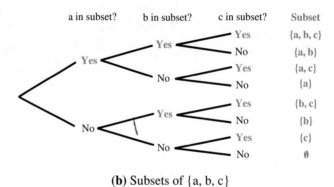

(b) Subsets of $\{a, b, c\}$

FIGURE 8.1

By using the fact that there are two possibilities for each element (either it is in the subset or it is not) we have found that a set with 2 elements has 4 ($= 2^2$) subsets and a set with 3 elements has 8 ($= 2^3$) subsets. Similar arguments work for any finite set and lead to this conclusion.

A set of n distinct elements has 2^n subsets.

▶ **EXAMPLE 4**   Find the number of subsets for each of the following sets.

(a) $\{3, 4, 5, 6, 7\}$

Since this set has 5 elements, it has 2^5 or 32 subsets.

Find the number of subsets for each of the following sets.

(a) $\{x \mid x$ is a season of the year$\}$

(b) $\{-6, -5, -4, -3, -2, -1, 0\}$

(c) $\{6\}$

Answers:

(a) 16

(b) 128

(c) 2

Refer to sets A, B, C, and U in the diagram.

(a) Is $A \subseteq B$?

(b) Is $C \subseteq B$?

(c) Is $C \subseteq U$?

(d) Is $\emptyset \subseteq A$?

Answers:

(a) No

(b) Yes

(c) Yes

(d) Yes

⑦

Let $U = \{a, b, c, d, e, f, g\}$, with $K = \{c, d, f, g\}$ and $R = \{a, c, d, e, g\}$. Find

(a) K';

(b) R'.

Answers:

(a) $\{a, b, e\}$

(b) $\{b, f\}$

(b) $\{x \mid x$ is a day of the week$\}$
This set has 7 elements and therefore has $2^7 = 128$ subsets.

(c) $\emptyset$
Since the empty set has 0 elements, it has $2^0 = 1$ subset, $\emptyset$ itself. ◀ ⑤

Figure 8.2 shows a set A, which is a subset of a set B, because A is entirely in B. (The areas of the regions are not meant to be proportional to the size of the corresponding sets.) The rectangle represents the universal set, U. Such diagrams, called **Venn diagrams,** are used to illustrate relationships among sets. ⑥

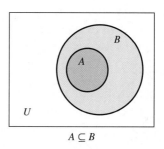

$A \subseteq B$

FIGURE 8.2

We can operate on sets to get other sets, just as operations on numbers (such as finding the negative, addition, and multiplication) produce other numbers. For example, given a set A and a universal set U, the set of all elements of U which do *not* belong to A is called the **complement** of set A. For example, if set A is the set of all the female students in your class and U is the set of all students in the class, then the complement of A would be the set of all male students in the class. The complement of set A is written A' (read "A-prime"). The Venn diagram of Figure 8.3 shows a set B. Its complement, B', is shown in color.

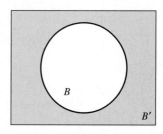

FIGURE 8.3

▶ **EXAMPLE 5** Let $U = \{1, 2, 3, 4, 5, 6, 7\}$, $A = \{1, 3, 5, 7\}$, and $B = \{3, 4, 6\}$. Find the following sets.

(a) A'
Set A' contains the elements of U that are not in A.

$$A' = \{2, 4, 6\}$$

(b) $B' = \{1, 2, 5, 7\}$

(c) $\emptyset' = U$ and $U' = \emptyset$ ◀ ⑦

Given two sets A and B, the set of all elements belonging to *both* set A and set B is called the **intersection** of the two sets, written $A \cap B$. For example, the elements that belong to both $A = \{1, 2, 4, 5, 7\}$ and $B = \{2, 4, 5, 7, 9, 11\}$ are 2, 4, 5, and 7, so

$$A \text{ and } B = A \cap B$$
$$= \{1, \mathbf{2}, \mathbf{4}, \mathbf{5}, \mathbf{7}\} \cap \{\mathbf{2}, \mathbf{4}, \mathbf{5}, \mathbf{7}, 9, 11\}$$
$$= \{2, 4, 5, 7\}.$$

The Venn diagram of Figure 8.4 shows two sets A and B with their intersection, $A \cap B$, shown in color.

(8)

Find the following.

(a) $\{1, 2, 3, 4\} \cap \{3, 5, 7, 9\}$

(b) Suppose set K is the set of all blue-eyed blondes in a class and J is the set of all blue-eyed brunettes in the class. Let $P = \{x \mid x \text{ is a brown-eyed redhead}\}$. If the class has only blondes or brunettes, find $K \cap P$.

Answers:

(a) $\{3\}$

(b) $\emptyset$

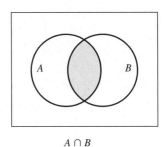

$A \cap B$

FIGURE 8.4

▶ **EXAMPLE 6** **(a)** $\{9, 15, 25, 36\} \cap \{15, 20, 25, 30, 35\} = \{15, 25\}$

The elements 15 and 25 are the only ones belonging to both sets.

(b) $\{x \mid x \text{ is a teen-ager}\} \cap \{x \mid x \text{ is a senior citizen}\}$ is an empty set. ◀ **(8)**

Two sets that have no elements in common are called **disjoint sets.** For example, there are no elements common to both $\{50, 51, 54\}$ and $\{52, 53, 55, 56\}$, so these two sets are disjoint, and

$$\{50, 51, 54\} \cap \{52, 53, 55, 56\} = \emptyset.$$

The result of this example can be generalized:

For any sets A and B,

if A and B are disjoint sets, then $A \cap B = \emptyset$.

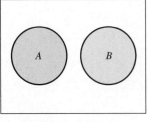

A and *B* are disjoint sets.

FIGURE 8.5

Figure 8.5 is a Venn diagram of disjoint sets.

The set of all elements belonging to set A or to set B, or to both sets, is called the **union** of the two sets, written $A \cup B$. For example, for sets $A = \{1, 3, 5\}$ and $B = \{3, 5, 7, 9\}$,

$$A \text{ or } B = A \cup B$$
$$= \{1, 3, 5\} \cup \{3, 5, 7, 9\}$$
$$= \{1, 3, 5, 7, 9\}.$$

The Venn diagram of Figure 8.6 shows two sets A and B, with their union $A \cup B$ shown in color.

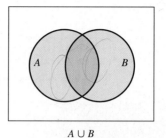

$A \cup B$

FIGURE 8.6

⑨

Find the following.

(a) $\{a, b, c\} \cup \{a, c, e\}$

(b) Describe $K \cup J$ in words for the sets given in side problem 8(b).

Answers:

(a) $\{a, b, c, e\}$

(b) All members of the class with blue eyes

▶ **EXAMPLE 7** **(a)** Find the union of $\{1, 2, 5, 9, 14\}$ and $\{1, 3, 4, 8\}$.

Begin by listing the elements of the first set, $\{1, 2, 5, 9, 14\}$. Then include any elements from the second set *that are not already listed.* Doing this gives

$$\{1, 2, 5, 9, 14\} \cup \{1, 3, 4, 8\} = \{1, 2, 3, 4, 5, 8, 9, 14\}.$$

(b) $\{$terriers, spaniels, chows, dalmatians$\} \cup \{$spaniels, collies, bulldogs$\} = \{$terriers, spaniels, chows, dalmatians, collies, bulldogs$\}$ ◀ ⑨

Finding the complement of a set, the intersection of two sets, or the union of two sets are examples of *set operations.* The set operations are summarized below.

Operations on Sets

Let A and B be any sets with U the universal set. Then

the **complement** of A, written A', is

$$A' = \{x \mid x \notin A \text{ and } x \in U\};$$

the **intersection** of A and B is

$$A \cap B = \{x \mid x \in A \text{ and } x \in B\};$$

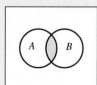

the **union** of A and B is

$$A \cup B = \{x \mid x \in A \text{ or } x \in B \text{ or both}\}.$$

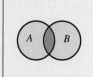

CAUTION As shown in the definitions given above, an element is in the *intersection* of sets A and B if it is in *both A and B*. On the other hand, an element is in the *union* of sets A and B if it is in *A or B or both*.

▶ **EXAMPLE 8** The table gives the annual high and low prices, the last price, and the weekly change for six stocks, as listed on the New York Stock Exchange at the end of a recent week.*

Stock	High	Low	Last	Change
BiotechT	206.00	92.51	138.20	+3.10
Cisco	70.00	13.19	20.49	+1.64
Disney	43.00	26.00	31.85	+.13
GenElec	60.50	36.42	48.14	−.86
Lucent	67.19	5.50	8.40	+.41

Let set A include all stocks with a high price greater than \$100; B all stocks with a last price between \$20 and \$40; C all stocks with a positive price change. Find the following.

Find all stocks with

(a) a last price between \$20 and \$40 and with a negative price change.

(b) a last price between \$20 and \$40 or with a negative price change.

Answers:

(a) ∅

(b) {Cisco, Disney, GenElec}

(a) B'

Set B' contains all the listed stocks that are not in set B, so

$$B' = \{\text{BiotechT, GenElec, Lucent}\}.$$

(b) $A \cap C$

The intersection, A and C, contains those stocks with a high price greater than \$100 and a positive price change.

$$A \cap C = \{\text{BiotechT}\}$$

(c) $A \cup B$

The union of A and B includes all stocks with a high price greater than \$100 or a last price between \$20 and \$40.

$$A \cup B = \{\text{BiotechT, Cisco, Disney}\} \quad ◀ \quad$$

Sacramento Bee, June 9, 2001.

8.1 Exercises

Write true *or* false *for each statement.*

1. $3 \in \{2, 5, 7, 9, 10\}$
2. $6 \in \{-2, 6, 9, 5\}$
3. $9 \notin \{2, 1, 5, 8\}$
4. $3 \notin \{7, 6, 5, 4\}$
5. $\{2, 5, 8, 9\} = \{2, 5, 9, 8\}$
6. $\{3, 7, 12, 14\} = \{3, 7, 12, 14, 0\}$
7. {all whole numbers greater than 7 and less than 10} $= \{8, 9\}$
8. {all counting numbers not greater than 3} $= \{0, 1, 2\}$
9. $\{x \mid x \text{ is an odd integer, } 6 \le x \le 18\} = \{7, 9, 11, 15, 17\}$

10. $\{x \mid x \text{ is a vowel}\} = \{a, e, i, o, u\}$

11. The elements of a set may be sets themselves, as in $\{1, \{1, 3\}, \{2\}, 4\}$. Explain why the set $\{\emptyset\}$ is not the same set as $\{0\}$.

12. What is set-builder notation? Give an example.

Let $A = \{-3, 0, 3\}$, $B = \{-2, -1, 0, 1, 2\}$, $C = \{-3, -1\}$, $D = \{0\}$, $E = \{-2\}$, *and* $U = \{-3, -2, -1, 0, 1, 2, 3\}$. *Insert* $\subseteq$ *or* $\not\subseteq$ *to make the following statements true. (See Example 2.)*

13. $A \underline{\hspace{1cm}} U$ **14.** $E \underline{\hspace{1cm}} A$

15. $A \underline{\hspace{1cm}} E$ **16.** $B \underline{\hspace{1cm}} C$

17. $\emptyset \underline{\hspace{1cm}} A$ **18.** $\{0, 2\} \underline{\hspace{1cm}} D$

19. $D \underline{\hspace{1cm}} B$ **20.** $A \underline{\hspace{1cm}} C$

Find the number of subsets for each set. (See Example 4.)

21. $\{A, B, C\}$

22. $\{\text{red, yellow, blue, black, white}\}$

23. $\{x \mid x \text{ is an integer strictly between 0 and 7}\}$

24. $\{x \mid x \text{ is a whole number less than 4}\}$

Find the complement of each set. (See Example 5.)

25. The set in Exercise 23 if U is the set of all integers.

26. The set in Exercise 24 if U is the set of all whole numbers.

27. Describe the intersection and union of sets. How do they differ?

Insert $\cap$ *or* $\cup$ *to make each statement true. (See Examples 6 and 7.)*

28. $\{5, 7, 9, 19\} \underline{\hspace{1cm}} \{7, 9, 11, 15\} = \{7, 9\}$

29. $\{8, 11, 15\} \underline{\hspace{1cm}} \{8, 11, 19, 20\} = \{8, 11\}$

30. $\{2, 1, 7\} \underline{\hspace{1cm}} \{1, 5, 9\} = \{1\}$

31. $\{6, 12, 14, 16\} \underline{\hspace{1cm}} \{6, 14, 19\} = \{6, 14\}$

32. $\{3, 5, 9, 10\} \underline{\hspace{1cm}} \emptyset = \emptyset$

33. $\{3, 5, 9, 10\} \underline{\hspace{1cm}} \emptyset = \{3, 5, 9, 10\}$

34. $\{1, 2, 4\} \underline{\hspace{1cm}} \{1, 2, 4\} = \{1, 2, 4\}$

35. $\{1, 2, 4\} \underline{\hspace{1cm}} \{1, 2\} = \{1, 2, 4\}$

36. Is it possible for two nonempty sets to have the same intersection and union? If so, give an example.

Let $U = \{2, 3, 4, 5, 7, 9\}$; $X = \{2, 3, 4, 5\}$; $Y = \{3, 5, 7, 9\}$; *and* $Z = \{2, 4, 5, 7, 9\}$.

List the members of each of the following sets, using set braces. (See Examples 5–7.)

37. $X \cap Y$ **38.** $X \cup Y$

39. X' **40.** Y'

41. $X' \cap Y'$ **42.** $X' \cap Z$

43. $X \cup (Y \cap Z)$ **44.** $Y \cap (X \cup Z)$

Let $U = \{\text{all students in this school}\}$; $M = \{\text{all students taking this course}\}$; $N = \{\text{all students taking accounting}\}$; $P = \{\text{all students taking zoology}\}$.

Describe each of the following sets in words.

45. M' **46.** $M \cup N$

47. $N \cap P$ **48.** $N' \cap P'$

49. Refer to the sets listed in the directions for Exercises 13–20. Which pairs of sets are disjoint?

50. Refer to the sets listed in the directions for Exercises 37–44. Which pairs are disjoint?

Refer to Example 8 in the text. Describe each of the sets in Exercises 51–54 in words; then list the elements of each set.

51. C' **52.** $A \cap B$

53. $(A \cup B)'$ **54.** $(B \cap C)'$

Business *An electronics store classifies credit applicants by sex, marital status, and employment status. Let the universal set be the set of all applicants, M be the set of male applicants, S be the set of all single applicants, and E be the set of employed applicants. Describe the following sets in words.*

55. $M \cap E$ **56.** $M' \cap S$ **57.** $M' \cup S'$

Business *The total U.S. advertising volume (in millions of dollars) spent by medium in 1998 and 1999 is shown in the table.**

Medium	1998	1999
Newspapers	44,292	46,648
Magazines	10,518	11,433
Television	47,720	50,440
Radio	15,073	17,215
Yellow Pages	11,990	12,652
Direct Mail	39,620	41,403
Internet	1,050	1,904

List the elements of each set.

58. The set of all media that collected more than $40,000 million in both 1998 and 1999.

59. The set of all media that collected less than $12,000 million in 1998 or 1999 (or both years).

60. The dollar amounts spent for magazine ads in each year shown.

*Advertising Age, May 22, 2000. www.adage.com.

The top five pay-cable services in 1999 are listed here.* Use this information for Exercises 61–66.

Network	Subscribers (in millions)	Content
The Disney Channel	42.9	Movies, cartoons
HBO/Cinemax	35.7	Movies, variety, sports, documentaries
Encore	23.1	On-demand movies
Showtime/The Movie Channel	22.3	Movies, variety, comedy, sports
Spice	11.0	Adult movies

List the elements of the following sets.

61. *F*, the set of networks with more than 25 million subscribers.

*National Cable Television Association, *Cable TV Developments* (Spring 2000).

62. *G*, the set of networks that show sports.

63. *H*, the set of networks that show comedy.

64. $F \cap G$ **65.** $H \cup F$ **66.** F'

The table below shows some symptoms of an overactive thyroid and an underactive thyroid.

Underactive Thyroid	Overactive Thyroid
Sleepiness, *s*	Insomnia, *i*
Dry hands, *d*	Moist hands, *m*
Intolerance of cold, *c*	Intolerance of heat, *h*
Goiter, *g*	Goiter, *g*

Let U be the smallest possible set that includes all the symptoms listed, N be the set of symptoms for an underactive thyroid, and O be the set of symptoms for an overactive thyroid. Use the lower case letters in the table to list the elements of each set.

67. O' **68.** N'

69. $N \cap O$ **70.** $N \cup O$

8.2 APPLICATIONS OF VENN DIAGRAMS

We used Venn diagrams in the last section to illustrate set union and intersection. The rectangular region in a Venn diagram represents the universal set, *U*. Including only a single set, *A*, inside the universal set, as in Figure 8.7, divides *U* into two nonoverlapping regions. Region 1 represents A', those elements outside set *A*, while region 2 represents those elements belonging to set *A*. (The numbering of these regions is arbitrary.)

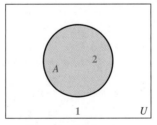

One set leads to 2 regions. (numbering is arbitrary).	Two sets lead to 4 regions. (numbering is arbitrary).
FIGURE 8.7	**FIGURE 8.8**

The Venn diagram of Figure 8.8 shows two sets inside *U*. These two sets divide the universal set into four nonoverlapping regions. As labeled in Figure 8.8, region 1 includes those elements outside both set *A* and set *B*. Region 2 includes those elements belonging to *A* and not to *B*. Region 3 includes those elements belonging

Draw Venn diagrams for the following.

(a) $A \cup B'$

(b) $A' \cap B'$

Answers:

(a)

$A' \cup B'$

(b)

$A' \cap B'$

to both A and B. Which elements belong to region 4? (Again, the numbering is arbitrary.)

▶ **EXAMPLE 1** Draw Venn diagrams similar to Figure 8.8 and shade the regions representing the following sets.

(a) $A' \cap B$

Set A' contains all the elements outside set A. As labeled in Figure 8.8, A' is represented by regions 1 and 4. Set B is represented by the elements in regions 3 and 4. The intersection of sets A' and B, the set $A' \cap B$, is given by the region common to regions 1 and 4 and regions 3 and 4. The result, region 4, is shaded in Figure 8.9.

(b) $A' \cup B'$

Again, set A' is represented by regions 1 and 4 and set B' by regions 1 and 2. To find $A' \cup B'$, identify the region that represents the set of all elements in A', B', or both. The result, which is shaded in Figure 8.10, includes regions 1, 2, and 4. ◀ ①

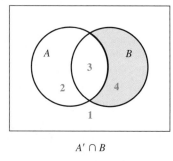

$A' \cap B$

FIGURE 8.9

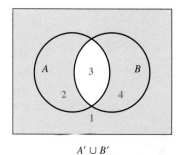

$A' \cup B'$

FIGURE 8.10

Venn diagrams also can be drawn with three sets inside U. These three sets divide the universal set into eight nonoverlapping regions which can be numbered (arbitrarily) as in Figure 8.11.

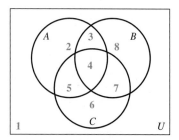

Three sets leads to 8 regions.

FIGURE 8.11

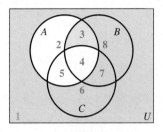

$A' \cup (B \cap C')$

FIGURE 8.12

(2)

Draw a Venn diagram for the following.

(a) $(B' \cap A) \cup C$

(b) $(A \cup B)' \cap C$

Answers:

(a)

$(B' \cap A) \cup C$

(b)

$(A \cup B)' \cap C$

▶ **EXAMPLE 2** Shade $A' \cup (B \cap C')$ in a Venn diagram.

First find $B \cap C'$. See Figure 8.12. Set B is represented by regions 3, 4, 7, and 8, and C' by regions 1, 2, 3, and 8. The overlap of these regions, regions 3 and 8, represents the set $B \cap C'$. Set A' is represented by regions 1, 6, 7, and 8. The union of regions 3 and 8 and regions 1, 6, 7, and 8 includes regions 1, 3, 6, 7, and 8, which are shaded in Figure 8.12. ◀ (2)

Venn diagrams can be used to solve problems that result from surveying groups of people. As an example, suppose a researcher collecting data on 100 households finds that

21 have a DVD player;

56 have a videocassette recorder (VCR); and

12 have both.

The researcher wants to answer the following questions.

(a) How many do not have a VCR?

(b) How many have neither a DVD player nor a VCR?

(c) How many have a DVD player but not a VCR?

A Venn diagram like the one in Figure 8.13 will help sort out the information. In Figure 8.13(a), we put the number 12 in the region common to both a VCR and a DVD player, because 12 households have both. Of the 21 with a DVD player, $21 - 12 = 9$ have no VCR, so in Figure 8.13(b) we put 9 in the region for a DVD player but no VCR. Similarly, $56 - 12 = 44$ households have a VCR but not a DVD player, so we put 44 in that region. Finally, the diagram shows that $100 - 44 - 12 - 9 = 35$ households have neither a VCR nor a DVD player. Now we can answer the questions:

(a) $35 + 9 = 44$ do not have a VCR;

(b) 35 have neither;

(c) 9 have a DVD player but not a VCR. (3)

(a)

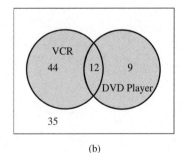

(b)

FIGURE 8.13

③

(a) Place numbers in the regions on a Venn diagram if the data on the 100 households showed

    29 DVD players;

    63 VCRs;

    20 with both.

(b) How many have a VCR but not a DVD player?

Answers:

(a)

(b) 43

▶ **EXAMPLE 3**   A group of 60 freshman business students at a large university was surveyed, with the following results.

    19 of the students read *Business Week;*

    18 read the *Wall Street Journal;*

    50 read *Fortune;*

    13 read *Business Week* and the *Journal;*

    11 read the *Journal* and *Fortune;*

    13 read *Business Week* and *Fortune;*

    9 read all three.

Use this data to answer the following questions.

    (a) How many students read none of the publications?

    (b) How many read only *Fortune?*

    (c) How many read *Business Week* and the *Journal* but not *Fortune?*

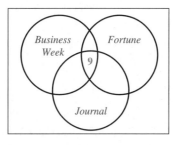

FIGURE 8.14(a)

Once again, use a Venn diagram to represent the data. Since 9 students read all three publications, begin by placing 9 in the area in Figure 8.14(a) that belongs to all three regions.

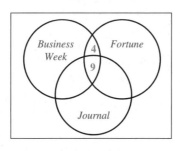

FIGURE 8.14(b)

Of the 13 students who read *Business Week* and *Fortune,* 9 also read the *Journal.* Therefore only 13 − 9 = 4 students read just *Business Week* and *Fortune.* So place a 4 in the region common only to *Business Week* and *Fortune* readers, as in Figure 8.14(b).

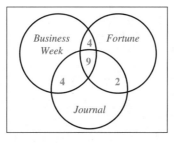

FIGURE 8.14(c)

In the same way, place a 4 in the region of Figure 8.14(c) common only to *Business Week* and the *Journal,* and 2 in the region common only to *Fortune* and the *Journal.*

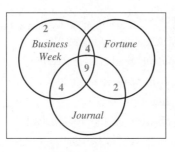

FIGURE 8.14(d)

The data shows that 19 students read *Business Week*. However, $4 + 9 + 4 = 17$ readers have already been placed in the *Business Week* region. The balance of this region in Figure 8.14(d) will contain only $19 - 17 = 2$ students. These 2 students read *Business Week* only—not *Fortune* and not the *Journal*.

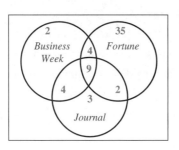

FIGURE 8.14(e)

In the same way, 3 students read only the *Journal* and 35 read only *Fortune*, as shown in Figure 8.14(e).

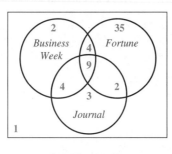

FIGURE 8.14(f)

A total of $2 + 4 + 3 + 4 + 9 + 2 + 35 = 59$ students are placed in the various regions of Figure 8.14(e). Since 60 students were surveyed, $60 - 59 = 1$ student reads none of the three publications and 1 is placed outside the other regions in Figure 8.14(f).

In the example about the three publications, how many students read exactly

(a) 1 of the publications?

(b) 2 of the publications?

Answers:

(a) 40

(b) 10

Figure 8.14(f) can now be used to answer the questions asked above.

(a) Only 1 student reads none of the publications.

(b) There are 35 students who read only *Fortune*.

(c) The overlap of the regions representing *Business Week* and the *Journal* shows that 4 students read *Business Week* and the *Journal* but not *Fortune*. ◀ ④

▶**EXAMPLE 4** Jeff Friedman is a section chief for an electric utility company. The employees in his section cut down tall trees, climb poles, and splice wire. Friedman reported the following information to the management of the utility.

Of the 100 employees in my section,

45 can cut tall trees;

50 can climb poles;

57 can splice wire;

28 can cut trees and climb poles;

⑤

In Example 4, suppose 46 employees can cut tall trees. Then how many

(a) can only cut tall trees?

(b) can cut trees or climb poles?

(c) can cut trees or climb poles or splice wire?

Answers:

(a) 4

(b) 68

(c) 91

20 can climb poles and splice wire;

25 can cut trees and splice wire;

11 can do all three;

  9 can't do any of the three (management trainees).

The data supplied by Friedman leads to the numbers shown in Figure 8.15. Add the numbers from all the regions to get the total number of Friedman's employees.

$$9 + 3 + 14 + 23 + 11 + 9 + 17 + 13 = 99$$

Friedman claimed to have 100 employees, but his data indicates only 99. The management decided that Friedman didn't qualify as a section chief and reassigned him as a nightshift meter reader in Guam. (Moral: He should have taken this course.) ◀ ⑤

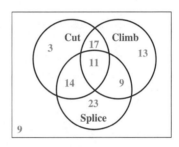

FIGURE 8.15

NOTE In all the examples above, we started in the innermost region with the intersection of the categories. This is usually the best way to begin solving problems of this type.

We use the symbol $n(A)$ to denote the *number* of elements in A. For instance, if $A = \{w, x, y, z\}$, then $n(A) = 4$. The following useful fact is proved below.

Union Rule for Counting

$$n(A \cup B) = n(A) + n(B) - n(A \cap B)$$

For example, if $A = \{r, s, t, u, v\}$ and $B = \{r, t, w\}$, then $A \cap B = \{r, t\}$, so that $n(A) = 5$, $n(B) = 3$, and $n(A \cap B) = 2$. By the formula in the box, $n(A \cup B) = 5 + 3 - 2 = 6$, which is certainly true since $A \cup B = \{r, s, t, u, v, w\}$.

Here is a proof of the statement in the box: let x be the number of elements in A that are not in B, y be the number of elements in $A \cap B$, and z be the number of elements in B that are not in A, as indicated in Figure 8.16. That diagram shows that $n(A \cup B) = x + y + z$. It also shows that $n(A) = x + y$ and $n(B) = y + z$, so that

$$n(A) + n(B) - n(A \cap B) = (x + y) + (z + y) - y$$
$$= x + y + z$$
$$= n(A \cup B).$$

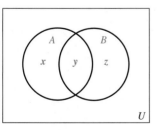

FIGURE 8.16

▶ **EXAMPLE 5** A group of 10 students meets to plan a school function. All are majoring in accounting or economics or both. Five of the students are economics majors and 7 are majors in accounting. How many major in both subjects?

Let A represent the set of accounting majors and B represent the set of economics majors. Use the union rule, with $n(A) = 5$, $n(B) = 7$, and $n(A \cup B) = 10$. We must find $n(A \cap B)$.

$$n(A \cup B) = n(A) + n(B) - n(A \cap B)$$
$$10 = 5 + 7 - n(A \cap B),$$

so

$$n(A \cap B) = 5 + 7 - 10 = 2. \quad ◀ \quad ⑥$$

⑥

If $n(A) = 10$, $n(B) = 7$, and $n(A \cap B) = 3$, find $n(A \cup B)$.

Answer:

14

▶ **EXAMPLE 6** Suppose a random sample of 200 voters was selected in the 1994 elections for the U.S. House of Representatives. The approximate numbers from the East, Midwest, South, or West who voted Democrat or Republican is given by the following table.*

	East (E)	Midwest (M)	South (S)	West (W)	Total
Democrat (D)	24	22	27	26	99
Republican (R)	22	28	33	18	101
Total	46	50	60	44	200

Using the letters given in the table, find the number of people in each of the following sets.

(a) $D \cap S$

The set $D \cap S$ consists of all those who voted Democrat *and* were from the South. From the table, we see that there were 27 such people.

(b) $D \cup S$

The set $D \cup S$ consists of all those who voted Democrat *or* were from the South. We include all 99 who voted Democrat, plus the 33 who were from the South and did not vote Democrat, for a total of 132. Alternatively, we could use the formula $n(D \cup S) = n(D) + n(S) - n(D \cap S) = 99 + 60 - 27 = 132$.

(c) $(E \cup W) \cap R'$

Begin with the set $E \cup W$, which is everyone from the East or the West. This consists of the four categories with 24, 22, 26, and 18 people. Of this set, take those who did *not* vote Republican, for a total of $24 + 26 = 50$ people. This is the number of people from the East or the West who did not vote Republican. ◀ ⑦

⑦

Find the number of people in each set.

(a) $R \cup M$

(b) $(R \cap W) \cup E'$

Answers:

(a) 123

(b) 154

New York Times, November 13, 1994, p. 24.

8.2 Exercises

Sketch a Venn diagram like the one below and use shading to show each of the following sets. (See Example 1.)

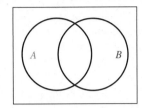

1. $B \cap A'$ 2. $A \cup B'$ 3. $A' \cup B$ 4. $A' \cap B'$
5. $B' \cup (A' \cap B')$ 6. $(A \cap B) \cup B'$
7. U' 8. $\emptyset'$
9. Three sets divide the universal set into at most _____ regions.
10. What does the notation $n(A)$ represent?

Sketch a Venn diagram like the one shown and use shading to show each of the following sets. (See Example 2.)

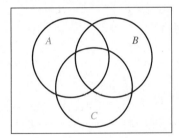

11. $(A \cap C') \cup B$ 12. $A \cap (B \cup C')$
13. $A' \cap (B \cap C)$ 14. $(A' \cap B') \cap C$
15. $(A \cap B') \cup C$ 16. $(A \cap B') \cap C$

Use Venn diagrams to answer the following questions. (See Examples 2 and 4.)

17. **Social Science** In 2000, the percent of white children under 18 years of age who lived with both parents was 75.3; the percent of white children under 18 years of age who lived with their father only was 4.3; and the percent of white children under 18 years of age who lived with neither parent was 3.1.* What percent of white children under age 18 lived with their mother only?

18. **Social Science** In 2000, there were 3842 (in thousands) children under the age of 18 living with their grandparents. Of these children, 531 had both parents also living

with them, 1732 had only their mother living with them, and 220 had only their father living with them.* How many children lived with their grandparents only?

19. **Social Science** A survey of people attending a Lunar New Year celebration in Chinatown yielded the following results:

   120 were women;
   150 spoke Cantonese;
   170 lit firecrackers;
   108 of the men spoke Cantonese;
   100 of the men did not light firecrackers;
   18 of the non-Cantonese-speaking women lit firecrackers;
   78 non-Cantonese-speaking men did not light firecrackers;
   30 of the women who spoke Cantonese lit firecrackers.

   (a) How many attended?
   (b) How many of those who attended did not speak Cantonese?
   (c) How many women did not light firecrackers?
   (d) How many of those who lit firecrackers were Cantonese-speaking men?

20. **Business** Jeff Friedman, of Example 4 in the text, was again reassigned, this time to the home economics department of the electric utility. He interviewed 140 people in a suburban shopping center to find out some of their cooking habits. He obtained the following results. Should he be reassigned yet one more time?

   58 use microwave ovens;
   63 use electric ranges;
   58 use gas ranges;
   19 use microwave ovens and electric ranges;
   17 use microwave ovens and gas ranges;
   4 use both gas and electric ranges;
   1 uses all three;
   2 cook only with solar energy.

21. **Natural Science** After a genetics experiment, the number of pea plants having certain characteristics was tallied, with the results as follows.

   22 were tall;
   25 had green peas;
   39 had smooth peas;
   9 were tall and had green peas;
   17 were tall and had smooth peas;
   20 had green peas and smooth peas;

---

*U.S. Bureau of the Census, *America's Families and Living Arrangements 2000*.

*U.S. Bureau of the Census.

6 had all three characteristics;

4 had none of the characteristics.

(a) Find the total number of plants counted.

(b) How many plants were tall and had peas that were neither smooth nor green?

(c) How many plants were not tall but had peas that were smooth and green?

22. Health   Human blood can contain either no antigens, the A antigen, the B antigen, or both the A and B antigens. A third antigen, called the Rh antigen, is important in human reproduction, and again may or may not be present in an individual. Blood is called type A-positive if the individual has the A and Rh antigens, but not the B antigen. A person having only the A and B antigens is said to have type AB-negative blood. A person having only the Rh antigen has type O-positive blood. Other blood types are defined in a similar manner. Identify the blood type of the individuals in regions (a)–(g) of the Venn diagram.

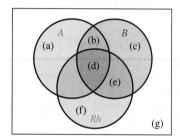

23. Natural Science   Use the diagram from Exercise 22. In a certain hospital, the following data was recorded.

25 patients had the A antigen;

17 had the A and B antigens;

27 had the B antigen;

22 had the B and Rh antigens;

30 had the Rh antigen;

12 had none of the antigens;

16 had the A and Rh antigens;

15 had all three antigens.

How many patients

(a) were represented?

(b) had exactly one antigen?

(c) had exactly two antigens?

(d) had O-positive blood?

(e) had AB-positive blood?

(f) had B-negative blood?

(g) had O-negative blood?

(h) had A-positive blood?

24. Social Science   At a pow-wow in Arizona, Native Americans from all over the Southwest came to participate in the ceremonies. A coordinator of the pow-wow took a survey and found that

15 families brought food, costumes, and crafts;

25 families brought food and crafts;

42 families brought food;

20 families brought costumes and food;

6 families brought costumes and crafts but not food;

4 families brought crafts but neither food nor costumes;

10 families brought none of the three items;

18 families brought costumes but not crafts.

(a) How many families were surveyed?

(b) How many families brought costumes?

(c) How many families brought crafts, but not costumes?

(d) How many families did not bring crafts?

(e) How many families brought food or costumes?

25. Social Science   The following table lists the number of head and neck injuries for 319 ice hockey players wearing either a full shield or half shield in the Canadian Inter-University Athletics Union during the 1997–98 season.* Using the letters given in the table, find the number of injuries in each set.

	Half Shield (H)	Full Shield (F)
Head and Face Injuries (A)	95	34
Concussions (B)	41	38
Neck Injuries (C)	9	7
Other Injuries (D)	202	150

(a) $A \cap F$ **(b)** $C \cap (H \cup F)$

(c) $D \cup F$ **(d)** $B' \cap C'$

26. Social Science   The number of female military personnel on active duty in September 1998 is given in the table.[†]

	Army (A)	Air Force (B)	Navy (C)	Marines (D)	Total
Officers (O)	10,367	11,971	7777	854	30,969
Enlisted (E)	60,787	53,542	42,261	8928	165,518
Cadets and Midshipmen (M)	624	653	656	0	1,933
Total	71,778	66,166	50,694	9782	198,420

Use this information and the letters given to find the number of female military personnel in each of the following sets.

(a) $A \cup B$ **(b)** $E \cup (C \cup D)$ **(c)** $O' \cap M'$

---

*Brian Benson, Nicholas Nohtaki, M. Sarah Rose, and Willem Meeuwisse, "Head and Neck Injuries Among Ice Hockey Players Wearing Full Face Shields vs. Half Face Shields," *JAMA* 282, no. 24 December 22/29, pp. 2328–2332.

†Borga Brunner, ed., *Time Almanac 2000*, p. 395.

27. Social Science The U.S. population by age and race in 2000 (in millions) is given in the table.*

	Non-Hispanic White (A)	Hispanic (B)	African-American (C)	Asian-American (D)	Totals
Under 45 (E)	120.1	25.7	24.4	7.6	177.8
45–64 (F)	47.5	4.9	6.3	2.2	60.9
65 and Over (G)	29.1	1.9	2.8	0.8	34.6
Totals	196.7	32.5	33.5	10.6	273.3

Using the letters given in the table, find the number of people in each set. (See Example 6.)
(a) $A \cap E$ **(b)** $F \cup B$ **(c)** $F \cup (C \cap G)$
(d) $E' \cap (B \cup G)$ **(e)** $G' \cup D$ **(f)** $F' \cap (A' \cap C')$

28. Health The table shows the numbers of breast cancers diagnosed among women in the California Teachers Study from 1996 through 1998.

Race (A)	No Breast Cancer (B)	Breast Cancer (C)	Total
White (D)	99,649	1236	100,885
Black (E)	3,083	36	3,119
Hispanic (F)	5,024	35	5,059
Asian/Pacific Islander (G)	4,167	45	4,212
Other (H)	2,500	21	2,521
Total	114,423	1373	115,796

Using the letters given in the table, find the number of women in each set. (See example 6.)
(a) $G \cap C$ **(b)** $B \cup F$
(c) $(D \cup E) \cap B$ **(d)** $(D' \cap C)'$

29. Restate the union rule in words.

Use Venn diagrams to answer the following questions. (See Example 5.)

30. If $n(A) = 5$, $n(B) = 8$, and $n(A \cap B) = 4$, what is $n(A \cup B)$?

31. If $n(A) = 12$, $n(B) = 27$, and $n(A \cup B) = 30$, what is $n(A \cap B)$?

32. Suppose $n(B) = 7$, $n(A \cap B) = 3$, and $n(A \cup B) = 20$. What is $n(A)$?

33. Suppose $n(A \cap B) = 5$, $n(A \cup B) = 35$, and $n(A) = 13$. What is $n(B)$?

Draw a Venn diagram and use the given information to fill in the number of elements for each region.

34. $n(U) = 38, n(A) = 16, n(A \cap B) = 12, n(B') = 20$

35. $n(A) = 26, n(B) = 10, n(A \cup B) = 30, n(A') = 17$

36. $n(A \cup B) = 17, n(A \cap B) = 3, n(A) = 8, n(A' \cup B') = 21$

37. $n(A') = 28, n(B) = 25, n(A' \cup B') = 45, n(A \cap B) = 12$

38. $n(A) = 28, n(B) = 34, n(C) = 25, n(A \cap B) = 14, n(B \cap C) = 15, n(A \cap C) = 11, n(A \cap B \cap C) = 9, n(U) = 59$

39. $n(A) = 54, n(A \cap B) = 22, n(A \cup B) = 85, n(A \cap B \cap C) = 4, n(A \cap C) = 15, n(B \cap C) = 16, n(C) = 44, n(B') = 63$

*In Exercises 40–43, show that the statements are true by drawing Venn diagrams and shading the regions representing the sets on each side of the equals signs.**

40. $(A \cup B)' = A' \cap B'$
41. $(A \cap B)' = A' \cup B'$
42. $A \cap (B \cup C) = (A \cap B) \cup (A \cap C)$
43. $A \cup (B \cap C) = (A \cup B) \cap (A \cup C)$

---

*U.S. Bureau of the Census, *Current Population Reports* (Projections).

*The statements in Exercises 40 and 41 are known as De Morgan's laws. They are named for the English mathematician Augustus De Morgan (1806–71).

8.3 INTRODUCTION TO PROBABILITY

If you go to a supermarket and buy 5 pounds of peaches at $1.26 per pound, you can easily find the *exact* price of your purchase: $6.30. On the other hand, the produce manager of the market is faced with the problem of ordering peaches. The manager may have a good estimate of the number of pounds of peaches that will be sold during the day, but it is impossible to predict the *exact* amount. The number of pounds that customers will purchase during a day is *random:* the quantity cannot be predicted exactly. A great many problems that come up in applications of mathematics involve random phenomena—those for which exact prediction is impossible. The best that we can do is determine the *probability* of the possible outcomes.

Sample Spaces

In probability, an **experiment** is an activity or occurrence with an observable result. Each repetition of an experiment is called a **trial.** The possible results of each trial are called **outcomes.** The set of all possible outcomes for an experiment is the **sample space** for that experiment. A sample space for the experiment of tossing a coin is made up of the outcomes heads (h) and tails (t). If S represents this sample space, then

$$S = \{h, t\}.$$

FIGURE 8.17

▶ **EXAMPLE 1**   Give the sample space for each experiment.

(a) A spinner like the one in Figure 8.17 is spun.
  The three outcomes are 1, 2, or 3, so the sample space is

$$\{1, 2, 3\}$$

(b) For the purposes of a public opinion poll, respondents are classified as young, middle-aged, or senior and as male or female.
  A sample space for this poll could be written as a set of orderend pairs:

$$\{(\text{young, male}), (\text{young, female}), (\text{middle-aged, male}),$$
$$(\text{middle-aged, female}), (\text{senior, male}), (\text{senior, female})\}$$

(c) An experiment consists of studying the numbers of boys and girls in families with exactly 3 children. Let b represent *boy* and g represent *girl.*
  A three-child family can have 3 boys, written *bbb,* 3 girls, *ggg,* or various combinations, such as *bgg.* A sample space with four outcomes (not equally likely) is

$$S_1 = \{3 \text{ boys}, 2 \text{ boys and } 1 \text{ girl}, 1 \text{ boy and } 2 \text{ girls}, 3 \text{ girls}\}.$$

Notice that a family with 3 boys or 3 girls can occur in just one way, but a family of two boys and 1 girl or 1 boy and 2 girls can occur in more than one way. If the order of the births is considered, so that *bgg* is different from *gbg* or *ggb,* for example, another sample space is

$$S_2 = \{bbb, bbg, bgb, gbb, bgg, gbg, ggb, ggg\}.$$

The second sample space, S_2, has equally likely outcomes if we assume that having a boy or a girl is equally likely. This assumption, while not quite true, is approximately true, so we will use it throughout this book. The outcomes in S_1 are not equally likely, since there is more than one way to get a family with 2 boys and 1 girl (*bbg, bgb,* or *gbb*) or a family with 2 girls and 1 boy (*ggb, gbg,* or *bgg*), but only one way to get 3 boys (*bbb*) or 3 girls (*ggg*). ◀

(1)

(a) Write an equally likely sample space for the experiment of tossing 2 fair coins.

(b) Write a sample space for the experiment of tossing 2 fair coins, if we are interested only in the number of heads. Are the outcomes in this sample space equally likely?

Answers:

(a) {hh, ht, th, tt}

(b) {0, 1, 2}; no

CAUTION An experiment may have more than one sample space, as shown in Example 1(c). The most convenient sample spaces have equally likely outcomes, but it is not always possible to choose such a sample space. (1)

Events

An **event** is a subset of a sample space. If the sample space for tossing a coin is $S = \{h, t\}$, then one event is $E = \{h\}$, which represents the outcome "heads."

An ordinary die is a cube whose six different faces show the following numbers of dots: 1, 2, 3, 4, 5, and 6. If the die is fair (not "loaded" to favor certain faces over others), then any one of the faces is equally likely to come up when the die is rolled. The sample space for the experiment of rolling a single fair die is $S = \{1, 2, 3, 4, 5, 6\}$. Some possible events are listed below.

The die shows an even number: $E_1 = \{2, 4, 6\}$.

The die shows a 1: $E_2 = \{1\}$.

The die shows a number less than 5: $E_3 = \{1, 2, 3, 4\}$.

The die shows a multiple of 3: $E_4 = \{3, 6\}$.

▶**EXAMPLE 2** For the sample space S_2 in Example 1(c), write the following events.

(a) Event H: the family has exactly 2 girls.
Families with three children can have exactly 2 girls with either *bgg*, *gbg*, or *ggb*, so that event H is

$$H = \{bgg, gbg, ggb\}.$$

(b) Event K: the three children are the same sex.
Two outcomes satisfy this condition, all boys or all girls.

$$K = \{bbb, ggg\}$$

(c) Event J: the family has three girls.
Only *ggg* satisfies this condition, so

$$J = \{ggg\}. \ ◀ \ (2)$$

(2)

Suppose a die is tossed. Write the following events.

(a) The number showing is less than 3.

(b) The number showing is 5.

(c) The number showing is 8.

Answers:

(a) {1, 2}

(b) {5}

(c) ∅

In Example 2(c), event J had only one possible outcome, *ggg*. Such an event, with only one possible outcome, is a **simple event.** If an event E equals the sample space S, then E is a **certain event.** If event $E = ∅$, then E is an **impossible event.**

▶**EXAMPLE 3** Suppose a fair die is rolled. As shown above, the sample space is $\{1, 2, 3, 4, 5, 6\}$.

(a) The event "the die shows a 4," {4}, has only one possible outcome. It is a simple event.

(b) The event "the number showing is less than 10" equals the sample space $S = \{1, 2, 3, 4, 5, 6\}$. This event is a certain event; if a die is rolled the number showing (either 1, 2, 3, 4, 5, or 6) must be less than 10.

Which of the events listed in side problem 2 is

(a) simple?

(b) certain?

(c) impossible?

Answers:

(a) Part (b)

(b) None

(c) Part (c)

Give the following events for the experiment of rolling a fair die if $E = \{1, 3\}$ and $F = \{2, 3, 4, 5\}$.

(a) $E \cap F$

(b) $E \cup F$

(c) E'

Answers:

(a) $\{3\}$

(b) $\{1, 2, 3, 4, 5\}$

(c) $\{2, 4, 5, 6\}$

(c) The event "the die shows a 7" is the empty set, ∅; this is an impossible event. ◀ ③

Since events are sets, we can use set operations to find unions, intersections, and complements of events. A summary of the set operations for events is given below.

Set Operations for Events

Let E and F be events for a sample space S.

$E \cap F$ occurs when both E **and** F occur;

$E \cup F$ occurs when E **or** F **or both** occur;

E' occurs when E does **not** occur.

▶ **EXAMPLE 4** A study of workers earning the minimum wage grouped such workers into various categories, which can be interpreted as events when a worker is selected at random.* Consider the following events:

E: worker is under 20;

F: worker is white;

G: worker is female.

Describe each of the following events in words.

(a) E'

E' is the event that the worker is 20 or over.

(b) $F \cap G'$

$F \cap G'$ is the event that the worker is white and not a female, that is, the worker is a white male.

(c) $E \cup G$

$E \cup G$ is the event that the worker is under 20 or is female. Note that this event includes all workers under 20, both male and female, and all female workers of any age. ④

Two events that cannot both occur at the same time, such as getting both a head and a tail on the same toss of a coin, are called **mutually exclusive events.**

Mutually Exclusive Events

Events E and F are mutually exclusive events if $E \cap F = \emptyset$.

For any event E, E and E' are mutually exclusive. By definition, mutually exclusive events are disjoint sets.

*New York Times, April 19, 1996, p. A26.

FIGURE 8.18

In Example 5, let $F = \{2, 4, 6\}$, $K = \{1, 3, 5\}$, and G remain the same. Are the following events mutually exclusive?

(a) F and K

(b) F and G

Answers:

(a) Yes

(b) No

▶ **EXAMPLE 5** Let $S = \{1, 2, 3, 4, 5, 6\}$, the sample space for tossing a die. Let $E = \{4, 5, 6\}$ and let $G = \{1, 2\}$. Then E and G are mutually exclusive events because they have no outcomes in common; $E \cap G = \emptyset$. See Figure 8.18. ◀ ⑤

Probability

For sample spaces with *equally likely* outcomes, the probability of an event is defined as follows.

> ### Basic Probability Principle
>
> Let S be a sample space of equally likely outcomes and let event E be a subset of S. Then the **probability that event E occurs** is
> $$P(E) = \frac{n(E)}{n(S)}.$$

By this definition, the **probability of an event** is a number that indicates the relative likelihood of the event.

CAUTION The basic probability principle only applies when the outcomes are equally likely.

▶ **EXAMPLE 6** Suppose a single fair die is rolled. Use the sample space $S = \{1, 2, 3, 4, 5, 6\}$ and give the probability of each of the following events.

(a) E: the die shows an even number.
Here, $E = \{2, 4, 6\}$, a set with three elements. Because S contains six elements,
$$P(E) = \frac{3}{6} = \frac{1}{2}.$$

(b) F: the die shows a number less than 10.
Event F is a certain event, with
$$F = \{1, 2, 3, 4, 5, 6\},$$
so that
$$P(F) = \frac{6}{6} = 1.$$

(c) G: the die shows an 8.
This event is impossible, so
$$P(G) = \frac{0}{6} = 0. \quad ◀ \quad ⑥$$

A fair die is rolled. Find the probability of rolling

(a) an odd number;

(b) 2, 4, 5, or 6;

(c) a number greater than 5;

(d) the number 7.

Answers:

(a) 1/2

(b) 2/3

(c) 1/6

(d) 0

A standard deck of 52 cards has four suits: hearts (♥), clubs (♣), diamonds (♦), and spades (♠), with 13 cards in each suit. The hearts and diamonds are red, and the spades and clubs are black. Each suit has an ace (A), a king (K), a queen (Q), a jack (J), and cards numbered from 2 to 10. The jack, queen, and king are called face cards and for many purposes can be thought of as having values 11, 12, and 13, respectively. The ace can be thought of as the low card (value 1) or the high card (value 14). See Figure 8.19. We will refer to this standard deck of cards often in our discussion of probability.

FIGURE 8.19

▶ **EXAMPLE 7** If a single card is drawn at random from a standard, well-shuffled, 52-card deck, find the probability of each of the following events.

(a) Drawing an ace

There are 4 aces in the deck. The event "drawing an ace" is

$$\{\text{heart ace, diamond ace, club ace, spade ace}\}.$$

Therefore,

$$P(\text{ace}) = \frac{4}{52} = \frac{1}{13}.$$

(b) Drawing a face card

Since there are 12 face cards,

$$P(\text{face card}) = \frac{12}{52} = \frac{3}{13}.$$

(c) Drawing a spade

The deck contains 13 spades, so

$$P(\text{spade}) = \frac{13}{52} = \frac{1}{4}.$$

(d) Drawing a spade or a heart

Besides the 13 spades, the deck contains 13 hearts, so

$$P(\text{spade or heart}) = \frac{26}{52} = \frac{1}{2}. \quad ◀ \quad ⑦$$

⑦

A single playing card is drawn at random from an ordinary 52-card deck. Find the probability of drawing

(a) a queen;

(b) a diamond;

(c) a red card.

Answers:

(a) 1/13

(b) 1/4

(c) 1/2

In the preceding examples, the probability of each event was a number between 0 and 1, inclusive. The same thing is true in general. Any event E is the subset of the sample space S, so $0 \le n(E) \le n(S)$. Since $P(E) = n(E)/n(S)$, it follows that $0 \le P(E) \le 1$.

For any event E,

$$0 \le P(E) \le 1.$$

▶**EXAMPLE 8**   The table gives the number of years of service of senators in the 107th Congress of the United States of America, which convened on January 3, 2001.*

Years of Service	Number of Senators
0–9	54
10–19	24
20–29	16
30–39	4
40 or more	2

Find the probability that a randomly selected senator of the 107th Congress had served 30 or more years when Congress convened.

Answer:

.06

Find the probability that a randomly selected senator of the 107th Congress served 20–29 years when Congress convened.

This probability is found by dividing the number of senators who served 20–29 years by the total number of senators. Thus,

$$P(\text{20–29 years}) = \frac{16}{100} = .16. \quad ◀ \quad ⑧$$

World Almanac and Book of Facts 2002, p. 83.

8.3 Exercises

1. What is meant by a "fair" coin or die?

2. What is the sample space for an experiment?

Write sample spaces for the experiments in Exercises 3–8. (See Example 1.)

3. A month of the year is chosen for a wedding.

4. A day in April is selected for a bicycle race.

5. A student is asked how many points she earned on a recent 80-point test.

6. A person is asked the number of hours (to the nearest hour) he watched television yesterday.

7. The management of an oil company must decide whether to go ahead with a new oil shale plant or to cancel it.

8. A coin is tossed and a die is rolled.

9. Define an event.

10. What is a simple event?

For the experiments in Exercises 11–16, write out an equally likely sample space, and then write the indicated events in set notation. (See Examples 2 and 3.)

11. A marble is drawn at random from a bowl containing 3 yellow, 4 white, and 8 blue marbles.
    (a) A yellow marble is drawn.
    (b) A blue marble is drawn.
    (c) A white marble is drawn.
    (d) A black marble is drawn.

12. Slips of paper marked with the numbers 1, 2, 3, 4, and 5 are placed in a box. After they are mixed, two slips are drawn.
    (a) Both slips are marked with even numbers.
    (b) One slip is marked with an odd number and the other is marked with an even number.
    (c) Both slips are marked with the same number.

13. An unprepared student takes a three-question true/false quiz in which he guesses the answers to all three questions.
    (a) The student gets three answers wrong.
    (b) The student gets exactly two answers correct.
    (c) The student gets only the first answer correct.

14. A die is tossed twice, with the tosses recorded as ordered pairs.
    (a) The first die shows a 3.
    (b) The sum of the numbers showing is 8.
    (c) The sum of the numbers showing is 13.

15. A jar contains four balls, labeled 1, 2, 3, and 4. A second jar contains five balls, labeled 1, 2, 3, 4, and 5. An experiment consists of taking one ball from the first jar, and then taking a ball from the second jar
    (a) The number on the first ball is even.
    (b) The number on the second ball is even.
    (c) The sum of the numbers on the two balls is 5.
    (d) The sum of the numbers on the two balls is 1.

16. From 5 employees, Strutz, Martin, Hampton, Williams, and Ewing, 2 are selected to attend a conference.
    (a) Hampton is selected.
    (b) Strutz and Martin are not both selected.
    (c) Both Williams and Ewing are selected.

A single fair die is rolled. Find the probabilities of the following events. (See Example 6.)

17. Getting a 5

18. Getting a number less than 4

19. Getting a number greater than 4

20. Getting a 2 or a 5

21. Getting a multiple of 3

22. Getting any number except 3

A card is drawn from a well-shuffled deck of 52 cards. Find the probability of drawing each of the following. (See Example 7.)

23. A 9

24. A black card

25. A black 9

26. A heart

27. The 9 of hearts

28. A 2 or a queen

29. A black 7 or a red 8

30. A red face card

A jar contains 5 red, 4 black, 7 purple, and 9 green marbles. If a marble is drawn at random, what is the probability that the marble is

31. red?

32. black?

33. green?

34. purple?

35. not black?

36. not purple?

37. red or black?

38. black or purple?

Solve each problem. (See Example 8.)

39. Finance As of July 2000, the Janus Mercury fund invested in equities throughout the world, as shown below.*

Region	% of Equities
Europe	18.69
Pacific Rim	6.44
United States	73.31
Latin America	1.56

Find the probability that a randomly selected equity would be from each of the following regions.
(a) The Pacific Rim
(b) Europe
(c) The United States

40. Health For a medical experiment, people are classified as to whether they smoke, have a family history of heart disease, or are overweight. Define events E, F, and G as follows.

    E: person smokes

    F: person has a family history of heart disease

    G: person is overweight

Describe each of the following events in words.
(a) G' (b) $F \cap G$ (c) $E \cup G'$

41. Health Refer to Exercise 40. Describe each of the following events in words.
(a) $E \cup F$ (b) $E' \cap F$ (c) $F' \cup G'$

*www.janus.com

42. Social Sciences   The population of the United States by race in 2000, and the projected population by race for the year 2025 are given in the table at the right (in thousands).* Find the probability of a randomly selected person being each of the following.

(a) Hispanic in 2000
(b) Hispanic in 2025
(c) African-American in 2000
(d) African-American in 2025

Race	*2000*	*2025*
White	196,700	209,900
Hispanic	32,500	56,900
African-American	33,500	44,700
Asian-American	10,600	24,000
Other	2,100	2,800

*"Projections of the Total Resident Population by 5-Year Age Groups, Race, and Hispanic Origin, 1999 to 2000," from U.S. Bureau of the Census, *Current Population Reports*.

8.4 BASIC CONCEPTS OF PROBABILITY

We determine the probability of more complex events in this section.

To find the probability of the union of two sets E and F in a sample space, we use the union rule for counting given in Section 8.2.

$$n(E \cup F) = n(E) + n(F) - n(E \cap F).$$

Dividing both sides by $n(S)$ shows that

$$\frac{n(E \cup F)}{n(S)} = \frac{n(E)}{n(S)} + \frac{n(F)}{n(S)} - \frac{n(E \cap F)}{n(S)}$$

$$P(E \cup F) = P(E) + P(F) - P(E \cap F).$$

This discussion is summarized below.

Union Rule for Probability

For any events E and F from a sample space S,

$$P(E \cup F) = P(E) + P(F) - P(E \cap F).$$

(Although the union rule applies to any events E and F from any sample space, the derivation we have given is valid only for sample spaces with equally likely simple events.)

▶ **EXAMPLE 1**   If a single card is drawn from an ordinary deck of cards, find the probability that it will be red or a face card.

Let R represent the event "red card" and F the event "face card." There are 26 red cards in the deck, so $P(R) = 26/52$. There are 12 face cards in the deck, so

A single card is drawn from an ordinary deck. Find the probability that it is black or a 9.

Answer:

7/13

$P(F) = 12/52$. Since there are 6 red face cards in the deck, $P(R \cap F) = 6/52$. By the union rule, the probability that the card will be red or a face card is

$$P(R \cup F) = P(R) + P(F) - P(R \cap F)$$

$$= \frac{26}{52} + \frac{12}{52} - \frac{6}{52} = \frac{32}{52} = \frac{8}{13}. \quad \blacktriangleleft ①$$

CAUTION Recall from Section 8.1, the word "or" always indicates a *union*.

▶ **EXAMPLE 2** Suppose two fair dice (plural of *die*) are rolled. Find each of the following probabilities.

(a) The first die shows a 2 or the sum of the results is 6 or 7.

The sample space for the throw of two dice is shown in Figure 8.20, where 1-1 represents the event "the first die shows a 1 and the second die shows a 1," 1-2 represents "the first die shows a 1 and the second die shows a 2," and so on. Let A represent the event "the first die shows a 2" and B represent the event "the sum of the results is 6 or 7." These events are indicated in Figure 8.20. From the diagram, event A has 6 elements, B has 11 elements, and the sample space has 36 elements. Thus,

$$P(A) = \frac{6}{36}, \quad P(B) = \frac{11}{36}, \quad \text{and} \quad P(A \cap B) = \frac{2}{36}.$$

By the union rule,

$$P(A \cup B) = P(A) + P(B) - P(A \cap B),$$

$$P(A \cup B) = \frac{6}{36} + \frac{11}{36} - \frac{2}{36} = \frac{15}{36} = \frac{5}{12}.$$

②

In the experiment of Example 2, find the following probabilities.

(a) The sum is 5 or the second die shows a 3.

(b) Both dice show the same number or the sum is at least 11.

Answers:

(a) 1/4

(b) 2/9

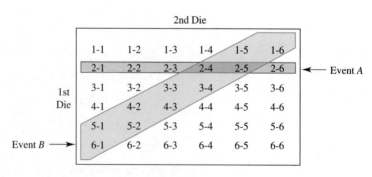

FIGURE 8.20

(b) The sum is 11 or the second die shows a 5.

$P(\text{sum is } 11) = 2/36$, $P(\text{second die shows } 5) = 6/36$, and $P(\text{sum is } 11 \text{ and second die shows } 5) = 1/36$, so

$$P(\text{sum is } 11 \text{ or second die shows } 5) = \frac{2}{36} + \frac{6}{36} - \frac{1}{36} = \frac{7}{36}. \quad \blacktriangleleft ②$$

CAUTION You may wonder why we did not use $S = \{2,3,4,5, \ldots, 12\}$ as the sample space in Example 2. Remember, we prefer to use a sample space with equally likely outcomes. The outcomes in set S above are not equally likely—a sum of 2 can occur in just one way, a sum of 3 in two ways, a sum of 4 in three ways, and so on, as shown in Figure 8.20.

If events E and F are mutually exclusive, then $E \cap F = \emptyset$ by definition; hence; $P(E \cap F) = 0$. Applying the union rule yields this useful fact.

Union Rule for Mutually Exclusive Events

For mutually exclusive events E and F,

$$P(E \cup F) = P(E) + P(F).$$

▶ **EXAMPLE 3**   Assume that the probability of a couple having a boy is the same as the probability of their having a girl. If the couple has 3 children, find the probability that at least 2 of them are girls.

The event of having at least 2 girls is the union of the mutually exclusive events E = "the family has exactly 2 girls" and F = "the family has exactly 3 girls." Using the equally likely sample space

$$\{\textbf{\textit{ggg}}, \textit{ggb}, \textit{gbg}, \textit{bgg}, \textit{gbb}, \textit{bgb}, \textit{bbg}, \textit{bbb}\},$$

we see that $P(2 \text{ girls}) = 3/8$ and $P(3 \text{ girls}) = 1/8$. Therefore,

$$P(\text{at least 2 girls}) = P(2 \text{ girls}) + P(3 \text{ girls})$$

$$= \frac{3}{8} + \frac{1}{8} = \frac{1}{2}. \quad ◀ \quad ③$$

③

In Example 3, find the probability of no more than 2 girls.

Answer:

7/8

By definition of E', for any event E from a sample space S,

$$E \cup E' = S \quad \text{and} \quad E \cap E' = \emptyset.$$

Because $E \cap E' = \emptyset$, events E and E' are mutually exclusive, so that

$$P(E \cup E') = P(E) + P(E').$$

However, $E \cup E' = S$, the sample space, and $P(S) = 1$. Thus

$$P(E \cup E') = P(E) + P(E') = 1.$$

Rearranging these terms gives the following useful rule.

Complement Rule

For any event E,

$$P(E') = 1 - P(E) \quad \text{and} \quad P(E) = 1 - P(E').$$

(a) Let $P(K) = 2/3$. Find $P(K')$.

(b) If $P(X') = 3/4$, find $P(X)$.

Answers:

(a) 1/3

(b) 1/4

▶ **EXAMPLE 4**   If a fair die is rolled, what is the probability that any number but 5 will come up?

If E is the event that 5 comes up, then E' is the event that any number but 5 comes up. $P(E) = 1/6$, so we have $P(E') = 1 - 1/6 = 5/6$. ◀ ④

In Example 5, find the probability that the sum of the numbers rolled is at least 5.

Answer:

5/6

▶ **EXAMPLE 5**   If two fair dice are rolled, find the probability that the sum of the numbers showing is greater than 3.

To calculate this probability directly, we must find the probabilities that the sum is 4, 5, 6, 7, 8, 9, 10, 11, or 12 and then add them. It is much simpler to first find the probability of the complement, the event that the sum is less than or equal to 3.

$$P(\text{sum} \leq 3) = P(\text{sum is } 2) + P(\text{sum is } 3)$$

$$= \frac{1}{36} + \frac{2}{36} = \frac{3}{36} = \frac{1}{12}$$

Now use the fact that $P(E) = 1 - P(E')$ to get

$$P(\text{sum} > 3) = 1 - P(\text{sum} \leq 3) = 1 - \frac{1}{12} = \frac{11}{12}. ◀ ⑤$$

Odds

Sometimes probability statements are given in terms of **odds**, a comparison of $P(E)$ with $P(E')$. For example, suppose $P(E) = \frac{4}{5}$. Then $P(E') = 1 - \frac{4}{5} = \frac{1}{5}$. These probabilities predict that E will occur 4 out of 5 times and E' will occur 1 out of 5 times. Then we say the **odds in favor** of E are 4 to 1 or 4:1.

Odds

The **odds in favor** of an event E are defined as the ratio of $P(E)$ to $P(E')$, or

$$\frac{P(E)}{P(E')}, \qquad P(E') \neq 0.$$

---

⑥

Suppose $P(E) = 9/10$. Find the odds

(a) in favor of E;

(b) against E.

Answers:

(a) 9 to 1

(b) 1 to 9

▶ **EXAMPLE 6**   Suppose the weather forecaster says that the probability of rain tomorrow is 1/3. Find the odds in favor of rain tomorrow.

Let E be the event "rain tomorrow." Then E' is the event "no rain tomorrow." Since $P(E) = 1/3$, $P(E') = 2/3$. By the definition of odds, the odds in favor of rain are

$$\frac{1/3}{2/3} = \frac{1}{2}, \quad \text{written 1 to 2} \quad \text{or} \quad 1:2.$$

On the other hand, the odds that it will *not* rain, or the odds *against* rain, are

$$\frac{2/3}{1/3} = \frac{2}{1}, \quad \text{written 2 to 1.} ◀ ⑥$$

If the odds in favor of an event are, say, 3 to 5, then the probability of the event is 3/8, while the probability of the complement of the event is 5/8. (Odds of 3 to 5 indicate 3 outcomes in favor of the event out of a total of 8 outcomes.) This example suggests the following generalization.

> If the odds favoring event E are m to n, then
>
> $$P(E) = \frac{m}{m + n} \quad \text{and} \quad P(E') = \frac{n}{m + n}.$$

▶ **EXAMPLE 7**   The odds that a particular bid will be the low bid are 4 to 5.

(a) Find the probability that the bid will be the low bid.
 Odds of 4 to 5 show 4 favorable chances out of $4 + 5 = 9$ chances altogether.

$$P(\text{bid will be low bid}) = \frac{4}{4 + 5} = \frac{4}{9}$$

(b) Find the odds against that bid being the low bid.
 There is a 5/9 chance that the bid will not be the low bid, so the odds against a low bid are

$$\frac{P(\text{bid will not be low})}{P(\text{bid will be low})} = \frac{5/9}{4/9} = \frac{5}{4}$$

or 5:4. ◀   ⑦

⑦

If the odds in favor of event E are 1 to 5, find

(a) $P(E)$;

(b) $P(E')$.

Answers:

(a) 1/6

(b) 5/6

Empirical Probability

In many real-life problems, it is not possible to establish exact probabilities for events. Instead, useful approximations are often found by drawing on past experience as a guide to the future. The next example shows one approach to such **empirical probabilities.**

▶ **EXAMPLE 8**   The table lists U.S. advertising volume in thousands of dollars by medium in 1998.*

Medium	Expenditures
Magazines	13,780,249
Newspapers	16,130,928
Network Television	16,271,972
Sport Television	15,486,766
Syndicated Television	2,691,648
Cable Television	6,671,978
Network Radio	824,007
Sunday Magazines	1,029,447

We can find the empirical probability that a dollar of advertising is spent on each medium by first finding the total spent and then dividing the amount spent on each medium by the total. Verify that the amounts in the table sum to 72,886,995.

*The World Almanac and Book of Facts 2000, p. 192.

The probability that a dollar is spent on newspapers, for example, is $P(\text{newspapers}) = 16,130,928/72,886,995 \approx .221$. Similarly, we could divide each amount by 72,886,995, with the results (rounded to three decimal places) shown below.

Medium	Probabilities
Magazines	.189
Newspapers	.221
Network Television	.223
Sport Television	.212
Syndicated Television	.037
Cable Television	.092
Network Radio	.011
Sunday Magazines	.014

⑧

A traffic engineer gathered the following data on the number of accidents at four busy intersections.

Intersection	Number of Accidents
1	25
2	32
3	28
4	35

If an accident occurs at one of these four intersections, what is the probability (to 2 decimal places) that it occurs at intersections 1 or 3?

Answer:

.44

The numbers in this table sum to .999. In theory, they should total 1.000, but this does not always occur with rounded numbers.

The categories in the table are mutually exclusive simple events. Thus, to find the probability that an advertising dollar is spent on newspapers or sport television, we use the union rule to calculate

$$P(\text{newspapers or sport television}) = .221 + .212 = .433.$$

We could get this same result by summing the amount spent on newspapers and sport television and dividing the total by 72,886,995.

Thus, nearly half of all advertising dollars are spent on these two media, a figure that should be of interest to both advertisers and the owners of the various media. ◀ ⑧

A table of probabilities, as in Example 8, sets up a probability distribution; that is, for each possible outcome of an experiment, a number, called the probability of that outcome, is assigned. This assignment may be done in any reasonable way (on a empirical basis, as in Example 8, or by theoretical reasoning, as in Section 8.3), provided that it satisfies the following conditions.

Properties of Probability

Let S be a sample space consisting of n distinct outcomes $s_1, s_2, \ldots, s_n$. An acceptable probability assignment consists of assigning to each outcome s_i a number p_i (the probability of s_i) according to these rules.

1. The probability of each outcome is a number between 0 and 1.

$$0 \le p_1 \le 1, \quad 0 \le p_2 \le 1, \ldots, \quad 0 \le p_n \le 1$$

2. The sum of the probabilities of all possible outcomes is 1.

$$p_1 + p_2 + p_3 + \cdots + p_n = 1$$

Probability distributions are discussed further in the next chapter.

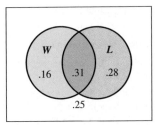

FIGURE 8.21

▶**EXAMPLE 9** Susan is a college student who receives heavy sweaters from her aunt at the first sign of cold weather. She estimates that the probability that a sweater is the wrong size is .47, the probability that it is a loud color is .59, and the probability that it is both the wrong size and a loud color is .31.

(a) Find the probability that the sweater is the correct size and not a loud color.

Let W be the event "wrong size" and L be the event "loud color." Place the given information on a Venn diagram, starting with .31 in the intersection of the regions W and L (see Figure 8.21). As stated earlier, event W has probability .47. Since .31 has already been placed inside the intersection of W and L,

$$.47 - .31 = .16$$

goes inside region W but outside the intersection of W and L. In the same way,

$$.59 - .31 = .28$$

goes inside the region for L and outside the overlap.

Using regions W and L, the event we want is $W' \cap L'$. From the Venn diagram in Figure 8.21, the labeled regions have a total probability of

$$.16 + .31 + .28 = .75$$

Since the entire region of the Venn diagram must have probability 1, the region outside W and L, or $W' \cap L'$, has probability

$$1 - .75 = .25.$$

The probability is .25 that the sweater is the correct size and not a loud color.

(b) Find the probability that the sweater is the correct size *or* not a loud color.

The corresponding region. $W' \cup L'$, has probability

$$.25 + .16 + .28 = .69. \quad ◀ \quad ⑨$$

Find the probability that the sweater in Example 9 is

(a) not a loud color;

(b) the correct size *and* a loud color.

Answers:

(a) .41

(b) .28

8.4 Exercises

1. Define mutually exclusive events in your own words.

Decide whether the events in Exercises 2–7 are mutually exclusive.

2. Owning a car and owning a truck

3. Wearing glasses and wearing sandals

4. Being married and being over 30 years old

5. Being a teenager and being over 30 years old

6. Rolling a die once and getting a 4 and an odd number

7. Being a male and being a postal worker

Two dice are rolled. Find the probabilities of rolling the given sums. (See Examples 2, 4, and 5.)

8. **(a)** 2 **(b)** 4 **(c)** 5 **(d)** 6

9. **(a)** 8 **(b)** 9 **(c)** 10 **(d)** 13

10. **(a)** 9 or more **(b)** Less than 7
    (c) Between 5 and 8 (exclusive)

11. **(a)** Not more than 5 **(b)** Not less than 8
    (c) Between 3 and 7 (exclusive)

Two dice are rolled. Find the probabilities of the following events.

12. The first die is 3 or the sum is 8.

13. The second die is 5 or the sum is 10.

14. Three unusual dice, A, B, and C, are constructed such that die A has the numbers 3, 3, 4, 4, 8, 8; die B has the numbers, 1, 1, 5, 5, 9, 9; and die C has the numbers 2, 2, 6, 6, 7, 7.

    (a) If dice A and B are rolled, find the probability that B beats A, that is, the number that appears on die B is greater than the number that appears on die A.

    (b) Which die is better? Explain.

One card is drawn from an ordinary deck of 52 cards. Find the probabilities of drawing the following cards. (See Example 1.)

15. (a) A 3 or a king
 (b) An ace or a spade
 (c) A red card or a 7

16. (a) A 9 or 10 (b) A red card or a 3
 (c) A 9 or a black 10 (d) A heart or a black card
 (e) A face card or a diamond

17. (a) Less than a 4 (count aces as ones)
 (b) A diamond or a 7 (c) A black card or an ace
 (d) A heart or a jack (e) A red card or a face card

Ms. Elliott invites 10 relatives to a party: her mother, 2 aunts, 3 uncles, 2 brothers, 1 male cousin, and 1 female cousin. If the chances of any 1 guest arriving first are equally likely, find the probabilities that the first guest to arrive is as follows.

18. (a) A brother or an uncle
 (b) A brother or a cousin
 (c) A brother or her mother

19. (a) An uncle or a cousin
 (b) A male or a cousin
 (c) A female or a cousin

The numbers 1, 2, 3, 4, and 5 are written on slips of paper, and two slips are drawn at random one at a time without replacement. Find each of the probabilities in Exercises 20 and 21.

20. (a) The sum of the numbers is 9.
 (b) The sum of the numbers is 5 or less.
 (c) The first number is 2 or the sum is 6.

21. (a) Both numbers are even.
 (b) One of the numbers is even or greater than 3.
 (c) The sum is 5 or the second number is 2.

Use Venn diagrams to work Exercises 22 and 23.

22. Suppose $P(E) = .26$, $P(F) = .41$, and $P(E \cap F) = .14$. Find each of the following.
 (a) $P(E \cup F)$ (b) $P(E' \cap F)$
 (c) $P(E \cap F')$ (d) $P(E' \cup F')$

23. Let $P(Z) = .42$, $P(Y) = .35$, and $P(Z \cup Y) = .61$. Find each of the following probabilities.
 (a) $P(Z' \cap Y')$ (b) $P(Z' \cup Y')$
 (c) $P(Z' \cup Y)$ (d) $P(Z \cap Y')$

24. Define what is meant by odds.

A single fair die is rolled. Find the odds in favor of getting the results in Exercises 25–28. (See Examples 6 and 7.)

25. 5 **26.** 3, 4, or 5
27. 1, 2, 3, or 4 **28.** Some number less than 2

29. A marble is drawn from a box containing 3 yellow, 4 white, and 8 blue marbles. Find the odds in favor of drawing the following.
 (a) A yellow marble (b) A blue marble
 (c) A white marble

30. Find the odds of *not* drawing a white marble in Exercise 29.

31. Two dice are rolled. Find the odds of rolling a 7 or 11.

32. In the "Ask Marilyn" column of *Parade* magazine, a reader wrote about the following game: You and I each roll a die. If your die is higher than mine, you win. Otherwise, I win. The reader thought that the probability that each player wins is 1/2. Is this correct? If not, what is the probability that each player wins?*

33. On page 134 of Roger Staubach's autobiography, *First Down, Lifetime to Go*. Staubach makes the following statement regarding his experience in Vietnam:

> Odds against a direct hit are very low but when your life is in danger, you don't worry too much about the odds.

Is this wording consistent with our definition of odds, for and against? How could it have been said so as to be technically correct?

34. The table gives the odds that a particular event will occur.[†] Convert each odds to the probability that the event will occur.

Event	Odds for the Event
You will eat out today.	1 to 2
The next bottled water you buy will be nothing more than tap water.	1 to 4
The earth will be struck by a huge meteor during your lifetime.	1 to 9000
You will go to Disney World this year.	1 to 9
You'll regain weight you lost by dieting.	9 to 10

Which of Exercises 35–42 are examples of empirical probability?

35. The probability of heads on 5 consecutive tosses of a coin

36. The probability that a freshman entering college will graduate with a degree

37. The probability that a person is allergic to penicillin

38. The probability of drawing an ace from a standard deck of 52 cards

39. The probability that a person will get lung cancer from smoking cigarettes

---

Parade Magazine, November 6, 1994, p. 10. Reprinted by permission of the William Morris Agency, Inc. on behalf of the author. Copyright © 1994 by Marilyn vos Savant.

†The Forum for Investor Advice: Krantz, Les, *What the Odds Are* (Harper Perennial, 1992); and Larry Laudan, *Danger Ahead: The Risks You Really Face on Life's Highway* (John Wiley & Sons, 1997).

40. A weather forecast that predicts a 70% chance of rain tomorrow

41. A gambler's claim that on a roll of a fair die, $P(\text{even}) = 1/2$

42. A surgeon's prediction that a patient has a 90% chance of full recovery

43. What is a probability distribution?

An experiment is conducted for which the sample space is $S = \{s_1, s_2, s_3, s_4, s_5\}$. *Which of the probability assignments in Exercises 44–49 is possible for this experiment? If an assignment is not possible, tell why.*

44.

Outcomes	s_1	s_2	s_3	s_4	s_5
Probabilities	.09	.32	.21	.25	.13

45.

Outcomes	s_1	s_2	s_3	s_4	s_5
Probabilities	.92	.03	0	.02	.03

46.

Outcomes	s_1	s_2	s_3	s_4	s_5
Probabilities	1/3	1/4	1/6	1/8	1/10

47.

Outcomes	s_1	s_2	s_3	s_4	s_5
Probabilities	1/5	1/3	1/4	1/5	1/10

48.

Outcomes	s_1	s_2	s_3	s_4	s_5
Probabilities	.64	−.08	.30	.12	.02

49.

Outcomes	s_1	s_2	s_3	s_4	s_5
Probabilities	.05	.35	.5	.2	−.3

One way to solve a probability problem is to repeat the experiment many times, keeping track of the results. Then the probability can be approximated using the basic definition of the probability of an event E: $P(E) = n(E)/n(S)$, where E occurs $n(E)$ times out of $n(S)$ trials of an experiment. This is called the Monte Carlo method of finding probabilities. If physically repeating the experiment is too tedious, it may be simulated using a random number generator, available on most computers and scientific or graphing calculators. To simulate a coin toss or the roll of a die on a graphing calculator, change the setting to fixed decimal mode with 0 digits displayed. To simulate multiple tosses of a coin, press RAND (or RANDOM or RND#) in the PROB submenu of the MATH (or OPTN) menu and then press ENTER repeatedly. Interpret 0 as a head and 1 as a tail. To simulate multiple rolls of a die, press RAND $\times$ 6 + .5 and then press ENTER repeatedly.

50. Suppose two dice are rolled. Use the Monte Carlo method with at least 50 repetitions to approximate the following probabilities. Compare with the results of Exercise 11.
  (a) $P(\text{the sum is not more than 5})$
  (b) $P(\text{the sum is not less than 8})$

51. Suppose two dice are rolled. Use the Monte Carlo method with at least 50 repetitions to approximate the following probabilities. Compare with the results of Exercise 10.
  (a) $P(\text{the sum is 9 or more})$
  (b) $P(\text{the sum is less than 7})$

52. Suppose three dice are rolled. Use the Monte Carlo method with at least 100 repetitions to approximate the following probabilities.
  (a) $P(\text{the sum is 5 or less})$
  (b) $P(\text{neither a 1 nor a 6 is rolled})$

53. Suppose a coin is tossed 5 times. Use the Monte Carlo method with at least 50 repetitions to approximate the following probabilities.
  (a) $P(\text{exactly 4 heads})$ **(b)** $P(\text{2 heads and 3 tails})$

54. Business Suppose that 8% of a certain batch of calculators have a defective case and that 11% have defective batteries. Also, 3% have both a defective case and defective batteries. A calculator is selected from the batch at random. Find the probability that the calculator has a good case and good batteries.

55. Business The table shows the probabilities of a person accumulating specific amounts of credit card charges over a 12-month period. Find the probabilities that a person's total charges during the period are the following.
  (a) $500 or more   **(b)** Less than $1000
  (c) $500 to $2999   **(d)** $3000 or more

Charges	Probability
Under $100	.31
$100–$499	.18
$500–$999	.18
$1000–$1999	.13
$2000–$2999	.08
$3000–$4999	.05
$5000–$9999	.06
$10,000 or more	.01

Business *The table below shows the probability that a customer of a department store will make a purchase in the indicated range.*

Amount Spent	Probability
Below $10	.07
$10–$24.99	.18
$25–$49.99	.21
$50–$74.99	.16
$75–$99.99	.11
$100–$199.99	.09
$200–$349.99	.07
$350–$499.99	.08
$500 or more	.03

Find the probabilities that a customer makes a purchase in the following ranges.

56. (a) Less than $25
    (b) More than $24.99
    (c) $50 to $199.99

57. (a) Less than $350
    (b) $75 or more
    (c) $200 or more

58. Business The probability that a company will make a profit this year is .74. Find the odds against the company making a profit.

59. Natural Science A study on body types gave the following results: 45% were short; 25% were short and overweight; and 24% were tall and not overweight. Find the probabilities that a person is the following.
    (a) Overweight
    (b) Short but not overweight
    (c) Tall and overweight

60. Natural Science Color blindness is an inherited characteristic that is more common in males than in females. If M represents male and C represents red-green color blindness, we use the relative frequencies of the incidences of males and red-green color blindness as probabilities to get

$$P(C) = .039, \quad P(M \cap C) = .035, \quad P(M \cup C) = .495.^*$$

Find the following probabilities.
    (a) $P(C')$ (b) $P(M)$ (c) $P(M')$
    (d) $P(M' \cap C')$ (e) $P(C \cap M')$ (f) $P(C \cup M')$

61. Natural Science Gregor Mendel, an Austrian monk, was the first to use probability in the study of genetics. In an effort to understand the mechanism of character transmittal from one generation to the next in plants, he counted the number of occurrences of various characteristics. Mendel found that the flower color in certain pea plants obeyed this scheme:

    Pure red crossed with pure white produces red.

From its parents, the red offspring received genes for both red (R) and white (W), but in this case red is *dominant* and white *recessive,* so the offspring exhibits the color red. However, the offspring still caries both genes, and when two such offspring are crossed, several things can happen in the third generation. The table in the next column, which is called a *Punnet square,* shows the equally likely outcomes.

		SECOND PARENT	
		R	W
FIRST PARENT	R W	RR WR	RW WW

Use the fact that red is dominant over white to find each of the following. Assume that there are an equal number of red and white genes in the population.
    (a) $P($a flower is red$)$ (b) $P($a flower is white$)$

62. Natural Science Mendel found no dominance in snapdragons, with one red gene and one white gene producing pink-flowered offspring. These second-generation pinks, however, still carry one red and one white gene, and when they are crossed, the next generation still yields the Punnet square shown above. Find each of the following probabilities.
    (a) $P($red$)$ (b) $P($pink$)$ (c) $P($white$)$

(Mendel verified these probability ratios experimentally and did the same for many characteristics other than flower color. His work, published in 1866, was not recognized until 1890.)

63. Natural Science In most animals and plants, it is very unusual for the number of main parts of the organism (such as arms, legs, toes, or flower petals) to vary from generation to generation. Some species, however, have *meristic variability,* in which the number of certain body parts varies from generation to generation. One researcher studied the front feet of certain guinea pigs and produced the following probabilities.*

$$P(\text{only four toes, all perfect}) = .77$$

$$P(\text{one imperfect toe and four good ones}) = .13$$

$$P(\text{exactly five good toes}) = .10$$

Find the probability of each of the following events.
    (a) No more than four good toes
    (b) Five toes, whether perfect or not

64. Business The following data was gathered for 130 adult U.S. workers: 55 were women; 3 women earned more than $40,000; and 62 men earned less than or equal $40,000. Find the probabilities that an individual is
    (a) a woman earning less than or equal $40,000;
    (b) a man earning more than $40,000;
    (c) a man or is earning more than $40,000;
    (d) a woman or is earning less than or equal $40,000.

---

*The probabilities of a person being male or female are from *The World Almanac and Book of Facts,* 2002. The probabilities of a male and female being color-blind are from *Parsons' Diseases of the Eye* (18th ed.) by Stephen J. H. Miller (Churchill Livingston, 1990), p. 269. This reference gives a range of 3 to 4% for the probability of gross color blindness in men; we used the midpoint of this range.

---

*J. R. Wright, "An Analysis of Variability in Guinea Pigs," *Genetics,* Vol. 19, pp. 506–536.

65. Business A survey of 100 people about their music expenditures gave the following information: 38 bought rock music; 20 were teenagers who bought rock music; and 26 were teenagers. Find the probabilities that a person is
(a) a teenager who buys nonrock music;
(b) someone who buys rock music or is a teenager;
(c) not a teenager;
(d) not a teenager but a buyer of rock music.

66. Social Science In a refugee camp in southern Mexico, it was found that 90% of the refugees came to escape political oppression, 80% came to escape abject poverty, and 70% came to escape both. What is the probability that a refugee in the camp was not poor nor seeking political asylum?

67. Social Science There were 198,420 female military personnel on active duty in September 1998 in various ranks and military branches, as listed in the table below.*

	Army (A)	Air Force (B)	Navy (C)	Marines (D)
Officers (O)	10,367	11,971	7,777	854
Enlisted (E)	60,787	53,542	42,261	8928
Cadets and Midshipmen (M)	624	653	656	0

(a) Convert the numbers in the table to probabilities.
(b) Find the probability that a randomly selected woman is enlisted in the Army.
(c) Find the probability that a randomly selected woman is an officer in the Navy or Marine Corps.
(d) $P(A \cup B)$
(e) $P(E \cup (C \cup D))$

*Borga Brunner, ed., *Time Almanac 2000*, p. 395.

68. Social Science Research has been carried out to measure the amount of intolerance that U.S. citizens have for left-wing Communists and right-wing Fascists, as indicated in the table below. Note that the numbers are given as percents and each row sums to 100 (except for rounding).*
(a) Find the probability that a randomly chosen U.S. citizen would have at least some intolerance of right-wing Fascists.
(b) Find the probability that a randomly chosen U.S. citizen would have at least some intolerance of left-wing Communists.
(c) Compare your answers to parts (a) and (b) and provide possible reasons for these numbers.

	None at All	Don't Know	Not Very Much	Somewhat	Extremely
Left-Wing Communists	13.0	2.7	33.0	34.2	17.1
Right-Wing Fascists	10.1	3.3	20.7	43.1	22.9

*J. L. Gibson, "Putting Up with Fellow Russians: An Analysis of Political Tolerance in the Fledgling Russian Democracy," *Political Research Quarterly*, vol. 51, no. 6, March 1998, pp. 37–38.

8.5 CONDITIONAL PROBABILITY AND INDEPENDENT EVENTS

The training manager for a large brokerage firm has noticed that some of the firm's brokers use the firm's research advice, while other brokers tend to go with their own beliefs about which stocks will go up. To see whether the research department performs better than the beliefs of the brokers, the manager conducted a survey of 100 brokers, with the results shown in the following table.

①

Use the data in the table to find

(a) $P(B)$,

(b) $P(A')$,

(c) $P(B')$.

Answers:

(a) .45

(b) .4

(c) .55

	Picked a Stock That Went Up	Didn't Pick A Stock That Went Up	Totals
Used Research	30	15	45
Didn't Use Research	30	25	55
Totals	60	40	100

Letting A be the event "picked a stock that went up" and letting B be the event "used research," $P(A)$, $P(A')$, $P(B)$, and $P(B')$ can be found. For example, the chart shows that a total of 60 brokers picked stocks that went up, so $P(A) = 60/100 = .6.$ ①

Suppose we want to find the probability that a broker using research will pick a stock that goes up. From the table above, of the 45 brokers who use research, there are 30 who picked stocks that went up, so

$$P(\text{broker who uses research picks stocks that go up}) = \frac{30}{45} \approx .667.$$

This is a different number than the probability that a broker picks a stock that goes up, .6, because *we have additional information* (the broker uses research) *that has reduced the sample space.* In other words, we found the probability that a broker picks a stock that goes up, A, given the additional information that the broker uses research, B. This is called the *conditional probability* of event A, given that event B has occurred, written $P(A \mid B)$. ($P(A \mid B)$ may also be read as "the probability of A given B.")

In the example above,

$$P(A \mid B) = \frac{30}{45}.$$

If we divide the numerator and denominator by 100 (the size of the sample space), this can be written as

$$P(A \mid B) = \frac{\frac{30}{100}}{\frac{45}{100}} = \frac{P(A \cap B)}{P(B)},$$

where $P(A \cap B)$ represents, as usual, the probability that both A and B will occur.

To generalize this result, assume that E and F are two events for a particular experiment. Assume that the sample space S for this experiment has n possible equally likely outcomes. Suppose event F has m elements and $E \cap F$ has k elements ($k \le m$). Using the fundamental principle of probability,

$$P(F) = \frac{m}{n} \quad \text{and} \quad P(E \cap F) = \frac{k}{n}.$$

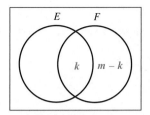

Event F has a total of m elements.

FIGURE 8.22

We now want to find $P(E \mid F)$, the probability that E occurs given that F has occurred. Since we assume F has occurred, reduce the sample space to F: look only at the m elements inside F. (See Figure 8.22.) Of these m elements, there are k elements where E also occurs, because $E \cap F$ has k elements. This makes

$$P(E \mid F) = \frac{k}{m}.$$

Divide numerator and denominator by n to get

$$P(E \mid F) = \frac{k/n}{m/n} = \frac{P(E \cap F)}{P(F)}.$$

The last result motivates the definition of conditional probability.

> The **conditional probability** of an event E given event F, written $P(E \mid F)$, is
>
> $$P(E \mid F) = \frac{P(E \cap F)}{P(F)}, \qquad P(F) \neq 0.$$

This definition tells us that, for equally likely outcomes, conditional probability is found by *reducing the sample space to event F,* and then finding the number of outcomes in F that are also in event E. Thus,

$$P(E \mid F) = \frac{n(E \cap F)}{n(F)}.$$

Although the definition of conditional probability was motivated by an example with equally likely outcomes, it is valid in all cases. For an intuitive explanation, think of the formula as giving the probability that both E and F occur compared with the entire probability of F.

▶ **EXAMPLE 1**   Use the information given in the table on the preceding page to find the following probabilities.

(a) $P(B \mid A)$

This represents the probability that the broker used research, given that the broker picked a stock that went up. Reduce the sample space to A. Then find $n(A \cap B)$ and $n(A)$.

$$P(B \mid A) = \frac{n(A \cap B)}{n(A)} = \frac{30}{60} = \frac{1}{2}$$

If a broker picked a stock that went up, then the probability is 1/2 that the broker used research.

(b) $P(A' \mid B)$

In words, this is the probability that a broker picks a stock that does not go up, even though he used research.

$$P(A' \mid B) = \frac{n(A' \cap B)}{n(B)} = \frac{15}{45} = \frac{1}{3}$$

The table shows the results of a survey of a buffalo herd.

	Males	Females	Totals
Adults	500	1300	1800
Calves	520	500	1020
Totals	1020	1800	2820

Let M represent "male" and A represent "adult." Find each of the following.

(a) $P(M \mid A)$

(b) $P(M' \mid A)$

(c) $P(A \mid M')$

(d) $P(A' \mid M)$

(e) State the probability in part (d) in words.

Answers:

(a) 5/18

(b) 13/18

(c) 13/18

(d) 26/51

(e) The probability that a buffalo is a calf given that it is a male

3

Find $P(F \mid E)$ if $P(E) = .3$, $P(F) = .4$, and $P(E \cup F) = .6$.

Answer:

1/3

(c) $P(B' \mid A')$

Here, we want the probability that a broker who picked a stock that did not go up did not use research.

$$P(B' \mid A') = \frac{n(B' \cap A')}{n(A')} = \frac{25}{40} = \frac{5}{8} \quad \blacktriangleleft \quad ②$$

Venn diagrams can be used to illustrate problems in conditional probability. A Venn diagram for Example 1, in which the probabilities are used to indicate the number in the set defined by each region, is shown in Figure 8.23. In the diagram, $P(B \mid A)$ is found by *reducing the sample space to just set A*. Then $P(B \mid A)$ is the ratio of the number in that part of set B which is also in A to the number in set A, or $.3/.6 = .5$.

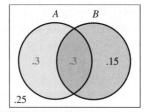

FIGURE 8.23   FIGURE 8.24

▶ **EXAMPLE 2**   Given $P(E) = .4$, $P(F) = .5$, and $P(E \cup F) = .7$, find $P(E \mid F)$.

Find $P(E \cap F)$ first. Then use a Venn diagram to find $P(E \mid F)$. By the union rule,

$$P(E \cup F) = P(E) + P(F) - P(E \cap F)$$
$$.7 = .4 + .5 - P(E \cap F)$$
$$P(E \cap F) = .2.$$

Now use the probabilities to indicate the number in each region of the Venn diagram in Figure 8.24. $P(E \mid F)$ is the ratio of the probability of that part of E which is in F to the probability of F or

$$P(E \mid F) = \frac{P(E \cap F)}{P(F)} = \frac{.2}{.5} = \frac{2}{5} = .4. \quad \blacktriangleleft \quad ③$$

▶ **EXAMPLE 3**   Two fair coins were tossed, and it is known that at least one was a head. Find the probability that both were heads.

The sample space has four equally likely outcomes, $S = \{hh, ht, th, tt\}$. Define two events:

$$E_1 = \text{at least 1 head} = \{hh, ht, th\}$$

and

$$E_2 = 2 \text{ heads} = \{hh\}.$$

Because there are four equally likely outcomes, $P(E_1) = 3/4$. Also, $P(E_1 \cap E_2) = 1/4$. We want the probability that both were heads, given that at least one was a head: that

In Example 3, find the probability that exactly one coin showed a head, given that at least one was a head.

Answer:

2/3

is, we want to find $P(E_2 \mid E_1)$. Because of the condition that at least one coin was a head, the reduced sample space is

$$\{hh, ht, th\}.$$

Since only one outcome in this reduced sample space is 2 heads,

$$P(E_2 \mid E_1) = \frac{1}{3}.$$

Alternatively, use the definition given above.

$$P(E_2 \mid E_1) = \frac{P(E_2 \cap E_1)}{P(E_1)} = \frac{1/4}{3/4} = \frac{1}{3} \quad \blacktriangleleft \quad \text{④}$$

It is important not to confuse $P(A \mid B)$ with $P(B \mid A)$. For example, in a criminal trial, a prosecutor may point out to the jury that the probability of the defendant's DNA profile matching that of a sample taken at the scene of the crime, given that the defendant is innocent, is very small. What the jury must decide, however, is the probability that the defendant is innocent, given that the defendant's DNA profile matches the sample. Confusing the two is an error sometimes called "the prosecutor's fallacy," and the 1990 conviction of a rape suspect in England was overturned by a panel of judges, who ordered a retrial, because the fallacy made the original trial unfair.*

In the next section, we will see how to compute $P(A \mid B)$ when we know $P(B \mid A)$.

Product Rule

If $P(E) \neq 0$ and $P(F) \neq 0$, then the definition of conditional probability shows that

$$P(E \mid F) = \frac{P(E \cap F)}{P(F)} \quad \text{and} \quad P(F \mid E) = \frac{P(F \cap E)}{P(E)}.$$

Using the fact that $P(E \cap F) = P(F \cap E)$, and solving each of these equations for $P(E \cap F)$, we obtain the following rule.

Product Rule of Probability

If E and F are events, then $P(E \cap F)$ may be found by either of these formulas.

$$P(E \cap F) = P(F) \cdot P(E \mid F) \quad \text{or} \quad P(E \cap F) = P(E) \cdot P(F \mid E).$$

The **product rule** gives a method for finding the probability that events E and F both occur. Here is a simple way to remember the ordering of E and F in the probability rule.

$$P(E \cap F) = P(F) \cdot P(E \mid F) \quad \text{or} \quad P(E \cap F) = P(E) \cdot P(F \mid E)$$

---

*David Pringle, "Who's the DNA Fingerprinting Pointing At?" *New Scientist*, January 29, 1994, pp. 51–52.

In a litter of puppies, 3 were female and 4 were male. Half the males were black. Find the probability that a puppy chosen at random from the litter would be a black male.

Answer:

2/7

▶**EXAMPLE 4** In a class with 2/5 women and 3/5 men, 25% of the women are business majors. Find the probability that a student chosen at random from the class is a female business major.

Let B and W represent the events "business major" and "woman," respectively. We want to find $P(B \cap W)$. By the product rule,

$$P(B \cap W) = P(W) \cdot P(B \mid W).$$

From the given information, $P(W) = 2/5 = .4$ and the probability that a woman is a business major is $P(B \mid W) = .25$. Then

$$P(B \cap W) = .4(.25) = .10. \quad ◀ \quad ⑤$$

In Section 8.1 we used a tree diagram to find the number of subsets of a given set. By including the probabilities for each branch of a tree diagram, we convert it to a **probability tree.** The next examples show how a probability tree is used with the product rule to find the probability of a sequence of events.

▶**EXAMPLE 5** A company needs to hire a new director of advertising. It has decided to try to hire either person A or person B, who are assistant advertising directors for its major competitor. To decide between A and B, the company does research on the campaigns managed by A or B (none are managed by both) and finds that A is in charge of twice as many advertising campaigns as B. Also, A's campaigns have satisfactory results three out of four times, while B's campaigns have satisfactory results only two out of five times. Suppose one of the competitor's advertising campaigns (managed by A or B) is selected randomly.

We can represent this situation schematically as follows. Let A denote the event "Person A does the job" and B the event "Person B does the job." Let S be the event "satisfactory results" and U the event "unsatisfactory results." Then the given information can be summarized in the probability tree in Figure 8.25. Since A does twice as many jobs as B, $P(A) = 2/3$ and $P(B) = 1/3$, as noted on the first-stage branches of the tree. When A does a job, the probability of satisfactory results is 3/4 and of unsatisfactory results 1/4, as noted on the second-stage branches. Similarly, the probabilities when B does the job are noted on the remaining second-stage branches. The composite branches labeled 1–4 represent the four mutually exclusive possibilities for the running and outcome of the campaign.

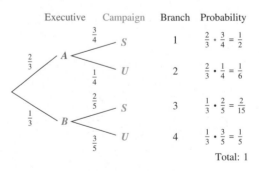

FIGURE 8.25

(a) Find the probability that A is in charge of a campaign that produces satisfactory results.

We are asked to find $P(A \cap S)$. We know that when A does the job, the probability of success is 3/4, that is, $P(S \mid A) = 3/4$. Hence, by the product rule,

$$P(A \cap S) = P(A) \cdot P(S \mid A) = \frac{2}{3} \cdot \frac{3}{4} = \frac{1}{2}.$$

The event $A \cap S$ is represented by branch 1 of the tree, and as we have just seen, its probability is the product of the probabilities that make up that branch.

(b) Find the probability that B runs a campaign that produces satisfactory results.

We must find $P(B \cap S)$. The event is represented by branch 3 of the tree and, as before, its probability is the product of the probabilities of the pieces of that branch:

$$P(B \cap S) = P(B) \cdot P(S \mid B) = \frac{1}{3} \cdot \frac{2}{5} = \frac{2}{15}.$$

(c) What is the probability that the selected campaign is satisfactory?

The event S is the union of the mutually exclusive events $A \cap S$ and $B \cap S$, which are represented by branches 1 and 3 of the tree diagram. By the union rule,

$$P(S) = P(A \cap S) + P(B \cap S) = \frac{1}{2} + \frac{2}{15} = \frac{19}{30}.$$

Thus, the probability of an event that appears on several branches is the sum of the probabilities of each of these branches.

(d) What is the probability that the selected campaign is unsatisfactory?

$P(U)$ can be read from branches 2 and 4 of the tree.

$$P(U) = \frac{1}{6} + \frac{1}{5} = \frac{11}{30}$$

Alternatively, because U is the complement of S,

$$P(U) = 1 - P(S) = 1 - \frac{19}{30} = \frac{11}{30}.$$

(e) Find the probability that either A runs the campaign or the results are satisfactory (or possibly both).

Event A combines branches 1 and 2, while event S combines branches 1 and 3, so use branches 1, 2, and 3.

$$P(A \cup S) = \frac{1}{2} + \frac{1}{6} + \frac{2}{15} = \frac{4}{5} \quad \blacktriangleleft \quad \text{⑥}$$

---

⑥

Find each of the following probabilities for Example 5.

(a) $P(U \mid A)$

(b) $P(U \mid B)$

Answers:

(a) 1/4

(b) 3/5

▶ **EXAMPLE 6**   From a box containing 1 red, 3 white, and 2 green marbles, two marbles are drawn one at a time without replacing the first before the second is drawn. Find the probability that one white and one green marble are drawn.

A probability tree showing the various possible outcomes is given in Figure 8.26 on the next page. In this diagram, W represents the event "drawing a white marble" and G represents "drawing a green marble." On the first draw, $P(W$ on the 1st$) = 3/6 = 1/2$ because three of the six marbles in the box are white. On the second draw, $P(G$ on the 2nd $\mid W$ on the 1st$) = 2/5$. One white marble has been removed, leaving 5, of which 2 are green.

Find the probability of drawing a green marble and then a white marble.

Answer:

1/5

In Example 6, find the probability of drawing 1 white and 1 red marble.

Answer:

1/5

We want to find the probability of drawing exactly one white marble and exactly one green marble. Two events satisfy this condition: drawing a white marble first and then a green one (branch 2 of the tree) or drawing a green marble first and then a white one (branch 4). For branch 2,

$$P(W \text{ on 1st}) \cdot P(G \text{ on 2nd} \mid W \text{ on 1st}) = \frac{1}{2} \cdot \frac{2}{5} = \frac{1}{5} \quad \text{⑦}$$

For branch 4, where the green marble is drawn first,

$$P(G \text{ first}) \cdot P(W \text{ second} \mid G \text{ first}) = \frac{1}{3} \cdot \frac{3}{5} = \frac{1}{5}.$$

Since these two events are mutually exclusive, the final probability is the sum of the two probabilities.

$$P(\text{one } W, \text{ one } G) = P(W \text{ on 1st}) \cdot P(G \text{ on 2nd} \mid W \text{ on 1st})$$

$$+ P(G \text{ on 1st}) \cdot P(W \text{ on 2nd} \mid G \text{ on 1st}) = \frac{2}{5} \quad ◀ \quad \text{⑧}$$

First Choice	Second Choice	Branch	Probability
	$\frac{2}{5}$ W	1	——
W	$\frac{2}{5}$ G	2	$\frac{1}{2} \cdot \frac{2}{5} = \frac{1}{5}$
	$\frac{1}{5}$ R	3	——
	$\frac{3}{5}$ W	4	$\frac{1}{3} \cdot \frac{3}{5} = \frac{1}{5}$
G	$\frac{1}{5}$ G	5	——
	$\frac{1}{5}$ R	6	——
	$\frac{3}{5}$ W	7	——
R	$\frac{2}{5}$ G	8	——

$\frac{1}{2}$, $\frac{1}{3}$, $\frac{1}{6}$

FIGURE 8.26

The product rule is often used when dealing with *stochastic processes,* which are mathematical models that evolve over time in a probabilistic manner. For example, drawing different colored marbles from a box (without replacing them) is such a process, in which the probabilities change with each successive draw. (Particular stochastic processes are studied further in Section 9.5.)

▶ **EXAMPLE 7** Two cards are drawn without replacement from an ordinary deck (52 cards). Find the probability that the first card is a heart and the second card is red.

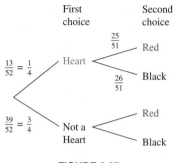

First
choice

Second
choice

$\frac{13}{52} = \frac{1}{4}$ Heart

$\frac{25}{51}$ Red

$\frac{26}{51}$ Black

$\frac{39}{52} = \frac{3}{4}$ Not a
Heart

Red

Black

FIGURE 8.27

Find the probability of drawing a heart on the first draw and a black card on the second, if two cards are drawn without replacement.

Answer:

$13/102 \approx .1275$

Use the tree in Example 8 to find the probability that exactly one of the cards is red.

Answer:

$13/34 \approx .382$

Start with the probability tree of Figure 8.27. (You may wish to refer to the deck of cards shown in Figure 8.19.) On the first draw, since there are 13 hearts in the 52 cards, the probability of drawing a heart first is $13/52 = 1/4$. On the second draw, since a (red) heart has been drawn already, there are 25 red cards in the remaining 51 cards. Thus the probability of drawing a red card on the second draw, given that the first is a heart, is $25/51$. By the product rule of probability,

P(heart on 1st and red on 2nd)

$= P$(heart on 1st) $\cdot P$(red on 2nd | heart on 1st)

$= \dfrac{1}{4} \cdot \dfrac{25}{51} = \dfrac{25}{204} \approx .1225.$ ◀ ⑨

▶**EXAMPLE 8** Three cards are drawn, without replacement, from an ordinary deck. Find the probability that exactly 2 of the cards are red.

Here we need a probability tree with three stages, as shown in Figure 8.28. The three branches indicated with arrows produce exactly 2 red cards from the draws. Multiply the probabilities along each of these branches and then add.

P(exactly 2 red cards) $= \dfrac{26}{52} \cdot \dfrac{25}{51} \cdot \dfrac{26}{50} + \dfrac{26}{52} \cdot \dfrac{26}{51} \cdot \dfrac{25}{50} + \dfrac{26}{52} \cdot \dfrac{26}{51} \cdot \dfrac{25}{50}$

$= \dfrac{50{,}700}{132{,}600} = \dfrac{13}{34} \approx .382$ ◀ ⑩

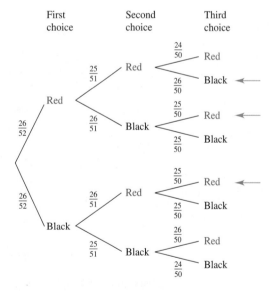

FIGURE 8.28

Independent Events

Suppose, in Example 7 that we draw the two cards *with* replacement rathen than without replacement (that is, we put the first card back before drawing the second card). If the first card is a heart, then the probability of drawing a red card on the second draw is 26/52, rather than 25/51 because there are still 52 cards in the deck, 26 of them red. In this case, P(red second | heart first) is the same as P(red second). The value of the second card is not affected by the value of the first card. We say that the event that the second card is red is *independent* of the event that the first card is a heart since the knowledge of the first card does not influence what happens to the second card. On the other hand, when we draw without replacement, the events that the first card is a heart and the second is red are *dependent* events. The fact that the first card is a heart means there is one less red card in the deck, influencing the probability that the second card is red.

As another example, consider tossing a fair coin twice. If the first toss shows heads, the probability that the next toss is heads is still 1/2. Coin tosses are independent events, since the outcome of one toss does not influence the outcome of the next toss. Similarly, rolls of a fair die are independent events. On the other hand, the events "the milk is old" and "the milk is sour" are dependent events; if the milk is old, there is an increased chance that it is sour. Also, in the example at the beginning of this section, the events A (broker picked a stock that went up) and B (broker used research) are dependent events, because information about the use of research affects the probability of picking stocks that go up. That is, $P(A \mid B)$ is different from $P(A)$.

If events E and F are independent, then the knowledge that E has occurred gives no (probability) information about the occurrence or nonoccurrence of event F. That is, $P(F)$ is exactly the same as $P(F \mid E)$, or

$$P(F \mid E) = P(F).$$

This, in fact, this is the formal definition of independent events.

> E and F are **independent events** if
>
> $$P(F \mid E) = P(F) \quad \text{or} \quad P(E \mid F) = P(E).$$

If the events are not independent, they are **dependent events.**

When E and F are independent events, then $P(F \mid E) = P(F)$ and the product rule becomes

$$P(E \cap F) = P(E) \cdot P(F \mid E) = P(E) \cdot P(F).$$

Conversely, if this equation holds, it follows that $P(F) = P(F \mid E)$. Consequently, we have this useful fact.

> ## Product Rule for Independent Events
>
> E and F are independent events if and only if
>
> $$P(E \cap F) = P(E) \cdot P(F).$$

▶ **EXAMPLE 9** A calculator requires a key-stroke assembly and a logic circuit. Assume that 99% of the key-stroke assemblies are satisfactory and 97% of the logic circuits are satisfactory. Find the probability that a finished calculator will be satisfactory.

Find the probability of getting 4 successive heads on 4 tosses of a fair coin.

Answer:

1/16

If the failure of a key-stroke assembly and the failure of a logic circuit are independent events, then

P(satisfactory calculator)

$\quad = P$(satisfactory key-stroke assembly) $\cdot P$(satisfactory logic circuit)

$\quad = (.99)(.97) \approx .96.$ ◄ (11)

CAUTION It is common for students to confuse the ideas of *mutually exclusive* events and *independent* events. Events E and F are mutually exclusive if $E \cap F = \emptyset$. For example, if a family has exactly one child, the only possible outcomes are $B = \{\text{boy}\}$ and $G = \{\text{girl}\}$. These two events are mutually exclusive. However, the events are *not* independent, since $P(G \mid B) = 0$ (if a family with only one child has a boy, the probability it has a girl is then 0). Since $P(G \mid B) \neq P(G)$, the events are not independent. Of all the families with exactly *two* children, the events $G_1 = \{\text{first child is a girl}\}$ and $G_2 = \{\text{second child is a girl}\}$ are independent, because $P(G_2 \mid G_1)$ equals $P(G_2)$. However, G_1 and G_2 are not mutually exclusive, since $G_1 \cap G_2 = \{\text{both children are girls}\} \neq \emptyset$.

To show that two events E and F are independent, we can show that $P(F \mid E) = P(F)$ or that $P(E \mid F) = P(E)$ or that $P(E \cap F) = P(E) \cdot P(F)$. Another way is to observe that knowledge of one outcome does not influence the probability of the other outcome, as we did for coin tosses.

NOTE In some cases, it may not be apparent from the physical description of the problem whether two events are independent or not. For example, it is not obvious whether the event that a baseball player gets a hit tomorrow is independent of the event that he got a hit today. In such cases, it is necessary to use the definition and calculate whether $P(F \mid E) = P(F)$ or, equivalently, whether $P(E \mid F) = P(E) \cdot P(F)$.

(12)

In the U.S. population, the probability of being Hispanic is .11, the probability of living in California is .12, and the probability of being a Hispanic living in California is .04. Are the events being Hispanic and living in California independent?

Answer:

No

► **EXAMPLE 10** On a typical January day in Manhattan the probability of snow is .10, the probability of a traffic jam is .80, and the probability of snow or a traffic jam (or both) is .82. Are the event "it snows" and the event "a traffic jam occurs" independent?

Let S represent the event "it snows" and T represent the event "a traffic jam occurs." We must determine whether

$$P(T \mid S) = P(T) \quad \text{or} \quad P(S \mid T) = P(S).$$

We know $P(S) = .10, P(T) = .8,$ and $P(S \cup T) = .82$. We can use the union rule (or a Venn diagram) to find $P(S \cap T) = .08, P(T \mid S) = .8,$ and $P(S \mid T) = .1$. Since

$$P(T \mid S) = P(T) = .8 \quad \text{and} \quad P(S \mid T) = P(S) = .1,$$

the events "it snows" and "a traffic jam occurs" are independent. ◄ (12)

Although we showed $P(T \mid S) = P(T)$ and $P(S \mid T) = P(S)$ in Example 10, only one of these results is needed to establish independence. It is also important to note that independence of events does not necessarily follow intuition; it is established from the mathematical definition of independence.

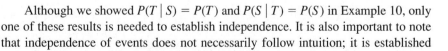

8.5 Exercises

If a single fair die is rolled, find the probability of rolling the following. (See Examples 1 and 2.)

1. 3, given that the number rolled was odd

2. 5, given that the number rolled was even

3. An odd number, given that the number rolled was 3

If two fair dice are rolled (recall the 36-outcome sample space), find the probability of rolling the following.

4. A sum of 8, given the sum was greater than 7

5. A sum of 6, given the roll was a "double" (two identical numbers)

6. A double, given that the sum was 9

If two cards are drawn without replacement from an ordinary deck (see Example 7), find the following probabilities.

7. The second is a heart, given that the first is a heart

8. The second is black, given that the first is a spade

9. The second is a face card, given that the first is a jack

10. The second is an ace, given that the first is not an ace

11. A jack and a 10 are drawn.

12. An ace and a 4 are drawn.

13. Two black cards are drawn.

14. Two hearts are drawn.

15. In your own words, explain how to find the conditional probability $P(E \mid F)$.

16. Your friend asks you to explain how the product rule for independent events differs from the product rule for dependent events. How would you respond?

17. Another friend asks you to explain how to tell whether two events are dependent or independent. How would you reply? (Use your own words.)

18. A student reasons that the probability in Example 3 of both coins being heads is just the probability that the other coin is a head, that is, 1/2. Explain why this reasoning is wrong.

19. In a two-child family, if we assume that the probabilities of a male child and a female child are each .5, are the events *each child is the same sex* and *at most one male* independent? Are they independent for a three-child family?

20. Let A and B be independent events with $P(A) = \dfrac{1}{4}$ and $P(B) = \dfrac{1}{5}$. Find $P(A \cap B)$ and $P(A \cup B)$.

Business *Use a probability tree or Venn diagram in Exercises 21–24 (See Examples 2 and 5–9. A shop that produces custom kitchen cabinets has two employees, Sitlington and Capek. 95% of Capek's work is satisfactory and 10% of Sitlington's work is unsatisfactory. 60% of the shop's cabinets are made by Capek (the rest by Sitlington). Find the following probabilities.*

21. An unsatisfactory cabinet was made by Capek. (*Hint:* Consider which event came first.)

22. A finished cabinet is unsatisfactory.

23. Social Science   Marriages between cousins are very common in some countries, where about 50% of marriages are consanguineous (between first cousins or people even more closely related). A recent study in Pakistan has shown that 16% of children from unrelated marriages died by age 10, while 21% of children from consanguineous marriages died by age 10. Find the following probabilities.

(a) a child survives

(b) a child from a consanguineous marriage survives

24. Business   Among users of automated teller machines (ATMs), 92% use ATMs to withdraw cash, and 32% use them to check their account balance.* Suppose that 96% use ATMs to either withdraw cash or check their account balance (or both). Find the probability that

(a) an ATM user withdraws funds if that person uses the ATM to check an account balance.

(b) an ATM user checks an account balance given that the ATM is used to withdraw funds.

25. Health   The table at the top of the next column, based on data from the Centers for Disease Control, gives the number of new cases of the AIDS virus for men and women in the United States in 1998 by method of transmission.[†]

*Chicago Tribune, Sec. 4, p. 1, December 18, 1995.

†Health United States, 1999, CDC, National Center for HIV, STD, and TB Prevention, Division of HIV/AIDS Prevention and The World Almanac and Book of Facts 2000, p. 903.

Method of Transmission	Male	Female	Total
Homosexual Contact	8388	0	8388
Heterosexual Contact	1146	1806	2952
Intravenous Drug Use	3652	1541	5193
Other	5237	2054	7291
Total	18,423	5401	23,824

(a) Find the probability that a male resident who was newly diagnosed with AIDS in 1998 contracted it via homosexual contact.

(b) Find the probability that a female resident who was newly diagnosed with AIDS in 1998 contracted it via intravenous drug use.

(c) Find the probability that a newly diagnosed person with AIDS is a female.

(d) Find the probability that a newly diagnosed person who contracted AIDS via heterosexual contact was a female.

Health *The table lists the number of head and neck injuries for 319 ice hockey players' exposures wearing either a full shield or half shield in the Canadian Inter-University Athletics Union during the 1997–1998 season.* [*]

	Half Shield (H)	Full Shield (F)	Total
Head and Face Injuries (A)	95	34	129
Concussions (B)	41	38	79
Neck Injuries (C)	9	7	16
Other Injuries (D)	202	150	352
Total	347	229	576

For a randomly selected injury, find the following probabilities.

26. $P(A)$

27. $P(C \mid F)$

28. $P(A \mid H)$

29. $P(B' \mid H')$

30. **Natural Science** In a letter to the journal *Nature*, Robert A. J. Matthews gives the following table of outcomes of forecast and weather over 1000 1-hour walks, based on the United Kingdom's Meteorological Office's 83% accuracy in 24-hour forecasts.[†]

	Rain	No Rain	Sum
Forecast of Rain	66	156	222
Forecast of No Rain	14	764	778
Sum	80	920	1000

(a) Verify that the probability that the forecast called for rain, given that there was rain, is indeed 83%. Also verify that the probability that the forecast called for no rain, given that there was no rain, is also 83%.

(b) Calculate the probability that there was rain, given that the forecast called for rain.

(c) Calculate the probability that there was no rain, given that the forecast called for no rain.

(d) Observe that your answer to part (c) is higher than 83% and that your answer to part (b) is much lower. Discuss which figure best describes the accuracy of the weather forecast in recommending whether or not you should carry an umbrella.

Business *The table shows 1999 employment figures (in thousands) in two occupational categories.* [*]

	Managerial/ Professional	Technical/Sales/ Adminis. Support	Totals
Women	14,100	24,800	38,900
Men	20,400	20,000	40,400
Totals	34,500	44,800	79,300

Let A represent managerial/professional workers, B represent Technical/Sales/Administrative Support workers, C represent women, and D represent men. Express each of the following probabilities in words and find its value. (See Example 1.)

31. $P(A \mid D)$

32. $P(C \mid A)$

33. $P(B \mid C)$

34. $P(D \cap A)$

Natural Science *The following table shows frequencies for red-green color blindness, where M represents male and C represents color-blind.*

	M	M'	Totals
C	.042	.007	.049
C'	.485	.466	.951
Totals	.527	.473	1.000

Use this table to find the following probabilities.

35. $P(M)$

36. $P(C)$

37. $P(M \cap C)$

38. $P(M \cup C)$

39. $P(M \mid C)$

40. $P(M' \mid C)$

---

*Brian Benson, Nicholas Nohtaki, M. Sarah Rose, and Willem Meeuwisse, "Head and Neck Injuries among Ice Hockey Players Wearing Full Face Shields vs. Half Face Shields," *JAMA* 282, no. 24 (December 22/29, 1999): 2328–2332.

†Robert A. J. Matthews, Nature 382, August 29, 1996: p 3.

*Bureau of Labor Statistics, U.S. Department of Labor.

41. Are the events C and M described above dependent? (Recall that two events E and F are dependent if $P(E \mid F) \neq P(E)$. (See Example 10.)

42. Natural Science A scientist wishes to determine if there is any dependence between color blindness (C) and deafness (D). Given the probabilities listed in the table below, what should his findings be? (See Example 10.)

	D	D'	Totals
C	.0004	.0796	.0800
C'	.0046	.9154	.9200
Totals	.0050	.9950	1.0000

Social Science *The Motor Vehicle Department has found that the probability of a person passing the test for a driver's license on the first try is .75. The probability that an individual who fails on the first test will pass on the second try is .80, and the probability that an individual who fails the first and second tests will pass the third time is .70. Find the probability that an individual*

43. fails both the first and second tests;

44. will fail three times in a row;

45. will require at least two tries to pass the test.

46. Social Science A smooth-talking young man has a 1/3 probability of talking a policeman out of giving him a speeding ticket. The probability that he is stopped for speeding during a given weekend is 1/2. Find the probabilities of the events in parts (a) and (b).
   (a) He will receive no speeding tickets on a given weekend.
   (b) He will receive no speeding tickets on 3 consecutive weekends.
   (c) We have assumed that what happens on the second or third weekend is the same as what happened on the first weekend. Is this realistic? Will driving habits remain the same after getting a ticket?

Natural Science *Four different medications, C, D, E, and F, may be used to control high blood pressure. A physician usually prescribes C first because it is least likely to cause side effects. If blood pressure remains high, the patient is switched to D. If this fails to work, the patient is switched to E, and if necessary to F. The probability that C will work is .7. If C fails, the probability that D will work is .8. If D fails, the probability that E will work is .62. If E fails, the probability that F will work is .45. Find the probability that*

47. A patient's blood pressure will not be reduced by any of the medications.

48. A patient will have to take at least two medications and will have his or her blood pressure reduced.

49. If medications C and D fail, what is the probability that a patient's blood pressure will be reduced by medication E or F?

Business *The number of new passenger cars in thousands from major worldwide producers in selected years is shown in the table.*

	U.S.	Canada	Germany	Japan
1993	10,898	1468	184	1597
1994	12,263	1591	188	1593
1995	11,985	1678	207	1387
1996	11,799	1688	234	1191

Find the following probabilities for a vehicle selected at random.

50. It was made in the United States.

51. It was made in 1993.

52. It was made in Japan in 1995.

53. It was made in 1996, given that it was made in Germany.

The probability that the first record by a singing group will be a hit is .32. If their first record is a hit, so are all their subsequent records. If their first record is not a hit, the probability of their second record and all subsequent ones being hits is .16. If the first two records are not hits, the probability that the third is a hit is .08. The probability of a hit continues to decrease by half with each successive nonhit record. Find the probability that

54. the group will have at least one hit in their first four records.

55. the group will have exactly one hit in their first three records.

56. the group will have a hit in their first six records if the first three are not hits.

Work the following problems on independent events. (See Examples 9 and 10.)

57. Business Corporations such as banks, where a computer is essential to day-to-day operations, often have a second, backup computer in case of failure by the main computer. Suppose that there is a .003 chance that the main computer will fail in a given time period and a .005 chance that the backup computer will fail while the main computer is being repaired. Assume these failures represent independent events and find the fraction of the time that the corporation can assume it will have computer service. How realistic is our assumption of independence?

58. Business According to a booklet put out by Eastwest Airlines, 98% of all scheduled Eastwest flights actually take place. (The other flights are canceled due to weather, equipment problems, and so on.) Assume that the event that a given flight takes place is independent of the event that another flight takes place.
   (a) Elisabeta Guervara plans to visit her company's branch offices; her journey requires 3 separate flights on Eastwest Airlines. What is the probability that all of these flights will take place?
   (b) Based on the reasons we gave for a flight to be canceled, how realistic is the assumption of independence that we made?

59. Social Science   The probability that a key component of a space rocket will fail is .03.

(a) How many such components must be used as backups to ensure that the probability of at least one of the components' working is .999999?

(b) Is it reasonable to assume independence here?

60. Natural Science   A medical experiment showed that the probability that a new medicine is effective is .75, the probability that a patient will have a certain side effect is

.4, and the probability that both events occur is .3. Decide whether these events are dependent or independent.

61. Social Science   A teacher has found that the probability that a student studies for a test is .6, the probability that a student gets a good grade on a test is .7, and the probability that both occur is .52. Are these events independent?

62. Business   Refer to Exercises 31–34. Are the events being a woman and being employed in the managerial/ professional category independent?

8.6 BAYES' FORMULA

Suppose the probability that a person gets lung cancer, given that the person smokes a pack or more of cigarettes daily, is known. For a research project, it might be necessary to know the probability that a person smokes a pack or more of cigarettes daily, given that the person has lung cancer. More generally, if $P(E \mid F)$ is known for two events E and F, can $P(F \mid E)$ be found? The answer is yes, we can find $P(F \mid E)$ using the formula to be developed in this section. To develop this formula, we can use a probability tree to find $P(F \mid E)$. Since $P(E \mid F)$ is known, the first outcome is either F or F'. Then for each of these outcomes, either E or E' occurs, as shown in Figure 8.29.

FIGURE 8.29

The four cases have the probabilities shown on the right. By the definition of conditional probability and the product rule,

$$P(E) = P(F \cap E) + P(F' \cap E),$$

$$P(F \cap E) = P(F) \cdot P(E \mid F), \quad \text{and} \quad P(F' \cap E) = P(F') \cdot P(E \mid F').$$

By substitution,

$$P(E) = P(F) \cdot P(E \mid F) + P(F') \cdot P(E \mid F'),$$

and

$$P(F \mid E) = \frac{P(F \cap E)}{P(E)} = \frac{P(F) \cdot P(E \mid F)}{P(F) \cdot P(E \mid F) + P(F') \cdot P(E \mid F')}.$$

We have proved a special case of Bayes' formula, which is generalized later in this section. ①

① Use the special case of Bayes' formula to find $P(F \mid E)$ if $P(F) = .2, P(E \mid F) = .1,$ and $P(E \mid F') = .3$. (*Hint:* $P(F') = 1 - P(F)$.)

Answer:

$1/13 \approx .077$

Bayes' Formula (Special Case)

$$P(F \mid E) = \frac{P(F) \cdot P(E \mid F)}{P(F) \cdot P(E \mid F) + P(F') \cdot P(E \mid F')}.$$

▶**EXAMPLE 1**   For a fixed length of time, the probability of worker error on a certain production line is .1, the probability that an accident will occur when there is a worker error is .3, and the probability that an accident will occur when there is no worker error is .2. Find the probability of a worker error if there is an accident.

Let E represent the event of an accident, and let F represent the event of worker error. From the information above,

$$P(F) = .1, \quad P(E \mid F) = .3, \qquad \text{and} \qquad P(E \mid F') = .2.$$

These probabilities are shown on the probability tree in Figure 8.30.

	Branch	Probability
$P(E \backslash F) = .3$ E	1	$P(F) \cdot P(E \mid F)$
$P(E' \mid F) = .7$ E'	2	$P(F) \cdot P(E' \mid F)$
$P(E \backslash F') = .2$ E	3	$P(F') \cdot P(E \mid F')$
$P(E' \mid F') = .8$ E'	4	$P(F') \cdot P(E' \mid F')$

$P(F) = .1$ F

$P(F') = .9$ F'

FIGURE 8.30

In Example 1, find $P(F' \mid E)$.

Answer:

$6/7 \approx .857$

Find $P(F \mid E)$ by dividing the probability that both E and F occur, given by branch 1, by the probability that E occurs, given by the sum of branches 1 and 3.

$$P(F \mid E) = \frac{P(F) \cdot P(E \mid F)}{P(F) \cdot P(E \mid F) + P(F') \cdot P(E \mid F')}$$

$$= \frac{(.1)(.3)}{(.1)(.3) + (.9)(.2)} \approx .143 \quad ◀ \quad ②$$

The special case of Bayes' formula can be generalized to more than two possibilities with the probability tree of Figure 8.31. This diagram shows the paths that can produce an event E. We assume that events $F_1, F_2, \ldots, F_n$ are pairwise mutually exclusive events (that is, events which, taken two at a time, are disjoint), whose union is the sample space, and E is an event that has occurred. See Figure 8.32.

Branch   Probability

FIGURE 8.31

FIGURE 8.32

The probability $P(F_i \mid E)$, where $1 \leq i \leq n$, can be found by dividing the probability for the branch containing $P(E \mid F_i)$ by the sum of the probabilities of all the branches producing event E.

Bayes' Formula

$$P(F_i \mid E) = \frac{P(F_i) \cdot P(E \mid F_i)}{P(F_1) \cdot P(E \mid F_1) + \cdots + P(F_n) \cdot P(E \mid F_n)}.$$

This result is known as **Bayes' formula,** after the Reverend Thomas Bayes (1702–61), whose paper on probability was published about two hundred years ago.

The statement of Bayes' formula can be daunting. Actually, it is easier to remember the formula by thinking of the probability tree that produced it. Go through the following steps.

Using Bayes' Formula

Step 1 Start a probability tree with branches representing events F_1, $F_2, \ldots, F_n$. Label each branch with its corresponding probability.

Step 2 From the end of each of these branches, draw a branch for event E. Label this branch with the probability of getting to it, or $P(E \mid F_i)$.

Step 3 There are now n different paths that result in event E. Next to each path, put its probability—the product of the probabilities that the first branch occurs, $P(F_i)$, and that the second branch occurs, $P(E \mid F_i)$: that is, $P(F_i) \cdot P(E \mid F_i)$.

Step 4 $P(F_i \mid E)$ is found by dividing the probability of the branch for F_i by the sum of the probabilities of all the branches producing event E.

▶ **EXAMPLE 2**   Based on past experience, a company knows that an experienced machine operator (one or more years of experience) will produce a defective item 1% of the time. Operators with some experience (up to one year) have a 2.5% defect rate, while new operators have a 6% defect rate. At any one time, the company has 60% experienced employees, 30% with some experience, and 10% new employees. Find the probability that a particular defective item was produced by a new operator.

Let E represent the event "item is defective," with F_1 representing "item was made by an experienced operator," F_2 "item was made by an operator with some experience," and F_3 "item was made by a new operator." Then

$$P(F_1) = .60 \qquad P(E \mid F_1) = .01$$
$$P(F_2) = .30 \qquad P(E \mid F_2) = .025$$
$$P(F_3) = .10 \qquad P(E \mid F_3) = .06.$$

We need to find $P(F_3 \mid E)$, the probability that an item was produced by a new operator, given that it is defective. First, draw a probability tree using the given information, as in Figure 8.33. The steps leading to event E are shown.

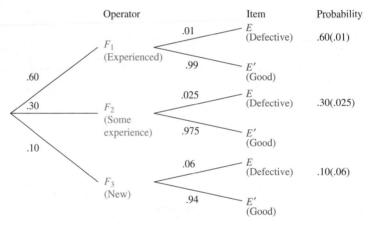

FIGURE 8.33

③

In Example 2, find

(a) $P(F_1 \mid E)$;

(b) $P(F_2 \mid E)$.

Answers:

(a) $4/13 \approx .3077$

(b) $5/13 \approx .3846$

Find $P(F_3 \mid E)$ using the bottom branch of the tree shown in Figure 8.33: divide the probability for this branch by the sum of the probabilities of all the branches leading to E.

$$P(F_3 \mid E) = \frac{.10(.06)}{.60(.01) + .30(.025) + .10(.06)} = \frac{.006}{.0195} \approx .3077 \quad ◀ \;③$$

After working problem 3 at the side, check that $P(F_1 \mid E) + P(F_2 \mid E) + P(F_3 \mid E) = 1$. (That is, the defective item was made by *someone*.)

▶ **EXAMPLE 3**   A manufacturer buys items from six different suppliers. The fraction of the total number of items obtained from each supplier, along with the probability that an item purchased from that supplier is defective, is shown in the following table.

Supplier	Fraction of Total Supplied	Probability of Defective
1	.05	.04
2	.12	.02
3	.16	.07
4	.23	.01
5	.35	.03
6	.09	.05

Find the probability that a defective item came from supplier 5.

Let F_1 be the event that an item came from supplier 1, with $F_2, F_3, F_4, F_5,$ and F_6 defined in a similar manner. Let E be the event that an item is defective. We want to find $P(F_5 \mid E)$. Use the probabilities in the table above to prepare a probability tree. By Bayes' formula,

$$P(F_5 \mid E)$$

$$= \frac{(.35)(.03)}{(.05)(.04) + (.12)(.02) + (.16)(.07) + (.23)(.01) + (.35)(.03) + (.09)(.05)}$$

$$= \frac{.0105}{.0329} \approx .319.$$

There is about a 32% chance that a defective item came from supplier 5. Even though supplier 5 has only 3% defectives, his probability of being "guilty" is relatively high, about 32%, because of the large fraction supplied by 5. ◀ ④

④

In Example 3, find the probability that the defective item came from

(a) supplier 3;

(b) supplier 6.

Answers:

(a) .340

(b) .137

8.6 Exercises

For two events, M and N, $P(M) = .4$, $P(N \mid M) = .3$, and $P(N \mid M') = .4$. Find each of the following. (See Example 1.)

1. $P(M \mid N)$ **2.** $P(M' \mid N)$

For mutually exclusive events R_1, R_2, R_3, $P(R_1) = .05$, $P(R_2) = .6$, and $P(R_3) = .35$. In addition, $P(Q \mid R_1) = .40$, $P(Q \mid R_2) = .30$, and $P(Q \mid R_3) = .60$. Find each of the following. (See Examples 2 and 3.)

3. $P(R_1 \mid Q)$ **4.** $P(R_2 \mid Q)$
5. $P(R_3 \mid Q)$ **6.** $P(R_1' \mid Q)$

Suppose three jars have the following contents: 2 black balls and 1 white ball in the first; 1 black ball and 2 white balls in the second; 1 black ball and 1 white ball in the third. If the probability of selecting one of the three jars is 1/2, 1/3, and 1/6, respectively, find the probability that if a white ball is drawn, it came from the

7. second jar;   **8.** third jar.

Social Science *In 1990, 1% of the U.S. population was Native American. The probability that a Native American lives in the West or South is .76. The probability is .56 that a person who is not Native American lives in the West or South. Find the following probabilities.*

9. A person living in the West or South is Native American.
10. A person who does not live in the West or South is not Native American.

Social Science *A federal study showed that in 1990, 49% of all those involved in a fatal car crash wore seat belts. Of those in a fatal crash who wore seat belts, 44% were injured and 27% were killed. For those not wearing seat belts, the comparable figures were 41% and 50%, respectively.*[*]

*National Highway Traffic Safety Administration, Office of Driver and Pedestrian Research: "Occupant Protection Trends in 19 Cities" (November 1989) and "Use of Automatic Safety Belt Systems in 19 Cities" (February 1991).

11. Find the probability that a randomly selected person who was killed in a car crash was wearing a seat belt.

12. Find the probability that a randomly selected person who was unharmed in a fatal crash was not wearing a seat belt.

Business *A building contractor buys 70% of his cement from supplier A, and 30% from supplier B. A total of 90% of the bags from A arrive undamaged, while 95% of the bags from B arrive undamaged. Give the probabilities that a damaged bag is from each of the following sources.*

13. Supplier A   14. Supplier B

Business *The probability that a customer of a local department store will be a "slow pay" is .02. The probability that a "slow pay" will make a large down payment when buying a refrigerator is .14. The probability that a person who is not a "slow pay" will make a large down payment when buying a refrigerator is .50. Suppose a customer makes a large down payment on a refrigerator. Find the probability that the customer is*

15. a "slow pay";   16. not a "slow pay."

17. Social Science A 1995 survey showed that of Baby Boomers (Americans born between 1946 and 1960), 56% have a family income of more than $40,000.* The probability that a Baby Boomer's family own their own home if their income is over $40,000 is .51, while the probability that they own their home if the family income is less than or equal to $40,000 is .27. Find the probability that a Baby Boomer has family income over $40,000, if it is known that the family owns its home.

18. Social Science During the murder trial of O. J. Simpson, Alan Dershowitz, an adviser to the defense team, stated on television that only about .1% of men who batter their wives actually murder them. Statistician I. J. Good observed that even if, given that a husband is a batterer, the probability he is guilty of murdering his wife is .001, what we really want to know is the probability that the husband is guilty, given that the wife was murdered.† Good estimates the probability of a battered wife being murdered, given that her husband is not guilty, as .001. The probability that she is murdered if her husband is guilty is 1, of course. Using these numbers and Dershowitz's .001 probability of the husband being guilty, find the probability that the husband is guilty, given that the wife was murdered.

19. Management The following information pertains to three shipping terminals operated by Krag Corp.:‡

Terminal	Percentage of Cargo Handled	Percentage of Error
Land	50	2
Air	40	4
Sea	10	14

Krag's internal auditor randomly selects one set of shipping documents, ascertaining that the set selected contains an error. Which of the following gives the probability that the error occurred in the Land Terminal?
(a) .02   (b) .10   (c) .25   (d) .50

Health *In a test for toxemia, a disease that affects pregnant women, the woman lies on her left side and then rolls over on her back. The test is considered positive if there is a 20 mm rise in her blood pressure within one minute. The results have produced the following probabilities, where T represents having toxemia at some time during the pregnancy and N represents a negative test.*

$$P(T' \mid N) = .90 \quad and \quad P(T \mid N') = .75$$

Assume that P(N') = .11, and find each of the following.

20. $P(N \mid T)$

21. $P(N' \mid T)$

22. Health In 2000, 7.3% of Americans were diabetic, 19.8% of Americans were obese, and 80% of diabetics were obese.* Find the probability that an American is diabetic given that person is obese.

23. Health In 1987, it was found that one woman in 10,000 was infected with the AIDS virus. The most commonly used test always detected an infected woman, but one healthy woman out of 20,000 was incorrectly identified as being infected with the AIDS virus.†
(a) Find the probability that a woman who tested positive in 1987 was not infected.
(b) It has been argued that everyone should be tested for AIDS. Based on the results of part (a), how useful would the results of such testing be?

24. Health The probability that a person with certain symptoms has hepatitis is .8. The blood test used to confirm this diagnosis gives positive results for 90% of people with the disease and 5% of those without the disease. What is the probability that an individual who has the symptoms and who reacts positively to the test actually has hepatitis?

---

*New York Times, sec. 3, p. 1, December 31, 1995.

†I. J. Good, "When Batterer Turns Murderer," *Nature* 375, no. 15 (June 15, 1995), p. 541.

‡Uniform CPA Examination, November 1989.

*The Sacramento Bee, September 12, 2001, p. A20.

†New York Times, September 5, 1987.

25. Health The sensitivity of a medical test is defined as the probability that a test will be positive given that a person has a disease, written $P(T^+ \mid D^+)$. The specificity of a test is defined as the probability that a test will be negative given that the person does not have the disease, written $P(T^- \mid D^-)$. For example, the sensitivity and specificity for breast cancer during a clinical breast examination by a trained expert are approximately .54 and .94, respectively.[*]

(a) If 2% of U.S. women have breast cancer,[†] find the probability that a woman who tests positive during a clinical breast examination actually has breast cancer.

(b) Given that a women tests negative during a clinical breast examination, find the probability that she does not have breast cancer.

(c) How many false positives could be expected for every 1000 clinical breast examinations? (A false positive refers to a person who does not have the disease, but tests positive.)

26. Health A new test for the virus that causes AIDS, developed by Octopus Diagnostics Research of Hantsport. Nova Scotia, shows the presence or absence of HIV in a drop of blood in two minutes, compared with five days for a conventional test.[‡] Preliminary results indicate a false positive rate (an indication that the HIV virus is present when it is not) of less than 2% and a false negative rate (a failure to detect the presence of the HIV virus) of up to 5%. Assume for this exercise that these rates are exactly 2% and 5%. In 1996, there were 780,000 people in North America with the HIV virus, out of a population of 295 million.[§] Suppose a resident of North America is chosen at random and given this test. If the result is positive, what is the probability that the person actually has the HIV virus?

27. Social Science A recent study by the Harvard School of Public Health reported that 86% of male students who live in a fraternity house are binge drinkers. The figure for fraternity members who are not residents of a fraternity house is 71%, while the figure for men who do not belong to a fraternity is 45%.[¶] Suppose that 10% of U.S. students live in a fraternity house, 15% belong to a fraternity, but do not live in a fraternity house, and 75% do not belong to a fraternity.

(a) What is the probability that a randomly selected male student is a binge drinker?

(b) If a randomly selected male student is a binge drinker, what is the probability that he lives in a fraternity house?

28. Finance A bank finds that the relationship between mortgage defaults and the size of the down payment is given by this table.

Down Payment (%)	5%	10%	20%	25%
Number of Mortgages with this Down Payment	1260	700	560	280
Probability of Default	.05	.03	.02	.01

If a default occurs, what is the probability that it is on a mortgage with a 5% down payment? (See Examples 2 and 3.)

Business *The table shows the percent of U.S. households in each region and the percent per region using central air conditioning.*

	Northeast	Midwest	South	West
Region	19.4	23.7	35.2	21.5
Uses Central AC	22.1	51.3	69.4	26.9

Find the probability that a household using central air conditioning is

29. in the South   **30.** in the Midwest

Social Science *The following tables give the proportions of men and women in the U.S. population, and the proportions of men and women who have never married, in a recent year.[*] Use them for Exercises 31–33.*

	MEN	
Age	Proportion in Population	Proportion Never Married
18–24	.133	.874
25–34	.205	.397
35–44	.232	.187
45–64	.287	.075
65 or Over	.142	.038

---

[*]Mary B. Barton, Russell Harris, and Suzanne Fletcher, "Does This Patient Have Breast Cancer? The Screening Clinical Breast Examinations: Should It Be Done? How?" *JAMA*, 282 no. 13 (October 6, 1999): 1270–1280.

[†]*The World Almanac and Book of Facts 2000*, p. 902.

[‡]*Maclean's*, February 17, 1997, p. 70.

[§]*The World Almanac Book of Facts 1997*.

[¶]*New York Times*, December 6, 1995, p. B16.

[*]"Marital Status and Living Arrangements of Adults 18 Years and Over," March, 1998, Table B. U.S. Bureau of the Census.

	WOMEN	
Age	*Proportion in Population*	*Proportion Never Married*
18–24	.123	.775
25–34	.194	.298
35–44	.219	.121
45–64	.284	.062
65 or Over	.181	.047

Social Science   *The following table gives the proportions of people over 18 who are in various age categories, along with the probabilities that a person in a given category will vote in a general election.*

Age	*Percent of Voting Age Population*	*Probability of a Person of This Age Voting*
18–21	11.0%	.48
22–24	7.6%	.53
25–44	37.6%	.68
45–64	28.3%	.64
65 or Over	15.5%	.74

31. Find the probability that a randomly selected man who has never married is between 35 and 44 years old (inclusive).

32. Find the probability that a randomly selected woman who has been married is between 18 and 24 (inclusive).

33. Find the probability that a randomly selected woman who has never been married is between 45 and 64 (inclusive).

Suppose a voter is picked at random. Find the probabilities that the voter is in the following age categories.

34. 18–21   **35.** 65 or over

36. Find the probability that a person who did not vote was in the age category 45–64.

CHAPTER 8   **SUMMARY**

Key Terms and Symbols

	empty set	simple event
	set-builder notation	certain event
	universal set	impossible event
	subset	mutually exclusive events
{ } set braces	set operations	basic probability principle
∈ is an element of	tree diagram	**8.4** union rule for probability
∉ is not an element of	Venn diagram	complement rule
∅ empty set	complement	odds
⊆ is a subset of	intersection	empirical probability
⊄ is not a subset of	disjoint sets	probability distribution
A' complement of set A	union	**8.5** conditional probability
∩ set intersection	**8.2** union rule for counting	product rule of probability
∪ set union	**8.3** experiment	probability tree
$P(E)$ probability of event E	trial	independent events
$P(F \mid E)$ probabilty of F, given that E has occurred	outcome	dependent events
	sample space	**8.6** Bayes' formula
8.1 set	event	
element (member)		

Key Concepts

Sets

Set A is a **subset** of set B if every element of A is also an element of B.

A set of n elements has 2^n subsets.

Let A and B be any sets with universal set U.

The **complement** of A is $A' = \{x \mid x \notin A \text{ and } x \in U\}$.

The **intersection** of A and B is $A \cap B = \{x \mid x \in A \text{ and } x \in B\}$.

The **union** of A and B is $A \cup B = \{x \mid x \in A \text{ or } x \in B \text{ or both}\}$.

$n(A \cup B) = n(A) + n(B) - n(A \cap B)$, where $n(X)$ is the number of elements in set X

Probability Summary

Basic Probability Principle

Let S be a sample space of equally likely outcomes and let event E be a subset of S. Then the probability that event E occurs is

$$P(E) = \frac{n(E)}{n(S)}.$$

Union Rule

For any events E and F from a sample space S.

$$P(E \cup F) = P(E) + P(F) - P(E \cap F).$$

For mutually exclusive events E and F,

$$P(E) \cup F) = P(E) + P(F).$$

Complement Rule

$P(E) = 1 - P(E) \quad \text{and } P(E') = 1 - P(E)$

Odds

The odds in favor of event E are $\dfrac{P(E)}{P(E')}$, $P(E') \neq 0$.

Properties of Probability

1. For any event E in sample space S, $0 \leq P(E) \leq 1$.

2. The sum of the probabilities of all possible distinct outcomes is 1.

Conditional Probability

The conditional probability of event E, given that event F has occurred, is

$$P(E \mid F) = \frac{P(E \cap F)}{P(F)}, \quad \text{where} \quad P(F) \neq 0.$$

For equally likely outcomes, conditional probability is found by reducing the sample space to event F; then

$$P(E \mid F) = \frac{n(E \cap F)}{n(F)}.$$

Product Rule of Probability

If E and F are events, then $P(E \cap F)$ may be found by either of these formulas.

$$P(E \cap F) = P(F) \cdot P(E \mid F) \text{ or } P(E \cap F) = P(E) \cdot P(F \mid E)$$

If E and F are independent events, then $P(E \cap F) = P(E) \cdot P(F)$.

Bayes' Formula

$$P(F_i \mid E) = \frac{P(F_i) \cdot P(E \mid F_i)}{P(F_1) \cdot P(E \mid F_1) + P(F_2) \cdot P(E \mid F_2) + \cdots + P(F_n) \cdot P(E \mid F_n)}$$

Chapter 8 Review Exercises

Write true *or* false *for each of the following.*

1. $9 \in \{8, 4, -3, -9, 6\}$

2. $4 \in \{3, 9, 7\}$

3. $2 \notin \{0, 1, 2, 3, 4\}$

4. $0 \notin \{0, 1, 2, 3, 4\}$

5. $\{3, 4, 5\} \subseteq \{2, 3, 4, 5, 6\}$

6. $\{1, 2, 5, 8\} \subseteq \{1, 2, 5, 10, 11\}$

7. $\emptyset \subseteq \{1\}$

8. $0 \subseteq \emptyset$

List the elements in the following sets.

9. $\{x \mid x$ is a national holiday$\}$

10. $\{x \mid x$ is an integer, $-3 \le x < 1\}$

11. $\{$all counting numbers less than 5$\}$

12. $\{x \mid x$ is a leap year between 1989 and 2003$\}$

Let $U = \{$ Vitamins A, B_1, B_2, B_3, B_6, B_{12}, C, D, E$\}$, $M = \{A, C, D, E\}$, *and* $N = \{A, B_1, B_2, C, E\}$. *Find the following.*

13. M'

14. N'

15. $M \cap N$

16. $M \cup N$

17. $M \cup N'$

18. $M' \cap N$

Let $U = \{$all students in a class$\}$, $A = \{$all male students$\}$, $B = \{$all A students$\}$, $C = \{$all students with red hair$\}$, *and* $D = \{$all students younger than 21$\}$. *Describe the following sets in words.*

19. $A \cap C$

20. $B \cap D$

21. $A \cup D$

22. $A' \cap D$

23. $B' \cap C'$

Draw a Venn diagram and shade the given set.

24. $B \cup A'$

25. $A' \cap B$

26. $A' \cap (B' \cap C)$

27. $(A \cup B)' \cap C$

Social Science   *A survey of a group of military personnel revealed the following information.*

   20 *officers*
   27 *minorities*
   19 *women*
   5 *women officers*
   8 *minority women*
   10 *minority officers*
   3 *women minority officers*
   6 *Caucasian male enlisted personnel*

28. How many were interviewed?

29. How many were enlisted minority women?

30. How many were male minority officers?

Write sample spaces for the following.

31. A die is rolled and the number of points showing is noted.

32. A card is drawn from a deck containing only 4 aces.

33. A color is selected from the set $\{$red, blue, green$\}$, and then a number is chosen from the set $\{10, 20, 30\}$.

A jar contains 5 discs labeled 2, 4, 6, 8, 10, *and another jar contains* 2 *blue and* 3 *yellow balls. One disc is drawn and then a ball is drawn. Give the following.*

34. The sample space

35. Event F, the ball is blue.

36. Event E, the disc shows a number greater than 5.

37. Are the outcomes in this sample space equally likely?

Business   *A company sells computers and copiers. Let E be the event "a customer buys a computer," and let F be the event "a customer buys a copier." In Exercises 38 and 39, write each of the following using* $\cap$, $\cup$, *or* ' *as necessary.*

38. A customer buys neither.

39. A customer buys at least one.

40. A student gives the answer to a probability problem as 6/5. Explain why this answer must be incorrect.

41. Describe what is meant by disjoint sets and give an example.

42. Describe what is meant by mutually exclusive events and give an example.

43. How are disjoint sets and mutually exclusive events related?

A single card is drawn from an ordinary deck. Find the probability of drawing each of the following.

44. A black king

45. A face card

46. A red card or a face card

47. A black card, given it is a 2

48. A jack, given it is a face card

49. A face card, given it is a jack

Find the odds in favor of drawing the following.

50. A spade

51. A red queen

52. A black face card or a 7

Business *A sample shipment of five electric motors is chosen at random. The probability of exactly 0, 1, 2, 3, 4, or 5 motors being defective is given in the following table.*

Number defective	0	1	2	3	4	5
Probability	.31	.25	.18	.12	.08	.06

Find the probability that

53. no more than 3 are defective.

54. at least 3 are defective.

Health *The partial table shows the four possible (equally likely) combinations when both parents are carriers of the sickle cell anemia trait. Each carrier parent has normal cells (N) and trait cells (T).*

		2nd Parent	
		N_2	T_2
1st Parent	N_1 T_1		N_1T_2

55. Complete the table.

56. If the disease occurs only when two trait cells combine, find the probability that a child born to these parents will have sickle cell anemia.

57. The child will carry the trait but not have the disease if a normal cell combines with a trait cell. Find this probability.

58. Find the probability that the child is neither a carrier nor has the disease.

Find the probabilities for the following sums when two fair dice are rolled.

59. 8

60. At least 10

61. No more than 5

62. Odd and greater than 8

63. 12, given the sum is greater than 10

64. 7, given that at least one die shows a 4

Suppose $P(E) = .51$, $P(F) = .37$, and $P(E \cap F) = .22$. Find each of the following probabilities.

65. $P(E \cup F)$

66. $P(E \cap F')$

67. $P(E' \cup F)$

68. $P(E' \cap F')$

69. For the events E and F, $P(E) = .2$, $P(E \mid F) = .3$, and $P(F) = .4$. Find each of the following.

(a) $P(E' \mid F)$ (b) $P(E \mid F')$

70. Define independent events and give an example.

71. Are independent events always mutually exclusive? Are they ever mutually exclusive? Give examples.

Business *Of the appliance repair shops listed in the phone book, 80% are competent and 20% are not. A competent shop can repair an appliance correctly 95% of the time; an incompetent shop can repair an appliance correctly 60% of the time. Suppose an appliance was repaired correctly. Find the probability that it was repaired by*

72. a competent shop;

73. an incompetent shop.

Suppose an appliance was repaired incorrectly. Find the probability that it was repaired by

74. a competent shop;

75. an incompetent shop.

76. **Business** A manufacturer buys items from four different suppliers. The fraction of the total number of items that is obtained from each supplier, along with the probability that an item purchased from that supplier is defective, is shown in the following table.

Supplier	Fraction of Total Supplied	Probability of Defective
1	.17	.04
2	.39	.02
3	.35	.07
4	.09	.03

(a) Find the probability that a defective item came from supplier 4.

(b) Find the probability that a defective item came from supplier 2.

77. **Social Science** The tables on the next page list the number of passengers who were on the Titanic and the number of passengers who survived, according to class of ticket.*

---

*Sandra L. Takis, "Titanic: A Statistical Exploration," *Mathematics Teacher* 92, no. 8, (November 1999): p. 660–664. Reprinted with permission. ©1999 by the National Council of Teachers of Mathematics. All rights reserved.

	CHILDREN		WOMEN	
	On	*Survived*	*On*	*Survived*
First Class	6	6	144	140
Second Class	24	24	165	76
Third Class	79	27	93	80
Total	109	57	402	296

	MEN		TOTALS	
	On	*Survived*	*On*	*Survived*
First Class	175	57	325	203
Second Class	168	14	357	114
Third Class	462	75	634	182
Total	805	146	1316	499

Use this information to determine the following (round answers to two decimal places).

(a) What is the probability that a randomly selected passenger was in second class?

(b) What is the overall probability of surviving?

(c) What is the probability of a first-class passenger surviving?

(d) What is the probability of a child who was also in third class surviving?

(e) Given that the survivor is from first class, what is the probability that she was a woman?

(f) Given that a male has survived, what is the probability that he was in third class?

(g) Are the events third-class survival and male survival independent events? What does this imply?

78. **Business** The table below shows the results of a survey of buyers of a certain model of car.

Car Type	Satisfied	Not Satisfied	Totals
New	300	100	
Used	450		600
Totals		250	

(a) Complete the table.

(b) How many buyers were surveyed?

(c) How many bought a new car and were satisfied?

(d) How many were not satisfied?

(e) How many bought used cars?

(f) How many of those who were not satisfied had bought a used car?

(g) Rewrite the event stated in part (f) using the expression "given that."

(h) Find the probability of the outcome in parts (f) and (g).

(i) Find the probability that a used-car buyer is not satisfied.

(j) You should have different answers in parts (h) and (i). Explain why.

Additional Probability Review Exercises

Use these exercises for practice deciding which rule, principle, or formula to apply.

1. Suppose $P(E) = .4$, $P(F) = .22$, and $P(E \cup F) = .52$. Find
(a) $P(E \cup F')$ (b) $P(E \cap F')$ (c) $P(E' \cup F)$

2. A jar contains 2 white, 3 orange, 5 yellow, and 8 black marbles. If a marble is drawn at random, find the probability that it is the following.
(a) White   (c) Not black
(b) Orange   (d) Orange or yellow

3. **Finance** GT Global Theme Funds has worldwide assets distributed as follows.*

Geographic Location	Percent
Africa and Middle East	4.2
Latin America	6.0
Asia-Pacific	18.7
Europe	23.1
United States and Canada	48.0

If an asset is chosen randomly, what is the probability that it is from the following regions?
(a) Asia-Pacific
(b) North or South America
(c) Not from Europe

4. **Finance** GT Global Health Care Fund has invested in the following sectors.*

Sector	Percent
Biotechnology	30.7
Pharmaceuticals	25.4
Medical Technology and Supplies	22.3
Health Care Services	8.2
Short-Term and Other	13.4

---

*Exercises 3 and 4 from GT Global Theme Funds Annual Report, October 31, 1996.

Find the probability that an investment selected at random from this fund is as follows.

(a) In biotechnology or health care services

(b) In pharmaceuticals or medical technology and supplies

(c) Not in biotechnology

5. A single fair die is rolled. Find the following probabilities, if the die shows

(a) a 2, given that the number was odd;

(b) a 4, given that the number was even;

(c) an even number given that the number was 6.

6. Finance According to an article in *Business Week*, in 1989 funding for university research in the United States totaled $15 billion. Support came from various sources, as shown in the table below.*

Source	Amount (in billions of dollars)
Federal government	9.0
State and local government	1.2
Institutional	2.7
Industry	1.0
Other	1.1

Find the probability that funds for a particular project came from each of the following sources.

(a) Federal government

(b) Industry

(c) The institution

7. Social Science The population of the United States by race in 1995, and the projected population by race for the year 2025 are given below.[†]

U.S. Population by Race (in thousands)		
Race	1995	2025
White	193,900	209,900
Hispanic	26,800	50,900
African-American	31,600	44,700
Asian-American	9,200	24,000
Other	1,900	2,800

Find the probability of a randomly selected person being each of the following.

(a) Hispanic in 1995

(b) Hispanic in 2025

(c) African-American in 1995

(d) African-American in 2025

8. Given $P(A) = .2$, $P(B \mid A) = .5$, $P(B \mid A') = .4$, find $P(A \mid B)$.

Business *Companies A, B, and C produce 15%, 40%, and 45%, respectively, of the major appliances sold in a certain area. In that area, 1% of the Company A appliances, $1\frac{1}{2}$% of the Company B appliances, and 2% of the Company C appliances need service within the first year. Suppose an appliance that needs service within the first year is chosen at random; find the probability that it was manufactured by Company*

9. A;   **10.** B.

11. Social Science In one area, 4% of the population drives luxury cars. However, 17% of the CPAs drive luxury cars. Are the events "person drives a luxury car" and "person is a CPA" independent?

12. Suppose $P(E) = .05$, $P(F) = .1$, and $P(E \cap F) = .02$. Find

(a) $P(E' \cap F)$ **(b)** $P(E' \cup F')$ **(c)** $P(E \cap F')$.

13. One orange and four red slips of paper are placed in a box. Two red and three orange slips are placed in a second box. A box is chosen at random, and a slip of paper is selected from it. The probability of choosing the first box is 3/8. If the selected slip of paper is orange, what is the probability that it came from the first box?

14. Find the probability that the slip of paper in Exercise 13 came from the second box, given that it is red.

15. Business A manufacturing firm finds that 70% of its new hires turn out to be good workers and 30% poor workers. All current workers are given a reasoning test. Of the good workers, 80% pass it; 40% of the poor workers pass it. Assume that these figures will hold true in the future. If the company makes the test part of its hiring procedure and only hires people who meet the previous requirements and pass the test, what percent of the new hires will turn out to be good workers?

Social Science *A survey has shown that 52% of the women in a certain community work outside the home. Of these women, 64% are married, while 86% of the women who do not work outside the home are married. Find the probability that a woman in that community is*

16. married;

17. a single woman working outside the home.

Social Science *A study showed that, in 1991, 31.6% of men and 35.0% of women were obese.* Given that 48.7% of

*"University Research: The Squeeze Is On," *Business Week*, May 20, 1991. Reprinted by special permission. Copyright © 1991 by McGraw-Hill, Inc.

†Projections of Hispanic and Non-Hispanic Populations by Age and Sex: 1995 to 2025, from U.S. Bureau of the Census, *Current Population Reports*.

*U.S. Bureau of the Census: Statistical Abstract of the United States, 1992.

Americans are men and 51.3% are women, find the probability that a randomly selected adult fits the following description.

18. An obese man

19. Obese

Business *On a given weekend in the fall, a tire company can buy television advertising time for a college football game, a baseball game, or a professional football game. If the company sponsors the college game, there is a 70% chance of a high rating, a 50% chance if they sponsor a baseball game, and a 60% chance if they sponsor a professional football game. The probability of the company sponsoring these various games is .5, .2, and .3, respectively. Suppose the company does get a high rating; find the probability that it sponsored*

20. a college game;

21. a professional football game.

22. **Social Science** A nationwide survey showed that the odds that an individual uses a seat belt when in a car's front seat are 17:8.* What is the probability that such an individual does not use a seat belt?

23. **Social Science** The survey mentioned in Exercise 22 also showed that the odds that a driver does not drink or does not drive after drinking are 4:1.* What is the probability that a driver was drinking?

24. If a marble is drawn from a bag containing 2 yellow, 5 red, and 3 blue marbles, what are the probabilities of the following?
(a) The marble is red.
(b) The marble is either yellow or blue.
(c) The marble is yellow or red.
(d) The marble is green.

25. The law firm of Alam, Bartolini, Chinn, Dickinson, and Ellsberg has two senior partners, Alam and Bartolini. Two of the attorneys are to be selected to attend a conference. Assuming that all are equally likely to be selected, find the following probabilities.
(a) Chinn is selected.
(b) Ellsberg is not selected.
(c) Alam and Dickinson are selected.
(d) At least one senior partner is selected.

*Exercises 22 and 23 based on an article in the *Sacramento Bee*, June 4, 1992.

Case 7

Medical Diagnosis

When a patient is examined, information, typically incomplete, is obtained about his or her state of health. Probability theory provides a mathematical model appropriate for this situation, as well as a procedure for quantitatively interpreting such partial information to arrive at a reasonable diagnosis.*

To do this, list the states of health that can be distinguished in such a way that the patient can be in one and only one state at the time of the examination. Each state of health H is associated with a number $P(H)$ between 0 and 1 such that the sum of all these numbers is 1. This number $P(H)$ represents the probability, before exam, that a patient is in the state of health H, and $P(H)$ may be chosen subjectively from medical experience, using any information available prior to the examination. The probability may be most conveniently established from clinical records; that is, a mean probability is established for patients in general, although the number would vary from patient to patient.

For example, limiting the discussion to the condition of a patient's heart, suppose there are exactly 3 states of health, with probabilities as follows.

	State of Health H	P(H)
H_1	patient has a normal heart	.8
H_2	patient has minor heart irregularities	.15
H_3	patient has a severe heart condition	.05

Having selected $P(H)$, the information from the examination is processed. First, the results of the examination must be classified. The examination itself consists of observing the state of a number of characteristics of the patient. Let us assume that the examination for a heart condition consists of a stethoscope examination and a cardiogram. The outcome of such an examination, C, might be one of the following:

$C_1 =$ stethoscope shows normal heart and cardiogram shows normal heart;

$C_2 =$ stethoscope shows normal heart and cardiogram shows minor irregularities;

and so on.

---

*From "Probabilistic Medical Diagnosis," Roger Wright, *Some Mathematical Models in Biology*, Robert M. Thrall, ed. (The University of Michigan, 1967), by permission of Robert M. Thrall.

It remains to assess for each state of health H the conditional probability $P(C \mid H)$ of each examination outcome C using only the knowledge that a patient is in a given state of health. (This may be based on the medical knowledge and clinical experience of the doctor.) The conditional probabilities $P(C \mid H)$ will not vary from patient to patient, so that they may be built into a diagnostic system, although they should be reviewed periodically.

Suppose the result of the examination is C_1. Let us assume the following probabilities.

$$P(C_1 \mid H_1) = .9$$
$$P(C_1 \mid H_2) = .4$$
$$P(C_1 \mid H_3) = .1$$

Now, for a given patient, the appropriate probability associated with each state of health H, after examination, is $P(H \mid C)$ where C is the outcome of the examination. This can be calculated by using Bayes' formula. For example, to find $P(H_1 \mid C_1)$—that is, the probability that the patient has a normal heart given that the examination showed a normal stethoscope examination and a normal cardiogram—we use Bayes' formula as follows.

$$P(H_1 \mid C_1)$$
$$= \frac{P(C_1 \mid H_1)P(H_1)}{P(C_1 \mid H_1)P(H_1) + P(C_1 \mid H_2)P(H_2) + P(C_1 \mid H_3)P(H_3)}$$
$$= \frac{(.9)(.8)}{(.9)(.8) + (.4)(.15) + (.1)(.05)} \approx .92$$

Hence, the probability is about .92 that the patient has a normal heart on the basis of the examination results. This means that in 8 out of 100 patients, some abnormality will be present and not be detected by the stethoscope or the cardiogram.

Exercises

1. Find $P(H_2 \mid C_1)$.
2. Assuming the following probabilities, find $P(H_1 \mid C_2)$:
   $P(C_2 \mid H_1) = .2, \quad P(C_2 \mid H_2) = .8, \quad P(C_2 \mid H_3) = .3.$
3. Assuming the probabilities of Exercise 2, find $P(H_3 \mid C_2)$.

Counting, Probability Distributions, and Further Topics in Probability

A survey by *Money* magazine found that supermarket scanners overcharged customers at 30% of stores. If you shop at 3 supermarkets that use scanners, what is the probability that you will be overcharged? In this chapter, we introduce additional techniques for finding the probabilities of events such as this.

To use the basic definition of probability, $P(E) = n(E)/n(S)$ (where S is a sample space with equally likely outcomes), up to now we have simply listed the outcomes in S and in E. However, when S has many outcomes, listing them all becomes very tedious. In the first section of this chapter, we introduce methods for counting the number of outcomes in a set without actually listing them, and then we use this approach in later sections to find probabilities.

9.1 PERMUTATIONS AND COMBINATIONS

We begin with a simple example. If there are three roads from town A to town B and 2 roads from town B to town C, in how many ways can someone travel from A to C by way of B? For each of the 3 roads from A there are 2 different routes leading from B to C, making $3 \cdot 2 = 6$ different trips, as shown in Figure 9.1 on the next page. This example illustrates a general principle of counting called the *multiplication principle*.

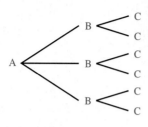

FIGURE 9.1

Multiplication Principle

Suppose n choices must be made, with

$$m_1 \text{ ways to make choice 1,}$$

and for each of these,

$$m_2 \text{ ways to make choice 2,}$$

and so on, with

$$m_n \text{ ways to make choice } n.$$

Then there are

$$m_1 \cdot m_2 \cdot \cdots \cdot m_n$$

different ways to make the entire sequence of choices.

(a) In how many ways can 6 business tycoons line up their golf carts at the country club?

(b) How many ways can 4 pupils be seated in a row with 4 seats?

Answers:

(a) $6 \cdot 5 \cdot 4 \cdot 3 \cdot 2 \cdot 1 = 720$

(b) $4 \cdot 3 \cdot 2 \cdot 1 = 24$

▶**EXAMPLE 1** A combination lock can be set to open to any 3-letter sequence. How many such sequences are possible?

Since there are 26 letters in the alphabet, there are 26 choices for each of the 3 letters, and, by the multiplication principle, $26 \cdot 26 \cdot 26 = 17,576$ different sequences. ◀

▶**EXAMPLE 2** A teacher has 5 different books to be arranged side by side. How many different arrangements are possible?

Five choices will be made, 1 for each space that will hold a book. Any of the 5 possible books could be chosen for the first space. There are 4 possible choices for the second space, since 1 book has already been placed in the first space, 3 possible choices for the third space, and so on. By the multiplication principle, the number of different possible arrangements (sequence of choices) is $5 \cdot 4 \cdot 3 \cdot 2 \cdot 1 = 120$. ◀

The use of the multiplication principle often leads to products such as $5 \cdot 4 \cdot 3 \cdot 2 \cdot 1$, the product of all the natural numbers from 5 down to 1. If n is a natural number, the symbol $n!$ (read "n factorial") denotes the product of all the natural numbers from n down to 1. If $n = 1$, this formula is understood to give $1! = 1$.

n-Factorial

For any natural number n,

$$n! = n(n-1)(n-2) \ldots (3)(2)(1).$$

Also, 0! is defined to be the number 1.

②

Evaluate:

(a) 4!

(b) 6!

(c) 1!

(d) 6!/4!

Answers:

(a) 24

(b) 720

(c) 1

(d) 30

With this symbol, the product $5 \cdot 4 \cdot 3 \cdot 2 \cdot 1$ can be written as 5!. Also, $3! = 3 \cdot 2 \cdot 1 = 6$. The definition of $n!$ could be used to show that $n! = n \cdot (n - 1)!$ for all natural numbers $n \geq 2$. It is helpful if this result also holds for $n = 1$. This can only happen if 0! equals 1, as defined above. ②

Almost all calculators have an $n!$ key. A calculator with a 10-digit display and scientific notation capability will usually give the exact value of $n!$ for $n \leq 13$ and approximate values of $n!$ for $14 \leq n \leq 69$. The value of 70! is approximately $1.198 \cdot 10^{100}$, which is too large for most calculators.

T E C H N O L O G Y   T I P   The factorial key on a graphing calculator is usually located in the PRB or PROB submenu of the MATH or OPTN menu.

▶ **EXAMPLE 3**   Morse code uses a sequence of dots and dashes to represent letters and words. How many sequences are possible with at most 3 symbols?

"At most 3" means "1 or 2 or 3." Each symbol may be either a dot or a dash. Thus, the following numbers of sequences are possible in each case.

③

How many sequences are possible with at most 4 symbols?

Answer:

30

Number of Symbols	Number of Sequences
1	2
2	$2 \cdot 2 = 4$
3	$2 \cdot 2 \cdot 2 = 8$

Altogether, $2 + 4 + 8 = 14$ different sequences of at most 3 symbols are possible. Because there are 26 letters in the alphabet, some letters must be represented by sequences of 4 symbols in Morse code. ◀ ③

④

In how many ways can 3 of 7 items be arranged?

Answer:

$7 \cdot 6 \cdot 5 = 210$

▶ **EXAMPLE 4**   Suppose the teacher in Example 2 wishes to place only 3 of the 5 books on his desk. How many arrangements of 3 books are possible?

The teacher again has 5 ways to fill the first space, 4 ways to fill the second space, and 3 ways to fill the third. Because he wants to use only 3 books, there are only 3 spaces to be filled giving $5 \cdot 4 \cdot 3 = 60$ arrangements. ◀ ④

Permutations

The answer 60 in Example 4 is called the number of *permutations* of 5 things taken 3 at a time. A **permutation** of r elements (where $r \geq 1$) from a set of n elements is any arrangement, *without repetition,* of the r elements. The number of permutations of n things taken r at a time (with $r \leq n$) is written $_nP_r$.* Based on the work in Example 4,

$$_5P_3 = 5 \cdot 4 \cdot 3 = 60,$$

which can be expressed as

$$5 \cdot 4 \cdot 3 = 5 \cdot 4 \cdot 3 \cdot \frac{2 \cdot 1}{2 \cdot 1} = \frac{5 \cdot 4 \cdot 3 \cdot 2 \cdot 1}{2 \cdot 1} = \frac{5!}{2!} = \frac{5!}{(5 - 3)!}$$

*Another notation that is sometimes used is $P(n, r)$.

This example illustrates the following general rule for permutations.

Permutations

If $_nP_r$ (where $r \leq n$) is the number of permutations of n elements taken r at a time, then

$$_nP_r = \frac{n!}{(n-r)!}.$$

TECHNOLOGY TIP The permutation function on a graphing calculator is in the same menu as the factorial key. As with $n!$, for large values of n and r, the calculator display for $_nP_r$ may be an approximation.

To find $_nP_r$, we can use either the rule above or direct application of the multiplication principle, as the following example shows.

▶ **EXAMPLE 5** Early in 1996, eight candidates sought the Republican nomination for president. In how many ways could voters rank their first, second, and third choices?

This is the same as finding the number of permutations of 8 elements taken 3 at a time. Since there are 3 choices to be made, the multiplication principle gives $_8P_3 = 8 \cdot 7 \cdot 6 = 336$. Alternatively, by the formula for $_nP_r$,

$$_8P_3 = \frac{8!}{(8-3)!} = \frac{8!}{5!} = \frac{8 \cdot 7 \cdot 6 \cdot 5 \cdot 4 \cdot 3 \cdot 2 \cdot 1}{5 \cdot 4 \cdot 3 \cdot 2 \cdot 1} = 8 \cdot 7 \cdot 6 = 336.$$

Figure 9.2 shows this result on a TI-83 graphing calculator. ◀ ⑤

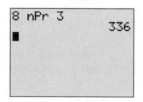

FIGURE 9.2

▶ **EXAMPLE 6** Find each of the following.

(a) The number of permutations of the letters A, B, and C
By the formula for $_nP_r$ with both n and r equal to 3,

$$_3P_3 = \frac{3!}{(3-3)!} = \frac{3!}{0!} = \frac{3!}{1} = 3! = 6.$$

The 6 permutations (or arrangements) are

$$ABC, \ ACB, \ BAC, \ BCA, \ CAB, \ CBA.$$

⑤

Find the number of permutations of

(a) 5 things taken 2 at a time;

(b) 9 things taken 3 at a time.

Find each of the following.

(c) $_3P_1$

(d) $_7P_3$

(e) $_{12}P_2$

Answers:

(a) 20

(b) 504

(c) 3

(d) 210

(e) 132

Find the number of permutations of the letters C, O, D, and E

(a) using all the letters;

(b) using 2 of the 4 letters;

(c) using 3 of the 4 letters.

Answers:

(a) 24

(b) 12

(c) 24

(b) The number of permutations using just 2 of the letters A, B, and C. Find $_3P_2$:

$$_3P_2 = \frac{3!}{(3-2)!} = \frac{3!}{1!} = 3! = 6.$$

This result is exactly the same answer as in part (a). This is because, in the case of $_3P_3$, after the first 2 choices are made, the third is already determined, as shown in the table below.

First Two Letters	AB	AC	BA	BC	CA	CB
Third Letter	C	B	C	A	B	A

◀ ⑥

▶ **EXAMPLE 7**   A televised talk show will include 4 women and 3 men as panelists.

(a) In how many ways can the panelists be seated in a row of 7 chairs?

Find $_7P_7$, the total number of ways to seat 7 panelists in 7 chairs.

$$_7P_7 = \frac{7!}{(7-7)!} = \frac{7!}{0!} = \frac{7!}{1} = 7 \cdot 6 \cdot 5 \cdot 4 \cdot 3 \cdot 2 \cdot 1 = 5040$$

There are 5040 ways to seat the 7 panelists.

⑦

A collection of 3 paintings by one artist and 2 by another is to be displayed. In how many ways can the paintings be shown

(a) in a row?

(b) if the works of the artists are to be alternated?

(c) if one painting by each artist is displayed?

Answers:

(a) 120

(b) 12

(c) 6

(b) In how many ways can the panelists be seated if the men and women are to be alternated?

Use the multiplication principle. In order to alternate men and women, a woman must be seated in the first chair (since there are 4 women and only 3 men), any of the men next, and so on. Thus, there are 4 ways to fill the first seat, 3 ways to fill the second seat, 3 ways to fill the third seat (with any of the 3 remaining women), and so on.

$$4 \cdot 3 \cdot 3 \cdot 2 \cdot 2 \cdot 1 \cdot 1 = 144$$

There are 144 ways to seat the panelists.

(c) In how many ways can the panelists be seated if the men must sit together and the women must sit together?

Use the multiplication principle. We first must decide how to arrange the two groups (men and women). There are 2! ways of doing this. Next, there are 4! ways of arranging the women and 3! ways of arranging the men, for a total of

$$2! \, 4! \, 3! = 2 \cdot 24 \cdot 6 = 288$$

ways.   ◀ ⑦

Combinations

In Example 4, we found there are 60 ways that a teacher can arrange 3 of 5 different books on a desk. That is, there are 60 permutations of 5 things taken 3 at a time. Suppose now that the teacher does not wish to arrange the books on his desk but rather wishes to choose, at random, any 3 of the 5 books to give to a book sale to raise money for his school. In how many ways can he do this?

At first glance, we might say 60 again, but this is incorrect. The number 60 counts all possible *arrangements* of 3 books chosen from 5. However, the following arrangements would all lead to the same set of 3 books being given to the book sale.

mystery-biography-textbook	biography-textbook-mystery
mystery-textbook-biography	textbook-biography-mystery
biography-mystery-textbook	textbook-mystery-biography

The list shows 6 different *arrangements* of 3 books but only one *subset* of 3 books. A subset of items selected *without regard to order* is called a **combination.**

The number of combinations of 5 things taken 3 at a time is written $\binom{5}{3}$ or $_5C_3$. Since they are subsets, combinations are *not ordered.*

To evaluate $\binom{5}{3}$, start with the $5 \cdot 4 \cdot 3$ *permutations* of 5 things taken 3 at a time. Combinations are unordered; therefore, we find the number of combinations by dividing the number of permutations by the number of ways each group of 3 can be ordered—that is, by 3!.

$$\binom{5}{3} = \frac{5 \cdot 4 \cdot 3}{3!} = \frac{5 \cdot 4 \cdot 3}{3 \cdot 2 \cdot 1} = 10$$

There are 10 ways that the teacher can choose 3 books at random for the book sale.

Generalizing this discussion gives the formula for the number of combinations of n elements taken r at a time, written $\binom{n}{r}$ or $_nC_r$. In general, a set of r elements can be ordered in $r!$ ways, so we divide $_nP_r$ by $r!$ to get $\binom{n}{r} = {}_nC_r$.

$$\binom{n}{r} = {}_nC_r = \frac{_nP_r}{r!}$$
$$= {}_nP_r \frac{1}{r!}$$
$$= \frac{n!}{(n-r)!} \cdot \frac{1}{r!} \qquad \text{Definition of } _nP_r$$
$$= \frac{n!}{(n-r)!r!}$$

This last form is the most useful for setting up the calculation. ◀ ⑧

⑧

Evaluate $\dfrac{_nP_r}{r!}$ for the following values.

(a) $n = 6, r = 2$
(b) $n = 8, r = 4$
(c) $n = 7, r = 0$

Answers:
(a) 15
(b) 70
(c) 1

Combinations

The number of combinations of n elements taken r at a time, where $r \leq n$, is

$$\binom{n}{r} = {}_nC_r = \frac{n!}{(n-r)!\,r!}.$$

Replacing r by $n - r$ in the combinations formula gives

$$\binom{n}{n-r} = \frac{n!}{(n - [n - r])! \, (n - r)!}$$

$$= \frac{n!}{r! \, (n - r)!} = \frac{n!}{(n - r)! \, r!}.$$

Therefore,

$$\binom{n}{r} = \binom{n}{n-r}, \quad \text{or equivalently,} \quad {}_nC_r = {}_nC_{n-r}.$$

For example, by this result,

$$\binom{5}{3} = \binom{5}{2} \quad \text{and} \quad \binom{10}{4} = \binom{10}{6}.$$

▶ **EXAMPLE 8** From a group of 8 professors, a committee is selected to develop a new curriculum. How many committees can be formed with the following numbers of members?

(a) 3

A committee is an unordered group, so find $\binom{8}{3}$. By the formula for combinations,

$$\binom{8}{3} = \frac{8!}{5! \, 3!} = \frac{8 \cdot 7 \cdot 6 \cdot 5 \cdot 4 \cdot 3 \cdot 2 \cdot 1}{5 \cdot 4 \cdot 3 \cdot 2 \cdot 1 \cdot 3 \cdot 2 \cdot 1} = \frac{8 \cdot 7 \cdot 6}{3 \cdot 2 \cdot 1} = 56.$$

(b) 5

We want $\binom{8}{5}$. By the result above,

$$\binom{8}{5} = \binom{8}{3} = 56.$$

This makes sense; since each group of 3 leaves a corresponding group of 5, the number of ways to choose 5 people from 8 is the same as the number of ways to choose 3 people from 8. ◀ ⑨

⑨

Use $\dfrac{n!}{(n - r)! \, r!}$ to evaluate $\dbinom{n}{r}$.

(a) $\dbinom{6}{2}$

(b) $\dbinom{8}{4}$

(c) $\dbinom{7}{0}$

Compare your answers with the answers for side problem 8.

Answers:

(a) 15

(b) 70

(c) 1

TECHNOLOGY TIP The combination function on a graphing calculator, designated ${}_nC_r$, is in the same menu as ${}_nP_r$. Figure 9.3 shows the result of computing ${}_8C_3$.

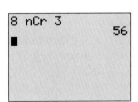

```
8 nCr 3
            56
■
```

FIGURE 9.3

▶**EXAMPLE 9**   Three managers are to be selected from a group of 30 to work on a special project.

(a) In how many different ways can the managers be selected?

Here we wish to know the number of 3-element combinations that can be formed from a set of 30 elements. (We want combinations, not permutations, since order within the group of 3 does not matter.)

$$\binom{30}{3} = \frac{30!}{27! \, 3!} = 4060$$

There are 4060 ways to select the project group.

(b) In how many ways can the group of 3 be selected if a certain manager must work on the project?

Since 1 manager has already been selected for the project, the problem is reduced to selecting 2 more from the remaining 29 managers.

$$\binom{29}{2} = \frac{29!}{27! \, 2!} = 406$$

In this case, the project group can be selected in 406 ways.

(c) In how many ways can a nonempty group of at most 3 managers be selected from these 30 managers?

The group is to be nonempty; therefore, "at most 3" means "1 or 2 or 3." Find the number of ways for each case.

Five orchids from a collection of 20 are to be selected for a flower show.

(a) In how many ways can this be done?

(b) In how many different ways can the group of 5 be selected if 2 particular orchids must be included?

(c) In how many ways can at least 1 and at most 5 orchids be selected? (*Hint:* Use a calculator.)

Answers:

(a) $\binom{20}{5} = 15{,}504$

(b) $\binom{18}{3} = 816$

(c) 21,699

Case	Number of Ways		
1	$\binom{30}{1} = \frac{30!}{29! \, 1!} = \frac{30 \cdot 29!}{29! \, 1!} = 30$		
2	$\binom{30}{2} = \frac{30!}{28! \, 2!} = \frac{30 \cdot 29 \cdot 28!}{28! \cdot 2 \cdot 1} = 435$		
3	$\binom{30}{3} = \frac{30!}{27! \, 3!} = \frac{30 \cdot 29 \cdot 28 \cdot 27!}{27! \cdot 3 \cdot 2 \cdot 1} = 4060$		

The total number of ways to select at most 3 managers will be the sum

$$30 + 435 + 4060 = 4525. \quad ◀ \quad \text{(10)}$$

The formulas for permutations and combinations given in this section will be very useful in solving probability problems in later sections. Any difficulty in using these formulas usually comes from being unable to differentiate between them. Both permutations and combinations give the number of ways to choose r objects from a set of n objects. The differences between permutations and combinations are outlined in the following summary.

Permutations	Combinations
Different orderings or arrangements of the r objects are different permutations.	Each choice or subset of r objects gives 1 combination. Order within the r objects does not matter.
$$_nP_r = \frac{n!}{(n-r)!}$$	$$\binom{n}{r} = {_nC_r} = \frac{n!}{(n-r)!\,r!}$$
Clue words: arrangement, schedule, order	Clue words: group, committee, set, sample
Order matters!	Order does not matter!

In the next examples, concentrate on recognizing which of the formulas should be applied.

▶ **EXAMPLE 10** For each of the following problems, tell whether permutations or combinations should be used to solve the problem.

(a) How many 4-digit code numbers are possible if no digits are repeated?

Since changing the order of the 4 digits results in a different code, we use permutations.

(b) A sample of 3 lightbulbs is randomly selected from a batch of 15 bulbs. How many different samples are possible?

The order in which the 3 lightbulbs are selected is not important. The sample is unchanged if the bulbs are rearranged, so combinations should be used.

(c) In a basketball conference with 8 teams, how many games must be played so that each team plays every other team exactly once?

Selection of 2 teams for a game is an *unordered* subset of 2 from the set of 8 teams. Use combinations again.

(d) In how many ways can 4 patients be assigned to 6 hospital rooms so that each patient has a private room?

The room assignments are an *ordered* selection of 4 rooms from the 6 rooms. Exchanging the rooms of any 2 patients within a selection of 4 rooms gives a different assignment, so permutations should be used. ◀ ⑪

⑪

Solve the problems in Example 10.

Answers:

(a) 5040

(b) 455

(c) 28

(d) 360

▶ **EXAMPLE 11** A manager must select 4 employees for promotion. 12 employees are eligible.

(a) In how many ways can the 4 be chosen?

Because there is no reason to consider the order in which the 4 are selected, we use combinations.

$$\binom{12}{4} = \frac{12!}{4!\,8!} = 495$$

A mailman has special delivery mail for 7 customers.

(a) In how many ways can he arrange his schedule to deliver to all 7?

(b) In how many ways can he schedule deliveries if he can deliver to only 4 of the 7?

Answers:

(a) 5040

(b) 840

In how many ways can 4 aces and any other card be dealt?

Answer:

48

(b) In how many ways can 4 employees be chosen (from 12) to be placed in 4 different jobs?

In this case, once a group of 4 is selected, they can be assigned in many different ways (or arrangements) to the 4 jobs. Therefore, this problem requires permutations.

$$_{12}P_4 = \frac{12!}{8!} = 11,880 \quad \blacktriangleleft \quad \text{⑫}$$

The following problems involve a standard deck of 52 playing cards, shown in Figure 8.19 in Chapter 8.

▶ **EXAMPLE 12** In how many ways can a full house of aces and eights (3 aces and 2 eights) be dealt in a 5-card poker hand?

The order of the 3 aces and the 2 eights does not matter, so use combinations and the multiplication principle. There are $\binom{4}{3}$ ways to get 3 aces from the 4 aces in the deck and $\binom{4}{2}$ ways to get 2 eights. By the multiplication principle, the number of ways to get 3 aces *and* 2 eights is

$$\binom{4}{3} \cdot \binom{4}{2} = 4 \cdot 6 = 24. \quad \blacktriangleleft \quad \text{⑬}$$

▶ **EXAMPLE 13** Five cards are dealt from a standard 52-card deck.

(a) How many such hands have all face cards?

The face cards are the king, queen, and jack of each suit. There are 4 suits, so there are 12 face cards. The order of the 5 cards is not important, so use combinations to get

$$\binom{12}{5} = \frac{12!}{5! \, 7!} = 792.$$

(b) How many such hands have exactly 2 hearts?

There are 13 hearts in the deck, so the 2 hearts will be selected from those 13 cards. The other 3 cards must come from the remaining 39 cards cards that are not hearts. Use combinations and the multiplication principle to get

$$\binom{13}{2}\binom{39}{3} = 78 \cdot 9139 = 712,842.$$

Notice that the two top numbers in the combinations add up to 52, the total number of cards, and the two bottom numbers add up to 5, the number of cards in a hand.

(c) How many 5-card hands have all cards of a single suit?

The arrangement of the 5 cards is not important, so use combinations. The total number of ways that 5 cards of a particular suit of 13 cards can be dealt is $\binom{13}{5}$.

Since there are 4 different suits, by the multiplication principle, there are

$$4 \cdot \binom{13}{5} = 4 \cdot 1287 = 5148$$

ways to deal 5 cards of the same suit. ◀ ⑭

---

⑭

In how many ways can 5 red cards be dealt?

Answer:

65,780

As Examples 12 and 13 show, often both combinations and the multiplication principle must be used in the same problem.

▶ **EXAMPLE 14** To illustrate the differences between permutations and combinations in another way, suppose 2 cans of soup are to be selected from 4 cans on a shelf: noodle (N), bean (B), mushroom (M), and tomato (T). As shown in Figure 9.4(a), there are 12 ways to select 2 cans from the 4 cans if the order matters (if noodle first and bean second is considered different from bean, then noodle, for example). On the other hand, if order is unimportant, then there are 6 ways to choose 2 cans of soup from the 4, as illustrated in Figure 9.4(b). ◀

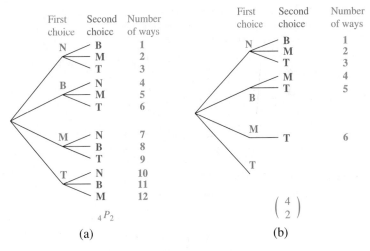

FIGURE 9.4

CAUTION It should be stressed that not all counting problems lend themselves to either permutations or combinations. Whenever a tree diagram or the multiplication principle can be used directly, use it.

9.1 Exercises

Evaluate the following factorials, permutations, and combinations.

1. $_4P_2$

2. $3!$

3. $\binom{8}{3}$

4. $7!$

5. $_8P_1$

6. $\binom{8}{1}$

7. $4!$

8. $_4P_4$

9. $_{12}C_5$

10. $\binom{10}{8}$

11. $_{13}P_2$

12. $_{12}P_3$

Use a calculator to find values for Exercises 13–20.

13. $_{25}P_5$

14. $_{38}P_4$

15. $_{14}P_5$

16. $_{17}P_8$

17. $\binom{21}{10}$

18. $\binom{34}{25}$

19. $_{25}C_{16}$

20. $\binom{30}{15}$

21. Some students find it puzzling that $0! = 1$, and they think that $0!$ should equal 0. If this were true, what would be the value of $P(4, 4)$ using the permutations formula?

22. If you already knew the value of $9!$, how could you find the value of $10!$ quickly?

Use the multiplication principle to solve the following problems. (See Examples 1–4.)

23. Social Science An ancient Chinese philosophical work known as the *I Ching* (Book of changes) is often used as an oracle from which people can seek and obtain advice. The philosophy describes the duality of the universe in terms of two primary forces: *yin* (passive, dark, receptive) and *yang* (active, light, creative). See the figure. The yin energy is represented by a broken line (– –) and the yang by a solid line (—). These lines are written on top of one another in groups of three, known as *trigrams*. For example, the trigram ☱ is called *Tui*, the Joyous, and has the image of a lake.

yin   yang

(a) How many trigrams are there altogether?

(b) The trigrams are grouped together, one on top of the other, in pairs known as hexagrams. Each hexagram represents one aspect of the *I Ching* philosophy. How many hexagrams are there?

24. Business How many different types of homes are available if a builder offers a choice of 5 basic plans, 3 roof styles, and 2 exterior finishes?

25. Business An auto manufacturer produces 7 models, each available in 6 different colors, with 4 different upholstery fabrics and 5 interior colors. How many varieties of the auto are available?

26. Business How many different 4-letter radio station call letters can be made
(a) if the first letter must be K or W and no letter may be repeated?
(b) if repeats are allowed (but the first letter is K or W)?
(c) How many of the 4-letter call letters (starting with K or W) with no repeats end in R?

27. Social Science A social security number has 9 digits. How many social security numbers are there? The U.S. population in 2000 was about 281 million. Is it possible for every U.S. resident to have a unique social security number? (Assume no restrictions.)

28. Social Science The United States Postal Service currently uses 5-digit zip codes in most areas. How many zip codes are possible if there are no restrictions on the digits used? How many would be possible if the first number could not be 0?

29. Social Science The Postal Service is encouraging the use of 9-digit zip codes in some areas, adding 4 digits after the usual 5-digit code. How many such zip codes are possible with no restrictions?

30. Social Science For many years, the state of California used 3 letters followed by 3 digits on its automobile license plates.
(a) How many different license plates are possible with this arrangement?
(b) When the state ran out of new numbers, the order was reversed to 3 digits followed by 3 letters. How many new license plate numbers were then possible?
(c) Several years ago, the numbers described in part (b) were also used up. The state then issued plates with 1 letter followed by 3 digits and then 3 letters. How many new license plate numbers will this provide?

31. Social Science A zydeco band from Louisiana will play 5 traditional and 3 original Cajun compositions at a concert. In how many ways can they arrange the program if

(a) they begin with a traditional piece?

(b) an original piece will be played last?

32. Business   A salesperson has the names of 6 prospects.

(a) In how many ways can she arrange her schedule if she calls on all 6?

(b) In how many ways can she arrange her schedule if she can call on only 4 of the 6?

Social Science   *The United States is rapidly running out of telephone numbers. In large cities, telephone companies have introduced new area codes as numbers are used up.*

33. **(a)** Until recently, all area codes had a 0 or a 1 as the middle digit and the first digit could not be 0 or 1. How many area codes were possible with this arrangement? How many telephone numbers does the current 7-digit sequence permit per area code? (The 3-digit sequence that follows the area code cannot start with 0 or 1. Assume there are no other restrictions.)

(b) The actual number of area codes under the previous system was 152. Explain the discrepancy between this number and your answer to part (a).

34. The shortage of area codes under the previous system was avoided by removing the restriction on the second digit. How many area codes are available under this new system?

35. A problem with the plan in Exercise 34 was that the second digit in the area code had been used to tell phone company equipment that a long-distance call was being made. To avoid changing all equipment, an alternative plan proposed a 4-digit area code and restricted the first and second digits as before. How many area codes would this plan have provided?

36. Still another alternative solution is to increase the local dialing sequence to 8 digits instead of 7. How many additional numbers would this plan create? (Assume the same restrictions.)

37. Define permutation in your own words.

Use permutations to solve each of the following problems. (See Examples 5–7.)

38. A baseball team has 20 players. How many 9-player batting orders are possible?

39. In a game of musical chairs, 12 children will sit in 11 chairs arranged in a row (one will be left out). In how many ways can the 11 children find seats?

40. Business   From a carton of 12 cans of a soft drink, 2 are selected for 2 people. In how many ways can this be done?

41. Social Science   In an election with 3 candidates for one office and 5 candidates for another office, how many different ballot orders may be used?

42. Business   From a pool of 7 secretaries, 3 are selected to be assigned to 3 managers. In how many ways can they be selected?

43. A chapter of union Local 715 has 35 members. In how many different ways can the chapter select a president, a vice president, a treasurer, and a secretary?

44. The television schedule for a certain evening shows 8 choices from 8 to 9 P.M., 5 choices from 9 to 10 P.M., and 6 choices from 10 to 11 P.M. In how many different ways could a person schedule that evening of television viewing from 8 to 11 P.M.? (Assume each program that is selected is watched for an entire hour.)

45. In a club with 15 members, how many ways can a slate of 3 officers consisting of president, vice-president, and secretary/treasurer be chosen?

Use combinations to solve each of the following problems. (See Examples 8, 9, 12, and 13.)

46. Business   Five items are to be randomly selected from the first 50 items on an assembly line to determine the defect rate. How many different samples of 5 items can be chosen?

47. Social Science   A group of 3 students is to be selected from a group of 12 students to take part in a class in cell biology.

(a) In how many ways can this be done?

(b) In how many ways can the group which will *not* take part be chosen?

48. Natural Science   From a group of 16 smokers and 20 nonsmokers, a researcher wants to randomly select 8 smokers and 8 nonsmokers for a study. In how many ways can the study group be selected?

49. Five cards are drawn from an ordinary deck. In how many ways is it possible to draw

(a) all queens;

(b) all face cards (face cards are the jack, queen, and king);

(c) no face card;

(d) exactly 2 face cards;

(e) 1 heart, 2 diamonds, and 2 clubs.

50. Business   A salesman has 10 accounts in a certain city.

(a) In how many ways can he select 3 accounts to call on?

(b) In how many ways can he select at least 8 of the 10 accounts to use in preparing a report?

51. Explain the difference between a permutation and a combination.

52. Padlocks with digit dials are often referred to as "combination locks." According to the mathematical definition of combination, is this an accurate description? Explain.

Exercises 53–70 are mixed problems that may require permutations, combinations, or the multiplication principle. (See Examples 10–12 and 14.)

53. Use a tree diagram to find the number of ways 2 letters can be chosen from the set {L, M, N} if order is important and
   (a) if repetition is allowed;
   (b) if no repeats are allowed.
   (c) Find the number of combinations of 3 elements taken 2 at a time. Does this answer differ from part (a) or (b)?

54. Repeat Exercise 53 using the set {L, M, N, P}.

55. Social Science   A legislative committee consists of 5 Democrats and 4 Republicans. A delegation of 3 is to be selected to visit a small Pacific Island republic.
   (a) How many different delegations are possible?
   (b) How many delegations would have all Democrats?
   (c) How many delegations would have 2 Democrats and 1 Republican?
   (d) How many delegations would include at least 1 Republican?

56. Natural Science   In an experiment on plant hardiness, a researcher gathers 6 wheat plants, 3 barley plants, and 2 rye plants. She wishes to select 4 plants at random.
   (a) In how many ways can this be done?
   (b) In how many ways can this be done if 2 wheat plants must be included?

57. Baskin-Robbins advertises that it has 31 flavors of ice cream.
   (a) How many different double-scoop cones can be made?
   (b) How many different triple-scoop cones can be made?

58. A concert to raise money for an economics prize is to consist of 5 works: 2 overtures, 2 sonatas, and a piano concerto.
   (a) In how many ways can the program be arranged?
   (b) In how many ways can the program be arranged if an overture must come first?

59. A lottery game requires that you pick 6 different numbers from 1 to 99. If you pick all 6 winning numbers, you win $4 million.
   (a) How many ways are there to choose 6 numbers if order is not important?
   (b) How many ways are there to choose 6 numbers if order matters?

60. In Exercise 59, if you pick 5 of the 6 numbers correctly, you win $5,000. In how many ways can you pick exactly 5 of the 6 winning numbers without regard to order?

61. The game of Sets* consists of a special deck of cards. Each card has on it either one, two, or three shapes. The shapes on each card are all the same color, either green, purple, or red. The shapes on each card are the same

style, either solid, shaded, or outline. There are three possible shapes: squiggle, diamond, and oval, and only one type of shape appears on a card. The deck consists of all possible combinations of shape, color, style, and number. How many cards are in a deck?

62. A bag contains 5 black, 1 red, and 3 yellow jelly beans; you take 3 at random. How many samples are possible in which the jelly beans are
   (a) all black;   **(b)** all red;
   (c) all yellow;   **(d)** 2 black, 1 red;
   (e) 2 black, 1 yellow;   **(f)** 2 yellow, 1 black;
   (g) 2 red, 1 yellow.

63. In Example 6, there are six 3-letter permutations of the letters A, B, and C. How many 3-letter subsets (unordered groups of letters) are there?

64. How many unordered 2-letter subsets of the letters A, B, and C are there? Compare with Example 6.

65. Natural Science   Eleven drugs have been found to be effective in the treatment of a disease. It is believed that the sequence in which the drugs are administered is important in the effectiveness of the treatment. In how many orders can 5 of the 11 drugs be administered?

66. Natural Science   A biologist is attempting to classify 52,000 species of insects by assigning 3 initials to each species. Is it possible to classify all the species in this way? If not, how many initials should be used?

67. One play in a state lottery consists of choosing 6 numbers from 1 to 44. If your 6 numbers are drawn (in any order), you win the jackpot.
   (a) How many possible ways are there to draw the 6 numbers?
   (b) If you get 2 plays for a dollar, how much would it cost to guarantee that 1 of your choices would be drawn?
   (c) Assuming that you work alone and can fill out a betting ticket (for 2 plays) every second and the lotto drawing will take place 3 days from now, can you place enough bets to guarantee that 1 of your choices will be drawn?

68. Powerball is a lottery game played in 15 states across the United States. For $1 a ticket, a player selects five numbers from 1 to 45 and one Powerball number from 1 to 45. A match of all six numbers wins the jackpot. How many different selections are possible?

69. In the game of bingo, each card has five columns. Column 1 has spaces for 5 numbers, chosen from 1 to 15. Column 2 similarly has 5 numbers, chosen from 16 to 30. Column 3 has a free space in the middle, plus 4 numbers chosen from 31 to 45. The 5 numbers in columns 4 and 5 are chosen from 46 to 60 and from 61 to 75, respectively. The numbers in each card can be in any order. How many different bingo cards are there?

*Copyright © Marsha J. Falco.

70. A television commercial for Little Caesars pizza announced that with the purchase of two pizzas, one could receive free any combination of up to five toppings on each pizza. The commercial shows a young child waiting in line at Little Caesars who calculates that there are 1,048,576 possibilities for the toppings on the two pizzas.* Verify the child's calculation. Use the fact that Little Caesar's has 11 toppings to choose from. Assume that the order of the two pizzas matters; that is, if the first pizza has combination 1 and the second pizza has combination 2, that is different from combination 2 on the first pizza and combination 1 on the second.

If the n objects in a permutations problem are not all distinguishable, that is, there are n_1 of type 1, n_2 of type 2, and so on for r different types, then the number of distinguishable permutations is

$$\frac{n!}{n_1! \, n_2! \, \ldots n_r!}.$$

Example   *In how many ways can you arrange the letters in the word* Mississippi?

This word contains 1 m, 4 i's, 4 s's, and 2 p's. To use the formula, let $n = 11$, $n_1 = 1$, $n_2 = 4$, $n_3 = 4$, *and* $n_4 = 2$ *to get*

$$\frac{11!}{1! \, 4! \, 4! \, 2!} = 34,650$$

arrangements. The letters in a word with 11 different letters can be arranged in $11! = 39,916,800$ *ways.*

---

*Joseph F. Heiser, "Pascal and Gauss meet Little Caesars," *Mathematics Teacher,* 87 (September 1994): 389. In a letter to *The Mathematics Teacher,* Heiser argued that the two combinations should be counted as the same, so the child has actually overcounted. In that case there would be 524,800 possibilities.

71. Find the number of distinguishable permutations of the letters in each of the following words.
(a) initial   (b) little   (c) decreed

72. A printer has 5 A's, 4 B's, 2 C's, and 2 D's. How many different "words" are possible which use all these letters? (A "word" does not have to have any meaning here.)

73. Mike has 4 blue, 3 green, and 2 red books to arrange on a shelf. In how many distinguishable ways can this be done
(a) if they can be arranged in any order?
(b) if books of the same color are identical and must be grouped together?
(c) if books of the same color are identical but need not be grouped together?

74. A child has a set of different shaped plastic objects. There are 3 pyramids, 4 cubes, and 7 spheres. In how many ways can she arrange them in a row
(a) if they are all different colors?
(b) if the same shapes must be grouped?
(c) In how many distinguishable ways can they be arranged in a row if objects of the same shape are also the same color but need not be grouped?

9.2 APPLICATIONS OF COUNTING

Many of the probability problems involving *dependent* events that were solved with probability trees in Chapter 8 can also be solved using counting principles, that is, permutations and combinations. Permutations and combinations are especially helpful when the number of choices is large. The use of counting rules to solve probability problems depends on the basic probability principle introduced in Section 8.3 and repeated here.

If event E is a subset of sample space S, then the probability that event E occurs, written $P(E)$, is

$$P(E) = \frac{n(E)}{n(S)}.$$

It is also helpful to keep in mind that in probability statements

"and" corresponds to multiplication,

"or" corresponds to addition.

To compare the methods of using permutations and combinations with the method of probability trees used in Section 8.5, the first example repeats Example 6 from that section.

▶ **EXAMPLE 1**   From a box containing 3 white, 2 green, and 1 red marble, 2 marbles are drawn one at a time without replacement. Find the probability that 1 white and 1 green marble are drawn.

In Example 6 of Section 8.5, it was necessary to consider the order in which the marbles were drawn. With combinations, it is not necessary. Simply count the number of ways in which 1 white and 1 green marble can be drawn from the given selection. The white marble can be drawn from the 3 white marbles in $\binom{3}{1}$ ways, and the green marble can be drawn from the 2 green marbles in $\binom{2}{1}$ ways. By the multiplication principle, both results can occur in

$$\binom{3}{1} \cdot \binom{2}{1} \text{ ways,}$$

giving the numerator of the probability fraction, $P(E) = m/n$. For the denominator, 2 marbles are to be drawn from a total of 6 marbles. This can occur in $\binom{6}{2}$ ways. The required probability is

$$P(1 \text{ white and } 1 \text{ green}) = \frac{\binom{3}{1}\binom{2}{1}}{\binom{6}{2}} = \frac{\dfrac{3!}{2!\,1!} \cdot \dfrac{2!}{1!\,1!}}{\dfrac{6!}{4!\,2!}} = \frac{6}{15} = \frac{2}{5}.$$

This agrees with the answer found earlier. ◀

▶ **EXAMPLE 2**   From a group of 22 nurses, 4 are to be selected to present a list of grievances to management.

(a) In how many ways can this be done?

Four nurses from a group of 22 can be selected in $\binom{22}{4}$ ways. (Use combinations, since the order in which the group of 4 is selected is unimportant.)

$$\binom{22}{4} = \frac{22!}{4!\,18!} = \frac{22(21)(20)(19)}{4(3)(2)(1)} = 7315$$

There are 7315 ways to choose 4 people from 22.

(b) One of the nurses is Michael Branson. Find the probability that Branson will be among the 4 selected.

The probability that Branson will be selected is the number of ways the chosen group includes him, divided by the total number of ways the group of 4 can be chosen. If Branson must be one of the 4 selected, the problem reduces to finding the number of ways that the 3 additional nurses can be chosen. The 3 are chosen from 21 nurses; this can be done in

$$\binom{21}{3} = \frac{21!}{3! \ 18!} = 1330$$

ways. The number of ways 4 nurses can be selected from 22 is

$$n = \binom{22}{4} = 7315.$$

The probability that Branson will be one of the 4 chosen is

$$P(\text{Branson is chosen}) = \frac{1330}{7315} \approx .182.$$

(c) Find the probability that Branson will not be selected.
The probability that he will not be chosen is $1 - .182 = .818.$ ◄ ①

▶ **EXAMPLE 3** When shipping diesel engines abroad, it is common to pack 12 engines in 1 container that is then loaded on a railcar and sent to a port. Suppose that a company has received complaints from its customers that many of the engines arrive in nonworking condition. To help solve this problem, the company decides to make a spot check of containers after loading. The company will test 3 engines from a container at random; if any of the 3 is nonworking, the container will not be shipped until each engine in it is checked. Suppose a given container has 2 nonworking engines. Find the probability that the container will not be shipped.

The container will not be shipped if the sample of 3 engines contains 1 or 2 defective engines. Letting $P(1 \text{ defective})$ represent the probability of exactly 1 defective engine in the sample,

$$P(\text{not shipping}) = P(1 \text{ defective}) + P(2 \text{ defectives}).$$

There are $\binom{12}{3}$ ways to choose the 3 engines for testing:

$$\binom{12}{3} = \frac{12!}{3! \ 9!} = \frac{12(11)(10)}{3(2)(1)} = 220.$$

There are $\binom{2}{1}$ ways of choosing 1 defective engine from the 2 in the container, and for each of these ways, there are $\binom{10}{2}$ ways of choosing 2 good engines from among the 10 in the container. By the multiplication principle, there are

$$\binom{2}{1}\binom{10}{2}$$

ways of choosing a sample of 3 engines containing 1 defective. ②

Using the result from side problem 2,

$$P(1 \text{ defective}) = \frac{90}{220}.$$

①

A jar contains 1 white and 4 red jelly beans.

(a) What is the probability that out of 2 jelly beans selected at random from the jar, 1 will be white?

(b) What is the probability of choosing 3 red jelly beans?

Answers:

(a) 2/5
(b) 2/5

②

Calculate $\binom{2}{1}\binom{10}{2}$.

Answer:

90

There are $\binom{2}{2}$ ways of choosing 2 defective engines from the 2 defective engines in the container, and $\binom{10}{1}$ ways of choosing 1 good engine from among the 10 good engines, giving

$$\binom{2}{2}\binom{10}{1}$$

ways of choosing a sample of 3 engines containing 2 defectives. ③

③

Calculate $\binom{2}{2}\binom{10}{1}$.

Answer:

10

Using the result from side problem 3,

$$P(2 \text{ defectives}) = \frac{10}{220}$$

and

$$P(\text{not shipping}) = P(1 \text{ defective}) + P(2 \text{ defectives})$$
$$= \frac{90}{220} + \frac{10}{220} = \frac{100}{220} = \frac{5}{11} \approx .455.$$

Notice that the probability is $1 - .455 = .545$ that the container *will* be shipped, even though it has 2 defective engines. The management must decide if this probability is acceptable; if not, it may be necessary to test more than 3 engines from a container. ◄

④

In Example 3, if a sample of 2 engines is tested, what is the probability that the container will not be shipped?

Answer:

.318

Instead of finding the sum $P(1 \text{ defective}) + P(2 \text{ defectives})$, the result in Example 3 could be found by calculating $1 - P(\text{no defectives})$.

$$P(\text{not shipping}) = 1 - P(\text{no defectives in sample})$$

$$= 1 - \frac{\binom{2}{0}\binom{10}{3}}{\binom{12}{3}} = 1 - \frac{1(120)}{220}$$

$$= 1 - \frac{120}{220} = \frac{100}{220} \approx .455 \quad ④$$

▶ **EXAMPLE 4**   In a common form of the card game poker, a hand of 5 cards is dealt to each player from a deck of 52 cards. There are a total of

$$\binom{52}{5} = \frac{52!}{5! \, 47!} = 2{,}598{,}960$$

such hands possible. Find each of the following probabilities.

(a) A hand containing only hearts, called a *heart flush*

There are 13 hearts in a deck; there are

$$\binom{13}{5} = \frac{13!}{5!8!} = \frac{13(12)(11)(10)(9)}{5(4)(3)(2)(1)} = 1287$$

different hands containing only hearts. The probability of a heart flush is

$$P(\text{heart flush}) = \frac{1287}{2{,}598{,}960} \approx .000495.$$

(b) A flush of any suit (5 cards, all from 1 suit)

There are 4 suits to a deck, so

$$P(\text{flush}) = 4 \cdot P(\text{heart flush}) = 4(.000495) \approx .00198.$$

(c) A full house of aces and eights (3 aces and 2 eights)

There are $\binom{4}{3}$ ways to choose 3 aces from among the 4 in the deck and $\binom{4}{2}$ ways to choose 2 eights.

$$P(3\ \text{aces},\ 2\ \text{eights}) = \frac{\binom{4}{3} \cdot \binom{4}{2}}{\binom{52}{5}} \approx .00000923$$

(d) Any full house (3 cards of one value, 2 of another)

There are 13 values in a deck, so there are **13** choices for the first value mentioned, leaving **12** choices for the second value (order *is* important here, since a full house of aces and eights, for example, is not the same as a full house of eights and aces).

$$P(\text{full house}) = \frac{13 \cdot \binom{4}{3} \cdot 12 \cdot \binom{4}{2}}{\binom{52}{5}}$$

Because, from part (c), the probability of any particular full house is .00000923,

$$P(\text{full house}) = 13 \cdot 12(.00000923) \approx .00144. \quad \blacktriangleleft \quad ⑤$$

Find the probability of being dealt a poker hand (5 cards) with 4 kings.

Answer:

.00001847

---

⑥

An office manager must select 4 employees, A, B, C, and D, to work on a project. Each employee will be responsible for a specific task and 1 will be the coordinator.

(a) If the manager assigns the tasks randomly, what is the probability that B is the coordinator?

(b) What is the probability of one specific assignment of tasks?

Answers:

(a) 1/4

(b) 1/24

▶ **EXAMPLE 5** A music teacher has 3 violin pupils, Fred, Carl, and Helen. For a recital, the teacher selects a first violinist and a second violinist. The third pupil will play with the others but will not solo. If the teacher selects randomly, what is the probability that Helen is first violinist, Carl is second violinist, and Fred does not solo?

Use *permutations* to find the number of arrangements in the sample space.

$$_3P_3 = 3! = 6$$

The 6 arrangements are equally likely, since the teacher will select randomly. Thus, the required probability is 1/6. ◀ ⑥

▶ **EXAMPLE 6** Suppose a group of 5 people is in a room. Find the probability that at least 2 of the people have the same birthday.

"Same birthday" refers to the month and the day, not necessarily the same year. Also, ignore leap years and assume that each day in the year is equally likely as a birthday. First find the probability that *no 2 people* among 5 people have the same birthday. There are 365 different birthdays possible for the first of the 5 people, 364 for the second (so that the people have different birthdays), 363 for the third, and so on. The number of ways the 5 people can have different birthdays is thus the number of permutations of 365 things (days) taken 5 at a time or

$$_{365}P_5 = 365 \cdot 364 \cdot 363 \cdot 362 \cdot 361.$$

The number of ways that the 5 people can have the same or different birthdays is

$$365 \cdot 365 \cdot 365 \cdot 365 \cdot 365 = (365)^5.$$

Finally, the *probability* that none of the 5 people have the same birthday is

$$\frac{_{365}P_5}{(365)^5} = \frac{365 \cdot 364 \cdot 363 \cdot 362 \cdot 361}{365 \cdot 365 \cdot 365 \cdot 365 \cdot 365} \approx .973.$$

The probability that at least 2 of the 5 people *do* have the same birthday is $1 - .973 = .027.$ ◀

Example 6 can be extended for more than 5 people. In general, the probability that no 2 people among n people have the same birthday is

$$\frac{_{365}P_n}{(365)^n}.$$

The probability that at least 2 of the n people *do* have the same birthday is

$$1 - \frac{_{365}P_n}{(365)^n}.\ \ ⑦$$

The following table shows this probability for various values of n.

Number of People, n	Probability That at Least 2 Have the Same Birthday
5	.027
10	.117
15	.253
20	.411
22	.476
23	.507
25	.569
30	.706
35	.814
40	.891
50	.970
365	1

The probability that 2 people among 23 have the same birthday is .507, a little more than half. Many people are surprised at this result—somehow it seems that a larger number of people should be required. ⑧

⑦

Evaluate $1 - \dfrac{_{365}P_n}{(365)^n}$ for

(a) $n = 3$;

(b) $n = 6$.

Answers:

(a) .008

(b) .040

⑧

Set up (do not calculate) the probability that at least 2 of the 9 members of the Supreme Court have the same birthday.

Answer:

$1 - _{365}P_9/365^9$

Using a graphing calculator, we can graph the probability formula in the previous example as a function of n, but the graphing calculator must be set to evaluate the function at integer points. Figure 9.5 was produced on a TI-83 by letting $Y_1 = 1 - (365 \text{ nPr } X)/365^X$ on $0 \le x \le 47$. (This domain ensures integer values for x.) Notice that the graph does not extend past $x = 39$. This is because $P(365, n)$ and 365^n are too large for the calculator when $n \ge 40$.

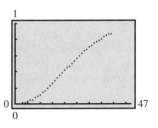

FIGURE 9.5

9.2 Exercises

Business *Refer to Example 3. The management feels that the probability of .545 that a container will be shipped even though it contains 2 defectives is too high. They decide to increase the sample size chosen. Find the probability that a container will be shipped even though it contains 2 defectives if the sample size is increased to*

1. 4.

2. 5.

Business *A shipment of 9 computers contains 2 with defects. Find the probability that a sample of the following size, drawn from the 9, will not contain a defective. (See Example 3.)*

3. 1

4. 2

5. 3

6. 4

A grab bag contains 10 $1 *prizes,* 6 $5 *prizes, and* 2 $20 *prizes. Three prizes are randomly drawn. Find the following probabilities. (See Examples 1 and 2.)*

7. All $1 prizes

8. All $5 prizes

9. Exactly two $20 prizes

10. One prize of each kind

11. At least one $20 prize

12. No $1 prize

Two cards are drawn at random from an ordinary deck of 52 cards. (See Example 4.)

13. How many 2-card hands are possible?

Find the probability that the two-card hand in Exercise 13 contains the following.

14. 2 queens

15. No aces

16. 2 face cards

17. Different suits

18. At least 1 black card

19. No more than 1 heart

20. Discuss the relative merits of using probability trees versus combinations to solve probability problems. When would each approach be most appropriate?

21. Several examples in this section used the rule $P(E') = 1 - P(E)$. Explain the advantage (especially in Example 6) of using this rule.

A bridge hand consists of 13 cards from a deck of 52. Set up the probabilities that a bridge hand includes each of the following.

22. 6 face cards

23. 2 aces and 3 kings

24. 7 cards of one suit and 6 of another

25. In Exercise 59 in Section 9.1, we found the number of ways to pick 6 different numbers from 1 to 99 in a state lottery.
 (a) Assuming order is unimportant, what is the probability of picking all 6 numbers correctly to win the big prize?
 (b) What is the probability, if order matters?

26. In Exercise 25, what is the probability of picking exactly 5 of the 6 numbers correctly?

27. In December 1993, Percy Ray Pridgen, a District of Columbia resident, and Charles Gill of Richmond, Virginia, shared a $90 million Powerball jackpot. The winning numbers were 1, 3, 13, 15, and 29. The Powerball number was 12. The answer to Exercise 68 in Section 9.1 gives the number of ways to win a Powerball jackpot as about 55 million. What is the probability of two individuals independently selecting the winning numbers?

28. Business A cellular phone manufacturer randomly selects 5 of every 50 phones from the assembly line and tests them. If at least 4 of the 5 pass inspection, the batch of 50 is considered acceptable. Find the probability that a batch is considered acceptable if it contains the following.

(a) 2 defective phones

(b) No defective phones

(c) 3 defective phones

29. Social Science Of the 16 members of President Clinton's first cabinet, 4 were women. Suppose the president randomly selected 4 advisers from the cabinet for a meeting. Find the probability that the group of 4 would be composed as follows.

(a) 2 women and 2 men

(b) All men

(c) At least 1 woman

30. Business A car dealer has 8 red, 11 gray, and 9 blue cars in stock. Ten cars are randomly chosen to be displayed in front of the dealership. Find the probability that

(a) 4 are red and the others are blue.

(b) 3 are red, 3 are blue, and 4 are gray.

(c) exactly 5 are gray and none are blue.

(d) all 10 are gray.

31. Social Science The teacher of a multicultural first-grade class on the first day of school found that in her class of 25 students, 10 spoke only English, 6 spoke only Spanish, 4 spoke only Russian, 3 spoke only Vietnamese, and 2 spoke only Hmong. The children were assigned randomly to groups of 5. Find the probability that a group included the following.

(a) 2 English-speaking and 3 Russian-speaking children

(b) All English-speaking children

(c) No English-speaking children

(d) At least two children who spoke Vietnamese or Hmong

For Exercises 32–34 refer to Example 6 in this section.

32. Set up the probability that at least 2 of the 42 men who have served as president of the United States have had the same birthday.*

33. Set up the probability that at least 2 of the 100 U.S. senators have the same birthday.

34. What is the probability that 2 of the 435 members of the House of Representatives have the same birthday?

35. An elevator has 4 passengers and stops at 7 floors. It is equally likely that a person will get off at any one of the 7 floors. Find the probability that no 2 passengers leave at the same floor.

36. Show that the probability that in a group of n people *exactly one pair* have the same birthday is

$$\binom{n}{2} \cdot \frac{_{365}P_{n-1}}{(365)^n}.$$

37. On National Public Radio, the "Weekend Edition" program on Sunday, September 7, 1991, posed the following probability problem: Given a certain number of balls, of which some are blue, pick 5 at random. The probability that all 5 are blue is 1/2. Determine the original number of balls and decide how many were blue.

38. The Big Eight Conference during the 1988 college football season ended the season in a "perfect progression," as shown in the table below.†

Won	Lost	Team
7	0	Nebraska (NU)
6	1	Oklahoma (OU)
5	2	Oklahoma State (OSU)
4	3	Colorado (CU)
3	4	Iowa State (ISU)
2	5	Missouri (MU)
1	6	Kansas (KU)
0	7	Kansas State (KSU)

Someone wondered what the probability of such an outcome might be.

(a) Assuming no ties and that each team had an equally likely probability of winning each game, find the probability of the perfect progression shown above.

(b) Find a general expression for the probability of a perfect progression in an n-team league with the same assumptions.

---

*In fact, James Polk and Warren Harding were both born on November 2. Although George W. Bush is the 43rd president, the 22nd and 24th presidents were the same man: Grover Cleveland.

†From Richard Madsen. "On the Probability of a Perfect Progression," *American Statistician* 45, no. 3 (August 1991): 214.

39. Use a computer or a graphing calculator and the Monte Carlo method with $n = 50$ to estimate the probabilities of the following hands at poker. (See the directions for Exercises 50–53 on page 449.) Assume aces are either high or low. Since each hand has 5 cards, you will need $50 \cdot 5 = 250$ random numbers to "look at" 50 hands. Compare these experimental results with the theoretical results.
  (a) A pair of aces
  (b) Any two cards of the same value
  (c) Three of a kind

40. Use a computer or a graphing calculator and the Monte Carlo method with $n = 20$ to estimate the probabilities of the following 13-card bridge hands. Since each hand has 13 cards, you will need $20 \cdot 13 = 260$ random numbers to "look at" 20 hands.
  (a) No aces
  (b) 2 kings and 2 aces
  (c) No cards of any one suit—that is, only 3 suits represented

9.3 BINOMIAL PROBABILITY

Many probability problems are concerned with repetitions of the same experiments. Some examples include finding the probability of getting 7 heads in 8 tosses of a coin, of hitting a target 6 times out of 6, and of finding 1 defective item in a sample of 15 items. Probability problems of this kind are called **Bernoulli trials** problems, or **Bernoulli processes,** and are named after the Swiss mathematician Jakob Bernoulli (1654–1705). In each case, some outcome is designated a success, and any other outcome is considered a failure. Thus, if the probability of a success in a single trial is p, the probability of failure will be $1 - p$. A Bernoulli trials problem, or **binomial experiment,** must satisfy the following conditions.

Binomial Experiment

1. The same experiment is repeated several times.
2. There are only two possible outcomes, success or failure.
3. The repeated trials are independent so that the probability of each outcome remains the same for each trial.

Consider a typical problem of this type: find the probability of getting 5 ones on 5 rolls of a die. Here, the experiment, rolling a die, is repeated 5 times. If getting a one is designated as a success, then getting any other outcome is a failure. The 5 trials are independent; the probability of success (getting a one) is 1/6 on each trial, while the probability of a failure (any other result) is 5/6. Thus, the required probability is

$$P(5 \text{ ones in } 5 \text{ rolls}) = P(1) \cdot P(1) \cdot P(1) \cdot P(1) \cdot P(1) = \left(\frac{1}{6}\right)^5$$

$$\approx .00013.$$

Now suppose the problem is changed to that of finding the probability of getting a one exactly 4 times in 5 rolls of the die. Again, a success on any roll of the die is defined as getting a one, and the trials are independent, with the same probability of success each time. The desired outcome for this experiment can occur in more than one way, as shown below, where s represents a success (getting a one), and f represents a failure (getting any other result), and each row represents a different outcome.

$$s \quad s \quad s \quad s \quad f$$
$$s \quad s \quad s \quad f \quad s$$
$$s \quad s \quad f \quad s \quad s$$
$$s \quad f \quad s \quad s \quad s$$
$$f \quad s \quad s \quad s \quad s$$

Using combinations, the total number of ways in which 4 successes (and 1 failure) can occur is $\binom{5}{4} = 5$. The probability of each of these 5 outcomes is

$$P(1) \cdot P(1) \cdot P(1) \cdot P(1) \cdot P(\text{not } 1) = \left(\frac{1}{6}\right)^4\left(\frac{5}{6}\right).$$

Since the 5 outcomes represent mutually exclusive alternative events, add the 5 probabilities, or multiply by $\binom{5}{4} = 5$.

$$P(4 \text{ ones in } 5 \text{ rolls}) = \binom{5}{4}\left(\frac{1}{6}\right)^4\left(\frac{5}{6}\right)^{5-4} = 5\left(\frac{1}{6}\right)^4\left(\frac{5}{6}\right)^1 = \frac{5^2}{6^5} \approx .0032$$

The probability of rolling a one exactly 3 times in 5 rolls of a die can be computed in the same way. The probability of any one way of achieving 3 successes and 2 failures will be

$$\left(\frac{1}{6}\right)^3\left(\frac{5}{6}\right)^2.$$

Again the desired outcome can occur in more than one way. The number of ways in which 3 successes (and 2 failures) can occur, using combinations, is $\binom{5}{3} = 10$.

$$P(3 \text{ ones in } 5 \text{ rolls}) = \binom{5}{3}\left(\frac{1}{6}\right)^3\left(\frac{5}{6}\right)^{5-3} = 10\left(\frac{1}{6}\right)^3\left(\frac{5}{6}\right)^2 = \frac{250}{6^5} \approx .032 \ \text{①}$$

A similar argument works in the general case.

Binomial Probability

If p is the probability of success in a single trial of a binomial experiment, the probability of x successes and $n - x$ failures in n independent repeated trials of the experiment is

$$\binom{n}{x}p^x(1 - p)^{n-x}.$$

This formula plays an important role in the study of *binomial distributions,* which are discussed in Section 10.4.

TECHNOLOGY TIP On the TI-83 calculator, use "binompdf(n,p,x)" in the DISTR menu to compute the probability of exactly x successes in n trials (where p is the probability of success in a single trial). Use "binomcdf(n,p,x)" to compute the probability of at most x successes in n trials. Figure 9.6 shows the probability of exactly three ones in five rolls of a die and the probability of at most three ones in five rolls of a die.

```
binompdf(5,1/6,3
)
        .0321502058
binomcdf(5,1/6,3
)
        .9966563786
■
```

FIGURE 9.6

▶ **EXAMPLE 1**   The advertising agency that handles the Diet Supercola account believes that 40% of all consumers prefer this product over its competitors. Suppose a random sample of 6 people is chosen. Assume that all responses are independent of each other. Find the probability of the following.

(a) Exactly 4 of the 6 people prefer Diet Supercola.

We can think of the 6 responses as 6 independent trials. A success occurs if a person prefers Diet Supercola. Then this is a binomial experiment with

$$p = P(\text{success}) = P(\text{prefer Diet Supercola}) = .4.$$

The sample is made up of 6 people, so $n = 6$. To find the probability that exactly 4 people prefer this drink, we let $x = 4$ and use the result in the box.

$$P(\text{exactly } 4) = \binom{6}{4}(.4)^4(1 - .4)^{6-4}$$

$$= 15(.4)^4(.6)^2$$

$$= 15(.0256)(.36) \qquad \text{Use a calculator.}$$

$$= .13824$$

(b) None of the 6 people prefers Diet Supercola.

Let $x = 0$.

$$P(\text{exactly } 0) = \binom{6}{0}(.4)^0(1 - .4)^6$$

$$= 1(1)(.6)^6$$

$$\approx .0467 \quad \blacktriangleleft \quad ②$$

---

②

Eighty percent of all students at a certain school ski. If a sample of 5 students at this school is selected, and if their responses are independent, find the probability that exactly

(a) 1 of the 5 students skis;

(b) 4 of the 5 students ski.

Answers:

(a) $\binom{5}{1}(.8)^1(.2)^4 = .0064$

(b) $\binom{5}{4}(.8)^4(.2)^1 = .4096$

Find the probability of getting 2 fours in 8 tosses of a die.

Answer:

$$\binom{8}{2}\left(\frac{1}{6}\right)^2\left(\frac{5}{6}\right)^6 \approx .2605$$

Five percent of the clay pots fired in a certain way are defective. Assume independence and find the probability of getting each result in a sample of 12 pots.

(a) 2 defective pots

(b) At most 2 defective pots

Answer:

(a) $\binom{12}{2}(.05)^2(.95)^{10} \approx .0988$

(b) .980

▶ **EXAMPLE 2** Find the probability of getting exactly 7 heads in 8 tosses of a fair coin.

The probability of success, getting a head in a single toss, is 1/2, so the probability of failure, getting a tail, is 1/2.

$$P(7 \text{ heads in 8 tosses}) = \binom{8}{7}\left(\frac{1}{2}\right)^7\left(\frac{1}{2}\right)^1 = 8\left(\frac{1}{2}\right)^8 = .03125 \quad \blacktriangleleft \; ③$$

▶ **EXAMPLE 3** The chance that an American falls asleep with the TV on at least three nights a week is 1/4.*

(a) Suppose a researcher selects 15 Americans at random and is interested in the probability that 3 are "TV sleepers."

The experiment, selecting a person, is repeated 15 times. If selecting a TV sleeper is labeled a success, then getting a non–TV sleeper is labeled a failure. Researchers consider such trials to be independent. Thus, the probability that 3 persons in the sample of 15 are sleepers is

$$\binom{15}{3}\left(\frac{1}{4}\right)^3\left(\frac{3}{4}\right)^{12} \approx .225.$$

(b) Find the probability that at most 3 of the 15 people in the sample are TV sleepers.

"At most 3" means 0 or 1 or 2 or 3. We must find the probability for each case and then use the union rule for mutually exclusive events.

$$P(0 \text{ TV sleepers}) = \binom{15}{0}\left(\frac{1}{4}\right)^0\left(\frac{3}{4}\right)^{15} \approx .0134$$

$$P(1 \text{ TV sleeper}) = \binom{15}{1}\left(\frac{1}{4}\right)^1\left(\frac{3}{4}\right)^{14} \approx .0668$$

$$P(2 \text{ TV sleepers}) = \binom{15}{2}\left(\frac{1}{4}\right)^2\left(\frac{3}{4}\right)^{13} \approx .156$$

By the union rule,

$$P(\text{at most 3 TV sleepers}) \approx .0134 + .0668 + .156 + .225 \approx .461. \quad \blacktriangleleft \; ④$$

In the next example we return to the problem posed at the beginning of this chapter.

▶ **EXAMPLE 4** A survey by *Money* magazine found that supermarket scanners are overcharging customers at 30% of stores.† Assume all such scanners consistently overcharge. If you shop at 3 supermarkets that use the scanners, what is the probability that you will be overcharged in at least one store?

Harper's Magazine, March 1996, p. 13.

†"Don't Get Cheated by Supermarket Scanners," pp. 132–138, by Vanessa O'Connell reprinted from the April 1993 issue of *Money* by special permission (reprinted in the *Chicago Tribune* as "Beware, When Scanners Err, You Pay Price"). Copyright © 1993, Time Inc.

In Example 4, find the probability that

(a) you are overcharged in more than 1 store.

(b) you are overcharged in less than 1 store.

Answers:

(a) .216

(b) .343

We can treat this as a binomial experiment, letting $n = 3$ and $p = .3$. At least 1 of 3 means 1 or 2 or 3. It will be simpler here to find the probability of being overcharged in none of the 3 stores, that is, $P(0 \text{ overcharges})$, and then find the probability of at least 1 overcharge, which is the number $= 1 - P(0 \text{ overcharges})$.

$$P(0 \text{ overcharges}) = \binom{3}{0}(.3)^0(.7)^3$$

$$= 1(1)(.343)$$

$$= .343$$

$$P(\text{at least } 1) = 1 - P(0 \text{ overcharges})$$

$$= 1 - .343$$

$$= .657 \quad \blacktriangleleft \quad \text{⑤}$$

▶ **EXAMPLE 5**   If each member of a 9-person jury acts independently of the other members and makes the correct determination of guilt or innocence with probability .65, find the probability that the majority of jurors will reach a correct verdict.*

Since the jurors in this particular situation act independently, we can treat this as a binomial experiment. Thus, the probability that the majority of the jurors will reach the correct verdict is given by

$$P(\text{at least } 5) = \binom{9}{5}(.65)^5(.35)^4 + \binom{9}{6}(.65)^6(.35)^3 + \binom{9}{7}(.65)^7(.35)^2$$

$$+ \binom{9}{8}(.65)^8(.35)^1 + \binom{9}{9}(.65)^9$$

$$\approx .2194 + .2716 + .2162 + .1004 + .0207$$

$$= .8283. \quad \blacktriangleleft$$

*Bernard Grofman, "A Preliminary Model of Jury Decision Making as a Function of Jury Size, Effective Jury Decision Rule, and Mean Juror Judgmental Competence," *Frontiers in Economics* (1979), pp. 98–110.

TECHNOLOGY TIP Some spreadsheets provide binomial probabilities. In Microsoft Excel, for example, the command "5BINOMDIST (5, 9, .65, 0)" gives .21939, which is the probability for $x = 5$ in Example 5. Alternatively, the command "5BINOMDIST (4, 9, .65, 1)" gives .17172 as the probability that 4 or fewer jurors will make the correct decision. Subtract .17172 from 1 to get .82828 as the probability that the majority of the jurors will make the correct decision. This value agrees with the value found in Example 5.

9.3 Exercises

In Exercises 1–39, see Examples 1–5.

Social Science   *In 1995 12.7% of the U.S. population was 65 or older, an increase from 4.1% in 1900.* Find the probabilities that the following number of persons selected at random from 20 U.S. residents in 1995 were 65 or older.*

1. Exactly 5   2. Exactly 1

3. None   4. All

5. At least 1   6. At most 3

A die is rolled 12 times. Find the probability of rolling

7. Exactly 12 ones;   8. Exactly 6 ones;

9. Exactly 1 one;   10. Exactly 2 ones;

11. No more than 3 ones;   12. No more than 1 one.

A coin is tossed 5 times. Find the probability of getting

13. All heads;   14. Exactly 3 heads;

15. No more than 3 heads;   16. At least 3 heads.

17. How do you identify a probability problem that involves a binomial experiment?

18. Why do combinations occur in the binomial probability formula?

Natural Science   *Alzheimer's disease affects an estimated 4 million Americans. The probability that an American aged 65 or older has Alzheimer's is .0127.* A group of 10 people aged 65 or older is selected randomly. Find the following probabilities.*

19. 2 people have the disease.

20. No one has the disease.

21. At most 1 person has the disease.

22. 8 people do not have the disease.

Business   *The survey discussed in Example 4 also found that customers overpay for 1 out of every 10 items, on average.*

Suppose a customer purchases 15 items. Find the following probabilities.

23. A customer overpays on 3 items.

24. A customer does not overpay for any item.

25. A customer overpays on at least one item.

26. A customer overpays on at least 2 items.

27. A customer overpays on at most 2 items.

Natural Science   *The probability that a birth will result in twins is .027.* Assuming independence (perhaps not a valid assumption), what are the probabilities that out of 100 births in a hospital, there will be the following numbers of sets of twins?*

28. Exactly 2 sets of twins   29. At most 2 sets of twins

30. Natural Science   In an experiment on the effects of a radiation dose on cells, a beam of radioactive particles is aimed at a group of 10 cells. Find the probability that 8 of the cells will be hit by the beam if the probability that any single cell will be hit is .6. (Assume independence.)

31. Natural Science   The probability of a mutation of a given gene under a dose of 1 roentgen of radiation is approximately 2.5×10^{-7}. What is the probability that in 10,000 genes, at least 1 mutation occurs?

32. Social Science   Refer to Example 3. The probability that an American falls asleep with the TV on at least three nights a week is 1/4. Suppose a researcher selects 5 Americans at random and is interested in the probability that one or more is a "TV sleeper." Find the following probabilities.

   (a) All 5 are TV sleepers.
   (b) Three of the 5 are TV sleepers.

33. Natural Science   Children in a certain school are monitored for a year and it is found that 30% of those who do not brush their teeth regularly don't have a cavity during the year. A child who brushes regularly has a 70% chance of having no cavities during the year.

*The World Almanac and Book of Facts, 2001, p. 721.

*The World Almanac and Book of Facts, 2001, p. 873.

(a) If 10 children brush regularly, what is the probability that at least 9 will have no cavities during the year?

(b) 150 children who brush regularly are divided into 15 groups of 10 each. What is the probability that in at least one group of 10, at least 9 will have no cavities during the year?

34. Social Science A recent study found that 33% of women would prefer to work part time rather than full time if money were not a concern.* Find the probability that if 10 women are selected at random, at least 3 of them would prefer to work part time.

35. Social Science The fatal accident involvement rate for drivers younger than 20 is .0006.[†] Find the probability that in 10 unrelated fatal accidents, the following number of accidents involved a driver younger than 20.

(a) 1   **(b)** at most 1

Social Science *Refer to Example 5. Find the probability of each result.*

36. All jurors reach a correct verdict.

37. No jurors reach a correct verdict.

38. At least 8 jurors reach a correct verdict.

39. 7 jurors reach a correct verdict.

40. Social Science According to a poll conducted by *U.S. News and World Report,* 84% of college students believe they need to cheat to get ahead in the world today.[‡]

(a) Do the results of this poll indicate that 84% of all college students cheat? Explain.

(b) If this result is accurate and 100 college students are asked if they believe that cheating is necessary to get ahead in the world, what is the probability that 90 or more of the students will answer affirmatively to the question?

41. Social Science In the "Numbers" section of a recent *Time* magazine, it was reported that 15.2% of low-birth-weight babies graduate from high school by age 19. On the other hand, it was reported that 57.5% of the normal-birth-weight siblings graduated from high school.[§]

(a) If 40 low-birth-weight babies were tracked through high school, what is the probability that fewer than 15 will graduate from high school by age 19?

(b) What are some of the factors that may contribute to the wide difference in high school success between these siblings? Do you believe that low birth weight is the primary cause of the difference? What other information do you need to better answer these questions?

42. Natural Science In placebo-controlled trials of Pravachol®, a drug that is prescribed to lower cholesterol, 7.3% of the patients who were taking the drug experienced nausea/vomiting, whereas 7.1% of the patients who were taking the placebo experienced nausea/vomiting.*

(a) If 100 patients who are taking Pravachol® are selected, what is the probability that 10 or more will experience nausea/vomiting?

(b) If a second group of 100 patients receives a placebo, what is the probability that 10 or more will experience nausea/vomiting?

(c) Since 7.3% is larger than 7.1%, do you believe that the Pravachol® causes more people to experience nausea/vomiting than a placebo? Explain.

43. Natural Science The use of DNA has become an integral part of many court cases. When DNA is extracted from cells and body fluids, genetic information is represented by bands of information, which look similar to a bar code at a grocery store. It is generally accepted that in unrelated people, the probability of a particular band matching is 1 in 4.[†]

(a) If 5 bands are compared in unrelated people, what is the probability that all 5 of the bands match? (Express your answer in terms of "1 chance in ?".)

(b) If 20 bands are compared in unrelated people, what is the probability that all 20 of the bands match? (Express your answer in terms of "1 chance in ?".)

(c) If 20 bands are compared in unrelated people, what is the probability that 16 or more bands match? (Express your answer in terms of "1 chance in ?".)

(d) If you were deciding paternity and there were 16 matches out of 20 bands compared, would you believe that the person being tested was the father? Explain.

44. Social Science In England, a woman was found guilty of smothering her two infant children. Much of the Crown's case against her was based on the testimony from a pediatrician who indicated that the chances of two crib deaths occurring in both siblings was only about 1 in 73 million. This number was calculated by assuming that the probability of a single crib death is 1 in 8500 and the probability of two crib deaths is 1 in 8500^2 (i.e., binomial).[‡] Why is the use of binomial probability not correct in this situation?

*Cathleen Ferraro, "Feelings of the Working Woman," *Sacramento Bee,* May 11, 1995, pp. Al, A22.

†National Safety Council; National Highway Traffic Safety Administration.

‡Carolyn Kleiner and Mary Lord, "The Cheating Game," *U.S. News and World Report,* November 22, 1999, pp. 55–66.

§"Numbers." *Time,* July 17, 2000, p. 21.

*Advertisement in *Time,* July 17, 2000, for Pravachol®, developed and marketed by Bristol-Myers Squibb Company.

†"Genetic Fingerprinting Worksheet." Centre for Innovation in Mathematics Teaching, http://www.ex.ac.uk/cimt/resource/fgrprnts.htm.

‡Stephen J. Watkins, "Conviction by Mathematical Error?" *British Medical Journal* 320. no. 7226 (January 1, 2000): 2–3.

9.4 PROBABILITY DISTRIBUTIONS AND EXPECTED VALUE

In this section we shall see that the *expected value* of a probability distribution is a type of average. Probability distributions were introduced briefly in Section 8.4. Now we take a more complete look at probability distributions. A probability distribution depends on the idea of a *random variable*, so we begin with that.

Random Variables

Suppose that the shipping manager at a company receives a package of one dozen computer monitors, of which, unknown to him, three are broken. He checks four of the monitors at random to see how many are broken in his sample of 4. The answer, which we will label x, is one of the numbers 0, 1, 2, or 3. Since the value of x is random, x is called a random variable.

> ### Random Variable
>
> A **random variable** is a function that assigns a real number to each outcome of an experiment.

In the example with the shipping manager, we can calculate the probability that 0, 1, 2, or 3 monitors in his sample of 4 are broken using the methods of Section 9.2. There are 3 broken monitors and 9 unbroken monitors, so the number of ways of choosing 0 broken monitors (which implies 4 unbroken monitors) is $\binom{3}{0}\binom{9}{4}$. The number of ways of choosing a sample of 4 monitors is $\binom{12}{4}$. Therefore, the probability of choosing 0 broken monitors is

$$P(x = 0) = \frac{\binom{3}{0}\binom{9}{4}}{\binom{12}{4}} = \frac{1\left(\dfrac{9 \cdot 8 \cdot 7 \cdot 6}{4 \cdot 3 \cdot 2 \cdot 1}\right)}{\left(\dfrac{12 \cdot 11 \cdot 10 \cdot 9}{4 \cdot 3 \cdot 2 \cdot 1}\right)} = \frac{126}{495} = \frac{14}{55}$$

Similary, the probability of choosing 1 broken monitor is

$$P(x = 1) = \frac{\binom{3}{1}\binom{9}{3}}{\binom{12}{4}} = \frac{3 \cdot 84}{495} = \frac{252}{495} = \frac{28}{55}$$

The probability of choosing 2 broken monitor is

$$P(x = 2) = \frac{\binom{3}{2}\binom{9}{2}}{\binom{12}{4}} = \frac{3 \cdot 36}{495} = \frac{108}{495} = \frac{12}{55}$$

The probability of choosing 3 broken monitors is

$$P(x = 3) = \frac{\binom{3}{3}\binom{9}{1}}{\binom{12}{4}} = \frac{1 \cdot 9}{495} = \frac{9}{495} = \frac{1}{55}$$

The results are entered in Table 1.

Table 1

x	0	1	2	3
$P(x)$	14/55	28/55	12/55	1/55

Such a table that lists the possible values of a random variable, together with the corresponding probabilities, is called a **probability distribution.** The sum of the probabilities in a probability distribution must always equal 1. (The sum in some distributions may vary slightly from 1 because of rounding.)

Instead of writing the probability distribution as a table, we could write the same information as a set of ordered pairs:

{(0, 14/55), (1, 28/55), (2, 12/55), (3, 1/55)}.

There is just one probability for each value of the random variable.

The information in a probability distribution is often displayed graphically as a special kind of bar graph called a **histogram.** The bars of a histogram all have the same width, usually 1. The heights of the bars are determined by the probabilities. A histogram for the data in Table 1 is given in Figure 9.7. A histogram shows important characteristics of a distribution that may not be readily apparent in tabular form, such as the relative sizes of the probabilities and any symmetry in the distribution.

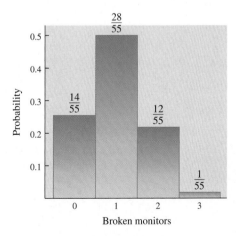

FIGURE 9.7

The area of the bar above $x = 0$ in Figure 9.7 is the product of 1 and 14/55, or $1 \cdot 14/55 = 14/55$. Since each bar has a width of 1, its area is equal to the probability that corresponds to that value of x. The probability that a particular value will occur is thus given by the area of the appropriate bar of the graph. For example, the probability that one or more monitors is broken is the sum of the areas for $x = 1$, $x = 2$, and $x = 3$. This area, shown in red in Figure 9.8, corresponds to 41/55 of the total area, since

$$P(x \geq 1) = P(x = 1) + P(x = 2) + P(x = 3)$$
$$= 28/55 + 12/55 + 1/55$$
$$= 41/55.$$

① **(a)** Give the probability distribution for the number of heads showing when 3 coins are tossed.

(b) Draw a histogram for the distribution in part (a). Find the probability that no more than 1 coin comes up heads.

Answers:

(a)

x	$P(x)$
0	1/8
1	3/8
2	3/8
3	1/8

(b) 1/2

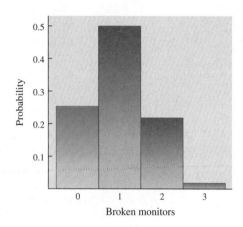

FIGURE 9.8

▶ EXAMPLE 1

(a) Give the probability distribution for the number of heads showing when two coins are tossed.

Let x represent the random variable "number of heads." Then x can take on the values 0, 1, or 2. Now find the probability of each outcome. The results are shown in Table 2.

Table 2

x	0	1	2
$P(x)$	1/4	1/2	1/4

(b) Draw a histogram for the distribution in Table 2. Find the probability that at least one coin comes up heads.

The histogram is shown in Figure 9.9. The portion in red represents

$$P(x \geq 1) = P(x = 1) + P(x = 2)$$
$$= \frac{3}{4}. \quad ◀ ①$$

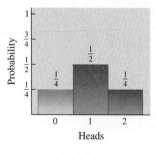

FIGURE 9.9

TECHNOLOGY TIP Virtually all graphing calculators can produce histograms. The procedures differ on various calculators, but it usually requires you to enter the outcomes in one list and the corresponding frequencies in a second list. For specific details, check your instruction manual under "statistics graphs" or "statistical plotting." To get the histogram in Figure 9.19 with a TI-83 calculator, we entered the outcomes 0, 1, and 2 in the first list and entered the probabilities .25, .5 and .25 in a second list. Two versions of the histogram are shown in Figure 9.10. They differ slightly because different viewing windows were used. With some calculators, the probabilities must be entered as integers, so make the entries in the second list 1, 2, and 1 (corresponding to 1/4, 2/4, and 1/4) and use a window with $0 \le y \le 4$.

 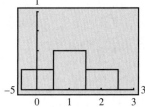

FIGURE 9.10

Expected Value

In working with probability distributions, it is useful to have a concept of the typical or average value that the random variable takes on. In Example 1, for instance, it seems reasonable that, on the average, one head shows when two coins are tossed. This does not tell what will happen the next time we toss two coins; we may get two heads, or we may get none. If we tossed two coins many times, however, we would expect that, in the long run, we would average about one head for each toss of two coins.

A way to solve such problems in general is to imagine flipping two coins 4 times. Based on the probability distribution in Example 1, we would expect that 1 of the 4 times we would get 0 heads, 2 of the 4 times we would get 1 head, and 1 of the 4 times we would get 2 heads. The total number of heads we would get, then, is

$$0 \cdot 1 + 1 \cdot 2 + 2 \cdot 1 = 4.$$

The expected number of heads per toss is found by dividing the total number of heads by the total number of tosses, or

$$\frac{0 \cdot 1 + 1 \cdot 2 + 2 \cdot 1}{4} = 0 \cdot \frac{1}{4} + 1 \cdot \frac{1}{2} + 2 \cdot \frac{1}{4} = 1.$$

Notice that the expected number of heads turns out to be the sum of the three values of the random variable x multiplied by their corresponding probabilities. We can use this idea to define the *expected value* of a random variable as follows.

Expected Value

Suppose the random variable x can take on the n values $x_1, x_2, x_3, \ldots, x_n$. Also, suppose the probabilities that these values occur are, respectively, $p_1, p_2, p_3, \ldots, p_n$. Then the **expected value** of the random variable is

$$E(x) = x_1 p_1 + x_2 p_2 + x_3 p_3 + \cdots + x_n p_n.$$

②

Find the expected value of the number of heads showing when four coins are tossed.

Answer:

2

EXAMPLE 2 In the example with the computer monitors, find the expected number of broken monitors that the shipping manager finds.

Multiply each outcome in Table 1 by its probability, and sum these products.

$$E(x) = 0 \cdot \frac{14}{55} + 1 \cdot \frac{28}{55} + 2 \cdot \frac{12}{55} + 3 \cdot \frac{1}{55} = 1$$

On the average, the shipping manager will find 1 broken monitor in the sample of 4. On reflection, this seems natural; 3 of the 12 monitors, or 1/4 of the total, are broken. We should expect, then, that 1/4 of the sample of 4 monitors are broken. ◀ ②

Physically, the expected value of a probability distribution represents a balance point. If we think of the histogram in Figure 9.7 as a series of weights with magnitudes represented by the heights of the bars, then the system would balance if supported at the point corresponding to the expected value.

EXAMPLE 3 Suppose a local symphony decides to raise money by raffling a microwave oven worth $400, a dinner for two worth $80, and 2 books worth $20 each. A total of 2000 tickets are sold at $1 each. Find the expected value of winning for a person who buys 1 ticket in the raffle.

Here the random variable represents the possible amounts of net winnings, where net winnings = amount won − cost of ticket. The net winnings of the person winning the oven are $400 (amount won) − $1 (cost of ticket) = $399. The net winnings for each losing ticket are $0 − $1 = −$1.

The net winnings of the various prizes, as well as their respective probabilities, are shown in Table 3. The probability of winning $19 is 2/2000 because there are 2 prizes worth $20. We have not reduced the fractions in order to keep all the denominators equal. Because there are 4 winning tickets, there are 1996 losing tickets, so the probability of winning −$1 is 1996/2000.

Table 3

x	$399	$79	$19	−$1
$P(x)$	1/2000	1/2000	2/2000	1996/2000

Suppose you buy 1 of 10,000 tickets at $1 each in a lottery where the prize is $5,000. What are your expected net winnings? What does this answer mean?

Answer:

−$ .50. On the average you lose $ .50 per ticket purchased.

The expected winnings for a person buying one ticket are

$$399\left(\frac{1}{2000}\right) + 79\left(\frac{1}{2000}\right) + 19\left(\frac{2}{2000}\right) + (-1)\left(\frac{1996}{2000}\right) = -\frac{1480}{2000}$$

$$= -.74.$$

On the average, a person buying 1 ticket in the raffle will lose $.74, or 74¢.

It is not possible to lose 74¢ in this raffle: either you lose $1, or you win a prize worth $400, $80, or $20, minus the $1 you pay to play. But if you bought tickets in many such raffles over a long period of time, you would lose 74¢ per ticket, on the average. It is important to note that the expected value of a random variable may be a number that can never occur in any one trial of the experiment. ◀ ③

NOTE An alternative way to compute expected value in this and other examples is to calculate the expected amount won and then subtract the cost of the ticket afterward. The amount won is either $400 (with probability 1/2000), $80 (with probability 1/2000), $20 (with probability 2/2000), or $0 (with probability 1996/2000). The expected winnings for a person buying one ticket are then

$$400\left(\frac{1}{2000}\right) + 80\left(\frac{1}{2000}\right) + 20\left(\frac{2}{2000}\right) + 0\left(\frac{1996}{2000}\right) - 1 = -\frac{1480}{2000}$$

$$= -.74.$$

▶ **EXAMPLE 4**   Each day Donna and Mary toss a coin to see who buys coffee (80 cents a cup). One tosses and the other calls the outcome. If the person who calls the outcome is correct, the other buys the coffee; otherwise the caller pays. Find Donna's expected winnings.

Assume that an honest coin is used, that Mary tosses the coin, and that Donna calls the outcome. The possible results and corresponding probabilities are shown below.

		Possible Results		
Result of Toss	Heads	Heads	Tails	Tails
Call	Heads	Tails	Heads	Tails
Caller Wins?	Yes	No	No	Yes
Probability	1/4	1/4	1/4	1/4

Find Mary's expected winnings.

Answer:

0

Donna wins an 80¢ cup of coffee whenever the results and calls match, and she loses an 80¢ cup when there is no match. Her expected winnings are

$$(.80)\left(\frac{1}{4}\right) + (-.80)\left(\frac{1}{4}\right) + (-.80)\left(\frac{1}{4}\right) + (.80)\left(\frac{1}{4}\right) = 0.$$

On the average, over the long run, Donna neither wins nor loses. ◀ ④

A game with an expected value of 0 (such as the one in Example 4) is called a **fair game.** Casinos do not offer fair games. If they did, they would win (on the average) $0 and have a hard time paying the help! Casino games have expected winnings for the house that vary from 1.5 cents per dollar to 60 cents per dollar. Exercises 30–35 at the end of the section ask you to find the expected winnings for certain games of chance.

The idea of expected value can be very useful in decision making, as shown by the next example.

⑤

A person can take one of two jobs. With job A, there is a 50% chance of making $60,000 per year after 5 years and a 50% chance of making $30,000. With job B, there is a 30% chance of making $90,000 per year after 5 years and a 70% chance of making $20,000. Based strictly on expected value, which job should be taken?

Answer:

Job A has an expected salary of $45,000; job B, $41,000. Take job A.

▶ **EXAMPLE 5**   At age 50, you receive a letter from Mutual of Mauritania Insurance Company. According to the letter, you must tell the company immediately which of the following two options you will choose: take $20,000 at age 60 (if you are alive, $0 otherwise) or $30,000 at age 70 (again, if you are alive, $0 otherwise). Based *only* on the idea of expected value, which should you choose?

Life insurance companies have constructed elaborate tables showing the probability of a person living a given number of years into the future. From a recent such table, the probability of living from age 50 to age 60 is .88, while the probability of living from age 50 to 70 is .64. The expected values of the two options are given below.

$$\text{First Option:}\quad (20{,}000)(.88) + (0)(.12) = 17{,}600$$
$$\text{Second Option:}\quad (30{,}000)(.64) + (0)(.36) = 19{,}200$$

Based strictly on expected values, choose the second option.   ◀ ⑤

▶ **EXAMPLE 6**   According to the National Center for Education Statistics, 79% of the U.S. holders of bachelor's degrees in education in 1992–1993 were women.* Suppose 5 holders of bachelor's degrees in education from 1992–93 are picked at random.

(a) Find the probability distribution for the number that are female.

We first note that each of the 5 people in the sample is either female (with probability .79) or male (with probability .21). As in the previous section, we may assume that the probability for each member of the sample is independent of that of any other. Such a situation is described by binomial probability with $n = 5$ and $p = .79$, for which we use the binomial probability formula

$$\binom{n}{x} \cdot p^x \cdot (1-p)^{n-x},$$

where x is the number of females in the sample. For example,

$$P(x = 0) = \binom{5}{0}(.79)^0(.21)^5 \approx .0004.$$

New York Times, January 7, 1996. Education Life, p. 24.

Similarly, we could calculate the probability that x is any value from 0 to 5, resulting in the probability distribution in Table 4 (with all probabilities rounded to four places).

Table 4

x	0	1	2	3	4	5
$P(x)$	.0004	.0077	.0578	.2174	.4090	.3077

(b) Find the expected number of females in the sample of 5 people.
   Using the formula for expected value, we have

$$E(x) = 0(.0004) + 1(.0077) + 2(.0578) + 3(.2174)$$
$$+ 4(.4090) + 5(.3077) = 3.95.$$

On the average, 3.95 of the people in the sample of 5 will be female. ◀

There is another way to get the answer in part (b) of the previous example. Because 79% of the U.S. holders of bachelor's degrees in education from 1992–93 are female, it is reasonable to expect 79% of our sample to be female. Thus, 79% of 5 is $5(.79) = 3.95$. Notice that what we have done is to multiply n by p. It can be shown that this method always gives the expected value for binomial probability.

Expected Value for Binomial Probability

For binomial probability, $E(x) = np$. In other words, the expected number of successes is the number of trials times the probability of success in each trial.

▶ **EXAMPLE 7**   Suppose a family has 3 children.

(a) Find the probability distribution for the number of girls.
   Assuming girls and boys are equally likely, the probability distribution is binomial with $n = 3$ and $p = 1/2$. Letting x be the number of girls in the formula for binomial probability, we find, for example,

$$P(x = 0) = \binom{3}{0}\left(\frac{1}{2}\right)^0\left(\frac{1}{2}\right)^3 = \frac{1}{8}.$$

The other values are found similarly, and the results are shown in Table 5.

Table 5

x	0	1	2	3
$P(x)$	1/8	3/8	3/8	1/8

We can verify this by noticing that in the sample space S of all 3-child families, there are eight equally likely outcomes: $S = \{ggg, ggb, gbg, gbb, bgg, bgb, bbg, bbb\}$. One of the outcomes has 0 girls, three have 1 girl, three have 2 girls, and one has 3 girls.

(b) Find the expected number of girls in a 3-child family using the distribution from part (a).

Using the formula for expected value, we have

$$\text{Expected number of girls} = 0\left(\frac{1}{8}\right) + 1\left(\frac{3}{8}\right) + 2\left(\frac{3}{8}\right) + 3\left(\frac{1}{8}\right)$$

$$= \frac{12}{8} = 1.5.$$

On average, a 3-child family will have 1.5 girls. This result agrees with our intuition that, on the average, half the children born will be girls.

(c) Find the expected number of girls in a 3-child family using the formula for expected value for binomial probability.

Using the formula $E(x) = np$ with $n = 3$ and $p = 1/2$, we have

$$\text{Expected number of girls} = 3\left(\frac{1}{2}\right) = 1.5.$$

This agrees with our answer from part (b), as it must. ◀ ⑥

We will return to the concepts in this section in Chapter 10.

---

⑥

The probability that a can of soda from a certain plant is defective is .005. A sample of 4 cans is selected at random. Write a distribution for the number of defective cans in the sample, and give its expected value.

Answers:

x	$P(x)$
0	.98
1	.0197
2	.00015
3	.0000005
4	6.25×10^{-10}

$E(x) = .02$

9.4 Exercises

For each of the experiments described below, let x determine a random variable and use your knowledge of probability to prepare a probability distribution. (See Example 1.)

1. Four coins are tossed, and the number of heads is noted.

2. Two dice are rolled, and the total number of points is recorded.

3. Three cards are drawn from a deck. The number of aces is counted.

4. Two balls are drawn from a bag in which there are 4 white balls and 2 black balls. The number of black balls is counted.

Draw a histogram for each of the following and shade the region that gives the indicated probability. (See Example 1.)

5. Exercise 1; $P(x \le 2)$

6. Exercise 2; $P(x \ge 11)$

7. Exercise 3; $P(\text{at least one ace})$

8. Exercise 4; $P(\text{at least one black ball})$

Find the expected value for each random variable. (See Example 2.)

9.

x	2	3	4	5
$P(x)$	.1	.4	.3	.2

10.

y	4	6	8	10
$P(y)$	.4	.4	.05	.15

11.

z	9	12	15	18	21
$P(z)$	.14	.22	.36	.18	.10

12.

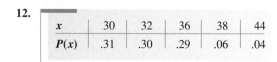

x	30	32	36	38	44
$P(x)$	.31	.30	.29	.06	.04

Find the expected values for the random variables x having the probability functions graphed below.

13.

14.

15.

16.

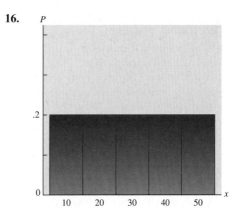

17. For the game in Example 4, suppose one day Mary brings a 2-headed coin and uses it to toss for the coffee. Since Mary tosses, Donna calls. Is this still a fair game?

18. In Exercise 17,
   (a) what is Donna's expected gain if she calls heads?
   (b) what is Donna's expected gain if she calls tails?

Solve each exercise. Many of these exercises require the use of combinations. (See Examples 3–5.)

19. Suppose 3 marbles are drawn from a bag containing 3 yellow and 4 white marbles.
   (a) Draw a histogram for the number of yellow marbles in the sample.
   (b) What is the expected number of yellow marbles in the sample?

20. Suppose 5 apples in a barrel of 25 apples are known to be rotten.
   (a) Draw a histogram for the number of rotten apples in a sample of 2 apples.
   (b) What is the expected number of rotten apples in a sample of 2 apples?

21. A delegation of 3 is selected from a city council made up of 5 liberals and 4 conservatives.
   (a) What is the expected number of liberals in the delegation?
   (b) What is the expected number of conservatives in the delegation?

22. From a group of 2 women and 5 men, a delegation of 2 is selected. Find the expected number of women in the delegation.

23. In a club with 20 senior and 10 junior members, what is the expected number of junior members on a 3-member committee?

24. If 2 cards are drawn at one time from a deck of 52 cards, what is the expected number of diamonds?

25. Suppose someone offers to pay you $5 if you draw 2 diamonds in the game in Exercise 24. He says that you should pay 50 cents for the chance to play. Is this a fair game?

26. Your friend missed class the day probability distributions were discussed. How would you explain a probability distribution to him?

27. Explain what expected value means in your own words.

28. A raffle offers a first prize of $100 and 2 second prizes of $40 each. One ticket costs $1, and 500 tickets are sold. Find the expected winnings for a person who buys 1 ticket. Is this a fair game?

29. A raffle offers a first prize of $1000, 2 second prizes of $300 each, and 20 third prizes of $10 each. If 10,000 tickets are sold at 50¢ each, find the expected winnings for a person buying 1 ticket. Is this a fair game?

Find the expected winnings for the games of chance described in Exercises 30–35.

30. A lottery requires you to choose 4 cards from an ordinary deck: 1 heart, 1 club, 1 diamond, and 1 spade in that order from the 13 cards in each suit. If all four choices are selected by the lottery, you win $5000. It costs $1 to play.

31. If exactly 3 of the 4 choices in Exercise 30 are selected, the player wins $200. (Ignore the possibility that all 4 choices are selected. It still costs $1 to play.)

32. In one form of roulette, you bet $1 on "even." If 1 of the 18 even numbers comes up, you get your dollar back, plus another one. If 1 of the 20 noneven (18 odd, 0, and 00) numbers comes up, you lose your dollar.

33. Repeat Exercise 32 if there are only 19 noneven numbers (no 00).

34. *Numbers* is a game in which you bet $1 on any three-digit number from 000 to 999. If your number comes up, you get $500.

35. In one form of the game Keno, the house has a pot containing 80 balls, each marked with a different number from 1 to 80. You buy a ticket for $1 and mark one of the 80 numbers on it. The house then selects 20 numbers at random. If your number is among the 20, you get $3.20 (for a net winning of $2.20).

36. **Business** A magazine distributor offers a first prize of $100,000, two second prizes of $40,000 each, and two third prizes of $10,000 each. A total of 2,000,000 entries are received in the contest. Find the expected winnings if you submit one entry to the contest. If it would cost you 50¢ in paper, and stamps to enter, would it be worth it?

37. **Business** A contest at a fast-food restaurant offered the following cash prizes and probabilities of winning on one visit.

Prize	Probability
$100,000	1/176,402,500
$25,000	1/39,200,556
$5000	1/17,640,250
$1000	1/1,568,022
$100	1/282,244
$5	1/7056
$1	1/588

Suppose you spend $1 to buy a bus pass that lets you go to 25 different restaurants in the chain and pick up entry forms. Find your expected value.

38. **Business** According to officials of Mars, the makers of M & M Plain Chocolate Candies, 20% of the candies in each bag are red.* 4 candies are selected from a bag and the number of red candies is recorded.
    (a) Give a probability distribution for the number of red candies.
    (b) Find the expected value.

39. **Business** A local used-car dealer gets complaints about his cars as shown in the following table.

Number of Complaints per Day	0	1	2	3	4	5	6
Probability	.01	.05	.15	.26	.33	.14	.06

Find the expected number of complaints per day.

40. **Finance** An insurance company has written 100 policies of $10,000, 500 of $5000, and 1000 of $1000 for people of age 20. If experience shows that the probability that a person will die at age 20 is .001, how much can the company expect to pay out during the year the policies were written?

41. **Business** Levi Strauss and Company uses expected value to help its salespeople rate their accounts.[†] For each account, a salesperson estimates potential additional vol-

---

NCTM News Bulletin, February 1995, p. 5.

[†]This example was supplied by James McDonald, Levi Strauss and Company, San Francisco.

Account Number	Existing Volume	Potential Additional Volume	Probability of Additional Volume	Expected Value of Potential	Existing Volume + Expected Value of Potential	Class
1	$15,000	$10,000	.25	$2,500	$17,500	C
2	40,000	0	—	—	40,000	C
3	20,000	10,000	.20			
4	50,000	10,000	.10			
5	5,000	50,000	.50			
6	0	100,000	.60			
7	30,000	20,000	.80			

ume and the probability of getting it. The product of these figures gives the expected value of the potential, which is added to the existing volume. The totals are then classified as A, B, or C as follows: $40,000 or below, class C; $40,000 up to and including $55,000, class B; above $55,000, class A. Complete the chart above for one salesperson.

42. Natural Science   In a certain animal species, the probability that a healthy adult female will have no offspring in a given year is .31, while the probabilities of 1, 2, 3, or 4 offspring are, respectively, .21, .19, .17, and .12. Find the expected number of offspring.

43. Natural Science   Otitis media, or middle ear infection, is initially treated with an antibiotic. Researchers have compared two antibiotics, amoxicillin and cefaclor, for their cost effectiveness. Amoxicillin is inexpensive, safe, and effective. Cefaclor is also safe. However, it is considerably more expensive and it is generally more effective. Use the tree diagram below (where the costs are estimated as the total cost of medication, office visit, ear check, and hours of lost work) to answer the following.*
   (a) Find the expected cost of using each antibiotic to treat a middle ear infection.
   (b) To minimize the total expected cost, which antibiotic should be chosen?

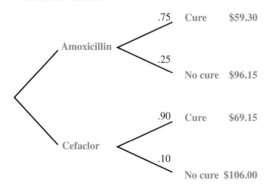

*Jeffrey Weiss and Shoshana Melman, "Cost Effectiveness in the Choice of Antibiotics for the Initial Treatment of Otitis Media in Children: A Decision Analysis Approach." *Journal of Pediatric Infectious Disease* vol. 7, no. 1 (1998): 23–26.

44. Physical Science   One of the few methods that can be used in an attempt to cut the severity of a hurricane is to *seed* the storm. In this process, silver iodide crystals are dropped into the storm. Unfortunately, silver iodide crystals sometimes cause the storm to *increase* its speed. Wind speeds may also increase or decrease even with no seeding. Use the tree diagram below to answer the following.*
   (a) Find the expected amount of damage under each option, "seed" and "do not seed."
   (b) To minimize total expected damage, what option should be chosen?

	Change in wind speed	Property damage (millions of dollars)
Seed 0.038	+32%	335.8
0.143	+16%	191.1
0.392	0	100.0
0.255		
0.172	−16%	46.7
	−34%	16.3
Do not seed 0.054	+32%	335.8
0.206	+16%	191.1
0.480	0	100.0
0.206		
0.054	−16%	46.7
	−34%	16.3

45. At the end of play in a major golf tournament, two players, an "old pro" and a "new kid," are tied. Suppose the first prize is $80,000 and second prize is $20,000. Find the expected winnings for the old pro if
   (a) both players are of equal ability;
   (b) the new kid will freeze up, giving the old pro a 3/4 chance of winning.

*Data from "The Decision to Seed Hurricanes," by R. A. Howard from *Science*, Vol. 176, pp. 1191–1202, Copyright 1972 by the American Association for the Advancement of Science.

46. Kimberly Workman has four cats: Riley, Abby, Beastie, and Sylvester. Each cat has a 30% probability of climbing into the chair in which Kimberly is sitting, independent of how many cats are already in the chair with Kimberly.
   (a) Find the probability distribution for the number of cats in the chair with Kimberly.
   (b) Find the expected number of cats in the chair with Kimberly using the probability distribution in part (a).
   (c) Find the expected number of cats in the chair with Kimberly using the formula for expected value of the binomial distribution.

47. Social Science   Mr. Statistics (a feature in *Fortune* magazine) investigated the claim of the United States Postal Service that 83% of first class mail in New York City arrives by the next day.* (The figure is 87% nationwide.) He mailed a letter to himself on 10 consecutive days; only 4 were delivered by the next day.
   (a) Find the probability distribution for the number of letters delivered by the next day if the overall probability of next-day delivery is 83%.
   (b) Using your answer to part (a), find the probability that 4 or fewer out of 10 letters would be delivered by the next day.
   (c) Based on your answer to part (b), do you think it is likely that the 83% figure is accurate? Explain.

(d) Find the number of letters out of 10 that you would expect to be delivered by the next day if the 83% figure is accurate.

48. After a team scores a touchdown, it can either attempt to kick an extra point or attempt a two-point conversion. During the 1999–2000 NFL season, two-point conversions were successful 37% of the time and the extra-point kicks were successful 94% of the time.*

   (a) Calculate the expected value of each strategy.
   (b) Which strategy, over the long run, will maximize the number of points scored?
   (c) Using this information, should a team always only use one strategy? Explain.

*Daniel Seligman, "Ask Mr. Statistics," *Fortune,* July 24, 1995, pp. 170–171.

*David Leonhardt, "In Football, 6 + 2 Often Equals 6," *New York Times,* Sunday, January 16, 2000. p. 4–2.

9.5 MARKOV CHAINS

In Section 8.5, we touched on **stochastic processes,** mathematical models that evolve over time in a probabilistic manner. In this section we study a special kind of stochastic process called a **Markov chain,** where the outcome of an experiment depends only on the outcome of the previous experiment. In other words, the next **state** of the system depends only on the present state, not on preceding states. Such experiments are common enough in applications to make their study worthwhile. Markov chains are named after the Russian mathematician A. A. Markov, 1856–1922, who started the theory of stochastic processes. To see how Markov chains work, we look at an example.

▶**EXAMPLE 1**   A small town has only two dry cleaners, Johnson and NorthClean. Johnson's manager hopes to increase the firm's market share by an extensive advertising campaign. After the campaign, a market research firm finds that there is a probability of .8 that a Johnson customer will bring his next batch of dirty items to Johnson and a .35 chance that a NorthClean customer will switch to Johnson for his next batch. Assume that the probability that a customer comes to a given cleaners depends only on where the last load of clothes was taken. If there is an .8 chance that a Johnson customer will return to Johnson, then there must be a 1 − .8 = .2 chance that the customer will switch to NorthClean. In the same way, there is a 1 − .35 = .65 chance that a NorthClean customer will return to NorthClean. If an individual bringing a load to Johnson is said to be in state 1 and an individual bringing a load to NorthClean is said to be in state 2, then these probabilities of change from one cleaner to the other are as shown in the following table.

		SECOND LOAD	
	State	*1*	*2*
FIRST LOAD	1	.8	.2
	2	.35	.65

The information from the table can be written in other forms. Figure 9.11 is a **transition diagram** that shows the two states and the probabilities of going from one state to another.

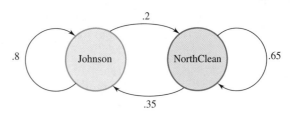

FIGURE 9.11

In a **transition matrix,** the states are indicated at the side and top, as follows.

$$
\begin{array}{c}
\quad\quad\quad\quad\quad \textit{Second Load} \\
\quad\quad\quad\quad \text{Johnson} \quad \text{NorthClean} \\
\textit{First Load}\;
\begin{array}{c} \text{Johnson} \\ \text{NorthClean} \end{array}
\left[\begin{array}{cc} .8 & .2 \\ .35 & .65 \end{array} \right]
\end{array}
$$ ①

A **transition matrix** has the following features.

1. It is square, since all possible states must be used both as rows and as columns.
2. All entries are between 0 and 1, inclusive, because all entries represent probabilities.
3. The sum of the entries in any row must be 1, because the numbers in the row give the probability of changing from the state at the left to one of the states indicated across the top.

▶**EXAMPLE 2** Suppose that when the new promotional campaign began, Johnson had 40% of the market and NorthClean had 60%. Use the probability tree in Figure 9.12 to find how these proportions would change after another week of advertising.

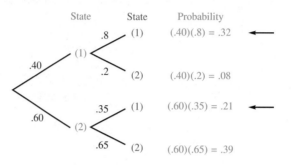

State 1: taking cleaning to Johnson
State 2: taking cleaning to NorthClean

FIGURE 9.12

Add the numbers indicated with arrows to find the proportion of people taking their cleaning to Johnson after 1 week.

$$.32 + .21 = .53$$

Similarly, the proportion taking their cleaning to NorthClean is

$$.08 + .39 = .47.$$

The initial distribution of 40% and 60% becomes, after 1 week, 53% and 47%. ◀

These distributions can be written as the *probability vectors*

$$[.40 \quad .60] \quad \text{and} \quad [.53 \quad .47].$$

A **probability vector** is a matrix of only one row, having nonnegative entries with the sum of the entries equal to 1.

The results from the probability tree above are exactly the same as the result of multiplying the initial probability vector by the transition matrix. (Multiplication of matrices was discussed in Section 6.4.)

$$[.4 \quad .6]\begin{bmatrix} .8 & .2 \\ .35 & .65 \end{bmatrix} = [.53 \quad .47]$$

If v denotes the original probability vector $[.4 \quad .6]$ and P denotes the transition matrix, then the market share vector after one week is $vP = [.53 \quad .47]$. To find the market share after 2 weeks multiply the vector $vP = [.53 \quad .47]$ by P, which amounts to finding vP^2. ②

Problem 2 at the side shows that after two weeks, the market share vector is $vP^2 = [.59 \quad .41]$. To get the market share after 3 weeks, multiply this vector by P, that is, find vP^3. Do not use the rounded answer from Problem 2. ③

② Find the product

$$[.53 \quad .47]\begin{bmatrix} .8 & .2 \\ .35 & .65 \end{bmatrix}.$$

Answer:

$[.59 \quad .41]$ (rounded)

Find each cleaner's market share after 3 weeks.

Answer:

$[.61 \quad .39]$ (rounded)

Continuing this process gives each cleaner's share of the market after additional weeks.

Weeks after Start	Johnson	NorthClean	
0	.4	.6	v
1	.53	.47	vP^1
2	.59	.41	vP^2
3	.61	.39	vP^3
4	.63	.37	vP^4
12	.64	.36	vP^{12}
13	.64	.36	vP^{13}

The results seem to approach the probability vector [.64 .36].

What happens if the initial probability vector is different from [.4 .6]? Suppose [.75 .25] is used; the same powers of the transition matrix as above give the following results.

Week after Start	Johnson	NorthClean	
0	.75	.25	v
1	.69	.31	vP^1
2	.66	.34	vP^2
3	.65	.35	vP^3
4	.64	.36	vP^4
5	.64	.36	vP^5
6	.64	.36	vP^6

The results again seem to be approaching the numbers in the probability vector [.64 .36], the same numbers approached with the initial probability vector [.4 .6]. In either case, the long-range trend is for a market share of about 64% for Johnson and 36% for NorthClean. The example above suggests that this long-range trend does not depend on the initial distribution of market shares. This means that if the initial market share for Johnson was less than 64%, the advertising campaign has paid off in terms of a greater long-range market share. If the initial share was more than 64%, the campaign did not pay off.

Regular Transition Matrices

One of the many applications of Markov chains is in finding these long-range predictions. It is not possible to make long-range predictions with all transition matrices, but for a large set of transition matrices, long-range predictions *are* possible. Such predictions are always possible with **regular transition matrices.** A transition matrix is **regular** if some power of the matrix contains all positive entries. A Markov chain is a **regular Markov chain** if its transition matrix is regular.

▶ **EXAMPLE 3** Decide whether the following transition matrices are regular.

(a) $A = \begin{bmatrix} .75 & .25 & 0 \\ 0 & .5 & .5 \\ .6 & .4 & 0 \end{bmatrix}$

Square A.

$$A^2 = \begin{bmatrix} .5625 & .3125 & .125 \\ .3 & .45 & .25 \\ .45 & .35 & .2 \end{bmatrix}$$

Since all entries in A^2 are positive, matrix A is regular.

(b) $B = \begin{bmatrix} .5 & 0 & .5 \\ 0 & 1 & 0 \\ 0 & 0 & 1 \end{bmatrix}$

Find various powers of B.

$$B^2 = \begin{bmatrix} .25 & 0 & .75 \\ 0 & 1 & 0 \\ 0 & 0 & 1 \end{bmatrix} \quad B^3 = \begin{bmatrix} .125 & 0 & .875 \\ 0 & 1 & 0 \\ 0 & 0 & 1 \end{bmatrix} \quad B^4 = \begin{bmatrix} .0625 & 0 & .9375 \\ 0 & 1 & 0 \\ 0 & 0 & 1 \end{bmatrix}$$

Notice that all of the powers of B shown here have zeros in the same locations. Thus, further powers of B will still give the same zero entries, so that no power of matrix B contains all positive entries. For this reason, B is not regular. ◀ ④

NOTE If a transition matrix P has some zero entries, and P^2 does as well, you may wonder how far you must compute P^k to be certain that the matrix is not regular. The answer is that if zeros occur in the identical places in both P^k and P^{k+1} for any k, they will appear in those places for all higher powers of P, so P is not regular.

Suppose that v is any probability vector. It can be shown that for a regular Markov chain with a transition matrix P, there exists a single vector V that does not depend on v, such that $v \cdot P^n$ gets closer and closer to V as n gets larger and larger.

Equilibrium Vector of a Markov Chain

If a Markov chain with transition matrix P is regular, then there is a unique vector V such that, for any probability vector v and for large values of n,

$$v \cdot P^n \approx V.$$

Vector V is called the equilibrium vector or the fixed vector of the Markov chain.

In the example with Johnson Cleaners, the equilibrium vector V is approximately $[.64 \quad .36]$. Vector V can be determined by finding P^n for larger and larger values of n and then looking for a vector that the product $v \cdot P^n$ approaches. Such an

④

Decide if the following transition matrices are regular.

(a) $\begin{bmatrix} 0 & 1 \\ 1 & 0 \end{bmatrix}$

(b) $\begin{bmatrix} .3 & .7 \\ 1 & 0 \end{bmatrix}$

Answers:

(a) No

(b) Yes

approach can be very tedious, however, and is prone to error. To find a better way, start with the fact that for a large value of n,

$$v \cdot P^n \approx V,$$

as mentioned above. From this result, $v \cdot P^n \cdot P \approx V \cdot P$, so that

$$v \cdot P^n \cdot P = v \cdot P^{n+1} \approx VP.$$

Since $v \cdot P^n \approx V$ for large values of n, it is also true that $v \cdot P^{n+1} \approx V$ for large values of n (the product $v \cdot P^n$ approaches V, so that $v \cdot P^{n+1}$ must also approach V). Thus, $v \cdot P^{n+1} \approx V$ and $v \cdot P^{n+1} \approx VP$, which suggests that

$$VP = V.$$

If a Markov chain with transition matrix P is regular, then the equilibrium vector V satisties

$$VP = V.$$

The equilibrium vector V can be found by solving a system of linear equations, as shown in the next examples.

▶ **EXAMPLE 4**   Find the long-range trend for the Markov chain in Examples 1 and 2, with transition matrix

$$P = \begin{bmatrix} .8 & .2 \\ .35 & .65 \end{bmatrix}.$$

This matrix is regular since all entries are positive. Let P represent this transition matrix and let V be the probability vector $[v_1 \quad v_2]$. We want to find V such that

$$VP = V,$$

or

$$[v_1 \quad v_2]\begin{bmatrix} .8 & .2 \\ .35 & .65 \end{bmatrix} = [v_1 \quad v_2].$$

Multiply on the left to get

$$[.8v_1 + .35v_2 \quad .2v_1 + .65v_2] = [v_1 \quad v_2].$$

Set corresponding entries from the two matrices equal to get

$$.8v_1 + .35v_2 = v_1 \qquad .2v_1 + .65v_2 = v_2.$$

Simplify each of these equations,

$$-.2v_1 + .35v_2 = 0 \qquad .2v_1 - .35v_2 = 0$$

These last two equations are really the same. (The equations in the system obtained from $VP = V$ are always dependent.) To find the values of v_1 and v_2, recall that $V = [v_1 \quad v_2]$ is a probability vector, so that

$$v_1 + v_2 = 1.$$

Find v_1 and v_2 by solving the system

$$-.2v_1 + .35v_2 = 0$$
$$v_1 + v_2 = 1.$$

From the second equation, $v_1 = 1 - v_2$. Substitute for v_1 in the first equation:

$$-.2(1 - v_2) + .35v_2 = 0. \; ⑤$$

Since $v_2 = .364$ (from problem 5 at the side) and $v_1 = 1 - v_2$, then $v_1 = 1 - .364 = .636$ and the equilibrium vector is $[.636 \quad .364] \approx [.64 \quad .36]$. ◀

⑤

Solve the equation for v_2.
Round to the nearest thousandth.

Answer:

.364

▶ **EXAMPLE 5** The probability that a complex assembly line works correctly depends on whether or not the line worked correctly the last time it was used. The various probabilities are as given in the following transition matrix.

	Works Properly Now	Does Not
Worked Properly Before	.9	.1
Did Not	.7	.3

Find the long-range probability that the assembly line will work properly.

Begin by finding the equilibrium vector $[v_1 \quad v_2]$, where

$$[v_1 \quad v_2]\begin{bmatrix} .9 & .1 \\ .7 & .3 \end{bmatrix} = [v_1 \quad v_2].$$

Multiplying on the left and setting corresponding entries equal gives the equations

$$.9v_1 + .7v_2 = v_1 \text{ and } \quad .1v_1 + .3v_2 = v_2$$

or

$$-.1v_1 + .7v_2 = 0 \quad \text{and} \quad .1v_1 - .7v_2 = 0.$$

Substitute $v_1 = 1 - v_2$ in the first of these equations to get

$$-.1(1 - v_2) + .7v_2 = 0$$
$$-.1 + .1v_2 + .7v_2 = 0$$
$$-.1 + .8v_2 = 0$$
$$.8v_2 = .1$$
$$v_2 = \frac{1}{8},$$

⑥

In Example 5, suppose the company modifies the line so that the transition matrix becomes

$$\begin{bmatrix} .95 & .05 \\ .8 & .2 \end{bmatrix}.$$

Find the long-range probability that the assembly line will work properly.

Answer:

16/17

and $v_1 = 1 - 1/8 = 7/8$. The equilibrium vector is $[7/8 \quad 1/8]$. In the long run, the company can expect the assembly line to run properly 7/8 of the time. ◀ ⑥

▶ **EXAMPLE 6**   Find the equilibrium vector for the transition matrix

$$K = \begin{bmatrix} .2 & .6 & .2 \\ .1 & .1 & .8 \\ .3 & .3 & .4 \end{bmatrix}.$$

Matrix K has all positive entries and thus is regular. For this reason, an equilibrium vector V must exist such that $VK = V$. Let $V = [v_1 \quad v_2 \quad v_3]$.

Then

$$[v_1 \quad v_2 \quad v_3] \begin{bmatrix} .2 & .6 & .2 \\ .1 & .1 & .8 \\ .3 & .3 & .4 \end{bmatrix} = [v_1 \quad v_2 \quad v_3].$$

Use matrix multiplication on the left.

$$[.2v_1 + .1v_2 + .3v_3 \quad .6v_1 + .1v_2 + .3v_3 \quad .2v_1 + .8v_2 + .4v_3] = [v_1 \quad v_2 \quad v_3]$$

Set corresponding entries equal.

$$.2v_1 + .1v_2 + .3v_3 = v_1$$
$$.6v_1 + .1v_2 + .3v_3 = v_2$$
$$.2v_1 + .8v_2 + .4v_3 = v_3$$

Simplify these equations.

$$-.8v_1 + .1v_2 + .3v_3 = 0$$
$$.6v_1 - .9v_2 + .3v_3 = 0$$
$$.2v_1 + .8v_2 - .6v_3 = 0$$

Since V is a probability vector,

$$v_1 + v_2 + v_3 = 1.$$

This gives a system of four equations in three unknowns:

$$v_1 + v_2 + v_3 = 1$$
$$-.8v_1 + .1v_2 + .3v_3 = 0$$
$$.6v_1 - .9v_2 + .3v_3 = 0$$
$$.2v_1 + .8v_2 - .6v_3 = 0.$$

This system can be solved with the Gauss-Jordan method in Section 6.2. Start with the augmented matrix

$$\begin{bmatrix} 1 & 1 & 1 & | & 1 \\ -.8 & .1 & .3 & | & 0 \\ .6 & -.9 & .3 & | & 0 \\ .2 & .8 & -.6 & | & 0 \end{bmatrix}.$$

The solution of this system is $v_1 = 5/23$, $v_2 = 7/23$, $v_3 = 11/23$, and

$$V = \begin{bmatrix} \dfrac{5}{23} & \dfrac{7}{23} & \dfrac{11}{23} \end{bmatrix} \approx [.22 \quad .30 \quad .48]. \quad ◀ \quad ⑦$$

⑦

Find the equilibrium vector for the transition matrix

$$P = \begin{bmatrix} .3 & .7 \\ .5 & .5 \end{bmatrix}.$$

Answer:

$[5/12 \quad 7/12]$

In Example 4, we found that [.64 .36] was the equilibrium vector for the regular transition matrix.

$$P = \begin{bmatrix} .8 & .2 \\ .35 & .65 \end{bmatrix}.$$

Observe what happens when you take powers of the matrix P (displayed enries have been rounded for easy reading, but the full decimals were used in the calculations).

$$P^2 = \begin{bmatrix} .71 & .29 \\ .51 & .49 \end{bmatrix} \qquad P^3 = \begin{bmatrix} .67 & .33 \\ .58 & .42 \end{bmatrix} \qquad P^4 = \begin{bmatrix} .65 & .35 \\ .61 & .39 \end{bmatrix}$$

$$P^5 = \begin{bmatrix} .64 & .36 \\ .62 & .38 \end{bmatrix} \qquad P^6 = \begin{bmatrix} .64 & .36 \\ .63 & .37 \end{bmatrix} \qquad P^{10} = \begin{bmatrix} .64 & .36 \\ .64 & .36 \end{bmatrix}$$

As these results suggest, higher and higher powers of the transition matrix P approach a matrix having all rows identical: these identical rows have as entries the entries of the equilibrium vector V.

If you have the technology to compute matrix powers easily (such as a graphing calculator), you can approximate the equilibrium vector by taking higher and higher powers of the transition matrix until all its rows are identical. Figure 9.13 shows part of this process for the transition matrix

$$B = \begin{bmatrix} .9 & .1 \\ .7 & .3 \end{bmatrix}$$

of Example 5.

FIGURE 9.13

Figure 9.13 indicates that the equilibrium vector is [.875   .125] = [7/8   1/8], which is what was found algebraically in Example 5.

The results of this section can be summarized as follows.

Properties of Regular Markov Chains

Suppose a regular Markov chain has a transition matrix P.

1. As n gets larger and larger, the product $v \cdot P^n$ approaches a unique vector V for any initial probability vector v. Vector V is called the *equilibrium vector or fixed vector*.
2. Vector V has the property that $VP = V$.
3. To find V, solve a system of equations obtained from the matrix equation $VP = V$ and from the fact that the sum of the entries of V is 1.
4. The powers P^n come closer and closer to a matrix whose rows are made up of the entries of the equilibrium vector V.

9.5 Exercises

Decide which of the following could be a probability vector.

1. $\begin{bmatrix} \frac{1}{3} & \frac{2}{3} \end{bmatrix}$

2. $\begin{bmatrix} \frac{3}{4} & \frac{1}{4} \end{bmatrix}$

3. $\begin{bmatrix} 0 & 1 \end{bmatrix}$

4. $\begin{bmatrix} .4 & .2 & 0 \end{bmatrix}$

5. $\begin{bmatrix} .3 & -.1 & .6 \end{bmatrix}$

6. $\begin{bmatrix} \frac{1}{4} & \frac{1}{8} & \frac{5}{8} \end{bmatrix}$

Decide which of the following could be a transition matrix. Sketch a transition diagram for any transition matrices.

7. $\begin{bmatrix} .7 & .1 \\ .5 & .5 \end{bmatrix}$

8. $\begin{bmatrix} \frac{1}{4} & \frac{3}{4} \\ 0 & 1 \end{bmatrix}$

9. $\begin{bmatrix} \frac{2}{3} & \frac{1}{3} \\ \frac{1}{5} & \frac{4}{5} \end{bmatrix}$

10. $\begin{bmatrix} 0 & 1 & 0 \\ \frac{1}{3} & \frac{1}{3} & \frac{1}{3} \\ 1 & 0 & 0 \end{bmatrix}$

11. $\begin{bmatrix} \frac{1}{2} & \frac{1}{4} & 1 \\ \frac{2}{3} & 0 & \frac{1}{3} \\ \frac{1}{3} & 1 & 0 \end{bmatrix}$

12. $\begin{bmatrix} .2 & .3 & .5 \\ 0 & 0 & 1 \\ .1 & .9 & 0 \end{bmatrix}$

In Exercises 13–15, write any transition diagrams as transition matrices.

13.

14.

15.

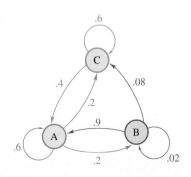

Decide whether the following transition matrices are regular. (See Example 3.)

16. $\begin{bmatrix} 1 & 0 \\ .6 & .4 \end{bmatrix}$

17. $\begin{bmatrix} .2 & .8 \\ .9 & .1 \end{bmatrix}$

18. $\begin{bmatrix} .3 & .5 & .2 \\ 1 & 0 & 0 \\ .5 & .1 & .4 \end{bmatrix}$

19. $\begin{bmatrix} 0 & 1 & 0 \\ .4 & .2 & .4 \\ 1 & 0 & 0 \end{bmatrix}$

20. $\begin{bmatrix} .25 & .40 & .30 & .05 \\ .18 & .23 & .59 & 0 \\ 0 & .15 & .36 & .49 \\ .28 & .32 & .24 & .16 \end{bmatrix}$

21. $\begin{bmatrix} .12 & .68 & 0 & .20 \\ 0 & .33 & .40 & .27 \\ 0 & 0 & 1 & 0 \\ .52 & 0 & .43 & .05 \end{bmatrix}$

Find the equilibrium vector for each of the following transition matrices. (See Examples 4 and 5.)

22. $\begin{bmatrix} .3 & .7 \\ .4 & .6 \end{bmatrix}$

23. $\begin{bmatrix} .8 & .2 \\ .1 & .9 \end{bmatrix}$

24. $\begin{bmatrix} \frac{1}{4} & \frac{3}{4} \\ \frac{1}{2} & \frac{1}{2} \end{bmatrix}$

25. $\begin{bmatrix} \frac{2}{3} & \frac{1}{3} \\ \frac{1}{8} & \frac{7}{8} \end{bmatrix}$

26. $\begin{bmatrix} .25 & .35 & .4 \\ .1 & .3 & .6 \\ .55 & .4 & .05 \end{bmatrix}$

27. $\begin{bmatrix} .16 & .28 & .56 \\ .43 & .12 & .45 \\ .86 & .05 & .09 \end{bmatrix}$

28. $\begin{bmatrix} .1 & .1 & .8 \\ .4 & .4 & .2 \\ .1 & .2 & .7 \end{bmatrix}$

29. $\begin{bmatrix} .5 & .2 & .3 \\ .1 & .4 & .5 \\ .2 & .2 & .6 \end{bmatrix}$

For each of the following transition matrices, use a graphing calculator or computer to find the first five powers of the matrix. Then find the probability that state 2 changes to state 4 after 5 repetitions of the experiment.

30. $\begin{bmatrix} .1 & .2 & .2 & .3 & .2 \\ .2 & .1 & .1 & .2 & .4 \\ .2 & .1 & .4 & .2 & .1 \\ .3 & .1 & .1 & .2 & .3 \\ .1 & .3 & .1 & .1 & .4 \end{bmatrix}$

31. $\begin{bmatrix} .3 & .2 & .3 & .1 & .1 \\ .4 & .2 & .1 & .2 & .1 \\ .1 & .3 & .2 & .2 & .2 \\ .2 & .1 & .3 & .2 & .2 \\ .1 & .1 & .4 & .2 & .2 \end{bmatrix}$

32. Social Science The chart on the next page shows the percent of the poor, middle class, and affluent that change into another class. The first graph shows the figures for 1967–1979, and the second those for 1980–1991.* For each time period, assume the number of people who move directly from poor to affluent or from affluent to poor is essentially 0.

---
*New York Times, June 4, 1995, p. E4.

(a) Find the long-range percent of poor, middle class, and affluent people if the 1967–1979 trends were to continue.

(b) Find the long-range percent of poor, middle class, and affluent people if the 1980–1991 trends were to continue.

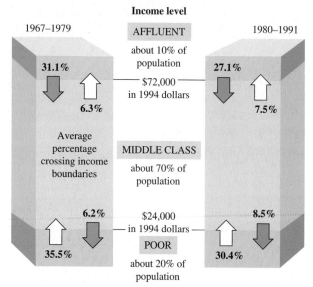

The poor are more likely to stay poor and the affluent are more likely to stay affluent.

Income level

1967–1979

AFFLUENT
about 10% of population

1980–1991

31.1%

6.3%

$72,000 in 1994 dollars

27.1%

7.5%

Average percentage crossing income boundaries

MIDDLE CLASS
about 70% of population

6.2%

35.5%

$24,000 in 1994 dollars

POOR
about 20% of population

8.5%

30.4%

All figures are for household after-tax income, including wages, salaries, and some Government assistance programs like food stamps.

Sources: Greg Duncan, Northwestern University; Timothy Smeeding, Syracuse University

33. Finance   In 1992, many homeowners refinanced their mortgages to take advantage of lower interest rates. Most of the mortgages could be classified into three groups: 30-year fixed-rate, 15-year fixed-rate, and adjustable rate. (For this exercise, the small number of loans that are not of those three types will be classified with the adjustable rate loans.) Sometimes when a homeowner refinanced, the new loan was the same type as the old loan, and sometimes it was different. The breakdown of the percent in each category is shown in the following table.*

| | NEW LOAN | | |
OLD LOAN	30-Year Fixed	15-Year Fixed	Adjustable
30-year fixed	.444	.479	.077
15-year fixed	.150	.802	.048
Adjustable	.463	.367	.170

New York Times, November 18, 1992, p. D2.

If these conversion rates were to persist, find the long-range trend for the percent of loans of each type.

34. Business   The probability that a complex assembly line works correctly depends on whether the line worked correctly the last time it was used. There is a .95 chance that the line will work correctly if it worked correctly the time before, and a .7 chance that it will work correctly if it did *not* work correctly the time before. Set up a transition matrix with this information and find the long-run probability that the line will work correctly. (See Example 5.)

35. Business   Suppose improvements are made in the assembly line of Exercise 34, so that the transition matrix becomes

$$\begin{array}{c} \\ \text{Works} \\ \text{Doesn't Work} \end{array} \begin{array}{cc} \text{Works} & \text{Doesn't Work} \\ \left[\begin{array}{cc} .95 & .05 \\ .85 & .15 \end{array} \right]. \end{array}$$

Find the new long-run probability that the line will work properly.

36. Natural Science   In Exercises 61 and 62 of Section 8.4 we discussed the effect on flower color of cross-pollinating pea plants. As shown there, since the gene for red is dominant and the gene for white is recessive, 75% of these pea plants have red flowers and 25% have white flowers, because plants with 1 red and 1 white gene appear red. If a red-flowered plant is crossed with a red-flowered plant known to have 1 red and 1 white gene, then 75% of the offspring will be red and 25% will be white. Crossing a red-flowered plant that has 1 red and 1 white gene with a white-flowered plant produces 50% red-flowered offspring and 50% white-flowered offspring.
(a) Write a transition matrix using this information.
(b) Write a probability vector for the initial distribution of colors.
(c) Find the distribution of colors after 4 generations.
(d) Find the long-range distribution of colors.

37. Natural Science   Snapdragons with 1 red gene and 1 white gene produce pink-flowered offspring. If a red snapdragon is crossed with a pink snapdragon, the probabilities that the offspring will be red, pink, or white are 1/2, 1/2, and 0, respectively. If 2 pink snapdragons are crossed, the probabilities of red, pink, or white offspring are 1/4, 1/2, and 1/4, respectively. For a cross between a white and a pink snapdragon, the corresponding probabilities are 0, 1/2, and 1/2. Set up a transition matrix and find the long-range prediction for the fraction of red, pink, and white snapdragons.

38. Many phenomena can be viewed as examples of a random walk. Consider the following simple example. A security guard can stand in front of any one of three doors 20 feet apart in front of a building, and every minute he decides whether to move to another door chosen at random. If he is at the middle door, he is equally likely to stay where he is, move to the door to the left, or move to the door to the right. If he is at the door on either end, he is equally likely to stay where he is or move to the middle door.

(a) Verify that the transition matrix is given by

$$\begin{array}{c} \\ 1 \\ 2 \\ 3 \end{array} \begin{array}{ccc} 1 & 2 & 3 \\ \left[\begin{array}{ccc} \frac{1}{2} & \frac{1}{2} & 0 \\ \frac{1}{3} & \frac{1}{3} & \frac{1}{3} \\ 0 & \frac{1}{2} & \frac{1}{2} \end{array} \right]. \end{array}$$

(b) Find the long-range trend for the fraction of time the guard spends in front of each door.

39. **Social Science** The probability that a homeowner will become a renter in five years is .03. The probability that a renter will become a homeowner in five years is .1. Suppose the proportions in the population are 64% homeowners (O), 35.5% renters (R), and .5% homeless (H), with the following transition matrix. Assume that these figures continue to apply.

$$\begin{array}{c} O \\ R \\ H \end{array} \begin{array}{ccc} O & R & H \\ \left[\begin{array}{ccc} .97 & .03 & 0 \\ .1 & .899 & .001 \\ 0 & .4 & .6 \end{array} \right] \end{array}$$

Find the long-range probabilities for the three categories.

40. **Business** An insurance company classifies its drivers into three groups: G_0 (no accidents), G_1 (one accident), and G_2 (more than one accident). The probability that a driver in G_0 will stay in G_0 after 1 year is .85, that he will become a G_1 is .10, and that he will become a G_2 is .05. A driver in G_1 cannot move to G_0 (this insurance company has a long memory). There is an .80 probability that a G_1 driver will stay in G_1 and a .20 probability that he will become a G_2. A driver in G_2 must stay in G_2.

(a) Write a transition matrix using this information.

Suppose that the company accepts 50,000 new policyholders, all of whom are in group G_0. Find the number in each group

(b) after 1 year; (c) after 2 years;

(d) after 3 years; (e) after 4 years.

(f) Find the equilibrium vector here. Interpret your result.

41. **Business** The difficulty with the mathematical model of Exercise 40 is that no "grace period" is provided; there should be a certain probability of moving from G_1 or G_2 back to G_0 (say, after 4 years with no accidents). A new system with this feature might produce the following transition matrix.

$$\left[\begin{array}{ccc} .85 & .10 & .05 \\ .15 & .75 & .10 \\ .10 & .30 & .60 \end{array} \right]$$

Suppose that when this new policy is adopted, the company has 50,000 policyholders in group G_0. Find the number of these in each group

(a) after 1 year; (b) after 2 years;

(c) after 3 years.

(d) Find the equilibrium vector here. Interpret your result.

42. **Business** Research done by the Gulf Oil Corporation produced the following transition matrix for the probability that a person with one form of home heating would switch to another.*

		Will Switch to		
		Oil	Gas	Electric
	Oil	.825	.175	0
Now Has	Gas	.060	.919	.021
	Electric	.049	0	.951

The current share of the market held by these three types of heat is given by [.26 .60 .14]. Find the share of the market held by each type of heat after

(a) 1 year; (b) 2 years; (c) 3 years.

(d) What is the long-range prediction?

43. The results of cricket matches between England and Australia have been found to be modeled by a Markov chain.[†] The probability that England wins, loses, or draws is based on the result of the previous game, with the following transition matrix.

	Wins	Loses	Draws
Wins	.443	.364	.193
Loses	.277	.436	.287
Draws	.266	.304	.430

(a) Compute the transition matrix for the game after the next one, based on the result of the last game.

(b) Use your answer from part (a) to find the probability that, if England won the last game, England will win the game after the next one.

(c) Use your answer from part (a) to find the probability that, if Australia won the last game, England will win the game after the next one.

44. **Social Science** At one liberal arts college, students are classified as humanities majors, science majors, or undecided. There is a 20% chance that a humanities major will change to a science major from one year to the next and a 45% chance that a humanities major will change to undecided. A science major will change to humanities with probability .15 and to undecided with probability .35. An undecided will switch to humanities or science with probabilities of .5 and .3, respectively. Find the long-range prediction for the fraction of students in each of these three majors.

---

*From "Forecasting Market Shares of Alternative Home-Heating Units by Markov Process Using Transition Probabilities Estimates from Aggregate Time Series Data" by Ali Ezzati, from *Management Science,* Vol. 21, No. 4, December 1974. Copyright © 1974 The Institute of Management Sciences. Reprinted by permission.

†From Derek Colwell, Brian Jones, and Jack Gillett, "A Markov Chain in Cricket," *Mathematical Gazette,* June 1991.

45. Business In the queuing chain, we assume that people are queuing up to be served by, say, a bank teller. For simplicity, let us assume that once two people are in line, no one else can enter the line. Let us further assume that one person is served every minute, as long as someone is in line. Assume further that in any minute, there is a probability of $\frac{1}{2}$ that no one enters the line, a probability of $\frac{1}{3}$ that exactly one person enters the line, and a probability of $\frac{1}{6}$ that exactly two people enter the line, assuming there is room. If there is not enough room for two people, then the probability that one person enters the line is $\frac{1}{2}$. Let the state be given by the number of people in line.

(a) Verify that the transition matrix is

$$
\begin{array}{c}
\ \ 0 \ \ \ 1 \ \ \ 2 \\
\begin{array}{c} 0 \\ 1 \\ 2 \end{array}
\begin{bmatrix}
\frac{1}{2} & \frac{1}{3} & \frac{1}{6} \\
\frac{1}{2} & \frac{1}{3} & \frac{1}{6} \\
0 & \frac{1}{2} & \frac{1}{2}
\end{bmatrix}.
\end{array}
$$

(b) Find the transition matrix for a two-minute period.

(c) Use your result from part (b) to find the probability that a queue with no one in line has two people in line two minutes later.

Use a graphing calculator or computer for Exercises 46 and 47.

46. Business A company with a new training program classified each employee in one of four states: s_1, never in the program; s_2, currently in the program; s_3, discharged; s_4, completed the program. The transition matrix for this company is given below.

$$
\begin{array}{c}
\ \ \ \ s_1 \ \ \ s_2 \ \ \ \ s_3 \ \ \ \ s_4 \\
\begin{array}{c} s_1 \\ s_2 \\ s_3 \\ s_4 \end{array}
\begin{bmatrix}
.4 & .2 & .05 & .35 \\
0 & .45 & .05 & .5 \\
0 & 0 & 1 & 0 \\
0 & 0 & 0 & 1
\end{bmatrix}
\end{array}
$$

(a) What percent of employees who had never been in the program (state s_1) completed the program (state s_4) after the program had been offered five times?

(b) If the initial percent of employees in each state was [.5 .5 0 0], find the corresponding percents after the program had been offered four times.

47. Business Find the long-range prediction for the percent of employees in each state for the company training program from Exercise 46.

9.6 DECISION MAKING

John F. Kennedy once remarked he had assumed that as president it would be difficult to choose between distinct, opposite alternatives when a decision needed to be made. Actually, however, he found that such decisions were easy to make; the hard decisions came when he was faced with choices that were not as clear-cut. Most decisions fall in this last category—decisions that must be made under conditions of uncertainty. In Section 9.4 we saw how to use expected values to help make a decision. These ideas are extended in this section, where we consider decision making in the face of uncertainty. We begin with an example.

► **EXAMPLE 1** Freezing temperatures are endangering the orange crop in central California. A farmer can protect his crop by burning smudge pots—the heat from the pots keeps the oranges from freezing. However, burning the pots is expensive; the cost is $4000. The farmer knows that if he burns smudge pots he will be able to sell his crop for a net profit (after smudge pot costs are deducted) of $10,000, provided that the freeze does develop and wipes out other orange crops in California. If he does nothing he will either lose $2000 already invested in the crop if it does freeze or make a profit of $9600 if it does not freeze. (If it does not freeze, there will be a large supply of oranges, and thus his profit will be lower than if there was a small supply.) What should the farmer do?

He should begin by carefully defining the problem. First, he must decide on the **states of nature,** the possible alternatives over which he has no control. Here there are two: freezing temperatures, or no freezing temperatures. Next, the farmer should list the things he can control—his actions or **strategies.** He has two possible strategies; to use smudge pots or not. The consequences of each action under each state

Explain how each of the following payoffs in the matrix were obtained.

(a) −$2000

(b) $10,000

Answers:

(a) If it freezes and smudge pots are not used, the farmer's profit is −$2000 for labor costs.

(b) If it freezes and smudge pots are used, the farmer makes a profit of $10,000.

of nature, called **payoffs,** are summarized in a **payoff matrix,** as shown below. The payoffs in this case are the profits for each possible combination of events.

$$
\begin{array}{cc}
 & \text{States of Nature} \\
\text{Strategies of Farmer} \begin{array}{l} \text{Use Smudge Pots} \\ \text{Do Not Use Pots} \end{array} & \begin{array}{cc} \text{Freeze} & \text{No Freeze} \\ \left[\begin{array}{cc} \$10,000 & \$5600 \\ -\$2000 & \$9600 \end{array}\right] \end{array}
\end{array}
$$

To get the $5600 entry in the payoff matrix, use the profit if there is no freeze, $9600, and subtract the $4000 cost of using the pots.

Once the farmer makes the payoff matrix, what then? The farmer might be an optimist (some might call him a gambler); in this case he might assume that the best will happen and go for the biggest number of the matrix ($10,000). For this profit, he must adopt the strategy "use smudge pots."

On the other hand, if the farmer is a pessimist, he would want to minimize the worst thing that could happen. If he uses smudge pots, the worst thing that could happen to him would be profit of $5600, which will result if there is no freeze. If he does not use smudge pots, he might face a loss of $2000. To minimize the worst, he once again should adopt the strategy "use smudge pots."

Suppose the farmer decides that he is neither an optimist nor a pessimist but would like further information before choosing a strategy. For example, he might call the weather forecaster and ask for the probability of a freeze. Suppose the forecaster says that this probability is only .1. What should the farmer do? He should recall the discussion of expected value and work out the expected profit for each of his two possible strategies. If the probability of a freeze is .1, then the probability that there is no freeze is .9. This information leads to the following expected values.

If smudge pots are used: $10,000(.1) + 5600(.9) = 6040$

If no smudge pots are used: $-2000(.1) + 9600(.9) = 8440$

Here the maximum expected profit, $8440, is obtained if smudge pots are not used.

What should the farmer do if the probability of a freeze is .6? What is his expected profit?

Answer:

Use smudge pots; $8240

As the example shows, the farmer's beliefs about the probabilities of a freeze affect his choice of strategies.

▶ **EXAMPLE 2**   A small manufacturer of Christmas cards must decide in February about the type of cards she should emphasize in her fall line. She has three possible strategies: emphasize modern cards, emphasize old-fashioned cards, or a mixture of the two. Her success is dependent on the state of the economy in December. If the economy is strong, she will do well with her modern cards, while in a weak economy people long for the old days and buy old-fashioned cards. In an in-between economy, her mixture of lines would do the best. She first prepares a payoff matrix for all three possibilities. The numbers in the matrix represent her profits in thousands of dollars.

$$
\begin{array}{cc}
 & \text{States of Nature} \\
\text{Strategies} \begin{array}{l} \text{Modern} \\ \text{Old-fashioned} \\ \text{Mixture} \end{array} & \begin{array}{ccc} \begin{array}{c} \text{Weak} \\ \text{Economy} \end{array} & \text{In-between} & \begin{array}{c} \text{Strong} \\ \text{Economy} \end{array} \\ \left[\begin{array}{ccc} 40 & 85 & 120 \\ 106 & 46 & 83 \\ 72 & 90 & 68 \end{array}\right] \end{array}
\end{array}
$$

(a) What would an optimist do?

If the manufacturer is an optimist, she should aim for the biggest number on the matrix, 120 (representing $120,000 in profit). Her strategy in this case would be to produce modern cards.

(b) How would a pessimist react?

A pessimistic manufacturer wants to find the best of the worst of all bad things that can happen. If she produces modern cards, the worst that can happen is a profit of $40,000. For old-fashioned cards, the worst is a profit of $46,000. From a mixture, the worst is a profit of $68,000. Her strategy here is to use a mixture.

(c) Suppose the manufacturer reads in a business magazine that leading experts believe there is a 50% chance of a weak economy at Christmas, a 20% chance of an in-between economy, and a 30% chance of a strong economy. How might she use this information?

The manufacturer can now find her expected profit for each possible strategy.

Modern: $40(.50) + 85(.20) + 120(.30) = 73$

Old-fashioned: $106(.50) + 46(.20) + 83(.30) = 87.1$

Mixture: $72(.50) + 90(.20) + 68(.30) = 74.4$

Here the best strategy is old-fashioned cards; the expected profit is 87.1, or $87,100. ◀ ③

③

Suppose the manufacturer reads another article which gives the following predictions: 35% chance of a weak economy, 25% chance of an in-between economy, and a 40% chance of a strong economy. What is the best strategy now? What is the expected profit?

Answer:

Modern; $83,250

9.6 Exercises

1. **Business** A developer has $100,000 to invest in land. He has a choice of two parcels (at the same price), one on the highway and one on the coast. With both parcels, his ultimate profit depends on whether he faces light opposition from environmental groups or heavy opposition. He estimates that the payoff matrix is as follows (the numbers represent his profit).

$$\begin{array}{c} \\ \text{Highway} \\ \text{Coast} \end{array} \begin{array}{cc} \textit{Opposition} & \\ \text{Light} & \text{Heavy} \\ \left[\begin{array}{cc} \$70{,}000 & \$30{,}000 \\ \$150{,}000 & -\$40{,}000 \end{array}\right] \end{array}$$

What should the developer do if he is
(a) an optimist?   **(b)** a pessimist?
(c) Suppose the probability of heavy opposition is .8. What is his best strategy? What is the expected profit?
(d) What is the best strategy if the probability of heavy opposition is only .4?

2. **Business** Hillsdale College has sold out all tickets for a jazz concert to be held in the stadium. If it rains, the show will have to be moved to the gym, which has a much smaller seating capacity. The dean must decide in ad-vance whether to set up the seats and the stage in the gym or in the stadium, or both, just in case. The payoff matrix below shows the net profit in each case.

$$\textit{Strategies} \begin{array}{c} \\ \text{Set up in Stadium} \\ \text{Set up in Gym} \\ \text{Set up Both} \end{array} \begin{array}{cc} \textit{States of Nature} & \\ \text{Rain} & \text{No Rain} \\ \left[\begin{array}{cc} -\$1550 & \$1500 \\ \$1000 & \$1000 \\ \$750 & \$1400 \end{array}\right] \end{array}$$

What strategy should the dean choose if she is
(a) an optimist?   **(b)** a pessimist?
(c) If the weather forecaster predicts rain with a probability of .6, what strategy should she choose to maximize expected profit? What is the maximum expected profit?

3. **Business** An analyst must decide what fraction of the items produced by a certain machine are defective. He has already decided that there are three possibilities for the fraction of defective items: .01, .10, and .20. He may recommend two courses of action: repair the machine or make no repairs. The payoff matrix below represents the *costs* to the company in each case, in hundreds of dollars.

$$\begin{array}{c} & & \textit{Defectives} \\ & & .01 \quad .10 \quad .20 \\ \textit{Strategies} & \begin{array}{l} \text{Repair} \\ \text{No Repair} \end{array} & \begin{bmatrix} 130 & 130 & 130 \\ 25 & 200 & 500 \end{bmatrix} \end{array}$$

What strategy should the analyst recommend if he is
(a) an optimist?   (b) a pessimist?
(c) Suppose the analyst is able to estimate probabilities for the three states of nature as follows.

Fraction of Defectives	Probability
.01	.70
.10	.20
.20	.10

Which strategy should he recommend? Find the expected cost to the company if this strategy is chosen.

4. **Business** The research department of the Allied Manufacturing Company has developed a new process that it believes will result in an improved product. Management must decide whether or not to go ahead and market the new product. The new product may be better than the old or it may not be better. If the new product is better and the company decides to market it, sales should increase by $50,000. If it is not better and they replace the old product with the new product on the market, they will lose $25,000 to competitors. If they decide not to market the new product they will lose $40,000 if it is better, and research costs of $10,000 if it is not.
(a) Prepare a payoff matrix.
(b) If management believes the probability that the new product is better to be .4, find the expected profits under each strategy and determine the best action.

5. **Business** A businessman is planning to ship a used machine to his plant in Nigeria. He would like to use it there for the next 4 years. He must decide whether or not to overhaul the machine before sending it. The cost of overhaul is $2600. If the machine fails when in operation in Nigeria, it will cost him $6000 in lost production and repairs. He estimates the probability that it will fail at .3 if he does not overhaul it and .1 if he does overhaul it. Neglect the possibility that the machine might fail more than once in the 4 years.
(a) Prepare a payoff matrix.
(b) What should the businessman do to minimize his expected costs?

6. **Business** A contractor prepares to bid on a job. If all goes well, his bid should be $30,000, which will cover his costs plus his usual profit margin of $4500. However, if a threatened labor strike actually occurs, his bid should be $40,000 to give him the same profit. If there is a strike

and he bids $30,000, he will lose $5500. If his bid is too high, he may lose the job entirely, while if it is too low, he may lose money.
(a) Prepare a payoff matrix.
(b) If the contractor believes that the probability of a strike is .6, how much should he bid?

7. **Natural Science** A community is considering an anti-smoking campaign.* The city council will choose one of three possible strategies: a campaign for everyone over age 10 in the community, a campaign for youths only, or no campaign at all. The two states of nature are a true cause-effect relationship between smoking and cancer and no cause-effect relationship. The costs to the community (including loss of life and productivity) in each case are as shown at the top of the next column.

	States of Nature	
Strategies	Cause-Effect Relationship	No Cause-Effect Relationship
Campaign for All	$100,000	$800,000
Campaign for Youth	$2,820,000	$20,000
No Campaign	$3,100,100	$0

What action should the city council choose if it is
(a) optimistic?   (b) pessimistic?
(c) If the Director of Public Health estimates that the probability of a true cause-effect relationship is .8, which strategy should the city council choose?

8. **Business** An investor has $20,000 to invest in stocks. She has two possible strategies: buy conservative blue-chip stocks or buy highly speculative stocks. There are two states of nature: the market goes up or the market goes down. The following payoff matrix shows the net amounts she will have under the various circumstances.

	Market Up	Market Down
Buy Blue-Chip	$25,000	$18,000
Buy Speculative	$30,000	$11,000

What should the investor do if she is
(a) an optimist?   (b) a pessimist?
(c) Suppose there is a .7 probability of the market going up. What is the best strategy? What is the expected profit?
(d) What is the best strategy if the probability of a market rise is .2?

Sometimes the numbers (or payoffs) in a payoff matrix do not represent money (profits or costs, for example), but utility. A utility is a number that measures the satisfaction (or lack of it) that results from a certain action. The numbers must be assigned by each individual, depending on how he or she feels about a

---

*This problem is based on an article by B. G. Greenberg in the September 1969 issue of the *Journal of the American Statistical Association.*

situation. For example, one person might assign a utility of $+20$ for a week's vacation in San Francisco, with -6 being assigned if the vacation were moved to Sacramento. Work the following problems in the same way as those above.

9. *Social Science* A politician must plan her reelection strategy. She can emphasize jobs or she can emphasize the environment. The voters can be concerned about jobs or about the environment. A payoff matrix showing the utility of each possible outcome is shown below.

$$
\begin{array}{c}
 & & \textit{Voters} \\
 & & \text{Jobs} \quad \text{Environment} \\
\textit{Candidate} & \begin{array}{c} \text{Jobs} \\ \text{Environment} \end{array} & \begin{bmatrix} +25 & -10 \\ -15 & +30 \end{bmatrix}
\end{array}
$$

The political analysts feel that there is a .35 chance that the voters will emphasize jobs. What strategy should the candidate adopt? What is its expected utility?

10. In an accounting class, the instructor permits the students to bring a calculator or a reference book (but not both) to an examination. The examination itself can emphasize either numbers or definitions. In trying to decide which aid to take to an examination, a student first decides on the utilities shown in the following payoff matrix.

$$
\begin{array}{c}
 & & \textit{Exam Emphasizes} \\
 & & \text{Numbers} \quad \text{Definitions} \\
\textit{Student Chooses} & \begin{array}{c} \text{Calculator} \\ \text{Book} \end{array} & \begin{bmatrix} +50 & 0 \\ +10 & +40 \end{bmatrix}
\end{array}
$$

(a) What strategy should the student choose if the probability that the examination will emphasize numbers is .6? What is the expected utility in this case?

(b) Suppose the probability that the examination emphasizes numbers is .4. What strategy should be chosen by the student?

CHAPTER 9   SUMMARY

Key Terms and Symbols

9.1 $n!$ n factorial
multiplication principle
permutations
combinations
9.3 Bernoulli trials (processes)
binomial experiment
binomial probability

9.4 random variable
probability distribution
histogram
expected value
fair game
9.5 stochastic processes
Markov chain
state
transition diagram

transition matrix
probability vector
regular transition matrix
regular Markov chain
equilibrium vector (fixed vector)
9.6 states of nature
strategies
payoffs
payoff matrix

Key Concepts

Multiplication Principle: If there are m_1 ways to make a first choice, m_2 ways to make a second choice, and so on, where none of the choices depend on any of the others, then there are $m_1 m_2 \cdots m_n$ different ways to make the entire sequence of choices.

The number of **permutations** of n elements taken r at a time is $_nP_r = \dfrac{n!}{(n-r)!}$.

The number of **combinations** of n elements taken r at a time is

$$
_nC_r = \binom{n}{r} = \frac{n!}{(n-r)!\,r!}.
$$

Binomial experiments have the following characteristics: The same experiment is repeated several times. There are only *two* outcomes, labeled success and failure. The trials are independent so that the probability of success is the same for each trial. If the probability of success in a single trial is p, then the probability of x successes in n trials is

$$\binom{n}{x} p^x (1 - p)^{n-x}.$$

Expected Value of a Probability Distribution: For a random variable x with values x_1, $x_2, \ldots, x_n$ and probabilities $p_1, p_2, \ldots, p_n$, respectively, the expected value is

$$E(x) = x_1 p_1 + x_2 p_2 + \cdots + x_n p_n.$$

Markov Chains: A **transition matrix** must be square, with all entries between 0 and 1 inclusive, and the sum of the entries in any row must be 1. A Markov chain is *regular* if some power of its transition matrix P contains all positive entries. The long-range probabilities for a regular Markov chain are given by the **equilibrium** or **fixed vector** V, where for any initial probability vector v, the products vP^n approach V as n gets larger and larger, and $VP = V$. To find V, solve the system of equations formed by $VP = V$ and the fact that the sum of the entries of V is 1.

Decision Making: A **payoff matrix** which includes all available strategies and states of nature is used in decision making to define the problem and the possible solutions. The expected value of each strategy can help to determine the best course of action.

Chapter 9 Review Exercises

1. In how many ways can 5 shuttle vans line up at the airport?

2. How many variations in first-, second-, and third-place finishes are possible in a 100-yard dash with 7 runners?

3. In how many ways can a sample of 3 pears be taken from a basket containing a dozen pears?

4. If 2 of the pears in Exercise 3 are spoiled, in how many ways can the sample of 3 include the following?
   (a) 1 spoiled pear   (b) No spoiled pears
   (c) At most 1 spoiled pear

5. In how many ways can 3 pictures, selected from a group of 6 pictures, be arranged in a row on a wall?

6. In how many ways can the 6 pictures in Exercise 5 be arranged in a row, if a certain one must be first?

7. In how many ways can the 6 pictures in Exercise 5 be arranged if 3 are landscapes and 3 are puppies, and if
   (a) like types must be kept together?
   (b) landscapes and puppies are alternated?

8. A representative is to be selected from each of 3 departments in a large company. There are 7 people in the first department, 5 in the second department, and 8 in the third department.
   (a) How many different groups of 3 representatives are possible?

   (b) How many groups are possible, if any number (at least 1) up to 3 representatives can form a group?

9. Explain under what circumstances a permutation should be used in a probability problem and under what circumstances a combination should be used.

10. Discuss under what circumstances the binomial probability formula should be used in a probability problem.

Suppose 2 cards are drawn without replacement from an ordinary deck of 52 cards. Find the probabilities of the following results.

11. Both cards are black.   12. Both cards are hearts.

13. Exactly 1 is a face card.   14. At most 1 is an ace.

A collection of golf balls contains 4 yellow, 2 blue, and 6 white balls. A golfer selects 3 balls at random. Find the probability that the selection includes the following.

15. All white balls   16. All yellow balls

17. At least 1 blue ball   18. 1 ball of each color

19. In this exercise we study the connection between sets (from Chapter 8) and combinations.
   (a) Given a set with n elements, what is the number of subsets of size 0? of size 1? of size 2? of size n?
   (b) Using your answer from part (a), give an expression for the total number of subsets of a set with n elements.

(c) Using your answer from part (b) and a result from Chapter 8, explain why the following equation must be true:

$$\binom{n}{0} + \binom{n}{1} + \binom{n}{2} + \cdots + \binom{n}{n} = 2^n.$$

(d) Verify the equation in part (c) for $n = 4$ and $n = 5$.

Business *A bottle capping machine has an error rate of .01. A random sample of 20 bottles is selected. Find the following probabilities.*

20. Exactly 4 bottles are improperly capped.

21. No more than 3 bottles are improperly capped.

22. At least 1 bottle is improperly capped.

Business *A certain machine that is used to manufacture screws produces a defect rate of .01. A random sample of 20 screws is selected. Find the probabilities that the sample contains the following.*

23. Exactly 4 defective screws

24. Exactly 3 defective screws

25. No more than 4 defective screws

26. Set up the probability that the sample has 12 or more defective screws. (Do not evaluate.)

27. **Business** A developer can buy a piece of property that will produce a profit of $16,000 with probability .7 or a loss of $9000 with probability .3. What is the expected profit?

In Exercises 28–32, (a) give a probability distribution, (b) *sketch its histogram, and (c)* find the expected value.

28. A pair of dice is rolled and the sum of the results for each roll is recorded.

29. A coin is tossed 4 times and the number of tails is recorded.

30. **Social Science** In 1992, the Big 10 collegiate sports conference moved to have women compose at least 40% of its athletes within 5 years.* Suppose they exactly achieved the 40% figure and that 5 athletes are picked at random from Big 10 universities. The number of women is recorded.

31. **Social Science** In the mathematics honors society at a college, 2 of the 8 members are African American. Three members are selected at random to be interviewed by the student newspaper, and the number of African Americans is noted.

32. **Social Science** In a small class of 10 students, 3 did not do their homework. The professor selects half of the class to present solutions to homework problems on the board, and records how many of those selected did not do their homework.

Solve the following problems.

33. Suppose someone offers to pay you $100 if you draw 3 cards from a standard deck of 52 cards and all the cards are clubs. What should you pay for the chance to win if it is a fair game?

34. You pay $6 to play in a game where you will roll a die, with payoffs as follows: $8 for a 6, $7 for a 5, and $4 for any other results. What are your expected winnings? Is the game fair?

35. A lottery has a first prize of $5000, two second prizes of $1000 each, and two $100 third prizes. Ten thousand tickets are sold, at $1 each. Find the expected winnings of a person buying 1 ticket.

36. Find the expected number of girls in a family of 5 children.

37. Three cards are drawn from a standard deck of 52 cards.
(a) What is the expected number of aces?
(b) What is the expected number of clubs?

Decide whether each of the following is a regular transition matrix.

38. $\begin{bmatrix} 0 & 1 \\ .8 & .2 \end{bmatrix}$ **39.** $\begin{bmatrix} -.1 & .4 \\ .3 & .7 \end{bmatrix}$

40. $\begin{bmatrix} .3 & 0 & .7 \\ .5 & .1 & .4 \\ 1 & 0 & 0 \end{bmatrix}$ **41.** $\begin{bmatrix} .2 & .3 & .5 \\ .4 & .4 & .2 \\ .5 & .1 & .4 \end{bmatrix}$

42. **Business** A credit card company classified its customers in three groups: nonusers in a given month, light users, and heavy users. The transition matrix for these states is

	Nonuser	Light	Heavy
Nonuser	.8	.15	.05
Light	.25	.55	.2
Heavy	.04	.21	.75

Suppose the initial distribution for the three states is [.4 .4 .2]. Find the distribution after
(a) 1 month;   **(b)** 2 months.
(c) What is the long-range prediction for the distribution of users?

43. **Business** Savmor Investments starts a heavy advertising campaign. At the start of the campaign, Savmor sells 35% of all mutual funds sold in the area, while 65% are sold by the Highrate Company. The campaigns produces the following transition matrix.

		After campaign	
		Savmor	Highrate
Before Campaign	Savmor	.8	.2
	Highrate	.4	.6

(a) Find the market share for each company after the campaign.
(b) Find the share of the market for each company after three such campaigns.
(c) Predict the long-range market share for Savmor.

*NCTM News Bulletin, February 1995, p. 5.

44. Business In labor-management relations, both labor and management can adopt either a friendly or a hostile attitude. The results are shown in the following payoff matrix. The numbers give the wage gains made by an average worker.

Management

		Friendly	Hostile
Labor	Friendly	$600	$800
	Hostile	$400	$950

(a) Suppose the chief negotiator for labor is an optimist. What strategy should he choose?

(b) What strategy should he choose if he is a pessimist?

(c) The chief negotiator for labor feels that there is a 70% chance that the company will be hostile. What strategy should he adopt? What is the expected payoff?

(d) Just before negotiations begin, a new management is installed in the company. There is only a 40% chance that the new management will be hostile. What strategy should be adopted by labor?

45. Social Science A candidate for city council can come out in favor of a new factory, be opposed to it, or waffle on the issue. The change in votes for the candidate depends on what her opponent does, with payoffs as shown.

Opponent

		Favors	Waffles	Opposes
Candidate	Favors	0	−1000	−4000
	Waffles	1000	0	−500
	Opposes	5000	2000	0

(a) What should the candidate do if she is an optimist?

(b) What should she do if she is a pessimist?

(c) Suppose the candidate's campaign manager feels there is a 40% chance that the opponent will favor the plant and a 35% chance that he will waffle. What strategy should the candidate adopt? What is the expected change in the number of votes?

(d) The opponent conducts a new poll that shows strong opposition to the new factory. This changes the probability he will favor the factory to 0 and the probability he will waffle to .7. What strategy should our candidate adopt? What is the expected change in the number of votes now?

Exercises 46 and 47 are taken from actuarial examinations given by the Society of Actuaries. *

46. Business A company is considering the introduction of a new product that is believed to have probability .5 of being successful and probability .5 of being unsuccessful.

Problem from "Course 130 Examination, Operations Research," of the Education and Examination Committee of the Society of Actuaries. Reprinted by permission of the Society of Actuaries.

Successful products pass quality control 80% of the time. Unsuccessful products pass quality control 25% of the time. If the product is successful, the net profit to the company will be $40 million; if unsuccessful, the net loss will be $15 million. Determine the expected net profit if the product passes quality control.

(a) $23 million   (b) $24 million

(c) $25 million   (d) $26 million

(e) $27 million

47. Business A merchant buys boxes of fruit from a grower and sells them. Each box of fruit is either Good or Bad. A Good box contains 80% excellent fruit and will earn $200 profit on the retail market. A Bad box contains 30% excellent fruit and will produce a loss of $1000. The a priori probability of receiving a Good box of fruit is .9. Before the merchant decides to put the box on the market, he can sample one piece of fruit to test whether it is excellent. Based on that sample, he has the option of rejecting the box without paying for it. Determine the expected value of the right to sample. (*Hint:* If the merchant samples the fruit, what are the probabilities of accepting a Good box, accepting a Bad box, and not accepting the box? What are these probabilities if he does not sample the fruit?)

(a) 0   (b) $16   (c) $34

(d) $72   (e) $80

48. Business The March 1982 issue of *Mathematics Teacher* included "Overbooking Airline Flights," an article by Joe Dan Austin. In this article, Austin developed a model for the expected income for an airline flight. With appropriate assumptions, the probability that exactly x of n people with reservations show up at the airport to buy a ticket is given by the binomial probability formula. Assume the following: 6 reservations have been accepted for 3 seats, $p = .6$ is the probability that a person with a reservation will show up, a ticket costs $100, and the airline must pay $100 to anyone with a reservation who does not get a ticket. Complete the following table.

Number Who Show Up (x)	0	1	2	3	4	5	6
Airline's Income							
P(x)							

(a) Use the table to find $E(I)$, the expected airline income from the 3 seats.

(b) Find $E(I)$ for $n = 3$, $n = 4$, and $n = 5$. Compare these answers with $E(I)$ for $n = 6$. For these values of n, how many reservations should the airline book for the 3 seats in order to maximize the expected revenue?

Case 8

Optimal Inventory for a Service Truck

For many different items it is difficult or impossible to take the item to a central repair facility when service is required. Washing machines, large television sets, office copiers, and computers are only a few examples of such items. Service for items of this type is commonly performed by sending a repair person to the item, with the person driving to the location in a truck containing various parts that might be required in repairing the item. Ideally, the truck should contain all the parts that might be required. However, most parts would be needed only infrequently, so that inventory costs for the parts would be high.

An optimum policy for deciding on which parts to stock on a truck would require that the probability of not being able to repair an item without a trip back to the warehouse for needed parts be as low as possible, consistent with minimum inventory costs. An analysis similar to the one below was developed at the Xerox Corporation.*

To set up a mathematical model for deciding on the optimum truck-stocking policy, let us assume that a broken machine might require one of 5 different parts (we could assume any number of different parts—we use 5 to simplify the notation). Suppose also that the probability that a particular machine requires part 1 is p_1; that it requires part 2 is p_2; and so on. Assume also that failures of different part types are independent, and that at most one part of each type is used on a given job.

Suppose that, on the average, a repair person makes N service calls per time period. If the repair person is unable to make a repair because at least one of the parts is unavailable, there is a penalty cost, L, corresponding to wasted time for the repair person, an extra trip to the parts depot, customer unhappiness, and so on. For each of the parts carried on the truck, an average inventory cost is incurred. Let H_i be the average inventory cost for part i, where $1 \leq i \leq 5$.

Let M_1 represent a policy of carrying only part 1 on the repair truck, M_{24} represent a policy of carrying only parts 2 and 4, with M_{12345} and M_0 representing policies of carrying all parts and no parts, respectively.

For policy M_{35}, carrying parts 3 and 5 only, the expected cost per time period per repair person, written $C(M_{35})$, is

$$C(M_{35}) = (H_3 + H_5) + NL[1 - (1 - p_1)(1 - p_2)(1 - p_4)].$$

(The expression in brackets represents the probability of needing at least one of the parts not carried, 1, 2, or 4 here.) As further examples,

$$C(M_{125}) = (H_1 + H_2 + H_5) + NL[1 - (1 - p_3)(1 - p_4)],$$

while

$$C(M_{12345}) = (H_1 + H_2 + H_3 + H_4 + H_5) + NL[1 - 1]$$
$$= H_1 + H_2 + H_3 + H_4 + H_5,$$

and

$$C(M_0) = NL[1 - (1 - p_1)(1 - p_2)(1 - p_3)(1 - p_4)(1 - p_5)].$$

To find the best policy, evaluate $C(M_0)$, $C(M_1)$, . . . , $C(M_{12345})$, and choose the smallest result. (A general solution method is in the *Management Science* paper.)

▶ **EXAMPLE 1** Suppose that for a particular item, only 3 possible parts might need to be replaced. By studying past records of failures of the item, and finding necessary inventory costs, suppose that the following values have been found.

p_1	p_2	p_3
.09	.24	.17

H_1	H_2	H_3
$15	$40	$9

*Reprinted by permission of Stephen Smith, John Chambers, and Eli Shlifer. "Optimal Inventories Based on Job Completion Rate for Repairs Requiring Multiple Items." *Management Science*, Vol. 26, No. 8, August 1980, copyright © 1980 by The Institute of Management Sciences.

Suppose $N = 3$ and L is \$54. Then, as an example,

$$
\begin{aligned}
C(M_1) &= H_1 + NL[1 - (1 - p_2)(1 - p_3)] \\
&= 15 + 3(54)[1 - (1 - .24)(1 - .17)] \\
&= 15 + 3(54)[1 - (.76)(.83)] \\
&\approx 15 + 59.81 \\
&= 74.81.
\end{aligned}
$$

Thus, if policy M_1 is followed (carrying only part 1 on the truck), the expected cost per repair person per time period is \$74.81. Also,

$$
\begin{aligned}
C(M_{23}) &= H_2 + H_3 + NL[1 - (1 - p_1)] \\
&= 40 + 9 + 3(54)(.09) \\
&= 63.58,
\end{aligned}
$$

so that M_{23} is a better policy than M_1. By finding the expected values for all other possible policies (see the exercises), the optimum policy may be chosen. ◀

Exercises

1. Refer to the example and find each of the following.
   (a) $C(M_0)$ (b) $C(M_2)$ (c) $C(M_3)$
   (d) $C(M_{12})$ (e) $C(M_{13})$ (f) $C(M_{123})$
2. Which policy leads to the lowest expected cost?
3. In Example 1, $p_1 + p_2 + p_3 = .09 + .24 + .17 = .50$. Why is it not necessary that the probabilities add up to 1?
4. Suppose an item to be repaired might need one of n different parts. How many different policies would then need to be evaluated?

Introduction to Statistics

Statistics is a branch of mathematics that deals with the collection and summarization of data. Methods of statistical analysis make it possible to draw conclusions about a population based on data from a sample of the population. Statistical models have become increasingly useful in manufacturing, government, agriculture, medicine, and the social sciences and in all types of research. An Indianapolis race-car team is using statistics to improve performance by gathering data on each run around the track. They sample data 300 times a second and use computers to process the data. In this chapter we give a brief introduction to some of the key topics from statistical theory.

10.1   FREQUENCY DISTRIBUTIONS AND MEASURES OF CENTRAL TENDENCY

Often, a researcher wishes to learn something about a characteristic of a population, but because the population is very large or mobile, it is not possible to examine all of its elements. Instead, a limited sample drawn from the population is studied to determine the characteristics of the population. For these inferences to be correct, the sample chosen must be a **random sample.** Random samples are representative of the population because they are chosen so that every element of the population is equally likely to be selected. For example, a hand dealt from a well-shuffled deck of cards is a random sample.

A random sample can be difficult to obtain in real life. For example, suppose you want to take a random sample of voters in your congressional district to see which candidate they prefer in the next election. If you do a telephone survey, you have a random sample of people who are at home to answer the telephone, under-representing those who work a lot of hours and are rarely home to answer the phone, or those who have an unlisted number, or those who cannot afford a telephone, or those who refuse to answer telephone surveys. Such people may have a different opinion than those you interview.

A famous example of an inaccurate poll was made by the *Literary Digest* in 1936. Their survey indicated that Alfred Landon would win the presidential election; in fact, Franklin Roosevelt won with 62% of the popular vote. The *Digest's* major error was mailing their surveys to a sample of those listed in telephone directories. During the depression, many poor people did not have telephones, and the poor voted overwhelmingly for Roosevelt. Modern pollsters use sophisticated techniques to ensure that their sample is as random as possible.

Once a sample has been chosen and all data of interest are collected, the data must be organized so that conclusions may be more easily drawn. One method of organization is to group the data into intervals; equal intervals are usually chosen.

▶ **EXAMPLE 1** A survey asked a random sample of 30 business executives for their recommendations as to the number of college units in management that a business major should have. The results are shown below. Group the data into intervals and find the frequency of each interval.

3	25	22	16	0	9	14	8	34	21
15	12	9	3	8	15	20	12	28	19
17	16	23	19	12	14	29	13	24	18

The highest number in the list is 34 and the lowest is 0; one convenient way to group the data is in intervals of size 5, starting with 0–4 and ending with 30–34. This gives an interval for each number in the list and results in seven equal intervals of a convenient size. Too many intervals of smaller size would not simplify the data enough, while too few intervals of larger size would conceal information that the data might provide. A rule of thumb is to use from six to fifteen intervals.

First tally the number of college units falling into each interval. Then total the tallies in each interval, as in the table below. This table is an example of a **grouped frequency distribution**.

An accounting firm selected 24 complex tax returns prepared by a certain tax preparer. The number of errors per return were as follows.

8	12	0	6	10	8	0	14
8	12	14	16	4	14	7	11
9	12	7	15	11	21	22	19

Prepare a grouped frequency distribution for this data. Use intervals 0–4, 5–9, and so on.

Answer:

Interval	Frequency
0–4	3
5–9	7
10–14	9
15–19	3
20–24	2
	Total: 24

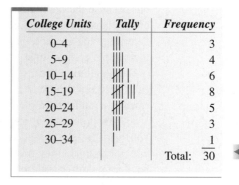

College Units	Tally	Frequency
0–4	|||	3
5–9	||||	4
10–14	ⅣⅢ		6
15–19	ⅣⅢ |||	8
20–24	ⅣⅢ	5
25–29	|||	3
30–34			1
		Total: 30

The frequency distribution in Example 1 shows information about the data that might not have been noticed before. For example, the interval with the largest number of units is 15–19, and 19 executives (more than half) had between 9 and 25 units. Also, the frequency in each interval increases rather evenly (up to 8) and then decreases at about the same pace. However, some information has been lost; for example, we no longer know how many executives had 12 units.

Picturing Data

The information in a grouped frequency distribution can be displayed in a histogram similar to the ones for probability distributions in the previous chapter. The intervals determine the widths of the bars; if equal intervals are used, all the bars have the same width. The heights of the bars are determined by the frequencies.

A **frequency polygon** is another form of graph that illustrates a grouped frequency distribution. The polygon is formed by joining consecutive midpoints of the tops of the histogram bars with straight line segments. Sometimes the midpoints of the first and last bars are joined to endpoints on the horizontal axis where the next midpoint would appear. (See Figure 10.1)

▶ **EXAMPLE 2**   A grouped frequency distribution of suggested college units was found in Example 1. Draw a histogram and a frequency polygon for this distribution.

First, draw a histogram, shown in blue in Figure 10.1. To get a frequency polygon, connect consecutive midpoints of the tops of the bars. The frequency polygon is shown in red.   ◀   ②

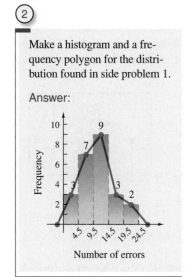

②

Make a histogram and a frequency polygon for the distribution found in side problem 1.

Answer:

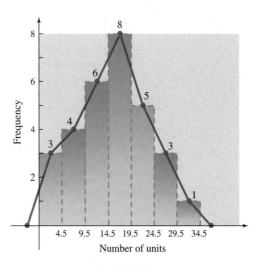

FIGURE 10.1

TECHNOLOGY TIP   As noted in Section 9.4, most graphing calculators can display histograms. Many will also display frequency polygons (which are usually labeled LINE or xyLINE in calculator menus). When dealing with grouped frequency distributions, however, certain adjustments must be made on a calculator.

1. *A calculator list of outcomes must consist of single numbers, not intervals.* The table in Example 1, for example, cannot be entered as shown. To convert the first column of the table for calculator use, choose one number in each interval, say 2 in the interval 0−4, 7 in the interval 5−9, 12 in the interval 10−14, etc. Then use 2, 7, 12, . . . as the list of outcomes to be entered into the calculator. The frequency list (the last column of the table) remains the same.

2. *The histogram bar width affects the shape of the graph.* If you use a bar width of 4 in Example 1, the calculator may produce a histogram with gaps in it. To avoid this, use the interval $0 \leq x < 5$ in place of $0 \leq x \leq 4$, and similarly for the other intervals and make 5 the bar width.

Following this procedure, we obtain the calculator-generated histogram and frequency polygon in Figure 10.2 for the data from Example 1. Note that the width of each histogram bar is 5. Some calculators cannot display both the histogram and the frequency polygon on the same screen as is done here.

FIGURE 10.2

Stem-and-leaf plots allow us to organize the data in a distribution without the disadvantage of losing the original information. In a **stem-and-leaf plot,** we separate the digits in each of the data into two parts consisting of the first one or two digits (the stem) and the other digits (the leaf).

▶ **EXAMPLE 3**   Construct a stem-and-leaf plot for the data in Example 1.

Since the data are two-digit numbers, we use the first digit for the stems: 0, 1, 2, and 3. The second digits provide the leaves. For example, as shown below, the first stem 0 has leaves 0, 3, 3 corresponding to the entries 3 and 0 in the first row of the data list and 3 in the second row. Here, each stem corresponds to an interval in the frequency table. The stems and leaves are separated by a vertical line.

Stem	Leaves
0	033
0	8899
1	222344
1	55667899
2	01234
2	589
3	4

Make a stem-and-leaf plot for the data in Example 1 using one stem for 0, 1, 2, and 3.

Answer:

Stem	Leaves
0	0338899
1	22234455667899
2	01234589
3	4

If we turn the page on its side, the distribution looks like a histogram but still retains each of the original values. We used each stem digit twice, because, as with a histogram, too few intervals conceal useful information about the shape of the distribution. ◀ ③

▶ **EXAMPLE 4** List the original data for the stem-and-leaf plot shown below.

Stem	Leaves
15	258
16	23469
17	145668
18	0137

④

List the original data for the following stem-and-leaf plot.

Stem	Leaves
4	36
5	2589
6	244678
7	1248
8	5

Answer:

43, 46, 52, 55, 58, 59, 62, 64, 64, 66, 67, 68, 71, 72, 74, 78, 85

The first stem and its leaves correspond to the data 152, 155, and 158. Similarly, the rest of the data are 162, 163, 164, 166, 169, 171, 174, 175, 176, 176, 178, 180, 181, 183, and 187. ◀ ④

NOTE The remainder of this section deals with topics that are generally referred to as "measures of central tendency." Computing these various measures is greatly simplified by the statistical capabilities of most scientific and graphing calculators. Calculators vary considerably in how data is entered, so read your instruction manual to learn how to enter lists of data and the corresponding frequencies. On scientific calculators with statistical capabilities, there are keys for finding most of the measures of central tendency discussed below. On graphing calculators, most or all of these measures can be obtained with a single keystroke (look for *one-variable statistics,* which is often labeled 1-VAR, in the STAT menu or its CALC submenu).

Mean

Three measures of central tendency, or "averages," are used with frequency distributions: the mean, the median, and the mode. The most important of these is the mean, which is similar to the expected value of a probability distribution. The **arithmetic mean** (the mean) of a set of numbers is the sum of the numbers, divided by the total number of numbers. We write the sum of n numbers $x_1, x_2, x_3, \ldots, x_n$ in a compact way using **summation notation,** also called **sigma notation.** With the Greek letter Σ (sigma), the sum

$$x_1 + x_2 + x_3 + \cdots + x_n$$

is written

$$x_1 + x_2 + x_3 + \cdots + x_n = \sum_{i=1}^{n} x_i.$$

In statistics, $\sum_{i=1}^{n} x_i$ is often abbreviated as just Σx. The symbol $\bar{x}$ (read "x-bar") is used to represent the mean of a sample.

Mean

The mean of the n numbers $x_1, x_2, x_3, \ldots, x_n$ is

$$\bar{x} = \frac{x_1 + x_2 + \cdots + x_n}{n} = \frac{\Sigma x}{n}.$$

▶ **EXAMPLE 5**   The numbers of bankruptcy petitions (in thousands) filed in the United States in the fiscal years (1994–1999) are given in the table below.* Find the mean number of bankruptcy petitions filed annually during this period.

Year	Petitions Filed (in thousands)
1994	845
1995	858
1996	1042
1997	1317
1998	1429
1999	1392

Let $x_1 = 845$, $x_2 = 858$, and so on. Here, $n = 6$, since there are 6 numbers in the list.

$$\bar{x} = \frac{845 + 858 + 1042 + 1317 + 1429 + 1392}{6} \approx 1147$$

The mean number of bankruptcy petitions filed during the given years is 1,147,000.

◀ ⑤

⑤

Find the mean of the following list of sales at a boutique.

$25.12	$42.58
$76.19	$32
$81.11	$26.41
$19.76	$59.32
$71.18	$21.03

Answer:

$45.47

TECHNOLOGY TIP The mean of the six numbers in Example 5 is easily found by using the $\bar{x}$ key on a scientific calculator or the one-variable statistics key on a graphing calculator. A graphing calculator will also display additional information, which will be discussed in the next section.

---

*Administrative Offices of the U.S. Courts, *Statistical Tables for the Federal Judiciary.*

The mean of data that have been arranged in a frequency distribution is found in a similar way. For example, suppose the following data is collected.

Value	Frequency
84	2
87	4
88	7
93	4
99	3
Total:	20

The value 84 appears twice, 87 four times, and so on. To find the mean, first add 84 two times, 87 four times, and so on; or get the same result faster by multiplying 84 by 2, 87 by 4, and so on, and then by adding the results. Dividing the sum by 20, the total of the frequencies, gives the mean.

$$\bar{x} = \frac{(84 \cdot 2) + (87 \cdot 4) + (88 \cdot 7) + (93 \cdot 4) + (99 \cdot 3)}{20}$$

$$= \frac{168 + 348 + 616 + 372 + 297}{20}$$

$$= \frac{1801}{20}$$

$$\bar{x} = 90.05$$

Verify that your calculator gives the same result.

▶ **EXAMPLE 6** Find the mean for the data shown in the following frequency distribution.

Value	Frequency	Value × Frequency
30	6	$30 \cdot 6 = 180$
32	9	$32 \cdot 9 = 288$
33	7	$33 \cdot 7 = 231$
37	12	$37 \cdot 12 = 444$
42	6	$42 \cdot 6 = 252$
Total:	40	Total: 1395

The value 30 appears six times, 32 nine times, and so on. To find the mean, first multiply 30 times 6, 32 times 9, and so on, to get the new column "Value × Frequency,"

Find $\bar{x}$ for the following frequency distribution.

Value	Frequency
7	2
9	3
11	6
13	4
15	1
17	4

Answer:

$\bar{x} = 12.1$

that has been added to the frequency distribution. Adding the products from this column gives a total of 1395. The total from the frequency column is 40. The mean is

$$\bar{x} = \frac{1395}{40} = 34.875. \quad \blacktriangleleft \quad \text{\textcircled{6}}$$

The mean of grouped data is found in a similar way. For grouped data, intervals are used, rather than single values. To calculate the mean, it is assumed that all these values are located at the midpoint of the interval. The letter x is used to represent the midpoints and f represents the frequencies, as shown in the next example.

▶ **EXAMPLE 7**   Listed below is the grouped frequency distribution for the 30 business executives described in Example 1. Find the mean from the grouped frequency distribution.

Interval	Midpoint, x	Frequency, f	Product, xf
0–4	2	3	6
5–9	7	4	28
10–14	12	6	72
15–19	17	8	136
20–24	22	5	110
25–29	27	3	81
30–34	32	1	32
		Total: 30	Total: 465

A column for the midpoint of each interval has been added. The numbers in this column are found by adding the endpoints of each interval and dividing by 2. For the interval 0–4, the midpoint is $(0 + 4)/2 = 2$. The numbers in the product column on the right are found by multiplying each frequency by its corresponding midpoint. Finally, we divide the total of the product column by the total of the frequency column to get

$$\bar{x} = \frac{465}{30} = 15.5$$

Find the mean of the following grouped frequency distribution.

Interval	Frequency
0–4	6
5–9	4
10–14	7
15–19	3

Answer:

8.75

Notice that this mean is slightly different from the earlier mean of 15.93. The reason for this difference is that we have acted as if each piece of data is at the midpoint, which is not true here and is not true in most cases. Information is always lost when the data are grouped. It is more accurate to use the original data, rather than the grouped frequency, when calculating the mean, but the original data might not be available. Furthermore, the mean based upon the grouped data is typically not too different from the mean based upon the original data, and there may be situations in which the extra accuracy is not worth the extra effort.   ◀   ⑦

NOTE   **1.** The midpoint of an interval in a grouped frequency distribution may be a value that the data does not assume. For example, if we grouped the data for the 30 business executives into the intervals 0–5, 6–11, 12–17, 18–23, 24–29, and 30–35, the midpoints would be 2.5, 8.5, 14.5, 20.5, 26.5, and 32.5, even though all the data are whole numbers.

2. If we used different intervals in Example 7, the mean would come out to be a slightly different number. Verify that with the intervals 0–5, 6–11, 12–17, 18–23, 24–29, and 30–35, the mean in Example 7 is 16.1.

The formula for the mean of a grouped frequency distribution is given below.

Mean of a Grouped Distribution

The mean of a distribution where x represents the midpoints, f the frequencies, and $n = \Sigma f$, is

$$\bar{x} = \frac{\Sigma(xf)}{n}.$$

The mean of a random sample is a random variable, and for this reason it is sometimes called the **sample mean.** The sample mean is a random variable because it assigns a number to the experiment of taking a random sample. If a different random sample were taken, the mean would probably have a different value, with some values more probable than others. If another set of 30 business executives were selected in Example 1, the mean number of college units in management recommended for a business major might be 13.22 or 17.69. It is unlikely that the mean would be as small as 1.21 or as large as 32.75, although these values are remotely possible.

We saw in Section 9.4 how to calculate the expected value of a random variable when we know its probability distribution. The expected value is sometimes called the **population mean,** denoted by the Greek letter μ. In other words,

$$E(x) = \mu.$$

Furthermore, it can be shown that the expected value of $\bar{x}$ is also equal to μ; that is,

$$E(\bar{x}) = \mu.$$

For instance, consider again the 30 business executives in Example 1. We found that $\bar{x} = 15.93$, but the value of μ, the average for all possible business executives, is unknown. If a good estimate of μ were needed, the best guess (based on this data) is 15.93. Although $E(x) = \mu$ and $E(\bar{x}) = \mu$, $E(x)$ and $E(\bar{x})$ are not the same. $E(x)$ is the population mean, while $E(\bar{x})$ is the expected value of all possible sample means.

Median

Asked by a reporter to give the average height of the players on his team, the Little League coach lined up his 15 players by increasing height. He picked out the player in the middle and pronounced this player to be of average height. This kind of average, called the **median,** is defined as the middle entry in a set of data arranged in either increasing or decreasing order. If there is an even number of entries, the median is defined to be the mean of the two middle entries. The following table shows how to find the median for two sets of data: {8, 7, 4, 3, 1} and {2, 3, 4, 7, 9, 12}.

Odd Number of Entries	*Even Number of Entries*
8	2
7	3
Median = 4	4 ⎫
3	7 ⎬ Median = $\frac{4 + 7}{2}$ = 5.5
1	9 ⎭
	12

NOTE   As shown in the table above, when there are an even number of entries, the median is not always equal to one of the data entries.

▶ **EXAMPLE 8**   Find the median for the following lists of numbers.

(a) 11,   12,   17,   **20**,   23,   28,   29

The median is the middle number, in this case 20. (Note that the numbers are already arranged in numerical order.) In this list, three numbers are smaller than 20 and three are larger.

(b) 15,   13,   7,   11,   19,   30,   39,   5,   10

First arrange the numbers in numerical order, from smallest to largest or vice versa.

$$5,\quad 7,\quad 10,\quad 11,\quad \mathbf{13},\quad 15,\quad 19,\quad 30,\quad 39$$

The middle number can now be determined; the median is 13.

(c) 47,   59,   32,   81,   74,   153

Write the numbers in numerical order.

$$32,\quad 47,\quad \mathbf{59},\quad \mathbf{74},\quad 81,\quad 153$$

There are six numbers here; the median is the mean of the two middle numbers, or

$$\text{Median} = \frac{59 + 74}{2} = \frac{133}{2} = 66.5. \quad ◀$$

CAUTION   Remember, the data must be arranged in numerical order before locating the median. ⑧

⑧

Find the median for each of the following lists of numbers.

(a)   12,   15,   17,   19, 35,   42,   58

(b)   28,   68,   7,   15,   47, 59,   13,   74,   32,   25

Answers:

(a)   19

(b)   30

TECHNOLOGY TIP   Many graphing calculators (including TI-82/83/86, HP-39, Casio 9850) display the median when doing one-variable statistics. You may have to scroll down to a second screen to find it.

Both the mean and the median are examples of a **statistic,** which is simply a number that gives information about a sample. In some situations, the median gives a truer representative or typical element of the data than the mean. For example, suppose in an office there are 10 salespersons, 4 secretaries, the sales manager, and Ms. Daly, who owns the business. Their annual salaries are as follows: support staff, $30,000 each; salespersons, $50,000 each; manager, $70,000; and owner, $400,000. The mean salary is

$$\bar{x} = \frac{(30{,}000)4 + (50{,}000)10 + 70{,}000 + 400{,}000}{16} = \$68{,}125.$$

However, since 14 people earn less than $68,125 and only 2 earn more, this does not seem very representative. The median salary is found by ranking the salaries by size: $30,000, $30,000, $30,000, $30,000, $50,000, $50,000, . . . , $400,000. There are 16 salaries (an even number) in the list, so the mean of the 8th and 9th entries will give the value of the median. The 8th and 9th entries are both $50,000, so the median is $50,000. In this example, the median is more representative of the distribution than the mean.

Mode

Find the mode for each of the following lists of numbers.

(a) 29, 35, 29, 18, 29, 56, 48

(b) 13, 17, 19, 20, 20, 13, 25, 27, 13, 20

(c) 512, 546, 318, 729, 854, 253

Answers:

(a) 29

(b) 13 and 20

(c) No mode

Sue's scores on ten class quizzes include one 7, two 8's, six 9's and one 10. She claims that her average grade on quizzes is 9, because most of her scores are 9's. This kind of "average," found by selecting the most frequent entry, is called the **mode.**

▶ **EXAMPLE 9**   Find the mode for each list of numbers.

(a) 57,   38,   55,   55,   80,   87,   98

The number 55 occurs more often than any other, so it is the mode. It is not necessary to place the numbers in numerical order when looking for the mode.

(b) 182,   185,   183,   185,   187,   187,   189

Both 185 and 187 occur twice. This list has *two* modes, so it is bimodal.

(c) 10,708,   11,519,   10,972,   17,546,   13,905,   12,182

No number occurs more than once. This list has no mode.   ◀ ⑨

The mode has the advantages of being easily found and not being influenced by data that are extreme values. It is often used in samples where the data to be "averaged" are not numerical. A major disadvantage of the mode is that there may be more than one, in case of ties, or there may be no mode at all when all entries occur with the same frequency.

The mean is the most commonly used measure of central tendency. Its advantages are that it is easy to compute, it takes all the data into consideration, and it is reliable—that is, repeated samples are likely to give very similar means. A disadvantage of the mean is that it is influenced by extreme values, as illustrated in the salary example above.

The median can be easy to compute and is influenced very little by extremes. Like the mode, the median can be found in situations where the data are not numerical. A disadvantage of the median is the need to rank the data in order; this can be tedious when the number of items is large and difficult or not possible when the data is not numerical.

▶ **EXAMPLE 10** Seeds that are dried, placed in an airtight container, and stored in a cool, dry place remain ready to be planted for a long time. The following table gives the amount of time that each type of seed can be stored and still remain viable for planting.*

Vegetable	Years
Beans	3
Cabbage	4
Carrots	1
Cauliflower	4
Corn	2
Cucumbers	5
Melons	4
Peppers	2
Pumpkin	4
Tomatoes	3

Find the mean, median, and mode of the information in the table.

The mean amount of time that the seeds can be stored is

$$\bar{x} = \frac{3 + 4 + 1 + 4 + 2 + 5 + 4 + 2 + 4 + 3}{10} = 3.2 \text{ years}$$

After the numbers are arranged in order from smallest to largest, the middle number, or median, is found; it is 3.5.

The number 4 occurs more often than any other, so it is the mode. ◀

TECHNOLOGY TIP Most scientific calculators have some statistical capability and can calculate the mean of a set of data; graphing calculators can often calculate the median as well. For example, Figure 10.3 shows the mean and the median for the data above calculated on a TI-83, where the data is stored in the list L_1. This calculator does not include a command for finding the mode.

FIGURE 10.3

Spreadsheets can also be used to find the mean, median, and mode. Using Microsoft Excel, place the data in cells A1 through A10. To find the mean of this data, type "=mean (A1 . . A10)" in cell A11, or any other unused cell, and then press Enter. The result of 3.2 will appear in cell A11. To find the median of this data, type "=median (A1 . . A10)" in cell A12, or any other unused cell, and press Enter. The result of 3.5 will appear in cell A12. To find the mode of this data, type "=mode (A1 . . A10)" in cell A13, or any other unused cell, and press Enter. The result of 4 will appear in cell A13.

*The Handy Science Answer Book, Second Edition, The Carnegie Library of Pittsburgh, p. 247.

10.1 Exercises

For Exercises 1–4, (a) group the data as indicated; (b) prepare a frequency distribution with columns for intervals and frequencies; (c) construct a histogram; (d) construct a frequency polygon. (See Examples 1 and 2.)

1. Use six intervals, starting with 0–24.

74	133	4	127	20	30
103	27	139	118	138	121
149	132	64	141	130	76
42	50	95	56	65	104
4	140	12	88	119	64

2. Use seven intervals, starting with 30–39.

79	71	78	87	69	50	63	51	60	46
65	65	56	88	94	56	74	63	87	62
84	76	82	67	59	66	57	81	93	93
54	88	55	69	78	63	63	48	89	81
98	42	91	66	60	70	64	70	61	75
82	65	68	39	77	81	67	62	73	49
51	76	94	54	83	71	94	45	73	95
72	66	71	77	48	51	54	57	69	87

3. Use 70–74 as the first interval.

79	84	88	96	102	104	110	108	106	106	
104	99	97	92	94	90	82	74	72	83	
84	92	100	99	101	107	111	102	97	94	92

4. Use 140–149 as the first interval.

174	190	172	182	179	186	171	152	174	185
180	170	160	173	163	177	165	157	149	167
169	182	178	158	182	169	181	173	183	176
170	162	159	147	150	192	179	165	148	188

Construct a stem-and-leaf plot for the data in the indicated exercise. (Use a 2-digit stem. See Example 3.)

5. Exercise 3 (Use Stems 07, 08, . . . ,11)

6. Exercise 4

7. How does a frequency polygon differ from a histogram?

8. Discuss the advantages and disadvantages of the mean as a measure of central tendency.

Find the mean for each list. Round to the nearest tenth. (See Example 5.)

9. 21,900, 22,850, 24,930, 29,710, 28,340, 40,000

10. 38,500, 39,720, 42,183, 21,982, 43,250

11. 9.4, 11.3, 10.5, 7.4, 9.1, 8.4, 9.7, 5.2, 1.1, 4.7

12. 30.1, 42.8, 91.6, 51.2, 88.3, 21.9, 43.7, 51.2

Find the mean for each distribution. Round to the nearest tenth. (See Examples 5, 6, and 7.)

13.

Value	Frequency
3	4
5	2
9	1
12	3

14.

Value	Frequency
9	3
12	5
15	1
18	1

15.

Value	Frequency
12	4
13	2
15	5
19	3
22	1
23	5

16.

Value	Frequency
25	1
26	2
29	5
30	4
32	3
33	5

Find the median for each of the following lists of numbers. (See Example 8.)

17. 12, 18, 32, 51, 58, 92, 106

18. 596, 604, 612, 683, 719

19. 100, 114, 125, 135, 150, 172

20. 1072, 1068, 1093, 1042, 1056, 1005, 1009

21. 28.4, 9.1, 3.4, 27.6, 59.8, 32.1, 47.6, 29.8

22. .6, .4, .9, 1.2, .3, 4.1, 2.2, .4, .7, .1

Find the mode or modes for each of the following lists of numbers. (See Example 9.)

23. 4, 9, 8, 6, 9, 2, 1, 3

24. 21, 32, 46, 32, 49, 32, 49

25. 74, 68, 68, 68, 75, 75, 74, 74, 70

26. 158, 162, 165, 162, 165, 157, 163

27. 6.8, 6.3, 6.3, 6.9, 6.7, 6.4, 6.1, 6.0

28. 12.75, 18.32, 19.41, 12.75, 18.30, 19.45, 18.33

29. When is the median the most appropriate measure of central tendency?

30. Under what circumstances would the mode be an appropriate measure of central tendency?

For grouped data, the modal class *is the interval containing the most data values. Give the mean and modal class for each of the following collections of grouped data. (See Example 7.)*

31. The distribution in Exercise 3.

32. The distribution in Exercise 4.

33. To predict the outcome of the next congressional election, you take a survey of your friends. Is this a random sample of the voters in your congressional district? Explain why or why not.

Work each problem. (See Example 10.)

34. Social Science   The number of nations participating in the winter Olympic games, from 1972 to 1998, is given below.*

Year	Nations Participating
1972	35
1976	37
1980	37
1984	49
1988	57
1992	64
1994	67
1998	72

Find the following statistics for the data.
(a) Mean   (b) Median   (c) Mode

35. Business   The total pay (in thousands of dollars) for the 10 highest paid CEOs in 1997 is given in the following table.[†]

Person, Company	Total Pay
Sanford Weill, Travelers Group	230,725
Roberto Goizueta, Coca-Cola	111,832
Richard Scrushy, Health South	106,790
Ray Irani, Occidental Petroleum	101,505
Eugene Isenberg, Nabors Industries	84,547
Joseph Costelo, Cadence Design Systems	66,842
Andrew Grove, Intel	52,214
Charles McCall, HBO Co.	51,409
Philip Purcell, Morgan Stanley Dean Witter	50,807
Robert Shapiro, Monsanto	49,326

(a) Find the mean compensation for this group of people.
(b) Find the median compensation for this group of people.
(c) What might account for the difference between these values?
(d) Write the salary for Shapiro.

36. Business   The top ten box-office receipts (in millions of dollars) for the weekend of 6/15/2001 through 6/17/2001 at the movies are given here.*

Movie (Weeks in release)	Weekend Receipts (millions)
Lara Croft: Tomb Raider (1)	47.7
Atlantis: The Lost Empire (2)	20.3
Shrek (5)	13.2
Swordfish (2)	12.7
Pearl Harbor (4)	9.9
Evolution (2)	6.6
The Animal (3)	5.8
Moulin Rouge (5)	5
What's the Worst Thing That Could Happen? (3)	3
The Mummy Returns (7)	2.7

(a) Find the mean weekend receipts in dollars for this group.
(b) Find the median weekend receipts in dollars for this group.

The New York Times Almanac, 2001.

†John W. Wright, *The American Almanac of Jobs and Salaries,* 2000–2001 ed., p. 321.

*Exhibitor Relations Co.

37. Natural Science The number of recognized blood types varies by species, as indicated by the following table.* Find the mean, median, and mode of this data.

Animal	Number of Blood Types
Pig	16
Cow	12
Chicken	11
Horse	9
Human	8
Sheep	7
Dog	7
Rhesus Monkey	6
Mink	5
Rabbit	5
Mouse	4
Rat	4
Cat	2

38. Social Science The number of weddings in each month of 1996 is given in the following table.[†]

Month	Number (in thousands)
January	110
February	155
March	118
April	172
May	241
June	242
July	235
August	239
September	225
October	231
November	171
December	184

(a) Calculate the mean and median for this data.
(b) Which month is closest to the mean?

Physical Science *The table below gives the average monthly temperatures in degrees Fahrenheit for a certain area.*

Month	Maximum	Minimum
January	39	16
February	39	18
March	44	21
April	50	26
May	60	32
June	69	37
July	79	43
August	78	42
September	70	37
October	51	31
November	47	24
December	40	20

Find the mean and median for each of the following.

39. The maximum temperature

40. The minimum temperature

Business *U.S. wheat prices and production figures for a recent decade are given below.**

Year	Price ($ per bushel)	Production (millions of bushels)
1989	3.72	2037
1990	2.61	2730
1991	3.00	1980
1992	3.24	2467
1993	3.26	2396
1994	3.45	2321
1995	4.55	2183
1996	4.30	2285
1997	3.38	2482
1998	2.65	2550

Find the mean and median for each of the following.

41. Price per bushel of wheat

42. Wheat production

---

*The Handy Science Answer Book, Carnegie Library of Pittsburgh, Pennsylvania, p. 264.

†The Amazing Almanac, Blackbirch Press, Inc., p. 245.

*The World Almanac and Book of Facts 2000, pp. 141–142.

43. Social Science The total household income for full-time African-American workers making under $100,000 in 1997 is given in the table below.* (See Example 7.)

Income Range	Midpoint Salary	Frequency (in thousands)
Under $5,000	$2500	923
$5,000–$9,999	$7500	1746
$10,000–$14,999	$12,500	1310
$15,000–$24,999	$20,000	2233
$25,000–$34,999	$30,000	1771
$35,000–$49,999	$42,500	1859
$50,000–$74,999	$62,500	1634
$75,000–$99,999	$87,500	574

Use this table to estimate the mean household income for full-time African-American workers in 1997.

44. Social Science The total household income for full-time white American workers making under $100,000 in 1997 is given in the following table.* (See Example 7.)

Income Range	Midpoint Salary	Frequency (in thousands)
Under $5,000	$2500	2411
$5,000–$9,999	$7500	5769
$10,000–$14,999	$12,500	6716
$15,000–$24,999	$20,000	12,571
$25,000–$34,999	$30,000	11,366
$35,000–$49,999	$42,500	14,207
$50,000–$74,999	$62,500	16,188
$75,000–$99,999	$87,500	8180

(a) Use this table to estimate the mean household income for full-time white American workers in 1997.

*Time Almanac 2000, Time Inc., p. 829.

(b) Compare this estimate with the estimate found in Exercise 43. Does this provide some evidence that full-time white American workers have higher household earnings than full-time African-American workers?

45. Social Science The histogram below shows estimates of the percent of the U.S. population in each age group in the year 2000.[†] What percent of the population is estimated to be in each of the following age groups?

(a) 10–19
(b) 60–69
(c) What age range has the largest percent of the population?

46. Social Science The histogram below shows estimates of the percent of the U.S. population in each age group in the year 2025.[†] What percent of the population is estimated to be in each of the following age groups then?

(a) 20–29
(b) 70+
(c) What age group will have the smallest percent of the population?
(d) Compare the histogram in Exercise 45 with the histogram below. What seems to be true of the U.S. population?

†U.S. Census Bureau, January 13, 2000.

10.2 MEASURES OF VARIATION

The mean, median, and mode are measures of central tendency of a list of numbers but tell nothing about the *spread* of the numbers in the list. For example, look at the following three lists.

I	3	5	6	3	3
II	4	4	4	4	4
III	10	1	0	0	9

Each of these three lists has a mean of 4, and yet they are quite different; the amount of dispersion or variation within the lists is different. Therefore, in addition to a measure of central tendency, another kind of measure is needed that describes how much the numbers vary.

The largest number in sample I is 6, while the smallest is 3, a difference of 3. In sample II this difference is 0; in sample III it is 10. The difference between the largest and smallest number in a sample is called the **range,** one instance of a measure of variation. The range of sample I is 3, of sample II is 0, and of sample III is 10. The range has the advantage of being very easy to compute and gives a rough estimate of the variation among the data in the sample. However, it depends only on the two extremes and tells nothing about how the other data are distributed between the extremes.

▶ **EXAMPLE 1** Find the range for each list of numbers.

(a) 12, 27, 6, 19, 38, 9, 42, 15

The highest number here is 42; the lowest is 6. The range is the difference of these numbers, or

$$42 - 6 = 36.$$

(b) 74, 112, 59, 88, 200, 73, 92, 175

$$\text{Range} = 200 - 59 = 141 \quad ◀ \quad ①$$

①

Find the range for the numbers 159, 283, 490, 390, 375, 297.

Answer:

331

TECHNOLOGY TIP Many graphing calculators list the largest and smallest numbers in a list when displaying one-variable statistics, usually on the second screen of the display.

To find a useful measure of variation, we begin by finding the **deviations from the mean,** the differences found by subtracting the mean from each number in a distribution.

▶ **EXAMPLE 2** Find the deviations from the mean for the numbers

32, 41, 47, 53, 57.

Adding these numbers and dividing by 5 gives a mean of 46. To find the deviations from the mean, subtract 46 from each number in the list. For example, the first deviation from the mean is $32 - 46 = -14$; the last is $57 - 46 = 11$.

Number	Deviation from Mean
32	−14
41	−5
47	1
53	7
57	11

Find the deviations from the mean for each set of numbers.

(a) 19, 25, 36, 41, 52, 61

(b) 6, 9, 5, 11, 3, 2

Answers:

(a) Mean is 39; deviations are −20, −14, −3, 2, 13, 22.

(b) Mean is 6; deviations from the mean are 0, 3, −1, 5, −3, −4.

To check your work, find the sum of these deviations. It should always equal 0. (The answer is always 0 because the positive and negative deviations cancel each other.) ◀ ②

To find a measure of variation, we might be tempted to use the mean of the deviations. However, as mentioned above, this number is always 0, no matter how widely the data are dispersed. To avoid the problem of the positive and negative deviations averaging 0, we take absolute values and find $\Sigma|x - \bar{x}|$, then divide it by n to get the *mean deviation*. Statisticians consider a better measure of variation to be squaring each deviation to get nonnegative numbers, then taking the square root of the mean of the squared variations to preserve the units of the original data (such as inches, pounds, and so on). The squared deviations for the data in Example 2 are shown in the table.

Number	Deviation from Mean	Square of Deviation
32	−14	196
41	−5	25
47	1	1
53	7	49
57	11	121

In this case the mean of the squared deviations is

$$\frac{196 + 25 + 1 + 49 + 121}{5} = \frac{392}{5} = 78.4.$$

This number is called the **population variance,** because the sum was divided by $n = 5$, the number of items in the original list.

Because the deviations from the mean must add up to 0, if we know any 4 of the 5 deviations, the 5th will be determined. That is, only $n - 1$ of the deviations are free to vary, so we really have only $n - 1$ independent pieces of information, or *degrees of freedom*. Using $n - 1$ as the divisor in the formula for the mean gives

$$\frac{196 + 25 + 1 + 49 + 121}{5 - 1} = \frac{392}{4} = 98.$$

This number 98 is called the **sample variance** of the distribution and is denoted s^2 because it is found by averaging a list of squares. In this case, the population and sample variances differ by quite a bit. But when n is relatively large, as is the case in real-life applications, the difference between them is rather small.

Sample Variance

The variance of a sample of n numbers $x_1, x_2, x_3, \ldots, x_n$, with mean $\bar{x}$, is

$$s^2 = \frac{\Sigma(x - \bar{x})^2}{n - 1}.$$

When computing the sample variance by hand, it is often convenient to use the following shortcut formula, which can be derived algebraically from the definition in the box above.

$$s^2 = \frac{\Sigma x^2 - n\bar{x}^2}{n - 1}$$

To find the sample variance, we square the deviations from the mean, so the variance is in squared units. To return to the same units as the data, we use the *square root* of the variance, called the **sample standard deviation,** denoted s.

Sample Standard Deviation

The standard deviation of a sample of n numbers $x_1, x_2, x_3, \ldots, x_n$, with mean $\bar{x}$, is

$$s = \sqrt{\frac{\Sigma(x - \bar{x})^2}{n - 1}}.$$

NOTE   The **population standard deviation** is

$$\sigma = \sqrt{\frac{\Sigma(x - \bar{x})^2}{n}},$$

where n is the population size.

TECHNOLOGY TIP When a graphing calculator computes one-variable statistics for a list of data, it usually displays the following information (not necessarily in this order and sometimes on two screens), and possibly other information as well.

Information	Notation
Number of data entries	n or $N\Sigma$
Mean	$\bar{x}$ or mean Σ
Sum of all data entries	Σx or TOT Σ
Sum of the squares of all data entries	Σx^2
Sample standard deviation	Sx or sx or $x\sigma_{n-1}$ or SSDEV
Population standard deviation	σx or $x\sigma_n$ or PSDEV
Largest/smallest data entries	maxX/minX or MAXΣ/MINΣ
Median	Med or MEDIAN

NOTE   In the rest of this section, we shall deal exclusively with the sample variance and the sample standard deviation. So whenever standard deviation is mentioned, it means "sample standard deviation," not population standard deviation.

As its name indicates, the standard deviation is the most commonly used measure of variation. The standard deviation is a measure of the variation from the mean. The size of the standard deviation indicates how spread out the data are from the mean.

▶ **EXAMPLE 3** Find the standard deviation of the numbers

$$7, \quad 9, \quad 18, \quad 22, \quad 27, \quad 29, \quad 32, \quad 40$$

by hand, using the shortcut variance formula at the top of page 563.
Arrange the work in columns, as shown in the table.

Number	Square of the Number
7	49
9	81
18	324
22	484
27	729
29	841
32	1024
40	1600
184	5132

Now find the mean.

$$\bar{x} = \frac{\Sigma x}{8} = \frac{184}{8} = 23$$

The total of the second column gives $\Sigma x^2 = 5132$. The variance is

$$s^2 = \frac{\Sigma x^2 - n\bar{x}^2}{n - 1}$$

$$= \frac{5132 - 8(23)^2}{8 - 1}$$

$$= 128.6 \quad \text{(rounded)},$$

and the standard deviation is

$$s \approx \sqrt{128.6} \approx 11.3. \quad \blacktriangleleft \quad ③$$

③

Find the standard deviation of each set of numbers. The deviations from the mean were found earlier in problem 2 at the side.

(a) 19, 25, 36, 41, 52, 61

(b) 6, 9, 5, 11, 3, 2

Answers:

(a) 15.9

(b) 3.5

TECHNOLOGY TIP The screens in Figure 10.5 show two ways to find variance and standard deviation on a TI-83 calculator, using the LIST menu and the STAT menu. The data are first entered in a list, here L_5. See your instruction book for details.

FIGURE 10.4

With a spreadsheet, the data are entered in cells A1 through A8. Then, in cell A9, type "=VAR (A1..A8)" and press Enter. The standard deviation can be calculated by either taking the square root of cell A9 or by typing "=STDEV (A1..A8)" in cell A10 and pressing Enter.

CAUTION We must be careful to divide by $n - 1$, not n, when calculating the standard deviation of a sample. Many calculators are equipped with statistical keys that compute the variance and standard deviation. Some of these calculators use $n - 1$ and others use n for these computations; some may have keys for both. Check your calculator's instruction book before using a statistical calculator for the exercises.

One way to interpret the standard deviation uses the fact that, for many populations, most of the data is within three standard deviations of the mean. (See Section 10.3.) This implies that, in Example 3, most of the population data from which this sample is taken are between

$$\bar{x} - 3s = 23 - 3(11.3) = -10.9$$

and

$$\bar{x} + 3s = 23 + 3(11.3) = 56.9.$$

This has important implications for quality control. If the sample in Example 3 represents measurements of a product that the manufacturer wants to be between 5 and 45, the standard deviation is too large, even though all the numbers are within these bounds.

For data in a grouped frequency distribution, a slightly different formula for the standard deviation is used.

Standard Deviation for a Grouped Distribution

The standard deviation for a sample distribution with mean $\bar{x}$, where x is an interval midpoint with frequency f, and $n = \Sigma f$, is

$$s = \sqrt{\frac{\Sigma fx^2 - n\bar{x}^2}{n-1}}.$$

The formula indicates that the product fx^2 is to be found for each interval. Then these products are summed, n times the square of the mean is subtracted, and the difference is divided by one less than the total frequency; that is, by $n - 1$. The square root of this result is s, the standard deviation. The standard deviation found by this formula may (probably will) differ somewhat from the standard deviation found from the original data.

CAUTION In calculating the standard deviation for either a grouped or ungrouped distribution, using a rounded value for the mean or variance may produce an inaccurate value.

▶ **EXAMPLE 4** Find s for the grouped data of Example 7, Section 10.1.

Begin by including columns for x^2 (where x is the midpoint of the interval) and fx^2.

Interval	x	x²	f	fx²
0–4	2	4	3	12
5–9	7	49	4	196
10–14	12	144	6	864
15–19	17	289	8	2312
20–24	22	484	5	2420
25–29	27	729	3	2187
30–34	32	1024	1	1024
			Total: 30	Total: 9015

Recall from Example 7 of Section 10.1 that $\bar{x} = 15.5$. Use the formula above with $n = 30$ to find s.

$$s = \sqrt{\frac{\Sigma fx^2 - n\bar{x}^2}{n - 1}}$$

$$= \sqrt{\frac{9015 - 30(15.5)^2}{30 - 1}}$$

$$\approx 7.89 \quad ◀ \; ④$$

④

Find the standard deviation for the grouped data that follows. (*Hint:* $\bar{x} = 28.5$)

Value	Frequency
20–24	3
25–29	2
30–34	4
35–39	1

Answer:

5.3

NOTE A calculator is almost a necessity for finding a standard deviation. With a non-graphing calculator, a good procedure to follow is to first calculate $\bar{x}$. Then for each x, square that number, and multiply the result by the appropriate frequency. If your calculator has a key that accumulates a sum, use it to accumulate the total in the last column of the table. With a graphing calculator, simply enter the midpoints and the frequencies, and then ask for the 1-variable statistics.

10.2 Exercises

1. How are the variance and the standard deviation related?

2. Why can't we use the sum of the deviations from the mean as a measure of dispersion of a distribution?

Find the range and standard deviation for each of the following sets of numbers. (See Examples 1 and 3.)

3. 6, 8, 9, 10, 12

4. 12, 15, 19, 23, 26

5. 7, 6, 12, 14, 18, 15

6. 4, 3, 8, 9, 7, 10, 1

7. 42, 38, 29, 74, 82, 71, 35

8. 122, 132, 141, 158, 162, 169, 180

9. 241, 248, 251, 257, 252, 287

10. 51, 58, 62, 64, 67, 71, 74, 78, 82, 93

Find the standard deviation for the grouped data in Exercises 11 and 12. (See Example 4.)

11. (From Exercise 1, Section 10.1)

College Units	Frequency
0–24	4
25–49	3
50–74	6
75–99	3
100–124	5
125–149	9

12. (From Exercise 2, Section 10.1)

Scores	Frequency
30–39	1
40–49	6
50–59	13
60–69	22
70–79	17
80–89	13
90–99	8

13. **Natural Science** Twenty-five laboratory rats, used in an experiment to test the food value of a new product, made the following weight gains in grams.

5.25	5.03	4.90	4.97	5.03
5.12	5.08	5.15	5.20	4.95
4.90	5.00	5.13	5.18	5.18
5.22	5.04	5.09	5.10	5.11
5.23	5.22	5.19	4.99	4.93

Find the mean gain and the standard deviation of the gains.

14. **Business** An assembly-line machine turns out washers with the following thicknesses (in millimeters).

1.20	1.01	1.25	2.20	2.58	2.19	1.29	1.15
2.05	1.46	1.90	2.03	2.13	1.86	1.65	2.27
1.64	2.19	2.25	2.08	1.96	1.83	1.17	2.24

Find the mean and standard deviation of these thicknesses.

An application of standard deviation is given by Chebyshev's Theorem. *(P. L. Chebyshev was a Russian mathematician who lived from 1821 to 1894.) This theorem applies to any distribution of numerical data. It states:*

For any distribution of numerical data, at least $1 - 1/k^2$ *of the numbers lie within k standard deviations of the mean.*

Example *For any distribution, at least*

$$1 - \frac{1}{3^2} = 1 - \frac{1}{9} = \frac{8}{9}$$

of the numbers lie within 3 standard deviations of the mean. Find the fraction of all the numbers of a data set lying within the following numbers of standard deviations from the mean.

15. 2   16. 4   17. 1.5

In a certain distribution of numbers, the mean is 50 with a standard deviation of 6. Use Chebyshev's theorem to tell what percent of the numbers are

18. between 32 and 68;

19. between 26 and 74;

20. less than 38 or more than 62;

21. less than 32 or more than 68;

22. less than 26 or more than 74.

23. **Business** The Britelite Company conducted tests on the life of its lightbulbs and those of a competitor (Brand X) with the following results for samples of 10 bulbs of each brand.

	Hours of Use (in 100s)									
Britelite	20	22	22	25	26	27	27	28	30	35
Brand X	15	18	19	23	25	25	28	30	34	38

(a) Compute the mean and standard deviation for each sample. Compare the means and standard deviations of the two brands and then answer the questions below.

(b) Which bulbs have a more uniform life in hours?

(c) Which bulbs have the highest average life in hours?

24. Business The weekly wages of the six employees of Harold's Hardware Store are $300, $320, $380, $420, $500, and $2000.

(a) Find the mean and standard deviation of this distribution.

(b) How many of the employees earn within one standard deviation of the mean? How many earn within two standard deviations of the mean?

25. Social Science The number of unemployed workers in the United States in recent years (in millions) is given below.*

Year	Number Unemployed
1989	6.53
1990	7.05
1991	8.63
1992	9.61
1993	8.94
1994	8.00
1995	7.40
1996	7.24
1997	6.74
1998	6.21

Use sample statistics to find the following.

(a) Find the mean number unemployed (in millions) in this period. Which year has unemployment closest to the mean?

(b) Find the standard deviation for the data.

(c) In how many of these years is unemployment within 1 standard deviation of the mean?

(d) In how many of these years is unemployment within 3 standard deviations of the mean?

26. Life Science A medical laboratory tested 21 samples of human blood for acidity on the pH scale, with the following results.

7.1	7.5	7.3	7.4	7.6	7.2	7.3
7.4	7.5	7.3	7.2	7.4	7.3	7.5
7.5	7.4	7.4	7.1	7.3	7.4	7.4

(a) Find the mean and standard deviation.

(b) What percent of the data is within 2 standard deviations of the mean?

27. Life Science The number of recognized blood types between species is given in the following table.* In Exercise 37 of the previous section, the mean was found to be 7.38.

Animal	Number of Blood Types
Pig	16
Cow	12
Chicken	11
Horse	9
Human	8
Sheep	7
Dog	7
Rhesus Monkey	6
Mink	5
Rabbit	5
Mouse	4
Rat	4
Cat	2

(a) Find the variance and the standard deviation of these data.

(b) How many of these animals have blood types that are within 1 standard deviation of the mean?

28. Social Science In 1999, fourteen state governors earned at least $115,000 annually (not counting expense allowances) as listed below. (Salaries are given in thousands of dollars and are rounded to the nearest $1000.)†

State	Salary
California	165
Georgia	116
Illinois	140
Maryland	120
Michigan	127
Minnesota	120
Nevada	117
New York	179
Ohio	119
Texas	115
Vermont	116
Virginia	125
Washington	132
Wisconsin	116

(a) Find the mean salary of these governors. Which state has the governor with the salary closest to the mean?

*The World Almanac and Book of Facts 2000, p. 145.

*The Handy Science Answer Book, Carnegie Library of Pittsburgh, Pennsylvania, p. 264.

†The World Almanac and Book of Facts 2000, p. 98.

(b) Find the standard deviation for the data.

(c) What percent of the governors have salaries within 1 standard deviation of the mean?

(d) What percent of the governors have salaries within 3 standard deviations of the mean?

29. Health   The amounts of time that it takes for various slow-growing tumors to double in size are listed in the table below.*

Type of Cancer	Doubling Time (days)
Breast cancer	84
Rectal cancer	91
Synovioma	128
Skin cancer	131
Lip cancer	143
Testicular cancer	153
Esophageal cancer	164

(a) Find the mean and standard deviation of these data.

(b) How many of these cancers have doubling times that are within 2 standard deviations of the mean?

(c) If a person had a nonspecified tumor that was doubling every 200 days, discuss if this particular tumor is growing at a rate that would be expected.

30. Business   The Quaker Oats Company conducted a survey to determine if a proposed premium, to be included in their cereal, was appealing enough to generate new sales.† Four cities were used as test markets, where the cereal was distributed with the premium, and four cities as control markets, where the cereal was distributed without the premium. The eight cities were chosen on the basis of their similarity in terms of population, per capita income, and total cereal purchase volume. The results were as follows.

		Percent Change in Average Market Shares per Month
Test Cities	1	+18
	2	+15
	3	+7
	4	+10
Control Cities	1	+1
	2	−8
	3	−5
	4	0

(a) Find the mean of the change in market share for the four test cities.

(b) Find the mean of the change in market share for the four control cities.

(c) Find the standard deviation of the change in market share for the test cities.

(d) Find the standard deviation of the change in market share for the control cities.

(e) Find the difference between the mean of part (a) and the mean of part (b). This represents the estimate of the percent change in sales due to the premium.

(f) The two standard deviations from part (c) and part (d) were used to calculate an "error" of ± 7.95 for the estimate in part (e). With this amount of error what is the smallest and largest estimate of the increase in sales? (*Hint:* Use the answer to part (e).)

On the basis of the results of Exercise 30, the company decided to mass produce the premium and distribute it nationally.

31. Business   The following table gives 10 samples of three measurements, made during a production run.

				SAMPLE NUMBER					
1	*2*	*3*	*4*	*5*	*6*	*7*	*8*	*9*	*10*
2	3	−2	−3	−1	3	0	−1	2	0
−2	−1	0	1	2	2	1	2	3	0
1	4	1	2	4	2	2	3	2	2

(a) Find the mean $\bar{x}$ for each sample of three measurements.

(b) Find the standard deviation s for each sample of three measurements.

(c) Find the mean $\bar{\bar{x}}$ of the sample means.

(d) Find the mean $\bar{s}$ of the sample standard deviations.

(e) The upper and lower control limits of the sample means here are $\bar{\bar{x}} \pm 1.954\bar{s}$. Find these limits. If any of the measurements are outside these limits, the process is out of control. Decide if this production process is out of control.

32. Discuss what the standard deviation tells us about a distribution.

Social Science   *The reading scores of a second-grade class given individualized instruction are shown below. The table also shows the reading scores of a second-grade class given traditional instruction in the same school.*

Scores	Individualized Instruction	Traditional Instruction
50–59	2	5
60–69	4	8
70–79	7	8
80–89	9	7
90–99	8	6

*Vincent Collins, R. Kenneth Lodffer, and Harold Tivey, "Observations on Growth Rates of Human Tumors," *American Journal of Roentgen*, 76, no. 5 (November 1956): 988–1000.

†This example was supplied by Jeffrey S. Berman, Senior Analyst, Marketing Information, Quaker Oats Company.

33. Find the mean and standard deviation for the individualized instruction scores.

34. Find the mean and standard deviation for the traditional instruction scores.

35. Discuss a possible interpretation of the differences in the means and the standard deviations in Exercises 33 and 34.

36. Business   Refer to Section 10.1. Exercise 35, which gives total pay for CEOs.
(a) Calculate the standard deviation of the data.
(b) What percent of the CEOs listed have compensation more than 2 standard deviations from the mean?
(c) What does your answer to part (b) suggest?

10.3 NORMAL DISTRIBUTIONS

Suppose a bank is interested in improving its services to customers. The manager decides to begin by finding the amount of time tellers spend on each transaction, rounded to the nearest minute. The times for 75 different transactions are recorded, with the results shown in the following table. The frequencies listed in the second column are divided by 75 to find the empirical probabilities.

Time	Frequency	Probability
1	3	$3/75 = .04$
2	5	$5/75 \approx .07$
3	9	$9/75 = .12$
4	12	$12/75 = .16$
5	15	$15/75 = .20$
6	11	$11/75 \approx .15$
7	10	$10/75 \approx .13$
8	6	$6/75 = .08$
9	3	$3/75 = .04$
10	1	$1/75 \approx .01$

Figure 10.5(a) shows a histogram and frequency polygon for the data. The heights of the bars are the empirical probabilities, rather than the frequencies. The transaction times are given to the nearest minute. Theoretically at least, they could have been timed to the nearest tenth of a minute, or hundredth of a minute, or even more precisely. In each case, a histogram and frequency polygon could be drawn. If the times are measured with smaller and smaller units, there are more bars in the histogram and the frequency polygon begins to look more and more like the curve in Figure 10.5(b) instead of a polygon. Actually, it is possible for the transaction times to take on any real number value greater than 0. A distribution in which the outcomes can take any real number value within some interval is a **continuous distribution.** The graph of a continuous distribution is a curve.

FIGURE 10.5

The distribution of heights (in inches) of college women is another example of a continuous distribution, since these heights include infinitely many possible measurements, such as 53, 58.5, 66.3, 72.666, . . . , and so on. Figure 10.6 shows the continuous distribution of heights of college women. Here the most frequent heights occur near the center of the interval shown.

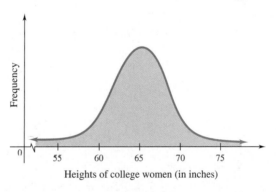

FIGURE 10.6

Another continuous curve, which approximates the distribution of yearly incomes in the United States, is given in Figure 10.7. The graph shows that the most frequent incomes are grouped near the low end of the interval. This kind of distribution, where the peak is not at the center, is called **skewed.**

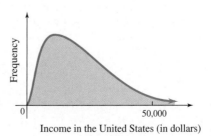

FIGURE 10.7

Normal Distributions

Many natural and social phenomena produce continuous probability distributions whose graphs are approximated very well by bell-shaped curves, such as those shown in Figure 10.8. Such distributions are called **normal distributions** and their graphs are called **normal curves.** Examples of distributions that are approximately normal are the heights of college women and the errors made in filling 1-pound cereal boxes. We use the Greek letters μ (mu) to denote the mean and σ (sigma) to denote the standard deviation of a normal distribution.

Three normal distributions

FIGURE 10.8

There are many normal distributions. Some of the corresponding normal curves are tall and thin and others short and wide, as shown in Figure 10.8. But every normal curve has the following properties.

1. Its peak occurs directly above the mean μ.
2. The curve is symmetric about the vertical line through the mean (that is, if you fold the graph along this line, the left half of the graph will fit exactly on the right half).
3. The curve never touches the x-axis—it extends indefinitely in both directions.
4. The area under the curve (and above the horizontal axis) is 1. (As can be shown with calculus, this is a consequence of the fact that the sum of the probabilities in any distribution is 1.)

A normal distribution is completely determined by its mean μ and standard deviation σ.* A small standard deviation leads to a tall, narrow curve like the one in the center of Figure 10.8, because most of the data are close to the mean. A large standard deviation means the data are very spread out, producing a flat wide curve like the one on the right in Figure 10.8.

Since the area under a normal curve is 1, parts of this area can be used to determine certain probabilities. For instance, Figure 10.9(a) is the probability distribution of the annual rainfall in a certain region. The probability that the annual rainfall will

---

*As shown in more advanced courses, its graph is the graph of the function

$$f(x) = \frac{1}{\sigma\sqrt{2\pi}}\,e^{-(x-\mu)^2/(2\sigma^2)},$$

where $e \approx 2.71828$ is the real number introduced in Section 4.1

be between 25 and 35 inches is the area under the curve from 25 to 35. The general case, shown in Figure 10.9(b), can be stated as follows.

The area of the shaded region under the normal curve from a to b is the probability that an observed data value will be between a and b.

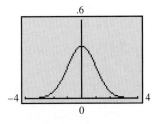

(a)   (b)

FIGURE 10.9

To use normal curves effectively we must be able to calculate areas under portions of these curves. These calculations have already been done for the normal curve with mean $\mu = 0$ and standard deviation $\sigma = 1$ (which is called the **standard normal curve**) and are available in Table 2 at the back of the book. The following examples demonstrate how to use Table 2 to find such areas. Later we shall see how the standard normal curve may be used to find areas under any normal curve.

The horizontal axis of the standard normal curve is usually labeled z. Since the standard deviation of the standard normal curve is 1, the numbers along the horizontal axis (the z-values) measure the number of standard deviations above or below the mean $z = 0$.

TECHNOLOGY TIP Some graphing calculators (such as the TI-83 and Casio 9850) have the ability to graph a normal distribution given its mean and standard deviation and to find areas under the curve between two x-values. For an area under the curve, some calculators will give the corresponding z-value. For details, see your instruction book. (Look for "distribution" or "probability distribution.") A calculator-generated graph of the standard normal curve is shown in Figure 10.10.

FIGURE 10.10

Find the percent of area between the mean and

(a) $z = 1.51$;

(b) $z = -2.04$.

(c) Find the percent of area in the shaded region.

Answers:

(a) 43.45%

(b) 47.93%

(c) 26.42%

If your calculator can graph probability distributions and find area, use it to find the areas required in Example 1.

Answers:

(a) 34.13%

(b) 49.25%

▶ **EXAMPLE 1** Find the following areas under the standard normal curve.

(a) The area between $z = 0$ and $z = 1$, the shaded region in Figure 10.11

Find the entry 1 in the z-column of Table 2. The entry next to it in the A-column is .3413, which means that the area between $z = 0$ and $z = 1$ is .3413. Since the total area under the curve is 1, the shaded area in Figure 10.11 is 34.13% of the total area under the normal curve.

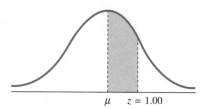

FIGURE 10.11

(b) The area between $z = -2.43$ and $z = 0$

Table 2 lists only positive values of z. But the normal curve is symmetric around the mean $z = 0$, so the area between $z = 0$ and $z = -2.43$ is the same as the area between $z = 0$ and $z = 2.43$. Find 2.43 in the z-column of Table 2. The entry next to it in the A-column shows that the area is .4925. Hence, the shaded area in Figure 10.12 is 49.25% of the total area under the curve. ◀ ① ②

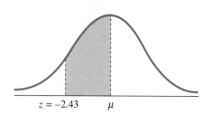

FIGURE 10.12

TECHNOLOGY TIP Because of convenience and accuracy, graphing calculators and computers have made normal curve tables less important. Figure 10.13 shows how part (b) of Example 1 can be done on a TI-83 calculator using a command from the DISTR menu. The second result in the calculator screen gives the area between $-\infty$ and $z = -2.43$; the entry -1E99 represents $-1 \cdot 10^{99}$, which is used to approximate $-\infty$.

FIGURE 10.13

Many statistical software packages are widely used today. All of these packages are set up in a way that is similar to a spreadsheet, and they all can be used to generate normal curve values. In addition, most spreadsheets can also perform a wide range of statistical calculations.

▶ EXAMPLE 2   Use technology or Table 2 to find the percent of the total area for the following areas under the standard normal curve.

(a) The area between 1.41 standard deviations *below* the mean and 2.25 standard deviations *above* the mean (that is, between $z = -1.41$ and $z = 2.25$)

First, draw a sketch showing the desired area, as in Figure 10.14. From Table 2, the area between the mean and 1.41 standard deviations below the mean is .4207. Also, the area from the mean to 2.25 standard deviations above the mean is .4878. As the figure shows, the total desired area can be found by *adding* these numbers.

$$
\begin{array}{r}
.4207 \\
+\ .4878 \\
\hline
.9085
\end{array}
$$

The shaded area in Figure 10.14 represents 90.85% of the total area under the normal curve.

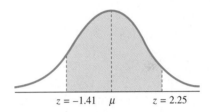

$z = -1.41 \quad \mu \qquad z = 2.25$

FIGURE 10.14

(b) The area between .58 standard deviation above the mean and 1.94 standard deviations above the mean

Figure 10.15 shows the desired area. The area between the mean and .58 standard deviation above the mean is .2190. The area between the mean and 1.94 standard deviations above the mean is .4738. As the figure shows, the desired area is found by *subtracting* one area from the other.

$$
\begin{array}{r}
.4738 \\
-\ .2190 \\
\hline
.2548
\end{array}
$$

The shaded area of Figure 10.15 represents 25.48% of the total area under the normal curve.

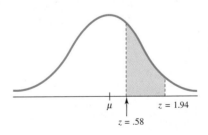

$\mu \qquad z = 1.94$
$z = .58$

FIGURE 10.15

(c) The area to the right of 2.09 standard deviations above the mean

The total area under a normal curve is 1. Thus, the total area to the right of the mean is 1/2, or .5000. From Table 2, the area from the mean to 2.09 standard

Find the following standard normal curve areas as percents of the total area.

(a) Between .31 standard deviations below the mean and 1.01 standard deviations above the mean

(b) Between .38 and 1.98 standard deviations below the mean

(c) To the right of 1.49 standard deviations above the mean

(d) What percent of the area is within 1 standard deviation of the mean? within 2 standard deviations of the mean? within 3 standard deviations of the mean? What can you conclude from the last answer?

Answers:

(a) 46.55%

(b) 32.82%

(c) 6.81%

(d) 68.3%, 95.47%, 99.7% Almost all the data lies within 3 standard deviations of the mean.

deviations above the mean is .4817. The area to the right of 2.09 standard deviations is found by subtracting .4817 from .5000.

$$\begin{array}{r} .5000 \\ - \ .4817 \\ \hline .0183 \end{array}$$

A total of 1.83% of the total area is to the right of 2.09 standard deviations above the mean. Figure 10.16 (which is not to scale) shows the desired area. ◀ ③

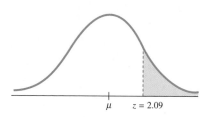

FIGURE 10.16

The key to finding areas under *any* normal curve is to express each number x on the horizontal axis in terms of standard deviations above or below the mean. The *z*-**score** for x is the number of standard deviations that x lies from the mean (positive if x is above the mean, negative if x is below the mean).

▶ **EXAMPLE 3** If a normal distribution has mean 50 and standard deviation 4, find the following *z*-scores.

(a) The *z*-score for $x = 46$
   Since 46 is 4 units below 50 and the standard deviation is 4, 46 is 1 standard deviation below the mean. So, its *z*-score is -1.

(b) The *z*-score for $x = 60$
   The *z*-score is 2.5 because 60 is 10 units above the mean (since $60 - 50 = 10$) and 10 units is 2.5 standard deviations (since $10/4 = 2.5$). ◀ ④

In Example 3(b) we found the *z*-score by taking the difference between 60 and the mean and dividing this difference by the standard deviation. The same procedure works in the general case.

If a normal distribution has mean μ and standard deviation σ, then the *z*-score for the number x is

$$z = \frac{x - \mu}{\sigma}.$$

The importance of *z*-scores is the following fact, whose proof is omitted.

④

Find each z-score using the information in Example 3.

(a) $x = 36$

(b) $x = 55$

Answers:

(a) -3.5

(b) 1.25

Area under a Normal Curve

The area under a normal curve between $x = a$ and $x = b$ is the same as the area under the standard normal curve between the z-score for a and the z-score for b.

Therefore, by converting to z-scores and using a graphing calculator or Table 2 for the standard normal curve, we can find areas under any normal curve. Since these areas are probabilities (as explained earlier), we can now handle a variety of applications.

Graphing calculators, computer programs, and CAS programs (such as DERIVE) can be used to find areas under the normal curve and hence, probabilities. The equation of the standard normal curve, with $\mu = 0$ and $\sigma = 1$, is

$$f(x) = (1/\sqrt{2\pi})e^{-x^2/2}.$$

A good approximation of the area under this curve (and above $y = 0$) can be found by using the x-interval $[-4, 4]$. See Chapter 13 for more information on finding such areas.

▶ **EXAMPLE 4**   Dixie Office Supplies finds that its sales force drives an average of 1200 miles per month per person, with a standard deviation of 150 miles. Assume that the number of miles driven by a salesperson is closely approximated by a normal distribution.

(a) Find the probability that a salesperson drives between 1200 and 1600 miles per month.

Here $\mu = 1200$ and $\sigma = 150$, and we must find the area under the normal distribution curve between $x = 1200$ and $x = 1600$. We begin by finding the z-score for $x = 1200$.

$$z = \frac{x - \mu}{\sigma} = \frac{1200 - 1200}{150} = \frac{0}{150} = 0$$

The z-score for $x = 1600$ is

$$z = \frac{x - \mu}{\sigma} = \frac{1600 - 1200}{150} = \frac{400}{150} = 2.67.*$$

So the area under the curve from $x = 1200$ to $x = 1600$ is the same as the area under the standard normal curve from $z = 0$ to $z = 2.67$, as indicated in Figure 10.17. A graphing calculator or Table 2 shows that this area is .4962. Therefore, the probability that a salesperson drives between 1200 and 1600 miles per month is .4962.

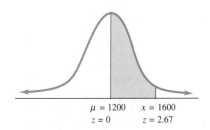

$\mu = 1200$ $x = 1600$
$z = 0$ $z = 2.67$

FIGURE 10.17

*All z-scores here are rounded to two decimal places.

(b) Find the probability that a salesperson drives between 1000 and 1500 miles per month.

As shown in Figure 10.18, z-scores for both $x = 1000$ and $x = 1500$ are needed.

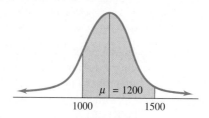

FIGURE 10.18

For $x = 1000$,

$$z = \frac{1000 - 1200}{150}$$

$$= \frac{-200}{150}$$

$$= -1.33.$$

For $x = 1500$,

$$z = \frac{1500 - 1200}{150}$$

$$= \frac{300}{150}$$

$$= 2.00.$$

From the table $z = 1.33$ leads to an area of .4082, while $z = 2.00$ corresponds to .4773. A total of $.4082 + .4773 = .8855$, or 88.55%, of all drivers travel between 1000 and 1500 miles per month. From this, the probability that a driver travels between 1000 and 1500 miles per month is .8855. ◀ ⑤

▶ **EXAMPLE 5**   A tire store finds that the tread life of its tires is normally distributed, with a mean of 26,640 miles and a standard deviation of 4000 miles. The store sold 9000 tires this month. How many of them can be expected to last more than 35,000 miles?

Here $\mu = 26,640$ and $\sigma = 4000$. The probability that a tire will last more than 35,000 miles is the area under the normal curve to the right of $x = 35,000$. The z-score for $x = 35,000$ is

$$z = \frac{x - \mu}{\sigma} = \frac{35,000 - 26,640}{4000} = \frac{8360}{4000} = 2.09.$$

Example 2(c) and Figure 10.16 show that the area to the right of $z = 2.09$ is .0183, which is 1.83% of the total area under the curve. Therefore, 1.83% of the tires can be expected to last more than 35,000 miles. Thus,

$$1.83\% \text{ of } 9000 = .0183 \cdot 9000 = 164.7,$$

or approximately 165 tires can be expected to last more than 35,000 miles. ◀

NOTE   Notice in Example 5 that $P(z \geq 2.09) = P(z > 2.09.)$. The area under the curve is the same, whether we include the endpoint or not. Notice also that $P(z = 2.09) = 0$, because no area is included.

CAUTION When calculating normal probability, it is wise to draw a normal curve with the mean and the z-scores every time. This will avoid confusion as to whether you should add or subtract probabilities.

As mentioned earlier, z-scores are standard deviations, so $z = 1$ corresponds to 1 standard deviation above the mean, and so on. As found in side problem 3(d) of this section, 68.3% of the area under a normal curve lies within 1 standard deviation of the mean. Also, 95.47% lies within 2 standard deviations of the mean, and 99.7% lies within 3 standard deviations of the mean. These results, summarized in Figure 10.19, can be used to get a quick estimate of results when working with normal curves.

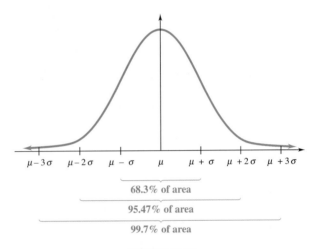

FIGURE 10.19

Boxplots

The normal curve is useful because you can easily read various characteristics of the data from the picture. Boxplots are another graphical means of presenting key characteristics of a data set. The idea is to arrange the data in increasing order and choose three numbers Q_1, Q_2, and Q_3 that divide it into four equal parts, as indicated schematically here.

The number Q_1 is called the **first quartile,** the median Q_2 is called the **second quartile,** and Q_3 is called the **third quartile.** These numbers are used to construct a boxplot, as illustrated in the following examples.

▶ **EXAMPLE 6** Construct a boxplot for the Britelite data in Exercise 23 of Section 10.2:

$$20, 22, 22, 25, \mathbf{26}, \mathbf{27}, 27, 28, 30, 35.$$

The data is already ranked. The minimum score is 20 and the maximum score is 35. Since there are an even number of scores, the median score Q_2 is 26.5 (halfway between the two center entries). To find Q_1, which separates the bottom 25% of the data from the rest, we first calculate

25% of n (rounded up to the nearest integer),

where n is the number of data. Here $n = 10$ and $.25(10) = 2.5$, which rounds to 3. Now count to the *third* score, to get $Q_1 = 22$.

Similarly, since Q_3 separates the bottom 75% of the data from the rest, we calculate

75% of n (rounded up to the nearest integer).

When $n = 10$, then $.75(10) = 7.5$, which rounds to 8. Count to the *eighth* score, getting $Q_3 = 28$.

The five key numbers for constructing the boxplot are

$$\text{Minimum,} \quad Q_1, \quad Q_2, \quad Q_3, \quad \text{Maximum}$$
$$\searrow \qquad \downarrow \quad \downarrow \quad \downarrow \qquad \swarrow$$
$$20, \quad 22, \quad \mathbf{26.5}, \quad 28, \quad 35$$

Draw a horizontal line parallel to the number axis, from the minimum to the maximum score. Around this line, construct a box whose ends are at Q_1 and Q_3 and mark the location of Q_2 by a vertical line in the box, as shown in Figure 10.20. ◀

FIGURE 10.20

Because they don't show the detail of a histogram or stem-and-leaf plots, boxplots are not as useful for analyzing a single data set. However, they can be useful for comparing two sets of data, as long as the same scale is used.

▶ **EXAMPLE 7** In Exercise 23 of Section 10.2, the following data was given for Brand X:

$$15, 18, 19, 23, 25, 25, 28, 30, 34, 38.$$

Construct a boxplot for this data and compare it to the one for Britelite in Example 6.

The minimum is 15, the maximum is 38, and the median is $Q_2 = 25$. Again $n = 10$ and

25% of $n = .25(10) = 2.5$, which rounds up to 3

so that $Q_1 = 19$ (the *third* data score). Similarly,

$$75\% \text{ of } n = .75(10) = 7.5, \text{ which rounds up to } 8,$$

so that $Q_3 = 30$ (the *eighth* data score). The minimum, maximum, and the three quartiles (namely, 15, 19, 25, 30, and 38) are used to get the boxplot in Figure 10.21.

FIGURE 10.21

⑥

Construct a box plot for the number of blood types data given in Section 10.2, Exercise 27.

Answer:

Comparing the graphs in Figures 10.20 and 10.21 shows that the range of hours in use is less for Brand X bulbs than for Britelite bulbs, the width of the boxplot indicates where the middle 50% of the distribution lie, and the median indicates how that 50% is distributed. The boxplot for Britelite bulbs is narrower than that for Brand X, showing that they are more consistent in number of hours of use, and the higher placement of the median indicates that they tend to last longer than brand X. ◀ ⑥

TECHNOLOGY   TIP   Many graphing calculators can graph boxplots for single-variable data. The procedure is similar to plotting data points using the STAT PLOT menu.

10.3 Exercises

1. The peak in a normal curve occurs directly above _____.
2. The total area under a normal curve (above the horizontal axis) is _____.
3. How are z-scores found for normal distributions where $\mu \neq 0$ or $\sigma \neq 1$?
4. How is the standard normal curve used to find probabilities for normal distributions?

Find the percent of the area under a normal curve between the mean and the following number of standard deviations from the mean. (See Example 2.)

5. 1.75
6. .23
7. $-.43$
8. -2.1

Find the percent of the total area under the standard normal curve between the following z-scores. (See Examples 1 and 2.)

9. $z = 1.41$ and $z = 2.83$
10. $z = .64$ and $z = 2.11$

11. $z = -2.48$ and $z = -.05$
12. $z = -1.74$ and $z = -1.02$
13. $z = -3.11$ and $z = 1.44$
14. $z = -2.94$ and $z = -.43$

Find a z-score satisfying each of the following conditions. (Hint: Use Table 2 backward or a graphing calculator.)

15. 5% of the total area is to the right of z.
16. 1% of the total area is to the left of z.
17. 15% of the total area is to the left of z.
18. 25% of the total area is to the right of z.
19. For any normal distribution, what is the value of $P(x \leq \mu)$? $P(x \geq \mu)$?
20. Compare the probability that a number will lie within 2 standard deviations of the mean of a probability distribution using Chebyshev's theorem and using the normal

distribution. (See Exercises 15–22, Section 10.2) Explain what you observe.

21. Repeat Exercise 20 using 3 standard deviations.

Assume the following distributions are all normal. (See Example 4.)

22. **Business** According to the label, a regular can of Campbell's soup holds an average of 305 grams with a standard deviation of 4.2 grams. What is the probability that a can will be sold that holds more than 306 grams?

23. **Business** A jar of Adams Old Fashioned Peanut Butter contains 453 grams with a standard deviation of 10.1 grams. Find the probability that one of these jars contains less than 450 grams.

24. **Business** A General Electric soft white 3-way bulb has an average life of 1200 hours with a standard deviation of 50 hours. Find the probability that the life of one of these bulbs will be between 1150 and 1300 hours.

25. **Business** A 100-watt lightbulb has an average brightness of 1640 lumens with a standard deviation of 62 lumens. What is the probability that a 100-watt bulb will have a brightness between 1600 and 1700 lumens?

26. **Social Science** The scores on a standardized test in a suburban high school have a mean of 80 with a standard deviation of 12. What is the probability that a student will have a score less than 60?

27. **Health** The mean clotting time of blood is 7.45 seconds with a standard deviation of 3.6 seconds. What is the probability that an individual's blood clotting time will be less than 7 seconds or greater than 8 seconds?

28. **Social Science** The average time to complete the test in Exercise 26 is 40 minutes with a standard deviation of 5.2 minutes. Find the probability that a student will require more than 50 minutes to finish the test.

29. **Social Science** A 120-minute videotape has a standard deviation of .7 minute. Find the probability that a 118-minute movie will not be entirely copied.

30. **Natural Science** The distribution of low temperatures for a city has an average of 44° with a standard deviation of 6.7°. What is the probability that the low temperature will be above 55°?

Business *A certain type of lightbulb has an average life of 500 hours, with a standard deviation of 100 hours. The length of life of the bulb can be closely approximated by a normal curve. An amusement park buys and installs 10,000 such bulbs. Find the total number that can be expected to last for each of the following periods of time. (See Example 5.)*

31. At least 500 hours

32. Less than 500 hours

33. Between 650 and 780 hours

34. Between 290 and 540 hours

35. Less than 740 hours

36. More than 300 hours

37. Find the shortest and longest lengths of life for the middle 80% of the bulbs.

Social Science *New studies by Federal Highway Administration traffic engineers suggest that speed limits on many thoroughfares are set arbitrarily and often are artificially low. According to traffic engineers, the ideal limit should be the "85th percentile speed." This means the speed at or below which 85 percent of the traffic moves. Assuming speeds are normally distributed, find the 85th percentile speed for roads with the following conditions.*

38. The mean speed is 50 mph with a standard deviation of 10 mph.

39. The mean speed is 30 mph with a standard deviation of 5 mph.

Social Science *One professor uses the following system for assigning letter grades in a course.*

Grade	Total Points
A	Greater than $\mu + \frac{3}{2}\sigma$
B	$\mu + \frac{1}{2}\sigma$ to $\mu + \frac{3}{2}\sigma$
C	$\mu - \frac{1}{2}\sigma$ to $\mu + \frac{1}{2}\sigma$
D	$\mu - \frac{3}{2}\sigma$ to $\mu - \frac{1}{2}\sigma$
F	Below $\mu - \frac{3}{2}\sigma$

What percent of the students receive the following grades?

40. A

41. B

42. C

43. Do you think the system in Exercises 40–42 would be more likely to be fair in a large freshman class in psychology or in a graduate seminar of five students? Why?

Health *In nutrition, the recommended daily allowance of vitamins is a number set by the government as a guide to an individual's daily vitamin intake. Actually, vitamin needs vary* drastically from person to person, but the needs are very closely approximated by a normal curve. To calculate the recommended daily allowance, the government first finds the average need for vitamins among people in the population and then the standard deviation. The recommended daily allowance is defined as the mean plus 2.5 times the standard deviation.

44. What percent of the population will receive adequate amounts of vitamins under this plan?

Find the recommended daily allowance for the following vitamins.

45. Mean = 500 units, standard deviation = 50 units

46. Mean = 1800 units, standard deviation = 140 units

47. Mean = 159 units, standard deviation = 12 units

48. Mean = 1200 units, standard deviation = 92 units

Social Science *The mean performance score of a large group of fifth-grade students on a math achievement test is 88. The scores are known to be normally distributed. What percent of the students had scores as follows?*

49. More than 1 standard deviation above the mean

50. More than 2 standard deviations above the mean

Social Science *Studies have shown that women are charged an average of $500 more than men for cars.* Assume a normal distribution of overcharges with a mean of $500 and a standard deviation of $60. Find the probability of a woman's paying the following additional amounts for a car.*

51. Less than $500

52. At least $600

53. Between $400 and $600

---

*"From repair shops to cleaners, women pay more," by Bob Dart as appeared in The Chicago Tribune, May 27, 1993. Reprinted by permission of the author.

Social Science *Women earn an average of $.73 for every $1 earned by a man, a difference of $.27.* Assume differences in pay are normally distributed with a mean of $.27 and a standard deviation of $.08. Find the probability that a woman earns the following amounts less than a man in the same job.*

54. Between $.20 and $.30   **55.** More than $.19

56. At most $.27   **57.** At least $.25

58. **Business** The table gives the percent of U.S. agricultural exports to selected nations.[†] Use this information in the following exercises. (See Example 6.)

	1990	*1998*
Japan	20.6	18.5
South Korea	6.7	5.4
Taiwan	4.2	4
China	2.1	1.8
Hong Kong	1.8	2.5
European Union	17.4	13.7
Mexico	6.5	11.7
Canada	10.7	14.6
Africa	4.9	4.5

(a) Construct a boxplot for the 1990 data.

(b) Construct a boxplot for the 1998 data.

(c) Use the graphs in parts (a) and (b) to compare the 1990 and 1998 distributions.

---

*Bureau of the Census, U.S. Department of Commerce.

†U.S. Department of Agriculture, Agricultural Statistics, annual.

10.4 NORMAL APPROXIMATION TO THE BINOMIAL DISTRIBUTION

As we saw in Section 9.3, many practical experiments have only two possible outcomes, sometimes referred to as success or failure. Such experiments are called Bernoulli trials or Bernoulli processes. Examples of Bernoulli trials include flipping a coin (with heads being a success, for instance, and tails a failure) or testing a computer chip coming off the assembly line to see whether or not it is defective. A binomial experiment consists of repeated independent Bernoulli trials, such as flipping a coin 10 times or taking a random sample of 20 computer chips from the assembly line. In Section 9.4 we found the probability distribution for several binomial experiments, such as sampling five people with bachelor's degrees in education and counting how many are women. The probability distribution for a binomial experiment is known as a **binomial distribution.**

As another example, it is reported that 29% of drivers on Interstate 10 in Texas exceed the 70 mph speed limit.* Suppose a state trooper wants to verify this statistic and records the speed of 10 randomly selected drivers. The trooper finds that 4 out of 10, or 40%, exceed the speed limit. How likely is this if the 29% figure is accurate? We can answer this question with the binomial probability formula

$$\binom{n}{x} \cdot p^x \cdot (1 - p)^{n-x},$$

where n is the size of the sample (10 in this case), x is the number of speeders (4 in this example), and p is the probability that a driver is a speeder (.29). This gives

$$P(x = 4) = \binom{10}{4} \cdot .29^4 \cdot (1 - .29)^6$$

$$= 210(.007073)(.1281) \approx .1903.$$

The probability is almost 20%, so this result is not unusual.

Suppose that the state trooper takes a larger random sample of 100 drivers. What is the probability that 40 or more drivers speed if the 29% figure is accurate? Calculating $P(x = 40) + P(x = 41) + \cdots + P(x = 100)$ is a formidable task. One solution is provided by graphing calculators or computers. On the TI-83, for example, we can first calculate the probability that 39 or fewer drivers exceed the speed limit using the DISTR menu command binomcdf (100, .29, 39). Subtracting the answer from 1 gives a probability of .0119. But this high-tech method fails as n becomes larger; the command binomcdf (1000, .29, 300) gives an error message. On the other hand, there is a low-tech method that works regardless of the size of n. It has further interest because it connects two different distributions: the normal and the binomial. The normal distribution is continuous, since the random variable can take on any real number. The binomial distribution is *discrete,* because the random variable can take on only integer values between 0 and n. Nevertheless, the normal distribution can be used to give a good approximation to binomial probability.

In order to use the normal approximation, we first need to know the mean and standard deviation of the binomial distribution. Recall from Section 9.4 that for the binomial distribution, $E(x) = np$. In Section 10.1, we referred to $E(x)$ as μ, and that notation will be used here. It is shown in more advanced courses in statistics that the standard deviation of the binomial distribution is given by $\sigma = \sqrt{np(1 - p)}$.

①

Find μ and σ for a binomial distribution having $n = 120$ and $p = 1/6$.

Answer:

$\mu = 20; \sigma = 4.08$

Mean and Standard Deviation for the Binomial Distribution

For the binomial distribution, the mean and standard deviation are given by

$$\mu = np \quad \text{and} \quad \sigma = \sqrt{np(1 - p)},$$

where n is the number of trials and p is the probability of success on a single trial. ①

Time, May 13, 1996, p. 34.

▶**EXAMPLE 1**   Suppose a fair coin is flipped 15 times.

(a) Find the mean and standard deviation for the number of heads.

Using $n = 15$ and $p = 1/2$, the mean is

$$\mu = np = 15\left(\frac{1}{2}\right) = 7.5.$$

The standard deviation is

$$\sigma = \sqrt{np(1 - p)} = \sqrt{15\left(\frac{1}{2}\right)\left(1 - \frac{1}{2}\right)}$$

$$= \sqrt{15\left(\frac{1}{2}\right)\left(\frac{1}{2}\right)} = \sqrt{3.75} \approx 1.94.$$

We expect, on average, to get 7.5 heads out of 15 tosses. Most of the time, the number of heads will be within three standard deviations of the mean, or between $7.5 - 3(1.94) = 1.68$ and $7.5 + 3(1.94) = 13.32$.

(b) Find the probability distribution for the number of heads and draw a histogram of the probabilities.

The probability distribution is found by putting $n = 15$ and $p = 1/2$ into the formula for binomial probability. For example, the probability of 9 heads is given by

$$P(x = 9) = \binom{15}{9}\left(\frac{1}{2}\right)^9\left(1 - \frac{1}{2}\right)^6 \approx .15274.$$

Probabilities for the other values of x between 0 and 15, as well as a histogram of the probabilities, are shown in Figure 10.22.

x	P(x)
0	.00003
1	.00046
2	.00320
3	.01389
4	.04166
5	.09164
6	.15274
7	.19638
8	.19638
9	.15274
10	.09164
11	.04166
12	.01389
13	.00320
14	.00046
15	.00003

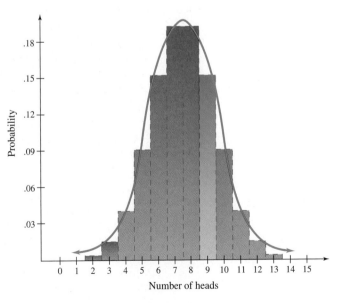

FIGURE 10.22

In Figure 10.22, we have superimposed the normal curve with $\mu = 7.5$ and $\sigma = 1.94$ over the histogram of the distribution. Notice how well the normal distribution fits the binomial distribution. This approximation was first discovered in 1718 by Abraham De Moivre (1667–1754) for the case $p = 1/2$. The result was generalized by the French mathematician Pierre-Simon Laplace (1749–1827) in a book published in 1812. As n becomes larger and larger, a histogram for the binomial distribution looks more and more like a normal curve. Figures 10.23 (a) and (b) show histograms of the binomial distribution with $p = .3$, using $n = 8$ and $n = 50$, respectively.

(a) $n = 8$

(b) $n = 50$

FIGURE 10.23

The probability of getting exactly 9 heads in the 15 tosses, or .15274, is the same as the area of the blue bar in Figure 10.22. As the graph suggests, the blue area is approximately equal to the area under the normal curve from $x = 8.5$ to $x = 9.5$. The normal curve is higher than the top of the bar in the left half but lower in the right half.

To find the area under the normal curve from $x = 8.5$ to $x = 9.5$, first find z-scores, as in the previous section. Use the mean and the standard deviation for the distribution, which we have already calculated, to get z-scores for $x = 8.5$ and $x = 9.5$.

For $x = 8.5$,
$$z = \frac{8.5 - 7.5}{1.94}$$
$$= \frac{1.00}{1.94}$$
$$z \approx .52.$$

For $x = 9.5$,
$$z = \frac{9.5 - 7.5}{1.94}$$
$$= \frac{2.00}{1.94}$$
$$z \approx 1.03.$$

(2)

Use the normal distribution to find the probability of getting exactly the following number of heads in 15 tosses of a coin.

(a) 7

(b) 10

Answers:

(a) .1985

(b) .0909

From Table 2, $z = .52$ gives an area of .1985, and $z = 1.03$ gives .3485. The difference between these two numbers is the desired result.

$$.3485 - .1985 = .1500$$

This answer (.1500) is not far from the more accurate answer of .15274 found above. ◀ (2)

CAUTION The normal curve approximation to a binomial distribution is quite accurate *provided that n* is large and *p* is not close to 0 or 1. As a rule of thumb, the normal curve approximation can be used as long as both *np* and $n(1 - p)$ are at least 5.

▶ **EXAMPLE 2** Consider the random sample discussed earlier of 100 drivers on Interstate 10 in Texas, where 29% of the drivers exceed the 70 mph speed limit.

(a) Use the normal distribution to approximate the probability that at least 40 drivers exceed the speed limit.

First find the mean and the standard deviation using $n = 100$ and $p = .29$.

$$\mu = 100(.29) \qquad \sigma = \sqrt{100(.29)(1 - .29)}$$
$$= 29 \qquad\qquad = \sqrt{100(.29)(.71)}$$
$$= \sqrt{20.59} \approx 4.54$$

As the graph in Figure 10.24 shows, we need to find the area to the right of $x = 39.5$ (since we want 40 or more speeders). The z-score corresponding to $x = 39.5$ is

$$z = \frac{39.5 - 29}{4.54} \approx 2.31.$$

From the table, $z = 2.31$ leads to an area of .4896, so

$$P(z > 2.31) = .5 - .4896 = .0104.$$

This value is close to the value of .0119 found earlier with the help of a graphing calculator. Either method tells us there is roughly a 1% chance of finding 40 or more speeders out of a random sample of 100. If the trooper found this many in his sample, he might suspect that either his sample is not truly random or the 29% figure is too low.

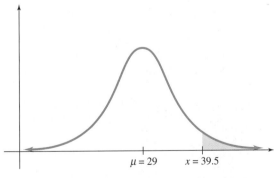

$\mu = 29$ $\quad$ $x = 39.5$

FIGURE 10.24

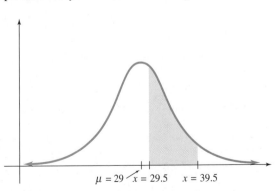

$\mu = 29$ $\quad \bar{x} = 29.5$ $\quad$ $x = 39.5$

FIGURE 10.25

About 9% of the transistors produced by a certain factory are defective. Find the approximate probability that in a sample of 200, the following numbers of transistors will be defective. (*Hint:* $\sigma = 4.05$)

(a) Exactly 11

(b) 16 or fewer

(c) More than 14

Answers:

(a) .0226

(b) .3557

(c) .8051

(b) Find the probability of finding between 30 and 35 speeders in a random sample of 100.

As Figure 10.25 shows, we need to find the area between $x = 29.5$ and $x = 35.5$.

$$\text{If } x = 29.5, \text{ then } z = \frac{29.5 - 29}{4.54} \approx .11.$$

$$\text{If } x = 35.5, \text{ then } z = \frac{35.5 - 29}{4.54} \approx 1.43.$$

Use the table to find that $z = .11$ gives an area of .0458 and $z = 1.43$ yields .4236. The final answer is the difference of these numbers.

$$P(.11 \leq z \leq 1.43) = .4236 - .0458 = .3798.$$

The probability of finding between 30 and 35 speeders is about .3798. ◀ ③

10.4 Exercises

1. What must be known to find the mean and standard deviation of a binomial distribution?

2. What is the rule of thumb for using the normal distribution to approximate a binomial distribution?

Suppose 16 coins are tossed. Find the probability of getting each of the following results (a) using the binomial probability formula, and (b) using the normal curve approximation. (See Examples 1 and 2.)

3. Exactly 8 heads

4. Exactly 7 heads

5. More than 12 tails

6. Fewer than 5 tails

For the remaining exercises in this section, use the normal curve approximation to the binomial distribution.

Suppose 1000 coins are tossed. Find the probability of getting each of the following results.

7. Exactly 500 heads

8. Exactly 510 heads

9. 480 heads or more

10. Fewer than 470 tails

A die is tossed 120 times. Find the probability of getting each of the following results.

11. Exactly twenty 5's

12. Exactly twenty-four 6's

13. More than eighteen 3's

14. Fewer than twenty-two 6's

15. A reader asked Mr. Statistics (a feature in *Fortune* magazine) about the game of 26 once played in the bars of

Chicago.* The player chooses a number between 1 and 6 and then rolls a cup full of 10 dice 13 times. Out of the 130 numbers rolled, if the number chosen appears at least 26 times, the player wins. Calculate the probability of winning.

16. Social Science According to the National Center for Education Statistics, 79% of the U.S. holders of bachelor's degrees in education in 1992–93 were women.[†] Suppose five holders of bachelor's degrees in education are picked at random.

(a) Find the probability that four of the five are women.

(b) Find the probability that at least two of the five are women.

17. Natural Science For certain bird species, with appropriate assumptions, the number of nests escaping predation has a binomial distribution.[‡] Suppose the probability of success (that is, a nest escaping predation) is .3. Find the probability that at least half of 26 nests escape predation.

18. Natural Science Under certain appropriate assumptions, the probability of a competing young animal eating

*Daniel Seligman and Patty De Llosa, "Ask Mr. Statistics," *Fortune,* May 1, 1995, p. 141.

†*The New York Times,* January 7, 1996, Education Life, p. 24.

‡From H. M. Wilbur, *American Naturalist,* vol. 111.

x units of food is binomially distributed, with *n* equal to the maximum number of food units the animal can acquire and *p* equal to the probability per time unit that an animal eats a unit of food.* Suppose $n = 120$ and $p = .6$.
 (a) Find the probability that an animal consumes 80 units of food.
 (b) Suppose the animal must consume at least 70 units of food to survive. What is the probability that this happens?

19. **Business**   A recent study of minimum wage earners found that 25.6% of them are teenagers.[†] Suppose a random sample of 600 minimum wage earners is selected. What is the probability that more than 160 of them are teenagers?

20. **Social Science**   In Exercise 19, what is the probability that less than 140 of them are teenagers?

21. **Health**   A study found that 23.6% of people on kidney dialysis in the United States die, more than in any other industrial country.[‡] A hospital surveyed 40 of its patients who had been on dialysis that year and found that 15 died so far. What is the probability that 15 or more people out of 40 would die if this hospital's patients are typical of those in the United States?

22. **Natural Science**   A flu vaccine has a probability of 80% of preventing a person who is inoculated from getting the flu. A county health office inoculates 134 people. Find the probabilities of the following.
 (a) Exactly 10 of the people inoculated get the flu.
 (b) No more than 10 of the people inoculated get the flu.
 (c) None of the people inoculated get the flu.

23. **Natural Science**   The probability that a male will be color blind is .042. Find the probabilities that in a group of 53 men, the following will be true.
 (a) Exactly 5 are color blind.
 (b) No more than 5 are color blind.
 (c) At least 1 is color blind.

24. **Business**   The probability that a certain machine turns out a defective item is .05. Find the probabilities that in a run of 75 items, the following results are obtained.
 (a) Exactly 5 defectives
 (b) No defectives
 (c) At least 1 defective

25. **Business**   A company is taking a survey to find out whether people like its product. Their last survey indicated that 70% of the population likes the product. Based on that, of a sample of 58 people, find the probabilities of the following.

 (a) All 58 like the product.
 (b) From 28 to 30 (inclusive) like the product.

26. **Natural Science**   The blood types B− and AB− are the rarest of the eight human blood types, representing 1.5% and .6% of the population, respectively.*
 (a) If the blood types of a random sample of 1000 blood donors are recorded, what is the probability that 10 or more of the samples are AB−?
 (b) If the blood types of a random sample of 1000 blood donors are recorded, what is the probability that 20 to 40 inclusive of the samples are B−?
 (c) If a particular city had a blood drive in which 500 people gave blood and 3% of the donations were B−, would we have reason to believe that this town has a higher than normal number of donors who are B−? (*Hint:* Calculate the probability of 15 or more donors being B− for a random sample of 500 and then consider the probability obtained.)

27. **Social Science**   In Washington State, there were 27.4 motorcycle rider fatalities per 1000 motorcycle crashes in 1989.[†] If this rate is a true reflection of the nationwide proportion of motorcycle crashes that result in death, find the probability that for a state that has 3000 motorcycle accidents in a given year, there will be between 80 and 100 fatalities.

28. **Social Science**   In one state, 55% of the voters expect to vote for Bill Poole. Suppose 1400 people are asked the name of the person for whom they expect to vote. Find the probability that at least 750 people will say that they expect to vote for Poole.

29. **Social Science**   A poll of 2000 teenagers found that 1 in 8 reported carrying a weapon for protection.[‡] In a typical high school with 1200 students, what is the probability that more than 120 students, but fewer than 180, carry a weapon?

30. **Social Science**   Only 1 out of 12 American parents requires that children do their homework before watching TV.[§] If your neighborhood is typical, what is the probability that that out of 51 parents, 5 or fewer require their children to do homework before watching TV?

31. In the 1989 U.S. Open, four golfers each made a hole in one on the same par-3 hole on the same day. *Sports Illustrated* writer R. Reilly stated the probability of a hole

*From G. deJong, *American Naturalist*, vol. 110.

†*Chicago Tribune*, December 28, 1995, pp. 1, 14.

‡*New York Times*, December 4, 1995, p. A1.

The Handy Science Answer Book, The Carnegie Library, Pittsburgh, Pennsylvania, 2000, p. 332.

†Jefferson Rowland, Frederick Rivara, Philip Salzberg, Robert Soderberg, Ronald Maier, and Thomas Keopsell, "Motorcycle Helmet Use and Injury Outcome and Hospitalization Costs from Crashes in Washington State," *American Journal of Public Health* 86, no. 1 (January 1996): 41–45.

‡*New York Times*, January 12, 1996, p. A6.

§"Harper's Index," *Harper's*, September 1996: 15.

in one for a given golf pro on a given par-3 hole to be 1/3709.*

(a) For a specific par-3 hole, use the binomial distribution to find the probability that 4 or more of the 156 golf pros in the tournament field shoot a hole in one.[†]

---

*R. Reilly, "King of the Hill," *Sports Illustrated,* June 1989, pp. 20–25.

[†]Bonnie Litwiller and David Duncan, "The Probability of a Hole in One," *School Science and Mathematics* 91, no. 1, (January 1991): 30.

(b) For a specific par-3 hole, use the normal approximation to the binomial distribution to find the probability that 4 or more of the 156 golf pros in the tournament field shoot a hole in one. Why must we be very cautious when using this approximation for this application?

(c) If the probability of a hole in one remains constant and is 1/3709 for any par-3 hole, find the probability that in 20,000 attempts by golf pros, there will be 4 or more hole in ones. Discuss whether this assumption is reasonable.

 CHAPTER 10 **SUMMARY**

Key Terms and Symbols

10.1 Σ summation (sigma) notation
   $\bar{x}$ sample mean
   μ population mean
   random sample
   grouped frequency distribution
   frequency polygon
   (arithmetic) mean
   median

statistic
mode
10.2 s^2 sample variance
   s sample standard deviation
   σ population standard deviation
   range
   deviations from the mean
   variance
   standard deviation
10.3 μ mean of a continuous distribution

   σ standard deviation of a normal distribution
continuous distribution
skewed distribution
normal distributions
normal curves
standard normal curve
z-score
boxplot
quartile
10.4   binomial distribution

Key Concepts

To organize the data from a sample, we use a **grouped frequency distribution,** a set of intervals with their corresponding frequencies. The same information can be displayed with a histogram, a type of bar graph with a bar for each interval. Each bar has width 1 and height equal to the probability of the corresponding interval. A **stem-and-leaf plot** presents the individual data in a similar form, so it can be viewed as bars. Another way to display this information is with a **frequency polygon,** which is formed by connecting the midpoints of consecutive bars of the histogram with straight line segments.

The **mean** $\bar{x}$ of a frequency distribution is the expected value.

For n numbers $x_1, x_2, \ldots , x_n,$

$$\bar{x} = \frac{\Sigma x}{n}.$$

For a grouped distribution,

$$\bar{x} = \frac{\Sigma(xf)}{n}.$$

The **median** is the middle entry in a set of data arranged in either increasing or decreasing order.

The **mode** is the most frequent entry in a set of numbers.

The **range** of a distribution is the difference between the largest and smallest numbers in the distribution.

The **sample standard deviation** s is the square root of the sample **variance.**

For n numbers,

$$s = \sqrt{\frac{\Sigma x^2 - n\bar{x}^2}{n - 1}}.$$

For a grouped distribution,

$$s = \sqrt{\frac{\Sigma fx^2 - n\bar{x}^2}{n - 1}}.$$

A **normal distribution** is a continuous distribution with the following properties: The highest frequency is at the mean; the graph is symmetric about a vertical line through the mean; the total area under the curve, above the x-axis, is 1. If a normal distribution has mean μ and standard deviation σ, then the z-score for the number x is $z = \dfrac{x - \mu}{\sigma}$. A **boxplot** organizes a list of data using the minimum and maximum values, the median, and the first and third quartiles to give a visual overview of the distribution.

Area under a Normal Curve   The area under a normal curve between $x = a$ and $x = b$ gives the probability that an observed data value will be between a and b.

The **binomial distribution** is a distribution with the following properties: For n independent repeated trials, where the probability of success in a single trial is p, the probability of x successes is $\dbinom{n}{x}p^x(1 - p)^{n-x}$. The mean is $\mu = np$ and the standard deviation is

$$\sigma = \sqrt{np(1 - p)}.$$

Chapter 10 Review Exercises

1. Discuss some reasons for organizing data into a grouped frequency distribution.

2. What is the rule of thumb for an appropriate interval in a grouped frequency distribution?

In Exercises 3 and 4, (a) write a frequency distribution; (b) draw a histogram; (c) draw a frequency polygon. (d) In Exercise 4, also construct a stem-and-leaf plot, using two separate intervals for 9 and 10.

3. The following numbers give the sales in dollars for the lunch hour at a local hamburger store for the last twenty Fridays. (Use intervals 450–474, 475–499, and so on.)

480 451 501 478 512 473 509 515 458 566
516 535 492 558 488 547 461 475 492 471

4. The number of units carried in one semester by the students in a business mathematics class was as follows. (Use intervals of 9–10, 11–12, 13–14, 15–16.)

10   9 16 12 13 15 13 16 15 11 13

12 12 15 12 14 10 12 14 15 15 13

Find the mean for each of the following.

5. 41, 60, 67, 68, 72, 74, 78, 83, 90, 97

6. 105, 108, 110, 115, 106, 110, 104, 113, 117

7.

Interval	Frequency
10–19	6
20–29	12
30–39	14
40–49	10
50–59	8

8.

Interval	Frequency
40–44	2
45–49	5
50–54	7
55–59	10
60–64	4
65–69	1

9. What do the mean, median, and mode of a distribution have in common? How do they differ? Describe each in a sentence or two.

Find the median and the mode (or modes) for each of the following.

10.   32, 35, 36, 44, 46, 46, 59

11.   38, 36, 42, 44, 38, 36, 48, 35

The modal class is the interval containing the most data values. Find the modal class for the distributions of

12.   Exercise 7 above.

13.   Exercise 8 above.

14.   What is meant by the range of a distribution?

15.   How are the variance and the standard deviation of a distribution related? What is measured by the standard deviation?

Find the range and standard deviation for each of the following distributions.

16.   14, 17, 18, 19, 32

17.   26, 43, 51, 29, 37, 56, 29, 82, 74, 93

Find the standard deviation for the following.

18.   Exercise 7 above.

19.   Exercise 8 above.

20.   Describe the characteristics of a normal distribution.

21.   What is meant by a skewed distribution?

Find the following areas under the standard normal curve.

22.   Between $z = 0$ and $z = 1.27$

23.   To the left of $z = .41$

24.   Between $z = -1.88$ and $z = 2.10$

25.   Between $z = 1.53$ and $z = 2.82$

26.   Find a z-score such that 8% of the area under the curve is to the right of z.

27.   Why is the normal distribution not a good approximation of a binomial distribution that has a value of p close to 0 or 1?

Business   The table gives the probability distribution of U.S. buyers of recorded music in 1999 by age.*

Age	Probability
10–14	.085
15–19	.126
20–24	.126
25–29	.105
30–34	.101
35–39	.104
40–44	.093
45 & up	.247

28.   **(a)** Draw a histogram for this distribution.
  (b) From the shape of the histogram, is this a normal distribution?

29.   As shown in the distribution above, 24.7% of buyers of recorded music are age 45 or over. Assume a binomial distribution with $n = 6$ and $p = .247$.
  (a) Find the probability that 4 of 6 randomly selected recorded music buyers are age 45 or over.
  (b) Find the mean and standard deviation of the binomial distribution. Interpret your results.

30.   Finance   The annual returns of two stocks for three years are given below.

	2000	2001	2002
Stock I	11%	−1%	14%
Stock II	9%	5%	10%

  (a) Find the mean and standard deviation for each stock over the three-year period.
  (b) If you are looking for security with an 8% return, which of these two stocks would you choose using your answers from part (a) to decide?

31.   Natural Science   The weight gains of two groups of 10 rats fed on two different experimental diets were as follows.

	Weight Gains									
Diet A	1	0	3	7	1	1	5	4	1	4
Diet B	2	1	1	2	3	2	1	0	1	0

Compute the mean and standard deviation for each group and compare them to answer the questions below.
  (a) Which diet produced the greatest mean gain?
  (b) Which diet produced the most consistent gain?

32.   Natural Science   Refer to Exercise 31.
  (a) Construct boxplots for each diet data.
  (b) Use the boxplots to compare weight gains for the two diets.

33.   Health   Between 1990 and 1999 HMO enrollment in the United States increased as shown below.*

Location	Gains
South	16.8%
Midwest	20.7%
Northeast	22.1%
West	18.2%

  (a) Compute the mean and standard deviation of the gains.
  (b) How many standard deviations from the mean is the greatest gain? the smallest gain?

34. Social Science   On standard IQ tests, the mean is 100, with a standard deviation of 15. The results are very close to fitting a normal curve. Suppose an IQ test is given to a very large group of people. Find the percent of people whose IQ score is
(a) more than 130;
(b) less than 85;
(c) between 85 and 115.

35. Business   A machine that fills quart orange juice cartons is set to fill them with 32.1 oz. If the actual contents of the cartons vary normally, with a standard deviation of .1 oz, what percent of the cartons contains less than a quart (32 oz)?

36. Business   About 60% of the bags of blended coffee produced by a certain company contain cheap coffee. Assume a binomial distribution and use the normal distribution to find the probability that, in a sample of 500 bags of coffee, the following is true.
(a) 310 or fewer contain cheap coffee.
(b) Exactly 300 contain cheap coffee.

37. Natural Science   An area infested with fruit flies is to be sprayed with a chemical that is known to be 98% effective for each application. A sample of 100 flies is checked. Assume a binomial distribution and use the normal distribution to approximate the following probabilities.
(a) 95% of the flies are killed in one application.
(b) At least 95% of the flies are killed in one application.
(c) At least 90% of the flies are killed in one application.
(d) All the flies are killed in one application. Compare your answer with the probability found using the binomial distribution.

38. Natural Science   Twenty percent of third-world women would like to avoid pregnancy but are not using birth control. For 50 such women, use the normal curve approximation to find the probability of each of the following. Assume a binomial distribution and use the normal distribution approximation.
(a) Exactly 10 are not using birth control.
(b) At least 20 are using birth control.

39. Health   A survey of injuries during practice or performance given to 313 Broadway performers produced the demographics in the table.*

	Mean	Standard Deviation
Dancer's Age (female)	28.0	5.5
Dancer's Age (male)	32.2	8.4
Height in m (female)	1.64	.08
Duration as Professional in Years (female)	9.5	6.0
Total No. of Injuries as Performer (female)	3.0	2.2

*Randolph Evans, Richard Evans, Scott Carvajal, and Susan Perry, "A Survey of Injuries among Broadway Performers," *American Journal of Public Health* 86, no.1 (January 1996): 77–80.

Assume that all of these demographics follow a normal distribution.
(a) Find the probability that a female dancer is 35 years old or older.
(b) Find the probability that a male dancer is 35 years old or older.
(c) Compare your answers to parts (a) and (b).
(d) Find the probability that a female performer is 1.4 meters tall or taller.
(e) Find the probability that a female performer has a career duration that is more than 1.5 standard deviations from the mean.
(f) Would a female who has more than 6 injuries during her career be considered a rare event? Explain.

40. Much of our work in Chapters 9 and 10 is interrelated. Note the similarities in the following parallel treatments of a frequency distribution and a probability distribution on the next page.

Frequency Distribution

Complete the table below for the following data.
(Recall that *x* is the midpoint of the interval.)

14, 7, 1, 11, 2, 3, 11, 6, 10, 13, 11, 11, 16, 12, 9, 11, 9, 10, 7, 12, 9, 6, 4, 5, 9, 16, 12, 12, 11, 10, 14, 9, 13, 10, 15, 11, 11, 1, 12, 12, 6, 7, 8, 2, 9, 12, 10, 15, 9, 3

Interval	Tally	x	f	x · f
1–3				
4–6				
7–9				
10–12				
13–15				
16–18				

Probability Distribution

A binomial distribution has $n = 10$ and $p = .5$. Complete the table to the right.

For each distribution, do the following for (a)–(c).

(a) Find the mean (or expected value).

(b) Find the standard deviation.

(c) Use Chebyshev's theorem to find the interval that contains 75% of the distribution. (See Section 10.2 exercises.)

(d) Use the normal approximation of the binomial probability distribution to find the interval that contains 95.44% of that distribution.

(e) Why can't we use the normal distribution to answer probability questions about the frequency distribution?

x	$P(x)$	$x \cdot P(x)$
0	.001	
1	.010	
2	.044	
3	.117	
4		
5		
6		
7		
8		
9		
10		

C a s e 9

Statistics in the Law—
The *Castaneda* Decision

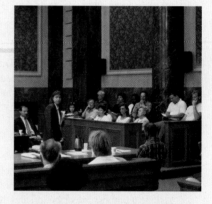

Statistical evidence is now routinely presented in both criminal and civil cases. In this application we'll look at a famous case that established use of the binomial distribution and measurement by standard deviation as an accepted procedure.*

Defendants who are convicted in criminal cases sometimes appeal their conviction on the grounds that the jury that indicted or convicted them was drawn from a pool of jurors that does not represent the population of the district in which they live. These appeals almost always cite the Supreme Court's decision in *Castaneda v. Partida* [430 U.S. 482], a case that dealt with the selection of grand juries in the state of Texas. The decision summarizes the facts this way:

*The Castaneda case and many other interesting applications of statistics in law are discussed in Michael O. Finkelstein and Bruce Levin, *Statistics for Lawyers*, New York, Springer-Verlag, 1990. U.S. Supreme Court decisions are online at http://www.findlaw.com/casecode/supreme.html, and most states now have important state court decisions online.

After respondent, a Mexican-American, had been convicted of a crime in a Texas District Court and had exhausted his state remedies on his claim of discrimination in the selection of the grand jury that had indicted him, he filed a habeas corpus petition in the Federal District Court, alleging a denial of due process and equal protection under the Fourteenth Amendment, because of gross underrepresentation of Mexican-Americans on the county grand juries.

The case went to the Appeals Court, which noted that "the county population was 79% Mexican-American, but, over an 11-year period, only 39% of those summoned for grand jury service were Mexican-American," and concluded that together with other testimony about the selection process, "the proof offered by respondent was sufficient to demonstrate a prima facie case of intentional discrimination in grand jury selection. . . ."

The state appealed to the Supreme Court, and the Supreme Court needed to decide whether the underrepresentation of Mexican-Americans on grand juries was indeed too extreme to be an effect of chance. To do so, they invoked the binomial distribution. Here is the argument:

> Given that 79.1% of the population is Mexican-American, the expected number of Mexican-Americans among the 870 persons summoned to serve as grand jurors over the 11-year period is approximately 688. The observed number is 339. Of course, in any given drawing some fluctuation from the expected number is predicted. The important point, however, is that the statistical model shows that the results of a random drawing are likely to fall in the vicinity of the expected value. . . .
>
> The measure of the predicted fluctuations from the expected value is the standard deviation, defined for the binomial distribution as the square root of the product of the total number in the sample (here 870) times the probability of selecting a Mexican-American (.791) times the probability of selecting a non-Mexican-American (.209). . . . Thus, in this case the standard deviation is approximately 12. As a general rule for such large samples, if the difference between the expected value and the observed number is greater than two or three standard deviations, then the hypothesis that the jury drawing was random would be suspect to a social scientist. The 11-year data here reflect a difference between the expected and observed number of Mexican-Americans of approximately 29 standard deviations. A detailed calculation reveals that the likelihood that such a substantial departure from the expected value would occur by chance is less than 1 in 10^{140}.

The Court decided that the statistical evidence supported the conclusion that jurors were not randomly selected, and that it was up to the state to show that its selection process did not discriminate against Mexican-Americans. The Court concluded:

> The proof offered by respondent was sufficient to demonstrate a prima facie case of discrimination in grand jury selection. Since the State failed to rebut the presumption of purposeful discrimination by competent testimony, despite two opportunities to do so, we affirm the Court of Appeals' holding of a denial of equal protection of the law in the grand jury selection process in respondent's case.

Exercises

1. Check the Court's calculation of 29 standard deviations as the difference between the expected number of Mexican-Americans and the number actually chosen.

2. Where do you think the Court's figure of 1 in 10^{140} came from?

3. The *Castaneda* decison also presents data from a $2\frac{1}{2}$-year period during which the state district judge supervised the selection process. During this period, 220 persons were called to serve as grand jurors, and only 100 of these were Mexican-American.

   (a) Considering the 220 jurors as a random selection from a large population, what is the expected number of Mexican-Americans, using the 79.1% population figure?

   (b) If we model the drawing of jurors as a sequence of 220 independent Bernoulli trials, what is the standard deviation of the number of Mexican-Americans?

   (c) About how many standard deviations is the actual number of Mexican-Americans drawn (100) from the expected number that you calculated in (a)?

   (d) What does the normal distribution table at the back of the book tell you about this result?

4. The following information is from an appeal brought by Hy-Vee stores before the Iowa Supreme Court, appealing a ruling by the Iowa Civil Rights Commission in favor of a female employee of one of their grocery stores.

   > In 1985, there were 112 managerial positions in the ten Hy-Vee stores located in Cedar Rapids. Only 6 of these managers were women. During that same year there were 294 employees; 206 were men and 88 were women.

   (a) How far from the expected number of women in management was the actual number, assuming that gender had nothing to do with promotion? Measure the difference in standard deviations.

   (b) Does this look like evidence of purposeful discrimination?

APPENDIX A

Graphing Calculators

A summary of the contents of this appendix appears below. The entire appendix is available at our Web site: www.aw.com/MWA8. You can print as much or as little of it as you want.

Part 1: Using Your Calculator

This is a brief introduction to the features of a graphing calculator that are relevant to the topics in this text, including the following.

> **Basics:** keyboard, edit/replay, scientific notation
> **Function Graphing:** trace, zoom, maximum/minimum finder, intersection finder
> **Function Evaluation:** tables, functional notation, eval key
> **Equation Solving:** root finder, polynomial solver, equation solver
> **Linear Regression and Other Regression Procedures**
> **Matrices:** matrix editor, matrix arithmetic, row operations

Part 2: Calculator Programs

The programs listed below are available for TI, Sharp 9600, and Casio 9850 (they work on some other Casio models as well). TI users may download them directly, if they have the appropriate link hardware and software. All programs may also be printed and then entered into your calculator by hand.

1. Fraction Conversion for Casio and Sharp 9600
2. Quadratic Formula for TI-82/83 and Sharp 9600
3. Table Maker for TI-85
4. Present and Future Value of an Annuity
5. Loan Payment
6. Loan Balance after n Payments
7. Amortization Table
8. RREF Program for TI-82 and Casio 9850
9. Simplex Method
10. Two-Stage Method
11. Rectangle Approximation of $\int_a^b f(x)\,dx$ (using left endpoints)

Programs 1–3 and 8 provide the specified calculators with programs that are built-in on most other models. Programs 7 and 9–11 are not built-in on any calculators. Programs 4–6 are included in the TI-83 Financial menu (and in similar menus that can be downloaded from TI for TI-86 and TI-89), though some students may find the versions here easier to use.

Tables

Table 1 Formulas from Geometry

CIRCLE
Area: $A = \pi r^2$
Circumference: $C = 2\pi r$

RECTANGLE
Area: $A = lw$
Perimeter: $P = 2l + 2w$

PARALLELOGRAM
Area: $A = bh$
Perimeter: $P = 2a + 2b$

TRIANGLE
Area: $A = \dfrac{1}{2} bh$

SPHERE
Volume: $V = \dfrac{4}{3} \pi r^3$
Surface area: $A = 4\pi r^2$

RECTANGULAR BOX
Volume: $V = lwh$
Surface area: $A = 2lh + 2wh + 2lw$

CIRCULAR CYLINDER
Volume: $V = \pi r^2 h$
Surface area: $A = 2\pi r^2 + 2\pi rh$

TRIANGULAR CYLINDER
Volume: $V = \dfrac{1}{2} bhl$

CONE
Volume: $V = \dfrac{1}{3} \pi r^2 h$

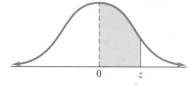

Table 2   Areas under the Normal Curve
The column under A gives the proportion of the area under the entire curve that is between $z = 0$ and a positive value of z.

z	A	z	A	z	A	z	A
.00	.0000	.48	.1844	.96	.3315	1.44	.4251
.01	.0040	.49	.1879	.97	.3340	1.45	.4265
.02	.0080	.50	.1915	.98	.3365	1.46	.4279
.03	.0120	.51	.1950	.99	.3389	1.47	.4292
.04	.0160	.52	.1985	1.00	.3413	1.48	.4306
.05	.0199	.53	.2019	1.01	.3438	1.49	.4319
.06	.0239	.54	.2054	1.02	.3461	1.50	.4332
.07	.0279	.55	.2088	1.03	.3485	1.51	.4345
.08	.0319	.56	.2123	1.04	.3508	1.52	.4357
.09	.0359	.57	.2157	1.05	.3531	1.53	.4370
.10	.0398	.58	.2190	1.06	.3554	1.54	.4382
.11	.0438	.59	.2224	1.07	.3577	1.55	.4394
.12	.0478	.60	.2258	1.08	.3599	1.56	.4406
.13	.0517	.61	.2291	1.09	.3621	1.57	.4418
.14	.0557	.62	.2324	1.10	.3643	1.58	.4430
.15	.0596	.63	.2357	1.11	.3665	1.59	.4441
.16	.0636	.64	.2389	1.12	.3686	1.60	.4452
.17	.0675	.65	.2422	1.13	.3708	1.61	.4463
.18	.0714	.66	.2454	1.14	.3729	1.62	.4474
.19	.0754	.67	.2486	1.15	.3749	1.63	.4485
.20	.0793	.68	.2518	1.16	.3770	1.64	.4495
.21	.0832	.69	.2549	1.17	.3790	1.65	.4505
.22	.0871	.70	.2580	1.18	.3810	1.66	.4515
.23	.0910	.71	.2612	1.19	.3830	1.67	.4525
.24	.0948	.72	.2642	1.20	.3849	1.68	.4535
.25	.0987	.73	.2673	1.21	.3869	1.69	.4545
.26	.1026	.74	.2704	1.22	.3888	1.70	.4554
.27	.1064	.75	.2734	1.23	.3907	1.71	.4564
.28	.1103	.76	.2764	1.24	.3925	1.72	.4573
.29	.1141	.77	.2794	1.25	.3944	1.73	.4582
.30	.1179	.78	.2823	1.26	.3962	1.74	.4591
.31	.1217	.79	.2852	1.27	.3980	1.75	.4599
.32	.1255	.80	.2881	1.28	.3997	1.76	.4608
.33	.1293	.81	.2910	1.29	.4015	1.77	.4616
.34	.1331	.82	.2939	1.30	.4032	1.78	.4625
.35	.1368	.83	.2967	1.31	.4049	1.79	.4633
.36	.1406	.84	.2996	1.32	.4066	1.80	.4641
.37	.1443	.85	.3023	1.33	.4082	1.81	.4649
.38	.1480	.86	.3051	1.34	.4099	1.82	.4656
.39	.1517	.87	.3079	1.35	.4115	1.83	.4664
.40	.1554	.88	.3106	1.36	.4131	1.84	.4671
.41	.1591	.89	.3133	1.37	.4147	1.85	.4678
.42	.1628	.90	.3159	1.38	.4162	1.86	.4686
.43	.1664	.91	.3186	1.39	.4177	1.87	.4693
.44	.1700	.92	.3212	1.40	.4192	1.88	.4700
.45	.1736	.93	.3238	1.41	.4207	1.89	.4706
.46	.1772	.94	.3264	1.42	.4222	1.90	.4713
.47	.1808	.95	.3289	1.43	.4236	1.91	.4719

Table 2   *(continued)*

z	A	z	A	z	A	z	A
1.92	.4726	2.42	.4922	2.92	.4983	3.42	.4997
1.93	.4732	2.43	.4925	2.93	.4983	3.43	.4997
1.94	.4738	2.44	.4927	2.94	.4984	3.44	.4997
1.95	.4744	2.45	.4929	2.95	.4984	3.45	.4997
1.96	.4750	2.46	.4931	2.96	.4985	3.46	.4997
1.97	.4756	2.47	.4932	2.97	.4985	3.47	.4997
1.98	.4762	2.48	.4934	2.98	.4986	3.48	.4998
1.99	.4767	2.49	.4936	2.99	.4986	3.49	.4998
2.00	.4773	2.50	.4938	3.00	.4987	3.50	.4998
2.01	.4778	2.51	.4940	3.01	.4987	3.51	.4998
2.02	.4783	2.52	.4941	3.02	.4987	3.52	.4998
2.03	.4788	2.53	.4943	3.03	.4988	3.53	.4998
2.04	.4793	2.54	.4945	3.04	.4988	3.54	.4998
2.05	.4798	2.55	.4946	3.05	.4989	3.55	.4998
2.06	.4803	2.56	.4948	3.06	.4989	3.56	.4998
2.07	.4808	2.57	.4949	3.07	.4989	3.57	.4998
2.08	.4812	2.58	.4951	3.08	.4990	3.58	.4998
2.09	.4817	2.59	.4952	3.09	.4990	3.59	.4998
2.10	.4821	2.60	.4953	3.10	.4990	3.60	.4998
2.11	.4826	2.61	.4955	3.11	.4991	3.61	.4999
2.12	.4830	2.62	.4956	3.12	.4991	3.62	.4999
2.13	.4834	2.63	.4957	3.13	.4991	3.63	.4999
2.14	.4838	2.64	.4959	3.14	.4992	3.64	.4999
2.15	.4842	2.65	.4960	3.15	.4992	3.65	.4999
2.16	.4846	2.66	.4961	3.16	.4992	3.66	.4999
2.17	.4850	2.67	.4962	3.17	.4992	3.67	.4999
2.18	.4854	2.68	.4963	3.18	.4993	3.68	.4999
2.19	.4857	2.69	.4964	3.19	.4993	3.69	.4999
2.20	.4861	2.70	.4965	3.20	.4993	3.70	.4999
2.21	.4865	2.71	.4966	3.21	.4993	3.71	.4999
2.22	.4868	2.72	.4967	3.22	.4994	3.72	.4999
2.23	.4871	2.73	.4968	3.23	.4994	3.73	.4999
2.24	.4875	2.74	.4969	3.24	.4994	3.74	.4999
2.25	.4878	2.75	.4970	3.25	.4994	3.75	.4999
2.26	.4881	2.76	.4971	3.26	.4994	3.76	.4999
2.27	.4884	2.77	.4972	3.27	.4995	3.77	.4999
2.28	.4887	2.78	.4973	3.28	.4995	3.78	.4999
2.29	.4890	2.79	.4974	3.29	.4995	3.79	.4999
2.30	.4893	2.80	.4974	3.30	.4995	3.80	.4999
2.31	.4896	2.81	.4975	3.31	.4995	3.81	.4999
2.32	.4898	2.82	.4976	3.32	.4996	3.82	.4999
2.33	.4901	2.83	.4977	3.33	.4996	3.83	.4999
2.34	.4904	2.84	.4977	3.34	.4996	3.84	.4999
2.35	.4906	2.85	.4978	3.35	.4996	3.85	.4999
2.36	.4909	2.86	.4979	3.36	.4996	3.86	.4999
2.37	.4911	2.87	.4980	3.37	.4996	3.87	.5000
2.38	.4913	2.88	.4980	3.38	.4996	3.88	.5000
2.39	.4916	2.89	.4981	3.39	.4997	3.89	.5000
2.40	.4918	2.90	.4981	3.40	.4997		
2.41	.4920	2.91	.4982	3.41	.4997		

Table 3   Integrals

(C is an arbitrary constant.)

1. $\displaystyle\int x^n\,dx = \frac{1}{n+1}x^{n+1} + C \quad (n \neq -1)$

2. $\displaystyle\int e^{kx}\,dx = \frac{1}{k}e^{kx} + C$

3. $\displaystyle\int \frac{a}{x}\,dx = a\ln|x| + C$

4. $\displaystyle\int \ln|ax|\,dx = x(\ln|ax| - 1) + C$

5. $\displaystyle\int \frac{1}{\sqrt{x^2 + a^2}}\,dx = \ln\left|x + \sqrt{x^2 + a^2}\right| + C$

6. $\displaystyle\int \frac{1}{\sqrt{x^2 - a^2}}\,dx = \ln\left|x + \sqrt{x^2 - a^2}\right| + C$

7. $\displaystyle\int \frac{1}{a^2 - x^2}\,dx = \frac{1}{2a}\cdot\ln\left|\frac{a+x}{a-x}\right| + C \quad (a \neq 0)$

8. $\displaystyle\int \frac{1}{x^2 - a^2}\,dx = \frac{1}{2a}\cdot\ln\left|\frac{x-a}{x+a}\right| + C \quad (a \neq 0)$

9. $\displaystyle\int \frac{1}{x\sqrt{a^2 - x^2}}\,dx = -\frac{1}{a}\cdot\ln\left|\frac{a + \sqrt{a^2 - x^2}}{x}\right| + C \quad (a \neq 0)$

10. $\displaystyle\int \frac{1}{x\sqrt{a^2 + x^2}}\,dx = -\frac{1}{a}\cdot\ln\left|\frac{a + \sqrt{a^2 + x^2}}{x}\right| + C \quad (a \neq 0)$

11. $\displaystyle\int \frac{x}{ax + b}\,dx = \frac{x}{a} - \frac{b}{a^2}\cdot\ln|ax + b| + C \quad (a \neq 0)$

12. $\displaystyle\int \frac{x}{(ax + b)^2}\,dx = \frac{b}{a^2(ax + b)} + \frac{1}{a^2}\cdot\ln|ax + b| + C \quad (a \neq 0)$

13. $\displaystyle\int \frac{1}{x(ax + b)}\,dx = \frac{1}{b}\cdot\ln\left|\frac{x}{ax + b}\right| + C \quad (b \neq 0)$

14. $\displaystyle\int \frac{1}{x(ax + b)^2}\,dx = \frac{1}{b(ax + b)} + \frac{1}{b^2}\cdot\ln\left|\frac{x}{ax + b}\right| \quad (b \neq 0)$

15. $\displaystyle\int \sqrt{x^2 + a^2}\,dx = \frac{x}{2}\sqrt{x^2 + a^2} + \frac{a^2}{2}\cdot\ln\left|x + \sqrt{x^2 + a^2}\right| + C$

16. $\displaystyle\int x^n\cdot\ln|x|\,dx = x^{n+1}\left[\frac{\ln|x|}{n+1} - \frac{1}{(n+1)^2}\right] + C \quad (n \neq -1)$

17. $\displaystyle\int x^n e^{ax}\,dx = \frac{x^n e^{ax}}{a} - \frac{n}{a}\cdot\int x^{n-1}e^{ax}\,dx + C \quad (a \neq 0)$

Answers to Selected Exercises

CHAPTER 1

Section 1.1 (Page 9)

1. true **3.** identity property of addition **5.** commutative property of addition **7.** associative property of multiplication
11. -54 **13.** $-\dfrac{1}{2}$ **15.** 18% **17.** .75% **19.** 4
21. -4 **23.** -1 **25.** irrational; 9.4248
27. irrational; 1.7321 **29.** $5 < 7$ **31.** $y \le 8.3$ **33.** $t > 0$
35.

$\qquad$ -8 $\qquad$ -1

37.

$\qquad$ -2 $\qquad$ 2

39.

$\qquad$ -4 -2 0 2 4

41.

$\qquad$ 3

43.

$\qquad$ -2

45. 4 **47.** 6 **49.** 0 **51.** (a) 17.6 (b) no
53. (a) 29.3 (b) no **55.** 47° **57.** 22° **59.** 4
61. -19 **63.** = **65.** = **67.** = **69.** =
71. $-(a - 7) = 7 - a$
77. Let x represent the number of dollars contributed annually by foreign students to the California economy. Then $x > 1,000,000,000$.
79. Let x represent the percent of foreign students now in the United States from Middle Eastern countries. Then $0 \le x < 7.5$.
81. Let x represent the percent of foreign students in the United States that are in California. Then $x > 13$.

Section 1.2 (Page 16)

1. 37,974,983.3583 **3.** 289.099133864 **7.** 2^7
9. $(-5)^7 = -5^7$ **11.** $(2z)^{11}$ **13.** $-x^3 + x^2 + 3x$
15. $-6y^2 + 3y + 10$ **17.** $-10x^2 + 4x - 2$
19. $-18m^3 - 27m^2 + 9m$ **21.** $12z^3 + 14z^2 - 7z + 5$
23. $12k^2 - 20k + 3$ **25.** $6y^2 + 7y - 5$
27. $18k^2 - 7kq - q^2$ **29.** $4.34m^2 + 5.68m - 4.42$
31. $-k + 3$ **33.** $R = 5000x$; $C = 150,000 + 2250x$;
$P = 2750x - 150,000$ **35.** $R = 7500x$;
$C = -3x^2 + 3480x + 299,675$; $P = 3x^2 + 4020x - 299,675$
37. (a) 9,400,000 (b) 8,900,000
39. (a) 32,700,000 (b) 32,036,000
41. 43,844,000 **43.** 49,076,000
45. (a) approximately 60,501,067 cu ft (b) The shape becomes a rectangular box with a square base, with volume b^2h. (c) yes
47. (a) 0, 1, 2, 3, or no degree (if one is the negative of the other)
(b) 0, 1, 2, 3, or no degree (if they are equal) (c) 6

Section 1.3 (Page 22)

1. $12x(x - 2)$ **3.** $r(r^2 - 5r + 1)$ **5.** $6z(z^2 - 2z + 3)$
7. $2(2y - 1)^2(5y - 1)$ **9.** $(x + 5)^4(x^2 + 10x + 28)$
11. $(x + 1)(x + 4)$ **13.** $(x + 3)(x + 4)$ **15.** $(z + 4)(z + 6)$
17. $(2x - 1)(x - 4)$ **19.** $(3p - 4)(5p - 1)$
21. $(2z - 5)(2z - 3)$ **23.** $(2x + 1)(3x - 4)$
25. $(5y - 2)(2y + 5)$ **27.** $(2x - 1)(3x + 4)$
29. $(2a + 5)(a - 1)$ **31.** $(x + 8)(x - 8)$ **33.** $(3p - 4)^2$
35. $(r + 2t)(r - 5t)$ **37.** $(m - 3n)^2$ **39.** $(2p + 3)(2p - 3)$
41. $3(x - 4z)^2$ **43.** cannot be factored
45. $(-x + 4)(x - 3)$ or $(x - 4)(-x + 3)$ **47.** $(3a + 5)(a - 6)$
49. $(7m + 2n)(3m + n)$ **51.** $(4y - x)(5y + 11x)$
53. $(y - 7z)(y + 3z)$ **55.** $(11x + 8)(11x - 8)$
57. $m^3(m - 1)^2(m^2 + m + 1)^2(5 + 3m^2 - 3m^5)$
59. $(a - 6)(a^2 + 6a + 36)$ **61.** $(2r - 3s)(4r^2 + 6rs + 9s^2)$
63. $(4m + 5)(16m^2 - 20m + 25)$
65. $(10y - z)(100y^2 + 10yz + z^2)$

Section 1.4 (Page 27)

1. $\dfrac{x}{5}$ **3.** $\dfrac{4}{7p}$ **5.** $\dfrac{5}{4}$ **7.** $\dfrac{4}{w + 3}$ **9.** $\dfrac{y - 4}{3y^2}$ **11.** $\dfrac{m - 2}{m + 3}$
13. $\dfrac{x + 4}{x + 1}$ **15.** $\dfrac{2p}{7}$ **17.** $\dfrac{9a^3}{4}$ **19.** $\dfrac{15}{8c}$ **21.** $\dfrac{3}{4}$ **23.** $\dfrac{3}{10}$
25. $\dfrac{2(a + 4)}{a - 3}$ **27.** $\dfrac{k + 2}{k + 3}$ **31.** $-\dfrac{1}{15z}$ **33.** $\dfrac{4}{3}$ **35.** $\dfrac{12 + x}{3x}$
37. $\dfrac{3m - 2}{m(m - 1)}$ **39.** $\dfrac{14}{3(a - 1)}$ **41.** $\dfrac{23}{20(k - 2)}$
43. $\dfrac{7x + 9}{(x - 3)(x + 1)(x + 2)}$ **45.** $\dfrac{y^2}{(y + 4)(y + 3)(y + 2)}$
47. $\dfrac{k^2 - 13k}{(2k - 1)(k + 2)(k - 3)}$ **49.** $\dfrac{x + 1}{x - 1}$ **51.** $\dfrac{-1}{x(x + h)}$
53. (a) $\dfrac{x - 5}{x^2 - 10x}$ (b) 32.7 sec; 8 sec; 4.5 sec

Section 1.5 (Page 37)

1. 1 **3.** $\dfrac{1}{6}$ **5.** $\dfrac{1}{32}$ **7.** $-\dfrac{1}{2401}$ **9.** 216 **11.** $\dfrac{9}{16}$
13. 7 **15.** 1.7 **17.** 9 **19.** -16 **21.** $\dfrac{81}{16}$ **23.** $\dfrac{1}{4^5}$
25. 4^3 **27.** 2 **29.** 8^5 **31.** $\dfrac{1}{9^{32/15}}$ **33.** z^3 **35.** $\dfrac{p}{4}$
37. $\dfrac{q^5}{r^2}$ **39.** $\dfrac{8}{25p^7}$ **41.** $2^{5/6}p^{3/2}$ **43.** $2p + 5p^{5/3}$ **45.** (f)
47. (h) **49.** (g) **51.** (c) **53.** 4 **55.** 5 **57.** -2
59. 9 **61.** $\sqrt{77}$ **63.** 7 **65.** 3 **67.** $\dfrac{\sqrt{5} - 1}{2}$

69. $-1 - \sqrt{3}$ **71.** $\dfrac{1}{7 + 4\sqrt{3}}$ **73.** (a) 14 (b) 85 (c) 58
75. (a) $-63.4°$ (b) $-46.7°$ (c) Approximation is very good.
77. 166.46 **79.** 175.36 **81.** about 9440 **83.** about 12,533
85. about 14,216 **87.** (a) about 2,069,800; about 3,159,800;
about 4,579,000 (b) about 2021; no

Section 1.6 (Page 48)

1. 5 **3.** 7 **5.** $-\dfrac{9}{8}$ **7.** 2 **9.** $\dfrac{40}{7}$ **11.** $\dfrac{26}{3}$

13. $-\dfrac{12}{5}$ **15.** $-\dfrac{59}{6}$ **17.** $-\dfrac{9}{4}$ **19.** $x = .72$

21. $r \approx -13.26$ **23.** $x = \dfrac{5a - b}{6}$ **25.** $x = \dfrac{3b}{a + 5}$

27. $V = \dfrac{k}{P}$ **29.** $g = \dfrac{V - V_0}{t}$ **31.** $B = \dfrac{2A}{h} - b$ or $B = \dfrac{2A - bh}{h}$

33. $2, -3$ **35.** $-2, 8$ **37.** $\dfrac{5}{2}, \dfrac{7}{2}$ **39.** $-7, -\dfrac{3}{7}$

41. 13% **43.** $4000 **45.** 1975 **47.** 1995 **49.** 1996
51. 2001 **53.** $205.41 **55.** (a) .0352 (b) about .015, or
1.5% (c) approximately one case **57.** $8000 **59.** $5000

61. about 838 mi **63.** $\dfrac{400}{3}$ L **65.** 142 mi **67.** 70 mph

69. 10.025 gal premium; 5.475 gal regular **71.** 6 cm **73.** 4 ft

Section 1.7 (Page 58)

1. $-3, 12$ **3.** $-5, 0$ **5.** $0, 2$ **7.** $-7, -8$ **9.** $\dfrac{3}{2}, 1$

11. $-\dfrac{1}{2}, \dfrac{1}{3}$ **13.** $\dfrac{5}{2}, 4$ **15.** $5, 2$ **17.** $\dfrac{4}{3}, -\dfrac{4}{3}$ **19.** $0, 1$

21. $2 \pm \sqrt{7}$ **23.** $\dfrac{1 \pm 2\sqrt{5}}{4}$ **25.** $\dfrac{-5 \pm \sqrt{17}}{4}$; $-.219, -2.281$

27. $\dfrac{-1 \pm \sqrt{5}}{4}$; $.309, -.809$ **29.** $\dfrac{-3 \pm \sqrt{19}}{5}$; $.272, -1.472$

31. no real number solutions **33.** $\dfrac{5}{2}, 1$ **35.** no real number

solutions **37.** $-5, 3/2$ **39.** 1 **41.** 2
43. $x \approx .4701$ or 1.8240 **45.** $x \approx -1.0376$ or $.6720$
47. (a) 1.5 sec (b) about 3.35 sec **49.** (a) about 32 and 62
(b) about 17 and 77 **51.** (a) 1985 (b) 2001 **53.** (a) $x + 20$
(b) northbound: $5x$; eastbound: $5(x + 20)$ or $5x + 100$
(c) $(5x)^2 + (5x + 100)^2 = 300^2$ (d) about 31.23 mph and 51.23 mph
55. (a) $150 - x$ (b) $x(150 - x) = 5000$ (c) length 100 m;
width 50 m **57.** 9 ft by 12 ft **59.** (a) 2 sec (b) $\dfrac{1}{2}$ sec or $\dfrac{7}{2}$sec
(c) It reaches the given height twice—once on the way up and once on
the way down. **61.** $t = \pm\dfrac{\sqrt{2Sg}}{g}$ **63.** $h = \dfrac{\pm d^2 \sqrt{kL}}{L}$

65. $R = \dfrac{-2Pr + E^2 \pm E\sqrt{E^2 - 4Pr}}{2P}$ **67.** (a) $x^2 - 2x = 15$

(b) $x = 5$ or $x = -3$ (c) $z = \pm\sqrt{5}$ **69.** $\pm\dfrac{\sqrt{6}}{2}$

71. $\pm\sqrt{\dfrac{3 + \sqrt{13}}{2}}$

Chapter 1 Review (Page 61)

1. $0, 6$ **3.** $-12, -6, -\dfrac{9}{10}, -\sqrt{4}, 0, \dfrac{1}{8}, 6$

5. commutative property of multiplication **7.** distributive property
9. $x \geq 6$ **11.** $-7, -3, -2, 0, \pi, 8$
13. $-|3 - (-2)|, -|-2|, |6 - 4|, |8 + 1|$ **15.** -3 **17.** -1
19.

```
          [
  ————————•————————————
       -3
```

21.

```
  ◄————————)———————————►
          -2
```

23. -8 **25.** $-\dfrac{3}{4}$

27. $4x^4 - 4x^2 + 13x$ **29.** $2q^4 + 14q^3 - 4q^2$ **31.** $15z^2 - 4z - 4$
33. $16k^2 - 9h^2$ **35.** $36x^2 + 36xy + 9y^2$ **37.** $k(2h^2 - 4h + 5)$
39. $a^2(3a + 1)(a + 4)$ **41.** $(2y - 1)(5y - 3)$ **43.** $(2a - 5)^2$
45. $(12p + 13q)(12p - 13q)$ **47.** $(2y - 1)(4y^2 + 2y + 1)$

49. $\dfrac{7x^2}{3}$ **51.** 4 **53.** $\dfrac{(y - 5)(4y - 5)}{6y}$ **55.** $\dfrac{(m - 1)^2}{3(m + 1)}$

57. $\dfrac{1}{6z}$ **59.** $\dfrac{64}{35q}$ **61.** $\dfrac{1}{5^3}$ or $\dfrac{1}{125}$ **63.** -1 **65.** $\dfrac{5^2}{6^2}$ or $\dfrac{25}{36}$

67. 4^3 **69.** $\dfrac{1}{9}$ **71.** 9^3 **73.** $\dfrac{3}{4}$ **75.** 25 **77.** $\dfrac{1}{3^5}$

79. $\dfrac{1}{5^{2/3}}$ **81.** $3^{7/2}\, a^{5/2}$ **83.** 3 **85.** $2\sqrt{6}$ **87.** $3pq\sqrt[3]{2q^2}$

89. $\dfrac{n\sqrt{30m}}{6m}$ **91.** $-8\sqrt{3}$ **93.** 4 **95.** $\dfrac{-\sqrt{2} + \sqrt{6}}{2}$

97. $64,000,000 **99.** $10,081,152 **101.** 3 **103.** -2

105. no solution **107.** $x = \dfrac{1}{5a - 1}$ **109.** $x = \dfrac{c - 3}{3a - ac - 2}$

111. $12, -6$ **113.** $-38, 42$ **115.** (a) 1999 (b) 1996
117. (a) 1993 (b) 2000 **119.** $60,000 at 8%; $40,000 at 5%
121. 2 **122.** 2 **123.** 1 **125.** 0 **127.** $\dfrac{-1 \pm \sqrt{7}}{2}$

129. $3, -\dfrac{5}{2}$ **131.** $-2 \pm \sqrt{5}$ **133.** $\dfrac{-1 \pm \sqrt{3}}{3}$ **135.** $\dfrac{1}{2}, \dfrac{1}{6}$

137. $-\dfrac{8}{3}, 2$ **139.** $\dfrac{\pm\sqrt{3}}{3}$ **141.** $\dfrac{\pm\sqrt{3}}{3}$

143. $r = \dfrac{-Rp \pm E\sqrt{Rp}}{p}$ **145.** $s = \dfrac{a \pm \sqrt{a^2 + 4K}}{2}$

147. (a) 2000 (b) 2006 **149.** 50 m by 225 m or 112.5 m by
100 m

Case 1 (Page 65)

1. $c = 700 + 85x$ **3.** The $700 refrigerator costs $300 more over
10 years.

CHAPTER 2

Section 2.1 (Page 73)

1. yes **3.** no
5.   **7.**

9.

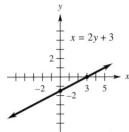

11. *x*-intercepts −1, 3.5; *y*-intercept 1   **13.** *x*-intercepts −2, 0, 2; *y*-intercept 0   **15.** *x*-intercept 8; *y*-intercept 6   **17.** *x*-intercept 3; *y*-intercept −2   **19.** *x*-intercepts 3 and −3; *y*-intercept −9

21.

23.

25.

27.

29.

31.

33.

35.

37.

39.

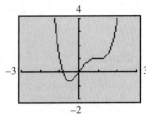

41. $x \approx -1.1038$ **43.** $x \approx 2.1017$ **45.** $x \approx -1.7521$ **47.** $r \approx 4.6580$ in.   **49.** 1989   **51.** In mid-1985 ($x \approx 10.46$) about 3,349,000 boys were involved in sports.   **53.** Fargo, about 2:00 P.M.; Seattle, about 5 P.M.   **55.** from 11 A.M. until 6 P.M.   **57.** **(a)** about \$1,250,000   **(b)** about \$1,750,000   **(c)** about \$4,250,000   **59.** **(a)** about \$500,000   **(b)** about \$1,000,000   **(c)** about \$1,500,000   **61.** approximately 12 sec   **63.** **(a)** about \$750, \$600, and \$300.   **(b)** during the 22nd year (that is, between $t = 21$ and $t = 22$)   **65.** from 1990 to the middle of 1993   **67.** It rose from about 117 at the beginning of 1997 to a high of about 138 in mid-1998, then dropped sharply to about 119, and then started up again, finishing the year around 127.   **69.** from June 1996 to approximately April 1997, briefly in fall 1998, and in the first two months of 2001

Section 2.2 (Page 86)

1. $-\dfrac{1}{2}$ **3.** -1 **5.** $-\dfrac{3}{2}$ **7.** undefined   **9.** $y = 3x + 5$ **11.** $y = -2.3x + 1.5$ **13.** $y = -\dfrac{3x}{4} + 4$ **15.** $m = 2; b = -7$ **17.** $m = 3; b = -2$ **19.** $m = \dfrac{2}{3}; b = -\dfrac{14}{9}$ **21.** $m = \dfrac{2}{3}; b = 0$ **23.** $m = 1; b = 5$ **25.** **(a)** C   **(b)** B   **(c)** B   **(d)** D

27.

29.

31.

33. perpendicular   **35.** parallel   **37.** neither   **39.** **(a)** $\dfrac{2}{5}, \dfrac{9}{8}, -\dfrac{5}{2}$ **(b)** yes   **41.** $3y = -2x + 4$ or $y = -\dfrac{2}{3}x + \dfrac{4}{3}$ **43.** $y = 2x + 2$ **45.** $y = 2$ **47.** $x = 6$ **49.** $3y = 4x + 7$ **51.** $2y = 5x - 1$ **53.** $y = 7x$ **55.** $x = 5$ **57.** $y = 2x - 2$ **59.** $y = x - 5$ **61.** $y = -x + 2$ **63.** \$1248.20

65. $5178.57   **67.** (a) 4,130,000   (b) 4,610,000   (c) 4,850,000
(d) 2004   **69.** (a) $6,928,230,000   (b) 2005
71. (a) (0, 54), (8, 63)   (b) $y = 1.125x + 54$ (c) 56,250,000
(d) 2005   **73.** (a) $y = .19x + 1.1$ (b) about 7.4 million
75. (a) $T = .03t + 15$ (b) 2103 approximately   **77.** (a) The
slope of $-.0221$ indicates that on the average, from 1912 to 1992, the
5000-meter run is being run .0221 second faster every year. It is negative because the times are generally decreasing as time progresses.
(b) World War II was going on in 1940 and 1944, and no Olympic
Games were held.

Section 2.3 (Page 96)

1. (a) $y = \dfrac{5}{9}(x - 32)$ (b) 10°C; 23.89°C   **3.** 462.2°C

5. $y = 4x + 130.7$; 150.7; 182.7   **7.** $y = 9x + 1$; $109 billion
9. $y = -.3125x + 14.25$; 42,500 sq mi
11. (a) For $y = .5x + 1.5$: 0, $-.5$, 0, $-.5$, 1; sum 0.
For $y = x$: 1, 0, 0, -1, 0; sum 0.   (b) 1.5 for $y = .5x + 1.5$ and 2 for
$y = x$ (c) $y = .5x + 1.5$ **13.** A straight line is a poor model
($r \approx .1$).   **15.** (a) two-point model: $y = -4.4615x + 522$; regression line model: $y = -4.8647x + 530.3210$ (b) two-point model:
about 321; regression line model: about 311   **17.** (a) two-point
model: $y = 15x + 2750$; regression line model: $y = 14.9x + 2822$
(b) two-point model: 5000, 6950, 9050; regression line model: 5055,
6992, 9078   (c) two-point model: 6275, 6500; regression line model:
6321.5, 6500   **19.** (a) (1, 4.8), (2, 5.8), (3, 8.8), (4, 11.5), (5, 16.2),
(6, 20.8), (7, 25.1), (8, 26.3), (9, 29.4), (10, 33.7)
(b) $y = 3.383x - .367$ (c) $43.6 billion; $50.4 billion
21. (a) $y = 1.5607x + 31.914$ (b) 2008
23. (a) $y = 56.4889x + 597.0404$ (b) Yes; coefficient of correlation is $r \approx .995$.   (c) about $1444.4 billion   (d) 2015
25. (a) $y = 50.9335x + 401.5133$ (b) Yes; coefficient of correlation is $r \approx .986$ (c) $1114.6 billion   (d) 2011

Section 2.4 (Page 106)

3. $[-4, \infty)$

5. $(-\infty, 0)$

7. $\left(-\infty, \dfrac{8}{3}\right]$

9. $(-\infty, -7]$

11. $(-\infty, 3)$ **13.** $(-1, \infty)$

15. $(-\infty, 1]$ **17.** $\left(\dfrac{1}{5}, \infty\right)$

19. $(-5, 6)$ **21.** $\left[\left(\dfrac{7}{3}, 4\right)\right]$

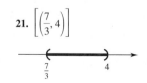

23. $\left[-\dfrac{11}{2}, \dfrac{7}{2}\right]$ **25.** $\left[-\dfrac{17}{7}, \infty\right)$

27. $x \geq 2$ **29.** $-2 < x \leq 5$ **31.** (a) Let x represent the number
of mg per L of lead in the water.   (b) $.038 \leq x \leq .042$ (c) yes
33. (a) $0 \leq x \leq 6000$; $6000 < x \leq 27,950$; $27,950 < x \leq 67,700$;
$67,700 < x \leq 141,250$; $141,250 < x \leq 307,050$; $x > 307,050$
(b) $0 \leq T \leq 600$; $600 < T \leq 3892.50$; $3892.50 < T \leq 14,625$;
$14,625 < T \leq 36,690$; $36,690 < T \leq 94,720$; $x > 94,720$
35. $(-1, 1)$ **37.** no solution

39. $(-4, -1)$

41. $\left(-\infty, -\dfrac{8}{3}\right]$ or $[2, \infty)$ **43.** $\left(-\dfrac{3}{2}, \dfrac{13}{10}\right)$

45. $29 \leq T \leq 69$ **47.** $43 \leq T \leq 81$
49. (a) $25.33 \leq R_L \leq 28.17$; $36.58 \leq R_E \leq 40.92$
(b) $5699.25 \leq T_L \leq 6338.25$ $8230.5 \leq T_E \leq 9207$
51. 1999 through 2003   **53.** 20 through 35   **55.** between 2 and
6 mi   **57.** $[500, \infty)$ **59.** $[45, \infty)$ **61.** impossible to break
even   **63.** $|x - 2| \leq 4$ **65.** $|z - 12| \geq 2$

Section 2.5 (Page 113)

1. $\left[-5, \dfrac{3}{2}\right]$ **3.** $(-\infty, -3)$ or $(-1, \infty)$

5. $\left[-2, \dfrac{1}{4}\right]$ **7.** $(-\infty, -1)$ or $\left(\dfrac{1}{4}, \infty\right)$

9. $[-5, 5]$ **11.** $(-\infty, 0)$ or $(16, \infty)$

13. $[-3, 0]$ or $[3, \infty)$ **15.** $[-6, -1]$ or $[4, \infty)$
17. $(-\infty, -4)$ or $(-1, 3)$ **19.** $\left(-\infty, -\dfrac{1}{2}\right)$ or $\left(0, \dfrac{4}{3}\right)$ **21.** no
23. $(-.0806, 2.4806)$ **25.** $[-2.2635, .7556]$ or $[3.5079, \infty)$
27. $(.5, .8393)$ **29.** $(-\infty, 1)$ or $[3, \infty)$ **31.** $\left(\dfrac{7}{2}, 5\right)$
33. $(-\infty, 2)$ or $(5, \infty)$ **35.** $(-\infty, -1)$ **37.** $(-\infty, -2)$ or $(0, 3)$
39. $[-1, .5]$ **41.** $\left(0, \dfrac{5}{3}\right)$ or $(10, \infty)$ **43.** $(100, 150)$
45. $[4, 9.75]$

Chapter 2 Review (Page 115)

1. $(-2, 3), (0, -5), (3, -2), (4, 3)$

3.

5.

7.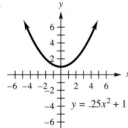

9. (a) about 11:30 A.M. to about 7:30 P.M. **(b)** from midnight until about 5 P.M. and after about 10:30 P.M. **13.** $-\dfrac{5}{6}$ **15.** $\dfrac{1}{4}$

17. 4 **19.** 0 **21.** -3

23.

25. $3y = 2x - 13$ **27.** $4y = -5x + 17$ **29.** $x = -1$
31. $3y = 5x + 15$ **33. (a)** $y = -.1625x + 25.3$ **(b)** Negative; smoking is decreasing. **(c)** 22.7% **35. (a)** $y = .365x + 9.955$
(b) $y = .3645x + 9.7975$ **(c)** two-point model: $12.15; regression model: $11.98; models high by $0.34 and $0.17, respectively
(d) two-point model: $15.07; regression model: $14.90
37. (a) $y = 695.88x + 14831.15$ **(b)** $r \approx .9878$; yes

(c) about $24,573 **39.** $\left(\dfrac{3}{8}, \infty\right)$ **41.** $\left(-\infty, \dfrac{1}{4}\right]$

43. $\left[-\dfrac{1}{2}, 2\right]$ **45.** $[-8, 8]$ **47.** $(-\infty, 2]$ or $[5, \infty)$

49. $\left[-\dfrac{9}{5}, 1\right]$ **51.** $5.5 \le y \le 9.5$ **53. (a)** $y = 23x + 415$

(b) until 1998 **(c)** after 2003 **55.** $(-3, 2)$

57. $(-\infty, -5]$ or $\left[\dfrac{3}{2}, \infty\right)$ **59.** $(-\infty, -5]$ or $[-2, 3]$ **61.** $[-2, 0)$

63. $\left(-1, \dfrac{3}{2}\right)$ **65.** $[-19, -5)$ or $(2, \infty)$ **67.** 1950–1992

Case 2 (Page 118)

3. The poor prediction isn't surprising, since we were extrapolating far beyond the range of the original data. **5.** It's not clear that any simple smooth function will fit this data—there seems to be a break in the pattern between 1970 and 1980. This will make it difficult to predict the life expectancy for females born in 2010. **7.** They used a regression equation of some kind to predict this value!

CHAPTER 3

Section 3.1 (Page 126)

1. function **3.** function **5.** not a function **7.** function
9. $(-\infty, \infty)$ **11.** $(-\infty, \infty)$ **13.** $(-\infty, 0]$ **15.** all real numbers except 1 **17.** $(-\infty, \infty)$ **19. (a)** 6 **(b)** 6 **(c)** 6 **(d)** 6
21. (a) 48 **(b)** 6 **(c)** 25.38 **(d)** 28.42
23. (a) $\sqrt{7} \approx 2.6458$ **(b)** 0 **(c)** $\sqrt{5.7} \approx 2.3875$
(d) not defined **25. (a)** -18.2 **(b)** -97.3 **(c)** -14.365
(d) -289.561 **27. (a)** 12 **(b)** 23 **(c)** 12.91 **(d)** 49.41
29. (a) $\dfrac{\sqrt{3}}{15} \approx .1155$ **(b)** not defined **(c)** $\dfrac{\sqrt{1.7}}{6.29} \approx .2073$

(d) not defined
31.

$Y_1 = 414.3125$

33. (a) $5 - p$ **(b)** $5 + r$ **(c)** $2 - m$
35. (a) $\sqrt{4 - p}\ (p \le 4)$ **(b)** $\sqrt{4 + r}\ (r \ge -4)$
(c) $\sqrt{1 - m}\ (m \le 1)$ **37. (a)** $p^3 + 1$ **(b)** $-r^3 + 1$
(c) $m^3 + 9m^2 + 27m + 28$

39. (a) $\dfrac{3}{p - 1}\ (p \ne 1)$ **(b)** $\dfrac{3}{-r - 1}\ (r \ne -1)$

(c) $\dfrac{3}{m + 2}\ (m \ne -2)$ **41.** 2 **43.** $2x + h$ **45. (a)** about

89% **(b)** about 47% **47. (a)** about 50,000; about 55,700
(b) about 157,300
49. (a)

X	Y1
4	32.536
5	36.05
6	38.004
7	39.022
8	39.728
9	40.746
10	42.7

$X = 10$

51. $f(t) = 200 - 475t$ **53. (a)** $C(x) = 125x + 36,000$ **(b)** $185

Section 3.2 (Page 137)

1.

$f(x) = -.5x + 2$

3.

$$f(x) = \begin{cases} x + 2 & \text{if } x \leq 1 \\ 3 & \text{if } x > 1 \end{cases}$$

5.

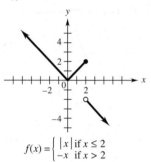

$$y = \begin{cases} 3 - x & \text{if } x \leq 0 \\ 2x + 3 & \text{if } x > 0 \end{cases}$$

7.

$$f(x) = \begin{cases} |x| & \text{if } x \leq 2 \\ -x & \text{if } x > 2 \end{cases}$$

9.

$f(x) = |x - 4|$

11.

$f(x) = |3 - 4x|$

13.

$y = -|x - 1|$

15.

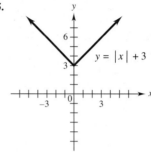

$y = |x| + 3$

17.

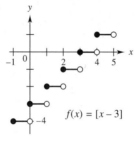

$f(x) = [x - 3]$

19.

$g(x) = [-x]$

21.

$f(x) = [x] + [-x]$

23.

25.

27.

29.

31. function   **33.** not a function   **35.** function
37. (a)

(b) When adjusted for inflation, the maximum yearly IRA contribution fell from $2000 to $1000 during this period.

39. (a)

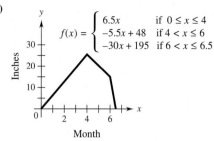

$$f(x) = \begin{cases} 6.5x & \text{if } 0 \le x \le 4 \\ -5.5x + 48 & \text{if } 4 < x \le 6 \\ -30x + 195 & \text{if } 6 < x \le 6.5 \end{cases}$$

(b) at the beginning of February; 26 in.   **(c)** begins in early October; ends in mid-April

41. (a)

$$f(x) = \begin{cases} 1.3x + 15.1 & \text{if } 0 \le x \le 25 \\ 8.5x - 164.1 & \text{if } 25 < x \le 50 \end{cases}$$

(b)

(c) 218.4   **(d)** 294.9

43. (a) No. The graph of the CPI for all items is always above the x-axis, so the percent of change is always positive; this means that the CPI is always increasing.   **(b)** Increasing 1980–1982, briefly in 1983–1984, 1987– late 1990, 1995–1997 and at the end of 1998. The percent of change was positive or zero during these years, which means the CPI was increasing.   **(c)** Decreasing 1982– late 1983, mid 1984–1987, late 1990–1995, 1997–1999. The percent of change was negative during these years, which means the CPI was decreasing.

45. (a)

(b) $y = .044x - 86.7$ **(c)** $f(x) = .044x - 86.7$ **(d)** 4.16; yes; yes
47. (a) 25; 33   **(b)** The figure has vertical line segments, which can't be part of the graph of a function (why?). To make the figure into the graph of f, delete the vertical line segments; then for each horizontal segment of the graph, put a closed dot on the left end and an open circle dot on the right end (as in Figure 3.9).
49. (a) $29   **(b)** $29   **(c)** $33   **(d)** $33

(e)

51. (a) There are many correct answers, including

Section 3.3 (Page 148)

1. Let $C(x)$ be the cost of renting a saw for x for hours; $C(x) = 12 + x$.
3. Let $P(x)$ be the cost (in cents) of parking for x half hours;
$P(x) = 30x + 35$.   **5.** $C(x) = 30x + 100$ **7.** $C(x) = 120x +$
3800   **9.** $48; $15.60; $13.80   **11.** $55.50; $11.40; $8.46
13. (a) $80,000   **(b)** $42.50   **(c)** $122,500; $1,440,000
(d) $122.50; $45   **15. (a)** $C(x) = .097x + 1.32$ **(b)** $98.32
(c) $98.417   **(d)** $.097, or 9.7¢   **(e)** $.097, or 9.7¢
17. (a) $C(x) = 10x + 500$ **(b)** $R(x) = 35x$ **(c)** $P(x) = 25x - 500$
(d) $2000   **19. (a)** $C(x) = 18x + 250$ **(b)** $R(x) = 28x$
(c) $P(x) = 10x - 250$ **(d)** $750   **21. (a)** $C(x) = 12.5x + 18,000$
(b) $R(x) = 25x$ **(c)** $P(x) = 12.5x - 18,000$ **(d)** $-$16,750$ (loss)

23. $(3, -1)$ **25.** $\left(-\dfrac{11}{4}, -\dfrac{61}{4}\right)$

27. (a) 200,000 policies $(x = 200)$
(b)

(c) revenue: $12,500; cost: $15,000   **29. (a)** $c = .126x + 1.5$
(b) $2.382 million   **(c)** about 17.857 units   **31.** Break-even point
is about 467 units; do not produce the item.   **33.** Break-even point
is about 1037 units; produce the item.   **35.** about (1988, $14.68)
37. (a) Ireland: $f(x) = 1.44x + 6.88$; Italy: $g(x) = .50x + 13.12$

(b)

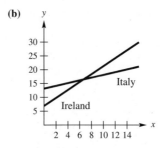

(c) About (6.64, 16.44), investment in both countries was about
$16.44 billion in late 1996.   **39.** $140   **41.** 10 items; 10 items
43. (a) $16   **(b)** $11   **(c)** $6   **(d)** 8 units   **(e)** 4 units
(f) 0 units   **(g)**

(h) 0 units   **(i)** $\dfrac{40}{3}$ units   **(j)** $\dfrac{80}{3}$ units   **(k)** See part (g).
(l) 8 units   **(m)** $6
45. (a)

(b) 125 units   **(c)** 50¢   **(d)** $[0, 125)$ **47.** Total cost increases
when more items are made (because it includes the cost of all previously made items), so the graph cannot move downward. No; the average cost can decrease as more items are made, so its graph may move
downward.

Section 3.4 (Page 158)

1. $(5, 2)$; upward   **3.** $(-6, -35)$; upward   **5.** $(-1, -1)$;
upward   **7.** x-intercepts 1, 3; y-intercept 9

9. x-intercepts $-1, -3$; y-intercept 6   **11.** $f(x) = \dfrac{1}{4}(x - 1)^2 + 2$

13. $f(x) = (x + 1)^2 - 2$
15. $(-2, 0), x = -2$ **17.** $(1, -3), x = 1$

19. $(2, 2)$, $x = 2$

$f(x) = x^2 - 4x + 6$
$f(x) = (x - 2)^2 + 2$

21. $(1, 3)$, $x = 1$

$f(x) = 2x^2 - 4x + 5$
$f(x) = 2(x - 1)^2 + 3$

23. about 47 **25.** 1995

27. (a)–(d)

(e) The graph of $f(x) = ax^2$ is the graph of $k(x) = x^2$ stretched away from the x-axis by a factor of a.

29. (a)–(d)

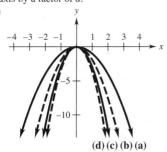

(e) Changing the sign of a reflects the graph in the x-axis.

31. (a)–(d)

(e) The graph of $f(x) = x^2 - c$ is the graph of $k(x) = x^2$ shifted c units downward. **33.** $f(x) = 3x^2$

Section 3.5 (Page 165)

1. (a) $30; $9; $105 (b)

(c) $(20, 5)$ (d) 20 boxes per day; $5 per box
3. (a) 10 milliseconds (b) 40 responses per millisecond
5. (a) 27 cases (c) 15 cases **7.** (a) about 12 books (b) 10 books (c) about 7 books (d) 0 books (e) 5 books
(f) about 7 books (g) 10 books (h) about 12 books
(i)

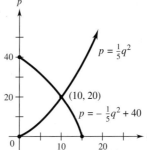

9. (a) $640 (b) $515 (c) $140
(d)

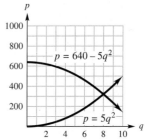

(e) 800 units (f) $320 **11.** equilibrium quantity 80; equilibrium price $3600 **13.** equilibrium quantity 30; equilibrium price $1500
15. 20 **17.** 10
19. (a) $R(x) = (100 - x)(200 + 4x)$
$= 20{,}000 + 200x - 4x^2$
(b)

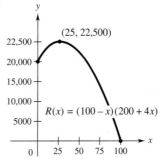

(c) 25 seats (d) $22,500 **21.** 13 weeks; $96.10/hog

23. (a) $f(x) = 1574.4023(x - 1)^2 + 159$
(b) $g(x) = 1738.7098x^2 - 4323.7917x - 4313.0848$

(c) $f(12) = 190,662$; $f(16) = 354,400$; $f(20) = 568,518$;
$g(12) = 194,176$; $g(16) = 371,616$; $g(20) = 604,695$
25. (a) $f(x) = .08677(x - 5)^2 + 109$
(b)

$g(x) \approx .060305x^2 - .0585x + 108.4373$; a very good model since
$r^2 \approx .998$, which is very close to 1
(c) $f(23) = 137.11$; $f(27) = 151$; $g(23) = 138.99$; $g(27) = 150.82$
(d) for $f(x)$, $= 163.23$ and 177.03; for $g(x)$, 160.96 and 172.18
27. 80 ft by 160 ft   **29. (a)** 11.3 and 88.7   **(b)** 50   **(c)** $3000
(d) $x < 11.3$ or $x > 88.7$ **(e)** $11.3 < x < 88.7$

Section 3.6 (Page 175)

1.

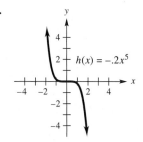

$f(x) = x^4$

3.

$h(x) = -.2x^5$

5. (a) yes **(b)** no **(c)** no **(d)** yes   **7. (a)** yes **(b)** no
(c) yes **(d)** no   **9.** D   **11.** B   **13.** E
15.

$f(x) = (x + 2)(x - 3)(x + 4)$

17.

$f(x) = x^2(x - 2)(x + 3)$

19.

$f(x) = x^3 + x^2 - 6x$

21.

$f(x) = x^3 + 3x^2 - 4x$

23. $-3 \le x \le 5$ and $-20 \le y \le 5$ **25.** $-3 \le x \le 4$ and
$-35 \le y \le 20$ **27. (a)** $933.33 billion   **(b)** $1200 billion
(c) $1145.8 billion   **(d)** $787.5 billion
(e)

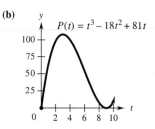

29. (a) 0; 108; 28; 10   **(b)**

$P(t) = t^3 - 18t^2 + 81t$

(c) increasing for years 0 to 3 and from the 9th year on; decreasing for
the years 3 to 9   **31. (a)** about 5880   **(b)** about 17,742
(c) about 47,134   **(d)** 2001
33. (a)

(b) $f(x) = -.037005x^4 + 1.51962x^3 - 18.4164x^2 + 63.5282x + 78.8951$ **(c)**

(d) a sharp increase, which seems unlikely

Section 3.7 (Page 185)

1. $x = -5, y = 0$

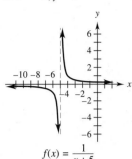

$$f(x) = \frac{1}{x+5}$$

3. $x = -\dfrac{5}{2}, y = 0$

$$f(x) = \frac{-3}{2x+5}$$

5. $x = 1, y = 3$

$$f(x) = \frac{3x}{x-1}$$

7. $x = 4; y = 1$

$$f(x) = \frac{x+1}{x-4}$$

9. $x = 3; y = -1$

$$f(x) = \frac{2-x}{x-3}$$

11. $x = -\dfrac{1}{2}; y = \dfrac{1}{2}$

$$f(x) = \frac{2x-1}{4x+2}$$

13. $x = -4, x = 1, y = 0$

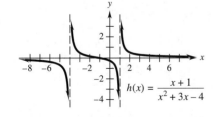

$$h(x) = \frac{x+1}{x^2+3x-4}$$

15. $x = -1, x = 1, y = 1$

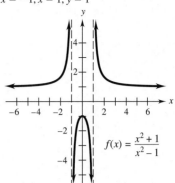

$$f(x) = \frac{x^2+1}{x^2-1}$$

17. $x = -2; x = 1$ **19.** $x = -1; x = 5$ **21.** **(a)** \$4300
(b) \$10,033.33 **(c)** \$17,200 **(d)** \$38,700 **(e)** \$81,700
(f) \$210,700 **(g)** \$425,700 **(h)** no
(i)

$$f(x) = \frac{4.3x}{100-x}$$

23. (a) $[0, \infty)$
(b)

$$f(x) = \frac{x}{1+x}$$

(c)

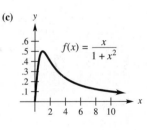

$$f(x) = \frac{x}{1+x^2}$$

(d) Increasing b makes the next generation smaller when this generation is larger.   **25. (a)** 6 min   **(b)** 1.5 min   **(c)** .6 min
(d) $A = 0$ **(e)**

$$W = \frac{S(S-A)}{A}$$

(f) W becomes negative. The waiting time approaches 0 as A approaches 3. The formula does not apply for $A > 3$ because there will be no waiting if people arrive more than 3 min apart.

27. (a) 30,000 reds; 10,000 blues

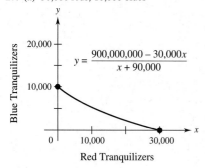

29. (a) $C(x) = 2.6x + 40,000$

(b) $\overline{C}(x) = \dfrac{2.6x + 40,000}{x} = 2.6 + \dfrac{40,000}{x}$

(c) $y = 2.6$; the average cost may get close to, but will never equal, $2.60.   **31.** about 73.9

33. (a)

(b) They appear almost identical because the parabola is an asymptote of the graph.

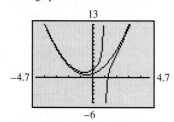

Chapter 3 Review (Page 188)

1. not a function   **3.** function   **5.** not a function
7. (a) 23   (b) -9 (c) $4p - 1$ (d) $4r + 3$
9. (a) -28 (b) -12 (c) $-p^2 + 2p - 4$ (d) $-r^2 - 3$
11. (a) -13 (b) 3   (c) $-k^2 - 4k$ (d) $-9m^2 + 12m$
(e) $-k^2 + 14k - 45$ (f) $12 - 5p$
13.

15.

17.

19.

21.

23.

25. (a)

(b) domain: $(0, \infty)$; range: $\{7, 11, 15, 19, \ldots\}$ **(c)** 3 days
27. These births appear to be leveling off.

29. (a) $C(x) = 30x + 60$ **(b)** $30 **(c)** $30.60   **31. (a)** $C(x) = 30x + 85$ **(b)** $30
(c) $30.85   **33. (a)** $18,000 **(b)** $R(x) = 28x$
(c) 4500 cartridges **(d)** $126,000   **35.** Equilibrium quantity is 36 million subscribers at a price of $12.95 per month.
37. upward; $(2, 6)$ **39.** downward; $(-1, 8)$
41.

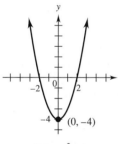

$f(x) = x^2 - 4$

43.

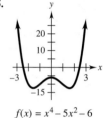

$f(x) = x^2 + 2x - 3$
$f(x) = (x + 1)^2 - 4$

45.

$f(x) = -x^2 - 4x + 1$
$f(x) = -(x + 2)^2 + 5$

47.

$f(x) = 2x^2 + 4x - 3$
$f(x) = 2(x + 1)^2 - 5$

49. minimum value; -11 **51.** maximum value; 7
53. 4 months after she began   **55.** 125 units
57. (a) $f(x) = \dfrac{11}{36}(x - 5)^2 + 39$ **(b)** $54 billion

59.

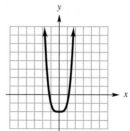

$f(x) = x^4 - 2$

61.

$f(x) = x(x - 2)(x + 3)$

63.

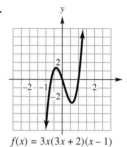

$f(x) = 3x(3x + 2)(x - 1)$

65.

$f(x) = x^4 - 5x^2 - 6$

67. about 313,152; about $690.72 per thousand
69. (a) $R(x) = 23x$; $P(x) = .000006x^3 - .07x^2 + 21x - 1200$
(b) about 76.54, which means that 77 must be made to earn a profit (you can't make .54 of a rack)   **(c)** 230   **(d)** 153; about $395.86

71. $x = 3, y = 0$ **73.** $x = 2, y = 0$

$f(x) = \dfrac{1}{x - 3}$

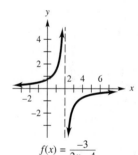

$f(x) = \dfrac{-3}{2x - 4}$

75. $x = -\dfrac{1}{2}, x = \dfrac{3}{2}, y = 0$

$g(x) = \dfrac{5x - 2}{4x^2 - 4x - 3}$

77. (a) about $10.83 **(b)** about $4.64 **(c)** about $3.61 **(d)** about $2.71 **(e)**

79. (a) $(10, 50)$

(b) $(10, \infty)$ **(c)** $(0, 10)$

CHAPTER 4

Section 4.1 (Page 197)

1. quadratic **3.** exponential **5. (a)** The graph lies entirely above the x-axis and falls from left to right. It falls relatively steeply until it reaches the y-intercept 1 and then falls slowly, with the positive x-axis as a horizontal asymptote. **(b)** $(0, 1), (1, .8)$
7. (a) The graph lies entirely above the x-axis and rises from left to right. The negative x-axis is a horizontal asymptote. The graph rises slowly until it reaches the y-intercept 1 and then rises quite steeply.
(b) $(0, 1), (1, 5^{.4})$

9.

11.

13.

15. (a)–(c)

17. 2.3 **19.** .75 **21.** .31 **23. (a)** $a > 1$
(b) domain: $(-\infty, \infty)$; range; $(0, \infty)$
(c)

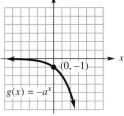

(d) domain: $(-\infty, \infty)$; range: $(-\infty, 0)$
(e)

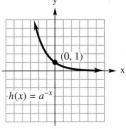

(f) domain: $(-\infty, \infty)$; range; $(0, \infty)$ **25. (a)** 3 **(b)** $\dfrac{1}{3}$ **(c)** 9
(d) 1

27.

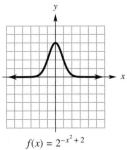

$$f(x) = 2^{-x^2 + 2}$$

29.

31. (a)

t	0	1	2	3	4	5	6	7	8	9	10
y	1	1.06	1.12	1.19	1.26	1.34	1.42	1.50	1.59	1.69	1.79

(b)

$y = (1.06)^t$

33. (a) about \$141,892 **(b)** about \$64.10 **35. (a)** about 113,210,000 **(b)** about 145,150,000 **(c)** No; the model predicts over 500 million cell phones in 2006, when the U.S. population is expected to be less than 300 million. **37. (a)** about .97 kg **(b)** about .75 kg **(c)** about .65 kg **(d)** 24,360 yr **39. (a)** about 6.2 billion **(b)** about 6.5 billion **(c)** about 8 billion **41.** \$19,420.26

Section 4.2 (Page 203)

1. (a) 100 g **(b)** about 70.9 g **(c)** about 50.3 g **(d)** about 11.6 g **3. (a)** about \$53.5 billion **(b)** about \$132.9 billion **(c)** about \$289.9 billion **5. (a)** $f(t) = 32.5(1.022362)^t$ **(b)** 40.5 million; 70.5 million **(c)** about 2023 **7. (a)** $f(t) = 800(1.01)^t$ **(b)** \$956.92; \$1015.79 **9. (a)** two-point: $f(t) = 1.003(.9684)^t$; regression: $g(t) = .9887(.96728)^t$ **(b)** $f(18) = .563$; $f(20) = .528$; $f(22) = .495$ and $g(18) = .543$; $g(20) = .508$; $g(22) = .476$ **(c)** in 2010 (regression model) or 2011 (two-point model) **11. (a)** two-point: $f(t) = 253.6(.9757)^t$; regression: $g(t) = 255(.9755)^t$ **(b)** $f(15) = 175.3$ and $f(30) = 121.2$; $g(15) = 175.8$ and $g(30) = 121.2$ **(c)** in 2007 **13. (a)** about 6 items **(b)** about 23 items **(c)** 25 items **15.** 6.25°C **17. (a)** .13 **(b)** .23 **(c)** about 2 weeks **19. (a)** about 49.7 million; about 73.5 million **(b)**

(c) in 2000 **(d)** No; the graph suggests that the number of subscribers will level off below the 80 million level. **21. (a)** about \$4695.1 billion; about \$5795.7 billion; about \$6453.4 billion **(b)**

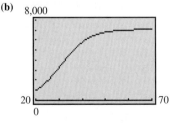

(c) in 2010 **(d)** yes, around 2020

Section 4.3 (Page 213)

1. a^y **3.** It is missing the value that equals b^y. If that value is x, the expression should read $y = \log_b x$.

5. $10^5 = 100,000$ **7.** $3^4 = 81$ **9.** $\log 75 = 1.8751$ **11.** $\log_3\left(\dfrac{1}{9}\right) = -2$ **13.** 3 **15.** 2 **17.** 3 **19.** -2 **21.** $\dfrac{1}{2}$ **23.** 3.78 **25.** 1.672 **27.** -1.047 **31.** $\log 16$ **33.** $\ln 5$ **35.** $\log\left(\dfrac{u^2 w^3}{v^6}\right)$ **37.** $\ln\left(\dfrac{(x+1)^2}{x+2}\right)$ **39.** $\dfrac{1}{2}\ln 6 + 2\ln m + \ln n$ **41.** $\dfrac{1}{2}\log x - \dfrac{5}{2}\log z$ **43.** $2u + 5v$ **45.** $3u - 2v$ **47.** 3.5145 **49.** 2.4851 **51.** There are many correct answers, including $b = 1$, $c = 2$. **53.**

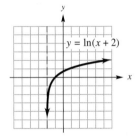

$y = \ln(x + 2)$

55.

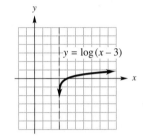

$y = \log(x - 3)$

59. $\ln 2.75 = 1.0116009$; $e^{1.0116009} = 2.75$ **61. (a)** 17.67 yr **(b)** 9.01 yr **(c)** 4.19 yr **(d)** 2.25 yr **(e)** $\dfrac{72}{4} = 18$; $\dfrac{72}{8} = 9$; $\dfrac{72}{18} = 4$; $\dfrac{72}{36} = 2$. So it takes about $\dfrac{72}{k}$ years for money to double at $k\%$ interest. **63. (a)** The percent of change in rents increased rapidly from 1992 to 1993, leveled off to a slow increase from 1993 to 1994, and then increased rapidly again from 1994 to 1996. From 1996 to 1999 (estimated) the percent of increase slowed and became constant at about 7%. **(b)** $f(92) \approx -3.4$; $f(99) \approx 7.1$; $f(92) \approx 3.4$ agrees closely with the y-value on the graph of about -3; $f(99) \approx 7.1$ is also reasonably close to the y-value on the graph of about 7.5.

Section 4.4 (Page 222)

1. 5 **3.** no solution **5.** 11 **7.** $\dfrac{8}{5}$ **9.** $\dfrac{1}{3}$ **11.** 10 **13.** 5.2378 **15.** 5 **17.** $\dfrac{4 + b}{4}$ **19.** $\dfrac{10^{2-b} - 5}{6}$ **23.** 3 **25.** $-\dfrac{5}{4}$ **27.** -2 **29.** -2 **31.** 1.465 **33.** 2.710 **35.** -1.825 **37.** .805 **39.** $-.123$ **41.** $\dfrac{\log d + 3}{4}$ **43.** $\dfrac{\ln b + 1}{2}$ **45.** 3 **47.** no solution **49.** $-2, 2$ **51.** 1, 10 **53.** 1 **55.** 4, -4 **57.** ± 2.141 **59.** 1.386 **63. (a)** about 3390; about 8692 **(b)** in 2005 **65.** 14.2 hr **67. (a)** 25 g **(b)** about 4.95 yr **69.** about 3689 yr old **71. (a)** approximately $3,981,072 i_0$ **(b)** approximately $3,162,278 i_0$ **(c)** about 1.26 times stronger. **73. (a)** 21 **(b)** 100 **(c)** 105 **(d)** 120 **(e)** 140 **75. (a)** 27.5% **(b)** \$130.14 **77. (a)**

(b) about 4717 ft   **79. (a)** $f(x) = 951.1369 + 4059.6835 \ln x$
(b) 13,500   **(c)** 2008

Chapter 4 Review (Page 226)

1. (c)   **3.** (d)   **5.** $0 < a < 1$ **7.** $(0, \infty)$
9.   **11.**

13. (a) about 2542.1 million   **(b)** 2002   **15.** $\log 47 = 1.6721$
17. $\ln 39 = 3.6636$ **19.** $10^3 = 1000$ **21.** $e^{4.5581} = 95.4$
23. 3   **25.** 7.4   **27.** $\dfrac{4}{3}$ **29.** $\log 20k^4$ **31.** $\log\left(\dfrac{b^2}{c^3}\right)$
33. 3   **35.** 8   **37.** 6   **39.** 2   **41.** -1 **43.** 2
45. 1.416   **47.** -2.807 **49.** -3.305 **51.** .747
53. 28.463   **55. (a)** \$15 million   **(b)** \$15.6 million
(c) \$16.4 million
57. (a) 10 g   **(b)** about 140 days   **(c)** about 243 days
59. 81.25° Celsius   **61. (a)** $f(x) = 345.4(.8142^x)$
(b) $g(x) = 336.8267(.818^x)$ **(c)** about 66.7 or 67.5 million, depending on the model   **(d)** in 2003   **63. (a)** $f(x) = \dfrac{73.27}{1 + 44e^{-.285x}}$
(b) in 2004

Case 3 (Page 230)

1. 23.6; 47.7; 58.4; the estimates are a bit low.

CHAPTER 5

Section 5.1 (Page 236)

1. time and interest rate   **3.** \$231.00   **5.** \$286.75
7. \$119.15   **9.** \$241.56   **11.** \$304.38   **13.** \$3654.10
17. \$46,265.06   **19.** \$28,026.37   **21.** \$8898.75
23. \$48,140.65   **25.** 10.7%   **27.** 11.8%   **29.** \$27,894.30
31. \$3056.25   **33.** 5.0%   **35.** 11.4%   **37.** \$1732.90
39. \$5778.50   **41.** \$3773; 13.58%   **43.** \$7278.68; no
45. (a) \$372.77   **(b)** \$26.94   **(c)** No, he will still owe
\$166.45.

Section 5.2 (Page 246)

1. r is the interest rate per year, while i is the interest rate per compounding period. t is number of years, while n is the number of
periods.   **3.** the interest rate and number of compounding
7. \$1593.85   **9.** \$1515.80   **11.** \$13,213.16   **13.** \$10,466.35
15. \$2307.95   **17.** \$1968.48   **19.** \$26,545.91
21. \$45,552.97   **23.** \$83,002.92   **25.** 4.04%   **27.** 8.16%
29. 12.36%   **31.** \$8954.58   **33.** \$3255.55   **35.** \$11,572.58
37. \$1000 now   **39.** \$30,611.30   **41.** about \$1.946 million
43. 7.11%   **45.** 4.91, 5.20, 5.34, 5.56, 5.63   **47.** \$7522.50
49. (a) \$16,659.95   **(b)** \$21,472.67   **51.** \$1000 now
53. about 14 yr   **55.** about 35 yr   **57.** 10.00%
59. (a) \$16,288.95   **(b)** \$16,436.19   **(c)** \$16,470.09
(d) \$16,486.65   **61. (a)**

Section 5.3 (Page 257)

3. (a) $a_1 = 1169$; $r = .916$ **(b)** $a_{10} = 531$; $a_{20} = 221$; This means
that a person who is 10 years from retirement should have savings of
531% of his or her annual salary; a person 20 years from retirement
should have savings of 221% of his or her annual salary.   **5.** 135
7. 6   **9.** 2315.25   **11.** 15   **13.** 6.24   **15.** 594.048
17. 15.91713   **19.** 21.82453   **21.** 22.01900
23. \$159,663.02   **25.** \$17,435.70   **27.** \$72,482.38
29. \$748.19   **31.** \$2038.01   **35.** \$7382.54   **37.** \$1463.16
39. \$258.73   **41.** \$6603.39   **43.** \$158,260.36
45. \$26,671.23   **47.** \$4421.89   **49.** \$144,872.14
51. \$4168.30   **53. (a)** \$149,850.69   **(b)** \$137,895.79
(c) \$11,954.90   **55.** \$130,159.72   **57.** \$284,527.35
59. (a) \$226.11   **(b)** \$245.77   **61.** \$522.85   **63.** \$7.397%
65. (a) \$1200   **(b)** \$3511.58

(c)

Payment Number	Amount of Deposit	Interest Earned	Total
1	\$3511.58	\$0.00	\$3511.58
2	3511.58	105.35	7128.51
3	3511.58	213.86	10,853.95
4	3511.58	325.62	14,691.15
5	3511.58	440.73	18,643.46
6	3511.58	559.30	22,714.34

Section 5.4 (Page 265)

1. (c)   **3.** 9.71225   **5.** 12.65930   **7.** 14.71787
11. \$8045.29   **13.** \$1,367,773.96   **15.** \$205,724.40
17. \$103,796.58   **21.** \$5309.69   **23.** \$11,727.32
25. \$260.19   **27.** \$1267.07   **29.** \$847.91   **31.** \$86.24
33. \$13.02   **35.** \$48,677.34   **37. (a)** \$158   **(b)** \$1584
39. (a) \$623,110.52   **(b)** \$456,427.28   **(c)** \$563,757.78
(d) \$392,903.18   **41.** \$280.46; \$32,310.74
43. (a) \$2349.51; \$197,911.80   **(b)** \$2097.30; \$278,352
(c) \$1965.82; \$364,746   **(d)** after 170 payments
45. (a) \$4025.90   **(b)** \$2981.93   **47.** \$432.53
49.

Payment Number	Amount of Payment	Interest for Period	Portion to Principal	Principal at End of Period
0	—	—	—	\$4000.00
1	\$1207.68	\$320.00	\$887.68	3112.32
2	1207.68	248.99	958.69	2153.63
3	1207.68	172.29	1035.39	1118.24
4	1207.70	89.46	1118.24	0

51.

Payment Number	Amount of Payment	Interest for Period	Portion to Principal	Principal at End of Period
0	—	—	—	\$7184.00
1	\$189.18	\$71.84	117.34	7066.66
2	189.18	70.67	118.51	6948.15
3	189.18	69.48	119.70	6828.45
4	189.18	68.28	120.90	6707.55

Chapter 5 Review (Page 269)

1. \$426.88   **3.** \$62.91   **7.** \$78,717.19   **9.** \$739.09
13. \$70,538.38; \$12,729.04   **15.** \$5282.19; \$604.96

17. \$12,857.07   **19.** \$1923.09   **21.** 4, 2, 1, $\dfrac{1}{2}$ **23.** -32

25. 5500   **29.** $137,925.91   **31.** $25,396.38
35. $2619.29   **37.** $916.12   **39.** $31,921.91
41. $14,222.42   **43.** $3628.00   **45.** $546.93   **47.** $896.06
49. $2696.12   **51.** $10,550.54   **53.** $8369.51
55. (a) about 37.44%   (b) about 24.74%   **57.** 3.6%
59. $24,818.76; $2418.76   **61.** $32.49   **63.** $5596.62
65. $3560.61   **67.** (d)   **69.** $5927.56

Case 4 (Page 274)

1.

$50(1 + i)^2 + 70(1 + i) = 127.40$; 4.3%

3. (a)

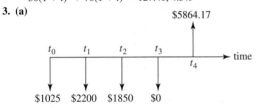

(b) 5.2%
5. (a)

$50(1 + i)^2 + 50(1 + i) - 90 = 0$; $-.068, -2.932$
(b) $-.068$ is reasonable, -2.932 is not.

CHAPTER 6

Section 6.1 (Page 287)

1. yes   **3.** $\left(\dfrac{11}{5}, -\dfrac{7}{5}\right)$ **5.** (28, 22)   **7.** (2, −1)

9. no solution   **11.** $(4y + 1, y)$ for any real number y
13. no solution   **15.** (a)   **17.** (5, 10)
19. (a) $r = 175,000$ and $b = 375,000$

21. $\begin{bmatrix} 3 & 4 & -2 & -3 & | & 0 \\ 1 & -3 & 7 & 4 & | & 9 \\ 2 & 0 & 5 & -6 & | & 0 \end{bmatrix}$

23. $3x + 2y + 6z = 18$
$2x - 2y + 5z = 7$
$x \quad\quad + 5z = 20$

25. $\begin{bmatrix} 1 & 5 & 2 & 0 & | & -1 \\ 8 & 5 & 4 & 6 & | & 6 \\ 0 & -15 & 1 & 1 & | & -1 \end{bmatrix}$ **27.** $\begin{bmatrix} 2 & 5 & 1 & | & -1 \\ -4 & 0 & 4 & | & 6 \\ \frac{3}{2} & 0 & 2 & | & -1 \end{bmatrix}$

29. dependent   **31.** inconsistent   **33.** independent
35. (2, 1, 4)   **37.** (−1, 4, 2)   **39.** no solution   **41.** (2, 0, 1)

43. $\left(\dfrac{(-7 - 2z)}{5}, \dfrac{(11z + 21)}{5}, z\right)$ for any real number z

45. (1, 2)   **47.** (a) $y = x + 1$ (b) $y = 3x + 4$ (c) $\left(-\dfrac{3}{2}, -\dfrac{1}{2}\right)$

51. $W_1 = \dfrac{300}{1 + \sqrt{3}} \approx 110$ lb

$W_2 = \dfrac{300 - W_1}{\sqrt{2}} \approx 134$ lb

53. 260 skirts; 270 blouses
55. (a) using the first equation, approximately 245 lb; using the second equation, approximately 253 lb   (b) for the first, 7.46 lb; and for the second, 7.93 lb   (c) at approximately 66 in. and 118 lb
57. .75 lb cashews; 2.25 lb peanuts
59. (a) $b + c$ (b) .35, .45, .2; A is tumorous, B is bone, C is healthy.
61. (a) $x_2 + x_3 = 700, x_3 + x_4 = 600$
(b) $(1000 - x_4, 100 + x_4, 600 - x_4, x_4)$ (c) 0, 600
(d) 400, 1000 for x_1; 100, 700 for x_2; 0, 600 for x_3 (e) If you know the number of cars entering or leaving three of the intersections, then the number entering or leaving the fourth is automatically determined, because the number leaving must equal the number entering.

Section 6.2 (Page 299)

1. $(3/2, 5, -2, 0)$ **3.** $(2 - w, -3 - 2w, 4, w)$ for any real number w
5. $(-3, 4, 0)$ **7.** $(1, 0, -1)$ **9.** $\left(\dfrac{-9z + 5}{23}, \dfrac{10z - 3}{23}, z\right)$ for any

real number z **11.** $(-1, 23, 16)$ **13.** no solution
15. $(-7, 5)$ **17.** $(-3, z - 17, z)$ for any real number z
19. $(-1, 1, -3, -2)$ **21.** no solution
23. Garcia 20 hr, Wong 15 hr   **25.** 40 lb pretzels, 20 lb dried fruit, 80 lb nuts   **27.** (a) .056057, 1.06657   (b) 228 ft
29. $15,000 in the mutual fund, $30,000 in bonds, and $25,000 in the food franchise   **31.** 6 cups of Roasted Chicken Rotini, 9 cups of Hearty Chicken, and 2 cups of Chunky Chicken Noodle; serving size 1.7 cups   **33.** (a) $12,000 at 6%, $7000 at 7%, and $6000 at 10%
(b) $10,000 at 6%, $15,000 at 7%, and $5000 at 10%
(c) $20,000 at 6%, $10,000 at 7%, and $10,000 at 10%
35. (a) $y = .01x^2 - .3x + 4.24$ (b) 15 platters; $1.99
37. (a) $C = .015t^2 + .7t + 315$ (b) 2081
39. (a) $L = 92.517t^2 - 1181.32t + 7053$ (b) 3296
(c) $L = -20.758t^3 + 362.38t^2 - 1928.6t + 7053$

Section 6.3 (Page 308)

1. 2×3; $\begin{bmatrix} -7 & 8 & -4 \\ 0 & -13 & -9 \end{bmatrix}$

3. 3×3; square matrix; $\begin{bmatrix} 3 & 0 & -11 \\ -1 & -\frac{1}{4} & 7 \\ -5 & 3 & -9 \end{bmatrix}$

5. 2×1; column matrix; $\begin{bmatrix} -7 \\ -11 \end{bmatrix}$ **7.** B is a 5×3 zero matrix.

9. $\begin{bmatrix} 9 & 12 & 0 & 2 \\ 1 & -1 & 2 & -4 \end{bmatrix}$ **11.** $\begin{bmatrix} 5 & 13 & 0 \\ 3 & 1 & 8 \end{bmatrix}$ **13.** $\begin{bmatrix} 3 & -7 & 7 \\ 4 & 4 & -8 \end{bmatrix}$

15. $\begin{bmatrix} -4 & 8 \\ 0 & 6 \end{bmatrix}$ **17.** $\begin{bmatrix} 24 & -8 \\ -16 & 0 \end{bmatrix}$ **19.** $\begin{bmatrix} -22 & -6 \\ 20 & -12 \end{bmatrix}$

21. $\begin{bmatrix} 4 & -\frac{7}{2} \\ 4 & \frac{21}{2} \end{bmatrix}$

23. $X + T = \begin{bmatrix} x & y \\ z & w \end{bmatrix} + \begin{bmatrix} r & s \\ t & u \end{bmatrix} = \begin{bmatrix} x+r & y+s \\ z+t & w+u \end{bmatrix}$ a 2×2 matrix

25. $X + (T + P) = \begin{bmatrix} x+(r+m) & y+(s+n) \\ z+(t+p) & w+(u+q) \end{bmatrix}$
$= \begin{bmatrix} (x+r)+m & (y+s)+n \\ (z+t)+p & (w+u)+q \end{bmatrix}$
$= (X + T) + P$

27. $P + O = \begin{bmatrix} m+0 & n+0 \\ p+0 & q+0 \end{bmatrix} = \begin{bmatrix} m & n \\ p & q \end{bmatrix} = P$

29. Several possible correct answers, including:

	basketball	hockey	football	baseball
percent of no shows	16	16	20	18
lost revenue per fan ($)	18.20	18.25	19	15.40
lost annual revenue (millions $)	22.7	35.8	51.9	96.3

31.

	1990	1998	2000
heart	1788	4121	4143
lung	308	3171	3614
liver	1237	12,070	15,359
kidney	17,883	38,270	45,273

; other answers are possible

33. (a) $A = \begin{bmatrix} 80 & 926 \\ 43 & 122 \\ 7 & 41 \end{bmatrix}$

(b) $B = \begin{bmatrix} 744 & 879 \\ 48 & 70 \\ 2 & 33 \end{bmatrix}$

(c) $A + B = \begin{bmatrix} 824 & 1805 \\ 91 & 192 \\ 9 & 74 \end{bmatrix}$; the matrix gives the total number of women's and men's teams in the three sports in 1981–82 and 1998–99.

(d) $C = \begin{bmatrix} 1019 \\ 134 \\ 45 \end{bmatrix}$

35. (a) $A = \begin{bmatrix} 2.61 & 4.39 & 6.29 & 9.08 \\ 1.63 & 2.77 & 4.61 & 6.92 \\ .92 & .75 & .62 & .54 \end{bmatrix}$

(b) $B = \begin{bmatrix} 1.38 & 1.72 & 1.94 & 3.31 \\ 1.26 & 1.48 & 2.82 & 2.28 \\ .41 & .33 & .27 & .40 \end{bmatrix}$

(c) $\begin{bmatrix} 1.23 & 2.67 & 4.35 & 5.77 \\ .37 & 1.29 & 1.79 & 4.64 \\ .51 & .42 & .35 & .14 \end{bmatrix}$

Section 6.4 (Page 321)

1. 2×2; 2×2 **3.** 3×2; BA does not exist

5. AB does not exist; 3×2 **7.** columns; rows **9.** $\begin{bmatrix} 13 \\ 25 \end{bmatrix}$

11. $\begin{bmatrix} -2 & 10 \\ 0 & 8 \end{bmatrix}$ **13.** $\begin{bmatrix} -4 & 1 \\ 2 & -3 \end{bmatrix}$ **15.** $\begin{bmatrix} 3 & -5 & 7 \\ -2 & 1 & 6 \\ 0 & -3 & 4 \end{bmatrix}$

17. $\begin{bmatrix} 16 & 11 \\ 37 & 32 \\ 58 & 53 \end{bmatrix}$ **19.** $AB = \begin{bmatrix} -30 & -45 \\ 20 & 30 \end{bmatrix}$, but $BA = \begin{bmatrix} 0 & 0 \\ 0 & 0 \end{bmatrix}$.

21. $(A + B)(A - B) = \begin{bmatrix} -7 & -24 \\ -28 & -33 \end{bmatrix}$, but $A^2 - B^2 = \begin{bmatrix} -37 & -69 \\ -8 & -3 \end{bmatrix}$.

23. $P(XT) =$
$\begin{bmatrix} (mx+nz)r+(my+nw)t & (mx+nz)s+(my+nw)u \\ (px+qz)r+(py+qw)t & (px+qz)s+(py+qw)u \end{bmatrix}$
$P(XT)$ is the same so $(PX)T = P(XT)$.

25. $k(X + T) = k\begin{bmatrix} x+r & y+s \\ z+t & w+u \end{bmatrix}$
$= \begin{bmatrix} k(x+r) & k(y+s) \\ k(z+t) & k(w+u) \end{bmatrix}$
$= \begin{bmatrix} kx+kr & ky+ks \\ kz+kt & kw+ku \end{bmatrix}$
$= \begin{bmatrix} kx & ky \\ kz & kw \end{bmatrix} + \begin{bmatrix} kr & ks \\ kt & ku \end{bmatrix} = kX + kT$

27. no **29.** yes **31.** no **33.** $\begin{bmatrix} 2 & -3 \\ -1 & 2 \end{bmatrix}$

35. no inverse **37.** $\begin{bmatrix} 2 & -3 \\ -\frac{1}{2} & 1 \end{bmatrix}$ **39.** $\begin{bmatrix} 3 & 3 & -1 \\ -2 & -2 & 1 \\ -4 & -5 & 2 \end{bmatrix}$

41. $\begin{bmatrix} 2 & 1 & -1 \\ 8 & 2 & -5 \\ -11 & -3 & 7 \end{bmatrix}$ **43.** no inverse

45. $\begin{bmatrix} \frac{7}{4} & \frac{5}{2} & 3 \\ -\frac{1}{4} & -\frac{1}{2} & 0 \\ -\frac{1}{4} & -\frac{1}{2} & -1 \end{bmatrix}$ **47.** $\begin{bmatrix} \frac{1}{2} & \frac{1}{2} & -\frac{1}{4} & \frac{1}{2} \\ -1 & 4 & -\frac{1}{2} & -2 \\ -\frac{1}{2} & \frac{5}{2} & -\frac{1}{4} & -\frac{3}{2} \\ \frac{1}{2} & -\frac{1}{2} & \frac{1}{4} & \frac{1}{2} \end{bmatrix}$

49. (a) $R = \begin{bmatrix} .024 & .008 \\ .025 & .007 \\ .015 & .009 \\ .011 & .011 \end{bmatrix}$

(b) $P = \begin{bmatrix} 1996 & 286 & 226 & 460 \\ 2440 & 365 & 252 & 484 \\ 2906 & 455 & 277 & 499 \\ 3683 & 519 & 310 & 729 \\ 4723 & 697 & 364 & 702 \end{bmatrix}$

(c) $PR = \begin{bmatrix} 63.504 & 25.064 \\ 76.789 & 29.667 \\ 90.763 & 34.415 \\ 114.036 & 43.906 \\ 143.959 & 53.661 \end{bmatrix}$

(d) Rows represent the years 1970, 1980, 1990, 2000, 2025. Column one gives the total births in those years; column two the total deaths. **(e)** 114,036,000; 53,661,000

51. (a) $C = \begin{bmatrix} 394 & 2753 \\ 595 & 3093 \\ 601 & 3085 \\ 610 & 3119 \end{bmatrix}$

(b) $D = \begin{bmatrix} 15.1 & 14.2 & 13.7 & 14.1 \\ 6.3 & 6.0 & 6.0 & 6.0 \end{bmatrix}$

(c) $DC = \begin{bmatrix} 31{,}233.1 & 171{,}733.3 \\ 13{,}318.2 & 73{,}125.9 \end{bmatrix}$

(d) row 1, column 1 **(e)** row 2, column 2

53. (a)

	A	B
Dept. 1	57	70
Dept. 2	41	54
Dept. 3	27	40
Dept. 4	39	40

(b) supplier A: $164; supplier B: $204

Section 6.5 (Page 332)

1. $\begin{bmatrix} 8 \\ 6 \end{bmatrix}$ **3.** $\begin{bmatrix} \frac{1}{2} & 1 \\ \frac{3}{2} & 1 \end{bmatrix}$ **5.** $\begin{bmatrix} 11 \\ -3 \\ 5 \end{bmatrix}$ **7.** $(-31, 24, -4)$

9. $(-31, -131, 181)$ **11.** $(15, -5, -1)$ **13.** $(-7, -34, -19, 7)$

15. $\begin{bmatrix} -6 \\ -14 \end{bmatrix}$ **17.** $\begin{bmatrix} \frac{32}{3} \\ \frac{25}{3} \end{bmatrix}$

19. 848 units of agriculture, 516 units of manufacturing, and 2970 units of households **21.** 195.5 million pounds of agriculture, 25.9 million pounds of manufacturing, and 13.6 million pounds of energy **23.** either 10 buffets, 5 chairs, no tables, or 11 buffets, 1 chair, 1 table **25.** 2340 of the first species, 10,128 of the second species, 224 of the third species **27.** jeans $34.50; jacket $72; sweater $44; shirt $21.75 **29.** about 1073 metric tons of wheat, about 1431 metric tons of oil **31.** gas $98 million, electric $123 million

33. (a) $\frac{7}{4}$ bushels of yams, $\frac{15}{8} \approx 2$ pigs **(b)** 167.5 bushels of yams, $153.75 \approx 154$ pigs

35. (a) $\begin{bmatrix} 1.67 & .56 & .56 \\ .19 & 1.17 & .06 \\ 3.15 & 3.27 & 4.38 \end{bmatrix}$

(b) These multipliers imply that if the demand for one community's output increases by $1 then the output in the other community will increase by the amount in the row and column of that matrix. For example, if the demand for Hermitage's output increases by $1 then output from Sharon will increase by $.56, from Farrell by $.06, and from Hermitage by $4.38.

37. $\begin{bmatrix} 23 \\ 51 \end{bmatrix}, \begin{bmatrix} 13 \\ 30 \end{bmatrix}, \begin{bmatrix} 45 \\ 96 \end{bmatrix}, \begin{bmatrix} 69 \\ 156 \end{bmatrix}, \begin{bmatrix} 87 \\ 194 \end{bmatrix}, \begin{bmatrix} 23 \\ 51 \end{bmatrix}, \begin{bmatrix} 51 \\ 110 \end{bmatrix}, \begin{bmatrix} 45 \\ 102 \end{bmatrix}, \begin{bmatrix} 69 \\ 157 \end{bmatrix}$

39. (a) 3 **(b)** 3 **(c)** 5 **(d)** 3

41. (a) $B = \begin{bmatrix} 0 & 2 & 3 \\ 2 & 0 & 4 \\ 3 & 4 & 0 \end{bmatrix}$ **(b)** $B^2 = \begin{bmatrix} 13 & 12 & 8 \\ 12 & 20 & 6 \\ 8 & 6 & 25 \end{bmatrix}$ **(c)** 12 **(d)** 14

43. (a)

	Dogs	Rats	Cats	Mice
Dogs	0	1	1	1
Rats	0	0	0	1
Cats	0	1	0	1
Mice	0	0	0	0

$C = $ (matrix above)

(b) $C^2 = \begin{bmatrix} 0 & 1 & 0 & 2 \\ 0 & 0 & 0 & 0 \\ 0 & 0 & 0 & 1 \\ 0 & 0 & 0 & 0 \end{bmatrix}$; C^2 gives the number of food sources once removed from the feeder. Thus, since dogs eat rats and rats eat mice, mice are an indirect as well as direct food source for dogs.

Chapter 6 Review (Page 337)

1. $(-5, 7)$ **3.** $(0, -2)$ **5.** no solution; inconsistent system.
7. $(-35, 140, 22)$ **9.** 8000 standard clips, 6000 extra large clips
11. $7000 in the first fund, $11,000 in the second fund
13. 50 from source I, 150 from source II, 100 from source III
15. $(0, 3, 3)$ **17.** no solution **19.** $(-79, 99, -8)$
21. 2×2; square **23.** 1×4; row **25.** 2×3

27. $\begin{bmatrix} 8 & 8 & 8 \\ 10 & 5 & 9 \\ 7 & 10 & 7 \\ 8 & 9 & 7 \end{bmatrix}$ **29.** $\begin{bmatrix} -2 & -3 & 2 \\ -2 & -4 & 0 \\ 0 & -1 & -2 \end{bmatrix}$ **31.** $\begin{bmatrix} 2 & 30 \\ -4 & -15 \\ 10 & 13 \end{bmatrix}$

33. not defined

35.

Next day		Two-day total	
2310	$-\frac{1}{4}$	4842	$-\frac{1}{2}$
1258	$-\frac{1}{4}$	2722	$-\frac{1}{8}$
5061	$\frac{1}{2}$	10,035	-1
1812	$\frac{1}{2}$	3566	1

37. $\begin{bmatrix} 18 & 80 \\ -7 & -28 \\ 21 & 84 \end{bmatrix}$ **39.** $\begin{bmatrix} 13 & 43 \\ 17 & 46 \end{bmatrix}$ **41.** $\begin{bmatrix} 222 & 632 \\ -77 & -224 \\ 231 & 672 \end{bmatrix}$

43. (a) $\begin{bmatrix} \frac{1}{4} & \frac{1}{2} \\ \frac{1}{4} & \frac{1}{3} \\ \frac{1}{3} & \frac{1}{3} \end{bmatrix}$ **(b)** cutting 34 hr; shaping 46 hr

45. There are many correct answers, including $\begin{bmatrix} 1 & 2 \\ 3 & 4 \end{bmatrix}$.

47. $\begin{bmatrix} -\frac{1}{4} & \frac{1}{6} \\ 0 & \frac{1}{3} \end{bmatrix}$ **49.** no inverse **51.** $\begin{bmatrix} \frac{1}{4} & \frac{1}{2} & \frac{1}{2} \\ \frac{1}{4} & -\frac{1}{2} & \frac{1}{2} \\ \frac{1}{8} & -\frac{1}{4} & -\frac{1}{4} \end{bmatrix}$

53. no inverse

55. $\begin{bmatrix} -\frac{2}{3} & -\frac{17}{3} & -\frac{14}{3} & -3 \\ \frac{1}{3} & \frac{1}{3} & \frac{1}{3} & 0 \\ -\frac{1}{3} & -\frac{10}{3} & -\frac{7}{3} & -2 \\ 0 & 2 & 1 & 1 \end{bmatrix}$ **57.** $\begin{bmatrix} -\frac{7}{19} & \frac{4}{19} \\ \frac{3}{19} & \frac{1}{19} \end{bmatrix}$

59. $\begin{bmatrix} 1 & 1 \\ -2 & -3 \end{bmatrix}$ **61.** no inverse **63.** $\begin{bmatrix} 18 \\ -7 \end{bmatrix}$ **65.** $\begin{bmatrix} -22 \\ -18 \\ 15 \end{bmatrix}$

67. $(2, 2)$ **69.** $(2, 1)$ **71.** $(-1, 0, 2)$ **73.** no inverse; no solution for the system **75.** 16 liters of the 9%, 24 liters of the 14% **77.** 30 liters of 40% solution, 10 liters of 60% solution

79. 80 bowls, 120 plates **81.** $12,750 at 8%, $27,250 at $8\frac{1}{2}$%, $10,000 at 11%

83. (a) $\begin{bmatrix} 1 & -\frac{1}{4} \\ -\frac{1}{2} & 1 \end{bmatrix}$ **(b)** $\begin{bmatrix} \frac{8}{7} & \frac{2}{7} \\ \frac{4}{7} & \frac{8}{7} \end{bmatrix}$ **(c)** $\begin{bmatrix} 2800 \\ 2800 \end{bmatrix}$

85. (a)
$$\begin{bmatrix} 1.30 & .045 & .567 & .012 & .068 & .020 \\ .204 & 1.03 & .183 & .004 & .022 & .006 \\ .155 & .038 & 1.12 & .020 & .114 & .034 \\ .018 & .021 & .028 & 1.08 & .016 & .033 \\ .537 & .525 & .483 & .279 & 1.73 & .419 \\ .573 & .346 & .497 & .536 & .087 & 1.94 \end{bmatrix}$$; every dollar of

increased demand for livestock will result in an increase of production demand of \$.204 in crops. **(b)** In millions of dollars make \$3855 of livestock, \$1476 in crops, \$2726 in food products, \$1338 in mining and manufacturing, \$8439 in households, and \$10,256 in other business sectors. (Solutions from nonrounded part (a) answer.)
87. agriculture \$140,909, manufacturing \$95,455

89. (a) $\begin{bmatrix} 54 \\ 32 \end{bmatrix}, \begin{bmatrix} 134 \\ 89 \end{bmatrix}, \begin{bmatrix} 172 \\ 113 \end{bmatrix}, \begin{bmatrix} 118 \\ 74 \end{bmatrix}, \begin{bmatrix} 208 \\ 131 \end{bmatrix}$ **(b)** $\begin{bmatrix} 2 & -3 \\ -\frac{1}{2} & 1 \end{bmatrix}$

Case 5 (Page 342)

1. Boston, Hyannis, Martha's Vineyard, Nantucket, New Bedford, and Providence may be reached by a two-flight sequence from New Bedford; all Cape Air cities may be reached by a three-flight sequence. **3.** The connection between Provincetown and Providence and the connection between Provincetown and New Bedford each take three flights. **5.** All Big Sky cities may be reached by a three-flight sequence from Helena. At least three flights must be used to get from Helena to Havre, Glendive, and Bismarck.

CHAPTER 7

Section 7.1 (Page 349)

1.

3.

5.

7.

9.

11.

13.

15.

17.

19.

23.

25.

27.

29.

31.

33.

35.

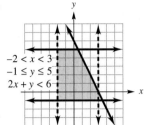

$$-2 < x < 3$$
$$-1 \leq y \leq 5$$
$$2x + y < 6$$

37.

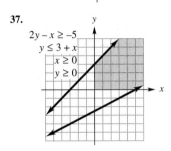

$$2y - x \geq -5$$
$$y \leq 3 + x$$
$$x \geq 0$$
$$y \geq 0$$

39.

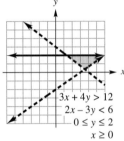

$$3x + 4y > 12$$
$$2x - 3y < 6$$
$$0 \leq y \leq 2$$
$$x \geq 0$$

41. $2 < x < 7$
$-1 < y < 3$

43. (a)

Number		Hours Spinning	Hours Dyeing	Hours Weaving
Shawls	x	1	1	1
Afghans	y	2	1	4
Maximum Number of Hours Available		8	6	14

(b) $x + 2y \leq 8$; $x + y \leq 6$; $x + 4y \leq 14$; $x \geq 0$; $y \geq 0$
(c)

$$x + 2y \leq 8$$
$$x + y \leq 6$$
$$x + 4y \leq 14$$
$$x \geq 0$$
$$y \geq 0$$

45. $x \geq 3000$; $y \geq 5000$; $x + y \leq 10{,}000$

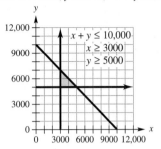

$$x + y \leq 10{,}000$$
$$x \geq 3000$$
$$y \geq 5000$$

47. $x + y \geq 3.2$; $.16x + .20y \leq .8$; $.5x + .3y \leq 1.8$; $x \geq 0$; $y \geq 0$

$$x + y \geq 3.2$$
$$.16x + .2y \leq .8$$
$$.5x + .3y \leq 1.8$$
$$x \geq 0$$
$$y \geq 0$$

Section 7.2 (Page 357)

1. maximum of 65 at (5, 10);   minimum of 8 at (1, 1)
3. maximum of 9 at (0, 12);   minimum of 0 at (0, 0)
5. (a) no maximum; minimum of 16 at (0, 8)
(b) no maximum; minimum of 18 at (3, 4)
(c) no maximum; minimum of 21 at $\left(\dfrac{13}{2}, 2\right)$
(d) no maximum; minimum of 12 at (12, 0)
7. maximum of $\dfrac{42}{5}$ at $x = \dfrac{6}{5}$, $y = \dfrac{6}{5}$ or $\left(\dfrac{6}{5}, \dfrac{6}{5}\right)$
9. minimum of 13 at $x = 5$, $y = 3$ or (5, 3)
11. maximum of $\dfrac{235}{4}$ at $x = \dfrac{105}{8}$, $y = \dfrac{25}{8}$ or $\left(\dfrac{105}{8}, \dfrac{25}{8}\right)$
13. no maximum; minimum of 9 at $\left(1, \dfrac{3}{2}\right)$
15. maximum of 22 at (2, 4); no minimum
17. (a) (18, 2)   **(b)** $\left(\dfrac{12}{5}, \dfrac{39}{5}\right)$ **(c)** no maximum

Section 7.3 (Page 363)

1. $6x + 4y \leq 90$, where x = number of canoes, y = number of rowboats   **3.** $150x + 750y \leq 8500$, where x = number of radio spots, y = number of TV ads   **5.** 12 chain saws; no chippers
7. 10 lb deluxe; 40 lb regular   **9. (a)** 6 units of policy A and 16 units of policy B for a minimum cost of \$940   **(b)** 30 units of policy A and 0 units of policy B for a minimum cost of \$750
11. $3\frac{3}{4}$ servings of A and $1\frac{7}{8}$ servings of B for a minimum cost of \$1.69
13. 800 type 1 and 1600 type 2 for a maximum revenue of \$272
15. 0 plants and 18 animals, for a minimum of 270 hr
17. From warehouse I ship 60 boxes to San Jose and 300 boxes to Memphis. From warehouse II ship 290 boxes to San Jose and none to Memphis, for a minimum cost of \$147.70.   **19.** \$20 million in bonds and \$20 million in mutual funds for a maximum annual interest of \$4 million   **21.** 5 humanities, 12 science   **23.** (b)   **25.** (c)

Section 7.4 (Page 378)

1. (a) 3

(b) x_3, x_4, x_5

(c) $\begin{aligned} 4x_1 + 2x_2 + x_3 \qquad\qquad &= 20 \\ 5x_1 + x_2 \qquad + x_4 \qquad &= 50 \\ 2x_1 + 3x_2 \qquad\qquad + x_5 &= 25 \end{aligned}$

3. (a) 3

(b) x_4, x_5, x_6

(c) $\begin{aligned} 3x_1 - x_2 + 4x_3 + x_4 \qquad\qquad\qquad &= 95 \\ 7x_1 + 6x_2 + 8x_3 \qquad + x_5 \qquad &= 118 \\ 4x_1 + 5x_2 + 10x_3 \qquad\qquad + x_6 &= 220 \end{aligned}$

5.

$$\begin{array}{ccccccc} x_1 & x_2 & x_3 & x_4 & x_5 & z & \\ \hline 2 & 3 & 1 & 0 & 0 & 0 & 6 \\ 4 & 1 & 0 & 1 & 0 & 0 & 6 \\ 5 & 2 & 0 & 0 & 1 & 0 & 15 \\ \hline -5 & -1 & 0 & 0 & 0 & 1 & 0 \end{array}$$

7.

$$\begin{array}{ccccccc} x_1 & x_2 & x_3 & x_4 & x_5 & x_6 & z & \\ \hline 1 & 2 & 3 & 1 & 0 & 0 & 0 & 10 \\ 2 & 1 & 1 & 0 & 1 & 0 & 0 & 8 \\ 3 & 0 & 2 & 0 & 0 & 1 & 0 & 6 \\ \hline -1 & -5 & -10 & 0 & 0 & 0 & 1 & 0 \end{array}$$

9. 3 in row 2, column 1   **11.** 6 in row 1, column 1

13.

$$\begin{array}{cccccc} x_1 & x_2 & x_3 & x_4 & x_5 & z & \\ \hline -1 & 0 & 3 & 1 & -1 & 0 & 16 \\ 1 & 1 & \tfrac{1}{2} & 0 & \tfrac{1}{2} & 0 & 20 \\ \hline 2 & 0 & -\tfrac{1}{2} & 0 & \tfrac{3}{2} & 1 & 60 \end{array}$$

15.

$$\begin{array}{ccccccc} x_1 & x_2 & x_3 & x_4 & x_5 & x_6 & z & \\ \hline -\tfrac{1}{2} & \tfrac{1}{2} & 0 & 1 & -\tfrac{1}{2} & 0 & 0 & 10 \\ \tfrac{3}{2} & \tfrac{1}{2} & 1 & 0 & \tfrac{1}{2} & 0 & 0 & 50 \\ -\tfrac{7}{2} & \tfrac{1}{2} & 0 & 0 & -\tfrac{3}{2} & 1 & 0 & 50 \\ \hline 2 & 0 & 0 & 0 & 1 & 0 & 1 & 100 \end{array}$$

17. (a) basic: x_3, x_5; nonbasic: x_1, x_2, x_4 **(b)** $x_1 = 0, x_2 = 0, x_3 = 16$, $x_4 = 0, x_5 = 29, z = 11$ **(c)** not maximum

19. (a) basic: x_1, x_2, x_5; nonbasic: x_3, x_4, x_6 **(b)** $x_1 = 6, x_2 = 13$, $x_3 = 0, x_4 = 0, x_5 = 21, x_6 = 0, z = 18$ **(c)** maximum

21. Maximum is 30 when $x_1 = 0, x_2 = 10, x_3 = 0, x_4 = 0, x_5 = 16$.

23. Maximum is 8 when $x_1 = 4, x_2 = 0, x_3 = 8, x_4 = 2, x_5 = 0$.

25. no maximum   **27.** no maximum   **29.** Maximum is 34 when $x_1 = 17, x_2 = 0, x_3 = 0, x_4 = 0, x_5 = 14$ or when $x_1 = 0$, $x_2 = 17, x_3 = 0, x_4 = 0, x_5 = 14$.   **31.** Maximum is 26,000 when $x_1 = 60, x_2 = 40, x_3 = 0, x_4 = 0, x_5 = 80, x_6 = 0$.   **33.** Maximum is 64 when $x_1 = 28, x_2 = 16, x_3 = 0, x_4 = 0, x_5 = 28, x_6 = 0$.

35. Maximum is 250 when $x_1 = 0, x_2 = 0, x_3 = 0, x_4 = 50, x_5 = 0$, $x_6 = 50$.   **37. (a)** Maximum is 24 when $x_1 = 12, x_2 = 0, x_3 = 0$, $x_4 = 0, x_5 = 6$.   **(b)** Maximum is 24 when $x_1 = 0, x_2 = 12, x_3 = 0$, $x_4 = 0, x_5 = 18$.   **(c)** The unique maximum value of z is 24, but this occurs at two different basic feasible solutions.

Section 7.5 (Page 384)

1.

$$\begin{array}{ccccc} x_1 & x_2 & x_3 & x_4 & x_5 & \\ \hline 2 & 1 & 1 & 0 & 0 & 90 \\ 1 & 2 & 0 & 1 & 0 & 80 \\ 1 & 1 & 0 & 0 & 1 & 50 \\ \hline -12 & -10 & 0 & 0 & 0 & 0 \end{array}$$

where x_1 is the number of Siamese cats and x_2 is the number of Persian cats

3.

$$\begin{array}{ccccccc} x_1 & x_2 & x_3 & x_4 & x_5 & x_6 & x_7 & \\ \hline 0 & 0 & .375 & .625 & 1 & 0 & 0 & 500 \\ 0 & .75 & .5 & .375 & 0 & 1 & 0 & 600 \\ 1 & .25 & .125 & 0 & 0 & 0 & 1 & 300 \\ \hline -90 & -70 & -60 & -50 & 0 & 0 & 0 & 0 \end{array}$$

where x_1 is the number of kg of P, x_2 is the number of kg of Q, x_3 is the number of kg of R, and x_4 is the number of kg of S

5. (a) Make no 1-speed or 3-speed bicycles; make 2295 10-speed bicycles; maximum profit is $55,080.   **(b)** 5280 units of aluminum are unused; all the steel is used.   **7. (a)** 12 min to the senator, 9 min to the congresswoman, and 6 min to the governor for a maximum of 1,320,000 viewers   **(b)** $x_4 = 0$ means that all of the 27 available min are used; $x_5 = 0$ means that the senator had *exactly* twice as much time as the governor; $x_6 = 0$ means that the senator and governor had a total time *exactly* twice the time of the congresswoman.

9. 4 radio ads, 6 TV ads, and no newspaper ads for a maximum exposure of 64,800 people   **11. (a)** 22 fund-raising parties, no mailings, and 3 dinner parties for a maximum of $6,200,000

13. (a) (3) **(b)** (4) **(c)** (3)   **15.** The breeder should raise 40 Siamese and 10 Persian cats for a maximum gross income of $580.

17. 163.6 kg of food P, none of Q, 1090.9 kg of R, 145.5 kg of S; maximum is 87,454.5.

Section 7.6 (Page 394)

1. $\begin{bmatrix} 3 & 1 & 0 \\ -4 & 10 & 3 \\ 5 & 7 & 6 \end{bmatrix}$ **3.** $\begin{bmatrix} 3 & 4 \\ 0 & 17 \\ 14 & 8 \\ -5 & -6 \\ 3 & 1 \end{bmatrix}$

5. Maximize $z = 4x_1 + 6x_2$
subject to: $3x_1 - x_2 \le 3$
$x_1 + 2x_2 \le 5$
$x_1 \ge 0, x_2 \ge 0.$

7. Maximize $z = 18x_1 + 15x_2 + 20x_3$
subject to: $x_1 + 4x_2 + 5x_3 \le 2$
$7x_1 + x_2 + 3x_3 \le 8$
$x_1 \ge 0, \ x_2 \ge 0, x_3 \ge 0.$

9. Maximize $z = 8x_1 + 12x_2$
subject to: $3x_1 + x_2 \le 1$
$4x_1 + 5x_2 \le 2$
$6x_1 + 2x_2 \le 6$
$x_1 \ge 0, x_2 \ge 0.$

11. Maximize $z = 5x_1 + 4x_2 + 15x_3$
subject to: $x_1 + x_2 + 2x_3 \le 8$
$x_1 + x_2 + x_3 \le 9$
$x_1 + 3x_3 \le 3$
$x_1 \ge 0, x_3 \ge 0, x_3 \ge 0.$

13. $y_1 = 0, y_2 = 100, y_3 = 0$; minimum is 100.   **15.** $y_1 = 0$, $y_2 = 12, y_3 = 0$; minimum is 12.   **17.** $y_1 = 0, y_2 = 0, y_3 = 2$; minimum is 4.   **19.** $y_1 = 10, y_2 = 10, y_3 = 0$; minimum is 320.

21. $y_1 = 0, y_2 = 0, y_3 = 4$; minimum is 4.   **23.** 4 servings of A, 2 servings of B for a minimum cost of $1.76   **25.** 28 units of regular beer and 14 units of light beer for a minimum cost of $1,680,000   **27. (a)**

29. (a) Minimize $w = 200y_1 + 600y_2 + 90y_3$
subject to: $y_1 + 4y_2 \ge 1$
$2y_1 + 3y_2 + y_3 \ge 1.5$
$y_1 \ge 0, y_2 \ge 0, y_3 \ge 0.$

(b) $y_1 = .6, y_2 = .1, y_3 = 0, w = 180$ **(c)** $186   **(d)** $179

Section 7.7 (Page 406)

1. (a) Maximize $z = 5x_1 + 2x_2 - x_3$
subject to:
$$2x_1 + 3x_2 + 5x_3 - x_4 \quad\;\; = 8$$
$$4x_1 - \;\; x_2 + 3x_3 + \;\;\; x_5 = 7$$
$$x_1 \geq 0, x_2 \geq 0, x_3 \geq 0, x_4 \geq 0, x_5 \geq 0.$$

(b)

x_1	x_2	x_3	x_4	x_5	
2	3	5	-1	0	8
4	-1	3	0	1	7
-5	-2	1	0	0	0

3. (a) Maximize $z = 2x_1 - 3x_2 + 4x_3$
subject to:
$$x_1 + x_2 + x_3 + x_4 \qquad\qquad = 100$$
$$x_1 + x_2 + x_3 \qquad - x_5 \qquad = 75$$
$$x_1 + x_2 \qquad\qquad\qquad - x_6 = 27$$
$$x_1 \geq 0, x_2 \geq 0, x_3 \geq 0, x_4 \geq 0, x_5 \geq 0, x_6 \geq 0.$$

(b)

x_1	x_2	x_3	x_4	x_5	x_6	
1	1	1	1	0	0	100
1	1	1	0	-1	0	75
1	1	0	0	0	-1	27
-2	3	-4	0	0	0	0

5. Maximize $z = -2y_1 - 5y_2 + 3y_3$
subject to:
$$y_1 + 2y_2 + 3y_3 \geq 115$$
$$2y_1 + y_2 + \;\; y_3 \leq 200$$
$$y_1 \qquad\quad + y_3 \geq 50$$
$$y_1 \geq 0, y_2 \geq 0, y_3 \geq 0.$$

y_1	y_2	y_3	y_4	y_5	y_6	
1	2	3	-1	0	0	115
2	1	1	0	1	0	200
1	0	1	0	0	-1	50
2	5	-3	0	0	0	0

7. Maximize $z = -y_1 + 4y_2 - 2y_3$
subject to: $7y_1 - 6y_2 + 8y_3 \geq 18$
$$4y_1 + 5y_2 + 10y_3 \geq 20$$
$$y_1 \geq 0, y_2 \geq 0, y_3 \geq 0$$

y_1	y_2	y_3	y_4	y_5	
7	-6	8	-1	0	18
4	5	10	0	-1	20
1	-4	2	0	0	0

9. Maximum is 480 when $x_1 = 40, x_2 = 0.$ **11.** Maximum is 114 when $x_1 = 38, x_2 = 0, x_3 = 0.$ **13.** Maximum is 90 when $x_1 = 12$, $x_2 = 3$ or when $x_1 = 0, x_2 = 9.$ **15.** Minimum is 40 when $y_1 = 0$, $y_2 = 20.$ **17.** Maximum is $133\frac{1}{3}$ when $x_1 = 33\frac{1}{3}, x_2 = 16\frac{2}{3}.$

19. Minimum is 512 when $y_1 = 6, y_2 = 8.$

21.

y_1	y_2	y_3	y_4	y_5	y_6	y_7	
1	1	1	-1	0	0	0	10
1	1	1	0	1	0	0	15
1	$-\frac{1}{4}$	0	0	0	-1	0	0
-1	0	1	0	0	0	-1	0
.30	.09	.27	0	0	0	0	0

23.

y_1	y_2	y_3	y_4	y_5	y_6	y_7	y_8	y_9	y_{10}	
1	1	0	0	-1	0	0	0	0	0	32
1	1	0	0	0	1	0	0	0	0	32
0	0	1	1	0	0	-1	0	0	0	20
0	0	1	1	0	0	0	1	0	0	20
1	0	1	0	0	0	0	0	1	0	25
0	1	0	1	0	0	0	0	0	1	30
14	12	22	10	0	0	0	0	0	0	0

25. Ship 200 barrels of oil from supplier S_1 to distributor D_1; ship 2800 barrels of oil from supplier S_2 to distributor D_1; ship 2800 barrels of oil from supplier S_1 to distributor D_2; ship 2200 barrels of oil from supplier S_2 to distributor D_2. Minimum cost is $180,400.

27. Use 1000 lb of bluegrass, 2400 lb of rye, and 1600 lb of Bermuda for a minimum cost of $560.

29. Allot $3,000,000 in commercial loans and $22,000,000 in home loans for a maximum return of $2,940,000.

31. Make 32 units of regular beer and 10 units of light beer for a minimum cost of $1,632,000.

33. $1\frac{2}{3}$ ounces of ingredient I , $6\frac{2}{3}$ ounces of ingredient II, and $1\frac{2}{3}$ ounces of ingredient III produce a minimum cost of $1.55 per barrel.

35. 22 from W_1 to D_1, 10 from W_2 to D_1, none from W_1 to D_2, and 20 from W_2 to D_2 for a minimum cost of $628.

Chapter 7 Review (Page 410)

1.

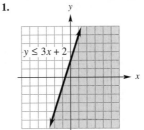

$y \leq 3x + 2$

3.

$3x + 4y \geq 12$

5.

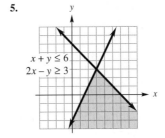

$x + y \leq 6$
$2x - y \geq 3$

7.

$2 \leq x \leq 5$
$1 \leq y \leq 7$
$x - y \leq 3$

9. Let x represent the number of batches of cakes and let y represent the number of batches of cookies. Then

$$2x + \left(\frac{3}{2}\right)y \le 15$$

$$3x + \left(\frac{2}{3}\right)y \le 13$$

$$x \ge 0$$

$$y \ge 0.$$

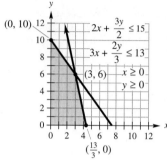

11. minimum of 8 at $(2, 1)$; maximum of 40 at $(6, 7)$
13. Maximum is 15 when $x = 5, y = 0$.   **15.** Minimum is 60 when $x = 0, y = 20$.   **17.** Make 3 batches of cakes and 6 batches of cookies for a maximum profit of $210.   **19. (a)** Let x_1 = number of item A; x_2 = number of item B; x_3 = number of item C.
(b) $z = 4x_1 + 3x_2 + 3x_3$
(c) $2x_1 + 3x_2 + 6x_3 \le 1200$
$\quad x_1 + 2x_2 + 2x_3 \le 800$
$\quad 2x_1 + 2x_2 + 4x_3 \le 500$
$\quad x_1 \ge 0, x_2 \ge 0, x_3 \ge 0$
21. (a) Let x_1 = number of gallons of Fruity wine and x_2 = number of gallons of Crystal wine.   **(b)** $z = 12x_1 + 15x_2$
(c) $2x_1 + x_2 \le 110$
$\quad 2x_1 + 3x_2 \le 125$
$\quad 2x_1 + x_2 \le 90$
$\quad x_1 \ge 0, x_2 \ge 0$
23. when there are more than 2 variables
25. any standard minimization problem
27. (a) $3x_1 + 5x_2 + x_3 \qquad\qquad = 47$
$\qquad x_1 + x_2 \qquad + x_4 \qquad = 25$
$\qquad 5x_1 + 2x_2 \qquad\quad + x_5 \qquad = 35$
$\qquad 2x_1 + x_2 \qquad\qquad\quad + x_6 = 30$

(b)

x_1	x_2	x_3	x_4	x_5	x_6	
3	5	1	0	0	0	47
1	1	0	1	0	0	25
5	2	0	0	1	0	35
2	1	0	0	0	1	30
-2	-7	0	0	0	0	0

29. (a) $x_1 + x_2 + x_3 + x_4 \qquad\qquad = 100$
$\qquad 2x_1 + 3x_2 \qquad\quad + x_5 \qquad = 500$
$\qquad x_1 \qquad + 2x_3 \qquad\qquad + x_6 = 350$

(b)

x_1	x_2	x_3	x_4	x_5	x_6	
1	1	1	1	0	0	100
2	3	0	0	1	0	500
1	0	2	0	0	1	350
-4	-6	-3	0	0	0	0

31. Maximum is 80 when $x_1 = 16, x_2 = 0, x_3 = 0, x_4 = 12, x_5 = 0$.
33. Maximum is 35 when $x_1 = 5, x_2 = 0, x_3 = 5, x_4 = 35$, $x_5 = 0, x_6 = 0$.   **35.** Maximize $z = -18y_1 - 10y_2$ with the same constraints.   **37.** Maximize $z = -6y_1 + 3y_2 - 4y_3$ with the same constraints.   **39.** Minimum is 40 when $y_1 = 10$ and $y_2 = 0$.
41. minimum of 1957 at $(47, 68, 0, 92, 35, 0, 0)$
43. $(9, 5, 8, 0, 0, 0)$; minimum is 62.   **45.** Get 250 of A, none of B or C for a maximum profit of $1000.   **47.** Make 17.5 gal of Crystal and 36.25 gal of Fruity for a maximum profit of $697.50.
49. Produce 660 cases of corn, no beans, and 340 cases of carrots for a minimum cost of $15,100.   **51.** Use 1,060,000 kg for whole tomatoes and 80,000 kg for sauce for a minimum cost of $4,500,000.

Case 6 (Page 414)

1. The answer in 100-gram units is 0.243037 units feta cheese, 2.35749 units lettuce, 0.3125 units salad dressing, and 1.08698 units tomato. Converting into kitchen units gives approximately 1/6 cup feta cheese, $4\frac{1}{4}$ cups lettuce, 1/8 cup salad dressing, and 7/8 of a tomato.

CHAPTER 8

Section 8.1 (Page 422)

1. false   **3.** true   **5.** true   **7.** true   **9.** false   **13.** $\subseteq$
15. $\not\subseteq$ **17.** $\subseteq$ **19.** $\subseteq$ **21.** 8   **23.** 64   **25.** $\{x \mid x$ is an integer less than or equal to 0 or greater than or equal to 7.$\}$
29. $\cap$ **31.** $\cap$ **33.** $\cup$ **35.** $\cup$ **37.** $\{3, 5\}$
39. $\{7, 9\}$ **41.** $\varnothing$ **43.** U or $\{2, 3, 4, 5, 7, 9\}$ **45.** all students in this school not taking this course   **47.** all students in this school taking accounting and zoology   **49.** C and D, A and E, C and E, D and E **51.** C' is the set of all stocks with a zero or negative price change; $C' = \{\text{GenElec}\}$.   **53.** $(A \cup B)'$ is the set of all stocks with a high price less than or equal to $100 and a last price either less than or equal to $20 or greater than or equal to $40; $(A \cup B)' = \{\text{GenElec, Lucent}\}$.   **55.** $M \cap E$ is the set of all male, employed applicants.   **57.** $M' \cup S'$ is the set of all female or married applicants.   **59.** $\{\text{Magazines, Yellow Pages, Internet}\}$
61. $F = \{\text{The Disney Channel, HBO/Cinemax}\}$
63. $H = \{\text{Showtime/The Movie Channel}\}$ **65.** $H \cup F = \{\text{The Disney Channel, HBO/Cinemax, Showtime/The Movie Channel}\}$
67. $\{s, d, c\}$ **69.** $\{g\}$

Section 8.2 (Page 431)

1.   **3.**

5. 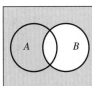   **7.** $\varnothing$ **9.** 8

11.   **13.**

15. 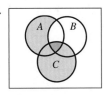   **17.** 17.3%

19. (a) 342 (b) 192 (c) 72 (d) 86   **21.** (a) 50 (b) 2
(c) 14   **23.** (a) 54 (b) 17 (c) 10 (d) 7 (e) 15 (f) 3
(g) 12 (h) 1   **25.** (a) 34 (b) 16 (c) 431 (d) 481
27. (a) 120.1 million (b) 88.5 million (c) 63.7 million
(d) 39.5 million (e) 239.5 million (f) 36.0 million   **31.** 9
33. 27

35.   **37.**

39.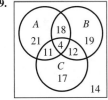

Section 8.3 (Page 439)

3. {January, February, March, . . ., December}
5. {0, 1, 2, . . . , 80}   **7.** {go ahead, cancel}
11. $\{y_1, y_2, y_3, w_1, w_2, w_3, w_4, b_1, b_2, b_3, b_4, b_5, b_6, b_7, b_8\}$
(a) $\{y_1, y_2, y_3\}$ (b) $\{b_1, b_2, b_3, b_4, b_5, b_6, b_7, b_8\}$
(c) $\{w_1, w_2, w_3, w_4\}$ (d) $\varnothing$
13. {ccc, ccw, cwc, cww, wcc, wcw, wwc, www}
(a) {www} (b) {ccw, cwc, wcc} (c) {cww}
15. {(1, 1), (1, 2), (1, 3), (1, 4), (1, 5), (2, 1), (2, 2), (2, 3), (2, 4), (2, 5),
(3, 1), (3, 2), (3, 3), (3, 4), (3, 5), (4, 1), (4, 2), (4, 3), (4, 4), (4, 5)}
(a) {(2, 1), (2, 2), (2, 3), (2, 4), (2, 5), (4, 1), (4, 2), (4, 3), (4, 4),
(4, 5)} (b) {(1, 2), (1, 4), (2, 2), (2, 4), (3, 2), (3, 4), (4, 2), (4, 4)}
(c) {(1, 4), (2, 3), (3, 2), (4, 1)} (d) $\varnothing$

17. $\dfrac{1}{6}$ **19.** $\dfrac{1}{3}$ **21.** $\dfrac{1}{3}$ **23.** $\dfrac{1}{13}$ **25.** $\dfrac{1}{26}$ **27.** $\dfrac{1}{52}$

29. $\dfrac{1}{13}$ **31.** $\dfrac{1}{5}$ **33.** $\dfrac{9}{25}$ **35.** $\dfrac{21}{25}$ **37.** $\dfrac{9}{25}$
39. (a) .0644 (b) .1869 (c) .7331   **41.** (a) The person
smokes or has a family history of heart disease. (b) The person does
not smoke and has a family history of heart disease. (c) The person
does not have a family history of heart disease or is not overweight.

Section 8.4 (Page 447)

3. no   **5.** yes   **7.** no   **9.** (a) $\dfrac{5}{36}$ (b) $\dfrac{1}{9}$ (c) $\dfrac{1}{12}$ (d) 0

11. (a) $\dfrac{5}{18}$ (b) $\dfrac{5}{12}$ (c) $\dfrac{1}{3}$ **13.** $\dfrac{2}{9}$

15. (a) $\dfrac{2}{13}$ (b) $\dfrac{4}{13}$ (c) $\dfrac{7}{13}$

17. (a) $\dfrac{3}{13}$ (b) $\dfrac{4}{13}$ (c) $\dfrac{7}{13}$ (d) $\dfrac{4}{13}$ (e) $\dfrac{8}{13}$

19. (a) $\dfrac{1}{2}$ (b) $\dfrac{7}{10}$ (c) $\dfrac{1}{2}$ **21.** (a) $\dfrac{1}{10}$ (b) $\dfrac{9}{10}$ (c) $\dfrac{7}{20}$

23. (a) .39 (b) .84 (c) .74 (d) .26   **25.** 1 to 5
27. 2 to 1   **29.** (a) 1 to 4 (b) 8 to 7 (c) 4 to 11   **31.** 2 to 7
35. not empirical   **37.** empirical   **39.** empirical   **41.** not
empirical   **45.** possible   **47.** Not possible; the sum of the probabilities is greater than 1.   **49.** Not possible; a probability cannot
be negative.
In Exercises 51 and 53, theoretical answers are given; actual answers
will vary.
51. (a) .2778 (b) .4167   **53.** (a) .15625 (b) .3125   **55.**
(a) .51 (b) .67 (c) .39 (d) .12
57. (a) .89 (b) .38 (c) .18   **59.** (a) .56 (b) .20 (c) .31

61. (a) $\dfrac{3}{4}$ (b) $\dfrac{1}{4}$ **63.** (a) .90 (b) .23

65. (a) .06 (b) .44 (c) .74 (d) .18
67. (a)

	A	B	C	D
O	.052	.060	.039	.004
E	.306	.270	.213	.045
M	.003	.003	.003	0

(b) about .31 (c) about .04 (d) about .7 (e) about .88

Section 8.5 (Page 462)

1. $\dfrac{1}{3}$ **3.** 1   **5.** $\dfrac{1}{6}$ **7.** $\dfrac{4}{17}$ **9.** $\dfrac{11}{51}$ **11.** .012

13. .245   **19.** No; yes   **21.** .43   **23.** (a) .815 (b) .79
25. (a) .46 (b) .29 (c) .23 (d) .61   **27.** .031   **29.** .834
31. The probability that a person in the group is employed in a man-
agerial/professional capacity, given that the person is male, is about
.505.   **33.** The probability that a person in the group is employed
in a technical/sales/administrative support capacity, given that the person is female, is .638.   **35.** .527   **37.** .042

39. $\dfrac{6}{7}$, or about .857   **41.** yes   **43.** .05   **45.** .25

47. .01254   **49.** .791   **51.** .236   **53.** .288   **55.** .0457
57. .999985   **59.** (a) 3 backups   **61.** no

Section 8.6 (Page 469)

1. $\dfrac{1}{3}$ **3.** $\dfrac{2}{41} \approx .0488$ **5.** $\dfrac{21}{41} \approx .5122$ **7.** $\dfrac{8}{17} \approx .4706$
9. .0135   **11.** .342   **13.** .824   **15.** .006   **17.** .706

19. (c)   **21.** .481   **23. (a)** .333   **25. (a)** .155   **(b)** .990
(c) about 59   **27. (a)** .53   **(b)** .1623   **29.** .524   **31.** .162
33. .086   **35.** .178

Chapter 8 Review (Page 474)

1. false   **3.** false   **5.** true   **7.** true   **9.** {New Year's Day,
Martin Luther King's Birthday, Presidents' Day, Memorial Day,
Independence Day, Labor Day, Columbus Day, Veterans' Day,
Thanksgiving, Christmas}   **11.** {1, 2, 3, 4}   **13.** {B_1, B_2, B_3, B_6,
B_{12}}   **15.** {A, C, E}   **17.** {$A, B_3, B_6, B_{12}, C, D, E$},
19. The set of all students in the class who are redheaded males
21. The set of all students in the class who are male or younger
than 21   **23.** The set of all students in the class who are not A
students and do not have red hair
25.

27.
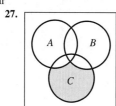

29. 5   **31.** {1, 2, 3, 4, 5, 6}   **33.** {(red, 10), (red, 20) (red, 30),
(blue, 10), (blue, 20), (blue, 30), (green, 10), (green, 20), (green, 30)}
35. {(2, blue), (4, blue), (6, blue), (8, blue), (10, blue)}   **37.** no

39. $E \cup F$ **45.** $\frac{3}{13}$ **47.** $\frac{1}{2}$ **49.** 1   **51.** 1 to 25

53. .86
55.

	N_2	T_2
N_1	N_1N_2	N_1T_2
T_1	T_1N_2	T_1T_2

57. $\frac{1}{2}$ **59.** $\frac{5}{36} \approx .139$ **61.** $\frac{5}{18} \approx .278$ **63.** $\frac{1}{3}$ **65.** .66

67. .71   **69. (a)** .7   **(b)** $\frac{2}{15} \approx .133$ **73.** $\frac{3}{22}$ **75.** $\frac{2}{3}$

77. (a) .271   **(b)** .379   **(c)** .625   **(d)** .342   **(e)** .690   **(f)** .514

Additional Probability Review (Page 476)

1. (a) .88   **(b)** .30   **(c)** .70   **3. (a)** .187   **(b)** .540   **(c)** .769
5. (a) 0   **(b)** 1/3   **(c)** 1   **7. (a)** .102   **(b)** .153   **(c)** .120

(d) .135   **9.** .091   **11.** no   **13.** $\frac{1}{6}$ **15.** 82.4%

17. .187   **19.** .333   **21.** .286   **23.** $\frac{1}{5}$

25. (a) $\frac{2}{5}$ **(b)** $\frac{3}{5}$ **(c)** $\frac{1}{10}$ **(d)** $\frac{7}{10}$

Case 7 (Page 479)

1. .076   **3.** .051

CHAPTER 9

Section 9.1 (Page 492)

1. 12   **3.** 56   **5.** 8   **7.** 24   **9.** 792   **11.** 156
13. 6,375,600   **15.** 240,240   **17.** 352,716   **19.** 2,042,975
21. Undefined   **23. (a)** 8   **(b)** 64   **25.** 840

27. $10^9 = 1,000,000,000$; yes   **29.** $10^9 = 1,000,000,000$
31. (a) 25,200   **(b)** 15,120   **33. (a)** 160; 8,000,000   **(b)** Some
numbers, such as 911, 800, and 900, are reserved for special purposes.
35. 1600   **39.** 479,001,600   **41.** 1440   **43.** 1,256,640
45. 2730   **47. (a)** 220   **(b)** 220   **49. (a)** 0   **(b)** 792
(c) 658,008   **(d)** 652,080   **(e)** 79,092   **53. (a)** 9   **(b)** 6
(c) 3; yes, from both   **55. (a)** 84   **(b)** 10   **(c)** 40   **(d)** 74
57. (a) 961   **(b)** 29,791   **59. (a)** 1,120,529,256
(b) 806,781,064,300   **61.** 81   **63.** 1   **65.** 55,440
67. (a) 7,059,052   **(b)** \$3,529,526   **(c)** no   **69.** 5.524×10^{26}
71. (a) 840   **(b)** 180   **(c)** 420   **73. (a)** 362,880   **(b)** 6
(c) 1260

Section 9.2 (Page 501)

1. .424   **3.** $\frac{7}{9}$ **5.** $\frac{5}{12}$ **7.** .147   **9.** .020   **11.** .314

13. 1326   **15.** .851   **17.** .765   **19.** .941

23. $\dfrac{\binom{4}{2}\binom{4}{3}\binom{44}{8}}{\binom{52}{13}}$ **25. (a)** 8.9×10^{-10} **(b)** 1.2×10^{-12}

27. 3.3×10^{-16} **29. (a)** .218   **(b)** .272   **(c)** .728

31. (a) .003   **(b)** .005   **(c)** .057   **(d)** .252   **33.** $1 - \dfrac{_{365}P_{100}}{(365)^{100}}$

35. .350   **37.** There were 10 balls with 9 of them blue.
39. Answers will vary. The theoretical answers are: **(a)** .0399
(b) .5191   **(c)** .0226

Section 9.3 (Page 508)

1. .067   **3.** .066   **5.** .934   **7.** .0000000005   **9.** .269

11. .875   **13.** $\frac{1}{32}$ **15.** $\frac{13}{16}$ **19.** .00655   **21.** .9932

23. .1285   **25.** .7941   **27.** .8159   **29.** .4913   **31.** .0025
33. (a) .1493   **(b)** .9116   **35. (a)** .0060   **(b)** .99998
37. .000079   **39.** .2162   **41. (a)** .9995
43. (a) 1 chance in 1024   **(b)** about 1 chance in 1.1×10^{12}
(c) about 1 chance in 2.6×10^6

Section 9.4 (Page 518)

1.

Number of Heads	0	1	2	3	4
Probability	1/16	1/4	3/8	1/4	1/16

3.

Number of Aces	0	1	2	3
Probability	.7826	.2042	.0130	.0002

5.   **7.**
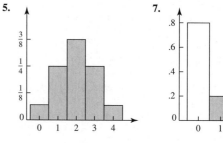

9. 3.6   **11.** 14.64   **13.** 2.7   **15.** 18   **17.** Yes, the
probability of a match is still $\frac{1}{2}$.

19. (a)

21. (a) $5/3 \approx 1.667$ **(b)** $4/3 \approx 1.333$ **23.** 1 **25.** No, the expected value is about $-21¢$. **29.** $-32¢$; no **31.** $-66¢$ **33.** $-2.7¢$ **35.** $-20¢$ **37.** $-87.8¢$ **39.** 3.51

41.

Account	Expected Value	Total	Class
3	$2000	$22,000	C
4	$1000	$51,000	B
5	$25,000	$30,000	C
6	$60,000	$60,000	A
7	$16,000	$46,000	B

43. (a) amoxicillin $68.51; cefaclor $72.84 **(b)** amoxicillin **45. (a)** $50,000 **(b)** $65,000 **47. (a)**

Number	0	1	2	3	4	5
Probability	.0000	.0000	.0000	.0003	.0024	.0141

Number	6	7	8	9	10
Probability	.0573	.1600	.2929	.3178	.1552

(b) .0027 **(d)** 8.3

Section 9.5 (Page 531)

1. yes **3.** yes **5.** no **7.** no
9. yes **11.** no

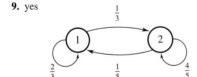

13. not a transition diagram
15.

$$\begin{array}{c} \\ A \\ B \\ C \end{array} \begin{array}{ccc} A & B & C \\ \begin{bmatrix} .6 & .2 & .2 \\ .9 & .02 & .08 \\ .4 & 0 & .6 \end{bmatrix} \end{array}$$

17. yes **19.** yes **21.** no

23. $\begin{bmatrix} \frac{1}{3} & \frac{2}{3} \end{bmatrix}$ **25.** $\begin{bmatrix} \frac{3}{11} & \frac{8}{11} \end{bmatrix}$ **27.** $[.4633 \quad .1683 \quad .3684]$

29. $\begin{bmatrix} \frac{1}{4} & \frac{1}{4} & \frac{1}{2} \end{bmatrix}$

31. The first power is the given transition matrix;

$$\begin{bmatrix} .23 & .21 & .24 & .17 & .15 \\ .26 & .18 & .26 & .16 & .14 \\ .23 & .18 & .24 & .19 & .16 \\ .19 & .19 & .27 & .18 & .17 \\ .17 & .20 & .26 & .19 & .18 \end{bmatrix} ; \begin{bmatrix} .226 & .192 & .249 & .177 & .156 \\ .222 & .196 & .252 & .174 & .156 \\ .219 & .189 & .256 & .177 & .159 \\ .213 & .192 & .252 & .181 & .162 \\ .213 & .189 & .252 & .183 & .163 \end{bmatrix} ;$$

$$\begin{bmatrix} .2205 & .1916 & .2523 & .1774 & .1582 \\ .2206 & .1922 & .2512 & .1778 & .1582 \\ .2182 & .1920 & .2525 & .1781 & .1592 \\ .2183 & .1909 & .2526 & .1787 & .1595 \\ .2176 & .1906 & .2533 & .1787 & .1598 \end{bmatrix} ;$$

$$\begin{bmatrix} .21932 & .19167 & .25227 & .17795 & .15879 \\ .21956 & .19152 & .25226 & .17794 & .15872 \\ .21905 & .19152 & .25227 & .17818 & .15898 \\ .21880 & .19144 & .25251 & .17817 & .15908 \\ .2185 & .19148 & .25253 & .17824 & .15918 \end{bmatrix} ; .17794$$

33. 24.02% will be 30-year fixed loans, 69.72% will be 15-year fixed loans and 6.26% will be adjustable loans. **35.** $\dfrac{17}{18}$

37.

$$\begin{array}{c} \\ R \\ P \\ W \end{array} \begin{array}{ccc} R & P & W \\ \begin{bmatrix} \frac{1}{2} & \frac{1}{2} & 0 \\ \frac{1}{4} & \frac{1}{2} & \frac{1}{4} \\ 0 & \frac{1}{2} & \frac{1}{2} \end{bmatrix} \end{array} ; \begin{bmatrix} \frac{1}{4} & \frac{1}{2} & \frac{1}{4} \end{bmatrix}$$

39. homeowners: .769; renters: .231; homeless: 0

41. (a) 42,500 in G_0; 5000 in G_1; 2500 in G_2 **(b)** 37,125 in G_0; 8750 in G_1; 4125 in G_2 **(c)** 33,281 in G_0; 11,513 in G_1; 5206 in G_2 **(d)** $\begin{bmatrix} \frac{28}{59} & \frac{22}{59} & \frac{9}{59} \end{bmatrix}$ or $[.475 \quad .373 \quad .152]$; in the long run, the probabilities of no accidents, one accident, and more than one accident are .475, .373, and .152, respectively.

43. (a) $\begin{bmatrix} .348 & .379 & .273 \\ .320 & .378 & .302 \\ .316 & .360 & .323 \end{bmatrix}$ **(b)** .348 **(c)** .320

45. (b) $\begin{bmatrix} \frac{5}{12} & \frac{13}{36} & \frac{2}{9} \\ \frac{5}{12} & \frac{13}{36} & \frac{2}{9} \\ \frac{1}{4} & \frac{5}{12} & \frac{1}{3} \end{bmatrix}$ **(c)** $\dfrac{2}{9}$

47. $[0 \quad 0 \quad .102273 \quad .897727]$

Section 9.6 (Page 536)

1. (a) choose the coast **(b)** choose the highway **(c)** choose the highway; $38,000 **(d)** choose the coast **3. (a)** make no repairs **(b)** make repairs **(c)** make no repairs; $10,750
5. (a)

$$\begin{array}{c} \\ \text{Overhaul} \\ \text{Don't} \end{array} \begin{array}{cc} \text{Fails} & \text{Doesn't} \\ \begin{bmatrix} -8600 & -2600 \\ -6000 & 0 \end{bmatrix} \end{array}$$

(b) do not overhaul before shipping **7. (a)** no campaign **(b)** campaign for all **(c)** campaign for all **9.** emphasize the environment; 14.25

Chapter 9 Review (Page 539)

1. 120 **3.** 220 **5.** 120 **7. (a)** 72 **(b)** 72 **11.** .245

13. .362 **15.** .091 **17.** .455 **19. (a)** $\binom{n}{0}$, or 1; $\binom{n}{1}$, or n;

$\binom{n}{2}$; $\binom{n}{n}$, or 1 **(b)** $\binom{n}{0} + \binom{n}{1} + \binom{n}{2} + \cdots + \binom{n}{n}$ **21.** .99996

23. .00004 **25.** 1.0000 **27.** $8500

29. (a)

0	1	2	3	4
$\frac{1}{16}$	$\frac{1}{4}$	$\frac{3}{8}$	$\frac{1}{4}$	$\frac{1}{16}$

(b) **(c)** 2

31. (a)

Number	0	1	2
Probability	10/28	15/28	3/28

(b) **(c)** 3/4

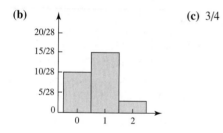

33. $1.29 **35.** -28¢ **37. (a)** $\frac{3}{13}$, or .23 **(b)** $\frac{3}{4}$, or .75

39. no **41.** yes **43. (a)** [.54 .46] **(b)** [.6464 .3536]

(c) $\frac{2}{3}$ **45. (a)** oppose **(b)** oppose **(c)** oppose; 2700

(d) oppose; 1400 **47. (c)**

Case 8 (Page 543)

1. (a) $69.01 **(b)** $79.64 **(c)** $58.96 **(d)** $82.54 **(e)** $62.88

(f) $64.00 **3.** The events of needing parts 1, 2, and 3 are not the only events in the sample space.

CHAPTER 10

Section 10.1 (Page 557)

1. (a)–(b)

Interval	Frequency
0–24	4
25–49	3
50–74	6
75–99	3
100–124	5
125–149	9

(c)–(d)

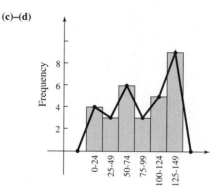

3. (a)–(b)

Interval	Frequency
70–74	2
75–79	1
80–84	3
85–89	2
90–94	6
95–99	5
100–104	6
105–109	4
110–114	2

(c)–(d)

5.

Stem	Leaves
07	249
08	23448
09	02224467799
10	0122446678
11	01

9. 27,955 **11.** 7.7 **13.** 6.7 **15.** 17.2 **17.** 51

19. 130 **21.** 29.1 **23.** 9 **25.** 68 and 74 **27.** 6.3

31. 94.9; 90–94 and 100–104

35. (a) $90,599,700 **(b)** $75,694,500 **(c)** There is one very large salary that affects the value of the mean. **(d)** $49,326,000

37. 7.38; 7; There are three modes: 7, 5, and 4. **39.** 55.5°F; 50.5°F **41.** $3.42; $3.32 **43.** $29,952.28

45. (a) 14 **(b)** 7 **(c)** 30–39 and 40–49

Section 10.2 (Page 567)

3. 6; 2.2 **5.** 12; 4.7 **7.** 53; 21.8 **9.** 46; 16.1 **11.** 45.2

13. $\bar{x} = 5.0876$; $s = .1087$ **15.** at least $\frac{3}{4}$ **17.** at least $\frac{5}{9}$

19. at least 93.75% **21.** no more than 11.1%

23. (a) Britelite: $\bar{x} = 26.2$, $s = 4.4$; Brand X: $\bar{x} = 25.5$, $s = 7.2$
(b) Britelite   (c) Britelite   **25.** (a) 7.635; 1995   (b) 1.122
(c) 7   (d) 10   **27.** (a) 14.8; 3.8   (b) 10
29. (a) 127.71 days; 30.16 days   (b) all of them
31. (a) $\frac{1}{3}$; 2; $-\frac{1}{3}$; 0; $\frac{5}{3}$, $\frac{7}{3}$; 1; $\frac{4}{3}$, $\frac{7}{3}$, $\frac{2}{3}$ (b) 2.1; 2.6; 1.5; 2.6; 2.5; .6;
1.0; 2.1; .6; 1.2   (c) 1.13   (d) 1.68   (e) 4.41; -2.15; the process is
out of control.   **33.** mean = 80.17; standard deviation = 12.2

Section 10.3 (Page 581)

1. The mean   **5.** 45.99%   **7.** 16.64%   **9.** 7.7%
11. 47.35%   **13.** 92.42%   **15.** 1.64 or 1.65   **17.** -1.04
19. .5; .5   **21.** .889; .997   **23.** .3821   **25.** .5762
27. .8887   **29.** .0021   **31.** 5000   **33.** 642   **35.** 9918
37. 372; 628   **39.** 35.2 mph   **41.** 24.17%   **45.** 625 units
47. 189 units   **49.** 15.87%   **51.** .5000   **53.** .9050
55. .8413   **57.** .5987

Section 10.4 (Page 588)

1. The number of trials and the probability of success on each trial
3. (a) .1964   (b) .1974   **5.** (a) .0106   (b) .0122   **7.** .0240
9. .9032   **11.** .0956   **13.** .6443   **15.** .1841   **17.** .0222
19. .2578   **21.** .0301   **23.** (a) .0472   (b) .9875   (c) .8814
25. (a) .0000   (b) .0018   **27.** .5977   **29.** .9898
31. (a) 1.2139×10^{-7} (b) essentially 0   (c) .7939

Chapter 10 Review (Page 591)

3. (a)

Interval	Frequency
450–474	5
475–499	6
500–524	5
525–549	2
550–574	2

(b)–(c)

5. 73   **7.** 34.9   **11.** 38; 36 and 38   **13.** 55–59
17. 67; 23.9   **19.** 6.2   **23.** .6591   **25.** .0606
29. (a) .0317   (b) 1.482; 1.056   **31.** (a) Diet A: (mean = 2.7)
(b) Diet B: ($s = .95$)   **33.** (a) $\bar{x} = 19.45\%$; $s = 2.39\%$
(b) 1.1; 1.1   **35.** 15.87%   **37.** (a) .0305   (b) .9938   (c) 1.000
(d) .1049; the binomial distribution gives .1326.
39. (a) .1190   (b) .3936   (d) .9987   (e) .1336

Case 9 (Page 595)

1. $z \approx 29.1$ **3.** (a) 174   (b) 6.03   (c) -12.3 (d) The probability is less than .004. (Actually, it is less than 10^{-30} which is the smallest probability in the table)

Index of Applications

Finance

Natural Science

Physical Science

Social Science

Transportation

Index

The **derivative** of the function f is the function denoted f' whose value at the number x is defined to be the number

$$f'(x) = \lim_{h \to 0} \frac{f(x + h) - f(x)}{h},$$

provided this limit exists.

RULES FOR DERIVATIVES

Assume all indicated derivatives exist.

Constant Function   If $f(x) = k$, where k is any real number, then

$$f'(x) = 0.$$

Power Rule   If $f(x) = x^n$, for any real number n, then

$$f'(x) = n \cdot x^{n-1}.$$

Constant Times a Function   Let k be a real number. Then the derivative of $y = k \cdot f(x)$ is

$$y' = k \cdot f'(x).$$

Sum or Difference Rule   If $y = f(x) \pm g(x)$, then

$$y' = f'(x) \pm g'(x).$$

Product Rule   If $f(x) = g(x) \cdot k(x)$, then

$$f'(x) = g(x) \cdot k'(x) + k(x) \cdot g'(x).$$

Quotient Rule   If $f(x) = \dfrac{g(x)}{k(x)}$, and $k(x) \neq 0$, then

$$f'(x) = \frac{k(x) \cdot g'(x) - g(x) \cdot k'(x)}{[k(x)]^2}.$$

Chain Rule   Let $y = f[g(x)]$. Then

$$y' = f'[g(x)] \cdot g'(x).$$

Chain Rule (Alternative Form)   If y is a function of u, say $y = f(u)$, and if u is a function of x, say $u = g(x)$, then $y = f[g(x)]$, and

$$\frac{dy}{dx} = \frac{dy}{du} \cdot \frac{du}{dx}.$$

Generalized Power Rule   Let u be a function of x, and let $y = u^n$ for any real number n. Then

$$y' = n \cdot u^{n-1} \cdot u'.$$

Exponential Function   If $y = e^{g(x)}$, then

$$y' = g'(x) \cdot e^{g(x)}.$$

Natural Logarithmic Function   If $y = \ln |g(x)|$, then

$$y' = \frac{g'(x)}{g(x)}.$$